MATERIALS SCIENCE RESEARCH
Volume 15

ADVANCES IN MATERIALS CHARACTERIZATION

MATERIALS SCIENCE RESEARCH

Recent volumes in the series:

Volume 5   CERAMICS IN SEVERE ENVIRONMENTS
            Edited by W. Wurth Kriegel and Hayne Palmour III

Volume 6   SINTERING AND RELATED PHENOMENA
            Edited by G. C. Kuczynski

Volume 7   SURFACES AND INTERFACES OF GLASS AND CERAMICS
            Edited by V. D. Fréchette, W. C. LaCourse, and V. L. Burdick

Volume 8   CERAMIC ENGINEERING AND SCIENCE: Emerging Priorities
            Edited by V. D. Fréchette, L. D. Pye, and J. S. Reed

Volume 9   MASS TRANSPORT PHENOMENA IN CERAMICS
            Edited by A. R. Cooper and A. H. Heuer

Volume 10   SINTERING AND CATALYSIS
            Edited by G. C. Kuczynski

Volume 11   PROCESSING OF CRYSTALLINE CERAMICS
            Edited by Hayne Palmour III, R. F. Davis, T. M. Hare

Volume 12   BORATE GLASSES: Structure, Properties, Applications
            Edited by L. D. Pye, V. D. Fréchette and N. J. Kreidl

Volume 13   SINTERING PROCESSES
            Edited by G. C. Kuczynski

Volume 14   SURFACES AND INTERFACES IN CERAMIC AND
            CERAMIC–METAL SYSTEMS
            Edited by Joseph Pask and Anthony Evans

Volume 15   ADVANCES IN MATERIALS CHARACTERIZATION
            Edited by David R. Rossington, Robert A. Condrate,
            and Robert L. Snyder

A Continuation Order Plan is available for this series. A continuation order will bring delivery of each new volume immediately upon publication. Volumes are billed only upon actual shipment. For further information please contact the publisher.

MATERIALS SCIENCE RESEARCH • Volume 15

ADVANCES IN MATERIALS CHARACTERIZATION

Edited by

David R. Rossington
Robert A. Condrate
and
Robert L. Snyder

New York State College of Ceramics
Alfred University
Alfred, New York

PLENUM PRESS • NEW YORK AND LONDON

Library of Congress Cataloging in Publication Data

Conference on Advances in Materials Characterization (1982: New York State College of Ceramics at Alfred University)
Advances in materials characterization.

(Materials science research; v. 15)
"Proceedings of a Conference on Advances in Materials Characterization, held August 15-18, 1982, at the New York State College of Ceramics at Alfred University, Alfred, New York"—P.
Includes bibliographical references and index.
1. Materials—Testing—Congresses. 2. Materials—Surfaces—Congresses. I. Rossington, David R., 1932-   . II. Condrate, Robert A., 1938-   . III. Snyder, Robert L., 1941-   IV. Title. V. Series.
TA410.C56 1982   620.1′1   83-4186
ISBN 0-306-41347-7

Proceedings of a Conference on
Advances in Materials Characterization,
held August 15-18, 1982,
at the New York State College of Ceramics
at Alfred University, Alfred, New York

©1983 Plenum Press, New York
A Division of Plenum Publishing Corporation
233 Spring Street, New York, N.Y. 10013

All rights reserved

No part of this book may be reproduced, stored in a retrieval system, or transmitted in any form or by any means, electronic, mechanical, photocopying, microfilming, recording, or otherwise, without written permission from the Publisher

Printed in the United States of America

PREFACE

The characterization of materials and phenomena has historically been the principal limitation to the development in each area of science. Once what we are observing is well defined, a theoretical analysis rapidly follows. Modern theories of chemical bonding did not evolve until the methods of analytical chemistry had progressed to a point where the bulk stoichiometry of chemical compounds was firmly established. The great progress made during this century in understanding chemistry has followed directly from the development of an analytical chemistry based on the Dalton assumption of multiple proportions.

It has only become apparent in recent years that the extension of our understanding of materials hinges on their non-stoichiometric nature. The world of non-Daltonian chemistry is very poorly understood at present because of our lack of ability to precisely characterize it. The emergence of materials science has only just occurred with our recognition of effects, which have been thought previously to be minor variations from ideality, as the principal phenomena controlling properties. The next step in the historical evolution of materials science must be the development of tools to characterize the often subtle phenomena which determine properties of materials.

The various discussions of instrumental techniques presented in this book are excellent summaries for the state-of-the-art of materials characterization at this rather critical stage of materials science. The application of the tools described here, and those yet to be developed, holds the key to the development of this infant into a mature science.

This volume constitutes the proceedings of the August 15-18, 1982 Conference on Advances in Materials Characterization, held at the New York State College of Ceramics at Alfred University. The conference was the 18th in the University Series on Ceramic Science, instituted in 1964 by Alfred University, the University of California at Berkeley, North Carolina State University and Notre Dame University.

PREFACE

The 49 papers included in these Proceedings cover all aspects of surface and bulk characterization and they have been divided into general areas according to different experimental techniques. However, as would be expected for such a broad topic as characterization, there is often considerable overlap among techniques.

The editors wish to thank all speakers and contributors to the conference, and especially to those who were invited to present overview papers in their field of expertise, namely: Dr. R. Conzemius, Dr. E. Etz, Dr. J. Ferraro, Dr. B. T. Khuri-Yakub, Dr. C. Pantano, Dr. R. L. Snyder, Dr. I.S.T. Tsong and Dr. K. Wefers.

The editors also gratefully acknowledge the conference sponsors listed below. Their financial support and commitment was an invaluable constituent to the success of the conference.

U.S. Dept. of the Interior - Bureau of Mines
Corning Glass Works Foundation
IBM Corporation
General Motors Corporation - AC Spark Plug Division
Leybold-Heraeus Vacuum Products
Perkin-Elmer Corporation
Siemens Corporation
Spex Industries, Inc.

The success of any conference depends not only upon the quality of papers presented and sponsorship, but also on the many people involved in the planning and preparation stages. It is impossible to list all the people at the New York State College of Ceramics who were responsible for the efficient running of the conference, but special thanks are due to Mrs. Darlene Brewer, Mrs. Josephine Schieder and Mrs. Doris Snowden for registration and secretarial services, Mr. William Emrick for projection facilities, Ms. Carol Binzer for accommodation facilities and Mr. Michael Linehan for food service. To all these people, and the many graduate students who performed the many last minute tasks that inevitably arise, the editors express their deep gratitude.

It is our sincere hope that this volume will prove to be of benefit to all people involved in some area of materials characterization.

Alfred, New York
December, 1982

David R. Rossington
Robert A. Condrate
Robert L. Snyder

CONTENTS

SURFACE SPECTROSCOPY

Surface Studies of Multicomponent Silicate Glasses:
    Quantitative Analysis, Sputtering Effects
    and the Atomic Arrangement. 1
    C. G. Pantano, J. F. Kelso and M. J. Suscavage

Depth-Profiling Studies of Glasses and Ceramics
    by Ion Beam Techniques. 39
    I. S. T. Tsong

Analysis of Solids by Spark Source and Laser
    Mass Spectroscopy 59
    Robert J. Conzemius and Harry J. Svec

Surface and Interface Studies of Metal
    Oxide/Glass Systems 71
    K. L. Smith

Atomic Structure at Electrode Surfaces 91
    Gerald A. Garwood Jr., and Arthur T. Hubbard

SURFACE TECHNIQUES

Surface Characterization of Certain Metal
    Oxides Determined by the Isothermal
    Adsorption and Desorption of Argon 109
    Richard G. Herman, Philip Pendleton,
    and John B. Bulko

Hysteresis in Mercury Porosimetry. 133
    S. Lowell and J. E. Shields

Pore Structure Characterization by
    Mercury Porosimetry 147
    O. J. Whittemore and G. D. Halsey

Surface Characteristics of Yttria Precursors
      in Relation to Their Sintering
         Behavior. 159
   Mufit Akinc and M. D. Rasmussen

VIBRATIONAL SPECTROSCOPIC TECHNIQUES

An Overview of Techniques Used in FT-IR
         Spectroscopy. 171
   John R. Ferraro

Raman Microprobe Spectroscopy of
         Polyphase Ceramics. 199
   D. R. Clarke and F. Adar

Characterization of Anodically Grown Native
      Oxide Films on $Hg_{0.7}Cd_{0.3}Te$ 215
   Fran Adar, R. E. Kvaas and D. R. Rhiger

The Raman Spectra of Potassium Boro-
      germanate Glasses 223
   I. N. Chakraborty and R. A. Condrate, Sr.

Characterization of the Structure and
      Nonstoichiometry of CaO-NiO
      Solid Solutions 239
   B. C. Cornilsen, E. F. Funkenbusch,
   C. P. Clarke, P. Singh and
   V. Lorprayoon

Vibrational Spectroscopies of Molecular
      Monolayers in Thin Film Geometries. 249
   J. R. Kirtley, J. C. Tsang, Ph. Avouris,
   and Y. Thefaine

Characterization of Rare Earth Sulfides. 267
   C. Lowe-Ma

ELECTRON OPTICAL METHODS

Applications of Analytical Microscopy
      to Ceramic Research 281
   Paul F. Johnson

Analysis of Second-Phase Particles in Al_2O_3. 297
   K. J. Morrissey and C. B. Carter

CONTENTS

Microstructural Characterization Abnormal
    Grain Growth Development in Al_2O_3. 309
    M. P. Harmer, S. J. Bennison and C. Narayan

Properties and Characterization of Surface
    Oxides on Aluminum Alloys. 321
    Karl Wefers

The Characterization of Microcracks
    in Brittle Solids. 323
    D. R. Clarke and D. J. Green

X-ray Energy Dispersive Spectroscopy of
    Intergranular Phases in
    $\beta 11$ and β' Sialons 339
    T. R. Dinger and G. Thomas

EM Study of the Structure and Composition
    of Grain Boundaries in $(Mn, Zn) Fe_2O_4$. 351
    I-Nan Lin, R. K. Mishra and G. Thomas

Cross-Sectional Transmission Electron
    Microscopy of Semiconductors 359
    D. K. Sadana

Microstructural Characterization of Nuclear
    Waste Ceramics 367
    F. J. Ryerson and D. R. Clarke

Semiautomatic Image Analysis for Microstructure
    and Powder Characterization. 387
    D. W. Readey, E. Bright, S. S. Campbell,
        J. H. Lee, R. S. Pan, T. Quadir, and
        K. A. Williams

ACOUSTIC AND MECHANICAL PROPERTIES

Acoustic Characterization of Structural
    Ceramics 401
    B. T. Khuri-Yakub

Characterization of Ceramics by
    Acoustic Microscopy 413
    D. E. Yuhas and L. W. Kessler

Determination of Slow Crack Growth Using
    an Automated Test Technique. 425
    Helen H. Moeller and Robert L. Farmer

Biaxial Compression Testing of
      Refractory Concretes. 437
   Albert H. Bremser and Oral Buyukozturk

Mechanical Testing of Glass Hollow
      Microspheres. 441
   Paul W. Bratt, J. P. Cunnion and Bruce D. Spivack

GENERAL CRYSTALLOGRAPHIC TECHNIQUES

The Renaissance of X-ray Powder Diffraction. 449
   Robert L. Snyder

Characterization of Imperfections in Plasma-
      Sprayed Titania 465
   C. C. Berndt, R. Korlipara, R. A. Zatorski,
   A. Jonca, T. Templeton and R. K. MacCrone

Characterization of the Mechanical Properties
      of Plasma-Sprayed Coatings. 473
   N. R. Shankar, C. C. Berndt and H. Herman

Analysis of Silicon Nitride. 491
   Gary Czupryna and Samuel Natansohn

Acid-Base Properties of Ceramic Powders. 449
   Alan Bleier

Recent Advances in Computerized High Temperature
      Differential Thermal Analysis 515
   Charles M. Earnest, W. P. Brennan and
   M. P. DiVito

Reflectance Technique for Measurement of
      Bottom Surface Tin Concentration
      on Clear Float Glass. 531
   Thomas O. LaFramboise

Characterization of Reinforcements for
      Inorganic Composites 535
   S. W. Bradstreet

GENERAL GLASS CHARACTERIZATION STUDIES

Nuclear Reaction Analysis of Glass Surfaces:
      The Study of the Reaction Between
      Water and Glass 549
   W. A. Lanford and C. Burman

CONTENTS

A Study of Water in Glass by an Autoradiographic
       Method that Utilizes Tritiated Water. 571
  S. H. Knickerbocker, S. B. Joshi, and
  S. D. Brown

Characterization of Borosilicate Glass Containing
       Savannah River Plant Radioactive Waste 591
  Ned E. Bibler and P. Kent Smith

Microstructure of Phase-Separated Sodium
       Borosilicate Glasses. 603
  Peter Taylor, Allan B. Campbell, Derrek G. Owen,
  David Simkin and Pierre Menassa

The Measurement of Thermal Diffusivity of
       Simulated Glass Forming Nuclear
       Waste Melts 615
  James U. Derby, L. David Pye and
  M. J. Plodinec

Volume-Temperature Relationships in
       Simulated Glass Forming Nuclear
       Waste Melts 627
  L. D. Pye, R. Locker and M. J. Plodinec

Property/Morphology Relationships in Glasses 639
  J. E. Shelby

The Characterization of Individual Redox
       Ions in Glasses 647
  Henry D. Schreiber

Author Index 659

Subject Index. 669

SURFACE STUDIES OF MULTICOMPONENT SILICATE GLASSES:

QUANTITATIVE ANALYSIS, SPUTTERING EFFECTS AND THE ATOMIC ARRANGEMENT

C. G. Pantano, J. F. Kelso and M. J. Suscavage

Department of Materials Science and Engineering
The Pennsylvania State University
University Park, PA 16802

INTRODUCTION

An 'ideal' glass surface can be created by fracturing bulk, homogeneous, microstructure-free glass in an ultra-high vacuum environment. In the absence of any atomic rearrangement ahead of or behind the advancing crack tip, the resulting fracture surface will exhibit the bulk composition of the glass. The surface monolayer will, of course, possess a high concentration of reactive, dangling bonds. This is in contrast to a 'real' glass surface which may have been created or treated in the presence of gaseous species, in contact with liquid and/or solid phases, and in many instances, while the glass itself is in the liquid state. Adsorption, contamination, volatization, thermal segregation, leaching, hydration, chemical reactions, and structural transformations can modify, significantly, the outermost atomic layers of the glass. Thus, the composition and structure of a 'real' glass surface can be very different from that of the bulk glass. In addition to the concentration gradients associated with the compositional and structural differences between the surface and bulk, the 'real' glass surface can be covered to varying degrees with adsorbed species and other reaction products. It is obvious, then, that a 'real' glass surface can extend far beyond the outermost monolayer which is characteristic of an 'ideal' surface. A 'real' glass surface is, perhaps, best defined as that region of a glass where the composition and structure can be clearly distinguished from the bulk.

Owing to the development and widespread availability of surface analysis techniques, it has become possible over the last ten years to characterize 'directly' the composition, structure, and extent of a 'real' glass surface. The most common techniques for this surface analysis are: Auger electron spectroscopy (AES), x-ray photoelectron spectroscopy (XPS), secondary ion mass spectroscopy (SIMS), ion scattering spectroscopy (ISS), and sputter induced photon spectroscopy (SIPS). These techniques can provide compositional, and in some cases structural, information about the outermost monolayer(s) of solid materials with a surface sensitivity of the order .5-5.0 nm. Through the combined use of inert ion sputtering, the material can be slowly eroded to permit an in-depth compositional analysis. Thus, these methods are used extensively for surface and in-depth analyses.

Unfortunately, the application of these methods to glasses and ceramics is often complicated by the insulating nature of these materials. There is no doubt that these complications—some of which are further discussed in this paper—severely limit the qualitative and quantitative analysis of insulating glass and ceramic materials. In general, these complications bring about a change in the composition or structure of the surface during the analysis. The most notable effects include unstable charging, beam-induced ion migration, electron stimulated desorption, ion bombardment induced redistribution of atomic species, and beam heating. These experimental perturbations will occur to varying degrees in all surface analysis methods involving charged particle excitation or detection.

A recent review [1] described the physical basis, analytical capabilities, experimental perturbations and some specific applications of these methods with regard to surface and in-depth analysis of 'real' glass and ceramic materials; many other references to surface analysis of glasses and ceramics are included therein.

It might also be noted that the high energy ion beam surface analysis techniques—particularly Rutherford backscattering (RBS) and nuclear reaction analysis (NRA)—have seen increasing applications for the characterization of 'real' glass surfaces [2-4]. Unfortunately, these methods require the use of a very high energy ion accelerator. Moreover, their surface sensitivity is limited to about 10.0-20.0 nm. For these reasons, RBS and NRA are not as widely applied as the five techniques listed above. Nonetheless, when a non-destructive, quantitative, in-depth profile in the range .1-2 µm is of primary interest, these techniques do, in fact, offer exceptional capabilities. Of course, they too are subject to complications when applied to insulating glass and ceramic materials.

In this paper, attention is focussed upon the study of multicomponent silicate glass surfaces prepared by the rapid fracture of bulk glasses. The objectives of the work are threefold:

(1) to establish the validity of using clean glass surfaces—which should exhibit the bulk composition—as calibration standards for quantitative AES and SIMS analysis of 'real' glass surfaces; of particular interest is a quantitative analysis of the surface composition and in-depth profile for tin-oxide in multicomponent silicate glasses.

(2) to use the clean glass surfaces—which have no in-depth concentration gradient—for measuring the degree to which argon ion bombardment can distort a 'real' concentration gradient during the acquisition of in-depth sputter profiles with AES and SIMS; again, specific reference is made to the measurement of tin-oxide diffusion profiles in multicomponent silicate glasses.

(3) to determine the degree of atomic rearrangement at the 'ideal' surfaces of alkali-silicate glasses using the monolayer sensitive ISS technique; in the absence of adsorbable species, atomic rearrangement may be the only available mechanism for elimination of the high energy states associated with the dangling bonds created at an 'ideal' glass surface.

EXPERIMENTAL PROCEDURE

The silicate glasses were prepared using standard glass melting techniques. The raw batch materials were SiO_2, Na_2CO_3, K_2CO_3, $CaCO_3$, $MgCO_3$ and $Na_2SnO_3 \cdot 3H_2O$. The use of sodium stannate, rather than pure tin oxide, was necessary in order to homogeneously introduce the tin into the melt. The batch formulations for glasses containing tin-oxide were calculated on the assumption that the tin would maintain its initial +4 oxidation state throughout melting and annealing. They were melted in air using platinum crucibles at 1400°C to 1500°C. Each melt was stirred at least three times and then allowed to fine and further homogenize overnight. The melts were cast into disks and square rods, or drawn into cylindrical rods. The cast samples were annealed for one hour at 550°C and were then slow cooled in the furnace. All of the tin-oxide doped glasses were subjected to an independent compositional analysis. The tin-oxide concentration of the glasses did not deviate from the batch composition by more than .1%.

The glass rods were prepared for fracture by introducing a deep notch with a file. The notched glass rods were mounted into specially designed sample holders which gripped the rod just below the notch. The rods were fractured in the vacuum system for AES or ISS analysis via a sharp blow of the extended rod against the side of the stainless steel chamber. In all cases, the mirror region extended entirely across the fractured rod. Thus, extremely smooth surfaces were available for the analyses.

Some fracture surfaces were created in air. These surfaces were not cleaned or handled in any way. The fracture surfaces created in air for AES analysis were under vacuum within one hour of fracture. The fracture surfaces which were created in air for SIMS analyses required further preparation because the sample holder in the SIMS instrument accepts specimens of a limited height. It proved impossible to prepare the required short rods by fracture. Thus, a somewhat longer rod was prepared by fracture. The short rod for SIMS analysis was then prepared by slicing off the fractured end with a diamond microsaw. Each sample was cut in fresh ethanol. The saw is kept very clean and used exclusively to prepare glass specimens for SIMS analysis; separate blades are kept for each compositional series of glasses.

The Auger analyses were performed with a Physical Electronics Industries, Inc. High Resolution Electron Energy Analyzer (Model 15-25G). Through the use of a specially designed sample tray, all of the sample surfaces made a 20° angle with respect to the incident electron beam. This angle of incidence was necessary in order to alleviate unstable charging of the surface. It resulted in a stable surface potential which uniformly shifted all the Auger peaks down in energy by 5 to 10 eV. The electron beam was operated routinely at 3 KeV and .72 mA/cm^2; the total beam current was ~7 µA and the elliptical beam spot was ~1.0 mm^2 in size. The rather low beam current density is necessary for minimizing any electron beam induced damage to the sample surface under study [5]. The beam current density was set as low as possible, while at the same time maintaining sufficient signal to noise to observe the Sn(M_4NN) Auger transitions in the glasses with $\geq$ 1 wt.% SnO_2. It was verified that these conditions did not result in any reduction of SiO_2, as evidenced by the stability of the Si(LMM) Auger peak at 78 eV. Of course, even under these conditions it was not possible to accurately measure the sodium Auger peak intensity due to its rapid migration and desorption under the beam. The Auger peak intensities reported herein are the Si(LMM), Ca(KLL), Sn(M_4NN), O(KLL) and Si(KLL). The Auger sputter profiling was performed with a Physical Electronics Industries, Inc. Sputter Ion Gun (Model 04-191). An argon ion beam was rastered over an area approximately 6 mm in diameter at the sample surface; the beam current density at 3 KeV was about 19 µA/cm^2 at the sample surface. The sputtering rate for all silicate glasses was assumed equal to the 1.0 nm/min rate measured for SiO_2 on this instrument. This rate was obtained by measuring the time required to sputter through SiO_2 films of known thickness which had been thermally grown on silicon substrates.

The ISS analyses were performed using a 3M Ion Scattering Spectrometer (Model 520). All ISS analyses were performed using $^4He^+$. The ion gun was operated routinely at 1500 eV with a total current of 0.17 µA; the nominal spot size was about 1 mm. The sputtering rate under these conditions was assumed to be .10 nm/min. The sample surface made a 70° angle with the incident beam, while the scattering angle was fixed at 90°. The charging was readily alleviated through the use of a charge neutralization system. A filament placed near the sample surface supplied electrons for neutralizing the build-up of any positive surface charge. A feedback loop, based upon the current measured through the sample, controlled the flood of electrons and thereby stabilized the sample surface potential. In order to avoid contamination, the neutralizing filament was always outgassed prior to preparing or placing a glass surface under the ion scattering spectrometer.

The AES and ISS analyses were performed in the same analytical vacuum chamber. The system is evacuated using cryo-sorption and metal ion pumping. The base pressure after an overnight pumping cycle was of the order 10^{-6} Pa ($\sim 10^{-8}$ torr). During the creation of a fracture surface and during the AES/ISS analyses, a liquid-nitrogen cooled titanium sublimation pump was used to minimize the partial pressure of residual gases such as H_2O, O_2 and H_2. Assuming, nonetheless, that all of the residual gas is H_2O (the most reactive gaseous species for these alkali silicate surfaces), and further assuming that its sticking probability is unity, one calculates that a monolayer of contamination will form on the 'ideal' glass surfaces after 20 to 30 minutes. The 'ideal' glass surfaces created in vacuum were always analyzed within 5 minutes. Thus, although a lower total base pressure is clearly desirable, conditions used here are adequate for these particular studies.

The SIMS analyses were performed using a Gatan Inc. SIPS-SIMS Scanning Ion Microprobe (Model 591C). This instrument routinely utilizes an ~ 50 µm diameter, 7 KeV argon ion beam. The primary ion beam passes through a velocity filter tuned for $^{40}Ar^+$. In addition, the beam is electrostatically deflected just prior to impinging upon the specimen in order to reject neutral species. A binocular microscope permits direct examination of the spot size and shape (using anodic Ta_2O_5 films), as well as of the samples during the analysis. The beam current can be conveniently measured using a Faraday cup. In this work, a .40 µA beam was rastered over an 850 x 850 µm area with 91% of the signal electronically gated off to reduce edge effects. The sputtering rate for a commercial soda-lime silicate glass under these conditions was found to be 18.0 nm/min. The sputtering rate was obtained by measuring crater depths with a profilometer. A range of crater depths were produced in a piece of commercial plate glass by sputtering for various times. The measured crater depths were plotted against the sputtering time and the slope of a least squares fit provided the 18.0 nm/min sputter rate.

The SIMS instrument utilizes an electrostatic energy filter and a quadrupole mass spectrometer for filtering and detection of the secondary ions. In order to successfully detect the $^{120}Sn^+$ isotope, it was necessary to use an rf tuning circuit whose minimum detectable mass is 3 amu; thus, it was not possible to analyze for $^1H^+$. In this work, a computerized digital data acquisition system controlled the spectrometer and thereby could repetitively measure the count rates every 10 seconds for $^{16}O^+$, $^{23}Na^+$, $^{26}Mg^+$, $^{28}Si^+$, $^{42}Ca^+$, $^{54}Fe^+$ and $^{120}Sn^+$. The band-pass of the energy filter and the resolution of the quadrupole were adjusted to maximize the $^{120}Sn^+$ count rate, and at the same time, maintain an $\sim$1 amu resolution.

An especially critical and difficult aspect of these SIMS analyses concerns charge neutralization. Here, a 1.5 KeV electron beam was directed, simultaneously, at the ion bombarded region. Since it is impossible to predict the conditions required for charge neutralization [6,7], the procedure is necessarily empirical. Thus, the neutralizing electron beam current and beam current density are adjusted to maximize the $^{28}Si^+$ secondary ion signal. The sample is then translated slightly to provide a new area of the sample surface for analysis. Most importantly, the charge neutralization is checked periodically during the acquisition of depth profiles. In the event that the secondary ion signals were not at their maximum values, the analysis is terminated and re-initiated on a new area of the specimen. While there is no reason, a priori, to believe that maximization of the signals is an indication of complete charge neutralization, it is the only procedure that has provided any measure of success. In this regard, it is significant that these maximized signals are reproducible to within 5% from sample to sample of the same composition. In fact, at least one glass standard (i.e., a fracture surface created in air) is now included with each loading of 'real' glass surfaces in order to verify the proper set-up and operation of the instrument. This procedure constitutes one of the practical outcomes of this study of 'ideal' glass surfaces.

AES AND SIMS STUDIES OF TIN-OXIDE DOPED SILICATE GLASSES

The chemical characterization of commercial float glass surfaces has been attempted by a number of investigators. Of

particular interest is the concentration profile of tin-oxide on
the 'tin-bath side' of the plate glass. The profile is determined
by the contact time between the glass ribbon and molten tin, by
the temperature gradient through the tin bath, and perhaps most
importantly, by the amount of oxygen dissolved in the molten tin.
It is probably influenced by the annealing process as well,
although this has not been verified. In general, the presence of
tin improves the scratch resistance and chemical durability of the
tin-bath side of the plate glass. In some instances, however, the
tin penetration can be detrimental. For example, the tin-side may
'bloom' during re-processing of the glass; that is, a white haze
will form on the tin-bath side when the float glass is re-heated
for thermal tempering or sagging. The propensity for 'tin bloom'
to occur is more related to the oxidation state and in-depth
concentration profile of the tin, than it is to the total tin
concentration in the surface; the details of this phenomena have
not been reported. It has also been observed that the chemical
durability of the tin-bath side can be drastically degraded under
some circumstances. An almost 'instant weathering' of the tin-bath
side of the glass plate can occur; again, a surface chemical
description of this phenomena has not been reported. These are but
two reasons for measuring, and understanding, the in-depth
concentration profile for tin-oxide in commercial float glass.
There also exists, of course, an academic interest in the tin-glass
interfacial reaction, the diffusion mechanism, and their role
in the commercial process.

The tin concentration profiles in commercial float glass have
been measured by virtually every available method; Colombin et al.
[8], as well as Kishi [9], have attempted to synthesize the results
obtained with various analytical techniques and by different
investigators. Naturally, each investigator examines a different
sample of float glass, obtained from a different manufacturer, and
perhaps processed under very different conditions. Thus, it should
not be surprising that the results of each analysis are not
identical. Nonetheless, some general characteristics of the tin-
side surface emerge regardless of the source of the float glass.
The outermost 10 to 100 nm of the tin-side surface exhibits an
exceedingly steep concentration profile for tin wherein the tin
concentration is anomalously high. Characterization of these
outermost 10-100 nm required the use of surface analysis techniques
such as AES [10,11], XPS [8,12,13], SIMS [14], and SIPS [15]. The
tin-oxide concentration was estimated from XPS [8,12] data to be of
the order 30 weight percent; these quantitative analyses were made
by correcting the measured photoelectron intensities with
calculated values of the photoionization cross-sections and
relative escape depths. The only other seemingly universal
observation is a very shallow tin concentration gradient extending
10-50 microns into the surface wherein the tin concentration is

significantly lower than in the outermost 10-100 nm. This region of the tin-side surface was analyzed with RBS [3,8] and electron microprobe analysis (EMPA) [16]; these analyses suggest tin-oxide concentrations of the order 5 wt.% and 2 wt.%, respectively, in the outermost 2 microns. The third important feature observed on the tin-side surface is a 'hump' in the tin concentration profile. While this anomalous hump has been of great academic interest, its existence and characteristics are not well founded. The hump has been observed anywhere from .5 to 10 µm below the geometric surface. In many instances, it is not observed at all, although this may be due to the limited depths examined in most published studies.

It is important to note that no one surface analysis technique can provide a complete measurement of the tin concentration profile in this 'real' surface. AES and XPS are very sensitive to the outermost .5-5.0 nm, but are impractical for in-depth profiling to any great extent. Conversely, SIMS and SIPS can provide a profile to depths of the order 1 µm but are sensitive to the outermost 0.5-10.0 nm. EMPA averages the composition in cross-section over 1.0-2.0 µm regions, and can measure the profile to any extent. RBS is perhaps best suited for quantitative analysis of the transition between the sharp near-surface gradient and the shallow sub-surface gradient; however, RBS is not especially sensitive to the outermost 5.0-10.0 nm. One should also be aware that using RBS, the tin profile deduced from the measured energy spectra requires some assumptions concerning the in-depth density gradient and the profiles of Ca, Na, Mg and Si [3]. It is also noteworthy that none of the ten or more surface analytical studies of commercial float glass have been able to measure the in-depth distribution of oxidation states for tin in the surface.

Of particular interest to us are the mechanisms associated with the tin penetration at elevated temperatures. Thus, laboratory diffusion specimens were prepared to examine the temperature and time dependence of the tin-oxide concentration profile in the outermost 20 to 200.0 nanometers. The complementary AES and SIMS techniques were utilized to provide a good measure of the tin-oxide concentration in the outermost 2.0 nm (via AES) and the in-depth profile through the outermost 500 nm (via SIMS). The results of these diffusion and interfacial reaction studies are presented elsewhere [17]. Here, attention is focussed upon the methods used to characterize the resulting surface concentration of tin-oxide and its concentration profile. Since the formation of an 'interphase' containing tin oxide is one explanation for the steep tin gradient observed at the surface of commercial float glass, a quantitative analysis was desired. A series of tin-oxide doped glasses with the commercial float glass composition were used to prepare clean glass surfaces for signal calibration. The use of

standards for signal calibration, rather than corrections of the
signal based upon published or calculated ionization cross-sections,
sputter ion yields, elemental sensitivity factors, etc., eliminates
the possibility of matrix effects. Most importantly, though, these
same clean surfaces can be used to investigate sputtering
artifacts which can distort the measured in-depth concentration
profiles. The importance of an accurate profile determination
relates to another hypothesis for explaining the sharp gradient
in the outermost 10-100 nm of commercial float glass based upon
diffusion control. The sharp transition in the tin profile within
the outer 10.0 to 100.0 nm cannot be associated with a simple
diffusion mechanism. The complexity of the profile is undoubtedly
related to the continuous temperature change during float glass
production, a compositional or structural modification in the
outermost 10 to 100 nm due to the tin penetration, a change in the
diffusion mechanism at temperatures below T_g, or some consequence
of multicomponent diffusion. In any case, it is necessary to make
an accurate measurement of the diffusion profile and its time and
temperature dependence under isothermal conditions. Thus, the
degree of distortion due to the sputter profiling analysis, itself,
must be determined.

It is especially important to point out that virtually all
reported surface analyses of 'real' glass surfaces with AES have
noted so-called electron beam damage [5]; that is, an electron
beam induced change in the composition or structure of the glass
surface during the analysis. The most notable effects involve
alkali-ion migration, electron stimulated desorption of halides,
water, oxygen, alkalis, and carbonaceous species, and finally,
electron beam heating. These authors are well aware of the
existence and consequences of these effects, and the methods
required to minimize them. These effects have already been
extensively reported for both real and ideal glass surfaces
[1,5,10,18-20]. Thus, they are not specifically addressed here,
except insofar as they affect the quantitative surface and in-depth
analysis of tin-oxide in these silicate glasses.

Calibration Curves

The calibration curve in Fig. 1 shows the dependence of the
$Sn(M_4NN)$ to $O(KLL)$ Auger signal ratio upon the bulk Sn to O atomic
ratio. The $Sn(M_4NN)$ signal has been normalized to the $O(KLL)$
signal for two reasons. First, this internal normalization
eliminates any signal variations due to unknown or unavoidable
fluctuations in data acquisition parameters such as electron beam
current, amplifier gains, scan rates, etc.; these instrumental
variables influence all the Auger peak intensities to the same
degree. Those instrumental variables which have selective effects
upon the Auger peak intensities are relatively easy to control and

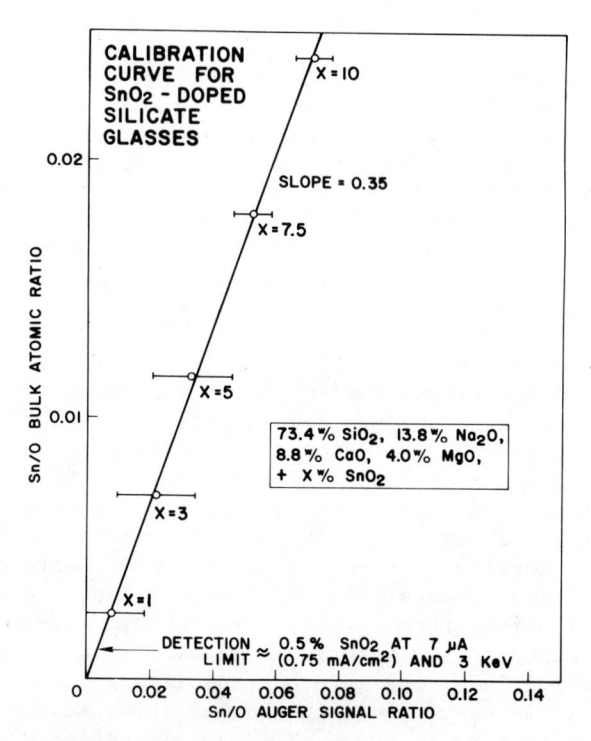

Fig. 1 The relationship between the bulk atomic ratio of Sn to O and the Auger signal ratio of Sn(M₄NN) to O(KLL); Auger signals obtained from standard glass fracture surfaces created in air and in vacuum.

thereby to keep constant; these include electron multiplier gain, energy resolution of the analyzer and electron beam energy. The second reason for normalizing to oxygen is that the atomic oxygen concentration in this series of glasses, and in most commercial silicate glasses, varies by less than 5%. Thus, one can expect the oxygen to provide an excellent internal standard for compositional analyses.

The scatter in the data originates, largely, from three related sources. First, low electron beam currents were used to minimize any electron beam induced modification of the surface. This reduces the signal to noise in the spectra and makes a precise measure of the Sn(M_4NN) signal, in particular, very difficult. Second, there was a measurable, albeit small, time-dependence of the Sn to O Auger signal ratio due to electron beam damage in spite of the low electron beam currents utilized (see below). Even though these data were obtained with less than 1.5 minutes of electron beam bombardment, any differences in the time dependence of the beam effect from sample to sample can introduce scatter. Third, the Auger data reported here were recorded with analog signal processing. It is our belief that through the use of digital data acquisition and signal averaging, the precision of the data can be improved significantly and the time required to record it can be reduced. Thus, even under the low signal to noise conditions, and in the presence of some electron beam damage, the scatter can, in principle, be considerably better than the data in Fig. 1 might suggest.

It is significant that the measured ratios did not depend upon whether the fracture surface was created in air or in vacuum. The fracture surfaces created in air seldom exhibited carbon or any other contamination; any adsorbed water on these surfaces was probably desorbed instantaneously via electron stimulated desorption (ESD) [5]. Thus, it is to be expected, at least for these glasses, that the fracture surfaces created in air and in vacuum exhibit comparable Sn to O Auger signal ratios. This is rather fortunate since the 'real' glass surfaces of interest will, of necessity, have been exposed to air. These results indicate that adsorbed water which is present on virtually all alkali silicate surfaces does not complicate the AES analysis. This is, perhaps, one situation where ESD is of benefit. One should be aware, though, that the presence of carbon contamination can significantly complicate the quantitative analysis. The results presented in the next section indicate that ion beam cleaning of contaminated surfaces can, in this case, further complicate the analysis.

Auger analysis of both the commercial float glasses, and the diffusion samples prepared in the laboratory, exhibited Sn to O

Auger signal ratios of the order .20 to .30. These are clearly
outside the range of the calibration provided by Fig. 1.
Unfortunately, it is not possible to make homogeneous glasses in
this system with more than 10-12.5 wt.% SnO_2. Thus, it becomes
necessary to consider an extrapolation to higher Sn/O ratios. In
an effort to establish the validity of the extrapolation, a single
crystal of naturally occurring SnO_2 (cassiterite) was cleaved in
air and the cleavage surface was analyzed. Figure 2 shows that the
extrapolation to pure SnO_2 is quite good. The slope of the
calibration curve in Fig. 1 is .35 with a correlation coefficient
greater than .999 while the slope in Fig. 2 is .32, also with a
correlation coefficient greater than .999 (using mean values only).

It is, perhaps, rather surprising that the calibration is
linear over the entire concentration range. However, the $Sn(M_4NN)$
and $O(KLL)$ Auger electrons have comparable kinetic energies. This
means that any influence of the matrix composition upon escape
depths will be nearly equal for the Sn and O Auger electrons.
Thus, a quantitative analysis using Sn/O is not complicated by the
escape depth parameter. Of course, the sampling volume for
quantitative analysis will be different if these escape depths are
altered significantly with the matrix composition. One parameter
that is expected to change with tin-oxide concentration is the
backscattering factor. Nonetheless, the data of Figs. 1 and 2
reveal that the combined effect of tin concentration, and the
associated increase in the backscattering function, is linear for
these oxides.

It is possible that matrix effects, due to the additions of
tin-oxide in the base glass composition, can influence the relative
signal intensities from the other glass components. Figure 3 shows
that the Ca/O Auger signal decreases with SnO_2 additions even
though the Ca/O bulk atomic ratio is effectively constant. There
is alot of scatter in the data, but the trend is evident; moreover,
the trend was present regardless of whether the fracture surface
was created in air or in vacuum. It is unlikely that the increases
in backscattering factor due to SnO_2 additions would influence the
ratio of Auger intensities. It is possible that the decrease in
the ratio is due to disproportionate changes in the Ca and O Auger
electron escape depths due to the SnO_2 additions; if so, it means
that the Ca(KLL) escape depth has decreased by more than the O(KLL)
escape depth with increasing SnO_2. This is inconsistent, though,
with the theory of inelastic mean free paths in inorganic solids
[21], and in any case, would not be of sufficient magnitude to
account for the observed change in the Ca/O Auger signal ratio. An
alternate interpretation is based upon the fundamental nature of
these surfaces and not with matrix effects in the analysis. Namely,
that the calcium ions have a tendency to relax upon creation of the

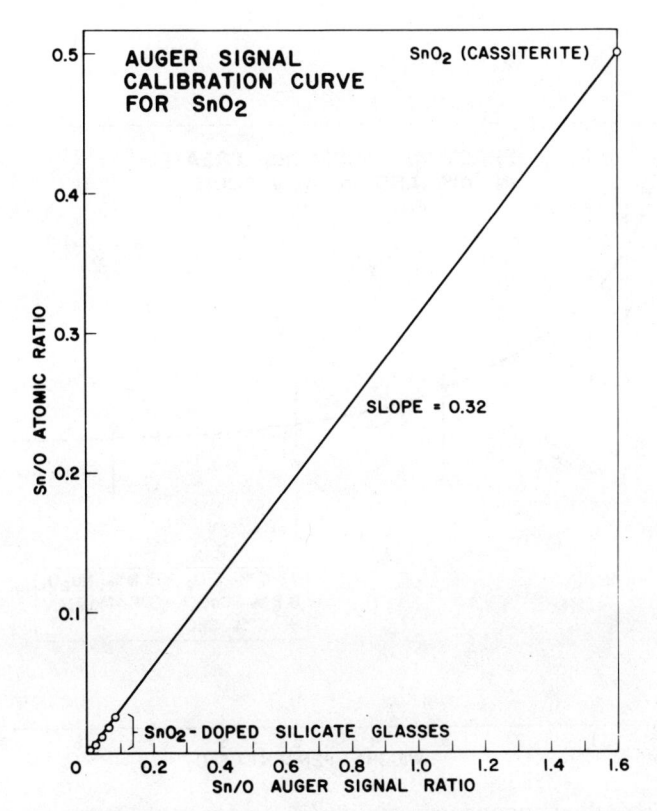

Fig. 2 The relationship between the bulk atomic ratio of Sn to O and the Auger signal ratio of Sn(M₄NN) to O(KLL); Auger signals obtained from the standard glass fracture surfaces and a cleavage surface of a natural cassiterite crystal (SnO_2).

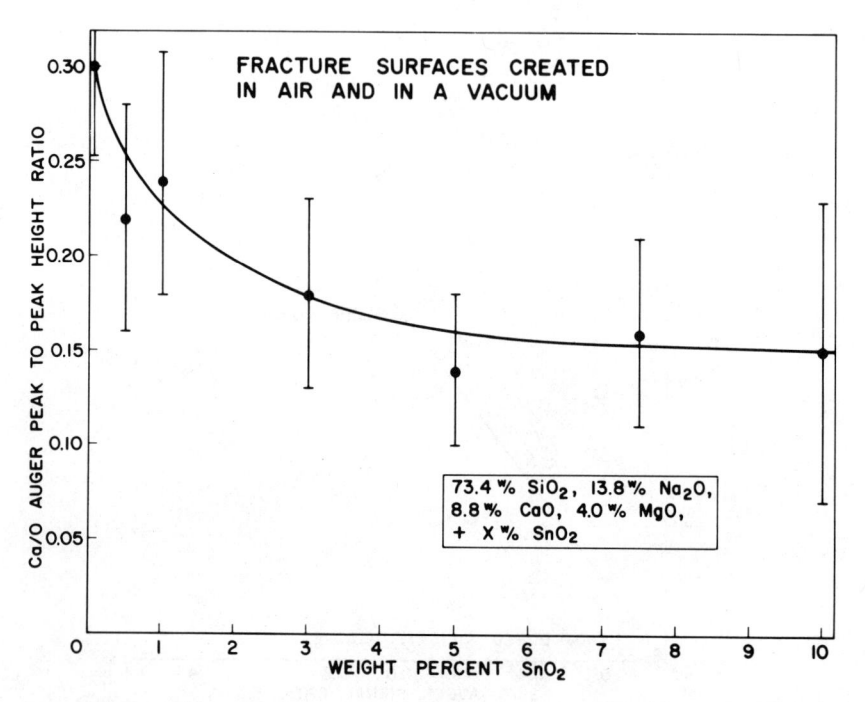

Fig. 3 The dependence of the Ca(KLL) to O(KLL) Auger signal ratio upon the SnO_2 content of standard glasses.

fracture surface, and thereby change their atomic position within the outermost monolayer of silicate tetrahedra. Since the Ca Auger electron has a relatively short escape depth, this atomic rearrangement is readily detected in spite of its confinement to the outermost monolayer of tetrahedra. Thus, Fig. 3 indicates that the addition of SnO_2 reduces the tendency for calcium ions to relax towards the surface. This phenomena is discussed further in the latter part of this paper where monolayer sensitive ISS analyses are presented to further substantiate the atomic rearrangement effect at 'ideal' surfaces of alkali silicate glasses.

The calibration curve in Fig. 4 was obtained from a SIMS analysis of the series of tin-oxide doped silicate glasses. These fracture surfaces were created in air because it was not possible to fracture the glass rods in the SIMS system. Thus, the $^{120}Sn^+$ to $^{16}O^+$ secondary ion signal ratios in Fig. 4 were measured after sputter removal of more than 50.0 to 100 nm of material, at which point all the secondary ion signals were constant with time. It is doubtful, even with 'ideal' surfaces, that the secondary ion signals would not show a transient at the onset of the SIMS analysis. This is due to the required attainment of the so-called equilibrium sputtered surface composition. This concept is discussed further in the section on sputtering effects; here it is emphasized only that the data in Fig. 4 were obtained after stabilization of all the secondary ion signals.

Although there is considerable scatter in the data, a linear dependence of the secondary ion signal ratios upon the bulk atomic ratios is evident. The acquisition of a standard calibration curve is most important for quantitative SIMS analysis. The matrix effects in SIMS are difficult, if not impossible, to handle with first principles corrections such as are used in AES and XPS. Here, the secondary ion signal for the species of interest is calibrated within the appropriate matrix. The only point of concern here is with regard to the possible effect of the tin oxidation state. In the commercial float glasses and the laboratory diffusion specimens, the equilibria may be shifted towards SnO; the standard glasses are expected to be largely SnO_2 since they were melted in air with SnO_2 batch additions [22].

The scatter in Fig. 4 is not due to any shortcomings in recording the data since a computerized digital data acquisition system has been used. It is more likely due to the very severe charging problems encountered, in general, during SIMS analyses of silicate glasses. It has already been pointed out that an electron beam neutralization system is used to minimize the charging. In all cases, the system is adjusted until the expected count rate for $^{28}Si^+$ is achieved. Nonetheless, the $^{28}Si^+$ count rate is considerably higher than that of the $^{120}Sn^+$ or $^{16}O^+$. Thus, undetectable differences in the $^{28}Si^+$ count rate due to charging

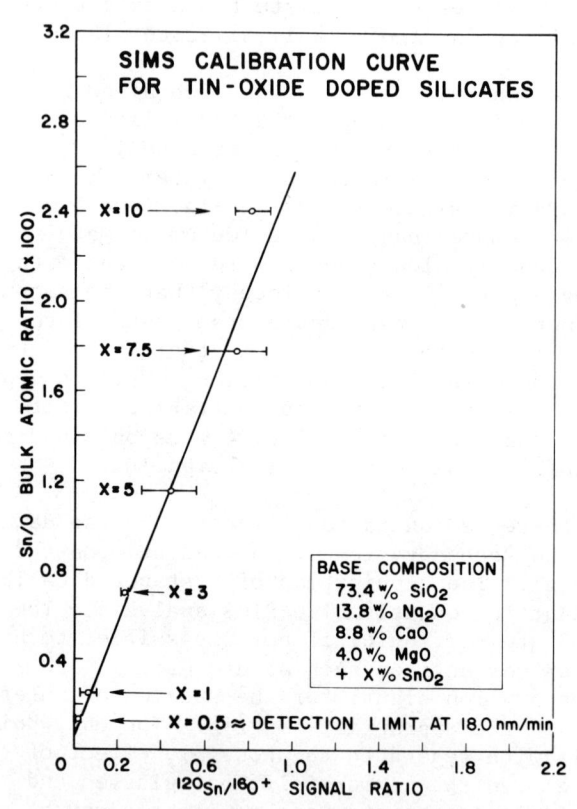

Fig. 4 The relationship between the bulk atomic ratio of Sn to O and the secondary ion intensity ratio of $^{120}Sn^+$ to $^{16}O^+$; secondary ion intensities obtained from standard glass fracture surfaces created in air.

can have measurable effects upon the $^{120}Sn^+$ and $^{16}O^+$. Most importantly, the charging is expected to influence these latter two secondary ions in a disproportionate fashion. Considering the difference in mass between Sn and O, these secondary ions can have very different kinetic energies. Thus, the $^{120}Sn^+$ to $^{16}O^+$ signal ratio can be very sensitive to minor fluctuations in the surface potential from sample to sample or from analysis to analysis on the same sample. Of course, any effects of the electron neutralizing beam itself, upon the surface under study, can also contribute to the scatter. Similarly, inhomogeneities in the bulk glasses may be further scattering the data, particularly since the standard deviations are greater in the glasses with the highest concentrations of SnO_2.

In view of the severe matrix effects associated with SIMS analysis, one is hesitant to attempt an extension of the calibration curve in Fig. 4 with anything other than SnO_2-doped silicate glasses. However, it has already been pointed out that the silicate glasses could not be prepared with greater than 10-12.5 wt.% SnO_2. Thus, a cleavage surface obtained from the single crystal of cassiterite was also examined by SIMS. The measured $^{120}Sn^+$ to $^{16}O^+$ secondary ion signal ratio was approximately equal to 10.0. A best fit line through the points in Fig. 4 gives a slope of .026 and a correlation coefficient greater than .999. A best fit line obtained by including the cassiterite data gives a slope of .051 and a correlation coefficient of .99. It is likely that this non-linearity is due to the use of the crystalline SnO_2 standard and might not occur in the glass matrix; however, these data do not permit a definitive conclusion. Thus, any extrapolation of the SIMS calibration curve in Fig. 4 is questionable.

Sputtering Effects

Figure 5 shows the changes observed in the Sn/O Auger signal ratio due to electron beam exposure (upper curve) and to 3 KeV argon ion sputtering (lower curve). Although the electron beam effect is not specifically addressed in this paper, it is nonetheless important to show the degree to which it influences the measured Sn/O Auger signal ratio. After all, electron beam exposure is required for Auger analysis of the argon sputtered surface. Thus, the upper curve in Fig. 5 was measured independent of the lower curve (in Fig. 5) by fracturing a glass rod in vacuum. The upper curve shows that even at these very low beam current densities, there is some electron beam induced change in the surface composition. Chappel and Stoddard [10] observed a much more dramatic electron beam effect upon their analysis of commercial float glass and a SnO_2-doped silicate glass. They used a considerably higher electron beam current density (~ 11 mA/cm^2) than the one used here (.72 mA/cm^2); they attributed the effect to an electric field induced migration of tin away from the surface.

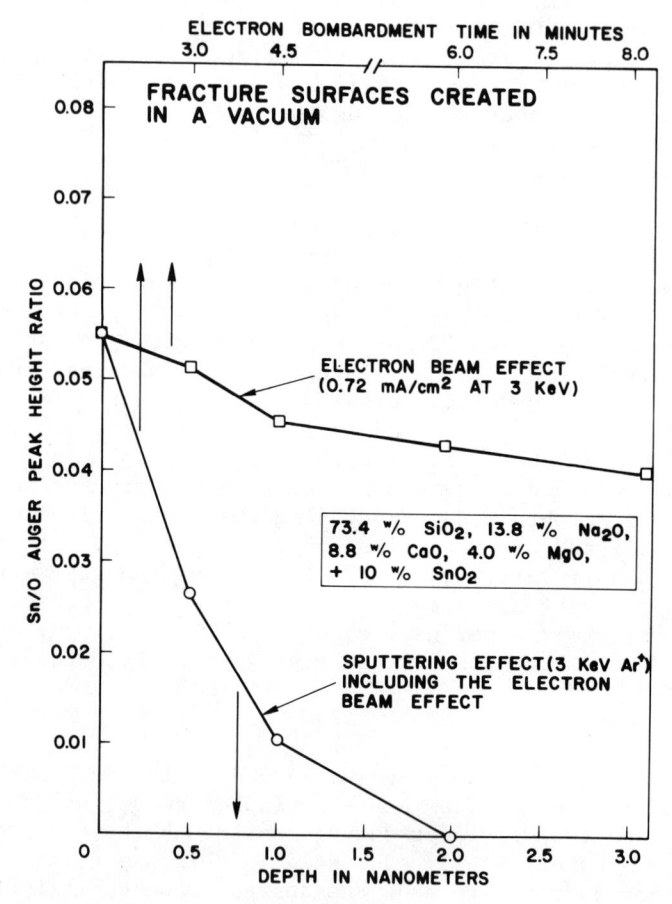

Fig. 5 The dependence of the Sn(M_4NN) to O(KLL) Auger signal ratio upon electron bombardment time and sputter etching.

The lower curve in Fig. 5 shows that a much more dramatic change occurs in the Sn/O Auger signal ratio with argon ion bombardment. It is emphasized that this is an 'ideal' glass surface; thus, it should not exhibit any in-depth concentration profile. The obvious conclusion here is that argon ion bombardment has significantly distorted the true in-depth concentration profile for tin. The data shown by the lower curve in Fig. 5 were obtained with minimal exposure to the electron beam. Moreover, the surface was not subjected to simultaneous electron and argon ion bombardment in order to eliminate interactive effects. The electron beam required for Auger analysis of the sputtered surface was turned on intermittently between each sputtering increment. The lower curve in Fig. 5 has been normalized to the electron bombardment time scale used for the upper curve in Fig. 5. Thus, direct comparison of the two curves reveals that most of the distortion in the true concentration profile is due to the argon ion bombsrdment. This means that even in a sputter profile analysis with XPS, where the electron beam effects are eliminated, the true in-depth profile will be subject to a significant distortion. The consequences of this effect upon SIMS depth profiles are discussed later.

Figure 6 shows the composition dependence of the ion-bombardment induced distortion of the tin-oxide concentration profiles. It is apparent that the Sn/O Auger signal ratio is reduced to below the detection limit in all of the SnO_2-doped silicate glasses. Although the exact shape of the profile varies somewhat from sample to sample, the Sn/O Auger ratio always goes to zero within 2-5 minutes of sputtering. No systematic difference in the profiles was observed regardless of whether the fracture surfaces were created in air or in vacuum.

Figure 7 shows the energy dependence of this ion bombardment-induced distortion on fragmentation of SnO_2; that is, a preferential removal of Sn relative to O within the sampling volume of the AES analysis. The sputter induced distortion was measured using 1 KeV, 3 KeV and 5 KeV argon ion bombardment. The data are plotted on a normalized time scale which accounts for the changes in ion current density (and therefore sputtering rate) that occur when the ion gun voltage is changed. Within the observed precision of these measurements, only the 1 KeV sputtering condition has any detectable effect upon the phenomena. This may be due to the fact that 1 KeV argon ion bombardment is insufficient to fragment the SnO_2 throughout the sampling volume of the Auger analysis. Thus, only those SnO_2 molecules within the outermost 1.0 nm are fragmented by 1 KeV argon ions, whereas the higher energy ions influence the SnO_2 to greater depths. Alternatively, the behavior of this glass under 1 KeV ion bombardment may be indicative of the chemical state of tin-oxide in these glasses. Kim and Winograd [23] reported that the SnO_2 oxide layer on polycrystalline tin-metal is stable to

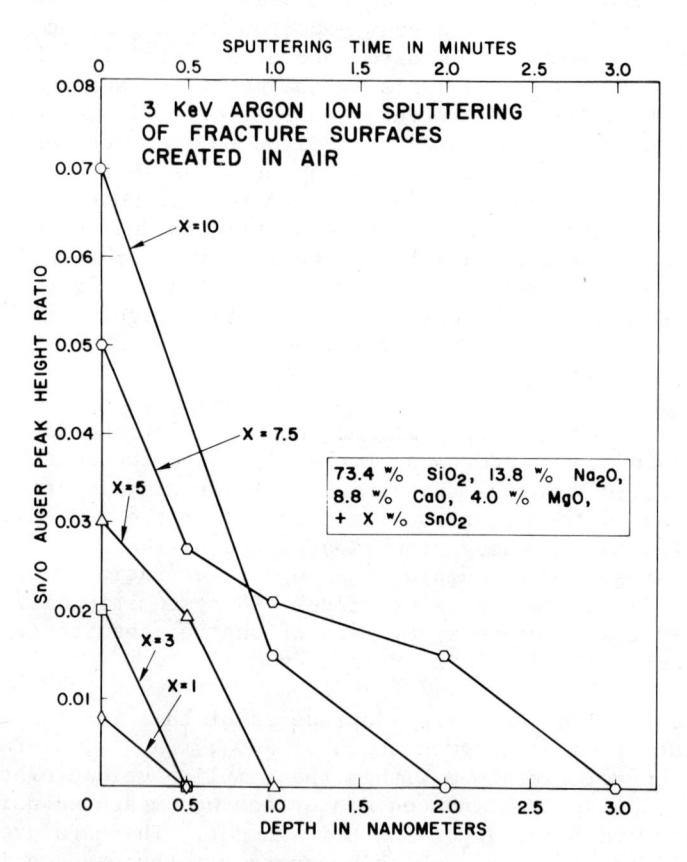

Fig. 6 The dependence of the Sn(M₄NN) to O(KLL) Auger signal ratio upon sputter etching for various glass compositions.

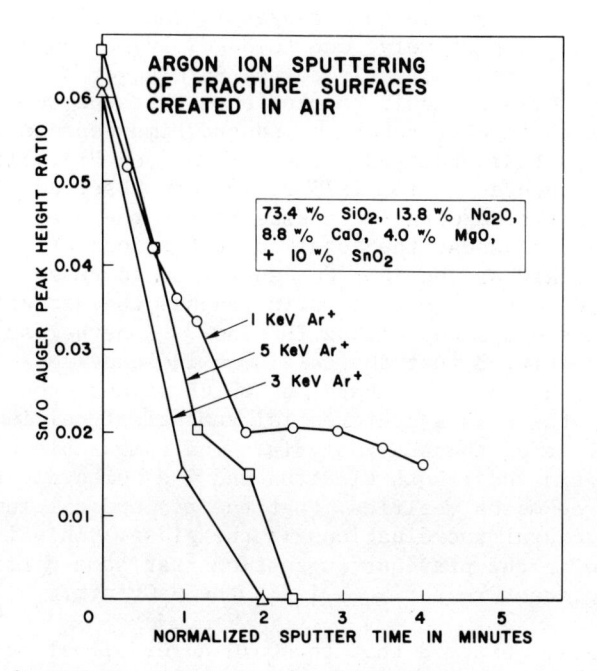

Fig. 7 The dependence of the Sn(M₄NN) to O(KLL) Auger signal ratio upon sputter etching at various ion energies.

1 KeV argon ion bombardment, but is subject to changes in its stoichiometry at energies in excess of 1 KeV. Thus, one interpretation of the data in Fig. 7 is that the tin-oxide in these glasses exists in both the SnO and SnO_2 forms. Under 1 KeV argon ion bombardment, the SnO may be fragmented, thereby changing the relative concentration of Sn and O in the sampling volume; however, the SnO_2 would be considerably more stable to stoichiometry changes. Under this hypothesis, Fig. 7 suggests a considerable percentage of the tin at these fracture surfaces is in the SnO form. Of course, at the 3 KeV and 5 KeV sputtering conditions, both SnO and SnO_2 undergo stoichiometry changes.

Figures 8 and 9 show the changes in the Ca/O and Si/O Auger signal ratios, respectively, due to 3 KeV argon ion bombardment; i.e., they correspond directly to the Sn/O Auger signal ratios shown in Fig. 6. It should be mentioned that the Ca/O Auger signal ratio exhibited an electron beam induced time-dependence to the same degree as that observed for the Sn/O (see Fig. 5); no electron beam induced changes in the Si/O ratio were observed. Also, note that although the Ca/O ratio is essentially the same in all of these standard glasses, the total concentrations of Na_2O, CaO, MgO and SiO_2 decrease as the SnO_2 concentration increases. For this reason, the Ca/O Auger signal ratio reaches the detection limit in Fig. 8 for the glass containing 10% SnO_2. Nonetheless, it is apparent from Fig. 8 that the calcium-oxide undergoes a stoichiometry change due to ion bombardment which is analogous to the tin oxide; i.e., there is a preferential removal of calcium within the sampling volume of the AES analysis. The comparable behavior of calcium and tin under both electron and ion bombardment suggests, although by no means verifies, that the tin and calcium oxides have similar structural coordinations in the glass. This is also consistent with the previous suggestion that some portion of the tin at these fracture surfaces is in the SnO form.

It is also observed that the Ca/O Auger signal ratios increase, although not to their original values, upon sputter etching for ten or more minutes; i.e., the Ca/O profile shows a depletion or 'dip' between 1 and 10 nanometers (not shown in Fig. 8). This distortion is especially noteworthy because it has been observed and reported for sodium and calcium in numerous surface studies of 'real' glass surfaces. Usually, it is attributed to a chemical reaction at the glass surface. In fact, the depletion zone for sodium has been shown to be exaggerated due to a chemical reaction with water vapor and other reactive gases [1]. Nonetheless, the calcium depletion zones observed for these clean glass surfaces, and their relative independence of whether the surface was created in vacuum or in air, indicates that they are an inherent artifact of the ion bombardment process. Any attempt to attribute them to a chemical reaction must be verified by systematic studies of the chemical

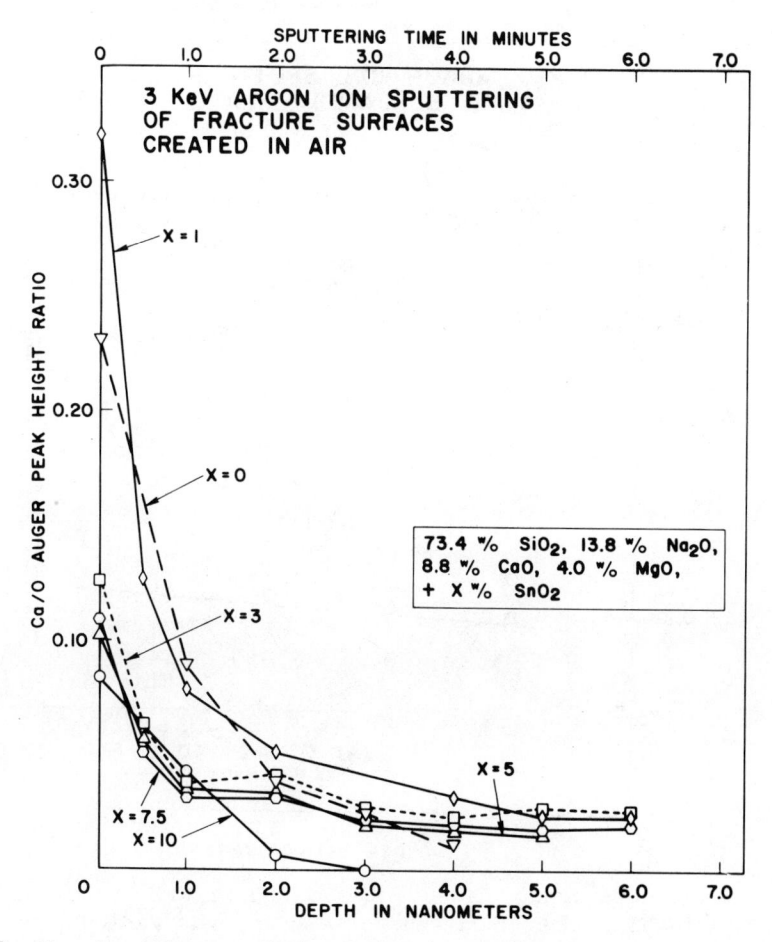

Fig. 8 The dependence of the Ca(KLL) to O(KLL) Auger signal ratio upon sputter etching for various glass compositions.

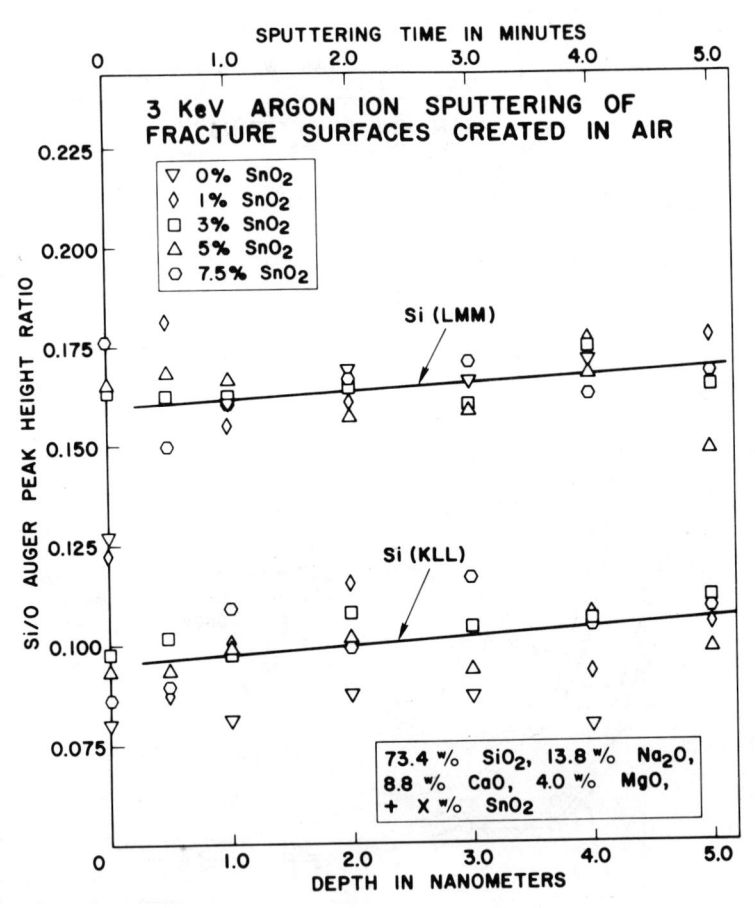

Fig. 9 The dependence of the Si(LMM) and Si(KLL) to O(KLL) Auger signal ratios upon sputter etching for various glass compositions.

reaction and also by a measure of the degree of the ion bombardment-induced depletion, and its extent, for an 'ideal' glass surface.

It is evident from Fig. 9 that the SiO_2 is hardly subject to these ion beam sputtering effects. The enhanced scatter in the data for the Si(KLL) curve is due to the poor signal to noise ratio obtained for that Auger peak intensity; this is not surprising because the Si(KLL) Auger transition is not especially prevalent under the 3 KeV electron beam excitation used here. The Si/O ratios are expected to be the same for all glass compositions; thus, the scatter in the data is random with respect to compositions. One exception to this is the Si(LMM)/O ratios observed for the X=0 and X=1 compositions; i.e., the initial ratios for these glasses are significantly lower than for the other compositions. It may be recalled that these same glass compositions exhibited statistically higher Ca/O Auger signal ratios at the surface; that observation was made whether the surfaces were created in air or in vacuum. Since the Si(LMM) Auger electron has an exceedingly small escape depth, it would be especially sensitive to the atomic arrangement in the outermost monolayer; i.e., this signal would be strongly attenuated by the presence of excess calcium ions on the surface. Thus, the hypothesis concerning a calcium ion relaxation at fracture surfaces, and its enhancement in glasses with little or no tin-oxide, is further substantiated.

Figure 10 shows the SIMS depth profiles obtained for a clean surface of the 7.5 wt.% SnO_2 glass standard. Although this surface is not quite as clean as the surfaces created in vacuum for the AES analyses, it is doubtful that the sample preparation has influenced the glass to the depths suggested by the $^{120}Sn^+$ and $^{42}Ca^+$ profiles. Thus, it is certain that the profiles for $^{120}Sn^+$, $^{42}Ca^+$, and to some extent, $^{26}Mg^+$, $^{16}O^+$ and $^{54}Fe^+$, are due to sputtering effects. However, the distortion of the expected flat concentration profiles appear less dramatic in these SIMS profiles than it did in the corresponding Auger sputter profiles. This is to be expected since AES analyzes the ion bombardment modified surface layers, whereas SIMS analyzes the flux of secondary ions ejected from the surface (see below). Nonetheless, it must be emphasized that these sputtering effects will have a significant influence upon a quantitative SIMS analysis of the SnO_2 concentration at the outermost atomic layers of a 'real' glass surface and within a steep concentration gradient. Similarly, the true representation of a short range diffusion profile (e.g., 1.0-100.0 nm) is questionable. However, they do not preclude quantitative SIMS analysis of SnO_2 in the bulk of a silicate glass or within a surface where the gradient is very shallow (i.e., extends over tens to hundreds of nanometers). The rationale behind these conclusions can be more easily explained using a simpler glass composition.

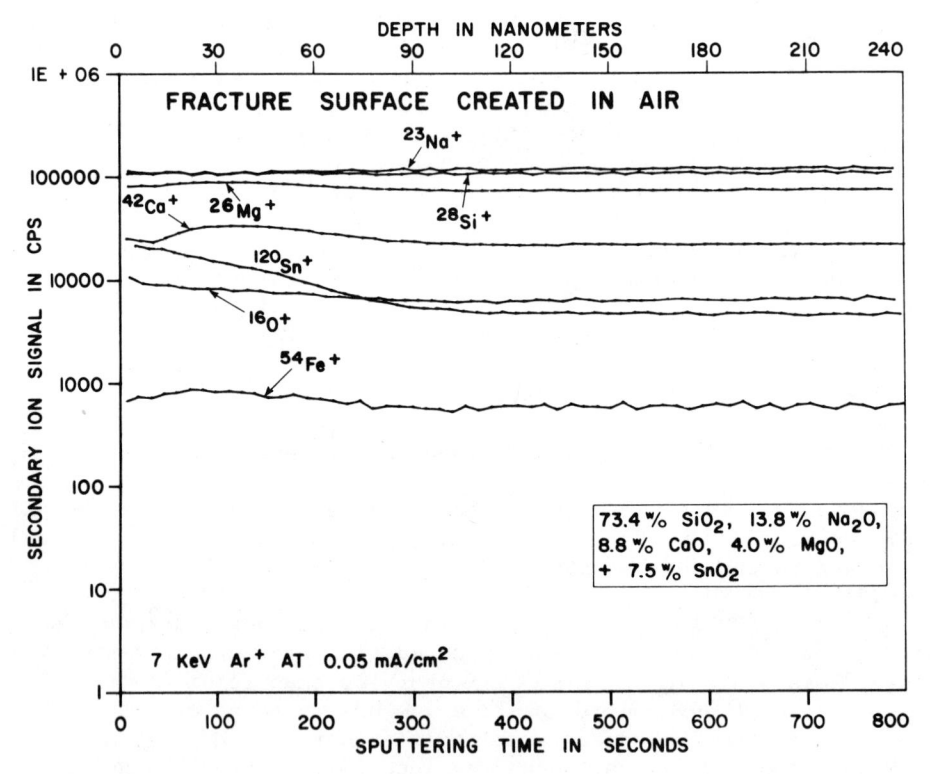

Fig. 10 The dependence of secondary ion intensities upon sputter etching for a multicomponent silicate glass.

Consider, for example, the 'ideal' surface of a two-component glass consisting of SiO_2 and SnO_2 where SnO_2 is the minor constitutent. Assume further that the sputter-induced changes in surface composition are due solely to preferential sputtering. That is, the sputtering yield of SnO_2 is greater than that of the SiO_2; a sputter rate for SnO_2 which is 4 times that of SiO_2 is a reasonable estimate. Thus, the initial flux of sputtered particles, and therefore the secondary ion signal, is enriched in SnO_2 relative to the true surface composition. At the same time, the outermost atomic layer(s) is very rapidly depleted in SnO_2. Eventually, the outermost atomic layers of the sputtered surface are sufficiently depleted in SnO_2 to compensate for the enhanced sputter yield of the SnO_2; i.e., the flux of sputtered particles is stoichiometric with the bulk composition of the material. At this point, the surface achieves the so-called steady-state (or equilibrium) sputtered surface composition. In the absence of other effects, it can be represented simply by:

$$\frac{C'_{SnO_2}}{C'_{SiO_2}} = \frac{C_{SnO_2}}{C_{SiO_2}} \cdot \frac{S_{SiO_2}}{S_{SnO_2}}$$

where C' is the steady state surface composition, C is the initial surface composition, and S is the sputtering rate. The approach to steady-state, again in the absence of other effects, will be exponential with time; the time constant for the process should be of the order S_{SiO_2}/S_{SnO_2} monolayers. In the present hypothetical case, the depletion of SnO_2, and corresponding reduction in the SIMS signal, should occur within 5.0 nm. (It may be worth noting that if the sputter rate of the minor constituent is less than that of the major constituent, the minor constituent will accumulate in the sputtered surface. Here, the approach to steady state would exhibit a somewhat longer time constant.)

This simplified model can also be used to interpret the ion bombardment induced surface composition changes exhibited in Auger sputter profiles. In this case, the first surface composition measured by AES—prior to any sputter etching—provides the most accurate measure of composition. Upon ion sputtering, the outermost atomic layers are rapidly depleted in SnO_2. Thus, the Auger spectra acquired as a function of sputtering time are a measure of the approach to the steady-state sputtered surface composition. The data in Fig. 5 suggests that the SnO_2 concentration is reduced to less than .5 wt.% within the ~3.0 nm sampling volume of the Auger analysis. The data in Fig. 8 suggests that the Ca concentration is also depleted significantly in the outermost monolayer. The fact that the depletion occurs within the expected 5.0 nm time constant is strong evidence for the existence of preferential sputtering effects in these multicomponent silicate glasses. However, preferential sputtering alone cannot account for the subsequent build-up of Ca with continued sputtering in the Auger and SIMS profiles, nor can it explain the fragmentation of CaO and SnO_2 (i.e., the removal of Ca and Sn preferential to O). Similarly, the distortions in the SIMS profiles of Fig. 10 extend to more than 100 nm; these too, suggest that other mechanisms in addition to preferential sputtering must be contributing to the sputtering effects observed overall.

Additional contributions to these sputtering effects include knock-on, diffusion, and especially in these insulating glasses, electromigration. Any attempt to further explain the detailed mechanisms of these sputtering effects would be beyond the scope of this presentation. This ion bombardment induced redistribution of atomic species—during the surface analysis—is, in general, a most complicated process. This is particularly true for these multi-component silicate systems which contain mobile ionic species. The reader is referred to excellent reviews of these phenomena by Kelly [24], Andersen [25] and also by Bach [26]; Coburn and Key have discussed these effects with more specific reference to composition profiling [27]. More recently, Smets and Lommen [28] commented on the origin of ion beam effects on 'ideal' glass surfaces on the basis of XPS measurements of the sputtered surface.

It should be obvious that a dynamic SIMS analysis cannot give an accurate measure of the composition at the outermost atomic layer(s) of a 'real' multicomponent glass surface. Whether or not it can accurately provide even a semi-quantitative in-depth analysis, for example a diffusion profile, depends upon the mechanisms responsible for the distortions. If preferential sputtering is the only mechanism contributing to the distortion, it should exhibit a short time constant. Thus, it could be distinguished, at least in principle, from the real profile. Unfortunately, the Sn profiles of interest with regard to float glass are within the time constant of preferential sputtering. Moreover, it has been shown that additional factors contribute to, and further complicate, the observed distortions in the SIMS profiles; specifically, the distortions shown in Fig. 10 extend over tens of nanometers and therefore can be easily confused with a real diffusion profile. The same words of caution apply to SIPS profiles of glass surfaces. Of course SIMS and SIPS can be used, in conjunction with a calibration curve such as the one presented in Fig. 4, to quantitatively measure the concentration of particular species within uniform films and bulk materials once the steady-state sputtered surface composition has been achieved.

In contrast, it is clear that AES, in conjunction with calibration curves of the sort presented in Figs. 1 and 2, can be used to provide a quantitative compositional analysis of the outermost atomic layers of 'ideal' and 'real' glass surfaces.

These will, of course, be subject to the well-known effects of electron beam damage. Auger sputter profiles, at least for minor species within a multicomponent silicate glass, are highly questionable. Since XPS is often utilized for glass surface analysis, it may be especially noteworthy that these same ion-bombardment induced effects can distort XPS sputter profiles.

These conclusions should not be taken to imply that AES, XPS, SIMS or SIPS profiles have no value for glass surface analyses. They can very successfully provide the composition profile of single component films or coatings applied to glass surfaces and of distinct reaction product layers at the surface. However, profiling the distribution of a minor constituent within a multicomponent glass surface must be approached with caution. The expected degree of distortion due to beam and sputtering effects should be established with clean glass surfaces. Subsequently, a systematic study of the surface chemical phenomenon of interest should be undertaken so that a meaningful description of the behavior of the 'real' glass surfaces can be obtained.

Altogether, the advantages of, and justification for, complementary surface analyses of the same samples is clearly indicated. Our combined use of AES for quantitative analysis of the tin oxide concentration in the outermost .50 to 5.0 nm of the surface, and SIMS for a quantitative analysis of the sub-surface tin oxide concentration, has provided a meaningful description of tin penetration in multicomponent silicate glasses [17]. Unfortunately, the diffusion coefficients deduced from the sharp, near-surface $^{120}Sn^+$ SIMS profiles are subject to further evaluation. Since the $^{120}Sn^+$ SIMS profiles generally observed in the laboratory diffusion specimens, and the commercial float glasses, are so distinct, it may be possible to develop correction methods for them based upon the degree of distortion exhibited by the 'ideal' surfaces.

ISS STUDIES OF ALKALI-SILICATE GLASSES

The atomic arrangement within the outermost atomic layers of a multicomponent silicate glass will have a profound influence upon physical and chemical surface phenomena. There is little doubt that accurate information about the surface structure—particularly the active sites for adsorption and the position of various modifier species with respect to the silicate tetrahedra and the geometric surface—can lead to a better understanding of the functionality and behavior of multicomponent silicate glass surfaces. Adsorption, catalysis, adhesion and friction are fundamentally dependent upon the atomic arrangement at glass surfaces; even the behavior of atomically sharp cracks in glass can depend upon the atomic arrangement at the crack-tip.

An 'ideal' silicate glass surface will possess a high concentration of reactive sites due to the presence of dangling oxygen and silicon bonds. Of course, exposure to the ambient atmosphere will lead to the rapid adsorption of water vapor. It is usually agreed that the water molecule dissociates to form silanol groups (SiOH) at the reactive $\equiv$Si-O$^-$ and $\equiv$Si$^+$ surface sites. In the absence of water vapor and/or other adsorbable species, however, the excess free energy associated with the 'ideal' surface may induce an atomic rearrangement within the outermost atomic layer(s). In the case of crystalline solids, atomic rearrangement, or reconstruction at 'ideal' surfaces has been observed in many non-metallic systems [29,30].

In contrast, 'real' glass surfaces are usually created in the presence of adsorbable species, and very often at elevated temperatures. Thus, there are a number of mechanisms through which a 'real' surface can be created with an exceedingly low excess free energy. These mechanisms include atomic, molecular and/or morphological rearrangement, Gibbsian segregation, adsorption, volatization and surface chemical reactions. Each of these phenomena exhibit strong temperature dependences. Since the majority of 'real' glass surfaces are created at elevated temperatures, and cooled to room temperature in various environments and at various rates, it is not surprising that the surface properties of multicomponent silicate glasses are so variable. The unique properties of fiber surfaces, for example, are surely the result of the high cooling rate and uniaxial stress under which these materials are fabricated. The mechanical creation of 'real' surfaces due to grinding, polishing, and fracture is equally complex. In this regard, the mechanochemical effects associated with the formation and propagation of atomically sharp cracks is, perhaps, the most challenging surface phenomena to explain.

Of particular interest to us is the atomic arrangement, or rearrangement, at the 'ideal' surfaces of alkali-silicate glasses, and its temperature dependence. Thus, ion scattering spectroscopy (ISS) is being used because of its unique sensitivity to the outermost monolayer. Of course, the concept of an 'atomic monolayer' has little meaning with regard to the tetrahedral coordination, network structure and interstitial modifiers associated with these multicomponent silicate glasses. It is for this very reason that the ISS spectra may provide a direct indication of the atomic arrangement at the surface, and in particular, the position of the alkali species relative to the geometric surface and the outermost layer of silicate tetrahedra. Here, attention is focused, specifically, upon the quantitative relationships between the ISS spectra and the bulk composition for a series of soda-silica and potassia-silica glasses. In addition, ternary and quaternary glasses of nearly a commercial composition were also examined.

Figure 11 shows the ISS spectra for SiO_2, $Na_2O \cdot 3SiO_2$ and $K_2O \cdot 3SiO_2$. All of these surfaces were created in vacuum and analyzed within 3-5 minutes. It is significant that relative to pure SiO_2, the oxygen signal is reduced by more than 50% in the soda-silica glass and is scarcely detected for the potassia-silica glass. The bulk atomic oxygen concentration of these alkali-trisilicate glasses is only 12% less than that of the pure silica. This provides strong evidence that the 'ideal' surface of these glasses is not simply a projection of the disordered bulk structure. Rather, the alkali-ions have assumed new positions within the outermost layer of silicate tetrahedra whereby they shield the terminal oxygens created by the fragmentation of the silicate network. It is important to emphasize that this atomic rearrangement need not involve any long range diffusion. At room temperature, at least, the rearrangement may require only a relaxation or jump of alkali species—already present in the outermost atomic layer(s)—but to a position which more effectively shields the outermost silicate tetrahedra. Whether this rearrangement occurs after the fracture event, or at the stressed surface of the advancing crack tip, is not yet known.

Figure 12 shows the dependence of this surface rearrangement upon bulk composition; it is represented in terms of the alkali ion to oxygen signal ratio. For the soda silica system, it is clear that the sodium to oxygen signal ratio is proportional to the bulk ratio, at least in this composition range. This appears to be true even in the case of the ternary and quaternary silicate glasses containing soda. However, the most significant observation is the fact that the calibration curve does not extrapolate through zero. Since the curve must pass through zero, but has a positive intercept in this case, it indicates that the sodium ions take a preferential position in the outermost atomic layer of these 'ideal' surfaces.

The potassia silica system exhibits a more complex behavior. A reduction in the bulk concentration of K_2O brings about an increase in its relative surface concentration. Nonetheless, the potassia silica calibration curve also shows a positive intercept. These observations indicate that the potassium is extremely surface active. The negative slope of the curve can be interpreted in one of two ways. It may imply that the rearrangement mechanism for these 'ideal' surfaces is impeded by an increased 'bulk' concentration of potassium. Alternatively, it can be due to the presence of a third surface active species, not detected in the ISS spectra, which is more favorably located in the surface at high K_2O concentrations in the bulk (or, the bulk concentration of the third component is increased at higher bulk K_2O concentrations). The most likely candidate for this additional surface active species is the proton, presumably derived from bulk or 'structural' water.

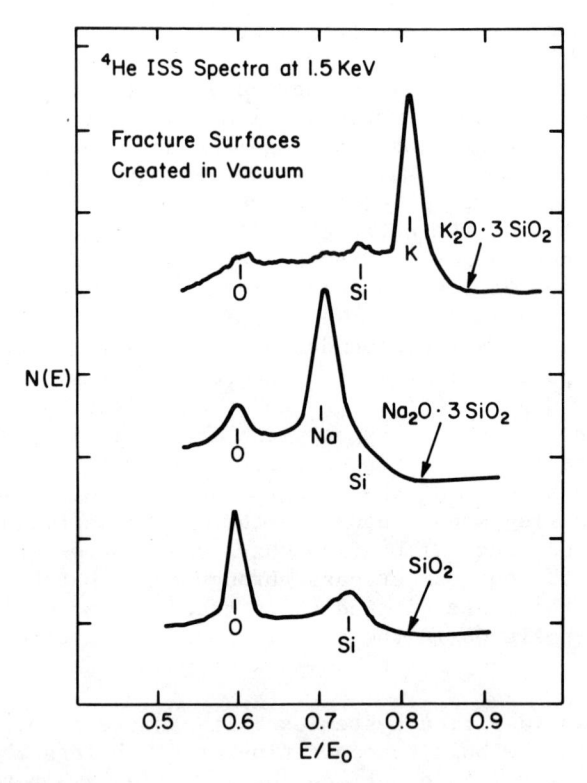

Fig. 11 The ISS spectra measured under identical conditions for three different glass compositions.

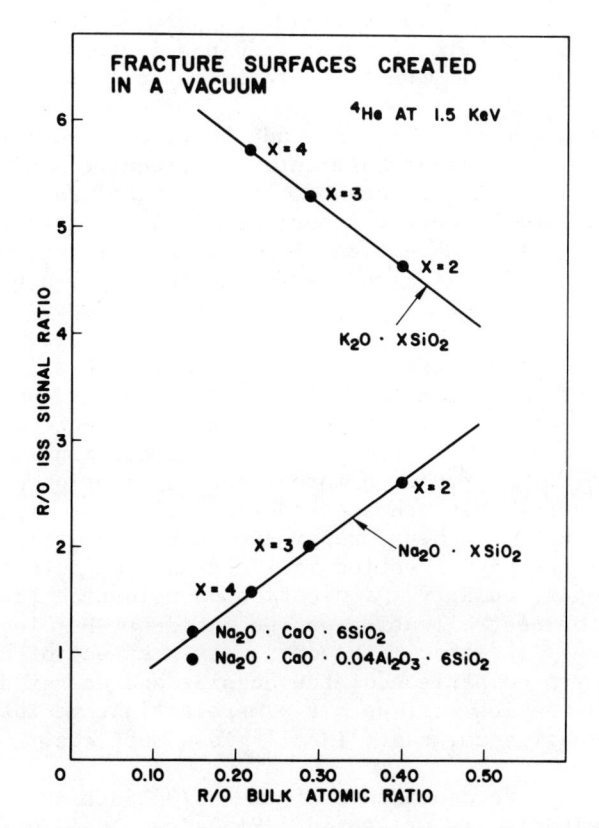

SURFACE STUDIES OF MULTICOMPONENT SILICATE GLASSES 33

Fig. 12 The relationship between the bulk atomic ratio of alkali to oxygen and the ratio of the alkali to oxygen ISS signals for various glass compositions (R = Na or K).

Unfortunately, hydrogen cannot be detected with conventional ISS analyses.

It may be especially significant that Lacharme et al. [31] also observed a local alkali-enrichment at the fracture surfaces of soda-silica and potassia-silica glasses. They fractured the glasses at a pressure of 2×10^{-8} Pa ($\approx 10^{-10}$ torr) and analyzed the surface composition using Auger electron spectroscopy; the Auger spectra were calibrated against a pure silica glass reference. Lacharme et al. interpreted their findings in terms of a local enrichment of alkali in the fracture propagation zone.

Of course, the observations made here, and the interpretation in terms of an atomic rearrangement, will require additional investigation. The dependence of the ISS signal ratios upon the angle of incidence between the $^4He^+$ beam and the fracture surface can provide a better indication of the atomic arrangement. In the event that these 'ideal' glass surfaces are simply a projection of the disordered bulk structure, no angular dependence of the signal ratios will be observed because shadowing and shielding effects will occur with equal probability at all relative orientations of the incident beam and sample surface. Any angular dependence observed, however, will further verify the existence of a preferred distribution of atomic species in the outermost monolayer, and at the same time, will provide a more direct indication of the atomic arrangement. Similarly, the dependence of the ISS signal ratios upon the incident $^4He^+$ beam energy should be examined; the energy dependence of the signal ratios will depend upon both the atomic arrangement—particularly anion-cation coordination [32]—and the probability for neutralization of the incident $^4He^+$ ions. Ion neutralization is, perhaps, the only matrix effect of concern in ISS analyses. A comparison of the angular and energy dependence of the ISS signal ratios will permit a more definitive interpretation of the atomic arrangement at 'ideal' glass surfaces.

In Fig. 13, the dependence of the Na/O ISS signal ratio upon ion bombardment time is presented. Since the $^4He^+$ ion bombardment required for ISS analysis is expected to sputter-etch the surface, albeit slowly, it is sometimes possible to acquire short-range depth profiles by sequential acquisition of ISS spectra. However, it has already been shown that these multicomponent silicate glasses are subject to preferential, and other, sputtering effects. Figure 13 demonstrates the preferential sputtering effect for a range of soda-silica and soda-calcia-silica glasses. The relative concentration of sodium in the sputtered surfaces is clearly independent of the bulk soda concentration. The effect is especially evident, and perhaps best studied, with ISS because the signal is sensitive to the outermost atomic layers. It is doubtful that anything quantitative can be obtained from ISS sputter profiles of multicomponent silicate glasses.

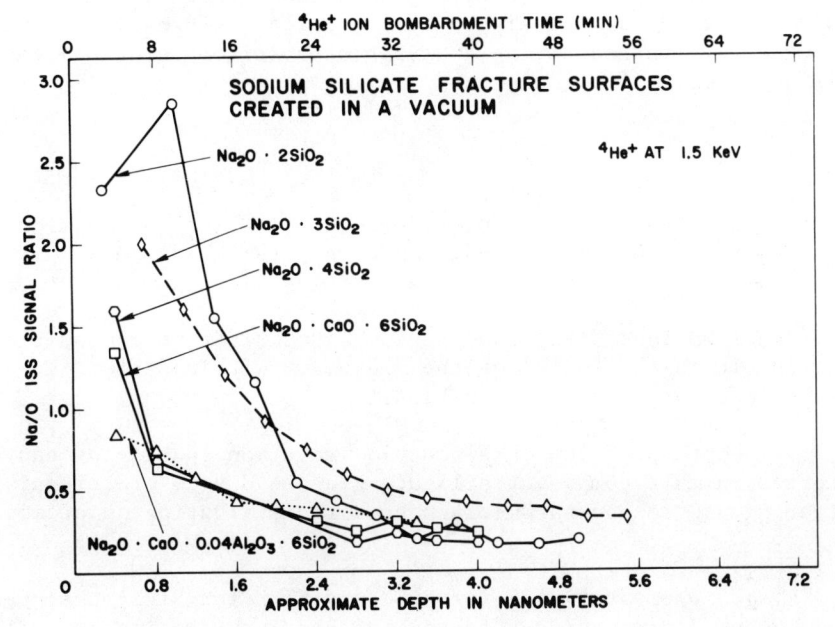

Fig. 13 The dependence of the ISS sodium to oxygen signal ratio upon ^{4}He$^+$ ion bombardment for various glass compositions.

It should be obvious that the acquisition of ISS spectra for multicomponent silicate glasses must be rapid in order to avoid this sputter-induced change in surface composition. The data presented in Figs. 11 and 12, for example, were recorded within five minutes of $^4He^+$ ion bombardment. Unfortunately, any subsequent attempt to follow changes in the atomic arrangement and/or surface composition of the 'ideal' surfaces due to elevated temperatures or exposure to reactive atmospheres, may require a new, unsputtered region of fracture surface (or a new specimen altogether). It is possible that elevated temperatures may well 'anneal-out' the sputter-induced surface composition change. On the other hand, adsorption phenomena will surely be influenced by the sputtered surface. These authors believe, nonetheless, that this inconvenience is far outweighed by the unique information provided by low energy ion scattering analysis of multicomponent silicate glass surfaces.

SUMMARY

(1) AES (and XPS) can provide a quantitative compositional analysis of the outermost .5-5.0 nm of a 'real' or 'ideal' glass surface.

(2) SIMS (and SIPS) can provide a quantitative compositional analysis within the 'bulk' of the glass, or within surface layers and coatings of uniform composition.

(3) Depth profiles obtained via inert ion sputtering can be distorted in multicomponent silicate glasses due to preferential sputtering and ion bombardment induced redistribution of atomic species.

(4) Complementary AES (or XPS) and SIMS (or SIPS) analyses should be used to provide a meaningful characterization of real glass surfaces.

(5) ISS is very sensitive to the atomic arrangement at clean surfaces of multicomponent silicate glasses.

(6) An atomic rearrangement appears to occur at the 'ideal' surfaces of alkali-silicate glasses due to an outward alkali-ion relaxation during or after the fracture event.

ACKNOWLEDGEMENT

The authors gratefully acknowledge the financial support of the National Science Foundation under DMR-8018473.

REFERENCES

1. C.G. Pantano, Am. Ceram. Soc. Bull. 60(11):1154 (1981).

2. J.A. Borders and G.W. Arnold, in: "Ion Beam Surface Layer Analysis," Volume 1, O. Meyer et al., eds., Plenum, New York (1976).

3. G. Della Mea, A.V. Drigo, V. Gottardi and F. Nicoletti, Silicates Industrials 9:179 (1978).

4. W.A. Lanford, IEEE Transactions on Nuclear Science, NS-26(1): 1795 (1979).

5. C.G. Pantano and T.E. Madey, Appl. Surf. Sci. 7:115 (1981).

6. K. Wittmack, J. Appl. Phys. 50:493 (1979).

7. H.W. Werner and A.E. Morgan, J. Appl. Phys. 47:1232 (1976).

8. L. Colombin, H. Charlier, A. Jelli, G. Debras, and J. Verbist, J. Non-Cryst. Sol. 38/39:551 (1980).

9. T. Kishii, Yogyo-Kyokai-Shi 89(1):56 (1981).

10. R.A. Chappell and C.T.H. Stoddart, Phys. Chem. Glasses 15(5): 130 (1974).

11. R.A. Chappell and C.T.H. Stoddart, J. Mater. Sci. 12:2001 (1977).

12. W.E. Baitinger, P.W. French and E.L. Swarts, J. Non-Cryst. Sol. 38/29:749 (1988).

13. K. Matsumoto, Yogyo-Kyokai-Shi 88(8):441 (1980).

14. B. Rauschenbach and W. Hinz, Silikattenchnik 30:151 (1979).

15. L. Colombin, A. Jelli, J. Riga, J.J. Pireaux, and J. Verbist, J. Non-Cryst. Sol. 253 (1977).

16. J.S. Sieger, J. Non-Cryst. Sol. 19:213 (1975).

17. M.J. Suscavage and C.G. Pantano, to be published.

18. B. Carriere and B. Lang, Surf. Sci. 64:209 (1977).

19. R.G. Gossink, H. Van Doveren and J.A.T. Verhoeven, J. Non-Cryst. Sol. 37:111 (1980).

20. F.M. Ohuchi, Ph.D. Dissertation, University of Florida (1981).

21. M.P. Seah and W.A. Dench, NPL Report Chem. 82 (1978).

22. V.H. Dannheim, H.J. Oel, and G. Tomandl, Glastechn. Ber. 49:170 (1976).

23. K.S. Kim and N. Winograd, Surf. Sci. 43:625 (1974).

24. R. Kelly, Nucl. Instr. Meth. 182/183:351 (1981).

25. H.H. Andersen, SPIG 1980 (B. Cobic, ed., Boris Kidric Institute of Nuclear Sciences, Beograd, Yugoslavia).

26. H. Bach, Radiation Effects 28:215 (1976).

27. J.W. Coburn and E. Kay, IBM Research Report No. RJ-1388, May 10, 1974.

28. B.M.J. Smets and T.P.A. Lommen, Comm. Am. Ceram. Soc. C-80 (June 1982).

29. M.A. Chesters and G.A. Somoyai, Ann. Rev. Mat. Sci. 5:99 (1975).

30. D.H. Lee and J.D. Joannopoulos, J. Vac. Sci. Technol. 21(2):351 (1982).

31. J.P. Lacharme, P. Champion and D. Leger, Scanning Elect. Micr. 237 (1981).

32. R.C. McCune, J. Vac. Sci. Technol. 18(3):700 (1981).

DEPTH-PROFILING STUDIES OF GLASSES AND CERAMICS

BY ION BEAM TECHNIQUES

I. S. T. Tsong

Department of Physics
Arizona State University
Tempe, AZ 85287

ABSTRACT

Various ion beam techniques are currently in use to measure the elemental concentration profile as a function of depth in a solid. At low energies, i.e. $\lesssim$ 10keV, the techniques include sputter-induced photon spectrometry (SIPS), secondary ion mass spectrometry (SIMS) and ion scattering spectrometry (ISS). When the primary ion energy increases to several MeV, the accelerator-based techniques such as Rutherford backscattering spectrometry (RBS), elastic recoil detection (ERD) and resonant nuclear reaction analysis (NRA) are very useful for depth profiling. Examples of the application of these techniques to study glass and ceramic materials are given in this report and the relative merits of the techniques are discussed.

INTRODUCTION

The interaction of a beam of heavy ions in the energy range of keV with a solid surface results in the sputtering of atomic and molecular species as well as the scattering of the incoming primary ions. This phenomenon forms the basis of the sputter-induced photon spectrometry (SIPS) and secondary ion mass spectrometry (SIMS) and ion scattering spectrometry (ISS) techniques for measuring the elemental concentration as a function of depth into the surface. At higher energies, i.e. MeV range, the probing ion can interact with a sample atom to produce either a scattering event or a nuclear reaction. Depending on the mass of the primary ion and the scattering geometry, the high energy scattering phenomenon has given rise to the Rutherford backscattering spectrometry (RBS) and elastic recoil detection (ERD) techniques. In the case of nuclear reaction, depth-profiling is accomplished by detecting one of the reaction products, usually an α-particle or a γ-ray.

In the present report, we will examine each technique in turn, citing examples of depth-profiling application to glass and ceramic materials. At the conclusion of the report, we will discuss the various strengths and weaknesses of these techniques.

SPUTTER-INDUCED PHOTON SPECTROMETRY (SIPS)

The SIPS technique detects the spectral lines emitted by sputtered atoms in excited states[1]. When the technique is used for depth-profiling, one simply monitors the most intense spectral line of the element of interest as a function of sputtering time. The time scale can be converted to a depth scale by measuring the depth of the sputter crater on the sample surface with a stylus profilometer.

The primary ion is usually Ar^+ and a suitable energy range is 5-10 keV. Like its companion technique SIMS, the SIPS signal is enhanced by the presence of oxygen on the surface[2]. However, since glass and ceramic materials usually contain oxygen, an O_2^+ primary beam is therefore not necessary. SIPS has the advantage of detecting neutral excited atoms, thereby circumventing to some extent surface charging problems which would otherwise influence the path of departing charged particles (secondary ions or electrons) from the surface to the detector.

The sensitivity of the SIPS technique is often limited by background radiation originated from continuum emission from excited molecules or clusters. SIPS is also not suitable for analysis of elements situated on the right-hand-side of the periodic table because the excitation is distributed over many weak lines rather than a few strong lines.

One of the most successful applications of SIPS is the study of the leaching or hydration of glasses. Tsong and his co-workers[3-6] have conducted a series of depth-profiling studies using SIPS on the diffusion of water into a variety of glasses by the ionic exchange process. Fig. 1 shows the hydrogen, sodium and potassium depth profiles for a number of commercial glasses listed in Table 1. Each of these glasses was hydrated for 48 hours at 90°C. The increasing hydrogen penetration due to higher alkali content in the glass is clearly demonstrated[3]. Ternary soda-lime glasses prepared in the laboratory also show this alkali effect[4] (Fig. 2). Increasing hydration times cause progressively deeper hydrogen penetration[4] as demonstrated in Fig. 3. Diffusion coefficients of hydrogen and sodium have been calculated for the sodium trisilicate glass and the $18Na_2O \cdot 12CaO \cdot 70SiO_2$ soda-lime glass from the SIPS depth-profiles[4,5] using Dosemus's interdiffusion model[7,8].

DEPTH-PROFILING STUDIES OF GLASSES AND CERAMICS

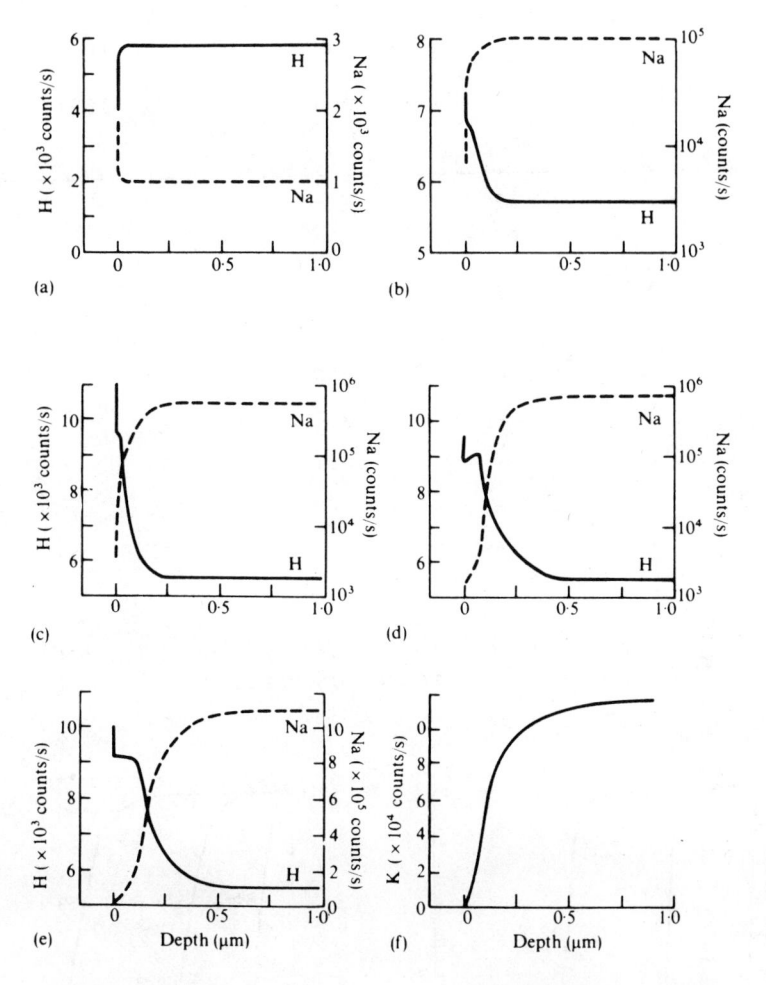

Fig. 1 H(———) and Na(---) depth profiles for (a) code 7910, (b) code 7740, (c) Penn Vernon sheet, (d) code 0160, and (e) code 0080 glasses determined by sputter-induces photon spectrometry; (f) shows the K profile for the code 0160 glass. The intensities of the H 6563Å, Na 5890Å, and K 7665Å lines were expressed in photon counts per second.

SECONDARY ION MASS SPECTROMETRY (SIMS)

Of all the surface analytical techniques available today, it is generally recognized that secondary ion mass spectrometry (SIMS) is one of the most sensitive. Detection limits in the range of one-part-per million or lower can be frequently achieved for all ele-

Table 1. Chemical compositions of the five glasses studied (mol%)

Glass*	Code 7910	Code 7740	Penn Vernon sheet	Code 0160	Code 0080
SiO_2	96.5	83.3	72.5	74.8	72.0
Al_2O_3	--	1.2	0.7	--	0.6
B_2O_3	3.5	11.5	--	0.5	--
Na_2O	Trace	4	12.8	2.6	16.2
MgO	--	--	5.2	--	5.9
CaO	--	--	8.7	--	5.3
K_2O	--	--	0.1	11.3	--
PbO	--	--	--	10.6	--
Sb_2O_3	--	--	--	0.1	--
Fe_2O_3	--	--	0.04	--	--

*Four-digit codes correspond to glasses made by Corning Glass Works and Penn Vernon sheet is a product of Pittsbrugh Plate Glass Co.

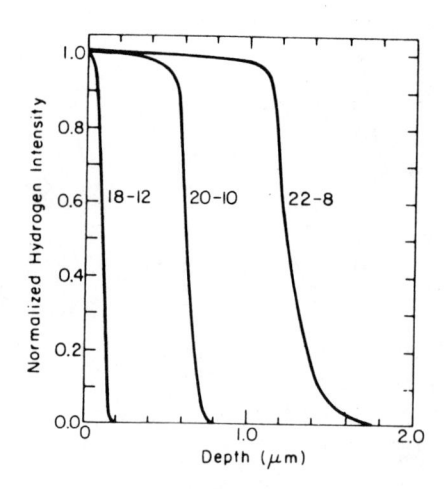

Fig. 2 The hydrogen depth profiles of $18Na_2O \cdot 70SiO_2$ (18-12), $20Na_2O \cdot 10CaO \cdot 70SiO_2$ (20-10) and $22Na_2O \cdot 8CaO \cdot 70SiO_2$ (22-8) glasses hydrated in deionized water at 90°C for 4 hours.

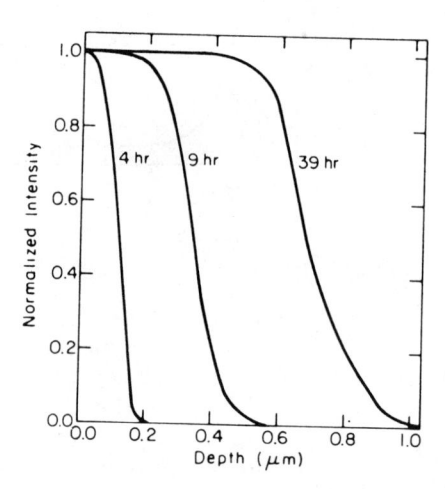

Fig. 3 The hydrogen depth-profiles of the $18Na_2O \cdot 12CaO \cdot 70SiO_2$ glass hydrated at 90°C for different times.

ments from H to U. By scanning the primary ion beam one can perform 4-dimensional analysis of a solid, i.e. area, depth and concentration. Scanning in the xy directions gives a secondary ion image of the monitored atomic species on the surface. Sputtering in the z direction provides information of concentration as a function of depth. Both ion-imaging and depth-profiling modes of operation have been extensively utilized in the characterization of surfaces and interfaces of materials.

In depth-profiling glasses and ceramics using SIMS, one usually comes across the surface charging problem. Fig. 4 illustrates how surface charging distorts the $^{16}O^+$ and $^{30}Si^+$ depth profiles when sputtering through a SiO_2/Si interface. Complete neutralization was achieved with a 2 keV electron beam bombardment of the surface at a current considerably larger than the ion current[9]. Magee and Harrington[10] have demonstrated the importance of surface neutralization using an energetic electron beam in SIMS depth-profiling of insulators, especially in arresting the charged-induced mobility of alkali ions.

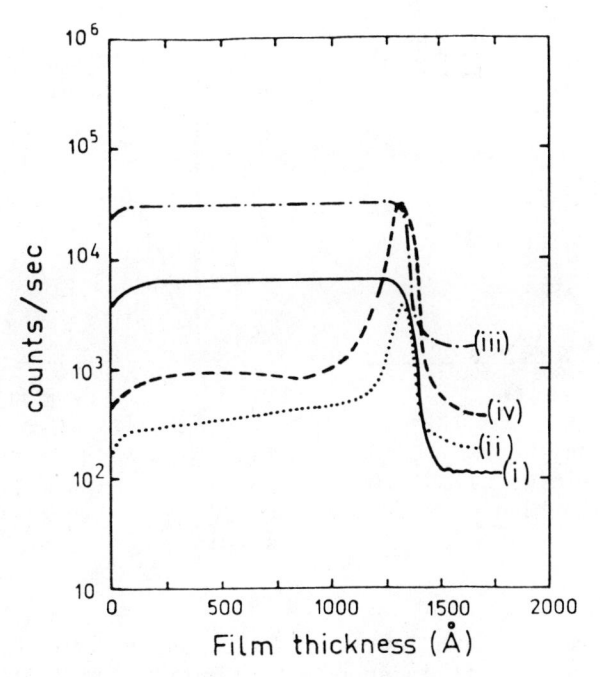

Fig. 4 SIMS depth profiles of $^{16}O^+$ in SiO_2 under (i) complete charge neutralization and (ii) incomplete charge neutralization. Depth profiles of $^{30}Si^+$ in SiO_2 under (iii) complete and (iv) incomplete charge neutralization.

As an example of the application of SIMS, we show our detection of hydrogen and chlorine in SiO$_2$ thermally grown on Si in different HCl/O$_2$ and Cl$_2$/O$_2$ ambients at 1100°C for 15 minutes[11]. This investigation is important because it has been well-documented[12] that the presence of chlorine during the oxidation of silicon produces several beneficial effects in the electrical characteristics of MOS devices. Fig. 5 clearly shows that the chlorine is found at or near the SiO$_2$/Si interface of the HCl oxides. Fig. 6 shows the distribution of hydrogen in various Cl$_2$ oxides. Our SIMS studies of hydrogen in SiO$_2$[9,11] provide conclusive proof for the first time that hydrogen exists at the SiO$_2$/Si interface although it has long been speculated that hydrogen played an important role in the interface properties by tying up the dangling bonds of Si and reducing interface states. Our success is in no small way due to the outstanding sensitivity for hydrogen in the SIMS technique.

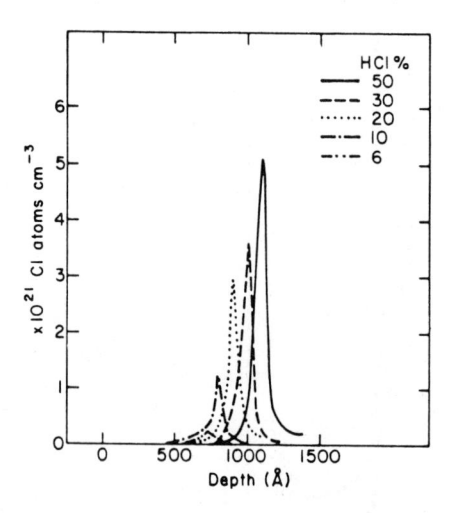

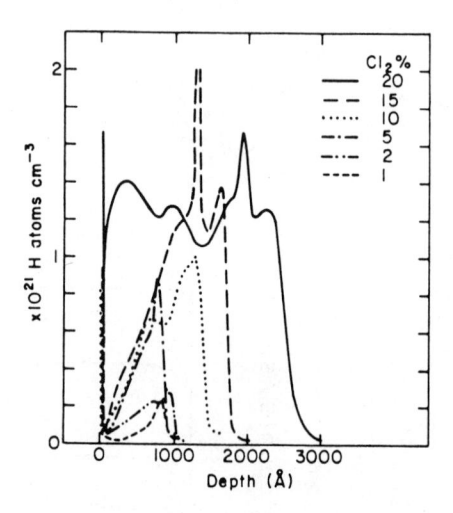

Fig. 5 SIMS depth profiles for ^{35}Cl$^+$ in five SiO$_2$ films grown in HCl/O$_2$ ambients at 1100°C for 15 min.

Fig. 6 SIMS depth profiles for ^{1}H$^+$ in six SiO$_2$ films grown in Cl$_2$/O$_2$ ambients at 1100°C for 15 min.

Because of the high levels of hydrogen in some of the Cl$_2$ oxides, we were able to independently confirm our H data shown in Fig. 6 using the ^{19}F resonant nuclear reaction analysis (details of which will be given in a later section). Fig. 7 shows the nuclear reaction profiles of hydrogen in oxides grown in 20, 10 and 5 volume percent of Cl$_2$. The agreement between the profiles in Fig. 6 and Fig. 7 is very striking considering the totally different experimental conditions.

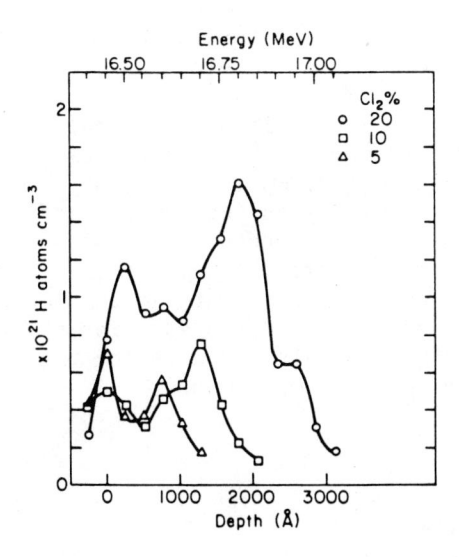

Fig. 7 ^{19}F nuclear reaction profiles in three SiO$_2$ films grown in Cl$_2$/O$_2$ ambients at 1100°C for 15 min. The data points were taken with the samples cooled to -130°C to prevent any beam-induced migration.

ION SCATTERING SPECTROMETRY (ISS)

In ISS, a beam of low energy ions, i.e. 0-10 keV, is scattered from the sample surface. If we measure the energy distribution of the ions scattered into a known angle θ and knowing the mass M_1, energy E_1 and current I_1 of the primary ions, we can obtain atomic composition and possibly concentration of the surface. Extensive reviews of the ISS technique have been given by several groups of workers, amongst whom are Neihus and Bauer[13] and Brongersma et al.[14] At these low energies, primary ions penetrating deeper than the first one or two atomic layers have a very high probability of being neutralized and these scattered neutrals are not detected in the conventional ISS technique using hemispherical electrostatic energy analyzers or cylindrical mirror analyzers. ISS is therefore an extremely sensitive technique for detecting atoms located in the outermost surface layer.

Usually He$^+$ ions are used in ISS giving a very slow sputtering rate. This is ideal if very shallow elemental depth-profile is required. For investigating deeper depth profiles, one has to resort to heavier ions, Ne$^+$ or Ar$^+$, to produce a faster sputtering rate. Using a heavier primary ion also has the advantage of improving the mass resolution for increasing masses of the surface atoms. For scattering angles less than 90° (measured between the incoming and outgoing trajectories), one can even use a heavy ion such as Ne$^+$ and Ar$^+$ to detect surface atoms of lighter masses than the probing ion. ISS is therefore a very useful technique for studying the top monolayer of a surface.

ISS can be used routinely to analyze insulators such as glasses and ceramics provided an electron flood gun or an electron beam is used to minimize surface charging. Fig. 8 shows an ISS analysis of ceramic powders by Brongersma et al.[14] The addition of a small amount of MoO_3 to the Bi_2MoO_6 catalysts strongly increases its catalytic activity for oxidation of propene $CH_3CH:CH_2$ to acrolein $CH_2:CHCHO$, whereas a small addition of Bi_2O_3 has the opposite effect. Fig. 8 shows that a small change in the bulk concentration of Mo or Bi has a pronounced effect on the surface composition of these catalysts. The Bi concentration is strongly depressed at the surface with 4% MoO_3 addition and strongly enhanced with 4% Bi_2O_3 addition. The dramatic change in surface composition probably accounts for the change in the catalytic process for the Bi_2MoO_6 powder.

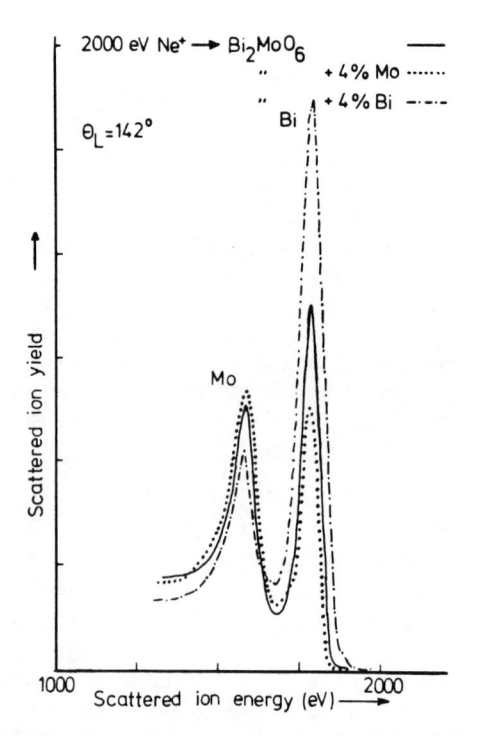

Fig. 8 ISS analysis of bismuth-molybdate catalysts having small differences in the bulk compositions.

RUTHERFORD BACKSCATTERING SPECTROMETRY (RBS)

RBS is probably the most established and widely used accelerator-based ion beam technique for surface analysis. Its principle is based on the elastic scattering of MeV energetic light ions from surface discovered by Rutherford and his co-workers at the beginning of this century. Several important factors account for the growing popularity of RBS: (1) it is an absolute quantitative technique and standards are not generally required; (2) it has the capability to produce depth-profiles without being destructive, i.e. no sputtering or erosion of sample surface; and (3) it can combine scattering and channeling effects to locate impurity atoms in single-crystalline samples. Many review articles have been written on the RBS technique, the most comprehensive being the book by Chu et al.[15] Other useful but shorter reviews have been given by Mitchell[16] and Mackintosh[17].

Although RBS has been tremendously successful in semiconductor investigations, it has not seen a great deal of use in glasses and ceramics. Our example is taken from the work of van der Meulen et al.[18] who used RBS to determine the concentration and distribution of chlorine in SiO_2 thermally grown as Si in HCl/O_2 and Cl_2/O_2 ambients. Fig. 9 shows an energy spectrum of the backscattered ^4He ions from such a SiO_2 film. The energy axis of Fig. 9 is both a mass scale and a depth scale. The location of the front edges of the peaks or plateaux provides information of the masses of the atoms while the depths can be derived by the widths of the peaks or plateaux. Referring to Fig. 9, the depth of the Cl_2 distribution in SiO_2 is given by

$$\delta x_{Cl} = \Delta x \, (\delta E_{Cl}/\Delta E_{Cl}) \quad [1]$$

where Δx is the SiO_2 film thickness determined by ellipsometry, δE_{Cl} is the energy difference between the Cl signal from the surface and that from a depth δx in the film and ΔE_{Cl} is the energy difference between the Cl signals from the surface and from the SiO_2/Si interface. It should be noted that the isotopes ^{35}Cl and ^{37}Cl can be resolved in this RBS analysis.

The chlorine concentration profiles unfolded from the RBS energy spectra are given in Figs. 10 and 11 for the HCl oxide and Cl_2 oxide respectively. In the HCl oxides, chlorine appears to be incorporated at the SiO_2/Si interface during oxidation whereas in the Cl_2 oxide, the chlorine appears to be less prominently peaked and more homogeneously distributed in the SiO_2. This finding is very similar to the SIMS analysis mentioned earlier. From the RBS data, the average concentration of the HCl oxides (1150°C, 9 vol.%) is 2.4×10^{20} at cm^{-3} while that for Cl_2 oxides (1000 C, 2 vol.%) is 4.4×10^{20} at cm^{-3}.

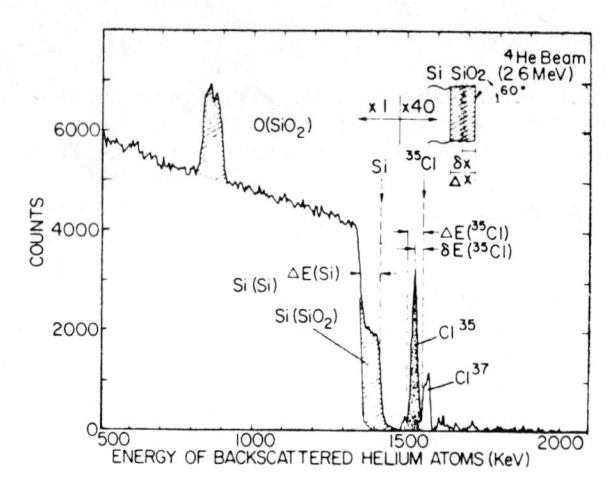

Fig. 9 Spectral distribution of ^{4}He ions backscattered from a thin SiO_2 film thermally grown in a chlorine containing ambient on a Si substrate.

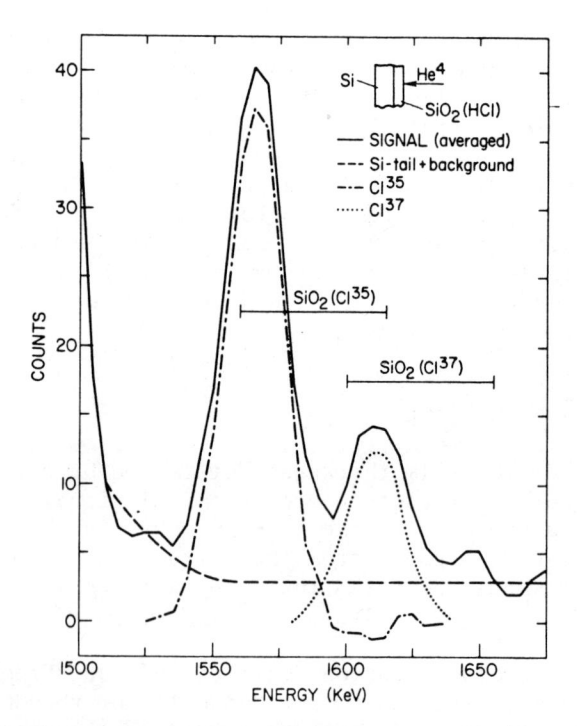

Fig. 10 Chlorine distribution as a function of depth in an HCl oxide (1150°C, 9 v/o HCl, 780Å). The measured signal has been separated into background and chlorine isotope profiles. The calculated energy range for both chlorine isotopes in the oxide film is indicated.

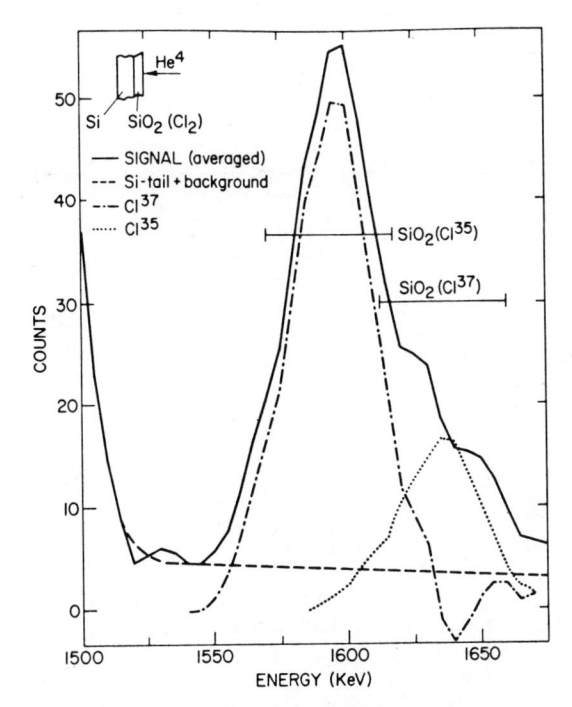

Fig. 11 Chlorine distribution as a function of depth in a Cl$_2$ oxide (1000°C, 2% Cl$_2$, 640Å). The measured signal has been separated into background and chlorine isotope profiles. Calculated energy ranges are indicated by horizontal bars.

ELASTIC RECOIL DETECTION (ERD)

The RBS technique is generally suitable to analysis of heavy impurity elements in a light matrix solid. If, however, one wishes to detect light elements in a heavy matrix, then one has to resort to the ERD technique. ERD was developed by Doyle and Peercy[19] of Sandia National Laboratories initially to obtain depth-profiles of hydrogen in solids. Since this technique is relatively new among ion beam techniques, a brief description of its principle and configuration is in order. The scattering geometry is illustrated in Fig. 12, in which the incident ion is forward-scattered. The basic principle is similar to RBS, except in this case instead of having a light ion, e.g. ^{1}H or ^{4}He, scattered from a heavy atom, e.g. ^{28}Si, we have a heavy ion, e.g. ^{28}Si, scattered from a light atom, e.g. ^{1}H. The recoil ion is detected the usual way by a surface-barrier detector. As in RBS, the energy loss of the recoil ion and the differential scattering cross section of the collision partners provide information on the mass of atom in the solid, its depth in the solid and its concentration.

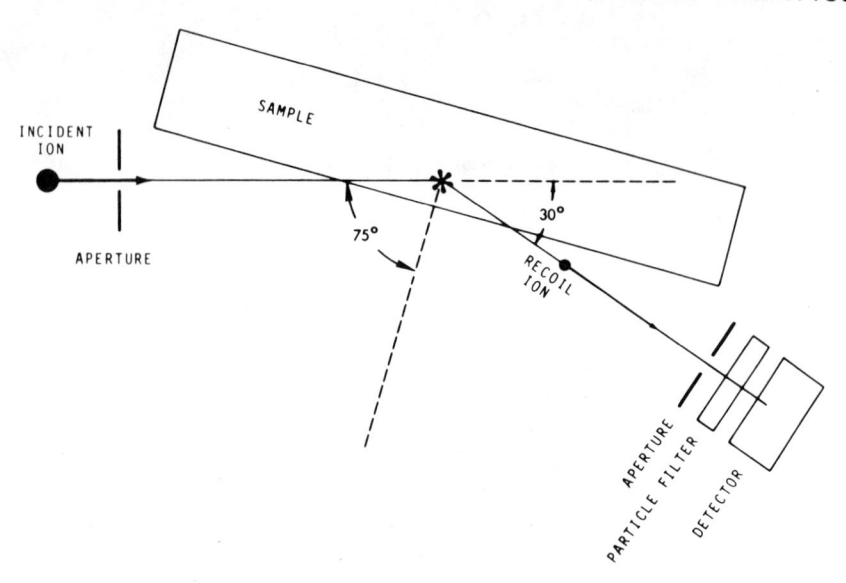

Fig. 12 Schematic illustration of the experimental arrangement for profiling H in solids using ERD.

An interesting example of the application of ERD in glasses has been given by Arnold et al.[20] A $Li_2O \cdot 2SiO_2$ glass sample was implanted with 2×10^{16} Xecm^{-2} at 250 keV. The Li profile was measured by ERD before implantation, after implantation and after implantation plus annealing (shown in Fig. 13). Note that the Li distribution is not uniform at the near surface before implantation and shows Li depletion in the first 3000Å. After Xe implantation, however, Li shows an increase in concentration near the surface. This enhanced Li concentration disappears after annealing at 500°C. The depletion of Li in the first 3000Å is apparently replaced by H as indicated in Fig. 14 which shows the H profile after implantation. Before implantation the H profile looks similar except the peak concentration is even higher, at 1.7×10^{22} atoms cm^{-3}.

In the same ERD study, Arnold et al.[20] implanted 10^{17} Li atoms cm^{-2} at 50 keV into fused silica. Fig. 15 shows the Li implant profiles as implanted and after annealing at 600°C. Further annealing at 700°C and 800°C shows migration of Li towards the surface (Fig. 16) of the SiO_2.

Arnold et al.[20] explain these interesting ERD observations in terms of crystallization of implanted regions upon annealing with the difference in Li behavior in the two samples attributed to phase separation in the lithia-silica system.

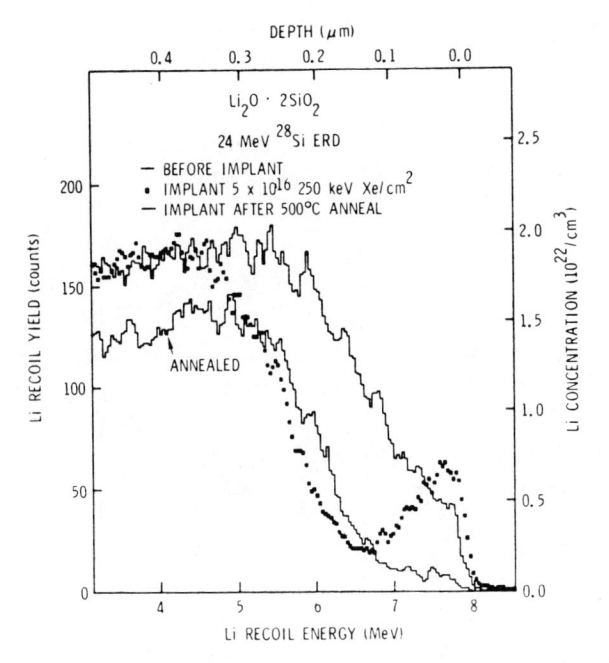

DEPTH-PROFILING STUDIES OF GLASSES AND CERAMICS 51

Fig. 13 Li depth distribution of $Li_2O \cdot 2\,SiO_2$ measured by ERD showing the effect of 5×10^{16} 250 keV Xe/cm^2 implantation and a subsequent thermal anneal at $500^{\circ}C$.

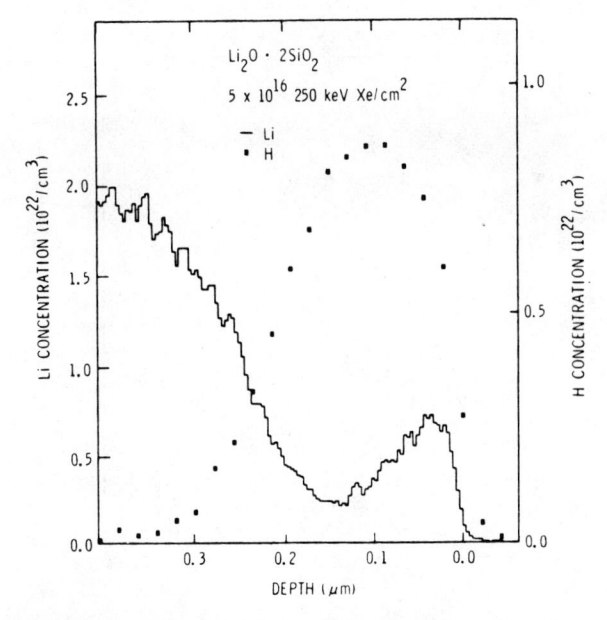

Fig. 14 Comparison of the Li and H profiles in the near-surface region of $Li_2O \cdot 2\,SiO_2$ implanted with 5×10^{16} 250 keV Xe/cm^2.

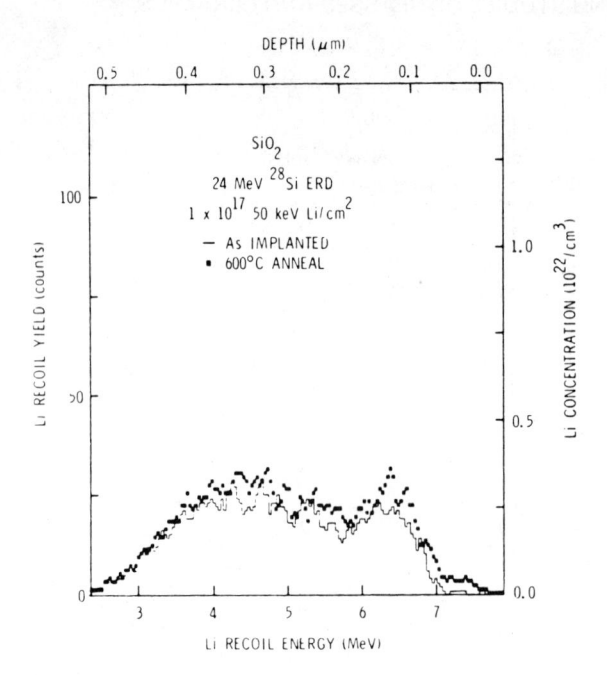

Fig. 15 Li depth distribution in CFS 7940 SiO_2 after implantation of 10^{17} 50 keV Li/cm^2 at room temperature and after annealing at 600°C.

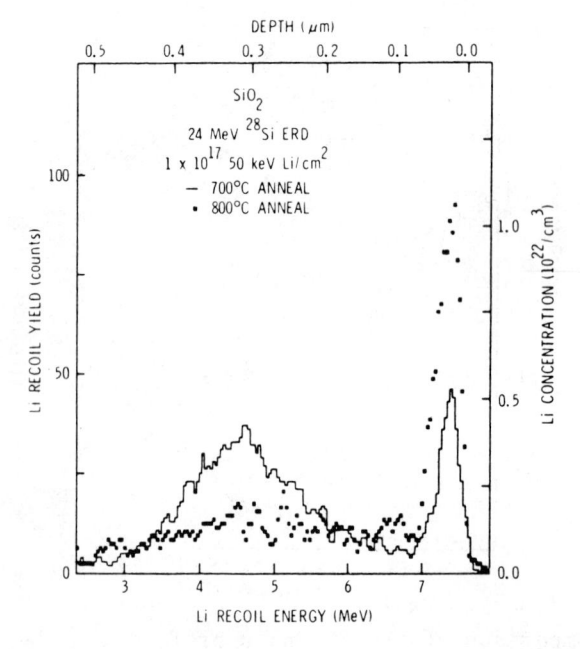

Fig. 16 Change in the Li profile in SiO_2 implanted with 10^{17} 50 keV Li/cm^2 after anneals at 700 and 800°C.

DEPTH-PROFILING STUDIES OF GLASSES AND CERAMICS

NUCLEAR REACTION ANALYSIS (NRA)

In NRA, MeV ions from an accelerator interacts with atoms in a solid to produce reaction products which can be detected and yield information about the distribution and concentration of the atoms in the solid. There are many nuclear reactions, but in this section we will only deal with resonant nuclear reactions which can be used to perform the depth-profiling operation. The two most well-known resonant nuclear reactions in the application to glass and ceramic materials are the ^{19}F and ^{15}N reactions for hydrogen depth profiling. They can be written as:

$$^{19}F + {}^{1}H = {}^{4}He(\alpha) + {}^{16}O^* = {}^{16}O + \alpha + \gamma \qquad [2]$$

and

$$^{15}N + {}^{1}H = {}^{4}He(\alpha) + {}^{12}C^* = {}^{12}C + \alpha + \gamma \qquad [3]$$

To utilize the $^{1}H(^{19}F,\alpha\gamma)^{16}O$ or $^{1}H(^{15}N,\alpha\gamma)^{12}C$ reaction as a probe for hydrogen, the sample is bombarded with ^{19}F (or ^{15}N) ions and the characteristic γ-ray yield is measured using a NaI scintillation detector. If there is hydrogen on the surface of the sample and the ^{19}F ions have energy 16.45 MeV in the lab frame (6.385 MeV for ^{15}N ions), the yield of the γ-rays is proportional to the amount of hydrogen on the surface.

To determine the concentration of hydrogen inside the sample, the ^{19}F (or ^{15}N) energy is raised so that it is above the resonant energy for hydrogen on the surface, but as the ^{19}F (or ^{15}N) slows down by electronic stopping inside the sample, it loses energy until it reaches the resonant energy at a certain depth inside the surface. The energy loss, dE/dx, in the target can be calculated with the aid of emperical tables of Northcliffe and Schilling[21] and Ziegler[22]. Therefore by raising the energy from 16.45 MeV for ^{19}F (or 6.385 MeV for ^{15}N) and measuring the yield of high energy γ-rays, the concentration of H versus depth can be determined.

The depth resolution is determined by the ratio of the resonance width to the electronic stopping power, dE/dx. This corresponds to ∼200Å for ^{19}F reaction and ∼75Å for the ^{15}N reaction in SiO_2.

The ^{15}N reaction is pioneered by Lanford and co-workers at SUNY, Albany and many interesting applications have originated from that laboratory. A review of the ^{15}N technique and its applications has been given by Lanford[23] and will not be repeated here.

An excellent example of NRA is the use of the ^{19}F reaction to test the controversy whether it is hydrogen or hydronium which is involved in the ionic exchange process in glass leaching[24], i.e. whether it is

$$H_2O + Na^+(glass) = H^+(glass) + NaOH \qquad [4]$$

or $2H_2O + Na^+(glass) = H_3O^+(glass) + NaOH.$ [5]

Three glasses of known composition (shown in Table 2) were hydrated in a large bath of deionized water at 88°C with the Corning 0080 and the Penn Vernon sheet glass for 48h and the less durable laboratory soda-lime glass for 1h. The hydrogen-depth profiles determined by the ^{19}F reaction with the samples cooled to -130°C to prevent loss of water to the vacuum and minimize beam-induced mobility is shown in Fig. 17. The hydrogen concentration $N_H(d)$ at depth d was determined by using the equation[25].

$$\frac{N_H(d)}{N_H(std)} = \frac{[counts(d) \cdot dE/dx]_{unknown}}{[counts \cdot dE/dx]_{std}} \quad [6]$$

The standard sample was a silicon wafer implanted with a known dose of hydrogen. The resulting H/Na ratios in the three hydrated glasses are shown in Table 3. Our results suggest that at least in the commercial glasses, it is the hydronium (H_3O^+) ions that are replacing the sodium ions, in agreement with a similar finding by Lanford et al.[8] The 2:1 ratio for the laboratory soda-lime glass may be due to unavoidable loss of hydrogen to the vacuum because the hydrogen is less strongly bound in this simpler glass.

Table 2. Chemical compositions of the three glasses studied (mol.%)

Glass	SiO_2	Al_2O_3	Na_2O	MgO	CaO	K_2O	Fe_2O_3
Corning 0080	72.0	0.6	16.2	5.9	5.3	...	...
Penn Vernon Sheet	72.5	0.7	12.8	5.2	8.7	0.1	0.04
Laboratory soda-lime	69.3	...	18.8	...	11.9	...	...

Table 3. H:Na atomic ratios in leached glasses.

Glass	N/Na
Corning 0080	3.2±0.4
Penn Vernon Sheet	2.6±0.4
Laboratory soda-lime	2.0±0.3

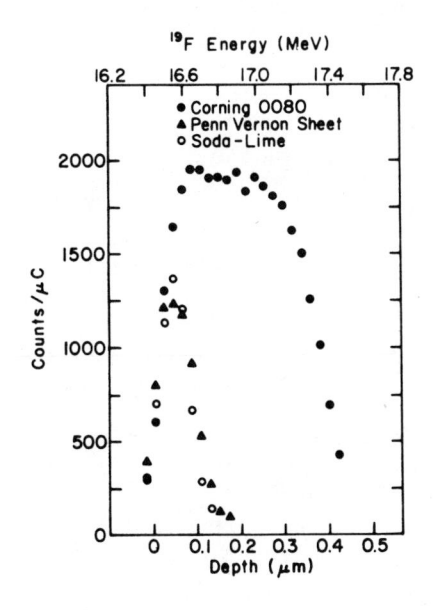

DEPTH-PROFILING STUDIES OF GLASSES AND CERAMICS 55

Fig. 17 Hydrogen depth profiles in three glasses determined by the $^1H(^{19}F,\alpha\gamma)^{16}O$ nuclear reaction.

ADVANTAGES AND DISADVANTAGES

Technique	Advantages	Disadvantages
SIPS	Reliable identification of chemical elements. Good for insulator samples. Good sensitivity for light elements, ~1ppm.	Sensitivity poor for some elements. Requires standards for quantitative analysis.
SIMS	Detection of all elements from H to U including isotopes. Both positive and negative ions can be used for detection, $\lesssim$ 1ppm sensitivity for all elements. Rapid multi-element analysis capability. Raster-gating technique produces excellent depth resolution, typically ~ 30Å.	Mass interference due to molecules and clusters. Surface charging requires careful neutralization for insulator samples. Ion yield enhancement due to chemical environment requires careful interpretation. Requires standards.

(continued)

(Continued)

Technique	Advantages	Disadvantages
ISS	Excellent sensitivity for outermost atomic layer. Good for insulator samples. Detection from H to U.	Mass resolution poor for heavier elements. Slow depth-profiling rate unless heavier ions are used. Standards difficult to prepare.
RBS	Absolute quantitative analysis is possible. Good depth resolution, ~ 100Å, for depth-profiling. Non-destructive. Good for thin film analysis. Good for insulators. Good sensitivity for heavy elements, $\lesssim 10$ppm.	Poor mass resolution for similar heavy masses. Ambiguity in spectra interpretation of multi-element samples, e.g. light element at the surface can be confused with heavy element within bulk. Poor sensitivity for light elements. Limited probing depth, 1-2µm.
ERD	Absolute quantitative analysis is possible. Non-destructive. Good for insulators. Good sensitivity for light elements, 10-100ppm.	Not suitable for heavy element analysis. Depth resolution ~ 300Å. Limited probing depth 1-2µm.
NRA	Good technique for hydrogen depth-profiling. Reasonably good sensitivity, 10-100ppm. Absolute quantification possible. Results are matrix independent. Good depth resolution for ^{15}N, ~ 70Å.	Other resonances limit depth range, ~ 0.4 m for ^{19}F and 3-4µm for ^{15}N. Not good for elements with Z>15. Not good for multi-element analysis.

REFERENCES

1. N. H. Tolk, I. S. T. Tsong and C. W. White, Anal. Chem. 49:16A (1977).
2. P. Williams, I. S. T. Tsong and S. Tsuji, Nucl. Instrum. Meth. 170:591 (1980).
3. I. S. T. Tsong, C. A. Houser and S. S. C. Tong, Phys. Chem. Glasses 21:197 (1980).
4. I. S. T. Tsong, C. A. Houser, W. B. White, G. L. Power and S. S. C. Tong, J. Non-Cryst. Solids 38/39:649 (1980).
5. C. A. Houser, J. S. Herman, I. S. T. Tsong and W. B. White, J. Non-Cryst. Solids 41:89 (1980).
6. C. A. Houser, I. S. T. Tsong and W. B. White, in "Scientific Basis for Nuclear Waste Management: Vol. 1, G. J. McCarthy, ed., Plenum, New York (1979) p. 131.
7. R. H. Doremus, J. Non-Cryst. Solids 19:137 (1975).
8. W. A. Lanford, K. Davis, P. LaMarche, T. Laursen and R. Groleau, J. Non-Cryst. Solids 33:249 (1979).
9. I. S. T. Tsong, M. D. Monkowski, J. R. Monkowski, P. D. Miller, C. D. Moak, B. R. Appleton and A. L. Wintenberg, in "The Physics of MOS Insulators", G. Lucovsky, S. T. Pantelides and F. L. Galeener, eds., Pergamon, New York (1980) p. 321.
10. C. W. Magee and W. L. Harrington, Appl. Phys. Letters 33:193 (1978).
11. I. S. T. Tsong, M. D. Monkowski, J. R. Monkowski, A. L. Wintenberg, P. D. Miller and C. D. Moak, Nucl. Instrum. Meth. 191:91 (1981).
12. J. R. Monkowski, Sol. State Tech. 22:58 and 113 (1979).
13. H. Niehus and E. Bauer, Surface Sci. 47:222 (1975).
14. H. H. Brongersma, L. C. M. Beirens and G. C. J. van der Ligt, in "Materials Characterization Using Ion Beams", J. P. Thomas and A. Cachard, eds., Plenum, New York (1978) p. 65.
15. W. K. Chu, J. W. Mayer and M. A. Nicolet, "Backscattering Spectrometry", Academic Press, New York (1978).
16. I. V. Mitchell, Phys. Bulletin 30:23 (1979).
17. W. D. Mackintosh, in "Characterization of Solid Surfaces", P. F. Kane and G. B. Larrabee, eds., Plenum, New York (1974) p. 403.
18. Y. J. van der Meulen, C. M. Osburn and J. F. Ziegler, J. Electrochem. Soc. 122:284 (1975).
19. B. L. Doyle and P. S. Peercy, Appl. Phys. Letters 34:811 (1979).
20. G. W. Arnold, P. S. Peercy and B. L. Doyle, Nucl. Instrum. Meth. 182/183:733 (1981).
21. L. C. Northcliffe and R. F. Schilling, Nucl. Data Tables A7:233 (1970).
22. J. F. Ziegler, "Handbook of Stopping Cross-Sections of Energetic Ions in All Elements", Pergamon, New York (1980).
23. W. A. Lanford, Nucl. Instrum. Meth. 149:1 (1978).
24. I. S. T. Tsong, C. A. Houser, W. B. White, A. L. Wintenberg, P. D. Miller and C. D. Moak, Appl. Phys. Letters 39:669 (1981).
25. J. F. Ziegler et al., Nucl. Instrum. Meth. 149:19 (1978).

ANALYSIS OF SOLIDS BY SPARK SOURCE AND LASER MASS SPECTROMETRY

Robert J. Conzemius and Harry J. Svec

Ames Laboratory and Dept. of Chemistry
Iowa State University
Ames, Iowa 50011

INTRODUCTION

In about 1960 two scientific developments occurred which have a direct bearing on this topic. One was the demonstration of the first laser (1) whose impact on this topic will be discussed later. The other was the scientific development in the late 50's and the commercial availability in the early 60's of the spark source mass spectrometer (SSMS). Actually the SSMS was first demonstrated back in 1935 by A. J. Dempster (2,3). Basically the development of SSMS as an analytical tool was a response to the great surge in solid state and high purity metallurgy accurring in the 1950's.

SPARK SOURCE MASS SPECTROMETRY

SSMS is very simple in concept but very complex in actual function (4). Specimens in the form of small rods ∿1 cm long and 1 mm sq. cross-section are inserted into a high vacuum chamber. A low duty cycle high frequency high voltage spark of a few µs in duration is made to cross a small gap (20 to 100 µm) between the rods. Ions formed in the spark which are representative of the specimens are accelerated with ∿20 kV and traverse electostatic and magnetic fields. Here the ions are separated into mass-to-charge ratios and a mass spectrum can be recorded. Subsequent interpretation of the mass spectrum permits direct identification of the elements present in the specimens with varying degrees of accuracy dependent upon the sophistication of the calibration process.

SSMS has been applied to virtually every known solid from frozen water to tungsten metal, from coal to ultrapure germanium, from organics to polymers. The initial promise of SSMS was to analyze any material without prior chemical treatment providing semiquantitative information without the need for standards for all elements in the periodic chart with a dynamic range from 100% to 1 ppb. Generally the instrumentation succeeded in this goal. However, technology soon demanded quantitative information and much effort was directed toward this goal. Some generalizations may be stated regarding what has transpired since 1960 and where the current state-of-the-art lies.

SSMS remains limited by the ability of the electronic excitation circuit to cause adequate sparking between the specimens. For example nonconducting matrices usually require crushing and mixing with a conducting medium such as graphite or powdered metals with subsequent pressing into a pelleted specimen. With respect to quantitation there has evolved three regimes of operation. The most common one utilizes semiquantitative measurements. This regime normally yields accuracies within a factor of 3 of the correct value for metallic elements and within a factor of 10 for nonmetallic elements. Standards are not used and elemental sensitivities are estimated from values observed from previous instrument calibrations. This regime is especially useful in the production of ultrapure materials. For example we at the Ames Laboratory have found SSMS to be highly useful in characterizing ultrapure rare earth metals (5).

In the second regime standards which are similar to the matrix are used to calibrate the SSMS. Here the accuracy is about 20 percent relative. In highly controlled experiments the accuracy may approach 5 percent relative but in less favorable cases may be as high as 50 percent. Specimen homogeniety can limit the accuracy since only a few mg of the specimen is sampled from an uncontrolled volume.

A third regime based upon isotopic dilution techniques has become quite popular and provides accuracies of one to three percent relative. An example of this type of analysis is that developed by Paul Paulson at the Bureau of Standards (6).

Some recent developments are significant. Most laboratories have small computers handling data acquisition. At the Ames Laboratory we are now controlling the entire instrument relieving the need for constant attention by a technical specialist. Considerable effort has occurred recently in Europe and the Soviet Union in studying more controlled spark breakdown with considerable improvement being reported for the damped, unidirectional spark (7,8). These developments are especially pertinent to scientists working with nonconducting materials since the unidirectional spark permits direct analysis of the nonconductor without grinding, mixing, and pressing into conducting specimens (9). Analysis of soils (10) and ashes (11) have indicated that confidence can be attained in SSMS analysis of these complex matrices by careful attention to sampling

ANALYSIS OF SOLIDS

and specimen preparation. An excellent series of descriptive papers on SSMS appeared in the 1981, Vol. 309, No. 4 issue of Fresenius Z. Anal. Chem.

LASER MASS SPECTROMETRY OF SOLIDS

Let us now turn to the laser. From its first demonstration in 1960 we followed its development closely especially when applied to mass spectrometry (12). Initial results were not very promising and indeed some very negative observations were made indicating severe selective volatilization and uncontrollably high instantaneous ion currents causing deterioration of mass resolution. Much of the early work emphasized correlations of molecular ions found in the mass spectra to short range ordering in the solid. Although very interesting, the results were always suspect due to recombination of particles in the ion source after leaving the solid and to a very high dependence of the ion signal upon the ionization potential of the ion species (13). Of course most of this early work was done using normal mode (non-Q switched) lasers in which the energy-per-pulse was quite high (joules or tenths of joules) and the duration of the pulse quite long (∼1 ms).

Table 1 shows how broad the ranges have been for parameters of laser beams used with the mass spectrometric study of solids. These ranges do not include work where continuous wave lasers have been used giving longer time periods per pulse and even lower power densities. Power density, the wattage per square cm, is an extremely important parameter. Figure 1 shows the ion fraction produced per total particle emitted as a general function of laser power density. At the lower power densities the ion fraction is very low — generally definable by the Saha-Langmuir equations for ordinary thermal vaporization. The range of ion fraction is thus very large due to elemental specificity caused by differences in volatility and ionization energy. Notice that as the power density approaches 10^9 W cm^{-2} the ion fraction rises to near unity and the range narrows dramatically. This is due to the transformation from ordinary thermal evaporation to the formation of a highly ionized micro plasma just above the specimen. Several studies have been made to characterize this micro plasma and many others are currently under way (12,14).

As the laser beam power density is increased the temperature of the plasma increases and the energy of the ions entering the mass spectrometer also increases (15). Figure 2 shows the general relationship between ion energy and power density. The average ion energy increases monotonically with laser power density. There is a threshold power density for each ion charge and the kinetic energy of ions obtained from such plasmas can be related directly to the ion charge state, the higher ion charge states possessing the higher energies. However even at higher power densities each

Table 1. Ranges of Laser Parameters Used in Mass Spectrometer Studies of Solids

Parameter	Range	Unit
Energy	$10^{-7} - 10^{1}$	Joules
Time	$10^{-9} - 10^{-3}$	s
Power	$10^{1} - 10^{8}$	Watts
Power Density	$10^{5} - 10^{12}$	Watts cm^{-2}
Wavelength	$0.22 - 10.6$	μm

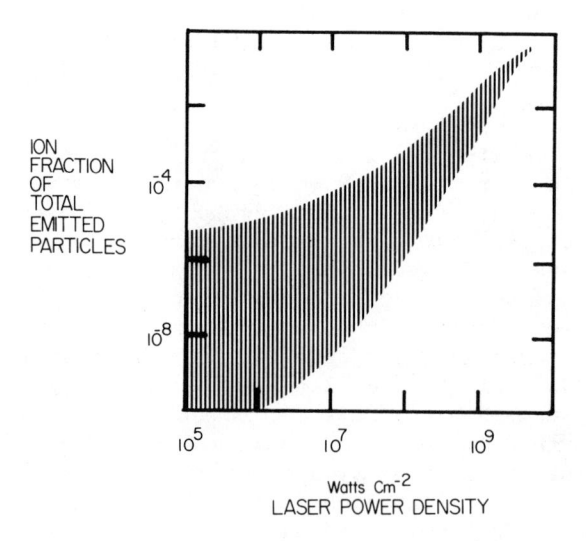

Figure 1. Range of ion fraction as a function of laser power density.

ion charge obtained from the plasma possesses an energy profile with maxima related to the mode of ion formation as indicated in Fig. 2 by the dashed lines for singly, doubly, and triply charged ions.

ANALYSIS OF SOLIDS

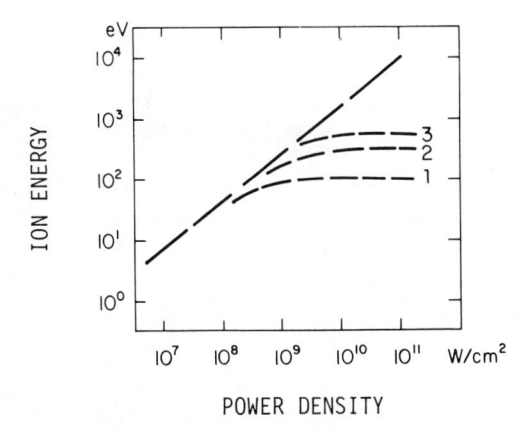

Figure 2. General relationship between ion energy and laser power density.

The pulsed nature of the ion beam along with the relatively high ion energy spread makes the choice of mass spectrometer non-trivial. Table 2 lists the basic types of mass spectrometers which have been used with laser excitation of solids. The single magnetic focusing, the TOF and the Quadrupole are all limited in their ability to provide useful mass resolution with high ion energy spreads. The two most common mass spectrometers used today, the quadrupole and the single magnetic focusing instruments, scan too slowly to handle the short ion pulses evolved by pulsed lasers. TOF not only does

Table 2. General Features of the Types of Mass Spectrometers Used in Studies of Laser Excitation of Solids

Instrument Type	Mass Resolution	Allowable Energy Spread	Mass Range	Scan Rate	General Cost
Double Focus	High	High	Medium	V. Slow	V. High
Single Focus	Low	Low	Low	V. Slow	High
Time-of-Flight	Low	Low*	High	V. High	High
Quadrupole	Low	Low	High	High-Limited	Low
Special Design	Medium	–	–	–	–

*Increased by use of ion reflecting mirror.

this fast scanning well but also appears to handle rather high energy spreads when the recent innovation of the ion reflecting mirror is utilized. The double focusing instruments generally provide for a wide mass range of simultaneous ion detection thus not requiring scanning. However they are very expensive.

Consider now some of the more practical aspects. The configuration of the laser beam and the specimen is shown in Fig. 3 to have six basic types. The configurations are based on the angle (ϕ) of laser illumination to the specimen surface and the angle (θ) of ion-take-off from the specimen surface to the ion optic axis of the mass spectrometer. Perhaps the only one needing clarification is the -90/90° orientation where the specimen is mounted on a thin film which is irradiated from the backside. The laser penetrates through the specimen and the ions are sampled from the front side of the specimen. This ion source configuration is used in a commercial laser mass spectrometer built by Leybold-Heraus. An excellent series of 26 articles describing this instrument and its applications may be found in the entire issue of Fresenius Z. Anal. Chem., 308 (1981) p. 193-320.

Since the ion source is inside a high vacuum chamber the laser created microplasma immediately begins to expand. The predominant

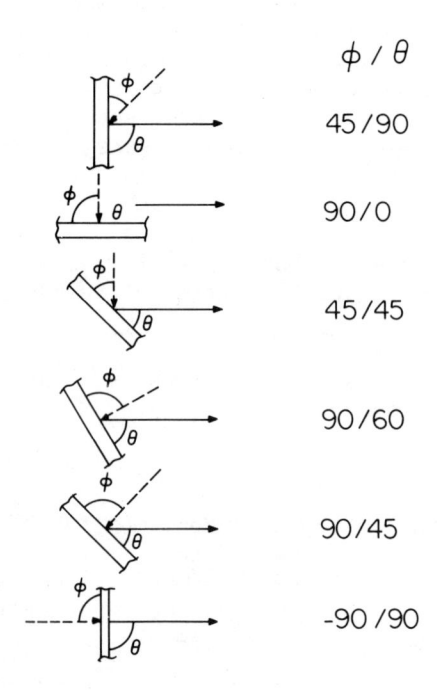

ϕ / θ

45/90

90/0

45/45

90/60

90/45

-90/90

Figure 3. Basic types of ion source configurations.

process is now recombination of ions and electrons until the plasma density reduces to the point where the probability of recombination goes to zero. Recombination is favored by high density, low plasma temperature, long plasma expansion time and large plasma dimensions. The directionality of the ions emitted by the microplasma follows generally a $\cos^2$ distribution with the higher charged particles tending to be accelerated more normal to the specimen surface.

At the Ames Laboratory, we have been studying the laser ion source using the ϕ/θ of 90/60 configuration shown in Fig. 3. The laser and the mass spectrometer which we employ has been described (16). Since the laser beam can be scanned along the x and y axes of the specimen during the analyses we call the system a scanning laser mass spectrometer (SLMS). Here we will show some results indicating the analytical usefulness and the potential for this analytical tool.

The laser we employ is a neodynium doped yttrium-aluminum-garnet which is continuously pumped and repetively pulsed by external control at rates up to 20,000 Hz using an acousto-optic Q-switch. We typically operate our laser at a 1000 Hz pulse rate in which the pulse duration is $\sim$100 ns long. The focused power density is on the order of gigawatts per cm^2. This provides a relatively stable average ion current signal of about a nA in the mass spectrometer which is a home designed and built instrument employing Mattauch Herzog Ion Optics.

A means of determining how well this instrument emulates an absolute detector can be shown by analyzing a standard specimen and observing the ratio of the observed signal to the known level of the element in the specimen. Ideally the ratio is unity. Table 3 lists ratios for 15 elements measured in a standard brass sample. In spite of the large differences in the physical characteristics of the elements listed in Table 3 the instrumental response is quite uniform indicating the laser initiated microplasma is ionizing virtually all the elements with similar probability. This allows semiquantitative analysis of specimens without the need for standards.

Another useful feature of the instrument lies in its ability to measure concentration profiles. Figure 4 is a plot of the silicon concentration measured in 25 µm steps across a welded junction of pure tantalum to tantalum containing ten ppmw silicon. The data indicates 25 µm spatial resolution for silicon at 1 to 10 ppmw. This analytical tool has been used for measuring numerous other concentration profiles in studies of chemical diffusion and electrotransport (17). Virtually any element of interest in any matrix can be measured.

The laser mass spectrometer can be used to measure elemental profiles of elements directly in rock samples (16). Both minor and

Table 3. Relative Elemental Response* of Laser Mass Spectrometer to Various Elements in NBS Brass Sample

Element	RSC	Element	RSC
Be	0.96	As	0.74
Al	1.16	Ag	0.48
Si	0.81	Cd	0.54
P	0.67	Sn	0.36
Mn	1.22	Sb	0.28
Fe	≡1.00	Pb	0.47
Ni	0.83	Zn	0.62
Cu	0.58		

*RSC = relative sensitivity coefficient = $\frac{\text{observed signal}}{\text{actual level in standard}}$

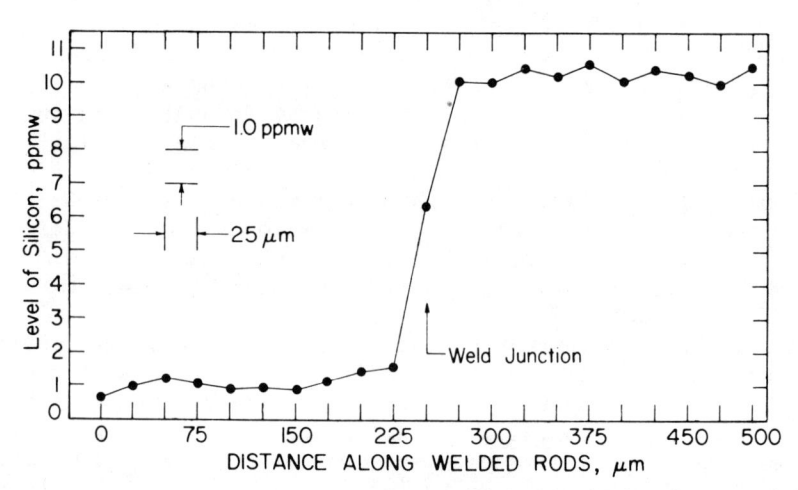

Figure 4. Plot of results from stepped scan of silicon in tantalum across a welded junction. "Reprinted with permission from (Anal. Chem. 50:1854). Copyright (1978) American Chemical Society."

ANALYSIS OF SOLIDS

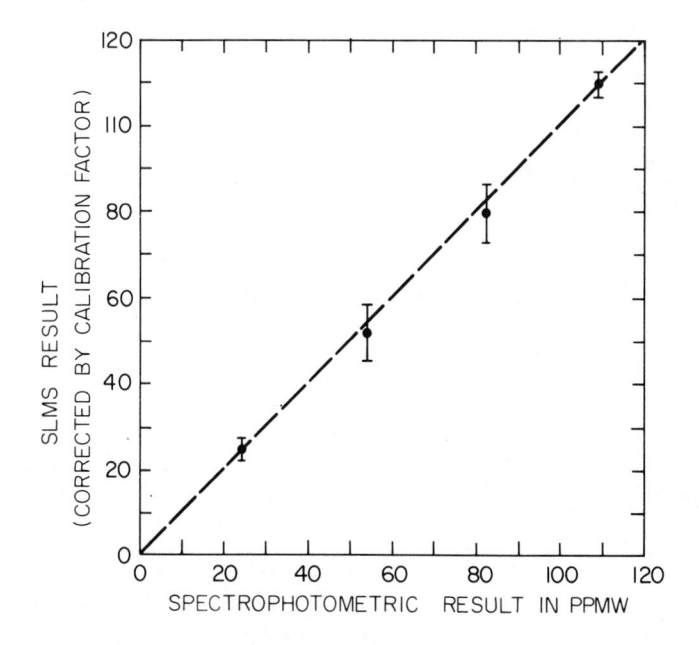

Figure 5. Plot of SLMS results vs spectrophotometric measurements for standards of molybdenum in thorium.

major constituents can be profiled simultaneously.

Quantitation is possible when standards are available. We prepared 4 specimens of thorium containing 24, 54, 82, and 109 ppmw molybdenum. These standards were analyzed using a classical technique, spectrophotometry, and also by SLMS. Figure 5 is a plot of the SLMS results vs the spectrophotometric results. The 45 degree straight line passing through zero indicates that a quantifiable response function is generated nicely. The Mo isotope at mass 98 was utilized for this measurement. At the lower standard level this nuclide is present in the specimen at ~6 ppm. This is measured only with a faraday cup and without electron multiplication indicating the magnitude of the ion currents delivered by this ion source. The uncertainties indicated on Fig. 5 are due to measurements with different ion source loadings on separate days.

Another example of the usefulness of this laser/mass spectrometer probe is demonstrated by the analysis of a 10 μm Si film on a sapphire substrate. The film could be analyzed directly on the substrate by scanning the laser over the surface during the analysis, effectively etching the film with 15 μm diameter microplasmas giving a mass spectrum of the film. Then the same technique was applied to

Table 4. Impurities Found in Silicon Film and in Sapphire (Al$_2$O$_3$) Substrate

Element*	~10 μm Si Film	Sapphire Substrate
Ta	18.	<2. (Ta hearth used)
Ba	1.4	<1.
Nb	2.7	<0.5
Zn	0.5	<0.5
Cu	24.	1.9
Ni	—	2.6
Fe	—	40.
Cr	14.	20.
Ti	3.	2.
Ca	60.	40.
K	90.	40.
Cl	3.	6.
Mg	6.	7.
Na	26.	32.
F	20.	40.
B	0.3	3.0
C	18.	17.
N	8.	2.
Si	10^6	—
Al	—	10^6

*Spectral interferences for Ce, S, P, Fe, Co, Ni, Rb, Ga, Mn in the Si matrix. Remaining elements not detected. Values are in ppma.

a pure sapphire substrate to provide the difference between the film and the film backing. Table 4 gives the listing of the impurities found. The results are in ppm atomic and for the film are relative to Si = 10^6 ppma and for the substrate relative to Al = 10^6 ppma.

The laser and the mass spectrometer have been interfaced to a small computer allowing control of experiments. We are studying effects of laser parameters such as focusing, power density, repetition rates, matrix effects, surface condition etc. These are interesting phenomena in themselves so that pertinent research and immediate applications are available within our solids mass spectrometry laboratory. The future appears to be bright for this direction in analytical chemistry and, fortunately for the scientific community especially interested in ceramics, this technique is suitable for direct analysis of nonconducting specimens.

REFERENCES

1. Maiman, T. H., Stimulated optical radiation in ruby, Nature, 187:493 (1960).
2. Dempster, A. J., New methods in mass spectroscopy, Proc. Am. Philos. Soc., 75:755 (1935).
3. Dempster, A. J., Ion sources for mass spectrometry, Rev. Sci. Instrum., 1:46 (1936).
4. Ahearn, A. J., "Trace Analysis by Mass Spectrometry", Academic Press, New York (1972).
5. Handbook on the Physics and Chemistry of Rare Earths, Vol. 4 Non-Metallic Compounds II. Chpt. 37C. "Analysis of Rare Earth Matrices by Spark Source Mass Spectrometry", K. A. Gschneidner, Jr. and L. Eyring, eds., North Holland Publishing Co., The Netherlands (1979).
6. Paulson, et al., Trace element determinations in low-alloy steel standard reference material by isotope dilution, spark source mass spectrometry, Appl. Spectrosc., 30:42 (1976).
7. Berthod, J., Polarity of breakdowns and properties of ion beams in the self-triggered mode of damped discharge, Adv. Mass Spectrom., 6:421 (1974).
8. Ramendik, et al., General, layer-by-layer, and local analysis of compact dielectrics on a spark source mass spectrometer, Zh. Anal. Khim., 34:1316 (1979).
9. Ramendik, et al., Real energy spread of ions produced in vacuum spark-discharge plasma, Int. J. Mass Spectrom. Ion Phys., 37:331 (1981).
10. Ure, A. M. and Bacon, J. R., Comprehensive analysis of soils and rocks by spark source mass spectrometry, Analyst 103:807, (1978).
11. Conzemius, et al., Elemental material balances by spark source mass spectrometry for a coal burning facility, To be published (1982).
12. Conzemius, R. J. and Capellen, J., A review of the applications to solids of the laser ion source in mass spectrometry, Int. J. Mass Spectrom. Ion Phys., 34:197 (1980).
13. Beahm, E. C., "An Investigation of the Laser Source Mass Spectrometer", Thesis, Pennsylvania State Univ., University Park, 87pp., University Microfilms Order No. 74-4214 (1973).
14. Haas, et al., A quantitative interpretation of LAMMA spectra based on a local thermodynamic equilibrium (LTE) model, Z. Anal. Chem., 308:270 (1981).
15. Bykovskii, et al., Mass Spectrometer study of laser plasma, Soviet Phys. JETP, 33:706 (1971).
16. Conzemius, R. J. and Svec, H. J., Scanning laser mass spectrometer milliprobe, Anal. Chem., 50:1854 (1978).
17. Conzemius, R. J., Schmidt, F. A., and Svec, H. J., Scanning laser mass spectrometry for trace level solute concentration profiles, Anal. Chem., 53:1899 (1981).

SURFACE AND INTERFACE STUDIES OF METAL OXIDE/GLASS SYSTEMS

K.L. Smith

Perkin-Elmer Corp., Physical Electronics Div.
6509 Flying Cloud Drive
Eden Prairie, Minnesota 55344

Optical displays for various devices often consist of a liquid crystal sandwiched between metal oxide layers deposited on glass substrates. Chemical interactions at any of the interfaces may affect the performance of the device. In this study, we have used the surface sensitive techniques of XPS (X-ray Photoelectron Spectroscopy) combined with Ar ion etching to probe the interfaces and the films in one type of these optical displays. Elemental concentration changes are found in some of the metal oxide films which could be correlated with the device performance.

INTRODUCTION

Liquid crystal displays (LCD's) are being used in increasingly varied applications including watches, gauges and TV's.[1] While LCD's present a fascinating materials problem, the part of the device of particular interest to a ceramist is the pattern of metal oxide lines deposited on the glass substrate. The patterns and coatings are similar to the coatings used in microelectronics or optics. As with many new products, there are problems and one of the scientist's objectives is to make the product more reliable. This implies that the scientist has a good technique to study the construction of the product. This work shows that Auger Electron Spectroscopy (AES) and X-ray Photoelectron Spectroscopy (XPS), also known as Electron Spectroscopy for Chemical Analysis (ESCA), can be successfully used to chemically characterize metal oxide/glass systems.

Experimental Procedure

Figure 1 shows the basic construction of the liquid crystal displays examined in this experiment. The tin doped indium electrical leads were sputter deposited on soda glass substrates in the required pattern and then oxidized to form the optically transparent metal oxides. Subsequent preparation included coating the surface with a polyimide and "brushing" it to promote alignment of the liquid crystal. After introduction of the liquid crystal the two glass plates were sealed.[1-3] When parallel polarizing coatings were adhered to the outer surfaces of the glass the device appeared dark. However, the device became transparent when a voltage was applied across the liquid crystal.

In this work the device was broken and the glass substrates separated. The liquid crystal was examined on the lead surface after a glass slide was drawn across the surface in a manner similar to preparation of a biological smear specimen. The remaining thin liquid crystal film could be admitted to high vacuum conditions but was still thick enough to be opaque to XPS. For the majority of the analysis, the liquid crystal was removed using a methanol ultrasonic clean which then allowed direct examination of the metal oxide and glass surfaces with XPS or AES.

The XPS analysis was performed on a Physical Electronics model 560 spectrometer using Mg K_α radiation. The data was charge referenced to C 1s at a binding energy of 284.6 eV (for C-C and C-H) and/or Si 2p at 103.4 eV (for SiO_2).[4] During XPS analysis, the surface is bombarded with X-rays and the photoelectrons generated at the surface are energy analyzed and counted in the spectrometer. As a result, elemental identification and chemical state information can be obtained from a near surface area approximately 3 mm in diameter and 30 Å deep.[5]

Auger Electron Spectroscopy usually involves bombarding the specimen with a finely focused beam of electrons. Again the spectrometer energy analyzes and counts the Auger electrons generated at the specimen surface. Because the kinetic energy of an Auger electron is unique to a specific element, chemical identification can be determined on an area as small as the electron beam.[6] For this experiment a Physical Electronics model 595 spectrometer was used. SEM pictures were taken on this system with an electron beam 500 Å in diameter. Auger data was obtained using a 3 kV electron beam and 30 nA which results in an analysis volume approximately 0.5 um in diameter and 30 Å deep.

When chemical analysis was required from a volume below the top 30 Å, an ion beam was used to physically remove the surface of the specimen. In this analysis a 4 kV Ar ion beam was used alternately with either XPS or AES. The sputter rate used was 20 Å/min during

METAL OXIDE/GLASS SYSTEMS 73

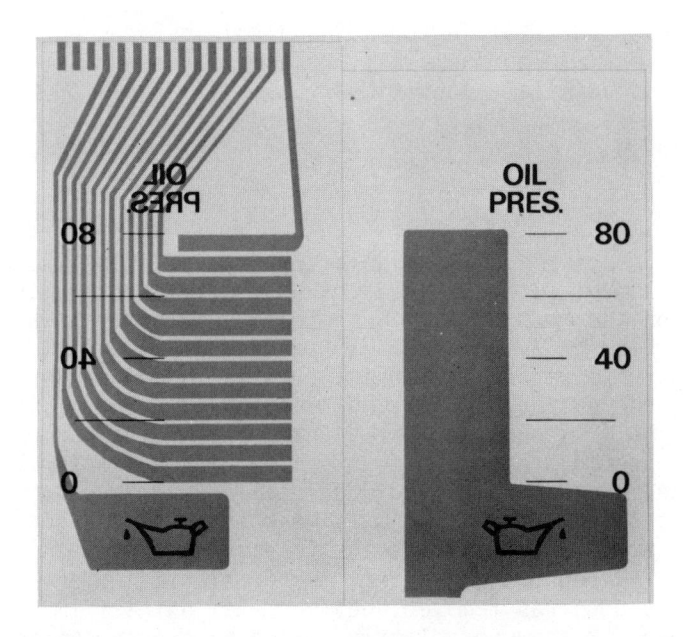

Fig. 1. Schematic of a liquid crystal device studied in this work.

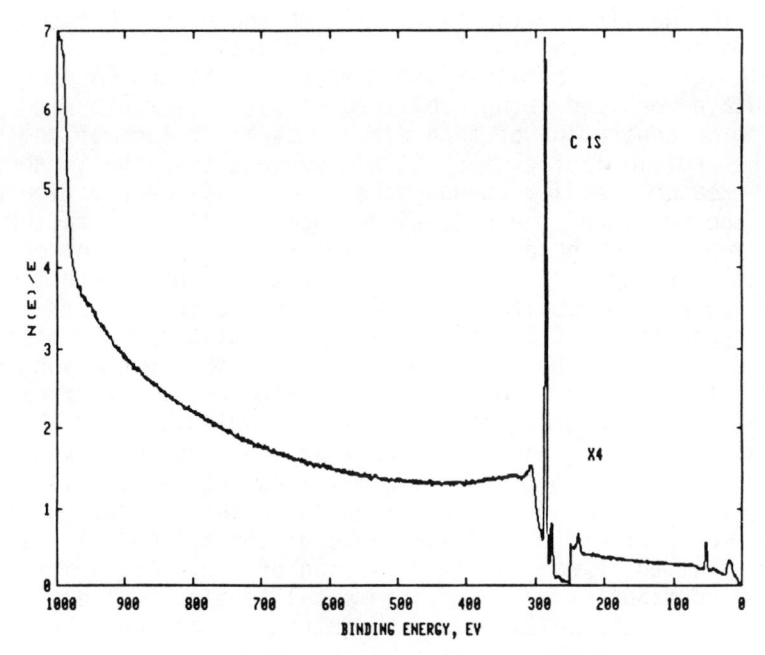

Fig. 2. XPS spectrum of the liquid crystal.

XPS analysis and 15 Å/min for AES analysis referenced to Ta_2O_5. Glass sputters at approximately the same rate as Ta_2O_5 and In_2O_3 sputters approximately four times faster.[7,8]

Results and Discussion

X-ray Photoelectron Spectroscopy

The XPS spectrum of the liquid crystal in figure 2 showed that the only element detectable was carbon. The crystal must contain only carbon and hydrogen in concentrations greater than 0.5 atomic percent. Figure 3A shows an XPS spectrum of the metal oxide coated glass substrate rinsed in methanol. While In and Sn were both detectable, the primary constituent was still carbon. Closer evaluation of the C 1s peak indicated that most of the carbon was bonded with hydrogen but some carbon-oxygen bonding was also observed.[9] This suggested that some liquid crystal and perhaps some methanol residue still remained. The O 1s peak in figure 3B showed that two chemical states were also present for oxygen; the metal oxide and the carbon-oxygen contamination. On sputter removing approximately 50 Å the residue was removed and the metal oxide film exposed. Figure 5A and B show very little carbon or oxygen contamination and figures 4C and D show single oxidation states for In and Sn.[10]

When XPS is coupled with ion sputtering, the concentration and chemical state of each element present can be monitored throughout the film. The mixed metal oxide film on the glass substrate was clearly seen, as shown in figure 5, by plotting the peak height of each element as a function of sputter time. Figure 6A shows the same data normalized using published sensitivity factors to yield the atom concentration of each element as a function of depth.[11] From the profile in figure 6B it was obvious that the Sn concentration decreased steadily through the film. This segregation was either the result of the oxidation process during manufacture or environmental testing on the LCD. Unfortunately no control display was available for verification. It might be expected that continued segregation could lead to a loss of desired electrical or optical properties of the film. While the distribution of the oxides did change, the oxidation state of both the tin and indium showed little change through the film. The reduction of oxides by sputtering reported for other metals was negligible for this materials system.[12] Figure 7A shows an XPS spectrum taken at the metal oxide/glass interface. The metal oxides as well as the glass components, Si, Na and Ca were detected. Since neither Na nor Ca were detected prior to the appearance of the silicon, it can be assumed that there was little diffusion of the cations into this film during manufacture or environmental testing. After 30 minutes of sputtering, the metal oxide concentration was nearly undetectable as seen in figure 7B where an XPS spectrum shows nearly pure glass.

METAL OXIDE/GLASS SYSTEMS

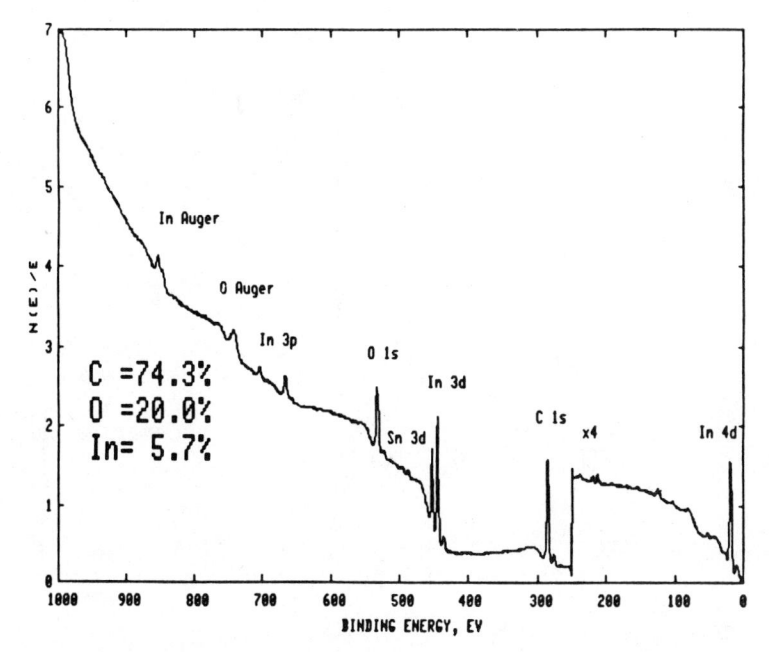

Fig. 3a. XPS spectrum of the metal oxide lead after rinsing with methanol.

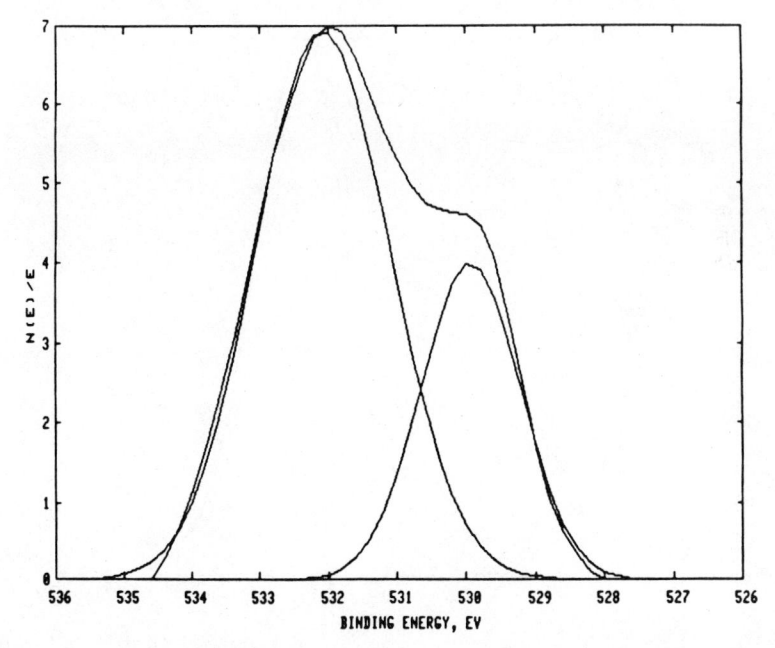

Fig. 3b. Expanded O1s peak showing two chemical states of oxygen; metal oxide (530 eV) and C-O bonding (532.5 eV).

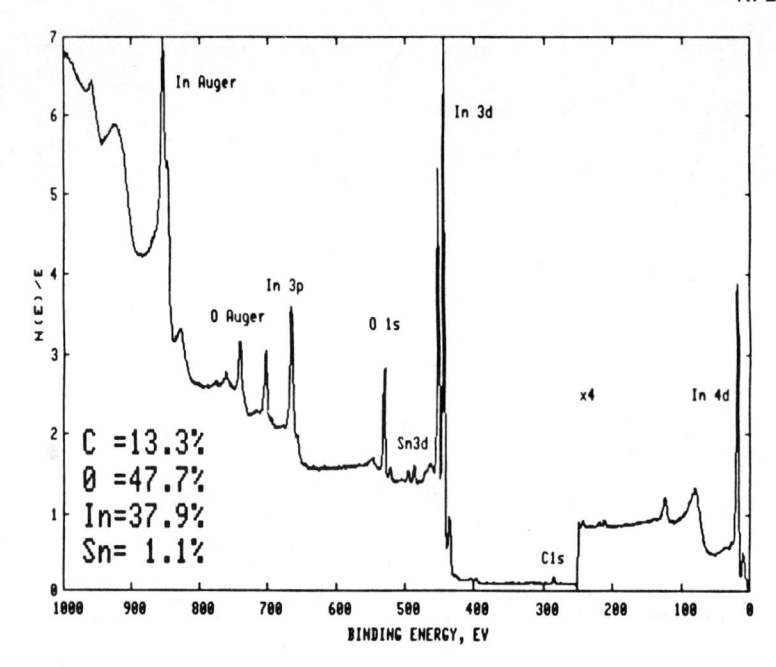

Fig. 4a. XPS spectrum of the metal oxide lead after removing 50 Å.

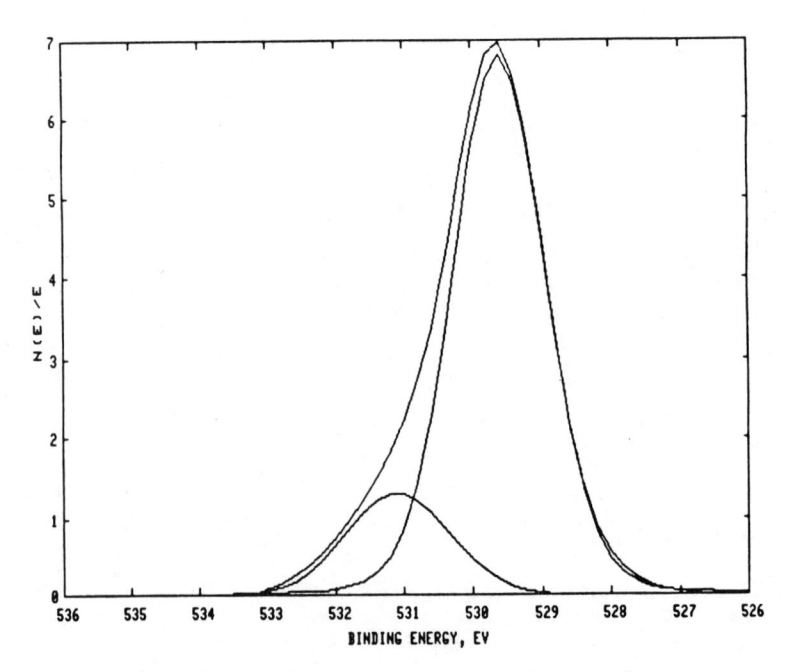

Fig. 4b. Expanded O1s peak showing metal oxide (530 eV) and less C-O bonding (532.5 eV) after removing 50 Å.

METAL OXIDE/GLASS SYSTEMS 77

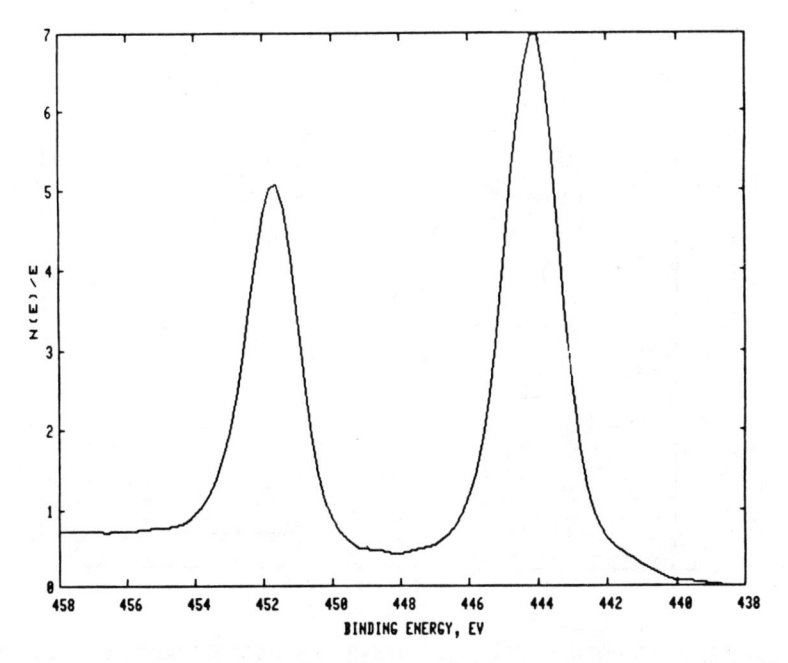

Fig. 4c. In3d peaks showing only the presence of In_2O_3 after removing 50 Å.

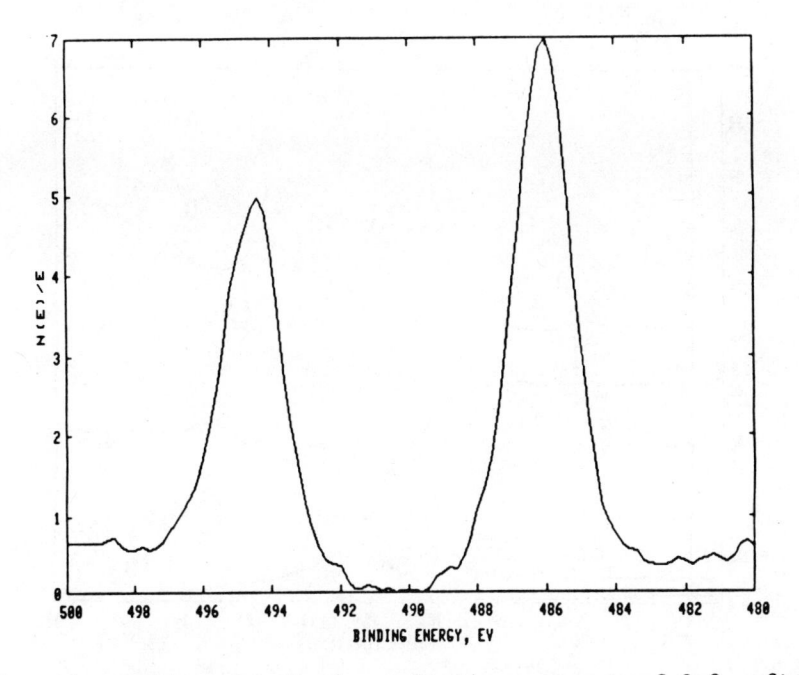

Fig. 4d. Sn3d peaks showing only the presence of SnO_2 after removing 50 Å.

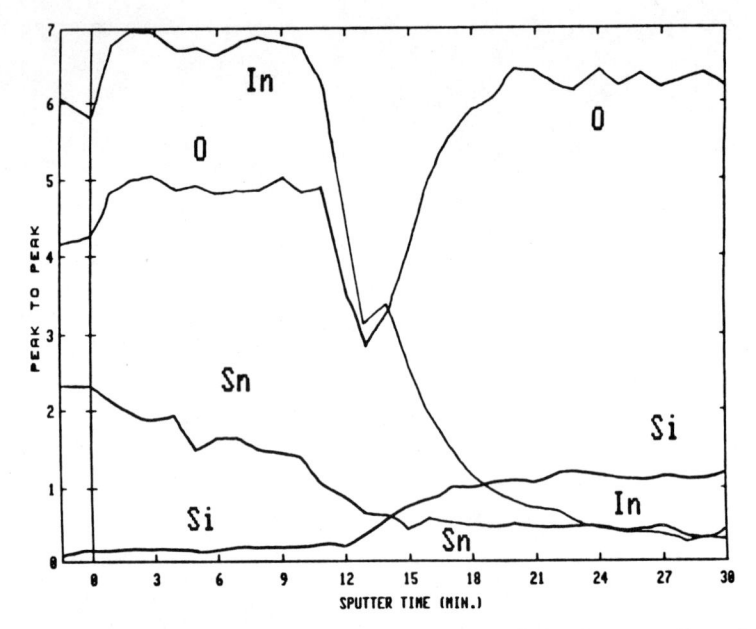

Fig. 5. The change of XPS peak intensities of the metal oxide lead as a function of depth. The sputter rate was 20 Å/min.

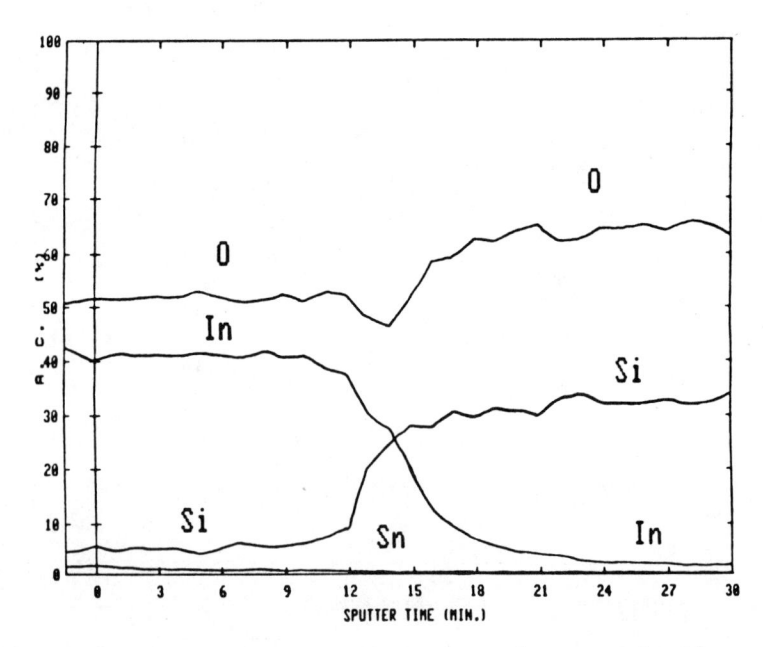

Fig. 6a. Data in Fig. 5 converted to atomic concentration as a function of depth.

Auger Electron Spectroscopy

While it is certainly useful to obtain chemical state information using XPS over a relatively large area, it is equally important to understand the uniformity of that surface. By virtue of the small size of the primary electron beam, scanning Auger Electron Spectroscopy reveals this information. SEM images of the edge of a metal-oxide lead and the adjacent glass, shown in figures 8A and B, clearly indicated that the surface was badly contaminated. In and Si Auger maps corresponding to the area in figure 8B are shown in figures 9A and B. Several of the features seen in figure 8B appear to be neither In or Si. Complimentary Auger maps shown in figures 10A and B indicate that calcium and carbon were the primary contaminants. With greater care during the methanol rinse, much of this nonuniformity can be eliminated.

Auger spectra at relatively clean points on the metal oxide and the glass are shown in figures 11A and B. Carbon was found to be the chief contaminant. Figures 12A and B show the composition of these two points as a function of depth after normalization to atomic concentration.[13] At both points the carbon residue was very thin. The compositions of the metal oxide layer, shown in figure 13, was free of contaminants of more than 1 atomic percent. It should be noted that Sn is not included in the AES profile shown in figure 12. Unfortunately there are small indium Auger peaks which overlap the major tin peaks, making quantitation difficult.

While AES was capable of determining the depth profile of a small area, 0.5 um diameter, it was not able to duplicate the information obtained using XPS. The two techniques are often complementary and are most effective when used together. In figure 14A the AES spectrum of the metal oxide approaching the metal-oxide/glass interface showed no contamination. While Na was not observed until well into the glass layer, its presence in the metal oxide layer is better determined with XPS data since Na is highly mobile in oxides when radiated by an electron beam.[8,14] Figure 14B shows the glass composition after sputtering. Here the Na was detectable but its mobility made quantitation unreliable.

Conclusions

This work has shown that surface analysis can be useful when examining liquid crystal displays. In particular this study illustrated that XPS and AES yield complementary information. The data showed that SnO_2 preferentially segregated to the surface of the SnO_2 doped In_2O_3 film. There also was no evidence of any chemical alteration of either of the metal oxides. That is, both indium and tin appeared only as saturated oxides throughout the film. Finally, there was no evidence of any alkali or other contaminant in the metal oxide film.

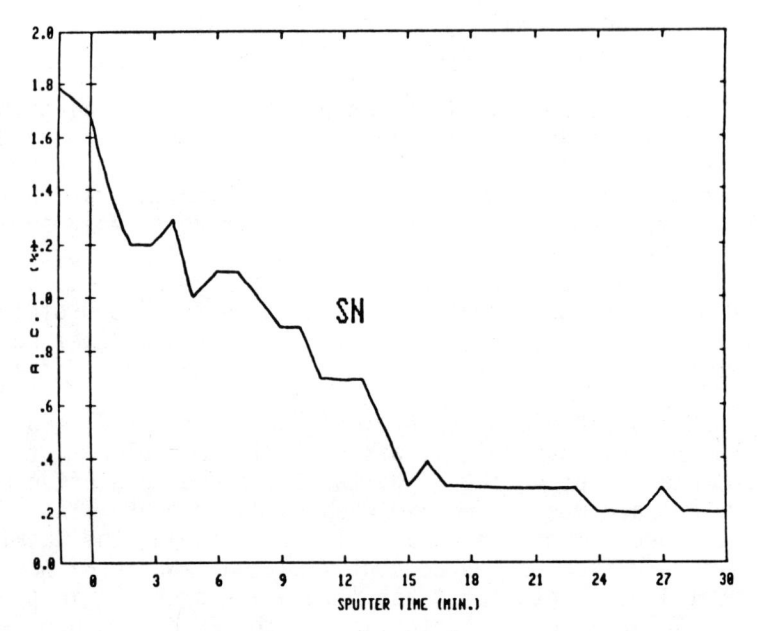

Fig. 6b. Expansion of the 0-2% range.

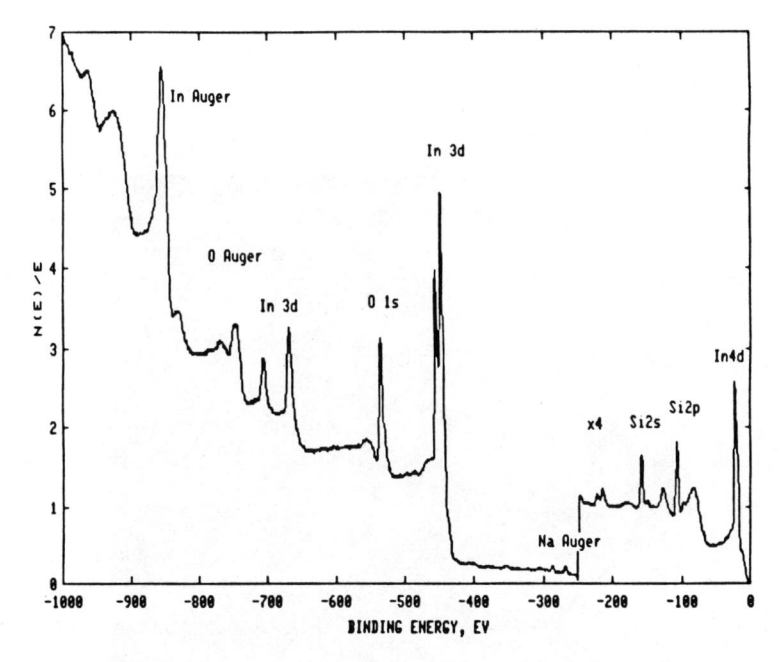

Fig. 7a. XPS spectrum of the metal oxide/glass interface.

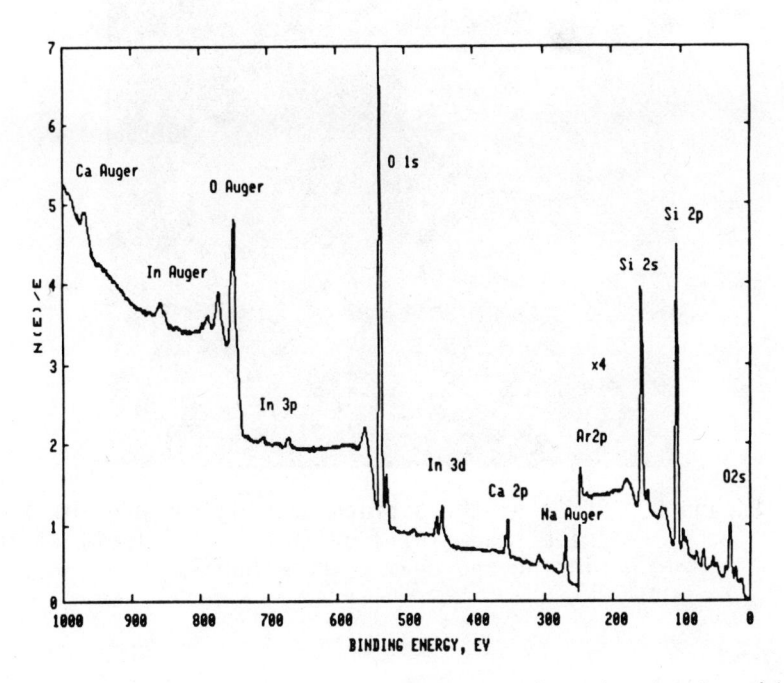

Fig. 7b. XPS spectrum after sputter removing the metal oxide layer.

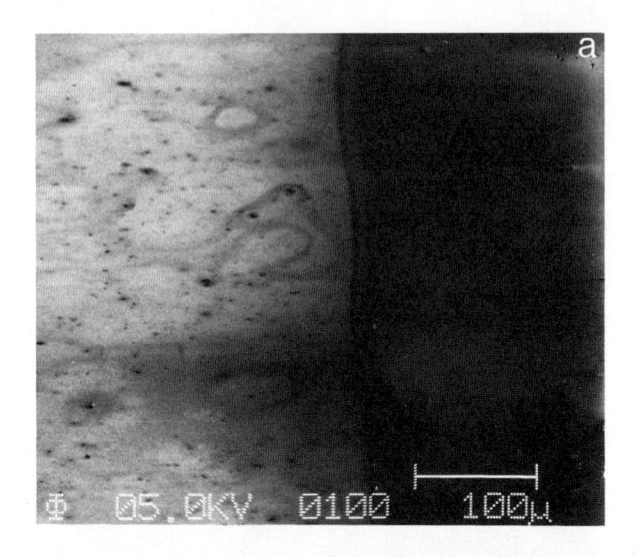

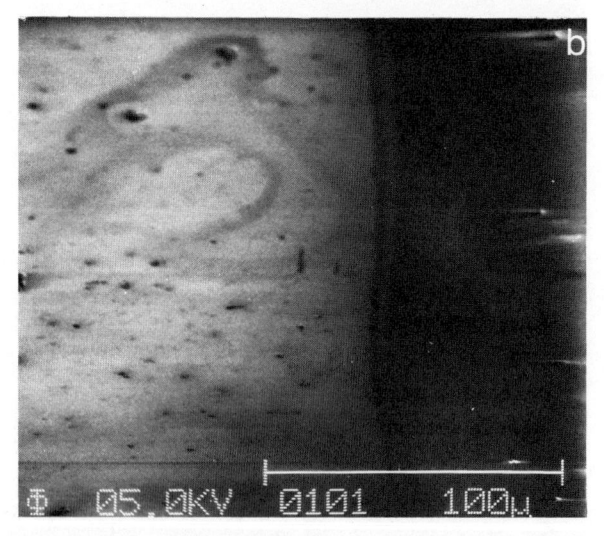

Fig. 8a and b. SEM's of the surface showing the edge of a metal oxide lead (left) on the glass (right) after rinsing the device in methanol.

METAL OXIDE/GLASS SYSTEMS

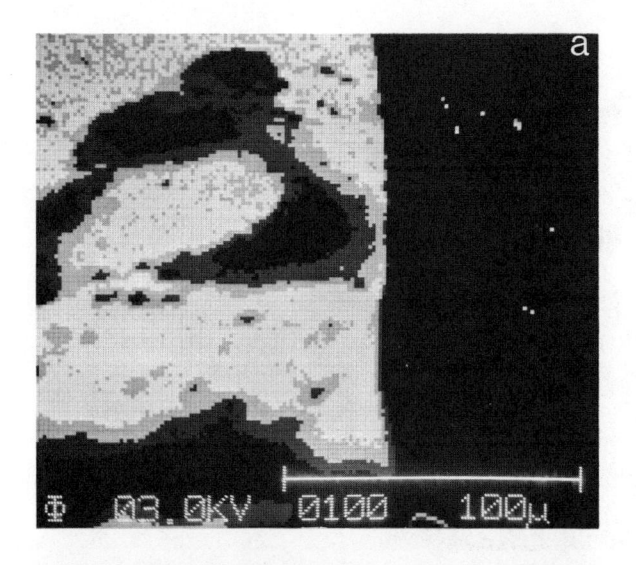

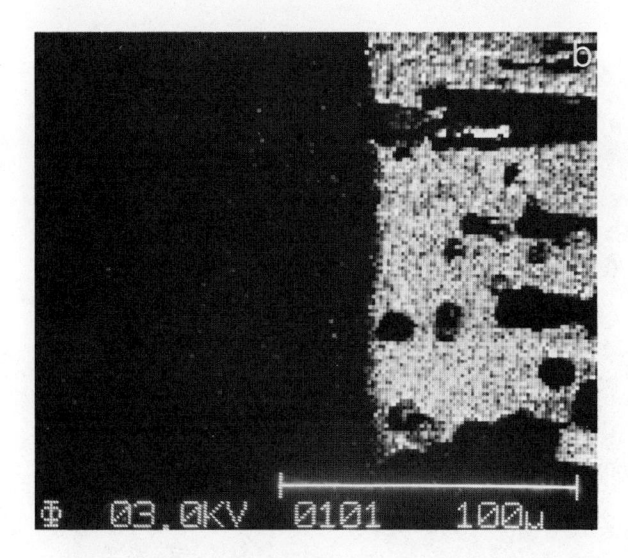

Fig. 9. a: Auger map of indium on the area shown in Fig. 8b.
b: Auger map of silicon.

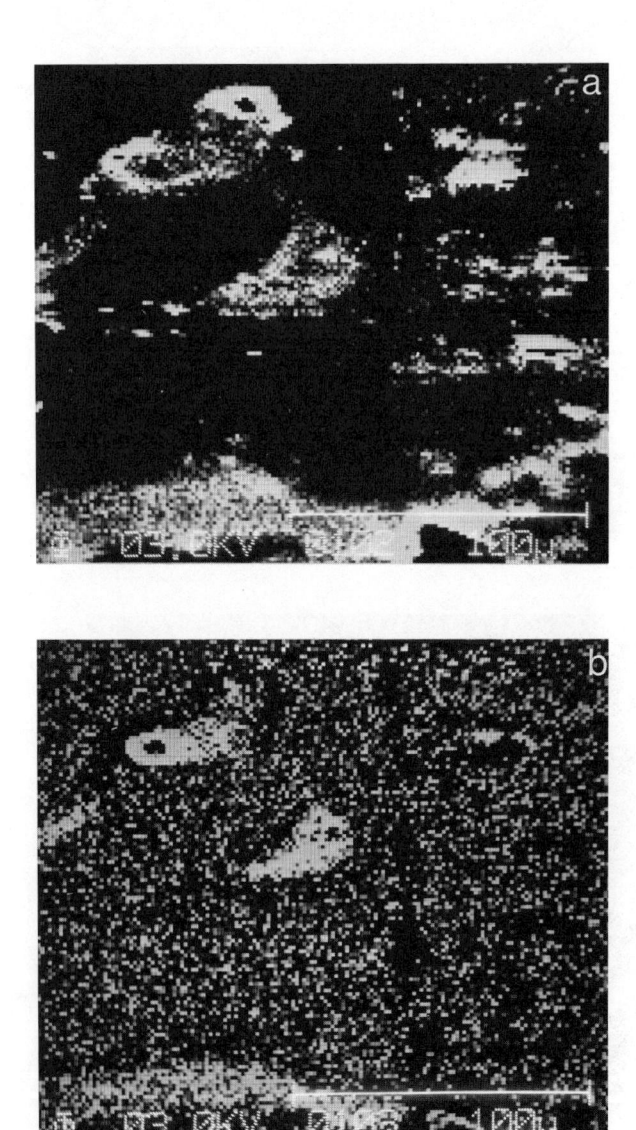

Fig. 10. Auger map of a: carbon and b: calcium of the area shown in Fig. 8b.

METAL OXIDE/GLASS SYSTEMS

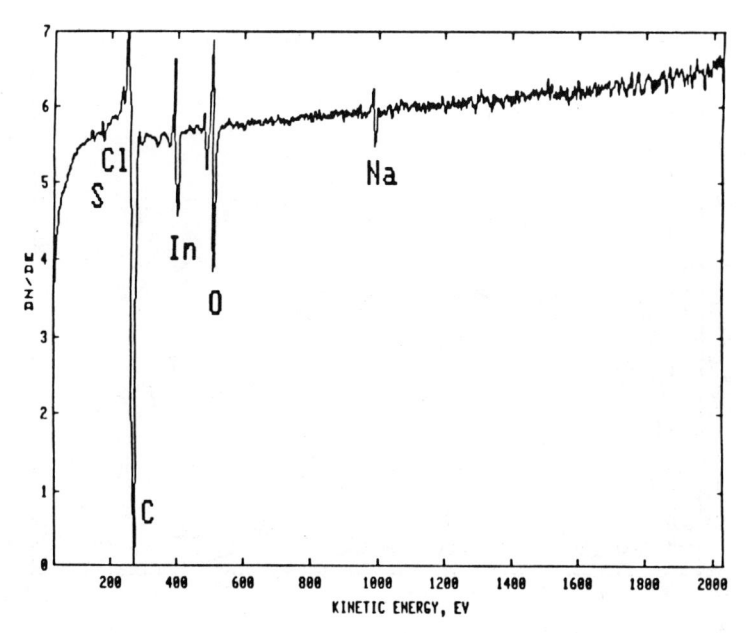

Fig. 11a. Auger spectrum at a relatively clean area on the metal oxide.

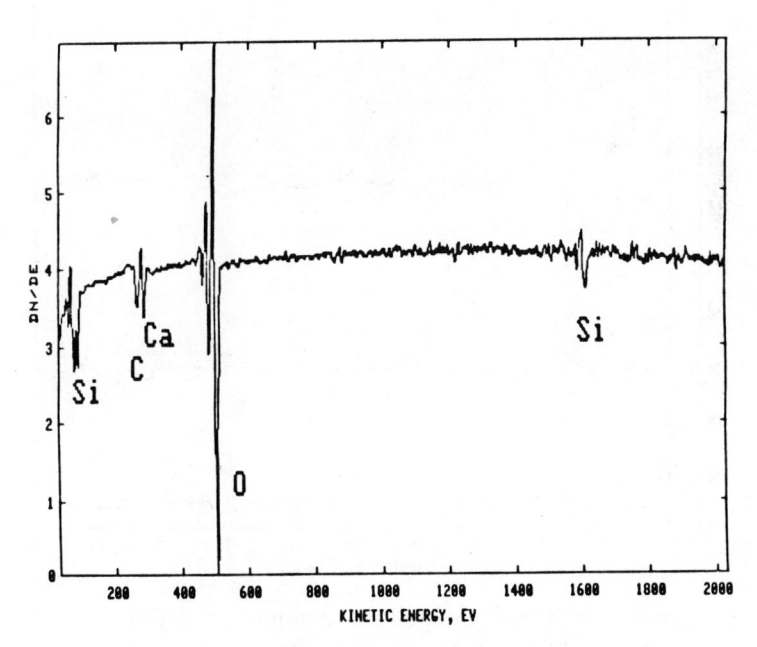

Fig. 11b. Auger spectrum on the glass.

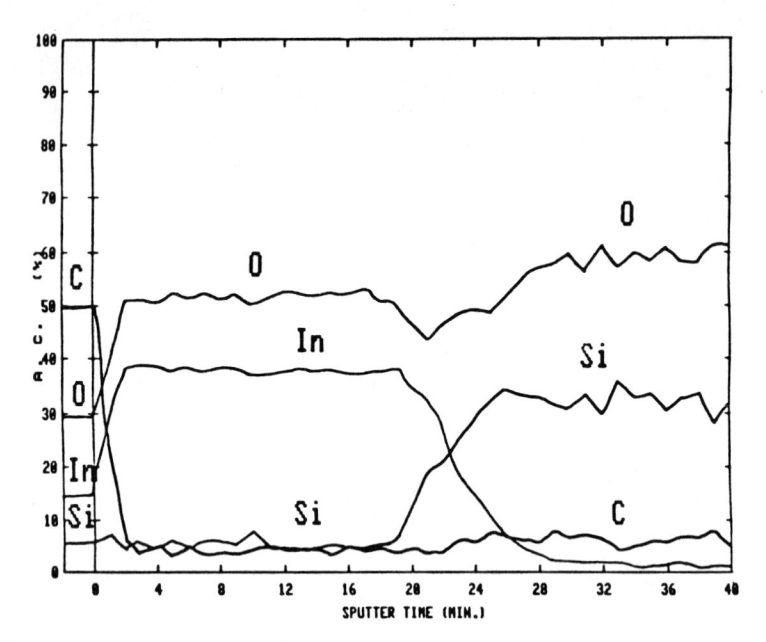

Fig. 12a. Auger profile showing the change in atomic concentration as a function of depth for the metal oxide lead.

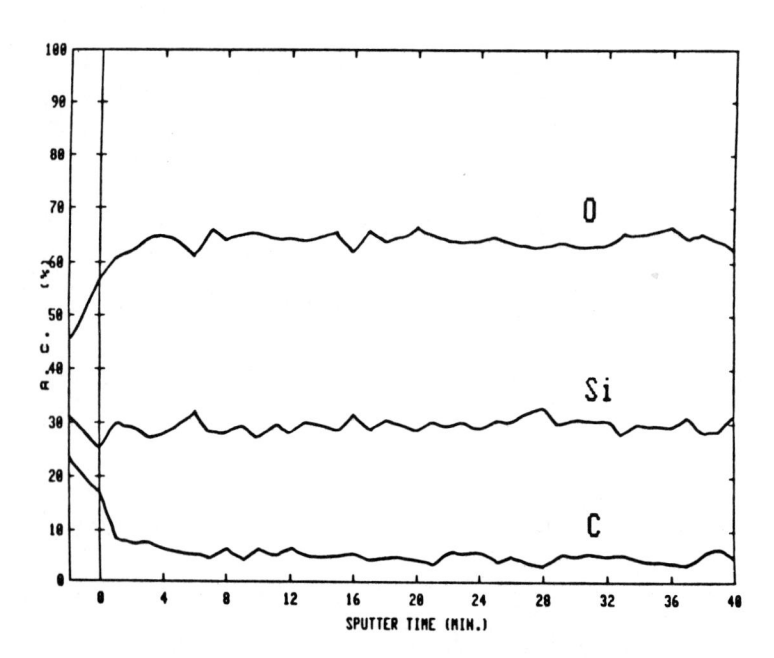

Fig. 12b. Auger profile showing the change in atomic concentration as a function of depth for the glass substrate.

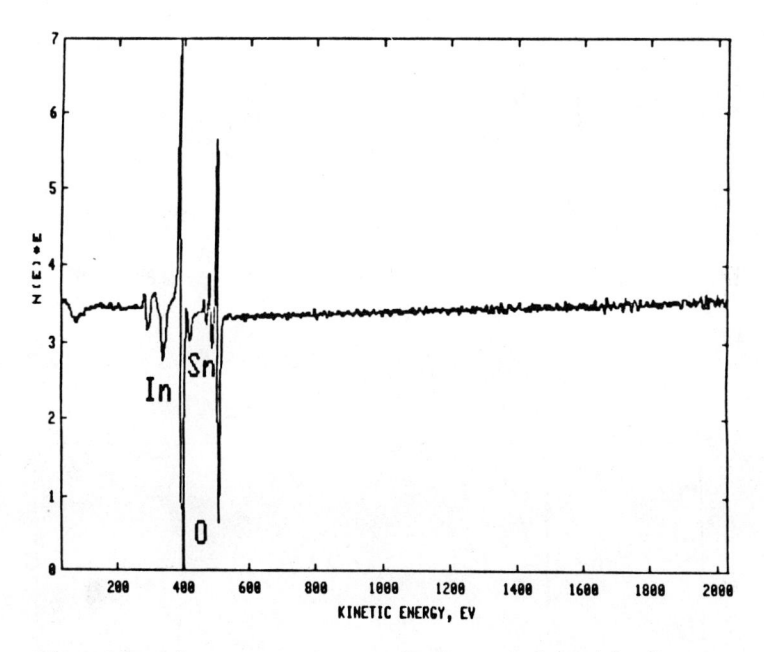

Fig. 13. Auger spectrum of the metal oxide layer.

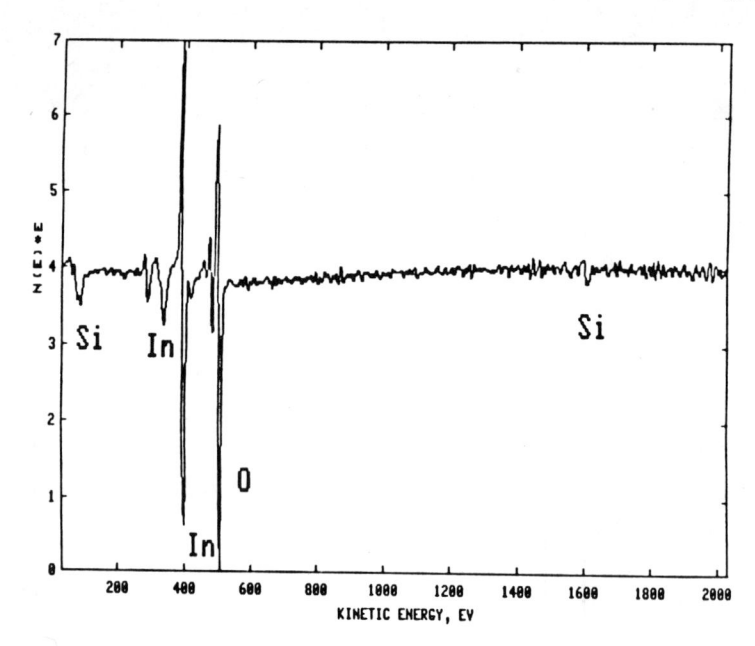

Fig. 14a. Auger spectrum of the metal oxide-glass interface.

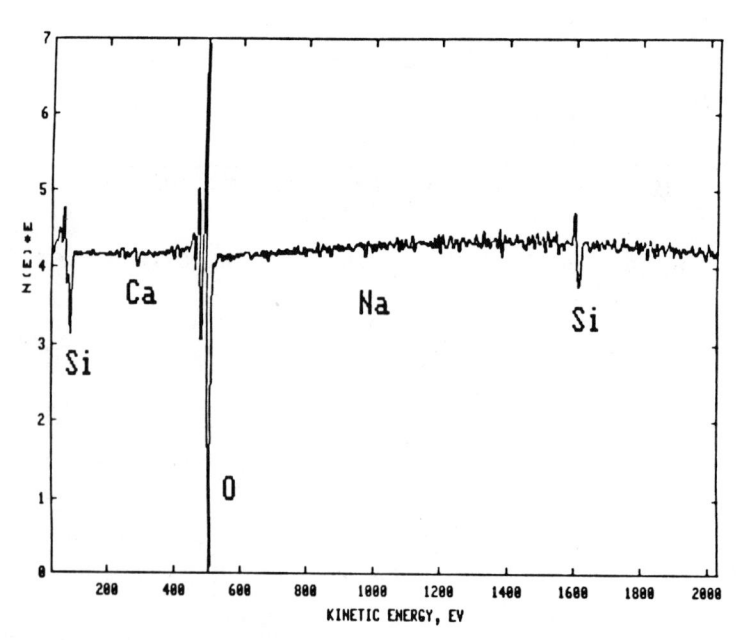

Fig. 14b. Auger spectrum of the sputtered glass.

REFERENCES

1. Frederic J. Kahn, Phys. Today $\underline{35}$, 66 (1982).

2. J. Davis Litster and Robert J. Birgeneau, Phys. Today $\underline{35}$, 26 (1982).

3. Liquid Crystal Displays in Japan, (Stanford Resources, Inc., San Jose, Calif. 1980), p. 124-140.

4. C. D. Wagner, H. A. Six, W. T. Jansen, J. A. Taylor, Appl. Surface Sci. $\underline{9}$, 203 (1981).

5. T. A. Carlson, Photoelectron and Auger Spectroscopy (Plenum, New York, 1975), p. 15-97.

6. A. Joshi, L. E. Davis and P. W. Palmberg in Methods of Surface Analysis, edited by A. W. Czanderna (Elsevier, New York, 1975), p. 158-169.

7. Bradway F. Phillips (unpublished).

8. C. G. Pantano and J. F. Kelso, "Spectroscopic Surface Studies of Glass and Ceramic Materials," same volume.

9. C. D. Wagner, W. M. Riggs, L. E. Davis, J. F. Moulder and G. E. Muilenberg, Handbook of X-ray Photoelectron Spectroscopy (Perkin-Elmer, Eden Prairie, MN 1979), p. 36-38.

10. A. W. C. Lin, N. R. Armstrong and T. Kuwana, Anal. Chem. $\underline{49}$, 1228 (1977).

11. C. D. Wagner, L. E. Davis, M. V. Zeller, J. A. Taylor, R. H. Raymond and L. H. Gale, Surf. Interface Anal. $\underline{3}$, 211 (1981).

12. N. S. McIntyre and F. W. Stanchell, J. Vac. Sci. Technol. $\underline{16}$, 798 (1979).

13. L. E. Davis, N. C. MacDonald, P. W. Palmberg, G. E. Riach and R. E. Weber, Handbook of Auger Electron Spectroscopy, (Physical Electronics Industries, Inc., Eden Prairie, MN, 1976), p. 11-17.

14. R. G. Gossink and T. P. A. Lommen, Appl. Phys. Lett. $\underline{34}$, 444 (1979).

ATOMIC STRUCTURE AT ELECTRODE SURFACES

Gerald A. Garwood, Jr.* and Arthur T. Hubbard

Department of Chemistry
University of California
Santa Barbara, CA 93106

INTRODUCTION

Modern techniques for preparation and electron spectroscopic characterization of surface structure and composition in ultra-high vacuum (UHV) have ushered in an era of rapid progress for gas-solid surface studies.[1-2] These advanced techniques have great potential usefulness in the science of electrochemistry.[3-4] Specifically, low-energy electron diffraction (LEED) provides data from which electrode surface and adsorbed layer crystallographic structure can be determined.[5] This represents a substantial advance for the science of electrochemistry in which surfaces of undefined structure on polycrystalline or bulk single-crystalline material have been the norm. Furthermore, Auger electron spectroscopy (AES) quantitates the elemental composition of the surface and adsorbed layer,[6] including interfacial impurities. Thermal desorption mass spectrometry (TDMS) provides data regarding the molecular constitution and thermal stability of the adsorbed material.[7] A growing number and variety of other techniques also highly useful for study of electrode surfaces have become available[8-13] although not exploited for the present study.

An understanding of how solvents and electrolytes interact with electrode surfaces is of both fundamental and practical importance.[14-15] A majority of polar compounds chemisorb on metal surfaces,[16] and electrode reactions are inherently sensitive to these chemisorption phenomena.[17] Adsorption of solvent can be controlled by surface pretreatment in solution.[18] Electrode

---

*Permanent Address: Santa Barbara Research Center, Goleta, California, 93117, U.S.A.

reactions in particular are strongly influenced by halogens and halides.[19] Accordingly, studies of adsorbed materials such as solvents and electrolytes by LEED and related methods will yield facts concerning stuctures, chemical composition and stability which will be informative to electrochemists.

The present article describes studies of the structures formed by typical polar solvents and other electrolytic materials such as I_2 and hydrogen halides on well-defined platinum single crystal surfaces in ultra-high vacuum. By combining effusion beam dosing with efficient pumping, dosages spanning the range from UHV to pressures approaching the vapor pressure of the pure liquid were included. The structure of the adsorbed layer was studied by LEED, elemental composition by Auger spectroscopy and thermal stability by thermal desorption mass spectrometry.

EXPERIMENTAL ASPECTS

Electrode surfaces were prepared and exposed to electrolyte vapor in a UHV system using a nozzle-beam doser at pressures low enough to allow use of electron techniques but high enough to approach the equilibrium vapor pressure of the electrolyte. A brief review of the UHV techniques employed in this study (LEED, AES and TDMS), together with a short description of the UHV system and material requirements are given below.

Brief Review of UHV Techniques

Low-energy electron diffraction provides data from which surface crystallographic structure can be determined, including the structure of the first few atomic layers of the substrate and any adsorbed material which might be present. The symmetry of the diffraction pattern is a direct consequence of the periodic arrangement of the atomic or molecular constituents in the surface of the crystal. In a diffraction experiment the elastic component of a scattered electron beam (which contains the primary diffraction information) can be separated from the inelastic components by a post-acceleration technique illustrated in Figure 1. The elastic component which has penetrated the cathodic grid system is accelerated (+7000 eV potential) onto a fluorescent screen causing light emission where the electron hits. In this study, LEED patterns were photographed by means of a Nikkormat camera on high-speed black-and-white film. An exposure of 30s at f5.6 was generally optimal at the beam currents stated in the figure captions. Patterns were observed throughout an incident energy range from 10 to 250 eV in order to locate all constituent spots.

Auger electron spectroscopy yields spectral data indicative of the identity and quantity of elements present in the surface

ATOMIC STRUCTURE AT ELECTRODE SURFACES

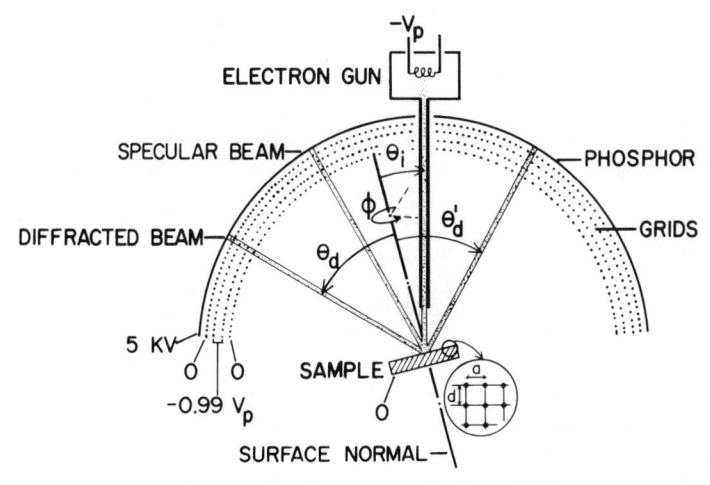

Figure 1. Schematic diagram of the LEED experiment.

region. In principle, all elements except H and He are determinable by means of AES provided that their abundance exceeds about 1%. Auger electron emission was induced by means of a 10 μA beam of 2000 eV electrons incident on the sample at a grazing angle (17°, measured from the surface plane). Auger electron energy distributions (spectra) were sinusoidally modulated (1000 Hz, 1.50 to 6.1V amplitude) and detected by using the four-grid LEED optics as a retarding-field kinetic energy analyzer,[20] Figure 2. Alternating current collected by the phosphorescent screen (operated at +250V) was converted to an alternating voltage by means of an audio-transformer (Ouncer UTC O-26). A lock-in amplifier supplied

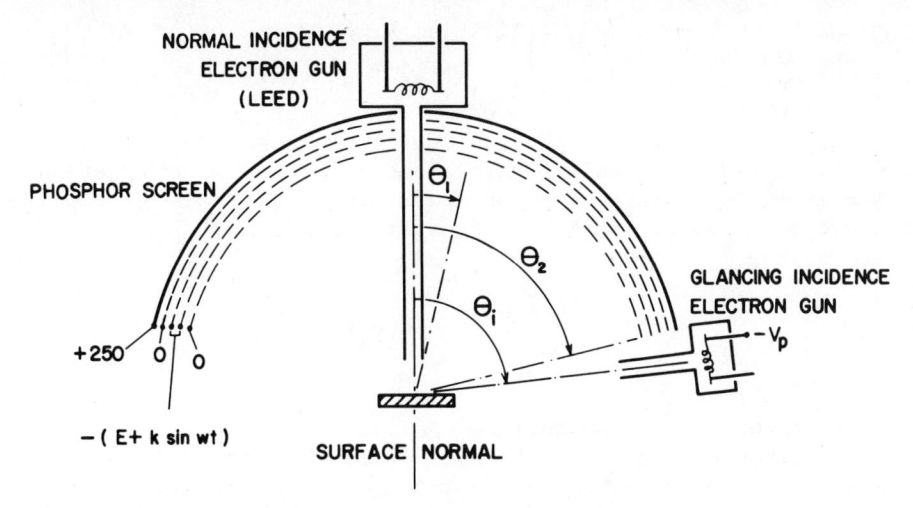

Figure 2. Schematic diagram of the AES experiment.

the modulating sine wave, and converted the second harmonic of the modulated Auger signal to a dc voltage (Model 128, Princeton Applied Research, Princeton, NJ 08540, USA). Spectra were scanned at a rate of 0.5 eV/s at the 50 eV sweep width and 1.5 eV/s otherwise. A time-constant setting on the lock-in-amplifier RC filter of 0.3s was selected. Integration of second-harmonic Auger spectra to obtain the Auger electron current due to the stated transitions of each element was obtained by Simpson's Rule numerical integration of the spectrum after correcting for background. The area enclosed under the resulting curve was determined either by graphical integration or by a second stage of Simpson's Rule numerical integration. Quantitative Auger spectroscopic theory and practice for adsorbed layers are reviewed in reference 21. A brief description will be given here. Total Auger electron current, I, for a given element (integrated over all transitions proceeding from the same initially ionized level) of an adsorbed atomic layer is given by Equation 1,

$$I = I° \phi_c G\Gamma \tag{1}$$

where $I°$ is the primary beam current, ϕ_c is the analyzer efficiency, and Γ is the atom density (atoms/cm^2). G is the proportionality factor, for which Equation 2 applies,

$$G = (Q \sec \theta_i + S) \tag{2}$$

where Q is the cross-section for the initial ionization, θ_i is the angle of incidence (measured from the surface normal) and S is the backscattering factor. Q was calculated from an equation derived by Gryzinski:[22]

$$Q = 6.56 \times 10^{-14} \frac{n}{E_p E_w} \left[\frac{(E_p/E_w)-1}{(E_p/E_w)+1}\right]^{3/2} \left\{1 + \frac{2}{3}\left(1 - \frac{E_w}{2E_p}\right) \ln\left[2.7 + \left(\frac{E_p}{E_w}-1\right)^{1/2}\right]\right\} \tag{3}$$

where E_p, the kinetic energy of electrons in the incident beam, was 2000 eV. E_w is the binding energy, and n the number of electrons of the level to be initially ionized; values of E_w were obtained from Siegbahn, et al.[23] S was found from Equation 4,

$$S = \frac{12.48}{kI_p} \int_{E_w}^{E_p} Q(E) A_1(E) \, dE \tag{4}$$

where k is the modulation amplitude (1.50 to 6.1V) and $A_1(E)$ is the observed first harmonic amplitude. Values of these experimental constants and integrals appear in Table 1. For simplicity, where quantitative values of Γ are not required, the ratio of Auger electron current to the maximum observed value is given, I/I_m.

Table 1. Values of Experimental Constants for Auger Spectroscopy of Adsorbed Layers.[21]

Element	ϕ_c	$G(10^{-19} cm^2)$	$S(10^{-19} cm^2)$	n	$E_w(eV)$[23]
C	0.098	17.0	6.68	2	284
N	0.100	8.70	2.79	2	399
S	0.094	140	71.3	6	164
Cl	0.095	97.4	45.4	6	201
I	0.096	15.9	3.66	10	685

Typically, Γ is used to calculate a parameter known as coverage, θ, which is given by

$$\theta = \Gamma / \Gamma_{substrate} \qquad (5)$$

where $\Gamma_{substrate}$ is 1.50×10^{15} and 1.30×10^{15} Pt atoms/cm² for the ideal Pt(111) and Pt(100) surfaces, respectively. Auger spectra and LEED patterns reported here did not change with time of exposure to the electron beams, ruling out beam damage as a factor in these experiments.

Thermal desorption mass spectrometry is a convenient source of valuable information regarding the stability and molecular constitution of the adsorbed layer. Thermal desorption mass spectra were obtained using a quadrupole mass filter (Model 1110A, R.M. Jordan Co., Mountain View, CA 94040, USA) mounted on a movable support permitting line-of-sight detection at very close proximity to the sample (1 mm), as depicted in Figure 3. The sample was heated by passing a constant current (5.5A) through the pair of fine wires supporting the sample (0.41 mm × ca. 10 mm). The limiting temperature at this power setting was 1075K, measured using an optical pyrometer (8622C, Leeds and Northrup, North Wales, PA 19454, USA). Temperatures in the range from 295 to 950K were estimated by means of Equation 6 which is consistent with the known power input and the heat equations (flow of heat in one dimension from an object at constant temperature to a second object initially at room temperature),[24]

$$T = 780[1 - \exp(-bt)] + 295 \qquad (6)$$

where T is the Kelvin temperature of the sample after heating for a time, t, in s and b = 0.010 to 0.030 for the various adsorbate studies. Equation 6 accurately represented the temperature in the range where it could be tested by the pyrometer (T ⩾ 950K), and presumably at the other temperatures, also.

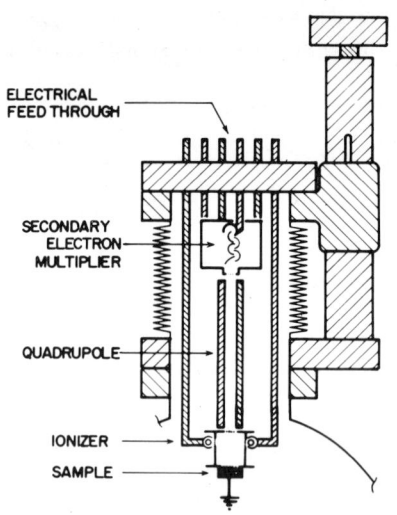

Figure 3. Schematic diagram of the TDMS experiment.

UHV System and Material Requirements

The instrument employed for this work was equipped for ultra-high vacuum LEED, Auger and thermal desorption mass spectroscopic characterization of the sample in rapid succession.[25] All internal components not constructed of stainless steel were electroplated with gold[26] (e.g., copper gaskets and electrical fittings) to alleviate problems encountered after introduction of corrosive gases. Prior to each experimental cycle, the entire vacuum enclosure was baked at 250°C for one to three days to facilitate removal of volatile materials. When the apparatus had cooled to room temperature, the residual pressure was below 10^{-9} Torr. The ion guage, mass spectrometer, electron guns and ion gun were operated at maximum temperature for at least 30 min prior to sample cleaning.

Platinum single crystals (purity > 99.99%) were obtained from the Materials Research Corp. (Orangeburg, NY 10962, USA). Each crystal was oriented to within 1° of angle by reflection X-ray diffraction (tungsten tube, operated at 40 kV and 25 mA). Reference 27 proved helpful in this regard. Each crystal was cut using a diamond wheel and polished with successively finer grades of alumina (Buehler, Ltd., Evanston, IL 60204, USA) so as to form a parallelopiped having all six faces equivalent to the stated crystallographic orientation. The finished sample was supported by a pair of fine (0.41 mm) Pt wires spot-welded to one face of the crystal; these also served for electrical heating of the sample. The sample was immersed in boiling, concentrated HNO_3 prior to installation in the vacuum system. At first use, each Pt crystal was cleaned by heating in O_2 for several hours (10^{-7} Torr O_2,

900°C), followed by Ar$^+$ ion-bombardment at 900K for 1h (5 × 10^{-5} Torr Ar, 10 μA/cm^2 Ar$^+$ ion current-density at 600 eV kinetic energy), and annealing to optimal sharpness of the LEED pattern (900K for about 5 min). During Ar$^+$ bombardment, the ion pumps were de-energized and a titanium sublimation unit with a cryogenically cooled baffle was employed to selectively remove reactive impurities from the residual gas. After this cleaning procedure Auger spectra were obtained from 50 to 2000 eV and if impurities were evident the cleaning procedure was continued. Brief ion-bombardment was adequate to remove adsorbed materials accumulated during the gas adsorption experiments. Preparation of clean Pt surfaces has been discussed.[28]

Pressurized gases (Ar, HBr, HCl, NH$_3$, HI, SO$_2$) were used as received from the supplier. Water was freed of hydrocarbons by pyrolytic distillation using O$_2$ at 900°C over a Pt gauze catalyst in a quartz glass apparatus.[29] Other chemicals were reagent grade, or better, and were used as received from the supplier (iodine, pyridine, DMSO, dimethylformamide, p-dioxane, dichloromethane, acetonitrile, acrylic acid, sulfolane, propylene carbonate, acetic acid). In order to realize a high level of purity in the material actually introduced onto the sample surface, the inlet manifolds were specially constructed of inert materials (Pyrex glass, Teflon) and the pressure regulators of Monel Alloy; a pair of Varian leak valves (modified by construction of its gasket and seat from platinum), each with a short length of tubing terminating in a one mm orifice[30] (Type 304 stainless steel), led to the sample. The movable mass analyzer permitted direct monitoring of the effusion beam (on line-of-flight). Preparation of the inlet manifold was continued until the mass spectrum showed that only pure solvent or electrolyte was present. Pressures of gases in the vacuum chamber were estimated by means of a Bayard-Alpert ionization gauge calibrated for nitrogen. Gauge reading at the onset of dosing was about 10^{-10} Torr, corrected for background, and was incremented gradually to about 10^{-5} Torr to verify that no further changes in adsorbed layer structure or composition occurred at extreme dosages. To prevent surface contamination by pyrolysis products, the gauge was employed merely for leak-valve calibration and was not operated during the actual dosing procedure. Pressure at the Pt crystal surface was estimated from the orifice diameter (1 mm) and the conductance to the pumping compartment to be a factor of 3 × 10^4 times that in the chamber. Accordingly, the initial dosing pressure was very approximately equivalent to 3 × 10^{-6} Torr; the maximum dosing pressure was about 0.3 Torr. Sticking coefficients and their variation with coverage were not documented quantitatively; however, the qualitative variation of coverage with dosage and time[31-35] indicates that the electrolyte and solvent compounds (except CH$_2$Cl$_2$, NH$_3$ and H$_2$O) are rapidly and efficiently adsorbed, reaching saturation after less than a few tens of Langmuirs. Care was taken to insure that the adsorbed material originated directly from the effusion beam;

the sample face opposite the one being dosed remained relatively free of adsorbed material.

STRUCTURE AT WELL-DEFINED PLATINUM SURFACES

Recent ultra-high vacuum-electrochemical studies have concentrated on Pt, although the techniques are broadly applicable.

Polar Solvents Adsorbed on Pt(111)

Well-defined Pt(111) surfaces were exposed in UHV conditions to a nozzle beam of a variety of typical polar solvent materials (I-XI):

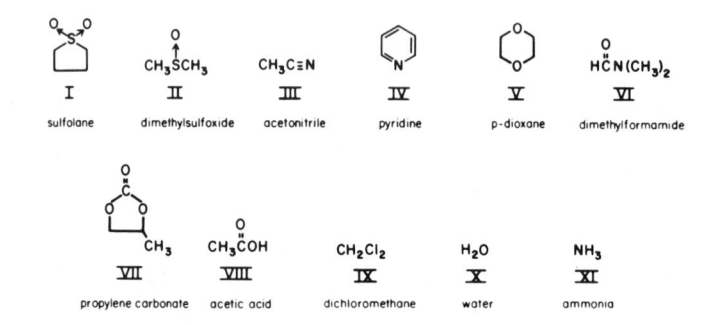

I sulfolane	II dimethylsulfoxide	III acetonitrile	IV pyridine	V p-dioxane	VI dimethylformamide
VII propylene carbonate	VIII acetic acid	IX dichloromethane	X water	XI ammonia	

The adsorbed layers were investigated by LEED, Auger spectroscopy and TDMS. The basal (hexagonal) plane, Pt(111), was employed for these studies since facets of this stable orientation constitute a major portion [about 55% vs. 45% for Pt(100)] of the surface of practical platinum electrodes. Interfacial concentration (coverage) was measured by AES.[21] The compounds are listed very approximately in order of decreasing coverage at comparable exposure. Compounds I-VIII were strongly adsorbed from vacuum, whereas compounds IX-XI were not adsorbed to an appreciable extent ($< 10^{-2}$ molecule/surface Pt atom) at nozzle-beam dosages equivalent to a pressure of 0.3 Torr. These UHV observations correlate nicely with the influence of strongly adsorbing solvents on the kinetics of electrode processes: compounds I-IV (as neat liquids and dilute aqueous solutions) reacted with polycrystalline Pt to produce a strongly adsorbed layer able to prevent electrolysis of dissolved reactants.[18] However, compounds V-XI did not interfere with electrolysis in keeping with their lesser affinity for Pt surfaces in UHV.

LEED pattens obtained for Pt(111) surfaces exposed in UHV to compounds I-VIII revealed a surprising trend: each of these compounds (except pyridine) displayed a sharply defined Pt(111)(2×2)

ATOMIC STRUCTURE AT ELECTRODE SURFACES   99

LEED pattern at approximately quarter coverage, Figure 4B. Higher coverages, up to a full monolayer, were reached without producing other distinct structures, except that sulfolane and DMSO each produced Pt(111)($\sqrt{3} \times \sqrt{3}$)R30° and Pt(111)(1×1) structures at about one-third and full monolayer coverage, respectively (Figures 4C, 4D). Heating of the adsorbed layer in each case produced a series of transitions in structure and composition, and caused cracking of the solvent molecule leading eventually to a refractory layer which persisted at red heat.[34] From these results one may learn that

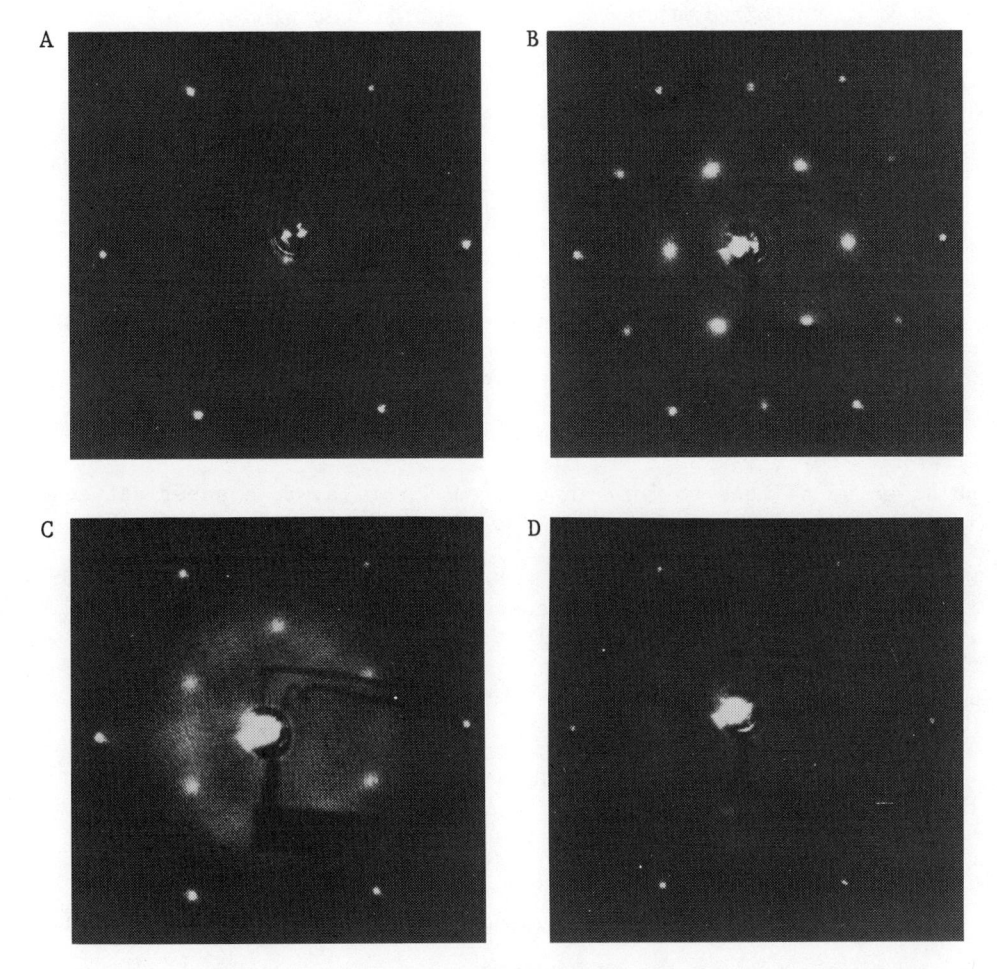

Figure 4. LEED patterns of Pt(111) after exposure to sulfolane: (A) clean Pt(111); (B) Pt(111)(2×2)-sulfolane, at an Auger intensity ratio for carbon of $(I/I_m)_C = 0.507$; (C) Pt(111)($\sqrt{3} \times \sqrt{3}$)R30°-sulfolane, at $(I/I_m)_C = 0.625$; (D) Pt(111)(1×1)-sulfolane, at $(I/I_m)_C = 1$. Incident beam current: 0.33 to 0.36 µA; kinetic energy: 60 eV.

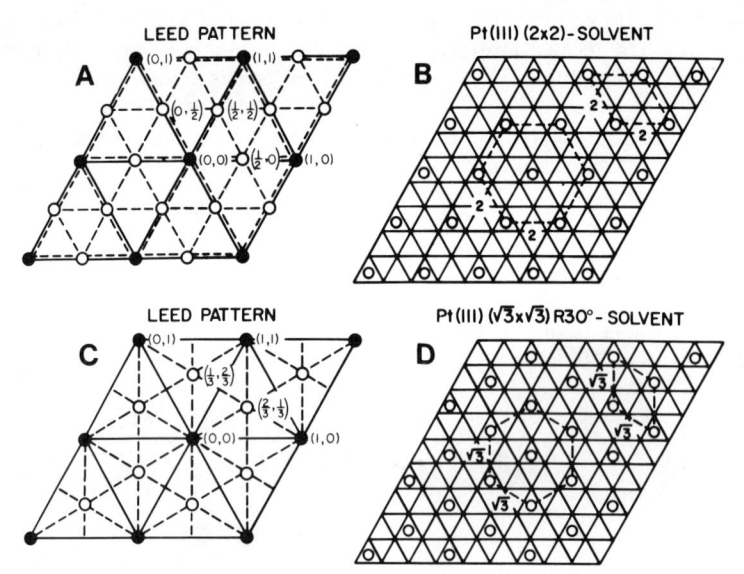

Figure 5. Nets of LEED patterns and structures formed by exposure of Pt(111) to polar solvents. LEED patterns: circles (o), fractional-index spots; dots (●), integral-index spots. Structures: circles (o), lattice points of adsorbed layer; intersections of solid lines, Pt atoms.

polar solvents tend to form an ordered first layer, a superlattice, in response to the structure of the electrode surface. Solvents adsorbed on the hexagonal (rhombic) Pt(111) surface are arranged hexagonally, in registry with the substrate, Figure 5. The size of the superlattice unit cell decreases as coverage increases, reaching an apparent limit of one solvent moiety per surface platinum atom.

Hydrogen Halides Adsorbed on Pt(111) and Pt(100)

HI and I_2 each react with Pt(111) and Pt(100) surfaces in UHV to form an interesting series of structurally related superlattices.[31-32] The reactions are evidently dissociative, Equations 7 and 8, as the structures produced by HI and I_2

$$HI \xrightarrow{Pt\ surface} I(adsorbed) + 1/2\ H_2(gaseous) \quad (7)$$

$$1/2\ I_2 \xrightarrow{Pt\ surface} I(adsorbed) \quad (8)$$

are identical and the adsorbed layer consists of I atoms, except that Pt(100) begins to retain HI in undissociated form when coverage exceeds $\theta = 0.50$. On the Pt(111) surface two superlattices are produced: Pt(111)($\sqrt{3} \times \sqrt{3}$)R30°-I ($\theta = 0.33$ I-atom per surface

ATOMIC STRUCTURE AT ELECTRODE SURFACES   101

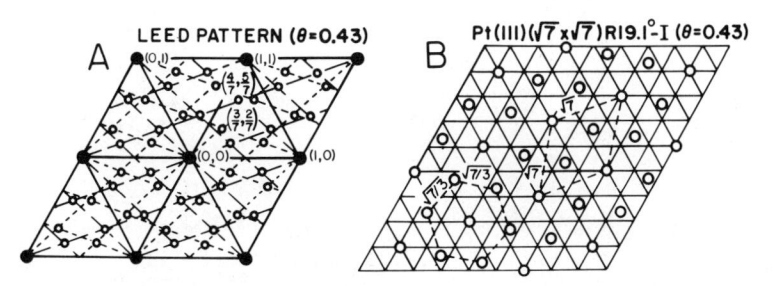

Figure 6. Nets of a LEED pattern and an I-atom structure formed by
           exposure of Pt(111) to HI: (A) LEED pattern, θ = 0.43;
           (B) structure: Pt(111)($\sqrt{7}$ × $\sqrt{7}$)R19.1°-I, θ = 0.43.
           LEED pattern: circles (o), fractional-index spots; dots
           (●), integral-index spots. Structure: circles (o), I
           atoms; intersections of solid lines, Pt atoms.

Pt atom) and Pt(111)($\sqrt{7}$ × $\sqrt{7}$)R19.1°-I (θ = 0.43), Figure 6. The
superlattices which form on Pt(100) are numerous and may be identified as Pt(100)[c(2×4)]-I (θ = 0.25), rings (0.34 ≤ θ < 0.43),
Pt(100)[c(2$\sqrt{2}$ × N$\sqrt{2}$)]R45°-I (0.43 ≤ θ < 0.50) and Pt(100)[c(2$\sqrt{2}$ ×
$\sqrt{2}$)]R45°-I (θ = 0.50), Figure 7. All of these structures consist
of hexagonal nets of I atoms, slightly distorted in some instances
to improve registry with the (100) substrate. Accordingly, these
structures are related by slight compression and/or rotations of
their hexagonal subunit. Quantitative Auger electron spectroscopy
confirms coverages inferred by LEED. Thermal desorption mass spectrometry reveals prominent rate maxima at 738 and 985K, associated
with desorption from the Pt(100)[c(2$\sqrt{2}$ × $\sqrt{2}$)]R45°-I and Pt(100)-
[c(2×4)]-I structures, respectively, while the Pt(111) structures
yield a small peak and a broad continuum. Atomic iodine is the
predominant thermal desorption product for θ ≤ 0.50. At coverages
above θ = 0.50, thermal desorption produces HI or I from adlayers
formed on Pt(100) by HI and I_2, respectively.

HBr and HCl react with Pt(111) and Pt(100) surfaces to form
adsorbed layers consisting of specific mixtures of halogen atoms
and hydrogen halide molecules.[33] Exposure of Pt(111) to HBr yields
a (3×3) pattern (Figure 8A) beginning at θ_{Br} = 2/9 and persisting
at the maximum coverage which consists of θ_{Br} = 1/3 plus θ_{HBr} = 1/9.
The most probable structure at maximum coverage, Pt(111)[c(3×3)]-
(3Br + HBr), has a rhombic unit cell encompassing nine surface Pt
atoms, and containing three Br atoms and one HBr molecule, Figure
8B. On Pt(100) the structure at maximum coverage appears to be
Pt(100)[c(2$\sqrt{2}$ × $\sqrt{2}$)]R45°-(Br + HBr), θ_{Br} = θ_{HBr} = 1/4; the rectangular unit cell involves four Pt atoms, one Br atom and one HBr
molecule, Figure 9A. Each of these structures consists of an hexagonal array of adsorbed atoms or molecules, excepting slight distortion for best fit with the substrate in the case of Pt(100).

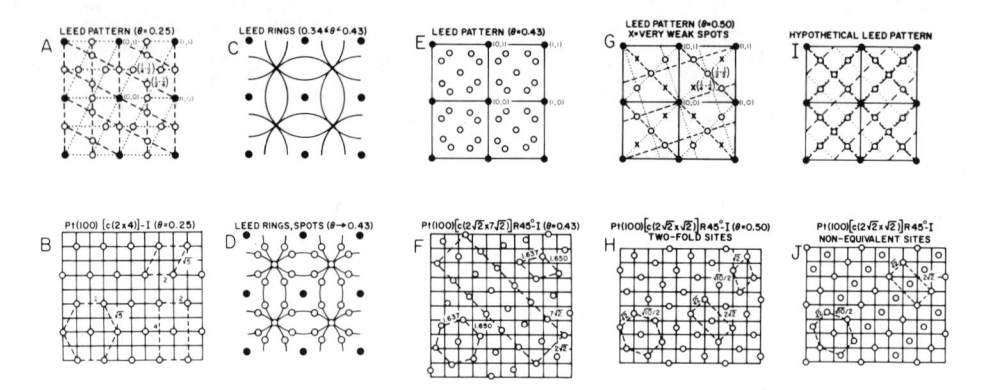

Figure 7. Nets of LEED patterns and I-atom structures formed by exposure of Pt(100) to HI: (A) LEED pattern, θ = 0.25; (B) structure: Pt(100)[c(2×4)]-I, θ = 0.25; (C) LEED rings, 0.34 ≤ θ < 0.43; (D) LEED rings, with spots, θ approaching 0.43; (E) LEED pattern, θ = 0.43; (F) structure: Pt(100)[c(2√2 × 7√2)]R45°-I, θ = 0.43; (G) LEED pattern, θ = 0.50, x = very weak spots; (H) structure: Pt(100)[c(2√2 × √2)]R45°-I, θ = 0.50, I atoms in two-fold sites; (I) LEED pattern expected from structure shown in Figure 8J; (J) structure: Pt(100)[c(2√2 × √2)]R45°-I, I atoms in non-equivalent sites. LEED patterns: circles (o), fractional-index spots; dots (•), integral-index spots. Structures: circles (o), I atoms; intersections of solid lines, Pt atoms.

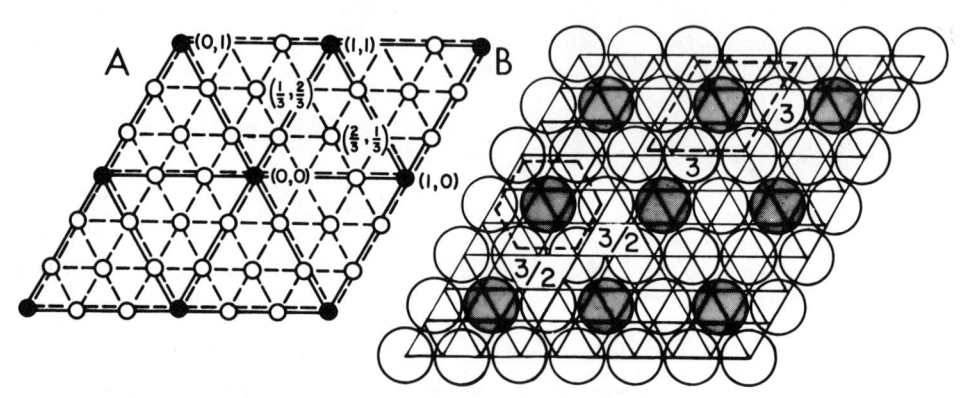

Figure 8. Nets of the LEED pattern and a Br/HBr structure formed by exposure of Pt(111) to HBr: (A) LEED pattern, θ = 2/9 to 4/9; (B) Pt(111)[c(3×3)]-(3Br + HBr) structure, θ_{Br} = 3/9, θ_{HBr} = 1/9. LEED pattern: dots, integral-index spots; circles, fractional-index spots. Structure: intersections of solid lines, Pt atoms; circles, Br atoms; shaded circles, HBr molecules.

ATOMIC STRUCTURE AT ELECTRODE SURFACES   103

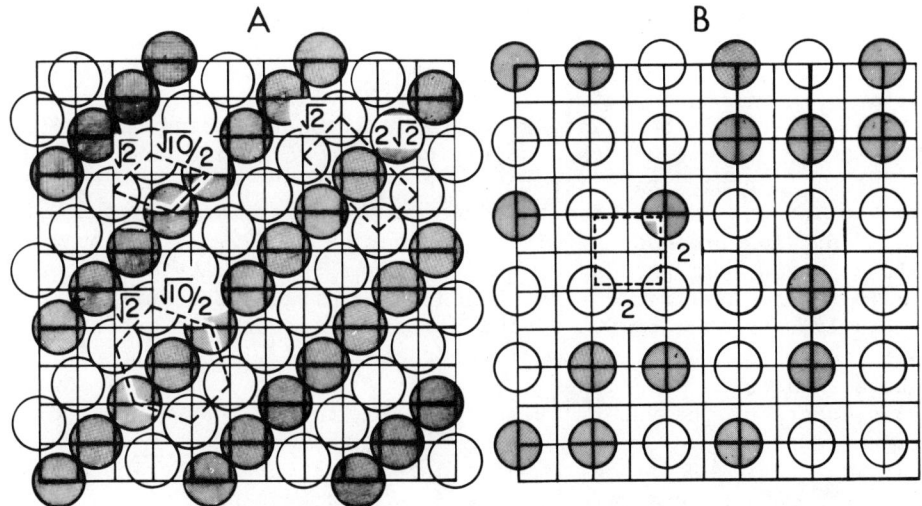

Figure 9. Structures formed by exposure of Pt(100) to HBr and HCl:
(A) Pt(100)$\left[c(2\sqrt{2} \times \sqrt{2})\right]$R45°-(Br + HBr) structure, θ_{Br} = θ_{HBr} = 1/4, Br and HBr at two-fold sites; (B) Pt(100) (2×2)-(Cl + HCl) structure, θ_{Cl} = 0.13, θ_{HCl} = 0.11. Intersections of solid lines, Pt atoms; circles, Br or Cl atoms; shaded circles, HBr or HCl molecules.

Treatment of Pt(100) with HCl produces a diffuse Pt(100)(2×2)-(Cl + HCl) structure at the maximum coverage of θ_{Cl} = 0.13, θ_{HCl} = 0.11, Figure 9B. Exposure of Pt(111) to HCl produces a disordered overlayer and a maximum coverage of $\theta_{(HCl + Cl)}$ = 0.32. Heating the adsorbed layers to 1075K causes the halogens to completely desorb, with Br and HBr, Cl and HCl as the predominant thermal desorption products. Thermal desorption data also reveal prominent rate maxima associated with the structural transitions observed by LEED.

Sequential Adsorption of Pairs of Compounds

Electrodes having Pt(111) or Pt(100) single-crystal surfaces were exposed to vapor of one substance followed by another, and the resulting adsorbed layers were characterized by LEED, AES and TDMS.[35] Pairs of compounds were chosen from among various common solvents, electrolytes, acids and bases: water, pyridine, acetonitrile, dimethylsulfoxide, hydrogen bromide, iodine, sulfur dioxide, acrylic acid ($CH_2 = CHO_2H$) and ammonia. Dosage was varied from low pressures (ca. 10^{-5} Torr) to pressures approaching the vapor pressure of the pure liquid. Specifically, the combinations investigated were: HBr and H_2O, HBr and NH_3, acryclic acid and NH_3, SO_2 and NH_3, I_2 and pyridine, I_2 and acetonitrile, I_2 and DMSO, DMSO and pyridine, pyridine and H_2O, I_2 and HBr.

A common characteristic of all pairs of compounds studied is that a chemisorbed atomic or molecular layer is not very reactive toward vapor, at least not at room temperature at well-defined Pt surfaces in pressures between 10^{-5} and 10^{-1} Torr. Even in the case of reagents having contrasting properties, such as acids (HBr, SO_2, $CH_2 = CHO_2H$) with bases (NH_3, pyridine), the reactions were very subdued. An explanation for the unreactivity of the chemisorption products of HBr can perhaps be found in the dissociative nature of the adsorption process: fully three-fourths on Pt(111) and one-half on Pt(100) is present as Br atoms, and the acidity of the remaining material would be weakened by bonding to the surface. Similarly, attachment of pyridine derivatives to the Pt surface most likely involves Pt-aromatic nitrogen bonding,[36] a type of interaction certain to lower the basicity of those compounds toward constituents of the vapor. Binding of SO_2 to the Pt surface would also be expected to diminish the acidic character of the molecule, and particularly so since comparison of the limiting packing density with molecular models suggests a flat orientation with all three atoms in contact with the surface. The lack of acidity of adsorbed acrylic acid toward ammonia vapor is particularly interesting. Aqueous studies of this compound and related ones at Pt/aqueous interfaces did indicate a specific interaction between the carboxylate group and the Pt surface, but retention of most or all of the acidic character was indicated in that case. Missing at the Pt/vacuum interface, however, is solvent to stabilize ions of a product salt, and a means of forming a three-dimensional ionic crystalline product. What remains then, is a simple competition for bonding to the Pt surface.

The result of exposing Pt(111) surfaces presaturated with polar solvents DMSO, acetonitrile and pyridine to iodine vapor is a matter of considerable significance for electrochemistry and surface science generally. Iodine appears to bring about a molecular reorientation within the adsorbed solvent layer. The first indication that this is occurring is that while modest desorption of solvent took place, the amount of iodine adsorbed considerably exceeded the amount of solvent displaced in each case: acetonitrile coverage decreased 23% while iodine reached 65% of saturation; pyridine decreased 13% as iodine increased to 32%; and DMSO decreased 10% while iodine increased to 50%. Thermal desorption data indicated that the desorption product, the I atom, was the same as for the clean surface; also, no iodine compounds were observed which might have indicated that the iodine had reacted with other than the Pt(111) surface. Data for DMSO and I_2 suggest the probable nature of this reorientation process. The limiting coverage of pure DMSO on Pt(111), $\theta = 0.25$, is exactly what one would expect if the DMSO molecules were arranged as shown in Figure 10B in which the "flat" orientation shown below is adopted:

ATOMIC STRUCTURE AT ELECTRODE SURFACES 105

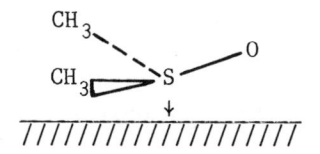

In this structure the sulfur atoms are located at the Pt surface lattice points, while the C and O atoms are located slightly to one side of the Pt lattice points (within about one-third of the Pt-Pt distance). This array of atoms resembles a (1×1) superlattice and would be expected to yield a (1×1) LEED pattern with slight diffuse scattering due to imperfect equivalency (placement and chemical identity) of the atomic scatterers, in agreement with experimental results. The DMSO-I mixed layer gives rise to a new LEED pattern (Figure 10C) which fixes the relative locations of the I atoms within the superlattice structure, Pt(111)[c(2×4√3/3)]-(DMSO + I), as illustrated in Figure 10D. The observed packing densities,

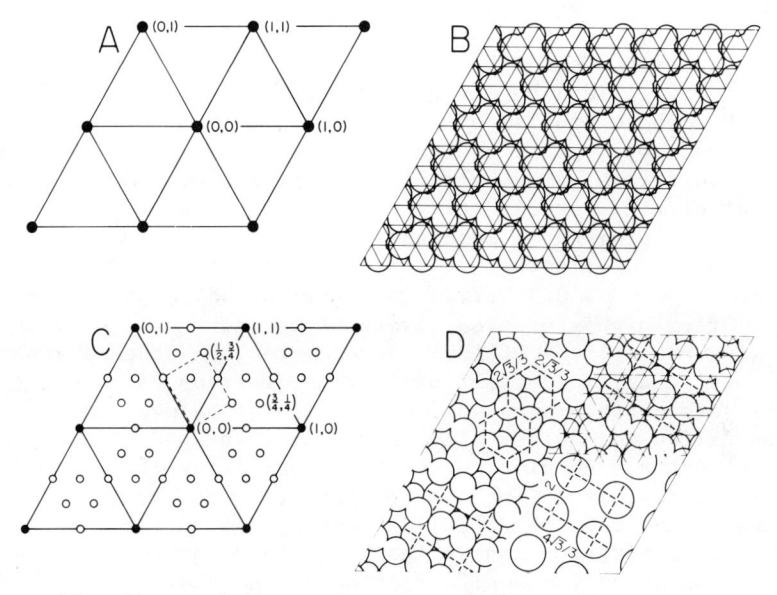

Figure 10. Nets of LEED patterns and structures on Pt(111): (A) LEED pattern after dosing clean Pt(111) surface with DMSO vapor to maximum coverage, $\theta = 1/4$; (B) structure, Pt(111)(1×1) - DMSO, $\theta_{DMSO} = 1/4$; (C) LEED pattern after dosing (B) with I_2 vapor; (D) structure, Pt(111)[c(2×4√3/3)] - (DMSO+I), $\theta_{DMSO} = \theta_I = 0.22$. LEED patterns: dots, integral-index spots; circles, fractional-index spots. Structures: intersections of solid lines, Pt atoms; circles, iodine atoms; ⊘ and ⊗, DMSO molecules in two different orientations.

based upon analysis[21] of the Auger spectra, $\theta_I = 0.21$, $\theta_{DMSO} = 0.21$, would not be spatially allowed apart from a reorientation of the DMSO molecule to a vertical orientation such as shown below, for which the calculated packing densities are $\theta_I = \theta_{DMSO} = 0.22$:

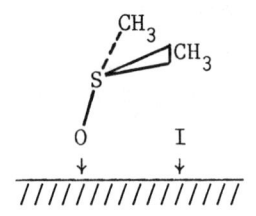

```
         CH₃
        /
       S───CH₃
       |
       O        I
       ↓        ↓
   /////////////////
```

This structure is illustrated in Figure 10D. Note that the I-atoms and methyl groups form an hexagonal network. This hexagonal network accounts for the observed packing densities and unique stability of the Pt(111)[c(2×4√3/3] - (DMSO+I) superlattice.

SUMMARY

Superlattices formed by the interaction of eleven polar solvents with Pt(111) and by HI, HBr, and HCl with Pt(111) and Pt(100) surfaces have been studied by LEED, Auger, and thermal desorption mass spectroscopy. Eight of the solvents chemisorb irreversibly on Pt(111): sulfolane, dimethylsulfoxide, pyridine, dimethylformamide, acetonitrile, acetic acid, propylene carbonate, and p-dioxane. The remaining three compounds (water, dichloromethane, and ammonia) do not chemisorb to an appreciable extent on Pt(111) at pressures as high as 0.3 Torr. The chemisorbed compounds (except pyridine) form highly ordered layers at certain coverages. HI reacts with Pt(111) and Pt(100) surfaces to form ordered adsorbed layers consisting exclusively of I atoms, whereas HBr and HCl produce adlayers containing specific mixtures of halogen atoms and hydrogen halide molecules. Each of the solvent and halide superlattice structures consists of an hexagonal array of adsorbed atoms and/or molecules, excepting slight distortion for best fit with the substrate in the case of Pt(100). Heating the surfaces to 1075K following chemisorption causes the complete desorption of the halides but removes only a small fraction of the chemisorbed solvent matter; the rest forms a very heat-resistant overlayer composed of most elements of the original adsorbed layer (C, N, O, S). Studies are also reported of the structure and composition of adsorbed layers formed by sequential exposure of Pt(100) and Pt(111) to pairs of compounds: solvents and electrolytic substances.

ACKNOWLEDGMENTS

We extend our thanks to the donors of the Petroleum Research Fund, administered by the American Chemical Society, and to the

National Science Foundation for support of this research. Dr. Robert Durand, Institut Nationale Polytechnic de Grenoble, France, assisted with the polar-solvent experiments. Acknowledgment is made to Santa Barbara Research Center for supporting Dr. Garwood in the presentation of this paper. Lastly, Dr. Garwood offers a personal and grateful acknowledgment to his "four-fathers."

REFERENCES

1. G.A. Somorjai, "Principles of Surface Chemistry," Prentice-Hall, Englewood Cliffs, NJ (1972).
2. M. Prutton, "Surface Physics," Clarendon, Oxford (1975).
3. A.T. Hubbard, Accounts Chem. Res. 13:177 (1980).
4. A.T. Hubbard, J. Vacuum Sci. Technol. 17:49 (1980).
5. J.B. Pendry, "Low-Energy Electron Diffraction," Academic Press, London (1974); G.A. Somorjai and H.H. Farrell, Advan. Chem. Phys. 20:215 (1971); M.A. Van Hove, Surface Sci. 80:1 (1979).
6. C.C. Chang, Surface Sci. 25:53 (1971); D.T. Hawkins, "Auger Electron Spectroscopy; A Bibliography, 1925-1975," Plenum, New York (1977); G.A. Somorjai and F.J. Szalkowski, Advan. High Temp. Chem. 4:137 (1971); N.J. Taylor, p. 117, in: "Techniques of Metals Research," Vol. 7, R.F. Bunshah, ed., Wiley-Interscience, New York (1971).
7. A. Benninghoven, Surface Sci. 35:427 (1973); R. Greenler, Critical Rev. Solid State Sci. 4:415 (1974); T.E. Madey and J.T. Yates, Jr., Surface Sci. 63:203 (1977); R.J. Madix, Catalysis Rev. Sci. Eng. 15:293 (1977); L.D. Schmidt, Catalysis Rev. Sci. Eng. 9:115 (1974).
8. B.G. Baker, p. 93, in: "Modern Aspects of Electrochemistry," Vol. 10, J.O'M. Bockris and B.E. Conway, ed., Plenum, New York (1975); T.A. Carlson, "Photoelectron and Auger Spectroscopy," Plenum, New York (1975); "Electron Spectroscopy for Surface Analysis," H. Ibach, ed., Springer-Verlag, Berlin (1977).
9. J.S. Hammond and N. Winograd, J. Electroanal. Chem. 80:123 (1977).
10. R.L. Gerlach, J.E. Houston and R.L. Park, Appl. Phys. Letters 16:179 (1970).
11. T.E. Madey and J.T. Yates, Jr., J. Vacuum Sci. Technol. 8:525 (1971).
12. C.A. Evans, Jr., Anal. Chem. 47(a):818A (1975).
13. P.E. Højiund-Nielsen, Surface Sci. 35:194 (1973).
14. C.K. Mann and K.K. Barnes, "Electrochemical Reactions in Non-Aqueous Systems," Marcel Dekker, New York (1970).
15. R.N. Adams, "Electrochemistry at Solid Electrodes," Marcel Dekker, New York (1969).
16. E. Gileadi, "Electrosorption," Plenum Press, New York (1967).
17. P. Delahay, "Double Layer and Electrode Kinetics," Interscience, New York (1965).

18. R.F. Lane and A.T. Hubbard, J. Phys. Chem. 81:734 (1977).
19. J.R. Cushing and A.T. Hubbard, J. Electroanal. Chem. 23:183 (1969); A.L.Y. Lau and A.T. Hubbard, J. Electroanal. Chem. 24:237 (1970); 33:77 (1971).
20. N.J. Taylor, Rev. Sci. Instr. 40:792 (1969).
21. J.A. Schoeffel and A.T. Hubbard, Anal. Chem. 49:2330 (1977).
22. M. Gryzinski, Phys. Rev. 138A:336 (1965).
23. K. Siegbahn, C. Nordling and A. Fahlman, "ESCA: Atomic, Molecular and Solid State Structure Studied by Means of Electron Spectroscopy," Amqvist-Wiksell, Uppsala (1967).
24. H.S. Carslaw and J.C. Jaeger, "Conduction of Heat in Solids," 2nd ed., London, Oxford (1969).
25. R.M. Ishikawa and A.T. Hubbard, J. Electroanal. Chem. 69:317 (1976); R.M. Ishikawa, J. Katekaru and A.T. Hubbard, J. Electroanal. Chem. 86:271 (1978).
26. "Modern Electroplating," 3rd ed., F.A. Lowenheim, ed., Wiley, New York (1974).
27. E.A. Wood, "Crystal Orientation Manual," Columbia Univ. Press, New York (1963).
28. P.W. Palmberg, in: "The Structure and Chemistry of Solid Surfaces," G.A. Somorjai, ed., Wiley, New York (1969); H.B. Lyon and G.A. Somorjai, J. Chem. Phys. 46:2539 (1967); J.T. Grant and T.W. Haas, Surface Sci. 18:457 (1969).
29. B.E. Conway, H. Angerstein-Kozlowska, W.B. Sharp and E.E. Criddle, Anal. Chem. 45:1331 (1973).
30. A.E. Morgan and G.A. Somorjai, Surface Sci. 12:405 (1968).
31. T.E. Felter and A.T. Hubbard, J. Electroanal. Chem. 100:473 (1979).
32. G.A. Garwood, Jr. and A.T. Hubbard, Surface Sci. 92:617 (1980).
33. G.A. Garwood, Jr. and A.T. Hubbard, Surface Sci. 112:281 (1981).
34. G.A. Garwood, Jr. and A.T. Hubbard, Surface Sci. 116:SSC652 (1982).
35. J.Y. Katekaru, G.A. Garwood, Jr., J.F. Hershberger and A.T. Hubbard, Surface Sci., in press.
36. M.P. Soriaga and A.T. Hubbard, J. Amer. Chem. Soc. 104:2735 (1982).

SURFACE CHARACTERIZATION OF CERTAIN METAL OXIDES DETERMINED BY THE
ISOTHERMAL ADSORPTION AND DESORPTION OF ARGON

Richard G. Herman, Phillip Pendleton, and John B. Bulko

Center for Surface and Coatings Research
Sinclair Laboratory, # 7
Lehigh University
Bethlehem, PA 18015

INTRODUCTION

The importance of surface areas and porosities of powdered solids are of vital interest to many scientific and technological disciplines. Knowledge of the magnitude of these properties is required when studying almost any interfacial phenomenon involving gas-solid or liquid-solid interactions. Among the techniques that have been utilized to determine the surface areas of oxide materials are electron microscopy, X-ray powder diffraction, small angle X-ray scattering, heat of immersion, adsorption from solution, and gas adsorption. The X-ray methods yield a volume-average crystallite diameter, while the adsorption techniques yield a surface-average value. On the other hand, electron microscopy yields a particle size distribution, from which a volume-average diameter and a surface-average value can be calculated by assuming a certain particle shape. Electron microscopy is often not readily accessible, while the X-ray techniques are applicable only to crystalline materials having crystallite sizes in the approximate diameter range of 3-60 nm.

The most frequently applied method to determine the surface area of metal oxides, as well as supported metals, is gas-phase physisorption. Among the procedures that can be employed are 1) static vacuum adsorption, 2) microbalance experiments, 3) temperature-programmed desorption, 4) continuous gas flow determinations, and 5) pulsed gas flow measurements. The static vacuum gas adsorption procedure is well developed for surface area determinations of porous adsorbents.[1-7] Instead of measuring pressures and volume changes, microbalance experiments measure gas adsorption by mass changes. Temperature-programmed desorption (TPD) techniques are still undergoing development since interpretation of the data can be difficult because the shape and position of the TPD curves can be affected by diffusional resistances, readsorption, flow rate changes of the carrier gas, and desorption kinetics.[8] Continuous flow measurements may be affected by diffusional limitations and by kinetically slow chemisorption, which is not readily detected. Pulsed flow determinations of surface areas can yield low results because weakly held gas is not maintained on the surface of the solid.[9]

The measurement of surface area by gas adsorption is achieved by accurately determining the amount of an inert gas that is adsorbed by van der Waals attraction on an accurately weighed solid sample at some constant temperature. The adsorption is carried out as a function of increasing gas pressure over the sample. The resultant adsorption isotherm data are then experimentally fit to equations that describe the adsorption process in terms of intermolecular forces operating at the gas-solid interface and the physicochemical parameters that control them. The best known theory of adsorption is that of Brunauer, Emmett, and Teller (BET),[1] and this theory provides an extension of the Langmuir monolayer model[10] to multilayer adsorption. Decreasing the gas pressure in small increments leads to sequential desorption of the gas from the solid and yields a desorption isotherm. With porous metal oxide adsorbents, adsorption-desorption isotherms generally exhibit hysteresis; that is, the desorption isotherm lies above the adsorption isotherm at the higher relative pressures of gas. Besides providing information on the surface area, the adsorption-desorption isotherms provide a wealth of information about the number, volume, diameter, and shape of pores in a porous solid.

The usual classification[11] of pores is according to their size (diameter or width): micropores (<2.0 nm), mesopores (2.0 - 50 nm), and macropores (>50 nm). Analysis of the microporosity of a solid can be carried out by comparing the shape of the experimental isotherm to the isotherm obtained with a nonporous reference material that is similar in composition. Differences between the two isotherms at low relative gas pressures are due to the filling of the micropores since this process proceeds by a volume-filling mechanism rather than a layer-by-layer adsorption mechanism.[12] This approach is commonly referred to as the comparison-plot method of microporosity evaluation, where the t-plot[13], α_s-plot[14], n-plot[15] and f-plot[16] methods have been utilized. The α_s technique is the most advantageous method to use for the determination of microporosity because the plot can be made independently of the BET calculation of the monolayer capacity of the sample. This is important when the monolayer coverage cannot be accurately determined from the BET analysis, e.g. when the adsorbate-adsorbent interaction energy is low. In addition, the α_s-plot method can be easily extended to microporosity analyses that utilize adsorbates other than inert gases that have well-established molecular cross-sectional areas.

Mesopores have dimensions that are much larger than the normally utilized adsorbate molecules, and normal monolayer-multilayer adsorption occurs on their surfaces. Mesopores are further characterized by their filling through a capillary condensation mechanism that usually is reflected by hysteresis in the adsorption-desorption isotherms. An upward deviation in the α_s- and t-plots in the high relative pressure region of the adsorption isotherm indicates the presence of mesopores. The approach employed here to determine the mesoporosity of metal oxides is based on Brunauer's corrected modelless method,[17] which can be contrasted with earlier methods that required an assumed pore geometry, e.g. cylindrical pores open on both ends[18] or closed at one end.[19] Brunauer's modelless method does not assume a particular pore shape for calculation of properties such as core volume and surface area. The core of a pore represents the empty space of a pore filled by capillary condensation and takes into account the thickness of the film of atoms or molecules adsorbed on the pore walls. From this, the core properties are "corrected" using a suitable pore geometry. A general outline of this method is the following:

A) The size of pores is expressed in terms of the volume to surface ratio, which is the hydraulic radius $r_h = V/S$.

B) The volume of a group of cores is taken as the difference between the volume adsorbed at two consecutive relative pressures on the adsorption or desorption isotherm.

C) The surface area of a core group is calculated through the integration of the Kiselev equation,[20] which is a generalized form of the Kelvin equation for capillary condensation, given in Equation 1 where S, γ, n, and P/P_o represent surface area, surface tension, amount of

$$S = -\frac{1}{\gamma} \int_{n_1}^{n_2} RT \ln (P/P_o) \, dn \qquad [1]$$

gas adsorbed (mol/g), and relative pressure with P_o = saturated vapor pressure, respectively. The integration limits correspond to the amounts adsorbed at the two consecutive relative pressures that form the group of cores. More recently, the graphical integration of

Equation 1 has been eliminated by assuming that if the chosen relative pressure intervals are small enough, the plot of log P/P_o vs V is linear.[21] Based on this analysis, a computer program was developed for mesopore evaluation by Brunauer's modelless method.[22]

D) From B and C, the hydraulic radii of core groups may be obtained, where the core is the part of the pore that fills up by capillary condensation or remains empty after capillary evaporation.

E) The correction for volume and surface of cores due to the adsorbed multilayer thickness requires an assumption of pore shape and the availability of a suitable t-curve on a non-porous reference solid. This results in a corrected core size distribution, which can be transformed into a corrected pore size distribution.

The total surface area of a metal oxide can be subdivided into microporous surface area and an external surface area that consists of the mesopore surface area and the non-porous surface area. In the α_s plot, the Y-intercept corresponds to the micropore volume of the sample, from which an equivalent monolayer surface area can be calculated. This surface area can be compared to the micropore surface area determined as the difference between the total surface area and the external surface area, which can be determined from the slope derived from the α_s plot. Knowledge of the surface properties of metal oxides can provide insight into their function and behavior as ceramic materials, adsorbants, and catalyst supports.

EXPERIMENTAL

Sample Preparation

The materials utilized in this research (CuO, ZnO, Al_2O_3, Cr_2O_3) were prepared from $Cu(NO_3)_2 \cdot 3H_2O$, $Zn(NO_3)_2 \cdot 6H_2O$, $Al(NO_3)_3 \cdot 9H_2O$, $Al(OH)_2(C_2H_3O_2)$, or $Cr(NO_3)_3 \cdot 9H_2O$ by an aqueous precipitation method. The metal salts were dissolved in distilled water (50 ml/g of final calcined product), which was stirred continuously and heated to 80-90°C. Precipitation was achieved by the dropwise addition of 1.0 M sodium carbonate until the pH of the solution increased to 6.8-7.0. This procedure typically took 1.5 hr and was followed by a 1 hr digestion period, during which the solution cooled to 30-40°C. After standing for a few minutes, the supernatant was decanted from the settled precipitate, which was then washed with added distilled water (100 ml/10 g of product). After standing, the clear solution was decanted, and this washing procedure was repeated two more times.

The precipitate was then filtered on Whatman #2 filters with the aid of suction, washed with 10 small portions of water, (total volume = 150 ml), and sucked as dry as possible in 10-15 min. The samples were dried at ambient conditions, and subsequent calcination was carried out in air by heating the materials in a furnace from 150 to 350°C in increments of 50°C every 30 min with the maximum temperature maintained for 3 hr. Preparation of binary and ternary combinations of these oxides has been described previously.[23]

When samples were desired that had been exposed to a reducing environment, portions of the oxides were placed in Pyrex or stainless steel tubes and were exposed to a H_2/N_2 = 2/98 vol% gas stream flowing at about 60 ml/min at atmospheric pressure. The temperature was increased from ambient to 250°C at a uniform rate of 3.5 $\pm$ 0.3°C/min. The desorption of water and any water produced by reduction of the metal oxide could be followed and determined by chromatographic analysis using an on-line Hewlett-Packard 573A gas chromatograph with an automated sampling valve and coupled with a Model 3388A integrator. The typical treatment, other than for CuO, was carried out for 3 hr/2.5 g of sample. CuO was treated until the disappearance of water from the exit gas was observed. The samples treated in this manner were transferred under nitrogen to cells utilized for surface area measurements.

Physical Measurements of Gas Adsorption

The surface areas of the materials were determined by adsorption of argon at liquid nitrogen temperature (77°K). A schematic of the apparatus used to obtain the volumetric adsorption-desorption data has been presented elsewhere.[24] The essential features of the Pyrex system are the following: 1) a vacuum system consisting of a mechanical pump, an oil diffusion pump, and a liquid nitrogen trap, 2) a gas manifold containing 5 l reservoirs of argon, nitrogen, and helium, 3) an electronic manometer equipped with a 1000 torr Barocel pressure sensor having a sensitivity of 2×10^{-4} torr and a null offset adaptor permitting a routine sensitivity of 0.001 torr over the entire range of the pressure unit, 4) an ion gauge and a control unit, and 5) a variable dosing volume section consisting of four independent and interconnected Pyrex bulbs of precalibrated and progressively larger volumes (V_1-V_4) that are housed in a cylindrical Pyrex thermostatting jacket filled with water. The sample holder, consisting of a 21 cm Pyrex tube with a standard 10/30 inner ground glass joint at one end and a 2 cm diameter bulb blown at the other end, was attached to the lower end of the latter section of the apparatus. The smallest volume, V_4 ($\sim$16 cm^3) served as a dosing volume in the low pressure region of the isotherm. The variable dosing volume was adequate for measuring the specific surface area in the range of 0.5 - 1000 m^2/g for a 0.5 g sample, although 0.1 - 0.2 g samples were normally used.

The sample bulb containing the accurately weighed sample was carefully evacuated at ambient temperature with this apparatus. The solid to be analyzed was then heated at approximately 115°C for at least 8 hr to remove any adsorbed matter, including water. After this activation treatment, the background pressure in the adsorption apparatus was typically 5 x 10^{-5} torr. The research grade gases utilized for the adsorption measurements had been previously purified and stored in the reservoirs. The purification unit consisted of a column containing clean copper turnings maintained at 400°C (to remove residual oxygen) and a column containing 3A zeolite maintained at 77°K (primarily to remove water vapor).

The isothermal adsorption measurements were begun by calibrating the dosing volume (V_d) and the sample bulb volume (V_s) with helium. After evacuation, the argon adsorbate gas was introduced into V_d, and its pressure was measured with a manometer. The contents of V_d were then expanded into V_s, and an equilibrium between the gas phase and the adsorbed phase was allowed to be established, at which time the pressure was measured (P_e). The decrease in pressure from the value given by the ideal gas law was used to calculate the amount of gas adsorbed on the surface of the sample. The amount of gas adsorbed was then converted to volume adsorbed (V_{ads}) at STP. Normally, 18-25 gas additions were made for the determination of the adsorption isotherm, while 8-10 data points were obtained down to 40-50 torr to generate the desorption isotherm. The latter was obtained by allowing V_s to expand into V_d by stepwise lowering the pressure in V_d by means of the pumping system.

The saturated vapor pressure (P_o) of argon was measured with a nitrogen gas thermometer in which the equilibrium pressure of the condensed nitrogen was measured with a differential mercury manometer readable to 1 torr. The condensation pressure of argon was determined from $P_o(N_2)$ by a graphical technique presented elsewhere,[25] in which $P_o(N_2)$, in the neighborhood of its boiling point, was plotted against data for $P_o(Ar)$.[25] Since the P_o value of liquid nitrogen changed by a few millimeters during the course of the adsorption measurements, a variable P_o was used in the calculation of relative pressures.

Surface Area Calculations

The specific surface area is determined from the adsorption isotherm by utilizing the BET equation,[1] which is expressed in the linear form

$$\frac{P}{V(P_o-P)} = \frac{1}{V_m C} + \frac{C-1}{V_m C} \frac{P}{P_o} , \qquad [2]$$

where P is the equilibrium pressure P_e, V is the quantity of gas adsorbed (V_{ads}) at the equilibrium relative pressure P/P_o, V_m is the monolayer capacity, and C is the BET constant that is related exponentially to the first layer heat of adsorption. The use of this equation assumes that adsorption of the first layer takes place on an array of surface sites of uniform energy, that molecules or atoms in the first layer act as sites for multilayer adsorption, that the evaporation-condensation constants for all layers above the first are identical, and that the corresponding heats of adsorption are equal to the heat of condensation of the adsorbate.[26] The linear relationship between $P/V(P_o-P)$ and P/P_o yields V_m, although the range of linearity of the BET plot is usually restricted to P/P_o values within 0.05 - 0.40. Multiplying V_m by the cross-sectional area of the adsorbate, argon in this case, yields the specific surface area of the solid. Assuming hexagonal close packing on the surface and a packing factor of 1.09, the calculated cross-sectional area of argon at 77°K is 0.142 nm^2. However, comparisons with geometric areas of some metal oxides have yielded values of 0.134, 0.147, and 0.182 nm^2 for the average area occupied by an argon atom.[27] Based on other research,[28] the value of 0.168 nm^2 for the cross-sectional area of argon was utilized in the present work.

The surface areas of the metal oxides were also determined by the B-point method,[14,29] in addition to the BET method of analysis. The B-point, taken as the indication of the completion of the adsorbed monolayer, can normally be located at the top of the knee in Type II adsorption isotherms,[2] which are characteristic of metal oxides. The onset of linearity at low relative pressure in the adsorption isotherm is typically chosen as the B-point, as will be shown in a later section of this paper. The B-point surface area determinations can usually be considered to be more reliable than the BET method when the sharpness of the isotherm knee provides for a reduction in the subjectiveness of determining the monolayer volume of the adsorbate. In addition, the usually observed rapid leveling off in uptake of the adsorbate at about 0.2 P/P_o allows a quick and precise determination of the monolayer coverage and of the total surface area.

Porosity Determinations

The mesopore distribution, and associated volumes and areas, was determined by Brunauer's corrected modelless method[17] using an available computer program,[22] in which the calculations were carried out for cylindrical pores. The input data were derived from the experimental desorption isotherms, and the output was used to plot $\Delta V/\Delta r$ vs pore radius.

The micropore volumes of the samples were determined by plotting α_s vs volume of argon adsorbed. α_s, in effect, compares the shape of two isotherms, one of which is determined with a non-porous reference solid adopted as a standard, powdered non-porous chromia gel. In the calculation of $\alpha_s = V/V_s$, V_s is the amount of argon adsorbed on the

TABLE I

Standard Adsorption Data for Argon at 77°K on Non-Porous Chromia Gel[30] and on Hydroxylated Silica[31]

P/P_o	Chromia Gel V_{ads}	α_s	Hydroxylated Silica V_{ads}	α_s
0.01			11.44	0.24
0.025	2.12	0.53	16.50	0.35
0.05	2.28	0.57	21.20	0.45
0.10	2.56	0.64	27.46	0.58
0.125	2.68	0.67	29.35	0.62
0.15	2.80	0.70	31.74	0.67
0.175	2.92	0.73	33.60	0.71
0.20	3.04	0.76	35.23	0.75
0.25	3.32	0.83	38.29	0.81
0.30	3.56	0.84	41.26	0.88
0.35	3.80	0.95	44.20	0.94
0.40	4.00	1.00	47.10	1.00
0.45	4.24	1.06	49.60	1.05
0.50	4.56	1.14	52.89	1.12
0.55	4.84	1.21	55.50	1.18
0.60	5.16	1.29	58.88	1.25
0.65	5.44	1.36	61.75	1.31
0.70	5.80	1.45	65.33	1.34
0.75	6.20	1.55	69.20	1.47
0.80	6.68	1.67		
0.85	7.28	1.82		
0.90	8.00	2.00		
0.95	8.92	2.23		
BET Area	11.7 m^2/g		149.5 m^2/g	
C	77.7		34.4	
k	0.342		0.315	

standard at a selected relative pressure, which is $P/P_o = 0.4$. Below this value, monolayer coverage and micropore filling occur, while any hysteresis in the isotherms should be observed at higher relative pressures. At the relative pressure of 0.4, α_s is set to unity and results in $V_s = 4.0$ cm^3/g at STP for the standard chromia gel as indicated in Table I. Plotting the resultant α_s values for a sample vs the volume of adsorbed argon yields an intercept that corresponds to the micropore volume of the sample.

Micropore surface area estimates were also obtained from the α_s plots. By determining the slopes of linear portions of the α_s plots and using a calibration constant (k) for the chosen standard (0.342 for chromia gel and 0.315 for hydroxylated silica) the external surface areas were estimated (slope/k). The surface areas attributable to the micropores can then be obtained by subtracting the external surface areas from the surface areas determined by the B-point method. The microporous surface area can also be derived from the micropore volume data determined from the α_s plot. As pointed out in the Introduction, the radii of micropores approach atomic dimensions, and vapor adsorption in micropores proceeds via a volume-filling mechanism rather than the layer-by-layer filling process observed in larger pores and on surfaces. The sorption capacity of a microporous material can be well characterized by its equivalent monolayer surface area (EMSA) as suggested previously,[32] where the EMSA is determined by visualizing the removal of all of the adsorbed contents of the micropores and the spreading of the contents as a close packed monolayer on a molecularly smooth surface. The area covered by the adsorbate is then considered to be the EMSA of the micropore adsorption capacity.

RESULTS

The argon adsorption-desorption isotherms that were obtained at 77°K with the metal oxides prepared via nitrate solutions are shown in Figures 1 and 2. Utilizing Equation 2, the adsorption data in the BET region (typically $P/P_o<0.3$) of the isotherms were used to construct the BET plots in Figure 3, which provides for the calculation of the surface areas since $1/V_m$ = slope + intercept. The derived surface areas of CuO, ZnO, Cr_2O_3, and Al_2O_3 are tabulated in Table II. The B-point, an example of which is shown in Figure 1, ideally represents the completion of the adsorbed monolayer and the beginning of multilayer adsorption of argon atoms. Using the V_m values determined at the B-points produced the B-point surface areas in Table II. The C constants in this table were calculated from the BET plots in Figure 3, C=1 + slope/intercept.

The α_s plots for Al_2O_3 and Cr_2O_3 prepared from the nitrates are shown in Figure 4, where the comparison standard was non-porous chromia. Linear portions of the curves that correspond to $P/P_o<0.3$ are extrapolated to the ordinate to yield the micropore volume, if any, in the oxides. The full range of α_s values are plotted to show that an upward deviation in the curves at high α_s, and therefore high P/P_o as indicated in Table I, is observed. This deviation to elevated V_{ads} values is evidence that capillary condensation (a volume phenomenon) in mesopores was occurring. The external surface areas, derived from the extrapolated slopes in Figure 4, and the micropore equivalent surface areas, calculated from the micropore volumes, are

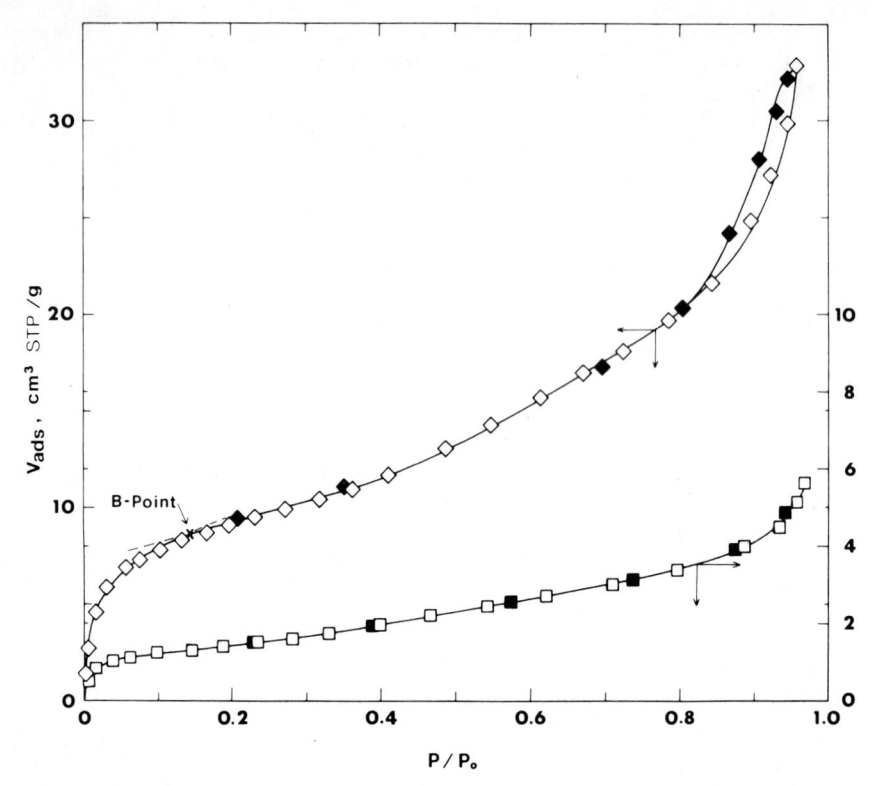

Fig. 1 Isotherms for the adsorption (open symbols) and desorption (solid symbols) of Argon at 77°K with calcined CuO (□,■) and ZnO (◇,◆) prepared via precipitation from nitrate solutions.

given in Table III. The sum of these areas yields the total α_s surface area. The mesopore equivalent surface area in each case was calculated from the desorption data.[22] The non-porous surface area is the difference between the external surface area and the mesopore equivalent surface area.

The α_s plots for ZnO and CuO are presented in Figures 5 and 6, respectively. Portions of each of these oxides were subjected to hydrogen reduction treatments and then were transferred to the gas adsorption apparatus without exposure to air. The physical appearance of ZnO was unaltered, but the black CuO was reduced to reddish metallic copper, as confirmed by subsequent analysis by X-ray powder diffraction.[23,33] A scanning electron micrograph of the latter sample indicated that the copper possessed a non-porous surface and that a decrease in surface area should be expected.[34] The specific surface areas determined by argon adsorption for these materials are tabulated in Tables II and III.

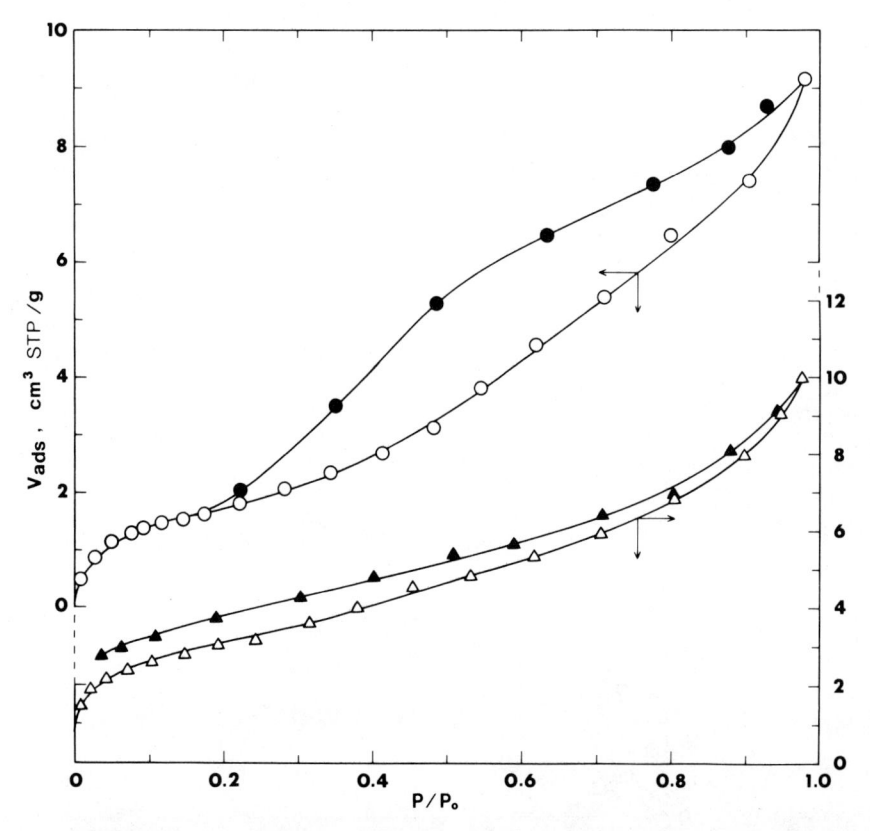

SURFACE CHARACTERIZATION OF CERTAIN METAL OXIDES

Fig. 2 Isotherms for the adsorption (open symbols) and desorption (solid symbols) of Argon at 77°K with calcined Cr_2O_3 ($\triangle$,$\blacktriangle$) and Al_2O_3 ($\bigcirc$,$\bullet$) prepared via precipitation from nitrate solutions.

An alumina was prepared by precipitation of a precursor compound from an acetate solution using sodium carbonate, and the precursor was calcined using the given stepwise procedure to 350°C. The resultant material consisted of hard granules having a glassy appearance. In comparison with the alumina prepared from nitrate solution, this glassy alumina was found to have a very high surface area, as shown in Table II. Figure 7 demonstrates that a pronounced hysteresis in the adsorption-desorption isotherms was observed. Figure 8 shows that the linear region of the BET plot extend only up to $P/P_o = 0.2$. The C constant in this region is approximately 35, while the C constant calculated for the upper portion of the curve is about 10.5. α_S analysis of the data using non-porous chromia gel as the standard would yield a negative intercept, while analysis using an hydroxylated silica standard produces a plot that can be extrapolated to the origin, as shown in Figure 9. The α_S values given in Table I for the

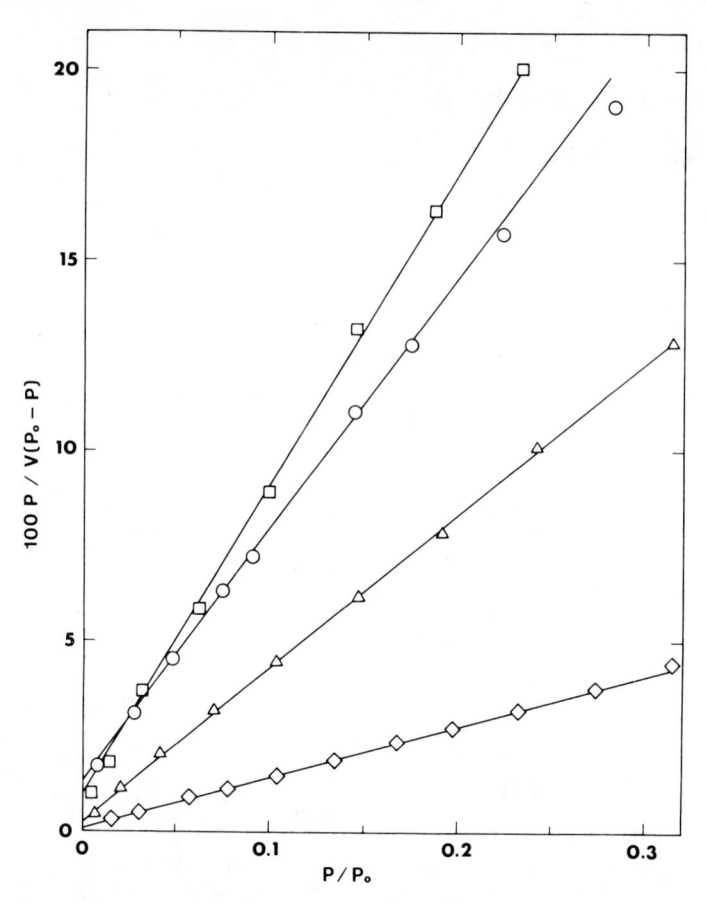

Fig. 3 BET plots utilized to determine the surface areas of ZnO($\diamond$), Cr_2O_3($\triangle$), Al_2O_3($\bigcirc$), and CuO($\square$).

hydroxylated silica standard have been calculated from data given elsewhere.[31] Figure 9 indicates that even with the very high surface area of the alumina, no microporous surface area is present in this sample. The enhanced argon adsorption V_{ads} at high α_s values suggests that this alumina is a mesoporous solid, and the mesoporosity calculations confirm this (Table III).

A commercial adsorption alumina was found to exhibit argon adsorption-desorption isotherms (Figure 10A) that were similar in shape to those obtained with alumina prepared from acetate solution (Figure 7), and the C constants were also similar (Table II). The 80-200 Mesh adsorption alumina was mechanically mixed with 70-325 Mesh CuO to yield a 25/75 wt% Al_2O_3/CuO mixture (3.66 g), which was then heated at 250°C for 15 hr in a flowing H_2/N_2 = 2/98 vol% gas stream. The resultant material would be expected to be Al_2O_3/Cu = 30/70 wt%,

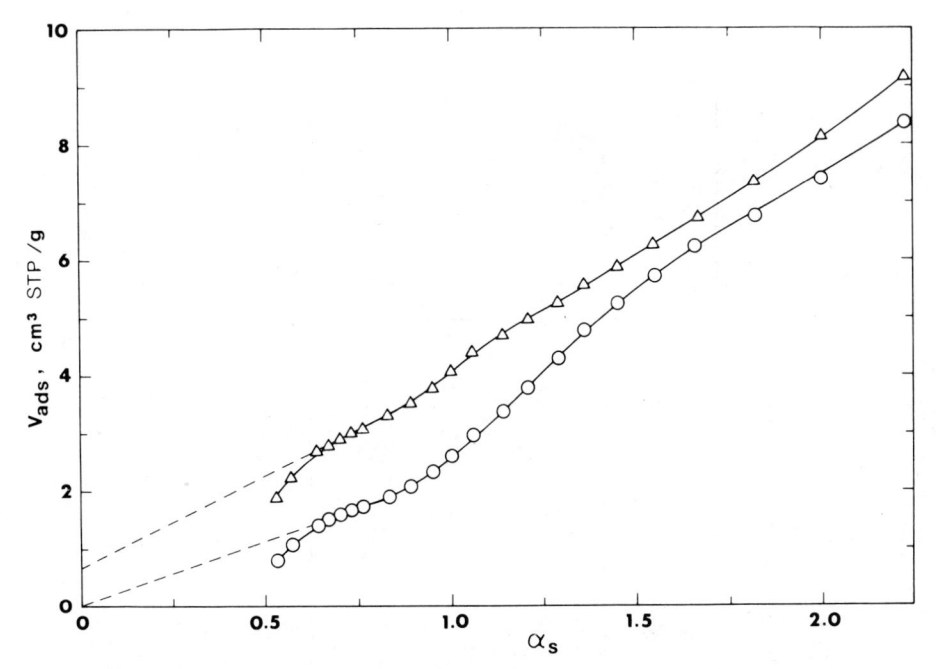

Fig. 4 α_s plots for calcined Al_2O_3 (O) and Cr_2O_3 ($\triangle$).

TABLE II

Specific Surface Areas and C Constants Derived from the Argon Adsorption Isotherms (77°K)

Sample[a]	BET Surface Area, m^2/g	B-Point Surface Area, m^2/g	C
CuO	6.0	5.6	236.
CuO-R	1.4	1.3	39.2
ZnO	33.6	38.8	137.
ZnO-R	22.2	23.7	116.
Cr_2O_3	11.2	12.4	135.
Al_2O_3	6.6	6.5	56.5
Al_2O_3-A	347.	343.	35.4
Al_2O_3-F	194.	167.	31.0
(Al_2O_3-F/CuO)-R	43.4	39.3	31.3

[a] Samples were prepared from nitrates except for the following: A, prepared from the acetate; F, adsorption alumina purchased from Fisher Scientific; R, samples subjected to hydrogen reduction treatments.

TABLE III

α_s Analyses of the Argon Adsorption Isotherms (77°K)

Sample[a]	External Surface Area m²/g	Mesopore Surface Area[b] m²/g	Non-porous Surface Area m²/g	Micropore Surface Area m²/g	Total α_s Surface Area[c] m²/g
CuO	5.1	2.7	2.4	0.5	5.6
CuO-R	1.4	0.8	0.6	0	1.4
ZnO	24.8	14.8	10.0	12.4	37.2
ZnO-R	18.4	10.6	7.8	5.3	23.7
Cr_2O_3	9.8	4.5	5.3	2.5	12.3
Al_2O_3	7.1	8.9	-	0	7.1
Al_2O_3-A	372.	348.	24.0	0	372.
Al_2O_3-F	172.	109.	63.0	0	172.
$(Al_2O_3$-F/CuO)-R	40.1			0	40.1

[a] See footnote to Table II

[b] Obtained by analysis of the Argon desorption isotherms

[c] External Surface Area + Micropore Surface Area

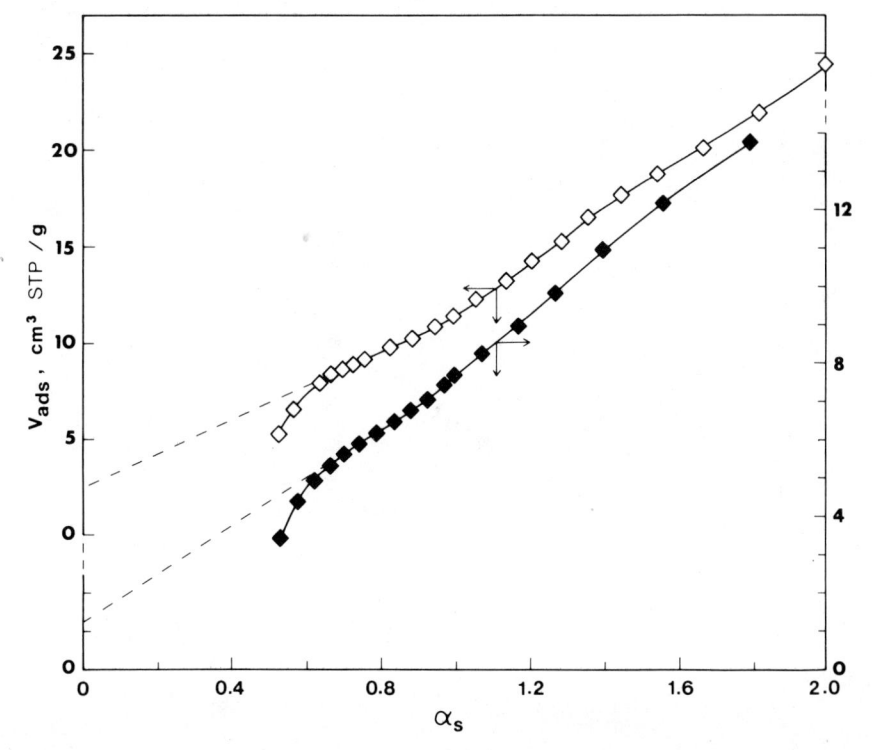

Fig. 5 α_s plots for calcined ZnO (◇) and ZnO subjected to a hydrogen reduction treatment at 250°C (◆).

and the argon adsorption-desorption isotherms are shown in Figure 10B. Subsequent characterization yielded the surface areas given in Tables II and III.

DISCUSSION

The curves shown in Figures 1 and 2 are typical adsorption-desorption isotherms that exhibit four different types of hysteresis loops. Argon adsorption-desorption is a reversible process on CuO, and no discernable hysteresis is observed. This indicates that the adsorbate interaction with the surface of the CuO particles is unhindered and that unrestricted monolayer-multilayer adsorption has occurred. Thus, the CuO can be characterized as consisting of external surface with few small pores, although mesoporosity analysis indicates the presence of a small number of mesopores with a radius of 6 nm, a Type II[2] isotherm. A Type A[35] hysteresis loop in a Type IV (Type II) isotherm is evident in Figure 1 for ZnO. Type A behavior is generally associated with capillary condensation in tubular mesopores that are open at both ends. The mesoporosity analysis

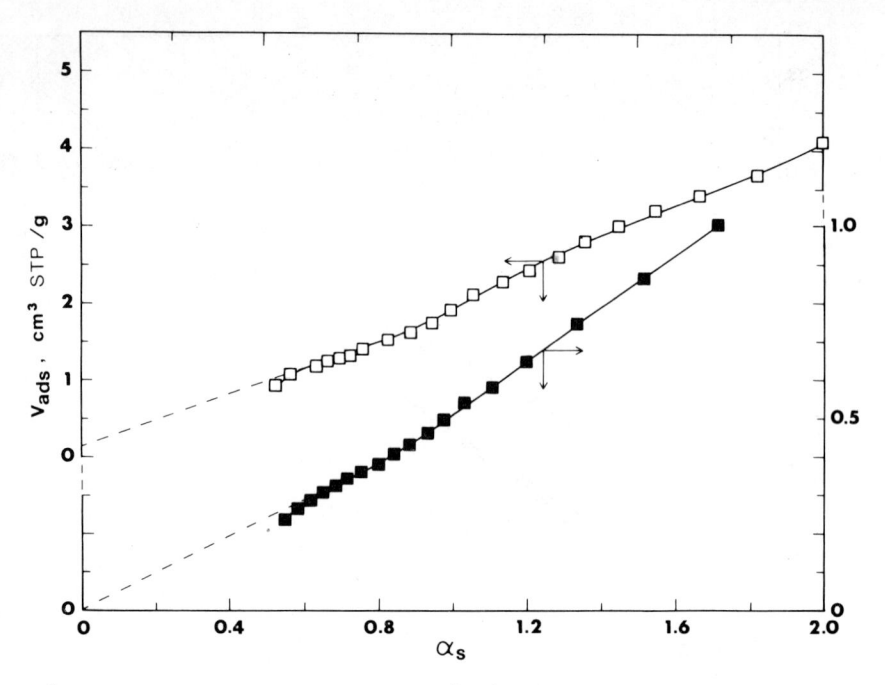

Fig. 6 α_s plots for calcined CuO (□) and CuO subjected to a hydrogen reduction treatment at 250°C (■).

indicated that nearly one-half of the total surface area of ZnO was associated with the mesopores (Table III), and the mesopore distribution depicted in Figure 11 shows that a dominant pore size is 6 ± 2 nm (radius).

The adsorption-desorption isotherms in Figure 2 for Al_2O_3 exhibit a large hysteresis loop that has the appearance of being a combination of Type B and Type E hysteresis. Type B loops are typical for slit-shaped pores, while Type E hysteresis is found for the adsorption-desorption processes with constricted pores that can be visualized as ink bottle shaped. Slit-shaped pores can arise from the packing of platelets. In this case, the desorption branch of the isotherm should resemble evaporation from a cylindrical meniscus, but mesoporosity analysis would yield a platelet separation distance rather than a cylindrical 'pore' size.[35,36] It has been pointed out that if the platelets are immobile, multilayer adsorption will proceed according to a non-porous (Type II) adsorption mechanism, but only up to a limiting relative pressure determined by the spacing between the platelets. The resultant shape of the isotherms will be a modified Type B hysteresis loop containing a Type E desorption branch. This is termed a Type B' hysteresis, and the Al_2O_3 isotherms in Figure 2 resemble this description. The large hysteresis indicates that the Al_2O_3 sample possesses high mesoporosity, and in fact the mesopore

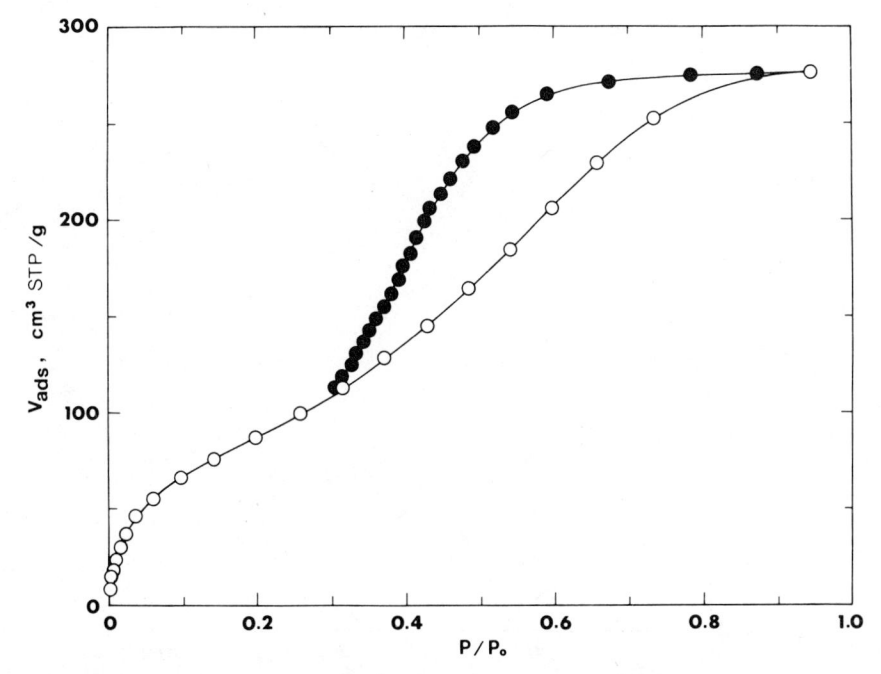

Fig. 7 Argon adsorption (○) and desorption (●) isotherms at 77°K obtained with calcined Al_2O_3 prepared via precipitation from acetate solution.

analysis attributed all of the surface area to that contained in the mesopores (Table III). The Cr_2O_3 hysteresis loop appears to be of Type B, but it did not close under the experimental conditions that were utilized. This might be due to an activated adsorption-desorption effect where the desorption rate is so slow that true equilibrium is not established. This effect has been observed with layer structures, e.g. exfoliated graphite,[38] that exhibit Type B hysteresis and can intercalate the adsorbed atoms. It would be expected that the observed hysteresis would be temperature and time dependent. When intercalation occurs, an activated entry effect could also be possible, and this has been observed with certain carbons[39] and microporous solids. In the case of activated entry, the surface area determined by heat of immersion measurements or by gravimetric gas adsorption should be higher than that obtained by the low temperature gas adsorption technique utilized in this research.

The Al_2O_3 prepared from acetate solution and the commercial Al_2O_3 exhibited classical Type E hysteresis in Figures 7 and 10. Desorption of argon from the pores of these materials did not begin to occur until the relative pressure was reduced to the neighborhood of 0.55. This indicates that the mouths of the pores are smaller than the bodies of the pores, and this feature maintains a capillary

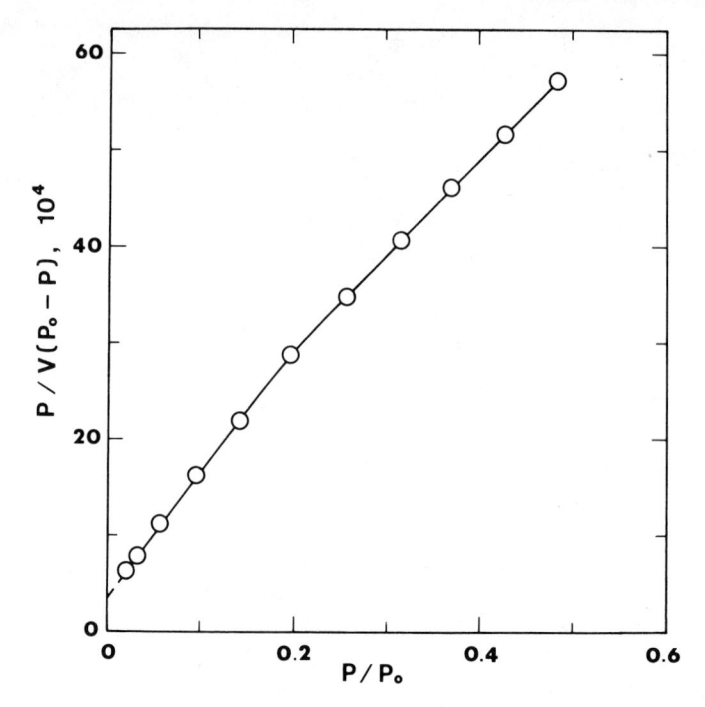

Fig. 8 BET plot of argon at 77°K on Al_2O_3 prepared from acetate solution.

condensed state until the relative pressure is reduced to a value appropriate for the evaporation of argon from pores having the dimensions of the pore mouths. The sharpness of the desorption break gives an indication of the uniformity of pore mouth size. The gradient of the desorption curve changes at the fastest rate in the area of $P/P_o = 0.4$, and this indicates that the dominant pore size will be about 2 nm (radius). Mesoporosity analysis of the desorption branch of the isotherm indicates a dominant pore dimension somewhat less than 2 nm. It has been shown that high surface area aluminas prepared by a wide variety of techniques generally exhibit the major pore radius of 2 - 2.5 nm.[40] It has been suggested that for Type E hysteresis, the adsorption branch of the isotherm more closely approaches true equilibrium and should be used for calculation of the real pore-size distribution.[14] Hydrogen reduction of the mechanically mixed CuO/Al_2O_3 sample hardly affected the isotherm hysteresis in the isotherms in Figure 10. However, a lower total surface area than expected (49.1 m^2/g) was obtained and the desorption break is somewhat sharper than for the untreated Al_2O_3 sample (Figure 10). This could be caused by narrowing of the pore mouths, and in some cases blockage of the pores, by the reduced copper. In general, the C constant decreases with increased sintering. The C constant in

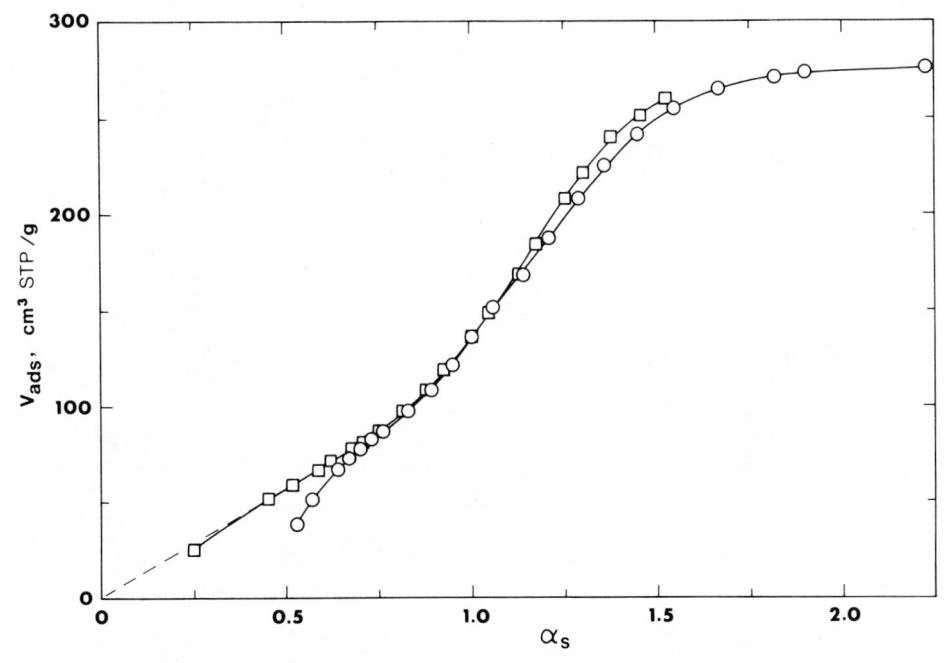

Fig. 9 α_s plots for calcined Al_2O_3, prepared via acetate solution, using non-porous chromia gel (○) and hydroxylated silica (□) as standards.

Table II for the hydrogen treated alumina sample compared with the constant for Al_2O_3-F would indicate that the lower surface area for the sample is not due to simple sintering.

The BET surface areas of all of the alumina samples were especially sensitive to the inclusion of the low partial pressure data points in the BET plots. A curvature in the line in Figure 8 is observed at $P/P_o = 0.2$. If the $P/P_o = 0.2 - 0.5$ portion of the plot is extrapolated to the axis and utilized to calculate V_m, a BET surface area of 455 m^2/g and a C constant of 10.5 is obtained for the Al_2O_3 sample. Similar analysis with the (Al_2O_3-F/CuO)-R sample yields a surface area of 57.8 m^2/g and a C constant of 10.1 Thus, using data beyond the 'BET region', which is restricted to lower P/P_o ranges for samples having lower true C constants, results in a very low calculated C constant. This reflects the fact that the higher adsorbed argon layers (the multilayers) begin to resemble the liquid state and the large entropy factor diminishes. This effect is more pronounced for samples exhibiting weaker physisorption; that is lower true C constants.

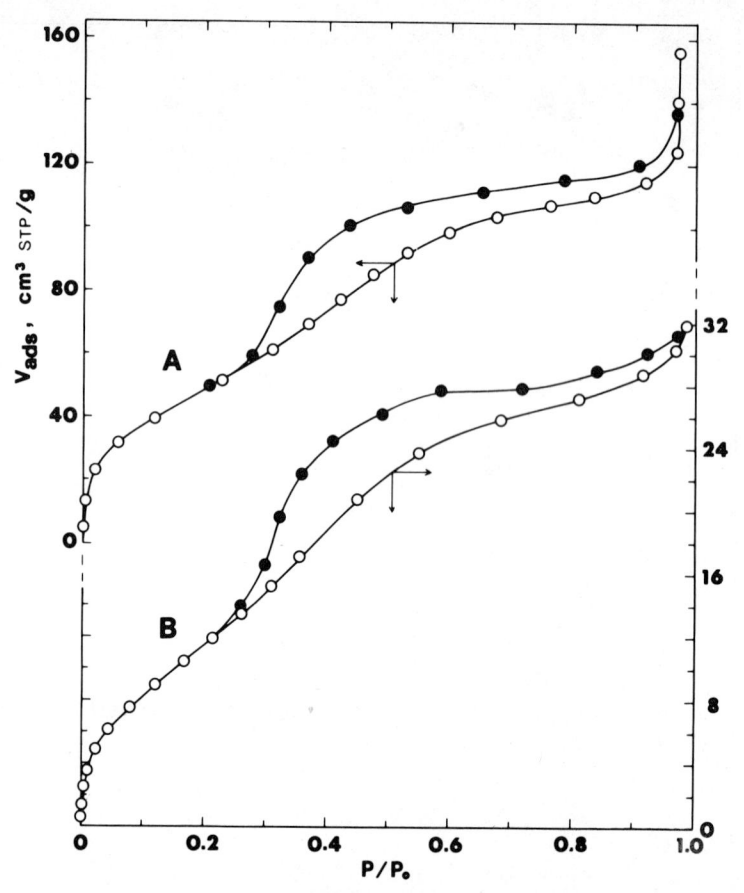

Fig. 10 Argon adsorption (O) and desorption (●) isotherms for (A) Fisher adsorption alumina and (B) for a CuO/Fisher Al$_2$O$_3$ mixture (67/33 wt%) that was subjected to a hydrogen reduction treatment at 250°C for 15 hr.

The CuO and ZnO samples that were subjected to hydrogen treatment at 250°C exhibited decreased surface areas. The CuO was sintered and reduced to metallic copper. The ZnO was apparently annealled, but a growth in crystallite size from approximately 16 nm was not detected by X-ray powder diffraction analysis. The calcined zinc oxide has been shown by electron microscopy to consist of small crystallites connected together to produce a lacelike network that is very porous (see Figure 3C in Reference 34). The crystallite size determined from the electron micrograph agrees with the size calculated from the X-ray diffraction pattern. The pore size appears in the micrograph to be somewhat smaller than the ZnO crystallite size, and this agrees with the mesopore diameter of about 12 nm shown in Figure 11. In agreement with mesopore size distribution analysis and X-ray

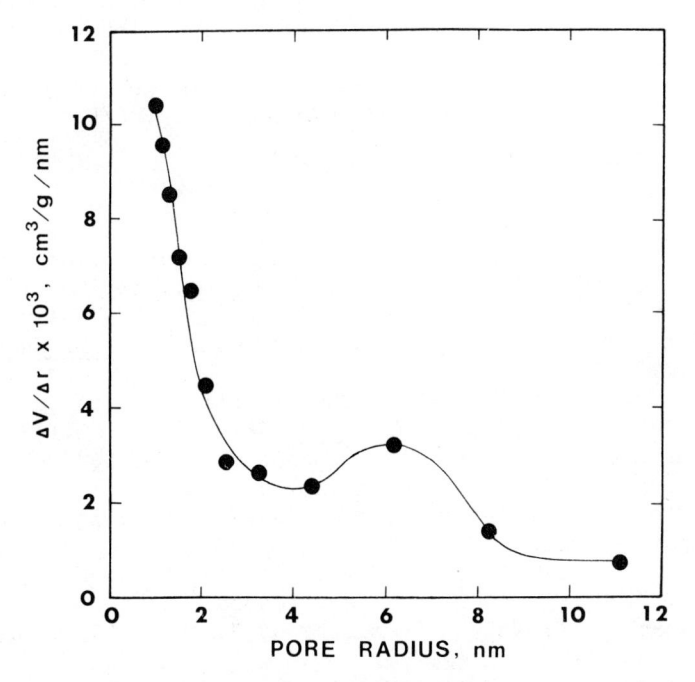

Fig. 11 Mesopore volume distribution for calcined ZnO.

crystallite size determination, electron microscopy showed no change in morphology or structure upon subjecting the ZnO to the hydrogen reduction treatment. Therefore, the mesopore surface area arises from the open porous structure formed by the association of the ZnO crystallites.

Figure 9 demonstrates that hydroxylated silica, rather than the chromia gel that has been selected as the initial standard because its surface area resembled those of the metal oxides prepared from nitrate solution and it was assumed that the adsorption process on those oxides would be similar to that on the chromia gel, is the better standard to use for the α_s plots of the Al_2O_3-A, Al_2O_3-F, and (Al_2O_3-F/CuO)-R samples. This results in better matching of surface areas and C constants between the standard and the high surface area analyzed samples. If a standard high surface area alumina[30] were available, it could also be utilized as a standard for these samples. It is evident from the tables that the total α_s surface areas generally approximate the B-point and BET surface areas quite well. It should be noted that one point surface areas are often utilized in routine screening of samples, prior to more complete analysis, and this method assumes that the BET plot goes through the origin and that the C constant is infinite. Generally, this will result in a surface area that is within 10% of the true surface area if an appropriate partial pressure is used, e.g. P/P_o = 0.2.

REFERENCES

1. S. Brunauer, P. H. Emmett, and E. Teller, Adsorption of Gases in Multimolecular Layers, J. Am. Chem. Soc., 60:309 (1938).
2. S. Brunauer, L. S. Deming, W. E. Deming, and E. Teller, On a Theory of the van der Waals Adsorption of Gases, J. Am. Chem. Soc., 62:1723 (1940).
3. S. Brunauer and P. H. Emmett, Chemisorptions of Bases on Iron Synthetic Ammonia Catalysts, J. Am. Chem. Soc., 62:1732 (1940).
4. S. Brunauer, "The Adsorption of Gases and Vapors," Princeton University Press, Princeton (1943).
5. W. A. Steele, "The Interaction of Gases with Solid Surfaces," Pergamon Press, Oxford (1974).
6. "Characterization of Powder Surfaces," ed. by G. D. Parfitt and K. S. W. Sing, Academic Press, London (1976).
7. A. W. Adamson, "Physical Chemistry of Surfaces," 3rd Ed., Wiley and Sons, New York (1976).
8. R. J. Gorte, Design Parameters for Temperature Programmed Desorption from Porous Catalysts, J. Catal., 75:164 (1982).
9. S. Parkash, Determining Surface Area, Chemtech, 10(9):572 (1980).
10. I. Langmuir, Chemical Reactions at Low Pressures, J. Am. Chem. Soc., 37:1139 (1915).
11. IUPAC Manual of Symbols and Terminology, Pure Appl. Chem., 31:579 (1972).
12. M. M. Dubinin, On Physical Feasibility of Brunauer's Micropore Analysis Method, J. Colloid Interface Sci., 46:351 (1974).
13. B. C. Lippens and J. H. DeBoer, Studies on Pore Systems in Catalysts. V. The t Method, J. Catal., 4:319 (1965).
14. S. J. Gregg and K. S. W. Sing, "Adsorption, Surface Area, and Porosity," Academic Press, London (1967).
15. R. S. Mikhail and D. A. Cadenhead, The Interaction of Methanol Vapor with Taurus-Littrow Orange Soil, J. Colloid Interface Sci., 55:462 (1976).
16. S. J. Gregg, A Simple Method for Comparing the Shapes of Closely Related Adsorption Isotherms, Chem. Commun., 699 (1975).
17. S. Brunauer, R. S. Mikhail, and E. E. Boder, Pore Structure Analysis Without a Pore Shape Model, J. Colloid Interface Sci., 24:451 (1967); and Some Remarks About Capillary Condensation and Pore Structure Analysis, J. Colloid Interface Sci., 25:353 (1967).
18. E. P. Barrett, L. G. Joyner, and P. P. Halenda, The Determination of Pore Volume and Area Determinations in Porous Substances. I. Computations from Nitrogen Isotherms, J. Am. Chem. Soc., 73:373 (1951).

19. R. W. Cranston and F. A. Inkley, The Determination of Pore Structures from Nitrogen Adsorption Isotherms, Advan. Catal., 9: 143 (1957).
20. A. V. Kiselev, Capillary Condensation Heat Maximums, Proc. 2nd Int. Congr. Catal. Act., 2:189 (1957).
21. E. E. Boder, I. Older, and J. P. Skalny, An Analytical Method for Pore Structure Analysis, J. Colloid Interface Sci., 32: 367 (1970).
22. J. C. Phillips and J. P. Skalny, Computer Program for Pore Structure Analysis, J. Colloid Interface Sci., 38:664 (1972).
23. R. G. Herman, K. Klier, G. W. Simmons, B. P. Finn, J. B. Bulko, and T. P. Kobylinski, Catalytic Synthesis of Methanol from CO/H_2. I. Phase Composition, Electronic Properties, and Activities of the $Cu/ZnO/M_2O_3$ Catalysts, J. Catal., 56:407 (1979).
24. A. C. Zettlemoyer, F. J. Micale, and K. Klier, Adsorption of Water on Well-characterized Solid Surfaces, in "Water-A Comprehensive Treatise," 5:249, ed. by F. Franks, Plenum Press, London (1975).
25. F. Din, "Thermodynamic Functions of Gases," Vol. 2, Butterworths, London (1956).
26. K. S. W. Sing, Surface Characterization: Physical, in "Characterization of Powder Surfaces," ed. by G. D. Parfitt and K. S. W. Sing, Academic Press, London, pp 1-56 (1976).
27. K. S. W. Sing, Adsorption at the Gas/Solid Interface, Colloid Sci., 1:48 (1973).
28. F. J. Micale, Determination of V_m and Effective Cross-sectional Area of Ar, N_2, and Kr for Determination of Specific Surface Areas, 49th Colloid and Surface Science Symposium of the American Chemical Society, Potsdam, NY (1974).
29. P. H. Emmett, Measurement of the Surface Area of Solid Catalysts, Catalysis, 1:31 (1954).
30. K. S. W. Sing, Utilisation of Adsorption Data in the BET Region, in "Surface Area Determination," IUPAC Symp. Proced., Butterworths, London, pp 25-34 (1969).
31. D. A. Payne, K. S. W. Sing, and D. H. Turk, Comparison of Argon and Nitrogen Adsorption Isotherms on Porous and Nonporous Hydroxylated Silica, J. Colloid Interface Sci., 43:287 (1973).
32. R. M. Barrer, Aspects of Sorption in Porous Crystals, in "The Structure and Properties of Porous Materials," ed. by D. H. Everett and F. S. Stone, Butterworths, London, pp 6-28 (1958).
33. J. B. Bulko, R. G. Herman, K. Klier, and G. W. Simmons, Optical Properties and Electronic Interactions of Microcrystalline Cu/ZnO Catalysts, J. Phys. Chem., 83:3118 (1979).
34. S. Mehta, G. W. Simmons, K. Klier, and R. G. Herman, Catalytic Synthesis of Methanol from CO/H_2. II. Electron Microscopy (TEM,STEM, Microdiffraction, and Energy Dispersive Analysis) of the Cu/ZnO and $Cu/ZnO/Cr_2O_3$ Catalysts, J. Catal., 57:339 (1979).

35. J. H. deBoer, The Shapes of Capillaries, in "The Structure and Properties of Porous Materials," ed. by D. H. Everett and F. S. Stone, Butterworths, London, pp 68-94 (1958).
36. R. M. Barrer and D. M. MacLeod, Intercalation and Sorption by Montmorillonite, Trans. Faraday Soc., 50:980 (1954).
37. B. C. Lippens, Ph.D. Dissertation, University of Delft (1961).
38. V. R. Dietz and E. Berlin, The Interaction of Krypton and an Exfoliated Graphite at 77.4°K, J. Colloid Interface Sci., 44:57 (1973).
39. F. A. P. Maggs, Reversal of Temperature Dependence for Physical Adsorption of Nitrogen, Research Correspondence, 6:135 (1953).
40. T. Kotanigawa, M. Yamamoto, M. Utiyama, H. Hattori, and K. Tanabe, The Influence of Preparation Methods on the Pore Structure of Alumina, Appl. Catal., 1:185 (1981).

HYSTERESIS IN MERCURY POROSIMETRY

S. Lowell and J. E. Shields

Quantachrome Corporation
Syosset, New York

INTRODUCTION

Since the development of the first commercial mercury porosimeter in 1945 [1], mercury porosimetry has become an important technique for the determination of pore size distributions in porous materials. The basic relationship of mercury porosimetry is the Washburn [2] equation (1) which gives the pressure, P, required to

$$Pr = -2\gamma \cos \theta \qquad (1)$$

intrude mercury into cylindrical pores of radius r. The surface tension and contact angle of mercury are given by γ and θ, respectively. A typical high pressure mercury intrusion-extrusion curve[3] is illustrated in Figure I.

A significant feature of all porosimetry curves is that they exhibit hysteresis, that is, the paths followed on intrusion and extrusion are not the same. At a given pressure the volume indicated on the extrusion curve is greater than that on the intrusion curve and for a given volume the pressure indicated on the intrusion curve is greater than that on the extrusion curve. A proposed explanation of intrusion-extrusion hysteresis is predicated on two factors: entrapment of mercury after a pressurization-depressurization cycle and a change in the contact angle of mercury on the pore walls between the processes of intrusion and extrusion. The concept of a pore potential has been proposed as an explanation for these factors.

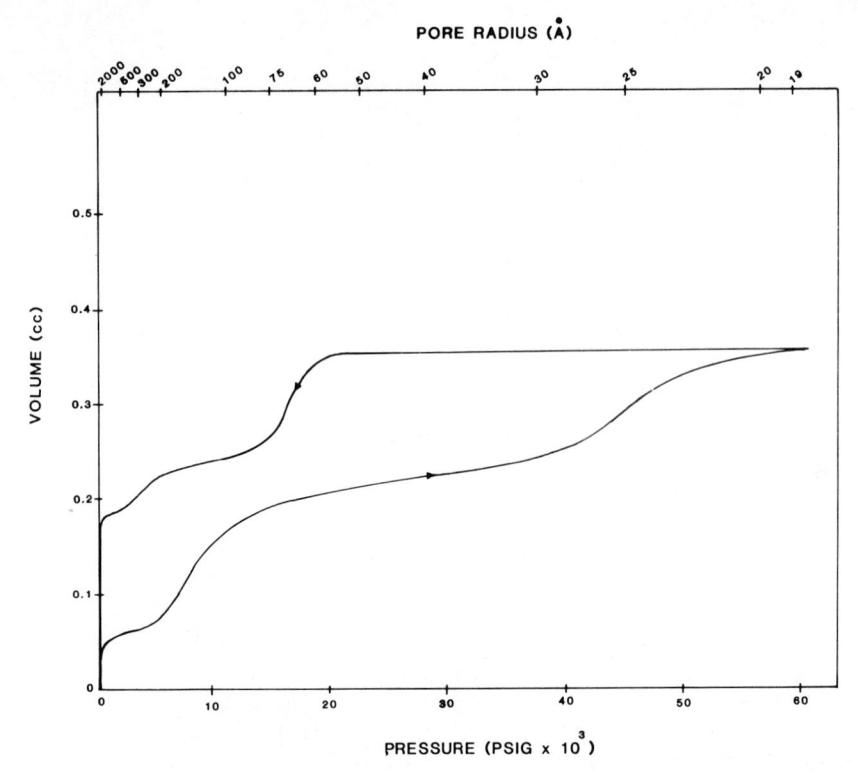

Fig. I. Mercury intrusion (→) and extrusion (←) curves on a sample of alumina/silica.

MERCURY ENTRAPMENT

Retention of mercury is recognized by the observation that, at the completion of an initial or first intrusion-extrusion cycle, some mercury always remains in the pores. Retained mercury is illustrated by V_t in Figure II in which a is the first intrusion curve and b the corresponding extrusion curve. Curve c is a second intrusion curve which joins the first intrusion curve at elevated pressures. Unlike intrusion curves, second and subsequent extrusion curves are always identical to the first extrusion curve (b). Usually after the sample is subjected to a first cycle, no additional entrapment is observed. However, in a few cases a third or even a fourth cycle is required before entrapment ceases.

CONTACT ANGLE

Contact angle hysteresis is a well documented phenomenon.[4] That is, the advancing (intrusion) contact angle differs from the receding (extrusion) angle. We have found[5] that if adjustments are made to the intrusion and extrusion contact angles, hysteresis

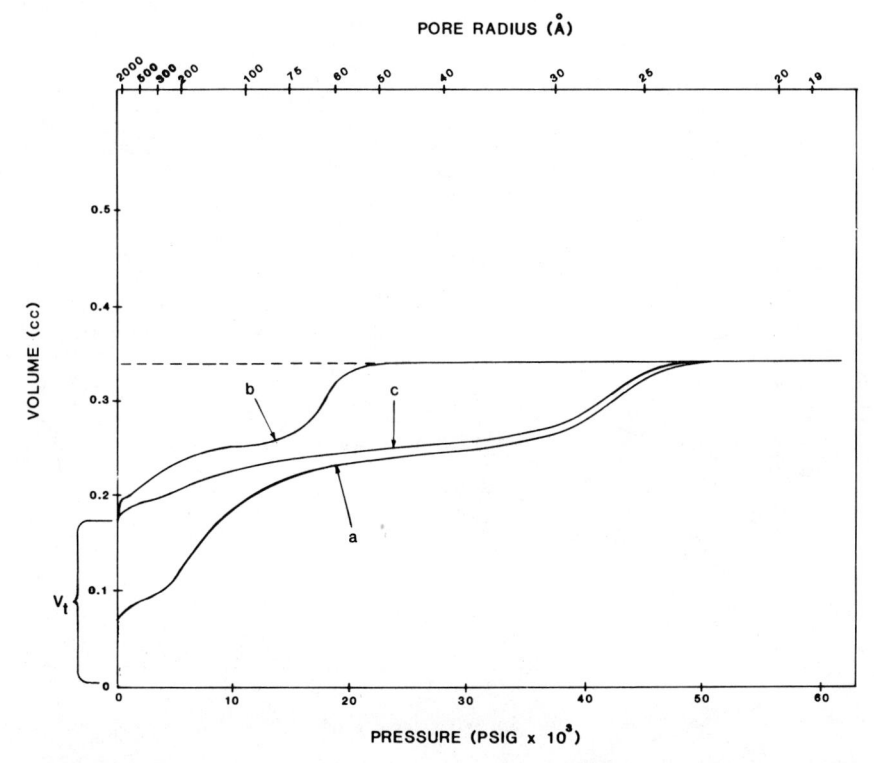

Fig. II. First and second intrusion-extrusion curves on a sample of alumina/silica.

between the intrusion and extrusion curves can be altered. Figure III illustrates an intrusion-extrusion cycle (A) using a contact angle of 140°. When the intrusion curve (B) was adjusted to a contact angle of 170° and the extrusion curve (C) adjusted to a contact angle of 110°, the two curves were brought into coincidence over only the high pressure portions. Since some mercury is permanently retained after the first depressurization, the intrusion and extrusion curve cannot be brought into coincidence along their entire path by means of contact angle adjustments. However, the effect of mercury entrapment can be eliminated by using second or subsequent scans where the hysteresis loop closes. Figure IV, curve B, represents a second run scan showing that the use of 170° for the intrusion curve and 110° for the extrusion curve results in exact superimposition of the intrusion and extrusion curves. Curves A and C in Figure IV are the corresponding curves obtained using a 140° contact angles for both intrusion and extrusion.

In order to determine the correct extrusion contact angle by superimposition of the intrusion and extrusion curves, it is necessary to know the intrusion contact angle. We have determined

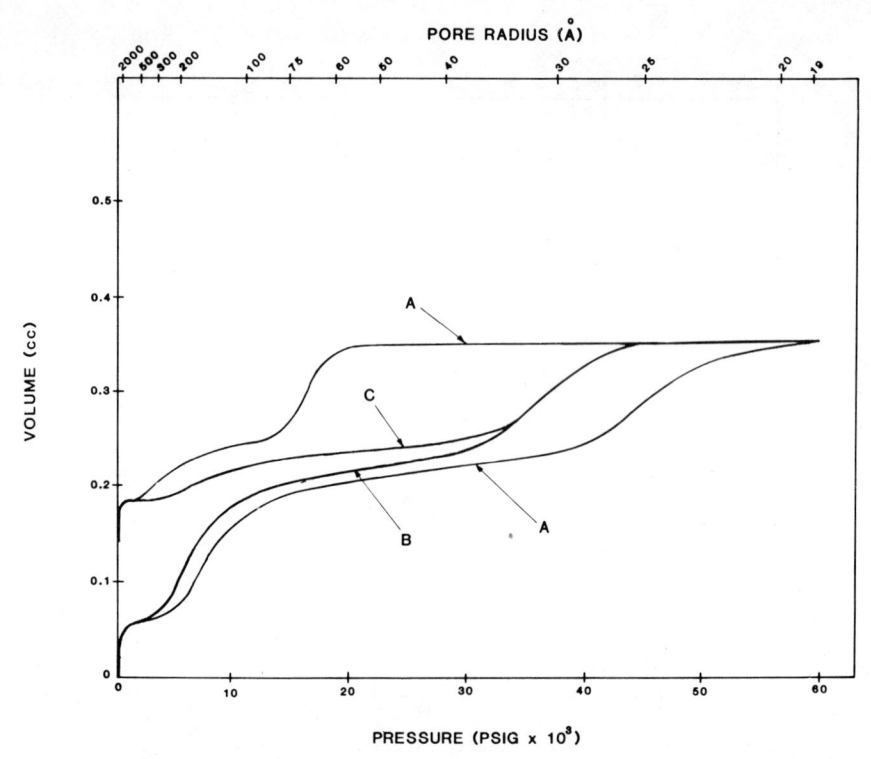

Fig. III. First intrusion-extrusion curves showing contact angle adjustments.

the contact angle of mercury on a variety of porous materials using a new type of mercury contact anglometer [6], which measures the contact angle by approximating the conditions used in mercury intrusion porosimetry, i.e., under vacuum and as the advancing or intrusion angle. In this technique mercury is forced by pressure through a hole of known diameter in an evacuated sample of compressed powder. Using 480 dynes/cm as the value for the surface tension of mercury, the contact angle can be calculated from the Washburn equation (1). The mean values of six measurements of the contact angles for ten powdered samples and the reproducibility of the measurement, expressed as standard deviation, are shown in Table I[6].

Figure V shows first and second cycle intrusion-extrusion scans for a sample of porous glass beads. Curve a is an initial intrusion cumulative volume distribution using the measured intrusion angle (136.8°) and curve c is the corresponding extrusion curve using the same contact angle. Curve b is a second intrusion curve using 136.8° for the contact angle and it is also the extrusion curve obtained by adjusting the contact angle to 103.5°. The pressure required for mercury to extrude from pores with a contact angle θ_e is less in all cases than that required to intrude

TABLE I

Material	Mean contact angle, θ	Standard deviation
Dimethylglyoxime	139.6°	0.45°
Galactose	140.3	0.43
Barium chromate	140.6	0.41
Titanium oxide	140.9	0.55
Zinc oxide	141.4	0.34
Dodecyl sodium sulfate	141.5	0.44
Antimony oxide	141.6	0.88
Fumaric acid	143.1	0.27
Starch	147.2	0.68
Carbon	154.9	1.2

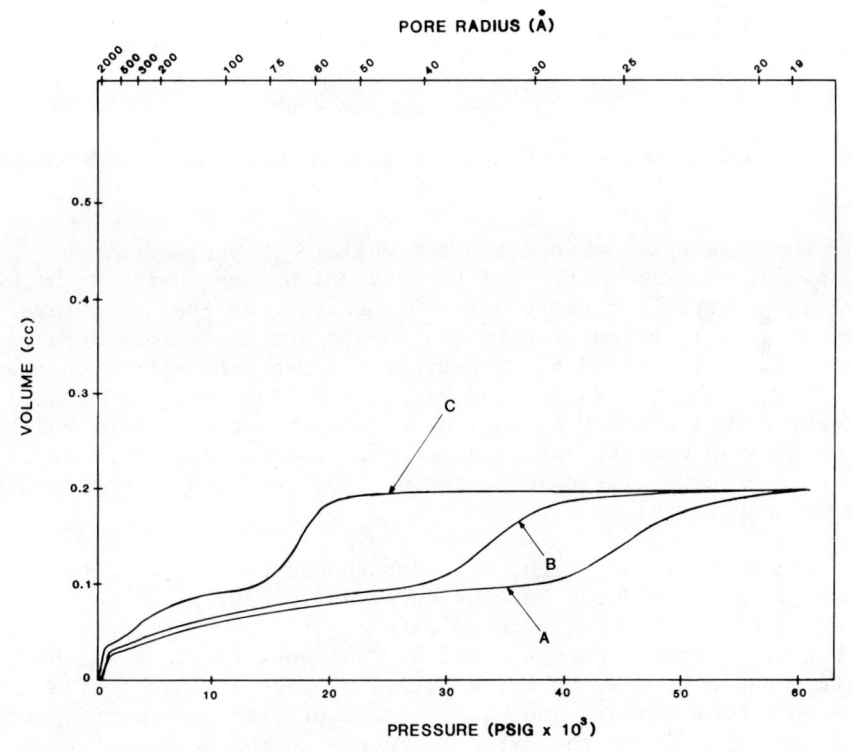

Fig. IV. Second intrusion-extrusion curves showing contact angle adjustments.

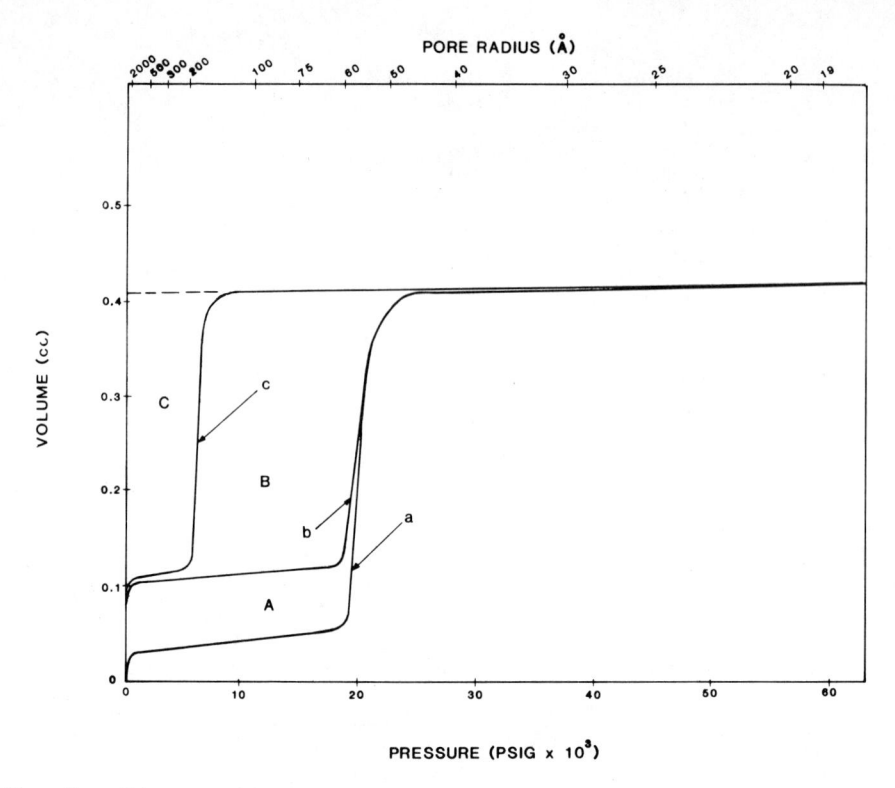

Fig. V. First and second intrusion-extrusion curves on a sample of porous glass.

into the same pores with a contact angle θ_i. Therefore, the successful superimposition of the extrusion curve using θ_e with the second intrusion curve using θ_i occurs when the curves are plotted as a function of pore radius and not pressure. When curve b (Figure V) is viewed as an extrusion curve, the extrusion pressure is incorrect by the factor $\cos\theta_e/\cos\theta_i$. However, the area between curves a and b as well as between b and c remains the same because curve b meets curves a and c at the maximum intruded volume and then extends off-scale at the maximum pressure without generating any additional area.

The work of intrusion, W_i, corresponds to the area above the intrusion curve and can be considered as consisting of three parts, shown in Figure V. The work of entrapment, W_t, is proportional to A, the area between curves a and b. The work of changing the contact angle from θ_i to θ_e, $W_{\Delta\theta}$, is proportional to B, the area between curves b and c and W_e, the work of extrusion corresponds to C, the area above the extrusion curve to the maximum intruded volume. These work terms have been obtained [7] from the graphical areas by using the following equations:

$$W_t^g = -A_t \gamma \cos \theta_i \tag{2}$$

$$W_{\Delta\theta}^g = -A_{\Delta\theta} \gamma (\cos \theta_i - \cos \theta_e) \tag{3}$$

$$W_e^g = A_e \gamma \cos \theta_i \tag{4}$$

In equations (2) - (4) A_t is the area of mercury entrapped in all the pores, $A_{\Delta\theta}$ is the area of mercury which undergoes a change in contact angle upon extrusion and is the same as A_e, the area of mercury which extrudes from all the pores. Assuming cylindrical pore geometry, the area, A, in any pore radius interval is given by

$$A = \frac{2}{\bar{r}} V \tag{5}$$

where $\bar{r}$ is the mean pore radius in a narrow pore radius interval and V is the volume of mercury intruded or extruded in that interval. The volume terms are determined as the volumetric difference between the desired curves less the volume differences for the previous interval. Thus, the three work terms can be expressed as

$$W_t = -2\gamma \cos \theta_i \sum_{1}^{n} \left(\frac{V_t}{\bar{r}}\right)_n \tag{6}$$

$$W_{\Delta\theta} = -2\gamma (\cos \theta_i - \cos \theta_e) \sum_{1}^{n} \left(\frac{V_{\Delta\theta}}{\bar{r}}\right)_n \tag{7}$$

$$W_e = 2\gamma \cos \theta_i \sum_{1}^{n} \left(\frac{V_e}{\bar{r}}\right)_n \tag{8}$$

The volume terms in equations (6), (7) and (8) can be obtained from mercury porosimetry data. The volume of mercury trapped (V_t) in a pore radius interval of mean radius $\bar{r}$ is obtained from the volume difference between curves a and b (Figure V) and the volume difference for the previous radius interval. Alternatively, the volume of mercury entrapped in the pores can be expressed as the difference between the volume intruded into a narrow radius interval and the volume extruded in the same interval, i.e.,

$$V_t = V_i - V_e \tag{9}$$

The term $V_{\Delta\theta}$ in equation (7) is the volume increment undergoing the contact angle change measured along curve b and V_e is obtained as small increments along the incorrect extrusion curve c. Thus, the terms $V_{\Delta\theta}$ and V_e are the same. An illustration of a calculation of W_t, $W_{\Delta\theta}$ and W_e from equations (6) through (8) for a porous glass sample is given Table II. The corresponding work terms obtained by graphical integration using equations (2) through (4) are given at the bottom of the table and show excellent agreement.

TABLE II

W_t, $W_{\Delta\theta}$, and W_e for Porous Glass Sample

Radius interval (Å)	$\bar{r}$ (Å)	$\sum_1^n (V_t)_n$ (cm³)	W_t (erg × 10⁻⁸)	$\sum_1^n (V_{\Delta\theta})_n$ (cm³)	$W_{\Delta\theta}$ (ergs × 10⁻⁸)	$\sum_1^n (V_e)_n$ (cm³)	W_e (ergs × 10⁻⁸)
17.1–47.5	32.3	0	0	0	0	0	0
47.5–52.3	49.9	0	0	0.211	2.01	0.004	0.0561
52.3–57.1	54.7	0.083	1.06	0.241	2.27	0.005	0.0689
57.1–61.9	59.9	0.083	1.06	0.241	2.27	0.005	0.0689
61.9–66.6	64.3	0.083	1.06	0.242	2.28	0.006	0.0798
66.6–76.1	71.4	0.084	1.07	0.245	2.30	0.006	0.0798
76.1–85.6	80.9	0.084	1.07	0.246	2.31	0.006	0.0798
85.6–104.7	95.2	0.084	1.07	0.248	2.32	0.007	0.0798
104.7–119.0	111.9	0.084	1.07	0.248	2.32	0.008	0.0871
119.0–166.5	142.5	0.084	1.07	0.259	2.35	0.220	0.0934
166.5–237.9	202.2	0.085	1.08	0.260	2.35	0.230	1.13
237.9–380.6	309.3	0.086	1.08	0.261	2.36	0.239	1.17
380.6–951.6	666.1	0.086	1.08	0.264	2.36	0.248	1.19
951.6–1903	1427	0.086	1.08	0.265	2.36	0.251	1.20
1903–9516	5710	0.086	1.08	0.265	2.36	0.255	1.20
9516–66612	38064	0.087	1.08	0.288	2.36	0.288	1.20

Note. $W_t = 1.08 \times 10^8$; $W_{\Delta\theta} = 2.36 \times 10^8$; $W_e = 1.20 \times 10^8$; $^gW_t = 1.09 \times 10^8$; $^gW_{\Delta\theta} = 2.37 \times 10^8$; $^gW_e = 1.22 \times 10^8$.

PORE POTENTIAL

The phenomena of hysteresis, mercury entrapment and contact angle changes in mercury porosimetry can be explained by means of a pore potential. The concept of a pore potential has been established[8] to account for capillary condensation of vapors at pressures substantially below the saturated equilibrium pressure.

Assuming cylindrical geometry, the force driving mercury out of a pore can be equated to the force associated with the pressure directing mercury into a pore. That is,

$$F = -2\pi r \gamma \cos\theta = P\pi r^2 \qquad (10)$$

If one postulates that the pore potential, U, is the difference between the interaction of mercury along the total length of all pores in a radius interval with radius $\bar{r}$ when the pores are filled at the intrusion pressure P_i and when partially emptied at the extrusion pressure P_e, then

$$-U = \int_0^{l_i} F_i \, dl_i - \int_0^{l_e} F_e \, dl_e \qquad (11)$$

HYSTERESIS IN MERCURY POROSIMETRY

F_i and F_e in equation (11) are the forces exerted on the mercury column within a pore at the pressures required for intrusion and extrusion, respectively, and l_i and l_e are the total lengths of the mercury column within a narrow band of pores of mean radius $\bar{r}$ when filled at P_i and when partially emptied at P_e. The combination of equation (10) and (11) leads to

$$U = 2\pi\bar{r}\,\gamma\,(\cos\theta_i \int_0^{l_i} dl_i - \cos\theta_e \int_0^{l_e} dl_e) \qquad (12)$$

or

$$U = \frac{2}{\bar{r}}\gamma\,(V_i \cos\theta_i - V_e \cos\theta_e) \qquad (13)$$

Extending equation (13) to a system of pores of various radii the total pore potential U can be expressed as

$$U = 2\gamma\left[\cos\theta_i \sum_i^n \left(\frac{V_i}{\bar{r}}\right)_n - \cos\theta_e \sum_i^n \left(\frac{V_e}{\bar{r}}\right)_n\right] \qquad (14)$$

Expressing V_t according to equation (9), recalling that $V_{\Delta\theta}$ is equal to V_e and combining equations (6) and (7) yields

$$W_t + W_{\Delta\theta} = -2\gamma\cos\theta_i\left[\sum_i^n \left(\frac{V_i}{\bar{r}}\right)_n - \sum_i^n \left(\frac{V_e}{\bar{r}}\right)_n\right]$$

$$-2\gamma\,(\cos\theta_i - \cos\theta_e)\sum_i^n \left(\frac{V_e}{\bar{r}}\right)_n \qquad (15)$$

which when combined with equation (14), gives

$$-U = W_t + W_{\Delta\theta} \qquad (16)$$

Equation (16) predicts that changes in U, the pore potential, will effect the magnitude of the entrapped mercury and/or the change in contact angle between extrusion and intrusion. Hence, changes in the pore potential will alter the size of the hysteresis loop.

The model developed here and expressed by equation (16) is that mercury, when intruded into a pore with a contact angle θ_i, acquires an increased interfacial free energy. Upon depressurization, mercury begins to extrude from the pore at pressure P_e reducing the interfacial area and simultaneously the contact angle. As the pressure decreases, mercury will continue extruding from the pore with the extrusion angle θ_e until the interfacial free energy is equal to the pore potential. At this point, extrusion ceases leaving some mercury trapped near the opening of the pore. This phenomenon is observed as discoloration of the sample due to finely divided mercury at the pore entrances on samples that have been penetrated.

Evidence for the pore potential was obtained from mercury intrusion-extrusion data of samples which had been impregnated with both polar and nonpolar materials in order to cause an alteration in the field intensity within a pore. Mercury porosimetry data for three samples which were impregnated with various ionic salts are given in Tables III- V.

TABLE III

Porosimetry Data for Al_2O_3 Impregnated with $CuSO_4$

%$CuSO_4$	V_i (cc/g)	V_t (cc/g)	P_i (psia)	P_e (psia)	%Hg retained	θ_i	θ_e
0	.347	.072	28500	8300	20.7	144.5°	103.7°
.5	.334	.075	28500	7700	22.5	142.5	102.4
1	.331	.093	26700	1800	28.1	145.4	93.2
2	.323	.128	28000	600	39.6	142.0	91.0
4	.313	.171	27400	600	54.6	143.1	91.0
8	.292	.229	27000	500	78.4	143.4	90.9
20	.294	.253	26600	500	86.1	143.6	90.8
30	.280	.255	26500	500	91.1	143.1	90.8

TABLE IV

Porosimetry Data for Al_2O_3 Impregnated with $NiCl_2$

%$NiCl_2$	V_i (cc/g)	V_t (cc/g)	P_i (psia)	P_e (psia)	%Hg retained	θ_i	θ_e
0	.348	.072	28300	8500	20.9	145.7°	104.4°
4	.322	.081	28500	6200	25.2	146.0	100.4
8	.315	.096	28500	5300	30.5	145.4	98.8
20	.288	.114	28500	2000	39.6	146.6	93.4
40	.244	.112	28200	1300	45.9	144.5	92.2

The percent salt indicated in these tables is the concentration of the salt solution used to coat the sample. The actual quantity of salt incorporated into the sample could not be determined since some of the substrate was lost in the process of removing the excess salt. The intrusion and extrusion pressures (P_i and P_e) were obtained from the maxima of the derivative curve

TABLE V

Porosimetry Data for Al_2O_3/SiO_2 Impregnated with $CoCl_2$

%$CoCl_2$	V_i (cc/g)	V_t (cc/g)	P_i (psia)	P_e (psia)	%Hg retained	θ_i	θ_e
0	.478	.259	6500	1600	54.2	145.1°	101.6°
4	.477	.264	7200	1800	55.3	142.8	101.5
8	.478	.282	7800	800	59.0	143.9	94.7
20	.459	.279	6750	500	60.8	142.8	93.4
30	.451	.271	7300	300	64.1	143.9	91.9

(dV/dP). Using the measured intrusion contact angle (θ_i) the extrusion contact angle (θ_e) was calculated from

$$\cos \theta_e = \cos \theta_i \frac{P_e}{P_i} \qquad (17)$$

and represents the extrusion contact angle required to superimpose the extrusion curve on a second run intrusion curve.

Tables III-V show that as the salt concentration is increased, the hysteresis increases, that is, the difference between P_i and P_e increases while the extrusion contact angle decreases. Thus, $W_{\Delta\theta}$, the work required to change the contact angle, increases with increasing amounts of ionic salt, in accordance with equation (7). Similarly, the work of entrapment, W_t, increases as the salt concentration is raised, as evidenced by the quantities of mercury entrapped [see equation (9)]. These effects can also be seen in Figure VI which illustrates the change in hysteresis resulting from impregnation of an alumina sample with varying amounts of copper sulfate. It should be noted that the intrusion curves for both untreated and treated samples are virtually identical, indicating that the process of impregnation does not significantly alter the radius of the pore opening. The same phenomenon was observed in all impregnated samples which were studied. However, in all cases the volume of mercury intruded, V_i, decreased with increasing salt concentration, an indication that precipitation of the salt occurred near the base of the pores. To eliminate the pore volume differences, the curves in Figure VI were normalized to the maximum volume intruded into the untreated sample.

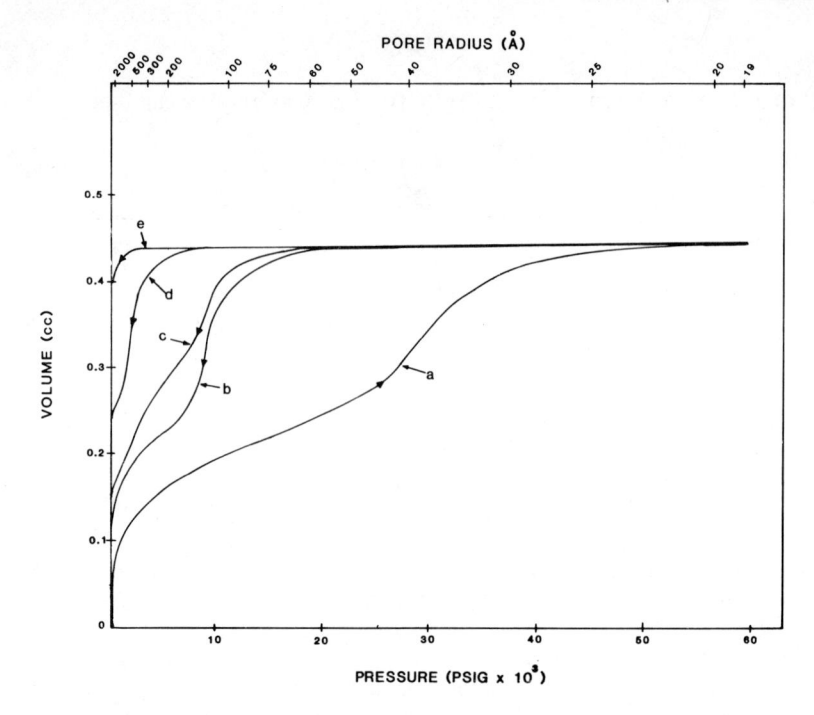

Fig. VI Intrusion-extrusion curves on an alumina sample coated with various amounts of copper sulfate; a) intrusion curve for all samples, b) extrusion curve for untreated alumina, c) extrusion curve for alumina treated with 0.5% $CuSO_4$, d) extrusion curve for alumina treated with 2% $CuSO_4$ and e) extrusion curve for alumina treated with 40% $CuSO_4$.

TABLE VI

Porosimetry Data for Various Porous Samples Impregnated with DCDMS[a]

Sample	V_i (cc/g)	V_t (cc/g)	P_i (psia)	P_e (psia)	%Hg retained	θ_i	θ_e
Porous Glass[b]	.656	.127	19750	5600	19.4	140°	102.5°
Porous Glass/DCDMS[c]	.585	.168	22000	8000	28.7	140	106.2
Al_2O_3[b]	.348	.072	28300	8500	20.9	145.7	104.4
Al_2O_3/DCDMS[c]	.293	.073	35000	15200	24.9	140.1	109.5
Al_2O_3-SiO_2[b]	.478	.259	6500	1600	54.2	145.1	101.6
Al_2O_3-SiO_2/DCDMS[c]	.454	.205	7800	3300	45.2	137.8	108.3

a, dichlorodimethylsilane; b, untreated sample; c, sample treated with dichlorodimethylsilane

The effect of pore impregnation with nonpolar material was studied by treating the sample with dichlorodimethylsilane (DCDMS). The results are shown in Table VI. In each case a decrease in the hysteresis, compared to the untreated material, was observed after impregnation with DCDMS. These increases in the extrusion contact angle resulted in decreases in $W_{\Delta\Theta}$. In some cases impregnation with DCDMS led to greater mercury retention or an increase in W_t oever the untreated material. However, this was always accompanied by a larger decrease in $W_{\Delta\Theta}$.

The hysteresis data shown above are consistent with the proposed pore potential as described by equation (16). Namely, the field intensity and therefore the potential within a pore is expected to increase upon addition of polar materials and decrease with nonpolar materials. Table VII gives the calculated pore potentials for a few typical examples. The work terms, W_t and $W_{\Delta\Theta}$, were obtained by graphical integration of the first and second intrusion-extrusion cycles. Impregnation of alumina with nickel chloride showed an increase in mercury entrapment and thus an increase in W_t. When added to the increase in $W_{\Delta\Theta}$, the expected increase in pore potential caused by ionic salts was observed. Table VII also shows the decrease in U resulting from coating two samples with DCDMS.

It is apparent that the shape and extent of hysteresis and mercury entrapment and thus the pore potential can be attributed to the nature of the material in which the pore exists and is subject to alterations by changing the nature of the pore wall.

TABLE VII

Pore Potential of Various Porous Samples

Sample	%Hg retained	$W_t (10^{-8})$ (ergs/cc Hg)	$W_{\Delta\Theta} (10^{-8})$ (ergs/cc Hg)	$U(10^{-8})$ (ergs/cc Hg)
Al_2O_3[a]	20.9	.63	6.51	7.14
$Al_2O_3/NiCl_2$[b]	45.9	1.94	7.04	8.98
Al_2O_3-SiO_2[a]	54.2	.25	1.95	2.20
Al_2O_3-SiO_2/DCDMS[c]	45.2	.23	1.32	1.55
Al_2O_3[a]	20.9	.63	6.51	7.14
Al_2O_3/DCDMS[c]	24.9	.89	5.37	6.25

a, untreated sample; b, alumina treated with 40% $NiCl_2$; c, samples treated with dichlorodimethylsilane.

REFERENCES

1. H. L. Ritter and L. C. Drake, Ind. Eng. Chem., Anal. Ed., **17**, 782 (1945).
2. E. W. Washburn, Phys. Rev., **17**, 273 (1921).
3. Obtained on a Quantachrome Autoscan-60.
4. L. S. Penn and B. Miller, J. Colloid Interface Sci., **77**, 574 (1980).
5. S. Lowell and J. E. Shields, J. Colloid Interface Sci., **80**, 192 (1981).
6. J. E. Shields and S. Lowell, Powder Technol., **31**, 227 (1982).
7. S. Lowell and J. E. Shields, J. Colloid and Interface Sci., **83**, 273 (1981).
8. A. W. Adamson, "Physical Chemistry of Surfaces", p. 335, Wiley-Interscience, New York, 1967.

PORE STRUCTURE CHARACTERIZATION BY

MERCURY POROSIMETRY

O. J. Whittemore and G. D. Halsey

University of Washington
Seattle, WA 98195

Ceramics when processed usually start as a highly porous compact and frequently after processing are porous, e.g. refractory and building brick. Therefore, porosity is a property of much interest and is often correlated with other properties, such as strength, permeability, electrical resistivity, etc. Measurements of apparent or total porosity are therefore usually determined on such materials. However, these are only single values and give no insight into the size and distribution of pores.

Mercury porosimetry is a recognized method for determination of open pore size distribution (PSD) from which also can be derived surface area, and pore size frequency. Proposed by Washburn in 1921[1], not until 1945 was the method applied by Ritter and Drake[2]. Use of mercury porosimetry has been rapidly developing. In the May/June 1981 issue of Powder Technology, which was devoted to the method, a bibliography[3] lists 998 references from 1921 to 1980 of which 75 were printed in 1979. Another paper[4] discusses characterization of porous ceramics including PSD specifications for tar bonded magnesia brick for steel making[5,6] and for frost resistant building brick[7]; the effects of sintering on plasma-sprayed alumina[8] and zircon[9] and on silica fiber compacts[10] used for the space shuttle insulation; the characterization of carbon/carbon composites[11], ceramic water filters[12], diatomites[13], and clays[14,15]. Ceramic processing studies included compaction studies on glass spheres[4], the gypsum plaster molds[4,16] used for slip casting. More studies have been achieved on firing of building brick and tile[7], of basic refractories[4,16], of sanitary ware[17].

Sintering studies have demonstrated by mercury porosimetry that many materials show pore growth during the initial stages of sintering rather than pore shrinkage which is assumed in many sintering models[18,19]. Pore size ranges are narrowed during sintering[19,20]. Also, work has shown the intermediate and final-stage shrinkage response to be dependent on pore structure in compacts fabricated by extrusion and by slip casting[21].

Mercury porosimetry can distinguish between materials with a narrow pore size distribution, such as a filter, or those with broad distributions, often as much as 3 to 4 orders of magnitude in pore size, like refractories.

Bimodal PSD often occurs in ceramic bodies before firing, caused by the presence of aggregates[18] or by a mixture of coarse particles with fine bond particles such as in refractories[4]. When these are fired or sintered, the smaller pore size disappears first. Cordierite exhaust catalyst supports have been shown to have bimodal PSD which changed to a trimodal PSD after gamma alumina was deposited in the pores[22].

PROCEDURE

Mercury porosimetry is conducted by first preparing a sample of a size dependent on the material porosity and with pore volume measurable by the equipment. The sample is placed in a holder called a penetrometer, evacuated, and then immersed in mercury at a low pressure. Increasing pressures are applied on the mercury and corresponding volumes of intruded mercury determined. The diameter, d, of a cylindrical pore being intruded by mercury of surface tension γ at pressure P at the contact angle θ with the sample, is determined from the Washburn equation:

$$d = \frac{-4 \gamma \cos \theta}{P} \qquad (1)$$

Since the surface tension and contact angle are considered to be constant (0.485 N/m and 130° to 140° are commonly used), the pore diameter is inversely proportional to the pressure. Equipment is available capable of pressures from 7kPa to 412MPa (1 to 60,000 psi) which thus will intrude pores from 200 μm to 3 nm or five orders of magnitude.

DATA PRESENTATION

The conventional method of PSD presentation is to plot pressure vs. cumulated volume of intruded mercury which, from equation (1), can also be a plot of pore diameter vs. cumulated pore volume. One such plot is shown in Figure 1 of an alumina compact with bimodal PSD together with PSD after sintering at 1000°C and at

PORE STRUCTURE CHARACTERIZATION

1200°C. The advantage of this cumulative function is that any fraction of pore volume in a pore size range can be quickly estimated.

A derivative function is often used as also shown in Figure 1 for d(vol.)/d(dia.). In this case, the bimodality is easily identified in the cumulative curve but some PSD curves can be better interpreted with derivative curves. However, the derivative function is distorted when pore sizes vary by orders of magnitude. Representation by the function d(vol)/d(log.dia) can also be used but in Fig. 1 would depress the peak at 0.02 μm.

Curve shape ratios have been used to describe PSD.[19] The ratios of the pore diameter at 5% total pore volume and at 95% total pore volume were calculated. When PSD is determined on equal sized spherical particles, an abrupt break-through of mercury occurs and the ratios were measured at 1.1 and 0.5, the same as theoretically calculated by Frevel and Kressley[23] for a log normal distribution for 20% of nested close-packing with 80% of close-packed layers stacked without nesting. Sintering was shown to narrow the PSD probably due to neck growth filling the space at particle contact.[19,20]

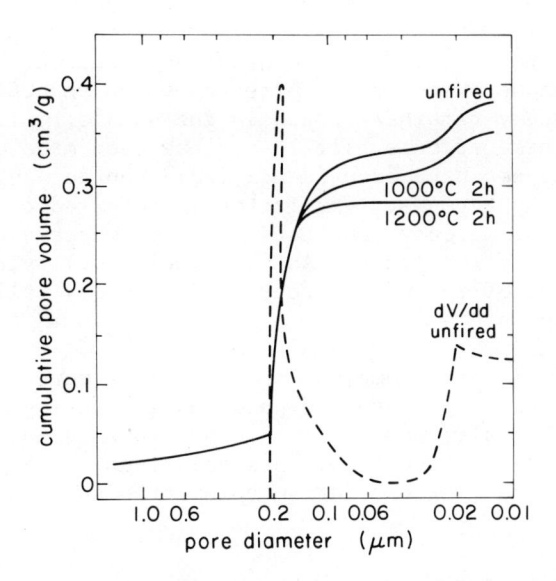

Fig. 1. PSD of compacted Linde A alumina before and after sintering. Solid lines are cumulative volumes. Dashed line is the derivative of the unfired sample curve.

The total pore volume can be calculated if all pores have been filled at the maximum pressure. In Fig. 1, it is obvious in the 1200°C. PSD that all pores are filled and larger than 0.06 μm but not so for the other two PSDs. If the true specific gravity of the sample is known, the porosity and apparent density, ρ_a, can be calculated from the total pore volume

$$\rho_a = 1/V_p + 1/\rho_t \qquad (2)$$

where V_p is the total pore volume per gram and ρ_t is the true specific gravity.

Surface area distribution can also be determined by integrating the volume-diameter curve, assuming that the pores are a series of cylinders of decreasing diameter. Assuming n number of cylindrical pores of length ℓ, these relationships follow:

$$\text{Pore Volume:} \quad V = n\pi d^2 \ell/4 \qquad (3)$$

$$\text{Surface Area:} \quad S = n\pi d\ell \qquad (4)$$

$$S = 4V/d \qquad (5)$$

A convenient single value from the PSD is the mid-pore diameter where half of the pore volume has been intruded by mercury.

When the pressure is reduced after penetration, the pressure-volume curve shows hysteresis. After complete reduction of pressure, some mercury remains trapped in the sample. Although many have studied these effects, little use has been made of them. However, two causes have been identified: contact angle hysteresis and structural hysteresis. The latter is referred to as the "ink-bottle" effect of mercury penetrating a pore whose entrance is smaller than the pore itself. As an example, well vitrified brick retain about 90% of intruded mercury and this may well be from "ink-bottle" pores.

As an example of a complex application of mercury porosimetry in ceramics, Varela[24,25] studied pore changes during sintering of MgO. Apparent density was determined by two methods: by mercury displacement and from the total pore volume using eq. (2). Good agreement was noted. The relative volumetric shrinkage was then obtained from:

$$\Delta V/V_o = \rho_o/\rho - 1 \qquad (6)$$

where $\Delta V/V_o$ is the relative volumetric shrinkage, ρ_o is the initial apparent density, and ρ is the apparent density of the sintered sample.

PORE STRUCTURE CHARACTERIZATION

To relate densification with the variation of mid-pore size, the relative volumetric shrinkage is plotted versus the relative mid-pore diameter cubed, $(\bar{d}/\bar{d}_o)^3-1$, in Figure 2 where $\bar{d}$ and $\bar{d}_o$ are the mid-pore diameters of the sintered sample and the initial state respectively.

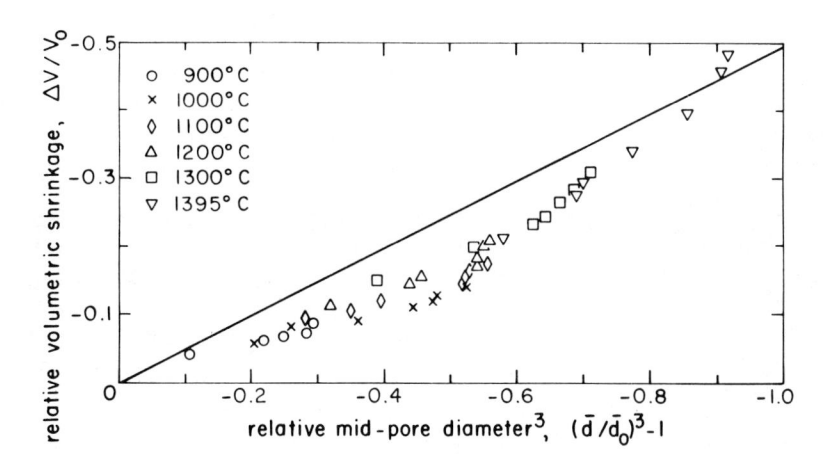

Fig. 2. Relative volumetric shrinkage vs. relative mid-pore size cubed of MgO compacts sintered in dry argon from 900 to 1395°C. and from 1 to 600 min. The straight line represents homogeneous shrinkage from eq. (7). All data are derived from mercury porosimetry.

The solid line in Fig. 2 represents the expected uniform shrinkage of the compact assuming uniform shrinkage of pores of the size of the mid-pore diameter. Equation (7) is thus derived:

$$\frac{\Delta V}{V_o} = \lambda \left[\left(\frac{\bar{d}}{\bar{d}_o}\right)^3 - 1 \right] \qquad (7)$$

where λ is a proportionality constant equal to the pore fraction or 0.49 for the compacts studied. The plot of eq. (7) is a straight line unless there is a wide distribution of pore sizes. These compacts had a narrow PSD so the deviation from a straight line would be small. The pore volume consisted of 80% of pores between 40 nm and 50 nm and less than 4% of pores smaller than 30 nm. If all of the pores smaller than 30 nm disappeared by shrinkage before the mid-pore size started to shrink, the plot on Fig. 2 would be only slightly higher.

However, nearly all points fall beneath the plot of eq. (7) showing more rapid decrease of the relative mid-pore volume with regard to the relative volume shrinkage than predicted up to 0.15 volumetric shrinkage (15%). At each temperature and up to 15% shrinkage, a straight line can be drawn, but the slopes of these lines increase with temperature, approaching the value of λ. In addition, it must be pointed out that as the shrinkage increases beyond 0.15, the slope must increase to approach the same end-point.

Thus, Varela showed a faster decrease of large pores during the initial stage contrary to an expected faster decrease of small pores. Rearrangement of particles during the early stage of sintering and extending through the initial stage is thus inferred.

Deviation from eq. (7) means nonuniform shrinkage and two processes could occur: rearrangement and coalescence. If coalescence occurred with shrinkage and particle growth, the points would be above the plot of eq. (7). With rearrangement, points would be below the plot as is shown. If pores grew with little or no compact shrinkage, as previous work[18,19] has shown, the points would occur on the left or positive side of Fig. 2.

The conventional method of studying sintering by measuring linear shrinkage would not give sufficient information to reach these conclusions. The particle size in these compacts was 43.5 nm which would be difficult to measure by electron microscopy. Mercury porosimetry appears to be the only characterization method by which distinction can be made between coalescence and rearrangement.

SURFACE AREA ESTIMATES

Any analysis of the pore structure of a rigid solid system involves an estimate of the total surface area of the structure as a primary characteristic of the system along with the pore geometry. Mercury intrusion itself provides information that leads to a surface area value, or that should be consistent with an independent estimate of this area.

Rootare and Prenzlow[26] have derived an equation for obtaining surface area without specific pore geometry:

$$S = -\frac{1}{\gamma \cos \theta} \int_0^V P dV \qquad (8)$$

where S is the specific surface expressed in area per volume pore space. The same expression is obtained for a cylinder or for an arrangement of parallel plates. The general case is not based on the specific structure of the pores; the final (integrated) energy of forming the mercury solid interface divided by the energy of forming unit area of this surface yields a surface area for the sample. There are severe conditions on the validity of this calculation however. Since the calculation is thermodynamic in its basis, the intrusion pressures must be assumed to be reversible or equilibrium values. Second, the upper limit of the intrusion pressure must fill the smallest pores in the surface inventory. Third, the composition and structure of the surface cannot change with pore size, or more formally, there must be no effect of pore radius on the interaction energy per unit area. These conditions, along with more practical considerations, rule out the meaningful analysis of intrusion data on solids containing micropores. Even the definition of the radius or interaction distance presents difficulties that in the definitions of Gibbs are characterized by the distinction between the dividing surface and the surface of tension at curved surfaces. The first is defined by the stoichiometry or amount of substances up to the surface, and the second by the energetics of the surface, and they in general can be made to coincide for (nearly) plane surfaces only.[27]

COMPARISON OF AREA ESTIMATES

If intrusion measurements are made on a porous solid with a crystallographically characterized regular pore structure, then in principle at least, a direct comparison of the area calculated from intrusion pressure and the "geometric" area could be made. Practically, such regular pores are too small to be in the applicable range of the intrusion measurements, and the regular cage-like voids with narrower approach channels are molecular size "ink bottles". The partially sintered aggregates which are the subject of our studies must of necessity have irregular structures that can be more readily characterized by equivalent cylindrical pores, but which have no readily estimated geometrical area.

For such solids the prime estimate of surface area has been provided by volumetric measurement of the physical adsorption of gases, in particular that of nitrogen in the classical BET method of surface area measurement. These estimates of surface area based on gas adsorption, although routine and reproducible, are subject to uncertainties that are discussed below. However, if we accept the BET area of a porous solid as correct the comparison with the area from mercury intrusion provides a check of consistency between the two methods of surface area determination. Reasonable agreement has been shown between surface areas determined by mercury porosimetry and by nitrogen adsorption (BET)[24]. It has been suggested[28] that the contact angle, θ, between the sample and mercury be calculated from equation (8) using the surface area determined by adsorption.

Knowledge of the whole gas adsorption isotherm provides auxiliary information. First, the shape of the isotherm provides information on the possible presence of pores too small to be included in the inventory provided by intrusion. Hysteresis in the adsorption isotherm at higher relative pressures is an indication of pore structure. For pores with a definite geometry, for example cylindrical pores, there is at least one simple and elegant explanation of this hysteresis; the filling radius of curvature in one direction is determined by the pore radius and is infinite in the other direction, whereas when the same pore empties from the outer end, the liquid-gas surface is spherical with both radii determined by the pore radius.

Second, the region of the isotherm used to determine the area is (except for the smallest pores) measured at lower pressure and thus independent of the region of the isotherm controlled by pore structure, for the class of systems to which mercury intrusion is a useful technique.

Fortunately, because of the high thermal instability of micropores, even partially sintered masses are unlikely to contain significant quantities of pores accessible to nitrogen but not to mercury. Our samples show no evidence of such pores. Thus, a lack of agreement between the two estimates of area must be explained by a discrepancy in the evaluation of the area, rather than the area accessible to the two experiments.

The area measured by mercury may be too high if pores fill irreversibly into "ink bottles". However, if the filling is reversible, or nearly so, any discrepancy between areas can be eliminated by a choice of interaction energy less than the surface tension of pure mercury, as long as the mercury surface area based on the surface tension of pure mercury is less than that based on gas measurements. This adjustment is usually expressed in terms of a suitable contact angle, that is maintained constant for a given class of surfaces or series of experiments.

Such a procedure is justified especially if it gives rise to a "standard" contact angle that can be used routinely for a given class of solids. If the underlying solid is an oxide of polyvalent cations, the contact angle is not very sensitive to the cations involved[29]. However, surface layers of contaminants, such as water, may have a greater effect on the angle than the solid itself, and furthermore may be labile under the conditions of sintering at various temperatures and humidities. In addition, hysteresis can be accommodated by the concept of advancing and retreating contact angles, but it is difficult to give a precise mechanism to the origin of this difference; that is, whether for example it is due to molecular or more macroscopic features of the surface structure.

VALIDITY OF THE BET SURFACE AREA ESTIMATES

If the BET area is used to establish a contact angle, one is implicitly relying on its absolute magnitude being correct. The model for this method is extremely simple in that it regards the monolayer as having the same structure as the bulk nitrogen adsorbate in order to calculate a surface area from a volume identified with the number of moles of nitrogen in a monolayer. This picture of the monolayer may drastically oversimplify its structure. In fact, it is strongly bound to a periodic solid surface, that may contain incidental features such as grain boundaries, steps, and foreign contaminants. The adsorbed layer may be in registry with the underlying solid, or it may have a structure influenced by its own intermolecular forces mediated by the solid. The only solid surface for which these possibilities have been at all thoroughly explored is the basal plane of graphite, which is not a high surface area material. Even the surface of high-surface area carbon black can only approximate the conditions explored for graphite in its regular crystallographic form.[30] There has never been any reason to believe that absolute BET surface areas are accurate to any better than ± 10%. (A general and concise overview of physical adsorption has been given by Pierotti and Thomas.[31])

These observations do not apply in such force to relative surface area measurements. In principle, these measurements are not based on any structural theory at all, but depend on the principle of similarity. That is if two isotherms are measured on the same type of surface at the same temperature and pressures, the ratio of the amount adsorbed for two samples should be a constant (equal to the surface area ratio) at any pressure. A measurement of the adsorption of argon on progressively sintered samples of SnO_2[32] fails to show this exact similarity; the conclusion reached was that the change in shape of the observed isotherms could be explained by a change in density at the surface as well as a reduction in surface area.

ACKNOWLEDGMENT

The authors wish to gratefully acknowledge support of this project through a grant, No. DMR 8111111, from the Division of Materials Research, Metallurgy Polymers and Ceramic Section of the National Science Foundation.

REFERENCES

1. E. W. Washburn, Note on a method of determining the distribution of pore sizes in a porous material, Proc. Nat. Acad. Sci., 7:115 (1921).

2. H. L. Ritter and L. C. Drake, Pore-size distribution in porous materials; pressure porosimeter and determinations of complete macropore-size distribution, Ind. Eng. Chem. Anal. Ed., 17,12:782 (1945).
3. S. Modry, M. Svata and J. Van Brakel, Thematic bibliography of mercury porosimetry, Powder Tech. 29:13 (1981).
4. O. J. Whittemore, Mercury porosimetry of ceramics, ibid, 29:167 (1981).
5. G. C. Ulmer and W. J. Smothers, Application of mercury porosimetry to refractory materials, Am. Ceram. Soc. Bull., 46:649 (1967).
6. W. C. Books, R. H. Herron and C. R. Beechan, Pore-size distribution in BOF refractories, ibid, 49:643 (1970).
7. A. Watson, J. O. May and B. Butterworth, "Studies of pore-size distribution. Part I. Apparatus and preliminary results, Trans. Brit. Ceram. Soc., 56:37 (1957).
8. V. S. Thompson and O. J. Whittemore, Structural changes on reheating plasma-sprayed alumina, Am. Ceram. Soc. Bull, 47:637 (1968).
9. O. J. Whittemore and D. A. Sullivan, Pore changes on reheating of plasma-sprayed zircon, J. Am. Ceram. Soc., 56:347 (1973).
10. T. J. Ormiston and O. J. Whittemore, Sintering of silica fiber compacts, Am. Ceram. Soc. Bull., 52:247 (1973).
11. D. B. Fischbach and C. S. Kucheria, Porosity of some laminar carbon-carbon composites, Proc. 4th London Int. Carbon Conf.; Soc. Chem. Ind., 422 (1976).
12. J. H. C. Castro, C. T. A. Suchicital and O. J. Whittemore, Characterization of porous ceramic materials, Ceramica (Sao Paulo) 25, 223-228 (1979) (in Portuguese).
13. O. J. Whittemore and J. H. C. Castro, Physical characteristics of Brasilian diatomites, Proc. Mat. Equip. Div., Am. Ceram. Soc., 48 (1976).
14. S. Diamond, Pore size distributions in clays, Clays, Clay Miner., 18:7 (1970).
15. S. Diamond, Microstructure and pore structure of impact-compacted clays, ibid, 19:239 (1971).
16. O. J. Whittemore, Pore morphography in ceramic processing, Mater. Sci. Res., 11:125 (1978).
17. O. J. Whittemore and J. H. C. Castro, Preliminary study of the physical properties of ceramic materials, Ceramica (Sao Paulo), 23:261 (1977) (in Portuguese).
18. O. J. Whittemore and J. J. Sipe, Pore growth during the initial stages of sintering ceramics, Powder Tech., 9:159 (1974).
19. O. J. Whittemore and J. A. Varela, Pore distributions and pore browth during the initial stages of sintering, in: "Sintering Processes," G. C. Kuczynski, ed., Mater. Sci. Res. 13:51 (1980).
20. H. M. Rootare and R. G. Craig, Characterization of the compaction and sintering of hydroxyapatite powders by mercury porosimetry, Powder Tech., 9:199 (1974).

21. T. G. Carbone and J. S. Reed, Dependence of sintering response with a constant rate of heating on processing-related pore distributions, Am. Ceram. Soc. Bull., 57:748 (1978).
22. A. J. Goodsel, Mercury penetration porosimetry in analysis of exhaust catalysts and catalyst supports, Powder Tech., 9:191 (1974).
23. L. K. Frevel and L. J. Kressley, Modifications in Mercury Porosimetry, Anal. Chem. 35:1492 (1963).
24. J. A. Varela and O. J. Whittemore, Structural rearrangement during the sintering of MgO, to be published in J. Am. Ceramic. Soc. (1983).
25. J. A. Varela, The initial stage of sintering MgO, Ph.D. dissertation, University of Washington (1981).
26. H. M. Rootare and C. F. Prenzlow, Surface areas from mercury porosimeter measurements, J. Phys. Chem. 71:2733 (1967).
27. J. W. Gibbs, "The Collected Works," Longmans Green and Co., New York (1928) Vol I, pp 219-237.
28. H. M. Rootare and A. C. Nyce, The use of porosimetry in the measurement of pore size distribution in porous materials, Int. J. Powder Met., 7:3 (1971).
29. A. T. Yeates and G. D. Halsey, The calculation of the contact angle of mercury on plane oxide surfaces, Powder Tech. (in press).
30. D. Heggarty, W. S. Ahn and G. D. Halsey, The validity of the BET equation near the monolayer volume, (in preparation).
31. R. A. Pierotti and H. E. Thomas, "Physical adsorption: the interaction of gases with solids," in Surface and Colloid Science, Vol. IV, E. Matijevic, ed., John Wiley and Sons, New York (1971), p 93.
32. D. C. Hinman and G. D. Halsey, Adsorption of argon on sintered tin dioxide analyzed by several methods, J. Phys. Chem. 81:739 (1979).

SURFACE CHARACTERISTICS OF YTTRIA PRECURSORS

IN RELATION TO THEIR SINTERING BEHAVIOR

Mufit Akinc and M. D. Rasmussen

Ames Laboratory, USDOE[†] Iowa State University, Ames
Iowa

ABSTRACT

Sintering experiments on yttria indicated that the most important factor in determining the sinterability of the oxide is the way in which the precursors are prepared, especially during the precipitation and drying steps. The powder retains the effects of these treatments even after calcining and isostatic pressing. The precursors are prepared by precipitating them in the form of yttriumhydroxynitrate hydrates. Physical characteristics of the surface of the precursors were determined and related to sinterability of the oxide.

INTRODUCTION

Among the various techniques used to prepare ceramic powder, precipitation of a precursor from a solution is probably the most widely known. A significant portion of ceramic powders are still produced[1,3] by this method. These precursors are then calcined to the desired form for further processing.

[†]Work was done at the Ames Laboratory, which is operated for the U.S. Department of Energy by Iowa State University under contract no. W-7405-ENG-82. This article was supported by the Director of Energy Research, Office of Basic Energy Sciences, Division of Materials Sciences.

The precipitation technique has several advantages over other techniques. For instance, reaction conditions are usually mild and easily controlled, the reaction is relatively rapid and quantitative, and, most importantly, components can be mixed at the atomic level relatively easily. One of the most critical steps in preparing a precursor is removing the solvent from the precipitated solid phase. If the precursor is dried properly, a very fine, reactive powder is obtained.[1,4,5]

In this paper, the effect of various drying techniques and the surface characteristics of the yttria precursors will be discussed and some correlations will be drawn between precursor characteristics and the sinterability of the oxide powders.

EXPERIMENTAL

Precipitation of the Precursors

Yttria precursors were precipitated as yttriumhydroxynitrate hydrate from acidic yttrium nitrate solution in an alkaline medium. It was previously reported that the addition sequence of reactants is important in determining the characteristics of the precursor.[6] Therefore, in this study, the precipitation procedure was kept constant. A 0.9 M solution of $Y(NO_3)_3$ was added at a rate of 24 ml/min to a 2.5 N solution of NH_4OH until the pH of the solution dropped to about 10.5. The solution was stirred continuously during the precipitation stage. The precipitate was filtered through a vacuum-assisted porous glass funnel. The filtered cake was washed with ammonium hydroxide solution (pH=10.5) and filtered again to remove excess salts. The precipitate was then dried in various ways as described below.

Drying of the Precursors

There are several different techniques used to dry the precipitate.[1,2] Among these, the simplest one is to dry the cake in an oven in an air atmosphere. A portion of the cake was dried according to this technique at 100 C overnight and the product is referred to as oven dried (OD) precursor. Another technique entails sequential washing with organic liquids to remove water; the technique developed earlier in our laboratory[7,8] involves the application of acetone-toluene-acetone (ATA) treatment. This technique was applied to another portion of the cake. The powder obtained by this technique is referred to as ATA powder. This third method involves aging of the precipitate in a controlled humidity chamber[a] at 90 C and 95% RH for five days.[9] This material was not dry when it was

---

[a]Blue M, Humid Flow (TM) Temperature & Humidity Chamber

removed from the chamber but no further attempt was made to dry it. Finally, some of the cake was dropped into liquid nitrogen and then freeze-dried[b]. The process took about two days.

Wet chemical analysis of the precursors gave a composition of $Y_2(OH)_{6-x}(NO_3)_x \cdot yH_2O$, where depending on the drying method, $0.2 < x < 1.0$ and $1.0 < y < 1.5$.

Sintering Experiments

All precursors are calcined in air at 1100 C for two hours. Powders were heated in Al_2O_3 crucibles at a rate of 8 C/min. Calcined powders were sieved through a 60 mesh screen. Oven dried (OD) and controlled humidity dried (CHD) powders had to be ground to pass through the 60 mesh screen. Using a tungsten carbide double-acting die, oxides were then pressed into 5-10 mm thick pellets which weighed 0.8 g each and had a diameter of 9.5 mm. Pellets were then isostatically pressed to 207 MPa. The pellets were sintered in yttria crucibles under vacuum for 1 hr at 1800 C. Heating and cooling rates were both 20 C/min. The bulk density of each specimen was determined by immersion technique. A schematic layout of the processing flow chart is given in Figure 1.

Characterization of Precursors and Oxides

Surface area determinations were done by N_2 adsorption at -195 C using a volumetric apparatus[c] and applying the BET method. The morphology of the precursors, oxide powders and pressed pellets was studied using the scanning transmission electron microscope[d] in the scanning mode. Micrographs of sintered pellets were taken by an optical microscope[e]. Potentiometric titrations were carried out on precursor powders to determine the point of zero charge, according to a method described by DeBruyn and his co-workers.[10,11] To obtain a 100 ml dispersion of known ionic strength, 75 ml of $NaNO_3$ solution of known concentration was added to 25 ml of dispersion (containing 1.0 g precursor). Titration was carried out by adding a known volume of 0.1 N NaOH to the dispersion and recording the pH of the solution. Zeta potential variation with pH was measured using a commercial apparatus[f]. Precursor powders were dispersed in solutions of constant ionic strength and pH was adjusted with 0.1 N NaOH solution.

---

[b]Labconco Model 75035 Bench Top Freeze Dryer
[c]Micrometritics AccuSorb 2100D
[d]JEOL 100 CX
[e]Zeiss Axiomat
[f]Lazer Zee-Meter, Model 500

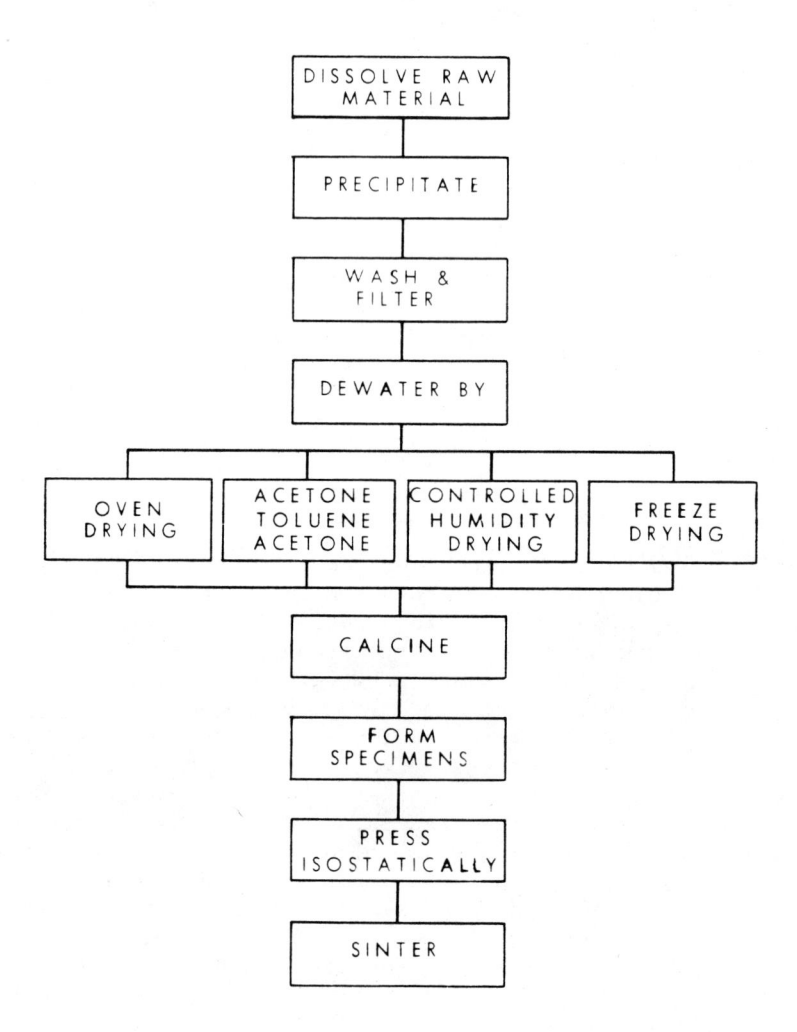

Fig. 1. Processing flow chart

RESULTS AND DISCUSSION

The specific surface areas of powders derived from different drying methods are given in Table 1. The mean equivalent spherical diameter (d_{BET}) determined from the specific surface area and the median particle size (d_{50}) determined from vibrating air column sieving are also given in the table.

Data in Table 1 indicate that at both the precursor and oxide stages, the specific surface area of the oven dried material is significantly lower than that of material dried by other methods. After calcination, the surface area of OD powders reduces to < 5 m^2/g while others approach 10-15 m^2/g, irrespective of the original precursor surface area. Median particle size, determined by sieving, essentially gives the extent of agglomeration of the oxide powders. Table 1 indicates that OD and CHD powders are highly agglomerated while ATA and FD powders are less agglomerated. Although CHD material seems to have larger agglomerate size, mean equivalent spherical diameter values suggest that original particles are comparable to both ATA and FD powders while the diameter of OD material is fivefold larger. One would probably visualize these two diameters as that of primary crystals (and/or aggregates) and agglomerates.

Densities and porosities of sintered pellets are summarized in Table 2. A comparison of specific area and sintered densities suggests a correlation between surface area of the powder and sinterability of the yttria. Again, oven drying is unique among the

Table 1. Specific Surface Area and Particle diameter of powders

Drying Method	Surface Area, m^2/g Precursor	Oxide	Particle Size, μm d_{BET}	d_{50}
OD	3.41 ± 0.05	2.78 ± 0.03	0.430	95
ATA	197 ± 1	13.9 ± 0.2	0.086	42
CHD	41	14	0.085	95
FD	32.31 ± 0.66	12.32 ± 0.25	0.097	<74[a]

[a]All of the powder passed through 200 mesh screen easily.

drying techniques in that sintered density is much lower, and the scatter in data is much larger than the rest. ATA, CHD and FD methods all give densities within 95% or better of the theoretical

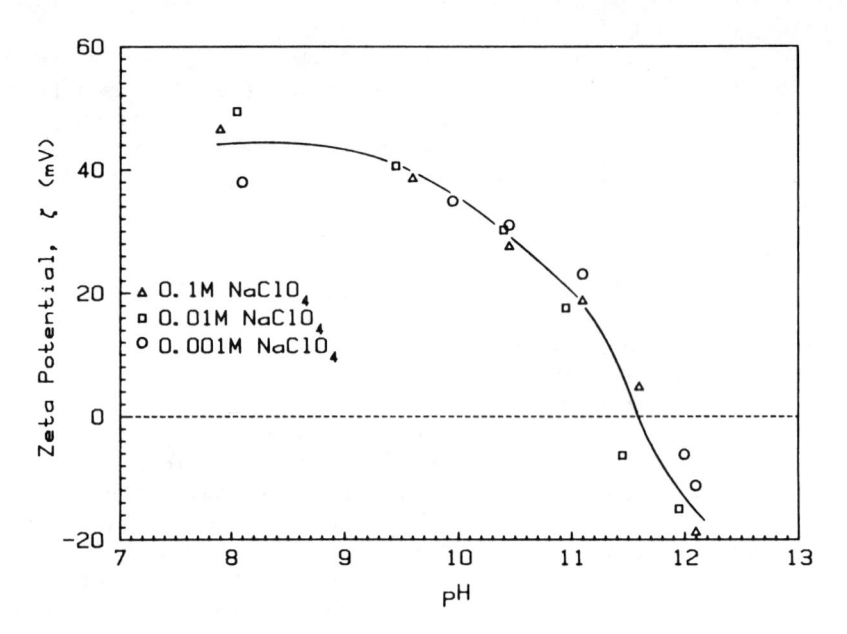

Figure 2. Variation of zeta potential with pH at different NaClO$_4$ concentrations

value. The majority of the pore volume is open in OD powder, whereas there is virtually no open porosity in other specimens.
Previously it was shown that as long as the final pH of the precipitate is less than 10, powder sinters to near theoretical densities, irrespective of the drying method. To understand the nature of the effect of drying technique at a pH greater than 10, electrochemical properties of the surface of the precursor were determined. Zeta potential of the freshly prepared yttriumhydroxynitrate hydrate precursors was measured as a function of pH and is presented in Figure 2. Zeta potential of the precursors was found to be highly positive (about 50 mV) in neutral solutions, to decrease gradually and approach zero around pH = 11.5, and to be negative in more alkaline solutions. The isoelectric point of the precursor was estimated to be at pH = 11.6 ± 0.3. This is an unusually high pH for the isoelectric point, compared to other trivalent hydroxides such as Al(OH)$_3$ and Fe(OH)$_3$. However, this may be due to the presence of nitrate ions in the precursors. For oxides and hydroxides, it is presumed that the isoelectric point coincides with point of zero charge and minimum solubility[10]. To confirm the assigned value of the isoelectric point, we measured the point of zero charge of precursors by potentiometric titration technique. The results of this experiment are given in Figure 3.

Determining the point of zero charge from potentiometric titrations is very difficult when the point of zero charge is at either very low or high pH values. The data are less accurate because of large amounts of titrant present in the solution. The determination also requires the use of higher concentrations of supporting electrolytes. In spite of these difficulties, a reasonable estimate for point of zero charge was made. It was found to be around pH = 11.6, in excellent agreement with the isoelectric point value.

It is interesting to note that when the final pH of the precipitate is higher than 10 (close to the isoelectric point), the precursor does not yield to sinterable powder when oven dried. On

Table 2. Sintered densities and porosities of Y_2O_3

Drying	DENSITY Bulk, g/cm^3	%Theoretical	% POROSITY Open	Closed
OD	3.49 ± 0.5	69.38 ± 10.0	30.2 ± 9.9	0.46 ± 0.15
ATA	4.78 ± 0.10	95.10 ± 1.95	0 ± 0.06	4.96 ± 1.97
CHD	4.95 ± 0.01	98.46 ± 0.19	0 ± 0.06	1.54 ± 0.16
FD	4.77 ± 0.01	94.83 ± 0.19	0.30 ± 0.09	4.87 ± 0.11

the other hand, if other drying techniques are used or the final pH of the precipitate is away from the isoelectric point, the powder is highly sinterable.

In order to explain what effect the drying technique has on the morphology of the powder and the sintering behavior, micrographs were taken at various stages of the processing. Figure 4 shows these stages for OD, ATA and FD materials. Figure 4a compares the precursor powders. The micrographs exhibit striking differences in morphologies of the various powders. Oven dried powder is irregular in shape with conchoidal fracture, has an average size of 100 μm and no surface detail is observed even at higher magnifications. The freeze-dried sample also has irregular particle shape but particles are considerably smaller in size. The ATA precursor is composed of very fine, loosely agglomerated, nearly spherical particles. Despite the loss of 35-40% of initial weight and the conversion to well-crystallized Y_2O_3, the original morphology seems to be preserved (Figure 4b). Some reduction in the size of the OD material is probably due to grinding. Fractured surfaces of pressed pellets are presented in Figure 4c. Again, the

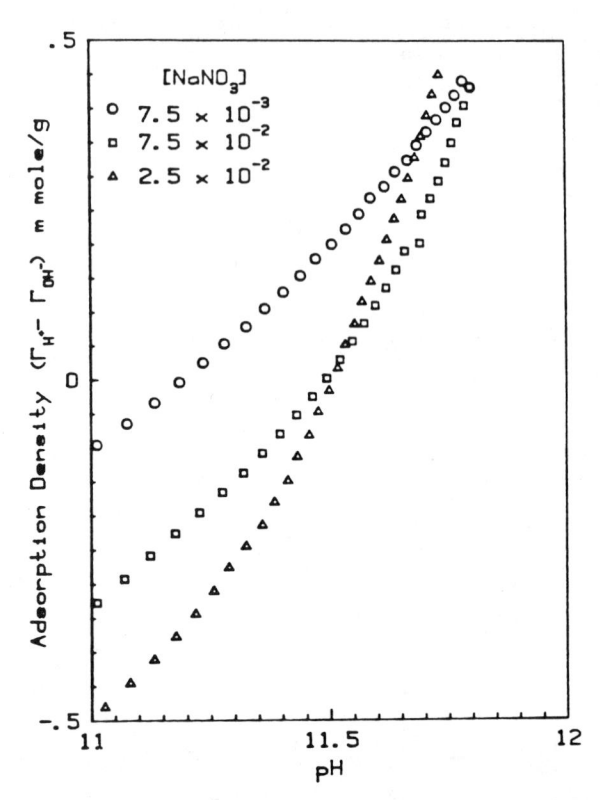

Figure 3. Plot of potentiometric titration. Adsorption density as a function of pH at different NaNO$_3$ concentrations

significant point is that pressing powders to about 200 MPa (or even higher pressures) does not change the original powder morphology. Fractured surfaces of pellets clearly indicate that oven dried material has a very irregular packing, whereas FD and ATA pellets show fine particles and pack much more uniformly.

Micrographs suggest that under similar sintering conditions both FD and ATA material would have higher densities than OD material. Sintered densities presented in Table 2 and optical micrographs of polished, etched surfaces of sintered pellets shown in Figure 5 substantiate these findings. As Figure 5 indicates, OD

PRECURSOR STAGE

OXIDE STAGE

PELLETS BEFORE FIRING

Figure 4. Micrographs of different stages of processing.

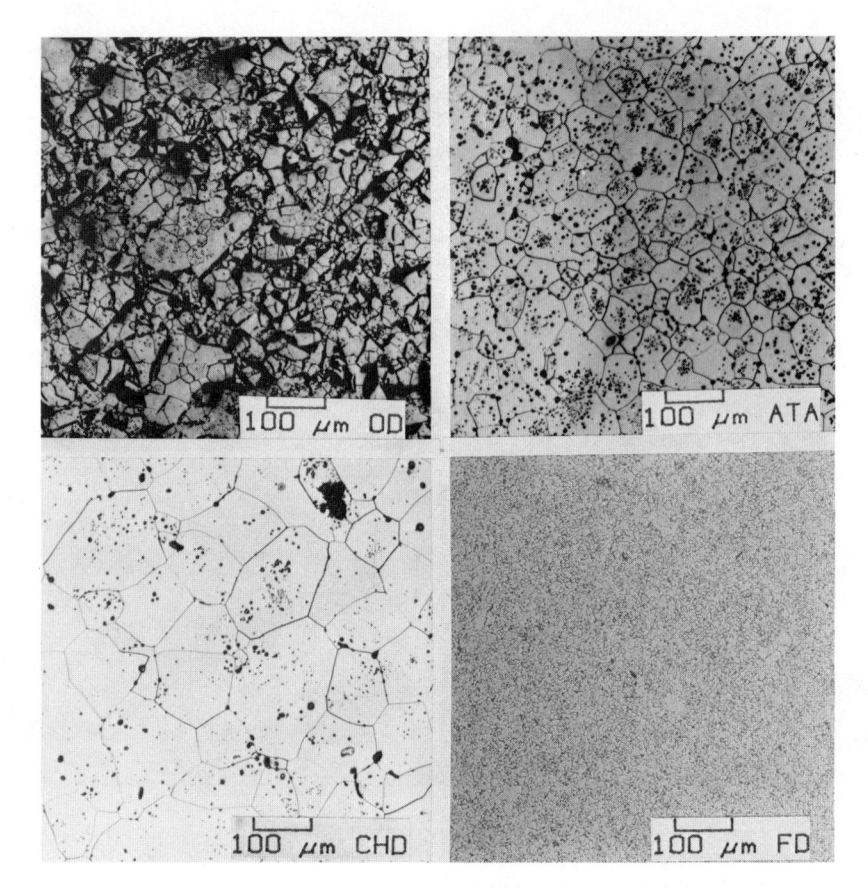

Figure 5. Optical micrographs of sintered pellets. Note: Unlike others, OD material does not sinter to high density. Also, there is a significant variation in grain size among the various drying methods.

material is very poorly sintered while others are sintered to high densities. Admittedly, they exhibit a wide range of grain sizes (about 10, 50 and 140 μm for FD, ATA and CHD powders respectively). A closer look at Figure 5a reveals regions of well-sintered aggregates of about 100 μm, composed of primary grains of about 20 μm. This suggests that the precursor particles in Figure 4a of OD material were aggregates of finer particles.

An explanation can be given for the presence of very tightly bound aggregates in OD material if one considers the precipitation and drying procedures. Precipitation around the isoelectric point causes coagulation of primary particles within aqueous media. During oven drying, removal of water causes a condensation-polymerization type bonding between primary particles. The end product is coarse gel-like material. This type of reaction is not likely to occur when material is freeze-dried or water is removed by organic liquids. Probably CHD aging causes slow and uniform growth of particles, which, in turn, prevents formation of gel-like structure as in the case of OD precursor.

CONCLUSIONS

We found that the method of precursor drying has a major effect (probably more than any other step) on the sinterability and microstructure of the yttria ceramics. A tentative value for the isoelectric point is assigned for freshly prepared hydroxide and an explanation is offered for effect of drying technique on the sinterability of yttria.

ACKNOWLEDGMENTS

The authors wish to express their appreciation to Connie Bailey and Lucille Kilmer for their assistance in the preparation of this paper.

REFERENCES

1. D. W. Johnson Jr., and P. K. Gallagher, In: "Ceramic Processing Before Firing," G. Y. Onoda and L. L. Hench, eds., pp 125-139, John Wiley & Sons, New York (1978).
2. R. Roy, Int. J. Powder Metall. 6(1):25 (1974).
3. P. E. D. Morgan, In: "Processing of Crystalling Ceramics," Hayne Palmour III. R. F. Davis and T. M. Hare, eds., pp 67-78, Materials Science Research Volume 11. Plenum Press, New York (1978).
4. Y. S. Kim and F. R. Monforte, Am. Ceram. Soc. Bull. 50(6):532 (1971).
5. R. E. Seager and T. J. Miller, Am. Ceram. Soc. Bull. 53(12): 855 (1974).
6. M. D. Rasmussen, G. Jordan, M. Akinc, O. Hunter Jr., and M. F. Berard, Influence of Precipitation Procedure on Sinterability of Y_2O_3 Prepared from Hydroxide Precursors, (to be published).
7. S. L. Dole, R. W. Scheidecker, L. E. Shiers, M. F. Berard and O. Hunter Jr., Mat. Sci. Eng. 32:277 (1978).

8. M. F. Berard, O. Hunter Jr., L. E. Shiers, S. L. Dole and R. W. Scheidecker, U.S. Patent #4140771, issued Feb. 20, 1978.
9. G. W. Jordan and M. F. Berard, Presented at the 83rd Annual Meeting of the American Ceramic Society, Washington, D.C., May 5, 1981. (Manuscript in preparation).
10. F. A. Parks and P. L. DeBruyn, J. Phys. Chem. 66:967 (1962).
11. G. Y. Onoda Jr., and P. L. DeBruyn, Surface Science 4:48 (1966).

AN OVERVIEW OF TECHNIQUES USED IN FT-IR SPECTROSCOPY

John R. Ferraro[*]

Chemistry Department
Loyola University
Chicago, IL. 60626

INTRODUCTION

The advent of commercial Fourier transform interferometry (FT-IR) instrumentation has provided the materials scientist, the analytical chemist, the research scientist and others, with another tool to solve their problems. It is our intention in this paper to present an overview of the techniques that can be used in FT-IR, and can be directed toward materials characterization.

Historical

About 90 years ago Michelson (1891,1892) discovered the device called an interferometer. Rayleigh (1892) recognized its potential when he identified that the interferogram that could be obtained was related to the spectrum of the radiation passing through the interferometer. A well-resolved interferogram was not measured until almost 20 years later by Rubens et al (1910,1911). However, the mathematical means of Fourier transformation to convert the interferogram into a meaningful spectrum was not available until 1949. Fellgett (1959) was the first to perform the Fourier transformation. The computations took many hours to complete, and this served to slow the interest in the technique. When Cooley and Tukey (1965) developed the fast Fourier transform algorithm, renewed interest took place. Computation time was reduced to minutes, and depending on resolution, even seconds. Simultaneously, computer technology advanced as solid-state devices and integrated circuits were developed. It became possible to interface the interferometer with

---

[*]Searle Professor of Chemistry

dedicated mini computers. Another purpose was also served. It compelled instrument manufacturers of dispersive IR to commence interfacing their instruments with computers.

FT-IR Advantages Over Dispersive IR

The theoretical advantages of interferometry (IF) over dispersive IR are recognized. Fellgett was interested in astronomical problems where the radiation intensity measured was weak. He realized that the interferometer measured all wavelengths at once compared to a dispersive grating instrument, which picks off one wavelength at a time. This advantage became known as the Fellgett Advantage or the Multiplex Advantage.

If N = number of resolution elements, ν_R = spectral range, $\Delta\nu_\ell$ = resolution, then $N = \nu_R/\Delta\nu_\ell$ and $\sqrt{N}$ is the analytical sensitivity of FT-IR over the dispersive technique for the same length of time to do the experiment. Consider the mid infrared region, where ν_R = 4000-400 = 3600cm^{-1} and $\Delta\nu_\ell$ = 2cm^{-1}, then N = 1800 and $\sqrt{N}$ = 42. This sensitivity decreases in the far infrared region (FIR) where ν_R is smaller. For example, $N = \frac{600}{2} = 300$ and $\sqrt{300}$ = 17.

Another inherent advantage of the Michelson interferometer is the greater throughput possible in contrast to a dispersive instrument using a monochromator involving slits to measure the same spectrum (Jacquinot, 1954). Throughput is defined as the product of area and solid angle of the beam from the source. Essentially more energy sees the sample and reaches the detector because of this advantage. The advantage is known as Jacquinots Advantage, and depends on the size or geometry of sample or source, and may be smaller than theoretically allowed. In the case of astronomy, the advantage is limited by the light grasp of the telescope. Griffiths (1975) pointed out that in the FIR the throughput is more a function of the "f" number of the source fore-optics of the interferometer, and therefore little advantage is obtained by an interferometer over the dispersive method in this case.

Two other advantages of the IF may be considered. These are the Connes Advantage (Connes and Connes 1966), which calibrates the wavelength for each scan, and the Stray-light Advantage. The former advantage is due to the use of a He-Ne laser to calibrate the wavelength for each scan, and is a requirement for accurate spectral manipulation of data. The latter advantage is due to the fact that each IR frequency is chopped by the IF at a different frequency, and the stray-light elimination is particularly evident in the FIR.

However, in actual practice the theoretical advantages (e.g. Fellgett's and Jacquinot's) are not entirely realized. For discussions on these topics see Sheppard et al (1977) and Griffiths (1977).

OVERVIEW OF TECHNIQUES USED IN FT-IR SPECTROSCOPY

INSTRUMENTATION

Fig. 1. shows the optical lay-out of a Michelson interferometer. This is the simplest IF and consists of 2 mirrors, a beam splitter, source and detector. One mirror is stationary while the other moves. The beam splitter transmits half of all of the incident radiation from a source to the moving mirror and reflects half to the stationary mirror. Each component reflected by the two mirrors returns to the beam splitter, where the amplitudes of the waves are combined to form an interferogram as seen by the detector. The interferogram is transformed by Fourier means into the familiar frequency spectrum.

Figures 2,3 present optical lay-outs for some of the commercial FT-IR instruments.

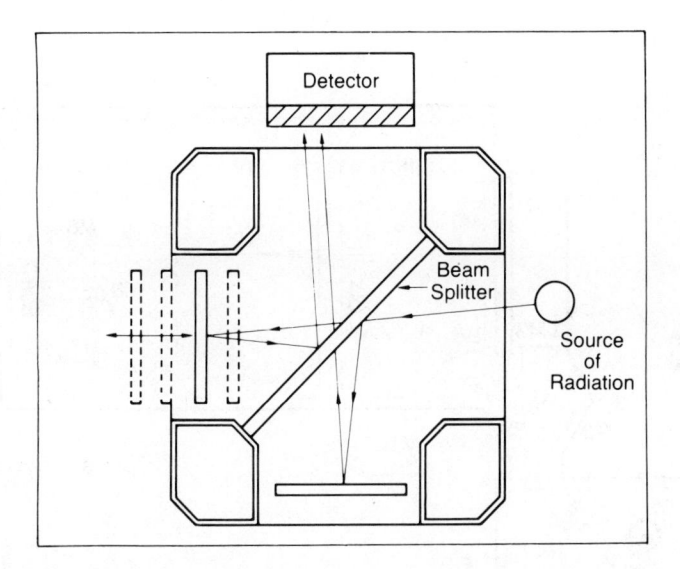

Fig. 1. The Michelson interferometer consists of two mirrors and a beam splitter. The beam splitter transmits half of all incident radiation from a source to a moving mirror and reflects half to a stationary mirror. Each component reflected by the two mirrors returns to the beam splitter, where the amplitudes of the waves are combined to form an interferogram as seen by the detector. It is the interferogram which is then Fourier transformed into the frequency spectrum.

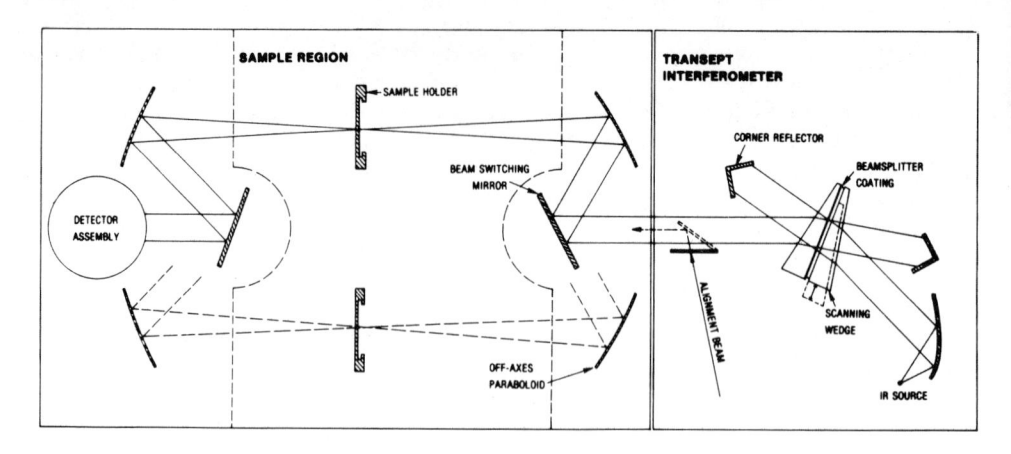

Fig. 2. Commercial Interferometer Manufactured by Analect Instruments, Irvine, CA.

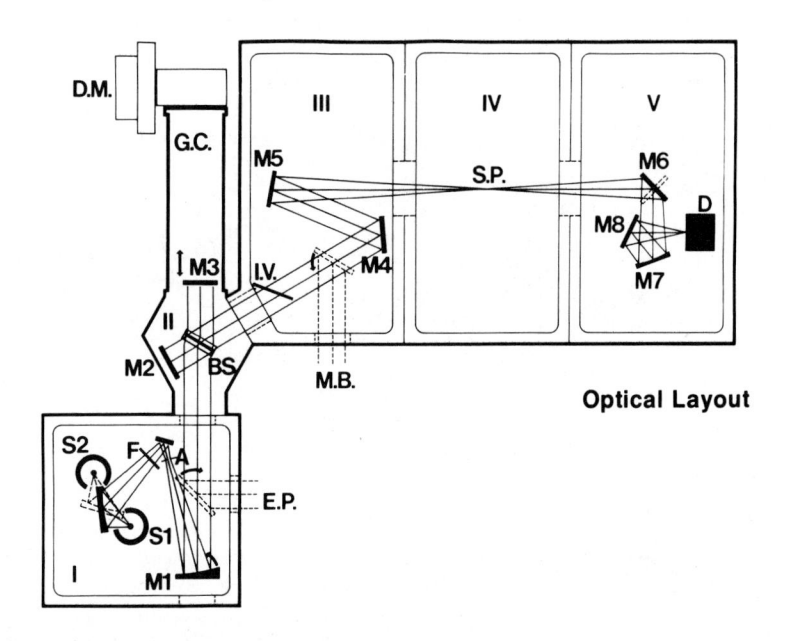

Fig. 3. Commercial Interferometer Manufactured by Bomem, Inc., Quebec, Canada.

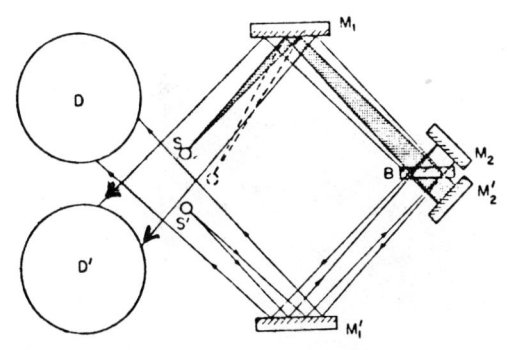

The diverging beam method of separating the input and output beams of a Michelson interferometer. Light from the telescope is focused on injection mirror S (or S') and then proceeds to diverge through the interferometer. By the time the light beam returns to S it has a much larger cross section and the injection mirror produces only a small obstruction. Two inputs S and S' and two detectors D and D' can thus be used. The spread of the beam (or the *f*-number) determines the maximum possible resolution. An alternate arrangement (shown by the dotted line) eliminates any obstruction to the beam at the expense of a larger off-axis angle and lower resolution.

Fig. 4. Interferometer Designed to Counteract Problems Incurred in Astronomical Measurements. (Fink et al. 1979), reprinted with permission of authors and Academic Press, N.Y.

Fig. 4 presents an interferometer designed to allow first-order cancellation of thermal background flux levels and some effects of source intensity fluctuations as encountered in astronomical problems. (see Fink et al., 1979).

It is not our intention to discuss the theory of FT-IR and for this the reader is referred to Griffiths (1975) and Bell (1972).

TECHNIQUES USED IN FT-IR

Several techniques have been used in FT-IR. For a summary see Krishnan and Ferraro (1982). The usual techniques of absorption (transmittance) are applicable depending on the nature of the sample. If not, other techniques are necessary. It is not intended in this overview to convey to the reader that these techniques are novel and unique only to FT-IR. However, in some cases, used with FT-IR instrumentation, they may offer some advantage.

Attenuated Total Reflectance

Attenuated total reflectance (ATR) is an excellent technique for the study of surfaces. Various types of samples may be studied, such as free-standing films, smooth surfaced solids, coatings on metals and liquids. The techniques and available accessories have been reviewed by Harrick (1967). Fig. 5 shows an ATR reflectance attachment and its optical lay-out. The accessory has a prism made

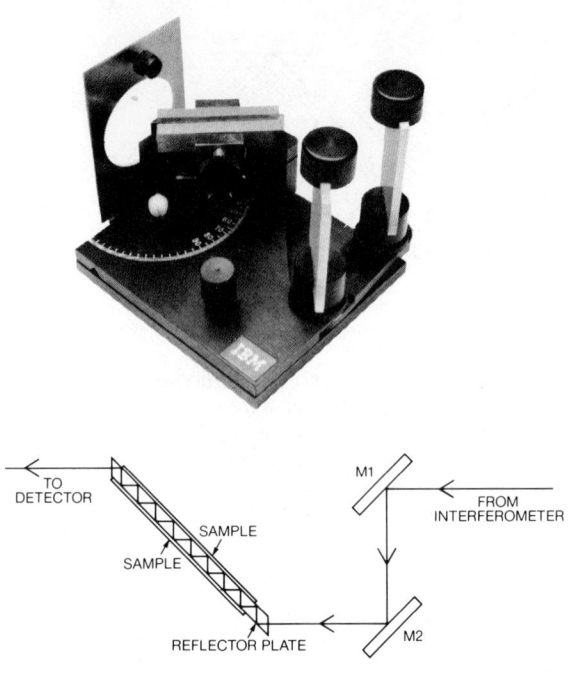

Fig. 5. Optical Layout of ATR Attachment for Use with FT-IR Instrumentation

of high-refractive-index material. When the IR beam is incident on the crystal at an angle larger than the critical angle, internal reflection takes place. If a sample is placed in intimate contact with the crystal face where the reflection takes place, a standing wave occurs at the crystal-sample interface. The reflected radiation has decreased intensity at those frequencies where the sample absorbs. The depth of penetration depends on the reflective indices of the crystal and sample, the effective angle of incidence of radiation, the wavelength, as well as the uniformity of the contacts between the sample and the prism.

Prism materials are generally KRS-5 and Ge. Table 1 shows penetration depths for polyethylene for different frequencies, prism materials and prism angles. The depth of penetration is lower for higher frequencies.

Crystal sizes are about 50 x 3 x 2 mm. With such crystals, depending on the effective angle, between 20 and 30 reflections will take place. Because of the significant energy losses occuring in the ATR experiment, the throughput advantage of FT-IR helps. The use of a mercury-cadmium telluride (MCT) detector is also beneficial. Figs. 6,7 illustrate typical ATR experiments made with an FT-IR.

Table 1[*]. Depth of Penetration for Polythene at Different Frequencies

Crystal	Angle(deg)	2000	1700	1000	500
KRS-5	45	1.00	1.18	2.00	4.00
	60	0.55	0.65	1.11	2.22
Ge	30	0.60	0.71	1.20	2.40
	45	0.33	0.39	0.66	1.32
	60	0.25	0.30	0.51	1.02

(units: cm^{-1})

[*]Smith, 1980

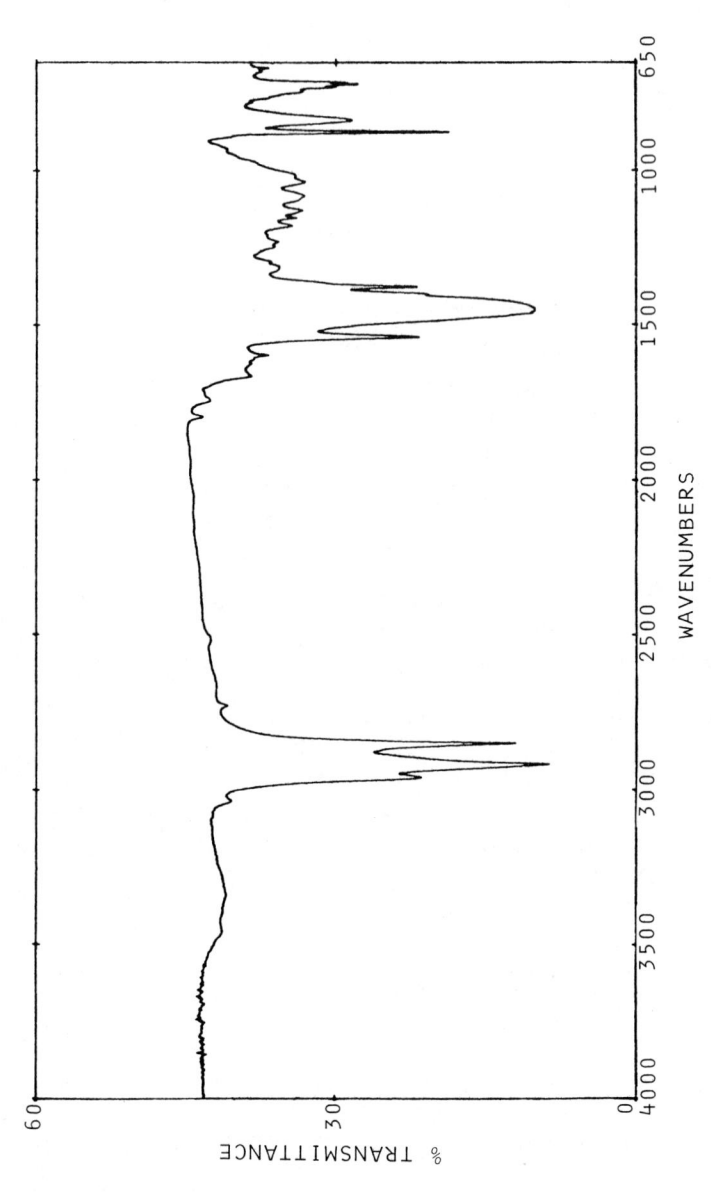

Fig. 6. ATR Spectrum of a Rubber Gasket from a Washing Machine Using a KRS-5 Crystal at 4cm^{-1} resolution and 30 sec. collection time. (Krishnan and Ferraro, 1982), reprinted with permission of authors and Academic Press, N.Y.

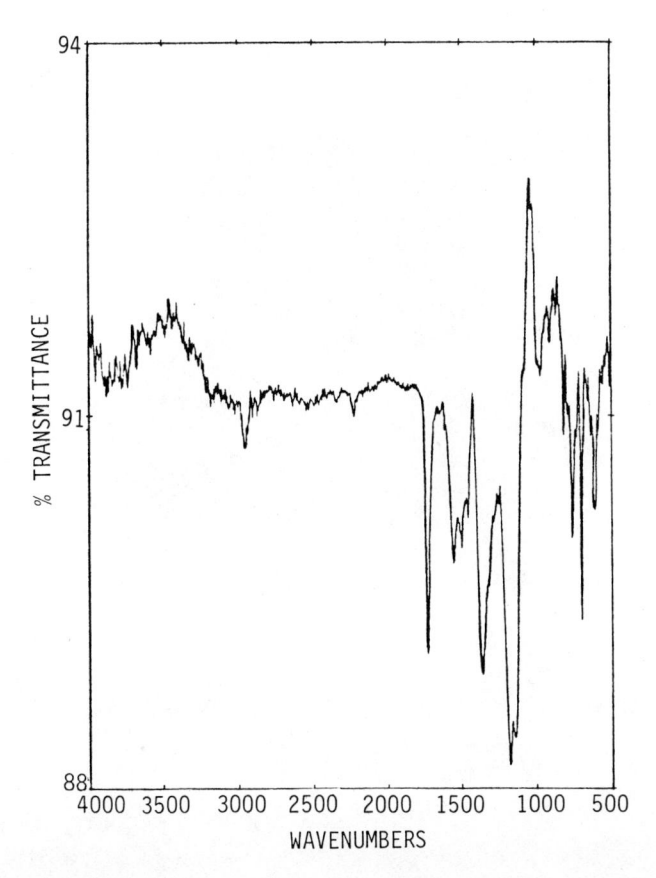

Fig. 7. ATR Spectrum of a 0.5 x 0.5 mm Paint Chip from a 1975 Plymouth Automobile. A KRS-5 crystal was used with 4 cm^{-1} resolution and 2 min. collection time. (Krishnan and Ferraro, 1982), reprinted with permission of authors and Academic Press, N.Y.

Diffuse Reflectance

The diffuse reflectance technique may be used with opaque powders, turbid liquids and to a limited extent, with coatings on flat surfaces. Fig. 8 illustrates the diffuse reflectance attachment. The common experimental device is an integrating sphere, the inside of which can be coated with a nonabsorbing material such as magnesium oxide.

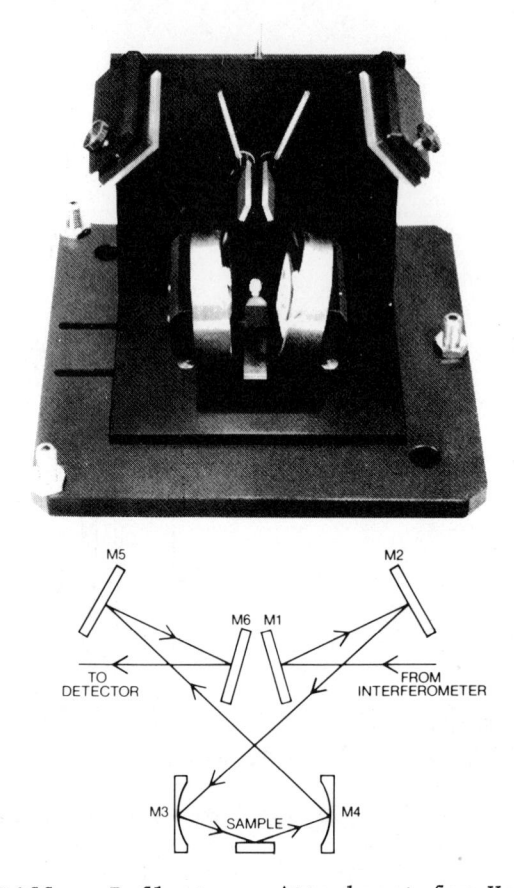

Fig. 8. Diffuse Reflectance Attachment for Use with an Interferometer

    Most of the success obtained by the diffuse reflectance technique with dispersive instruments was made in the UV-visible regions, with only limited success in the mid-infrared region. This was due to the limited energy throughput in a dispersive instrument. New accessories have made the technique feasible for both dispersive and FT-IR instruments. A MCT detector is recommended since the method is inefficient and the fraction of incident energy reaching the detector is small. Fig. 9 compares a diffuse reflectance spectrum of a coal sample in KBr powder with that of a KBr pellet.

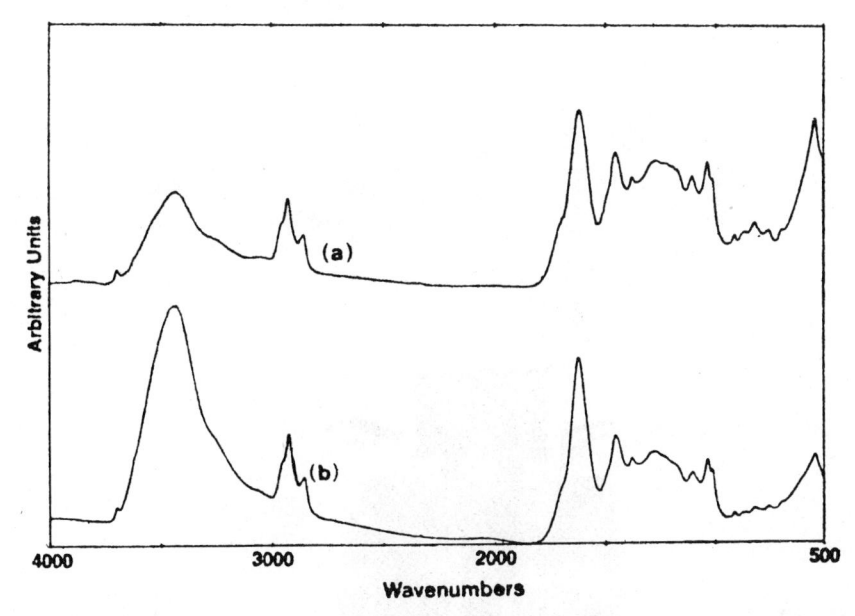

Fig. 9. Diffuse Reflectance Spectra - (a) KBr pellet spectrum of a coal sample, (b) diffuse reflectance spectrum of coal in KBr powder. (Krishnan and Ferraro, 1982), reprinted from Am. Lab. 12, (3) 104 (1980). Copyright 1980 by International Scientific Communications, Inc.

Specular Reflectance

   This method has been found to be useful for probing the surfaces of reflecting materials. Typical applications include the study of polymer coatings on metal surfaces and for the analysis of epitaxial layers on semi-conductor materials. The method can be used to study monomolecular layers, coatings and corrosion products on metal surfaces. Fig. 10 illustrates a specular reflectance attachment, and Fig. 11 shows a typical spectrum

Emission Spectroscopy

   Bates (1978) has discussed the technique of infrared emission spectroscopy samples which present a problem in obtaining an infrared spectra may be resolved by the emission technique. In particular, samples at high temperature (Bates and Boyd, 1973), remote samples, such as smokestack emissions (Low and Clancy, 1967), or using thin films such as monolayers or corrosion products on metals. The reference spectrum is that of a blackbody. Fig. 12 illustrates a furnace with a blackbody reference that can be used for emission spectroscopy. Fig. 13 shows typical emission spectra of sodium beta-alumina (Bates and Frech, 1976).

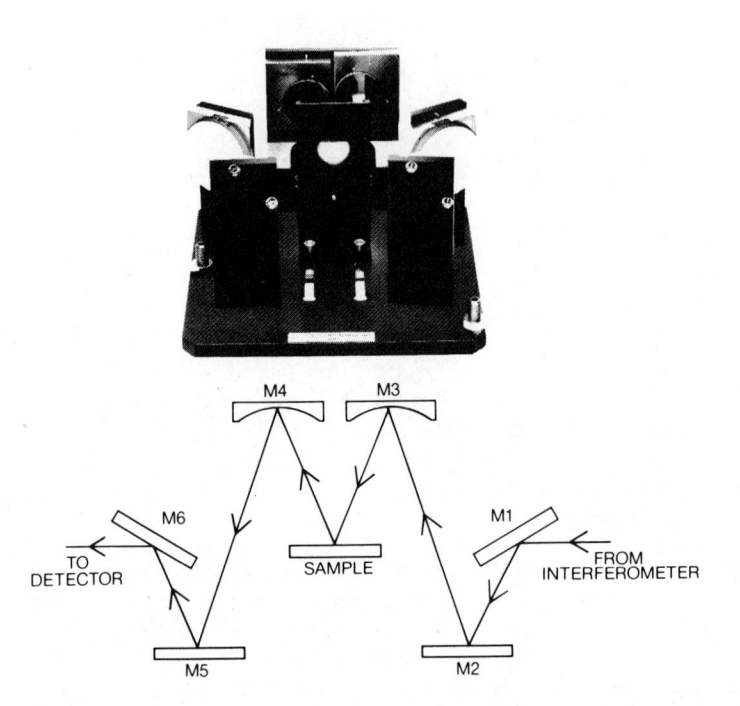

Fig. 10. Specular Reflectance Attachment for Use with an Interferometer

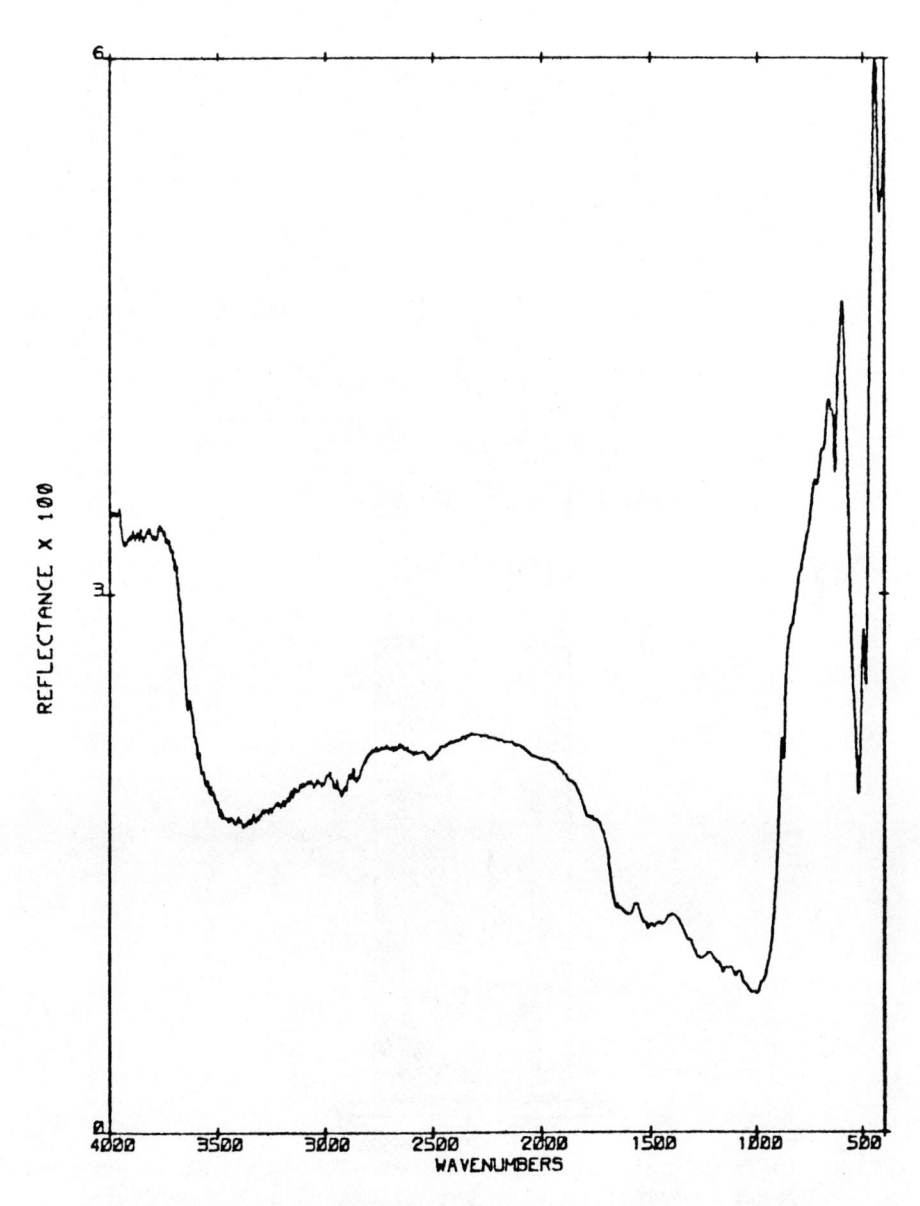

Fig. 11. Specular Reflectance Spectra of a Piece of Grinding Wheel Showing Al$_2$O$_3$ Bands and Bands of Organic Binder (Krishnan and Ferraro, 1982), reprinted with permission of authors and Academic Press, N.Y.

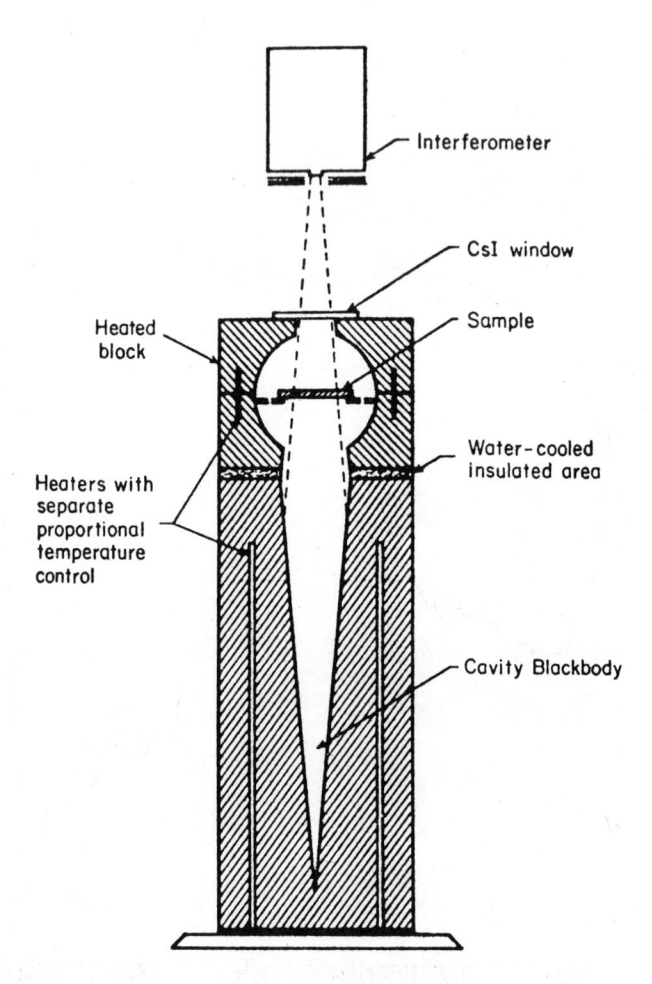

Fig. 12. Infrared Emission Furnace Used with an FT-IR Spectrophotometer. Reprinted with permission of Academic Press, N.Y.

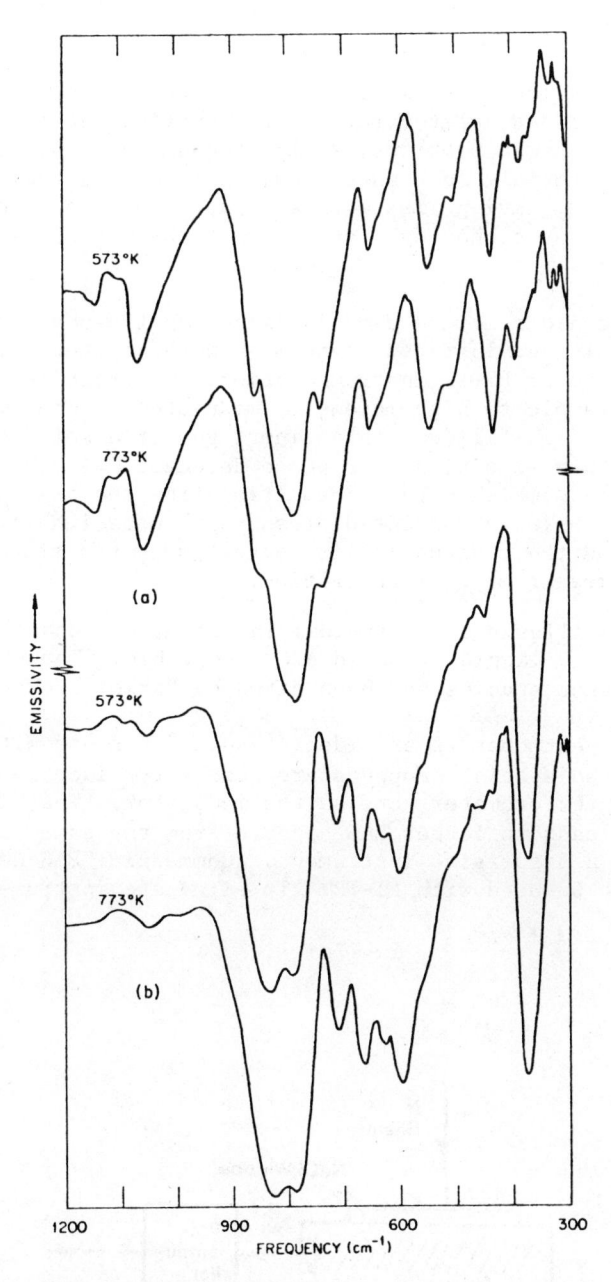

Fig. 13. Polarized Emission Spectra of a Single Crystal of Sodium
beta-alumina.
(a) Polarized along c axis.
(b) Polarized along a axis.
(Bates, 1978), reprinted with permission of author and
Academic Press, N.Y.

Photoacoustic FT-IR

There are many materials which are not adequately studied by the methods listed heretofore. Photoacoustic spectroscopy (PAS) may solve some of these problems, as the technique is useful for a variety of materials such as solids, semi-solids, powders, gels, adsorbed films, living tissue, highly absorbent colored and opaque substances. For a thorough treatment of the subject see Rosencwaig (1975, 1980).

The technique may be used in Raman spectroscopy as well as in the UV-visible and infrared regions. In the latter domain one can use dispersive or FT-IR instrumentation. The technique involves exposing a sample to high intensity modulated monochromatic radiation in a sealed cell (filled with an inert gas if a solid sample is used) equipped with a sensitive microphone detector. A temperature change occurs at the sample surface, resulting from the absorption of the radiation, causing a modulated pressure fluctuation in the gas in the cell, and the fluctuation is converted by the microphone to an electrical measurement of absorption.

Fig. 14 illustrates a typical layout of a photoacoustic detector. Samples of gas, liquid or solid are acceptable. The use of PAS for colored glasses studies has been cited by Parke and Warman (1980).

FT-IR spectrometers are ideally suited for use with the PAS technique. No special choppers are necessary since in FT-IR technique the interferometer acts as the modulator. The higher FT-IR throughput leads to larger PAS signals from the same sample when compared to a dispersive instrument. Commercial PAS cells are now available to be used with FT-IR. To eliminate instrumental background

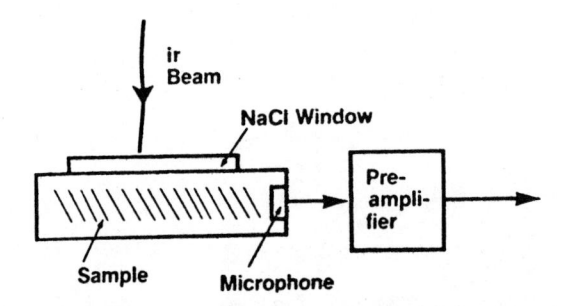

Fig. 14. Typical Layout of a Photoacoustic Detector. (Krishnan and Ferraro, 1982), reprinted permission of authors and Academic Press, N.Y.

OVERVIEW OF TECHNIQUES USED IN FT-IR SPECTROSCOPY 187

activated charcoal powder is used. Fig. 15 shows PAS spectrum for coal obtained by FT-IR means. (Rockley and Devlin, 1980). Fig. 16 records a PAS spectrum of cab-o-sil, a form of glass. Clearly observable are the bands at 3750 cm^{-1} due to non-bonded OH groups and at 3500-3600 cm^{-1} due to various hydrogen-bonded OH species (spectrum b). Lower frequencies (<2000 cm^{-1}) due to silica are also visible. For comparison purposes a spectrum obtained by despersive IR (a) is shown (Low and Parodi, 1979) and more features are clearly indicated in spectrum (b).

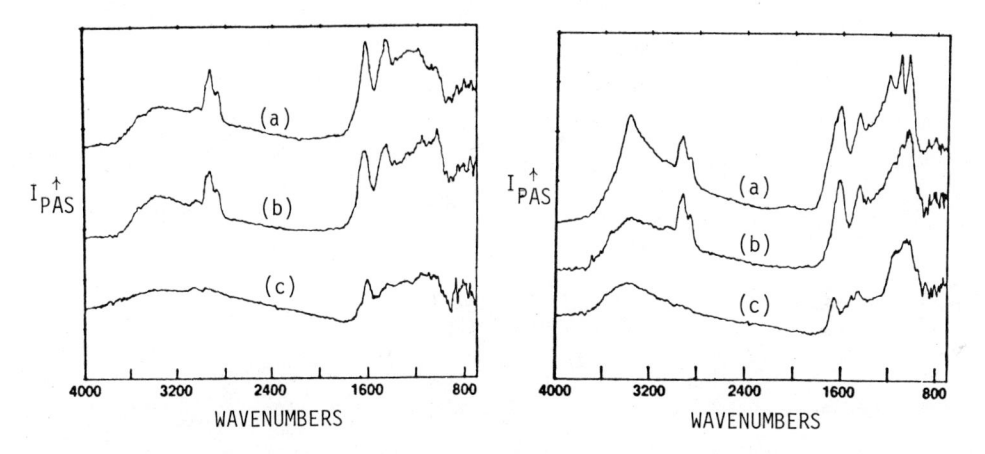

Fig. 15. PAS Spectra of Coal
  (a) Illinois Coal #6
  (b) Pittsburgh bituminous coal
  (c) Reading anthracite coal
  Left - freshly cleaved; Right - aged for a few days
  (Rockley and Devlin, 1980), reprinted with permission of authors and Academic Press, N.Y.

Microtechnique (The Diamond Anvil Cell)

The diamond anvil cell (DAC) has demonstrated its versatility during the past 20 years (Lippincott et al., 1960, 1961; Weir et al., 1959; Ferraro, 1971; Ferraro and Basile, 1974, 1979a,b; Adams and Payne, 1972). It has been used in combination with optical and vibrational spectroscopy, both as a microanalytical tool (Ferraro and Basile, 1979b) as well as a pressure cell (Lippincott et al., 1960, 1961; Ferraro, 1971; Ferraro and Basile 1974, 1979b; Adams and Payne, 1972). The first use of the DAC for sampling purposes was made by Weir et al. (1959), who interfaced the cell to a dispersive spectrophotometer in the mid-infrared region. The use of the

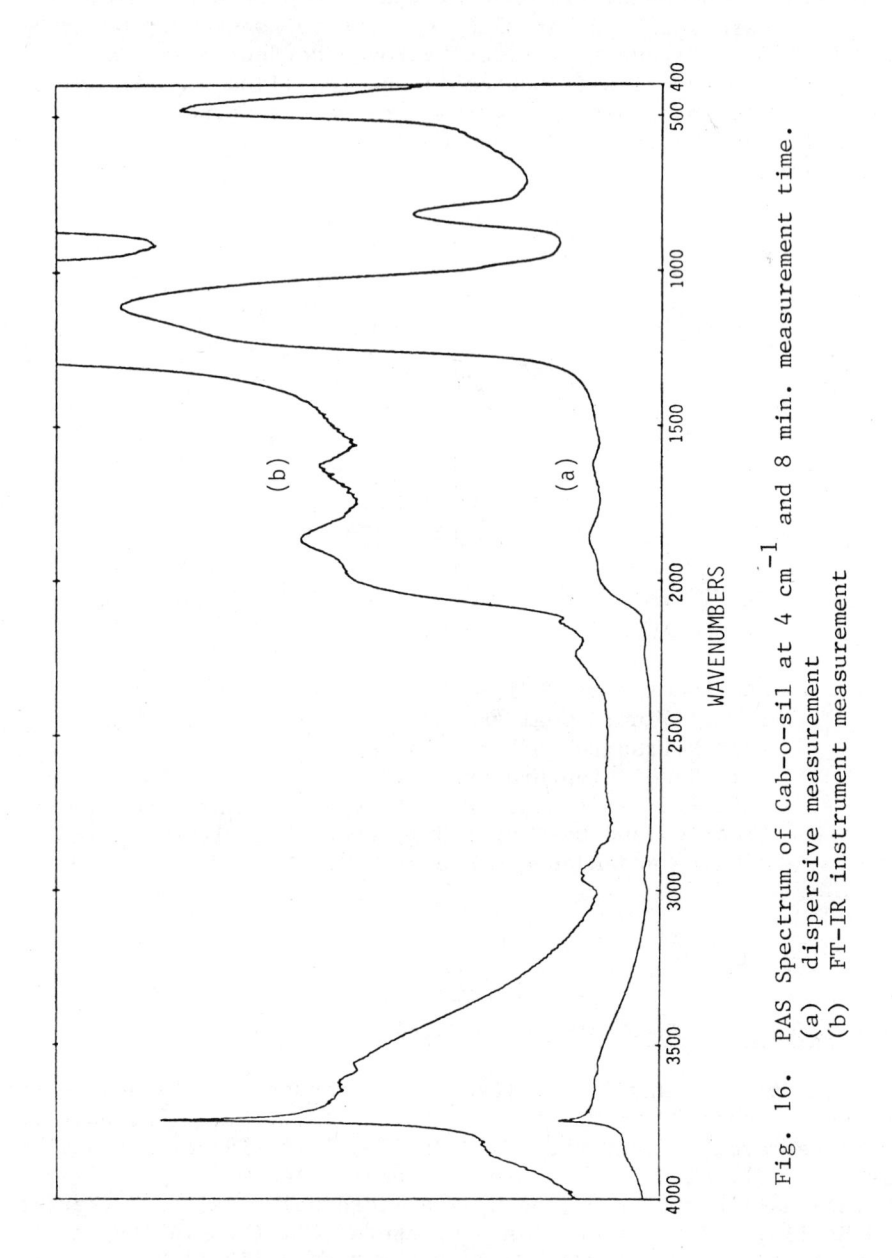

Fig. 16. PAS Spectrum of Cab-o-sil at 4 cm^{-1} and 8 min. measurement time.
(a) dispersive measurement
(b) FT-IR instrument measurement

cell in the far-infrared region (<200 cm^{-1}) was first demonstrated by Ferraro and co-workers (Ferraro et al., 1966, and Postmus et al., 1968).

Due to the high scattering caused by the diamonds, and the small aperture in the DAC (smallest diamonds can be ∿0.1 mm^2 in area) and the critical angle of diamonds, it becomes necessary to condense the source beam for infrared studies. For a discussion of the problems involved in the use of diamond windows in the DAC, see Adams and Sharma (1977). For dispersive instruments, a 4-to-6 X beam condenser has been used by Weir et al., 1959, Brasch, 1965, Jakobsen et al., 1970, Ferraro et al., 1966, and Postmus et al., 1968. In the case of interferometric measurements, a light pipe has been used successfully (McDevitt and co-workers, 1967). Additionally, a Perkin-Elmer 4 X beam condenser has been coupled with the Digilab model 14 (R.J. Jakobsen, Private communication), as well as the Bruker 1FS 114e (Klaeboe and Woldback, 1978). Recently, a 6 X Harrick beam condenser was interfaced with the Digilab FTS-20A and FTS-20 interferometers (Ferraro and Basile, 1980; Krishnan et al., 1980). Recently, lenses have been used to condense the light (Adams and Sharma, 1977, and Hirschfeld, 1981). For the measurement of emission interferometry at high pressure, the reader is referred to the review article by Lauer (1978), which deals with high-pressure infrared interferometry.

In obtaining spectra in the DAC, the transmission spectrum of the sample was recorded by ratioing the spectrum of the diamond cell with the sample against that of the blank cell. Figs. 17,18 show the comparison between the single-beam spectra with and without the diamond cell in the infrared beam. Fig. 18 shows an expanded view of the single-beam spectrum through the diamond cell. The strong absorptions between 2600 and 2000 cm^{-1} are plainly visible and limit the use of the cell in this region.

Krishnan and co-workers (1980) demonstrated the use of the DAC with the Digilab FTS-20 in the mid-infrared region. Very practical applications were demonstrated.

Fig. 19 show the diamond cell spectra of a paint chip. The spectrum was recorded at 4 cm^{-1} resolution for a measurement time of about 4 min/sample. Fig. 20 shows the use of absorbance-subtraction techniques using the diamond cell. The spectra presented are those of stock paper, a different region of the paper with a spot of ink on it, and the difference spectrum. The difference spectrum shows only the features caused by the ink. These examples illustrate the use of the diamond cell as a microsampling analytical device in the mid-infrared region using an interferometer.

The application of this technique in an energy-starved region, such as the far infrared (FIR), is demonstrated in the following figures. The work is that of (Ferraro et al.,1980, Ferraro and Basile,

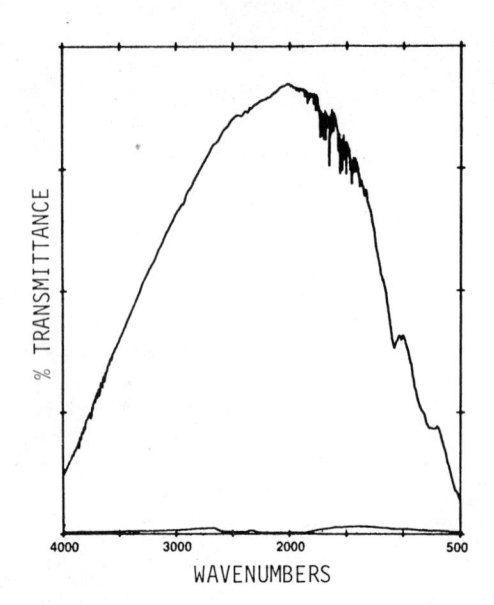

Fig. 17. Comparison of Single-beam Spectra With and Without the DAC.
(Krishnan and Ferraro, 1982), reprinted from Am. Lab., 12
(3), 104 (1980). Copyright 1980 by International Scientific
Communications, Inc.

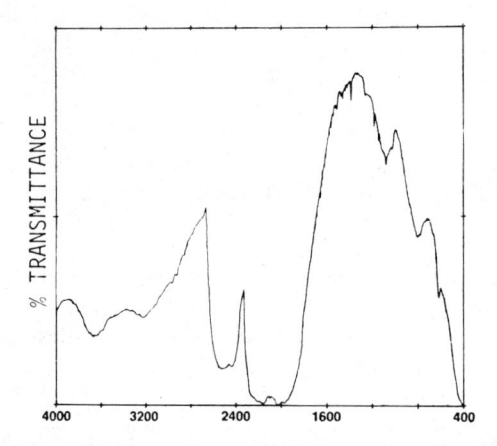

Fig. 18. Single-beam Spectrum through the DAC.
(Krishnan and Ferraro, 1982), reprinted from Am. Lab., 12
(3), 104 (1980). Copyright 1980 by International Scientific
Communications, Inc.

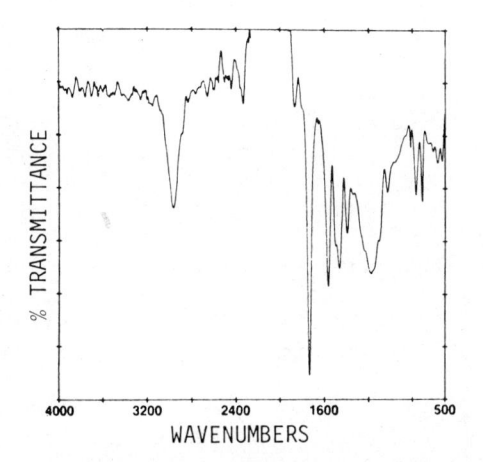

OVERVIEW OF TECHNIQUES USED IN FT-IR SPECTROSCOPY 191

Fig. 19. Spectra of Water-based Acrylic Enamel Paint Chip in DAC.
(Krishnan and Ferraro, 1982), reprinted from Am. Lab. _12_
(3), 104 (1980). Copyright 1980 by International Scientific
Communications, Inc.

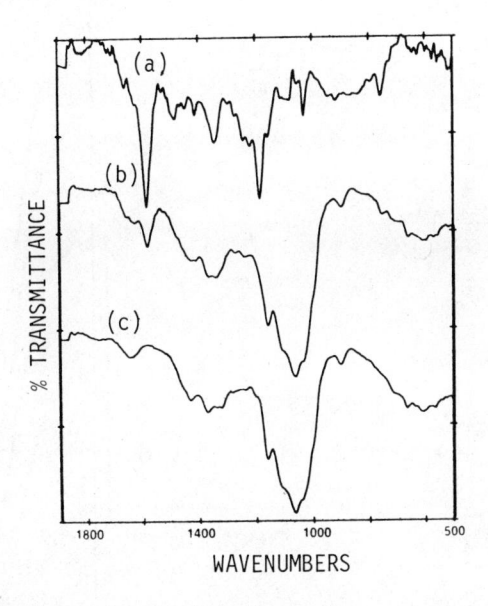

Fig. 20. Absorbance-subtraction Techniques Using the DAC.
(a) Spectrum of stock paper
(b) spectrum of stock paper with spot of ink
(c) spectrum of difference
(Krishnan and Ferraro, 1982), reprinted from Am. Lab, _12_
(3), 104 (1980). Copyright 1980 by International Scientific
Communications, Inc.

1980). Using 12.5- and 25-μm Mylar beam splitters, the useful region of the FIR was extended to 25 cm^{-1}. This accessibility allows new applications to surface. For example, AgI-type ionic conductors have ν_{AgI} modes at about 100 cm^{-1}, and these can be readily studied. It may also be possible to examine soft modes in other conductors in the region less than 50 cm^{-1} as well as other very low energy modes in other molecules with pressure. It should be cited that Klaiboe and Woldback (1978) have used similar techniques and obtained spectra of trans-1,4-diiodocyclohexane at 41 and 57 cm^{-1}. The spectrum of β-AgI is depicted in Fig. 21. Fig. 22 shows the spectrum of yellow HgO with absorption at 60 cm^{-1}. Fig. 23 illustrates the spectrum of the AgI mode at 107 cm^{-1} in pyridinium Ag$_5$I$_6$ as a function of pressure (Ferraro et al., 1980).

In conclusion, the spectra presented in this section illustrate the ease with which the infrared spectra of a variety of difficult samples may be obtained throughout the infrared region using an interferometer-DAC linkup. The reader should be aware of the fact that the present commercial dispersive infrared spectra are not amemable for use below ∿ 200 cm^{-1}.

Fig. 21. Spectrum of β-AgI in DAC.

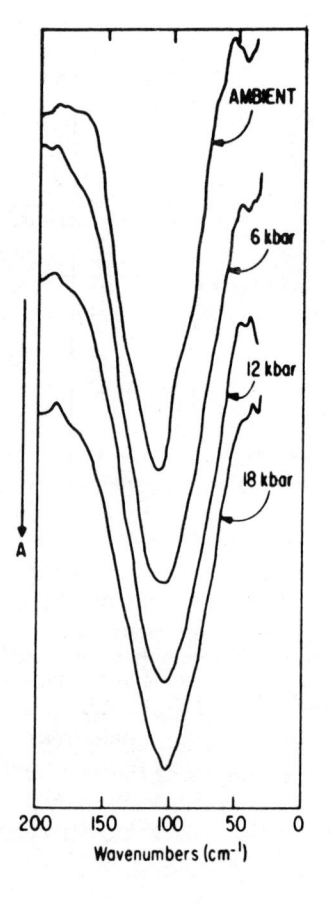

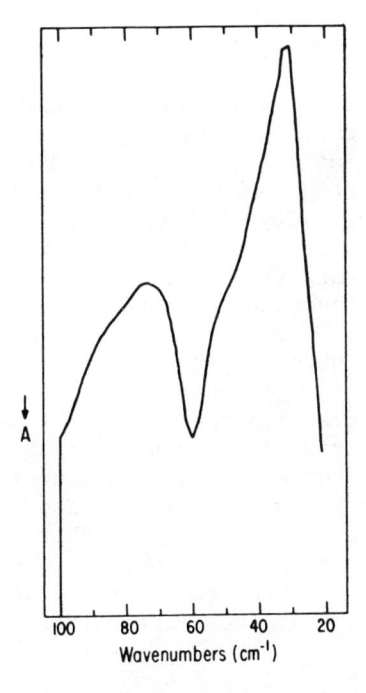

Fig. 22. Spectrum of Yellow HgO in DAC.

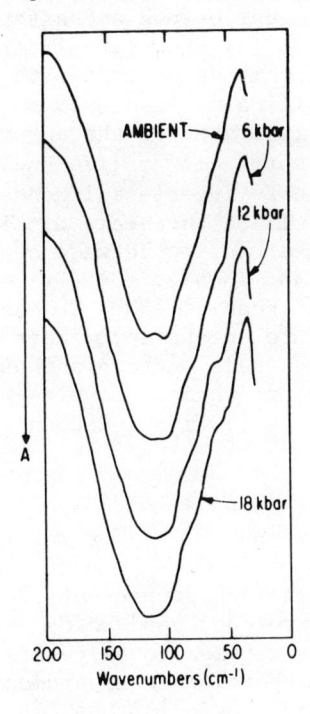

Fig. 23. Spectrum of Pyridinium Ag_5I_6 in DAC.

Studies of Glasses by FT-IR

Perhaps the most extensive research made on glass using FT-IR instrumentation are those of Koenig at Case Western Reserve University. For sometime interest has developed to determine the nature of the interaction between coupling agents on glass surfaces. The sensitivity and selectivity of dispersion instruments is insufficient to obtain infrared spectral information about the coupling agent inasmuch as its concentration is small compared to the bulk glass. FT-IR is a powerful technique for such studies.

Surface studies were made by FT-IR of the nature of coupling agents on glass surfaces (Ishida and Koenig, 1978a,b, Koenig, 1981). The coupling agents are used to enhance the adhesion between the fiber and polymer matrix in the construction of glass-reinforced composites. The mechanism of reinforcement is of utmost interest. Fig. 24 shows a spectrum of modified silica and that of the control. The absorption of the coupling agent is barely visible at 1411 cm^{-1} (A). However, when spectral subtraction is used, the difference spectrum (C) reveals the spectrum of the coupling agent; and further that it has polymerized on the glass surface.

The nature of the bond of the coupling agent to the glass surface is of importance as well. Fig. 25 shows the spectra of silica with a silane coupling agent before and after heat treatment. Spectrum (A) and spectrum (B) show very little difference. However, (C) the difference spectrum, demonstrates the condensation reaction of SiOH groups of the silica surface (970 cm^{-1}) and the coupling agent (893 cm^{-1}), as illustrated by the negative absorbances of these bands. The positive absorbances at 1170 and 1080 cm^{-1} of Si-O-Si bonds have been assigned to interfacial bonds formed between the coupling agent and the silica surface. Similar studies have been made on the nature of the bonding between coupling agents and E-glass fibers (Ishida and Koenig, 1978b). Other related studies have appeared Chiang and Koenig, 1981, Chiang et al., 1980). Studies of the effects of moisture on the epoxy matrix in glass reinforced composites have also been made (Antoon and Koenig, 1981). The hydrothermal degradation of various coupling agents on E-glass fibers has also been investigated (Ishida and Koenig, 1980).

SUMMARY

The attempt has been made to present some techniques, which are adaptable to FT-IR spectroscopy, and which can be used for materials characterization. Again we wish to reiterate that these techniques can also be used with dispersive IR instruments. The materials scientist and others are blessed in that two infrared techniques exist, and one has a choice that can be made dictated by the problem

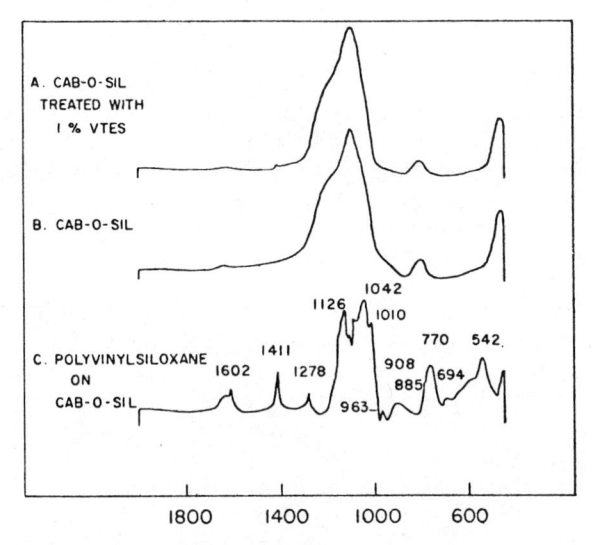

OVERVIEW OF TECHNIQUES USED IN FT-IR SPECTROSCOPY 195

Fig. 24. Spectrum of Modified Silica (Cab-o-Sil) and Control, reprinted with permission from Koenig (1981). Copyright 1981 American Chemical Society.

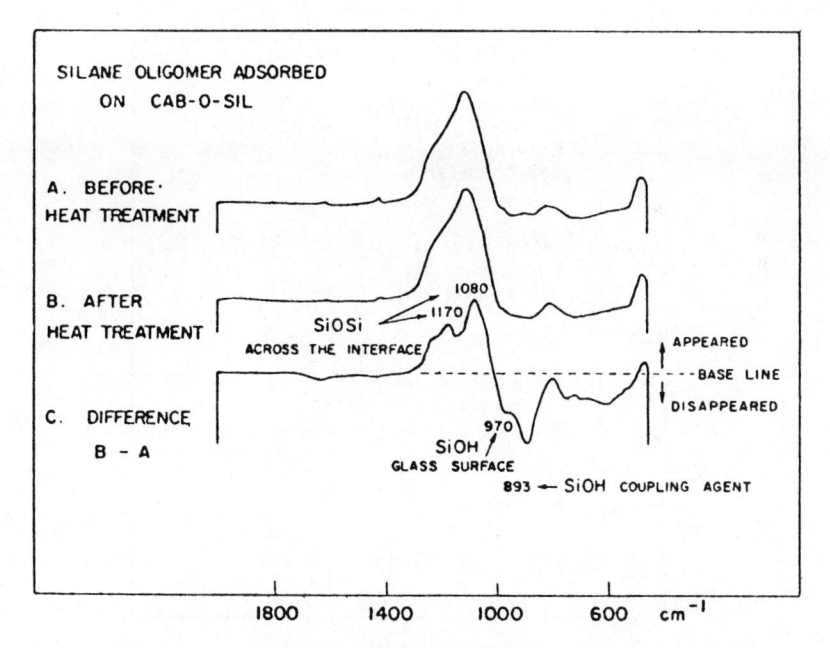

Fig. 25. Spectra of Modified Silica (Cab-o-Sil) Treated with a Silane Oligmer, reprinted with permission from Koenig (1981). Copyright 1981 American Chemical Society.

at hand. In the reflectance studies, because of serious energy losses the throughput advantage of FT-IR may be helpful. For the emission experiments certain difficulties arise. For example, the heated cell or furnace may serve as an IR source. To eliminate this unwanted energy one must use a ratio-recording double-beam dispersive IR spectrometer that pre-chops the IR beam prior to passing through the sample. In the case of a computer-dispersive IR and and FT-IR this unwanted source of energy can be compensated for by background measurements and data manipulation.

Griffiths (1974) has listed 3 situations where FT-IR is significantly superior to dispersive spectroscopy and these are:
1. High-resolution spectroscopy over a wide-spectra range. Fig. 26 illustrates the relationship of high resolution in comparison to other methods.
2. Rapid scan applications.
3. Applications where weak signals are incurred and an unacceptably long time is needed to measure the spectrum conventionally.
If the interest is in far infrared spectroscopy, the experimenter by necessity is committed to an FT-IR, as dispersive instruments have cut-offs at ~ 200 cm^{-1}.

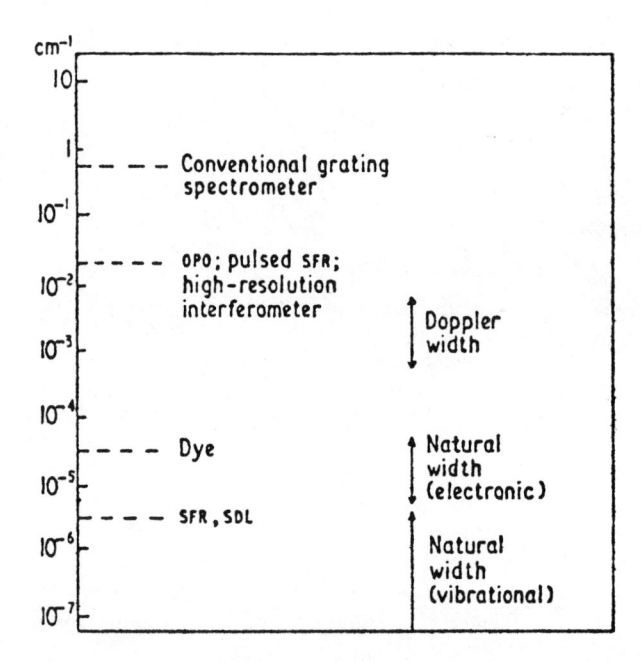

Fig. 26. Comparison of Resolution Obtained with Various Techniques. Taken in part from J.M. Colles and C.R. Pidgeon, Reports in Progress in Physics, <u>38</u>, 329 (1975).

ACKNOWLEDGEMENTS

The author wishes to thank the Searle Foundation for the support of this work.

REFERENCES

Adams, D. M., and Payne, S. J., 1972, Annual Reports A pp. 3-17, The Chemical Society, London.
Adams, D. M., and Sharma, S. K., 1977, J. Phys. E., 10:838.
Antoon, M. K. and Koenig, J. L., 1981, J. Polymer Sc., 19:197.
Bates, J. B., 1978, in: "Fourier Transform Infrared Spectroscopy," J. R. Ferraro and L. J. Basile, eds., Vol. 1, p. 99, Academic Press, N.Y.
Bates, J. B., and Boyd, G. E., 1973, Appl. Spectrosc., 27:204.
Bates, J. B., and Frech, R., 1976, unpublished work.
Bell, R. J., 1972, in: "Introductory Fourier Transform Spectroscopy," Academic Press, N.Y.
Brasch, J. W., 1965, J. Chem. Phys., 43:3473.
Chiang, C-H., and Koenig, J. L. 1981, Polymer Composites, 2:192.
Chiang, C-H., Ishida, H., and Koenig, J. L. 1980, J. Colloid and Interface Sc., 74:396.
Cooley, J. W., and Tukey, J. W., 1965, Math. Comput., 19:297.
Connes, J., and Connes, P., 1966, J. Opt. Soc. Am., 56:896.
Fellgett, P.B., 1959, J. Phys. Radium, 19:187.
Ferraro, J. R., and Basile, L. J., 1974, Appl. Spectrosc., 28:505.
Ferraro, J. R., and Basile, L. J., 1979a, in: "Fourier Transform Infrared Spectroscopy", Vol. 1, pp. 275-302, Academic Press, N.Y.
Ferraro, J. R., and Basile, L. J., 1979b, Am. Lab., 11:31.
Ferraro, J. R., 1971, in: "Spectroscopy in Inorganic Chemistry", C.N.R. Rao, and J.R. Ferraro, eds., pp. 55-77, Academic Press, N.Y.
Ferraro, J. R., and Basile, L. J., 1980, Appl. Spectrosc., 34:217.
Ferraro, J. R., and Mitra, S. S., and Postmus, C., 1966, Inorganic Nucl. Chem. Lett., 2:269.
Ferraro, J. R., Walling, P. L., and Sherren, A. T., 1980, Appl. Spectrosc., 34:570.
Fink, W., 1979, in: "Fourier Transform Infrared Spectroscopy", J. R. Ferraro and L. J. Basile, eds., Vol. 2, pp. 243-314, Academic Press, N.Y.
Griffiths, P. R., 1974, Anal. Chem., 46:645A.
Griffiths, P. R., 1975, "Chemical Infrared Fourier Transform Spectroscopy", J. Wiley and Sons, N.Y.
Griffiths, P. R., Sloane, H. J., and Hannah, R. W., 1977, Appl. Spectrosc., 31:485.
Harrick, N. J., 1967, "Internal Reflection Spectroscopy," Interscience, N.Y.

Hirschfeld, T., Pacific Conference on Chem. & Spectroscopy, Anaheim, CA., 1981.
Ishida, H., and Koenig, J. L.,1978a,J. Colloid Interface Sci., 64:555.
Ishida, H., and Koenig, J. L., 1978b, ibid, 64:565.
Ishida, H., and Koenig, J. L., 1980, J. Polymer Sc., 18:1931.
Jacquinot, P., 1954, 17 Congres du GANS, Paris.
Jakobsen, R. J., Mikawa, Y., and Brasch, J. W. 1970, Appl. Spectrosc., 24:33.
Klaeboe, P., and Woldback, T., 1978, Appl. Spectrosc., 32:588.
Koenig, J. L., 1981, Acc. Chem. Res., 14:171.
Krishnan, K., and Ferraro, J. R., 1982, in: "Fourier Transform Spectroscopy", Vol. 3, pp. 149-209, Academic Press, N.Y.
Krishnan, K., Hill, S. L., and Brown, R. H., 1980, Amer. Lab., 12:104.
Lauer, J. L., 1978, in: "Fourier Transform Infrared Spectroscopy", J. R. Ferraro and L. J. Basile, eds., Vol. 1, pp. 169-213. Academic Press, N.Y.
Lippincott, E. R., Weir, C. E., Van Valkenburg, H., and Bunting, E. N., 1960, Spectrochimica Acta, 16:58.
Lippincott, E. R., Welsh, F. E., and Weir, C. E., 1961, Anal. Chem., 33:137.
Low, M. J. D., and Clancy, F. K., 1967, Environm. Sci. Technol., 1:73.
Low, M. J. D., and Parodi, G. A., 1979, Appl. Spectrosc., 34:76.
Parke, S., and Warman, G. P., 1980, Phys. & Chem. of Glasses, 21:91.
Michelson, A. A., 1891, 31:256, 1892, 34:280, Phil. Mag. Ser. 5.
Piriou, B., and Arashi, H., 1980, High Temp. Sc., 13:299.
Postmus, C., Ferraro, J. R. and Mitra, S. S. 1981, Inorg. Nucl. Chem. 4:155.
Rayleigh, Lord, 1892, Phil. Mag. Ser., 5, 34:407.
Rockley, M. G. and Devlin, J. P., 1980, Appl. Spectrosc., 34:407.
Rosencwaig, A., 1975, Anal. Chem., 47:592A.
Rosencwaig, A., 1980, in: "Photoacoustics and Photoacoustic Spectroscopy, Wiley, N.Y.
Rubens, H., and Hollmagel, H., 1910, Phil. Mag. Ser., 6, 19:761.
Rubens, H., and Wood, R. W., 1911, Phil. Mag. Ser., 6, 21:249.
Sheppard, N., Greenler, R. G., and Griffiths, P. R., 1977, Appl. Spectrosc. 31:448.
Smith, F., 1980, Digilab Users Conference.
Weir, C. E., Lippincott, E. R., Van Valkenburg, A. and Bunting, E. N., 1959, J. Res. Nat. Bur. Std. U.S. Sect.A. 63:55.

RAMAN MICROPROBE SPECTROSCOPY OF POLYPHASE CERAMICS

D.R. Clarke* and F. Adar**

*Structural Ceramics Group
Rockwell International Science Center
Thousand Oaks, California 91360
**Applications Laboratory
Instruments Sa, Inc.
New Jersey 08840

INTRODUCTION

   One of the principal reasons that Raman spectroscopy has not been widely applied to problems in the characterization of ceramics is that until the recent advent of the Raman microprobe spectrometer it was not possible to localize the analysis volume to a dimension commensurate with the grain size typical of ceramic microstructures. This has proved to be an unfortunate limitation since many of the phases common to ceramic materials are strongly Raman active and generate quite distinct and recognizable Raman spectra. However, the ability to investigate regions with an optical probe as small as 1 µm in diameter, as can now be achieved in the Raman microprobe, finally enables the technique of Raman spectroscopy to complement existing microanalysis tools. To put the technique in perspective, it extends the capabilities of the optical microscope to include microanalysis just as x-ray microanalysis techniques (e.g., EDAX) have extended the capabilities of the scanning electron microscope and as x-ray microanalysis and electron energy loss spectroscopy have made the transmission electron microscope an analaytical instrument of exceptional power.

The small microanalysis volume and the associated high spatial resolution of the Raman microprobe are a direct consequence of the optical systems used to form the micron-sized probe and collect the scattered light. In the Dhamelincourt microprobe configuration[1,2] the illuminating laser beam is focused to a fine probe by an objective lens of high numerical aperture and the same lens is employed in collecting and directing the scattered photons into the monochrometer system. The axial spatial resolution (depth of field) is then determined by the depth of focus of the illuminating beam and the depth of field of the optics transferring the Raman scattered radiation to the monochrometer. The latter, as discussed by Dhamelincourt[1,2] and Rosasco,[3] can be limited by an aperture placed at an intermediate image plane between the Raman collecting lens and the monochrometer entrance slit. Under such optical conditions, fluorescence, Rayleigh scattering and Raman emission outside of the laser focal volume are optically blocked by the intermediate aperture. According to Dhamelincourt's calculations, which have recently been given preliminary experimental credence,[4] when the aperture size is equal to the lateral size of the image of the laser probe, the normalized depth (depth scaled by the refractive index) from which the Raman signal is collected corresponds to about 2-3 μm above and below the plane of focus.

As with the axial spatial resolution, the lateral spatial resolution of the microprobe is determined by both the illuminating optics and the collector angle subtended by the intermediate aperture of the Raman transfer optics. Consideration of these optics[1-4] suggests that an objective lens having a numerical aperture (n.a.) of 0.9 would focus about 84% of the illuminating laser energy into a diffraction limited spot (1.22 λ/n.a.) of about 0.7 μm diameter and that an intermediate aperture matched to the image size of the illuminating spot should confer a lateral resolution in the focal plane of the objective corresponding to the probe size. Thus, an analysis volume equal to the product of the probe size and the depth of field, which can be as small as ~ 6 μm^3, is attainable in the Dhamelincourt microprobe configuration.

One inherent advantage of the small microanalysis volume is that very small fractions of a phase may be identified, provided they can be distinguished in a microstructure and the probe focused onto it. This is in contrast to conventional x-ray diffraction where the analysis volume cannot be localized and as a result volume fraction of phases below 1-5% are difficult to detect. The advantage is especially striking in practice since micro phases and inclusions can often be seen in the optical microscope (on account of differences in reflectivity or color) and yet are indistinguishable in, for instance, the scanning electron microscope where different contrast mechanisms pertain.

At the time of writing few results of the use of Raman microprobe spectroscopy for the study of ceramics have been reported in the open literature.[4,5] In this work we present new data on a number of phases present in structural ceramics that is of general interest for identification of these phases. In the following sections the Raman spectra of the polymorphs of silicon nitride, boron nitride and zirconia are presented. In addition, spectra of silicon oxynitride and silicon carbide, phases that can co-exist in silicon nitride based ceramics, are presented to illustrate the ability to discriminate between these phases.

EXPERIMENTAL DETAILS

The Raman spectra reproduced here were obtained at room temperature from either polished sections of ceramic, fracture surfaces or powders using a commercially available Raman microprobe* based on the Dhamelincourt configuration. The samples were illuminated with a argon-ion laser beam (λ = 514.5 nm) focused down by the objective lens of the microprobe. The prevailing optical conditions typically employed were: a probe diameter of ~ 1 µm; objective lens: 40x or 80x; numerical aperture: 0.85 or 0.9, respectively; laser power at sample: 1 to 20 mW.

SILICON NITRIDE

Two polymorphs of silicon nitride are known to exist, the alpha variant and the beta variant having the space groups of P31c and P6$_{3/m}$, respectively. The stability of the alpha and beta forms is still a matter of some controversy[6,7] but it is widely recognized that α-Si$_3$N$_4$ is produced by processes such as chemical vapor deposition, reaction of silane and ammonia, and the low temperature nitridation of silicon. These findings and the results of "seeding" experiments[7] have been used to argue[6,7] that the alpha form is the low temperature modification. Its primary importance lies in the fact that it is usually the starting powder employed in the manufacture of hot-pressed or injection molded silicon nitride alloys. Beta-Si$_3$N$_4$ is formed during the liquid phase consolidation occuring during hot press-

---

*Micro-Ramanor, Instruments SA, Inc., Metuchen, NJ.

ing and is now generally accepted to grow at the expense of the alpha form by a solution-reprecipitation process through the liquid phase. In fact, the α → β phase transformation is often used as a monitor of the extent of the hot pressing process.

The Raman active modes of alpha silicon nitride have been investigated by Kuzuba et al,[8] who reported Raman spectra of both powdered and single crystal samples. According to their work the material exhibits strong bands at 262 and 365 cm^{-1} (whose strength is greater in the X(ZZ)Y configuration than in the X(ZX)Y configuration) with other relatively strong bands at 515, 668, 868, 915, 976, and 1034 cm^{-1}. Our findings (Fig. 1a) for CVD α-Si$_3$N$_4$ are in substantial agreement but with additional strong bands at 300 and 850 cm^{-1}. Far fewer bands are excited in the Raman spectrum of the beta-phase (shown in Fig. 1b) obtained from a powdered sample. The spectrum is in substantial agreement with that recently reported by Wada et al[9] in their study of the vibrational properties of the alpha- and beta-polymorphs.

Hot pressed silicon nitride ceramics also commonly contain silicon oxynitride as an additional crystalline phase. No previous reports of the Raman spectrum of this phase have been published. However, silicon oxynitride is a relatively strong Raman scatterer as shown in Fig. 2 and is easily detectable in polyphase samples containing low volume fractions of the phase. The spectrum is characterized by particularly strong bands at 184 and 250 cm^{-1}. A group analysis of the Raman and infrared active modes is presently underway and will be reported at a later date.

BORON NITRIDE

The discovery in 1957 by Wentdorf[10,11] that a cubic (zincblende) modification of boron nitride could be produced by high pressure-high temperature technology and that the phase (borazon) was exceptionally hard has spurred interest in the temperature-pressure phase diagram of boron nitride. Three phases are now recognized (hexagonal, cubic and wurtzite) and there is evidence to suggest that a rhombohedral variant[12] may also form under certain circumstances.

The common, low pressure, hexagonal form of boron nitride is similar to the crystalline structure of graphite and in consequence has a Raman spectrum that is in a number of respects similar to that of graphite. The spectrum (Fig. 3a) has a band at 1368 cm^{-1} corresponding to the nonpolar interlayer vibration

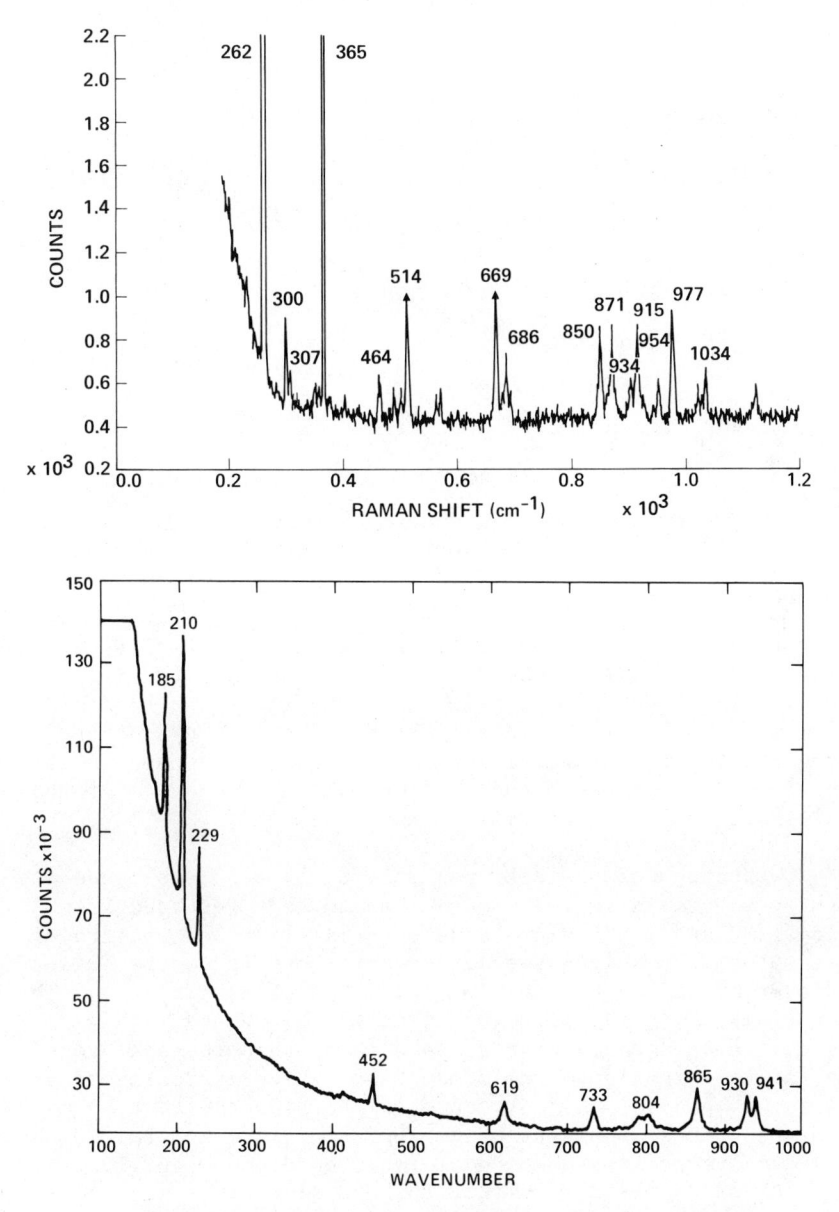

RAMAN MICROPROBE SPECTROSCOPY OF POLYPHASE CERAMICS 203

Fig. 1 A comparison of the Stokes-Raman spectra of the β and α polymorphs of silicon nitride. The top spectrum was recorded from a polycrystalline sample of alpha-Si_3N_4 prepared by chemical vapor deposition and the lower spectrum from high purity beta-powder. Excitation with 514.5 nm line of an argon-ion laser at room temperature.

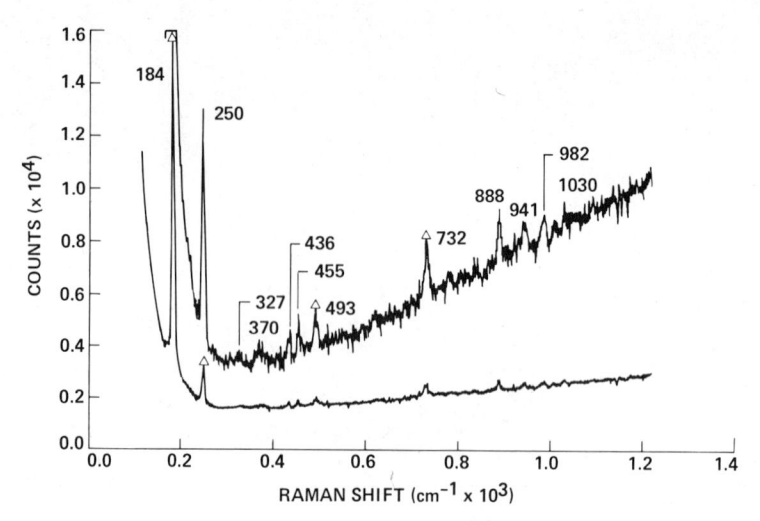

Fig. 2 The Stokes-Raman spectrum of silicon oxynitride powder.

of the hexagonal planes and is in agreement with earlier publications.[13,14] The frequency of the Raman band of BN at 1368 cm^{-1} is lower in comparison with the analogous band of graphite at 1575 cm^{-1}. Since the reduced masses of carbon-carbon and boron-nitrogen molecular units are the same as the lower frequency of the boron nitride band is a measure of the smaller atomic force constant in that material.

No spectrum of cubic BN has previously been published but that reproduced in Fig. 3b is similar to that recorded from cubic III-V and II-VI zincblende materials.[15] The spectra of these materials exhibit two lines corresponding to the TO and LO (transverse and longitudinal optic) modes of the lattice. The Raman lines of the cubic boron nitride at 1065 and 1368 cm^{-1} appear at considerably higher frequencies than the analogous lines of the zincblende semiconductors, which typically occur between 100 and 500 cm^{-1}, because of the higher effective masses and higher force constants of the boron nitride system. When the Raman spectrum is compared to that of diamond the TO frequency of boron nitride is seen to be at 1065 cm^{-1} whereas that of diamond is at 1332 cm^{-1}. As with the comparison with the hexagonal form of boron nitride, the lower frequency of the boron nitride line is attributed to weaker interatomic forces (the reduced masses being the same).

No Raman spectra have, to the author's knowledge, yet been published of the wurtzite form of boron nitride. Attempts to record spectra from the relatively recently discovered phase

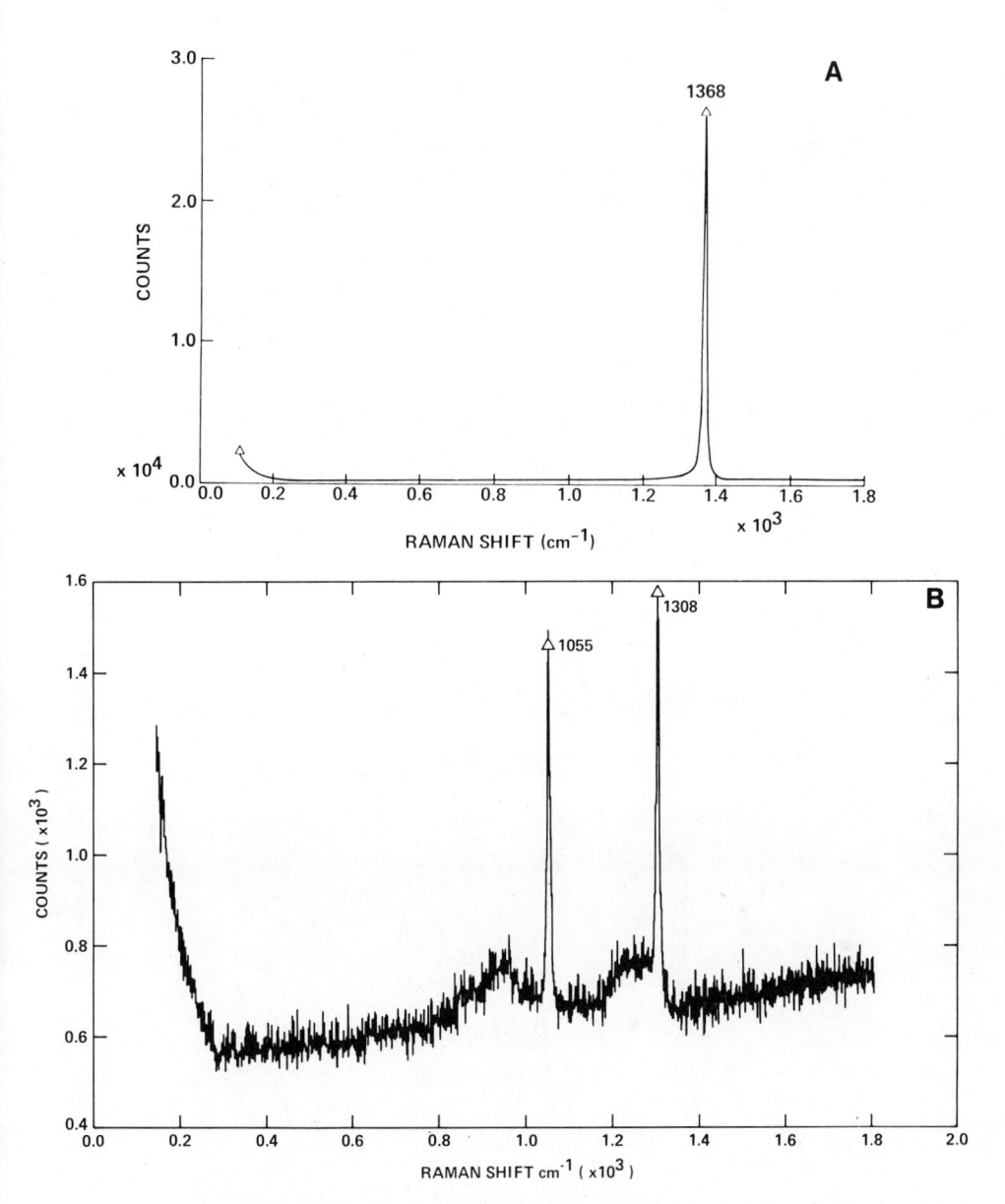

Fig. 3   The Stokes-Raman spectra recorded from (a) hexagonal and (b) cubic forms of boron nitride.

using samples manufactured by explosive compaction techniques and kindly supplied by Prof. Sawaoka have so far been unsuccessful due to fluorescence phenomena.

ZIRCONIA

In contrast to silicon nitride and boron nitride the principal feature of the Raman spectra of the polymorphs of zirconia have been investigated previously in some detail, dating from the work of Phillippi and Mazdiyasni.[16] The most recent high resolution spectra of tetragonal and monoclinic zirconia are reproduced in Fig. 4. The data are taken from a recent publication[5] by the authors in determining the size of the tetragonal → monoclinic transformation zone in a series of transformation toughened zirconia alloys having a grain size of approximately 1 µm.

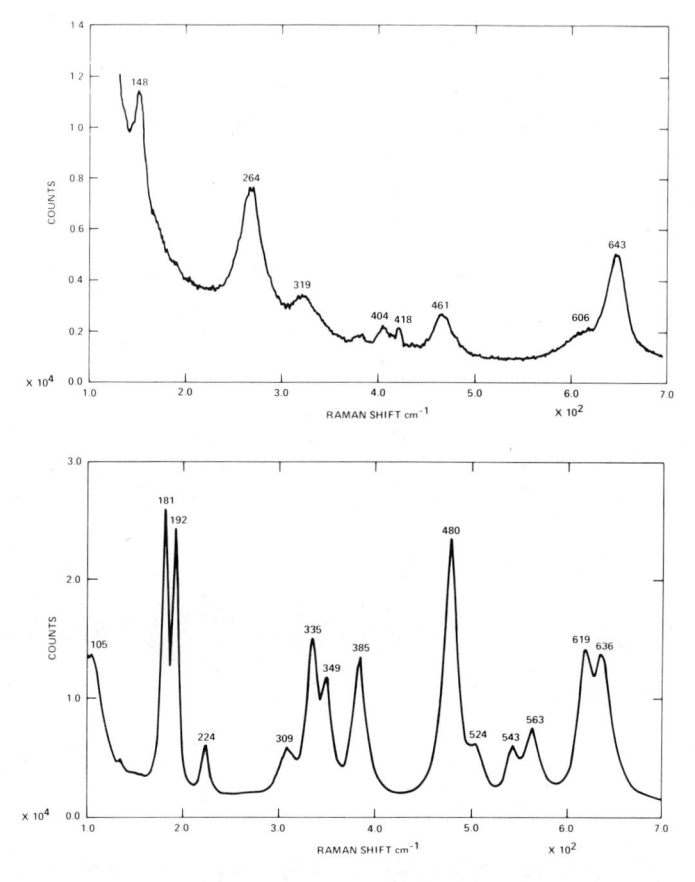

Fig. 4   A comparison of the spectra of the tetragonal (top) and monoclinic (bottom) crystal structures of zirconium dioxide. The principal distinguishing bands are the doublet at 181 and 192 cm^{-1} characteristic of the monoclinic phase and the bands at 148 and 264 cm^{-1} of the tetragonal form.

Despite the previous studies there has been some uncertainty concerning the assignment of Raman bands to the various polymorphs. As the tetragonal form is normally stable only above ~ 1150°C, the spectral lines recorded at these temperatures (e.g., in Ref. 17) are broad and relatively ill-defined. The other spectra[16,18] on which the Raman spectrum of tetragonal have previously been assigned were recorded from mixed powder so there has been some doubt due to the overlap with the monoclinic lines. By virtue of the processing techniques employed the material from which the spectrum of Fig. 4 was recorded was purely tetragonal and as the spectrum contains no lines characteristic of the monoclinic phase it can be attributed completely to the tetragonal polymorph. The spectra indicate that there is some similarity of the monoclinic and tetragonal lines for Raman shifts in excess of 300 cm^{-1}. However, over the range 100-300 cm^{-1} the characteristic monoclinic doublet (at 181 and 192 cm^{-1}) and the tetragonal bands, at 148 and 264 cm^{-1}, are well separated. The additional monoclinic bands at 105 and 224 cm^{-1} although weaker, are distinct from the tetragonal bands.

APPLICATIONS

The Raman microprobe has found application in a range of ceramic problems varying from simple minor phase identification to the investigation of phase transformations. In the following sub-sections, a number of examples in which the microprobe has made valuable contributions to the characterization of ceramic microstructures are described.

Inclusion Analysis

One of the simplest uses of the Raman microprobe to date has been in the identification of inclusions. This is illustrated by the inclusion shown in the optical and scanning electron micrographs of Figs. 5a,b revealed on the fracture surface of a silicon nitride tensile test bar tested as part of the DARPA/AFOSR Quantitative Nondestructive Evaluation Program. Despite being less than 100 μm across the presence of the inclusion was obvious to the naked eye in reflected light. The nature of the fracture markings (Fig. 5a) indicate that the inclusion acted as the origin for fracture and hence the importance in identifying it. The inclusion proved insufficiently large to contribute to an x-ray diffraction pattern and in any case this approach was probably precluded by the curvature of the fracture surface. When examined by x-ray microanalysis only silicon was detectable in the vicinity of the inclusion as it was away from the inclusion.* The Raman spectrum recorded by focussing the optical beam onto the inclusion is reproduced in

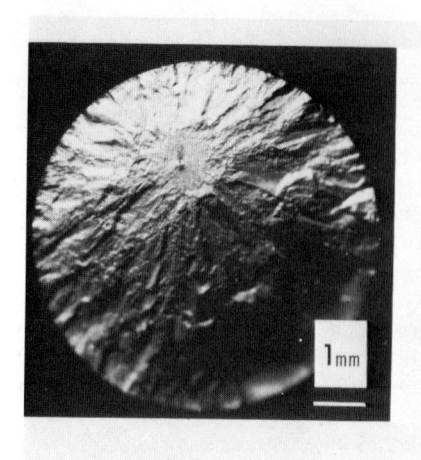

Fig. 5 Optical photomicrograph (left) of the fracture surface of a silicon test bar indicating that fracture originated at a volume defect. On examination (right) in the scanning electron microscope, the fracture origin is seen to be an inclusion ~100 μm in size.

Fig. 6 and comparison with the spectrum from a single crystal of 6H SiC demonstrates that the inclusion is SiC. This unequivocal identification is consistent with the EDS finding of Si and distinguishes between the possibilities of Si, Si_3N_4 and SiC.

ZrO_2-ThO_2

Another example of phase analysis that illustrates the significant advantages of Raman microprobe spectroscopy over more conventional techniques is the identification of phases in the ZrO_2-ThO_2 system, a ceramic that has come under scrutiny as a possible high toughness electromagnetic window material. In order to exhibit high values of fracture toughness, it is

---

*From a ceramics point of view, it would be unlikely that the inclusion would be silicon as it would be expected to have melted at the temperatures used to fabricate hot pressed silicon nitride. Likewise, the size and smooth fracture surface of the inclusion probably preclude it being silicon nitride.

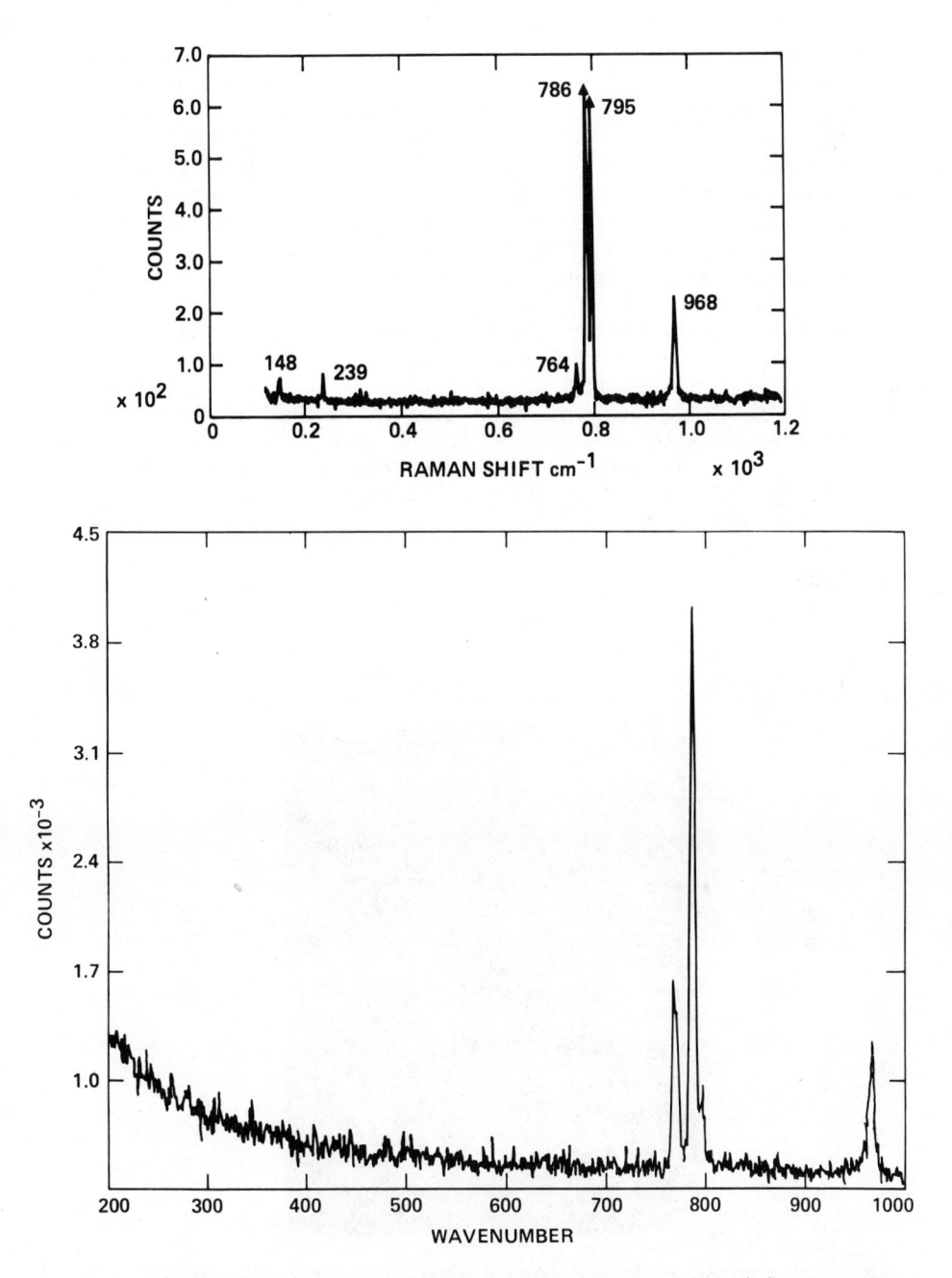

Fig. 6 A comparison of the Raman spectra recorded (a) from the inclusion of Fig. 5 and that recorded (b) from a single crystal of 6H silicon carbide.

necessary that the zirconia be in its tetragonal state but inadequate processing can result in a material containing either cubic or monoclinic polymorphs. Characterization of the phases present in the materials developed to date has proved difficult by x-ray diffraction methods since the tetragonality of the tetragonal form is close to unity and the reflections from the monoclinic, tetragonal and cubic forms either coincide or closely overlap. The problem is compounded by the similarity of the reflections from the ThO_2 to those of ZrO_2. Figure 7 is a scanning electron micrograph of a ZrO_2-ThO_2 sample including an atypical, abnormal grain. The Raman spectra recorded from the area of the photomicrograph is reproduced in Fig. 8 indicating that the region consists of mostly tetragonal ZrO_2 with a proportion (calculated from the relative intensities to be ~40%) of monoclinic ZrO_2 and including ThO_2. The ThO_2 phase is readily identifiable by its very strong line at 469 cm^{-1}. (Spectra from oxides with the fluorite structure, e.g., ThO_2, CeO_2 and UO_2 are characterized[19] by having a strong single line between 460 and 470 cm^{-1}.) By focussing the optical probe onto the grain, it was possible to determine that the abnormal grain was ThO_2.

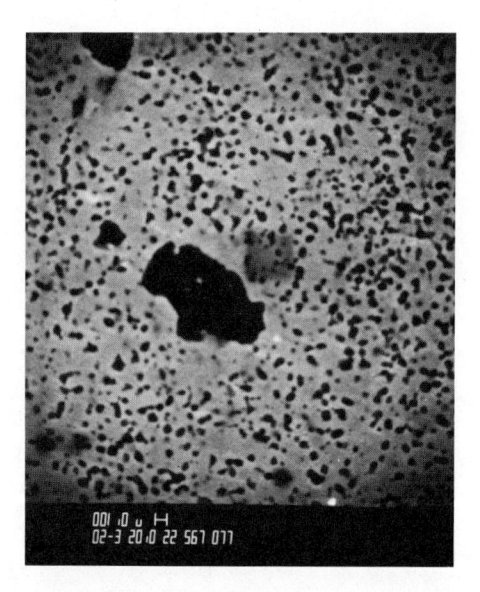

Fig. 7   Scanning electron micrograph of a ZrO_2-20 v/o ThO_2 sample illustrating a region including a large ThO_2 grain. The ThO_2 grains appear dark in contrast to the ZrO_2 grains.

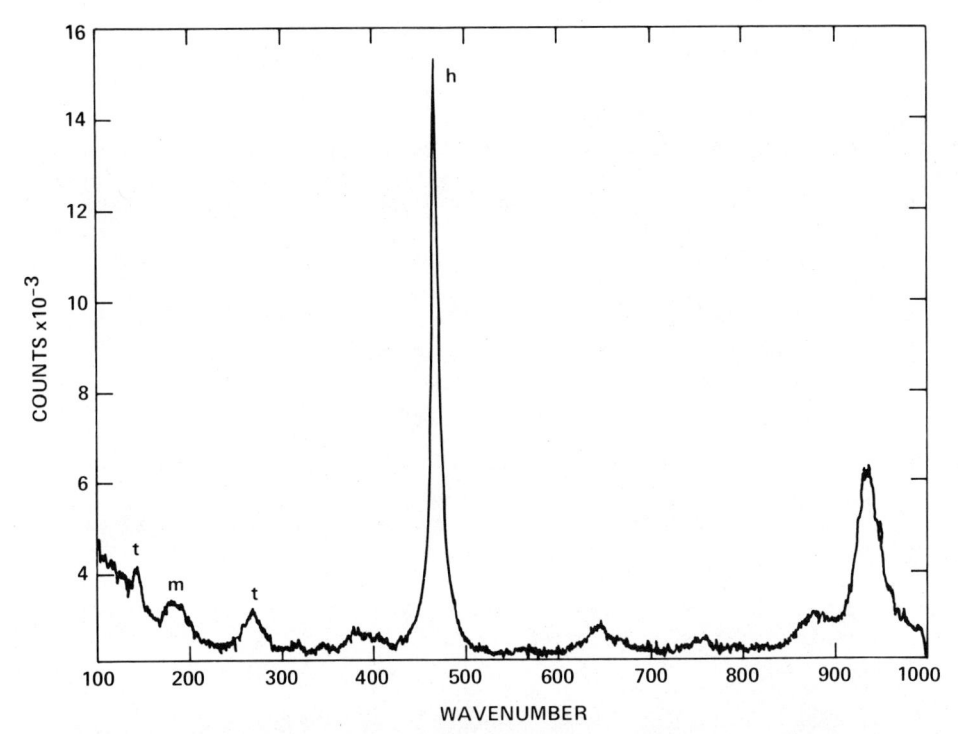

Fig. 8 Raman spectrum from the region of Fig. 7. The strong band (h) at 469 cm^{-1} in characteristic of ThO$_2$ whereas the bands (t) at 269 and 146 are indicative of the presence of tetragonal zirconia.

SiC/Si$_3$N$_4$ Refractory

Silicon nitride bonded silicon carbide refractory materials were originally developed several decades ago. They have been finding increased useage in recent years on account of both improved properties and as a lower cost alternative to silicon carbide bonded silicon carbide refractories. As illustrated by the photomicrographs of Fig. 9 the microstructure consists of large monolithic grains in a finer grain matrix. According to x-ray diffraction, the material typically consists of Si$_3$N$_4$, Si$_2$N$_2$O with some free silicon in addition to the SiC phase. The Raman spectrum (Fig. 10a) from the large grains clearly indicates that they are SiC. Other regions in the matrix can be identified, some containing Si$_3$N$_4$ and others, principally Si$_2$N$_2$O, as shown by the spectra of Fig. 10b. Pockets of needle-shaped grains in more porous regions give rise to Raman spectra characteristics of α-Si_3N_4, although additional spectral lines from other phases lining the pocket can also be discerned. The finding of needles of α-Si_3N_4 is consistent with the view that the phase has grown by a vapor phase mechanism in the porous region. One unusual feature of the spectrum of Fig. 10b is the broad band at ~ 825 cm^{-1}. Such broad bands are normally interpreted as originating from noncrystalline phases suggesting that a proportion of glass is present in the SiC/Si_3N_4 refractory material. These regions appear to coincide with regions found by x-ray microanalysis to contain relatively high concentrations of Ca in addition to Si, implying that the phase may be a calcium silicate based glass.

SUMMARY

The recent development of the Raman Microprobe has made it possible to characterize ceramic microstructives using the techniques of Raman spectroscopy. The capabilities of this new characterization instrument are illustrated with examples drawn from ceramics of contemporary interest.

Fig. 9   Scanning micrograph of part of the fracture surface of a SiC-Si_3N_4 refractory.

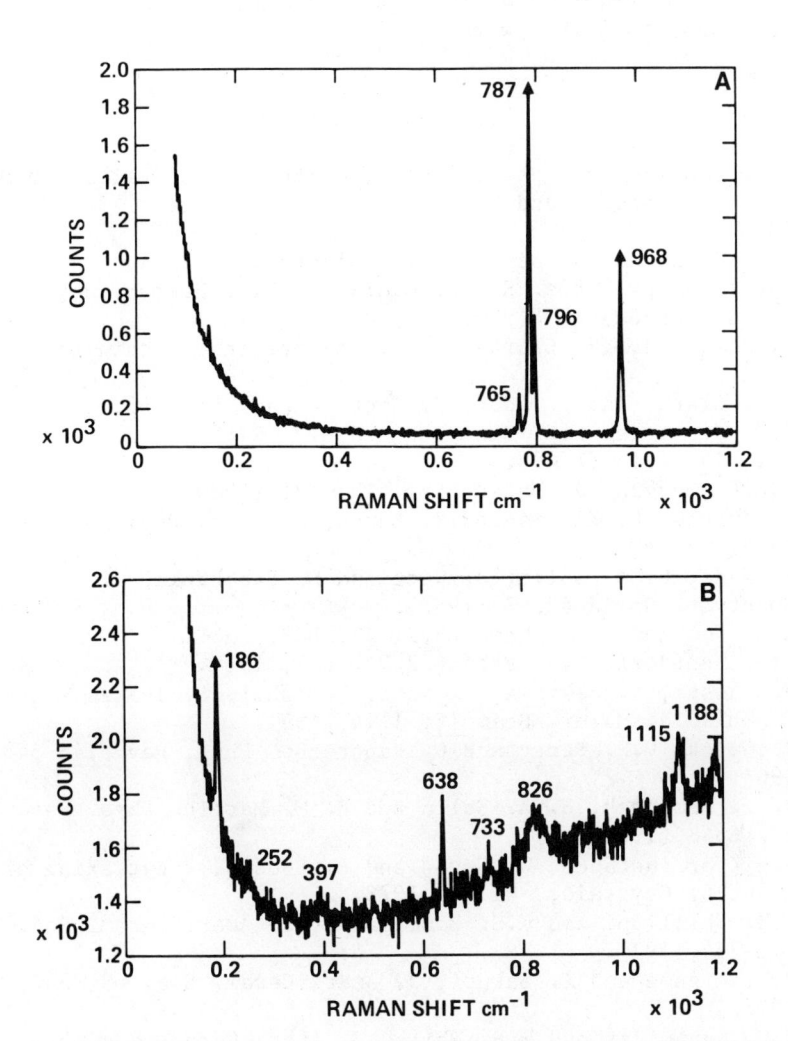

Fig. 10 (a) Raman spectrum from the large grained regions of Fig. 9. (b) Spectrum recorded from dark second phase. The band at 186 cm^{-1} is from silicon oxynitride and the broadband at ~826 cm^{-1} is indicative of anoncrystalline phase.

ACKNOWLEDGEMENT

The authors are indepted to the National Science Foundation under grant No. DMR-8007445 (DRC) and to Instruments, SA Inc. (FA) for support of this work.

REFERENCES

1. P. Dhamelincourt, Ph.D. Thesis, University of Lille, 1979.
2. P. Dhamelincourt and M. Delhaye, J. Raman Spect. 3, 33, 1975.
3. G.J. Rosasco, in "Advances in Infrared and Raman Spectroscopy," Eds. R.J.H. Clark and R.E. Hester, Heyden Press (London), 7, 1980.
4. F. Adar and D.R. Clarke, Proc. MAS Meeting, Washington, 1982.
5. D.R. Clarke and F. Adar, J. Amer. Ceram. Soc. 65, 1982.
6. H.M. Jennings, S.C. Danforth and M.H. Richman, J. Mater. Sci. 14, 1013 (1749).
7. P.E.D. Morgan, J. Mater. Sci. 15, 791 (1980).
8. T. Kuzuba, K. Kijima and Y. Bando, J. Chem. Phys. 69, 40, 1978.
9. N. Wada, S. A. Solin, J. Wong and S. Prochazka, J. Noncryst. Solid 43, 7 (1981).
10. R.H. Wentdorf, J. Chem. Phys. 34, 809, 1961.
11. R.H. Wentdorf, U.S. Patent 2,947,627, 1960.
12. Y. Matsui, Y. Sekiawa, T. Sato, T. Ishii, S. Isakasaa and K. Shii, J. Mater. Sci. 16, 1114, 1981.
13. R. Geick, C.H. Perry and G. Rapprecht, Phys. Rev. 146 543, 1966.
14. R. J. Nemarich, S. A. Solin and R. M. Martin, Phys. Rev. B 23, 6348 (1981).
15. See, for instance, W. Hayes and R. Loudon, "Scattering of Light by Crystals," Wiley, 1978.
16. G.M. Phillippi and K.S. Mazdiyasni, J. Amer. Ceram. Soc. 54, 254, 1971.
17. M. Ishigame and T. Sakuri, J. Amer. Ceram. Soc. 60, 367, 1977.
18. V.G. Keramidas and W.B. White, J. Amer. Ceram. Soc. 57, 22, 1974.
19. V. G. Keramidas and W. B. White, J. Chem. Phys. 59, 1561 (1973).

CHARACTERIZATION OF ANODICALLY GROWN NATIVE OXIDE FILMS ON

$Hg_{0.7}Cd_{0.3}Te$

Fran Adar

R.E. Kvaas and D.R. Rhiger

Instruments SA, Inc.
173 Essex Avenue
Metuchen, N.J. 08840

Santa Barbara Research Center
75 Coromar Drive
Goleta, CA 93117

ABSTRACT

The Raman spectrum of an anodically prepared 700Å native oxide film of $Hg_{0.7}Cd_{0.3}Te$ has been measured on the Raman microprobe. In order to model the native oxide this spectrum was compared to the spectra of oxide particles (mixed in the ratio determined to be present in the films) that had been sintered under vacuum. The spectra of the oxides indicate an amorphous phase and represents a lowest order approximation to the structure of the native oxide film. The Raman microprobe data complements that of XPS and Auger which assign oxidation states of +2, +2 and +4 to Hg, Cd, and Te respectively but which provide no information on intermediate or long range order. The identification of an amorphous phase in the oxides is also consistent with the known glass former property of TeO_2 and the tendency of HgO and CdO to be glass modifiers. The Raman technique is unique in its ability to differentiate between Cd/Hg TeO_3 and a mixture of CdO/HgO and TeO_2.

EXPERIMENTAL CONDITIONS

Raman microparticle spectra were recorded on the MOLE (Molecular Optical Laser Examiner) or on the Ramanor U1000 optically coupled to a research mircroscope for Raman microprobe analysis. The particles were dispersed on standard microscope slides and examined without further preparation. Particles of well characterized oxides (TeO_2, HgO, Hg_2O, $CdTeO_3$, $HgTeO_3$, TeO_3, $CdTeO_4$, H_6TeO_6) as well as sintered oxides (to model the native anaodic oxide film) were examined. In order to minimize the potential laser-induced evolution of mercury from Hg_2O, particles of this oxide were examined under water (which serves as a heat sink), as well as in normal atmospheric conditions.

The spectrum of the oxide film was recorded on the microscope accessory on the U1000. Because of the fragility of the films to high laser irradiance, the spectrum was recorded by averaging overnight signals generated by 0.5 to 1.0 mW of laser intensity at the sample. In addition, it was ncessary to defocus the laser beam to about 15 μm. Independent scans of different spots on the same sample produced the same spectrum shown here.

RESULTS

Reference Spectra. Spectra of microparticles of the oxides are presented in Figs. 1-4. One sees that the Raman spectra provide a clear fingerprint for composition. Several comments regarding these spectra are useful.

Examination of various particles of TeO_2 indicated some variability in the spectra, which can probably be attributed to excitation of infrared-active extraordinary phonons with wavevectors along axes between z and the x-y plane.[1]

The spectrum of Hg_2O is identical to that of HgO (Fig. 2). Since laser illumination may produce evolution of mercury from Hg_2O, we attempted to minimize this effect (if it were occuring) by examining particles under water. The spectra of Hg_2O under water and in air are essentially identical; the principal difference is the narrower linewidths of the bands of the former. We may thus conclude that either we could not prevent evolution of mercury from the lattice, or that the spectra of the two compounds are identical. In general, the second possibility would not be worthy of consideration. However, in this instance, there is independent evidence that in Hg_2O the second atom of mercury resides in the lattice of HgO without any chemical interaction.[2]

Samples of $CdTeO_3$ from two vendors were examined. The Raman spectra shown in Fig. 3 are quite distinguishable. This result implies the presence of two crystalline polymorphs for this material and was confirmed by x-ray diffraction. In addition, there is considerable variation of lineshape in the spectra of particles in both specimens, which is consistent with the known property of TeO_2 to be a glass former.

Sintered Oxides. Oxides of $CdTeO_3$, $HgTeO$ and TeO_3 were mixed in molar ratios of .54: .12: .35 and sintered in a quartz ampoule for 70 hours at 570°C. The powder had not fused into a solid so spectra of several particles were examined in order to assess the

ANODICALLY GROWN NATIVE OXIDE FILMS

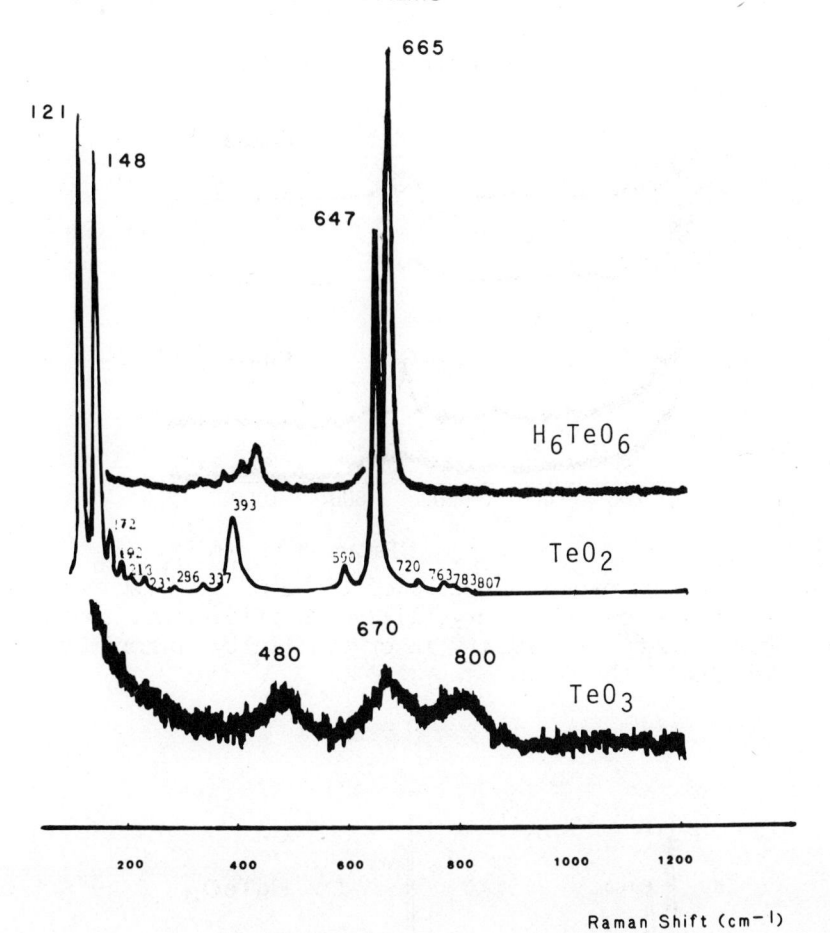

Fig. 1 Raman microparticle spectra of H$_6$TeO$_6$, TeO$_2$, and TeO$_3$

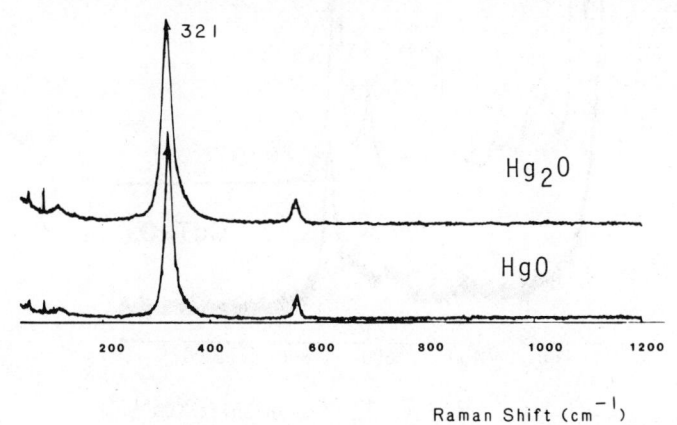

Fig. 2 Raman microparticle spectra of Hg$_2$O and HgO

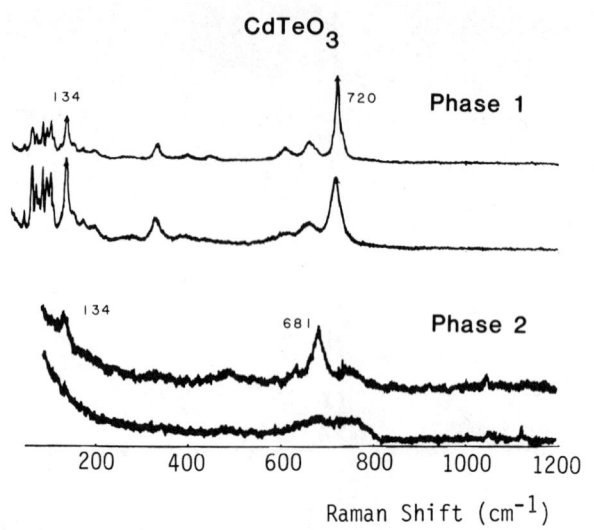

Fig. 3. Raman microparticle pectra of Cd Te O$_3$ from two vendors which represent two crystalline modifications. Both samples showed variability in crystallinity as monitored by widths of the bands.

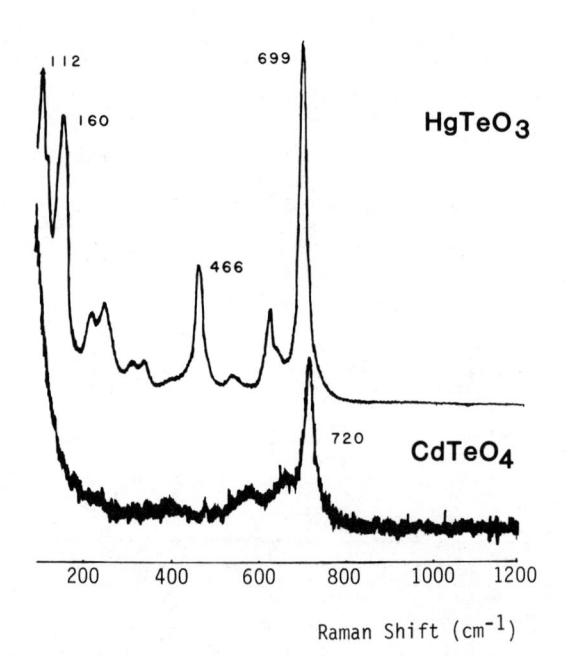

Fig. 4. Raman microparticle spectra of HgTeO$_3$ and CdTeO$_4$.

ANODICALLY GROWN NATIVE OXIDE FILMS 219

homogeneity of the reacted material. Raman microprobe spectra of two particles are shown in Fig. 6. The principle features are an intense band at 110 cm^{-1} and two broader features centered at ca 640 and 780 cm^{-1}. Examination of these spectra indicates considerable variability in lineshape and line width.

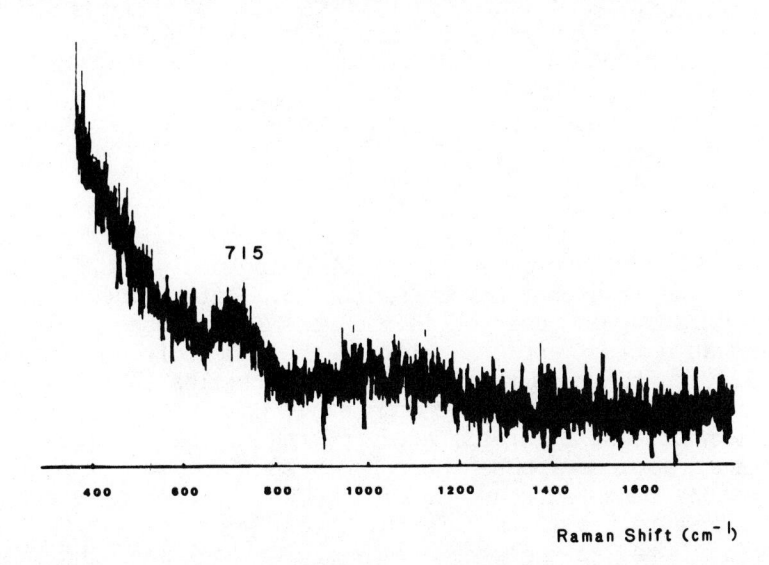

Fig.5 Raman spectrum of a 700Å film of the native oxide of $Hg_{0.7}Cd_{0.3}Te$.

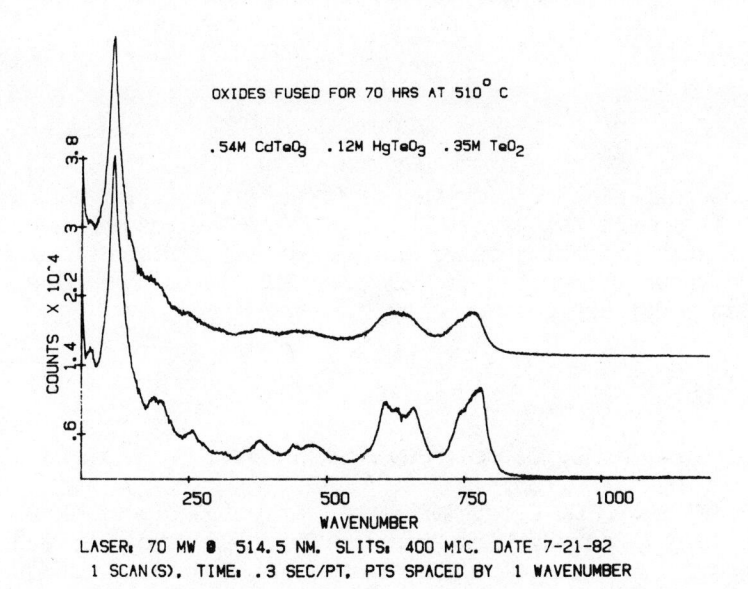

Fig.6 Raman spectra of two microparticles of sintered oxides.

DISCUSSION

The spectrum generated from the native oxide film certainly does not match that of any of the crystalline references oxides. Based on published Raman data,[4] it might contain substantial glassy TeO_2. The spectra of $CdTeO_3$ and $CdTeO_4$ also have bands which are not too different from those of the oxide. XPS and Auger spectrocopy eliminate CdO as a possible component of the native oxide[5]; this is the one possibility that is Raman inactive. However, the elemental techniques cannot identify more complicated phases of the oxides containing cadmium.

The sintered oxide sample was prepared as a first attempt to model the native oxide film. The composition of the film determinded by XPS and Auger was used in choosing the ratio of oxides that were mixed together before sintering. After sintering, some metallic mercury was observed in the vessel so these results must be viewed as a first approximation to the model of the native oxide. Variability in the spectra of different particles further show that formation of the glass was not complete. However, it is not difficult to expect that a careful study of the sintering conditions will produce a glassy phase whose Raman spectrum will contain a broad feature centered near 700 cm^{-1}. The failure to detect the stronger feature at 110 cm^{-1} in the spectrum of the native oxide can be attributed to the presence of a high Rayleigh wing when examing the Hg Cd Te platelet which is highly reflective; - ie, the ratio of the intensity of the Rayleigh wing at 100 cm^{-1} to the intensity of the Raman bands from a transparent 700Å film will be very high.

CONCLUSIONS

These experiments provide the first evidence for a glassy oxide film in the native oxide of HgCdTe. The results are consistent with the oxidation state assignments of XPS and Auger and with the known properties of TeO_2 and CdO/HgO to be a glass former and glass modifiers respectively.

REFERENCES

1. A. S. Pine and G. Dresselhaus, "Raman scattering in paratellurite, TeO ," Phys. Rev. B5: 4087-4093, 1972
2. R. E. Kirk and D. F. Othmer, Eds., Encyclopedia of Chemical Technology, Easton, Mack Printing Co., 1952, vol. 8, 888
3. S. Neov, I. Gerassimora, K. Krezhov, B. Sydzhimov, and V. Kozkukhavov, Pys. Status Solidi A47: 743, 1977

4. V. P. Cheremisinov and V. P. Zlomanov, "Vibrational spectra and the structure of cellurium dioxide in the crystalline and glassy states," 1961
5. G. D. Davis, T. S. Sun, S. P. Buchner and N. E. Byer, "Anodic oxide composition and Hg depletion at the oxide-semiconductor interface of $Hg_{1-x}Cd_xTe$," J. Vac. Sci. Technol.

THE RAMAN SPECTRA OF POTASSIUM BOROGERMANATE GLASSES

I. N. Chakraborty and R. A. Condrate, Sr.

New York State College of Ceramics
Alfred University
Alfred, N.Y.

INTRODUCTION

Properties such as density and refractive index are critical properties which must be optimized for glass materials that can be used in wave-guide systems.[1] Currently, some commercial groups are using or suggesting borogermanate or borogermanosilicate glasses for wave-guide applications because of their physical and chemical properties[2]. The above-mentioned properties intimately depend upon the structure of the glass. In this paper, Raman spectroscopy will be used to investigate the structure of potassium borogermanate glasses. This investigation will involve empirical comparison of the vibrational spectra of these glasses on a structural basis with those of related borate, germanate and borosilicate glasses and crystals.

EXPERIMENTAL PROCEDURE

Glasses were prepared from mixtures of ultrapure grade K_2CO_3, GeO_2 and H_3BO_3. H_3BO_3 was melted in a platinum crucible for an hour and quenched into a glass, which was used as a source of B_2O_3. The required amounts of starting materials were melted in a platinum crucible for 45 minutes in a globar furnace. The melting temperatures of the glasses were in the 900–1200°C range with the exact temperatures depending upon their compositions. The glasses were annealed in the 350–450°C range with the exact temperatures depending upon the melting temperatures, and were cooled slowly to room temperature. Six different series of glasses were prepared during this investigation. In all of the series, one of the components was held at a constant mole % concentration while the others were varied on a five mole % basis. All glasses were stored in desiccators over P_2O_5 after annealing in order to minimize moisture attack by the atmosphere.

Raman spectra were excited for the glass samples using a Spex model 1401 double spectrometer with a CRL model 54 argon ion laser. The glass samples were polished so that they had two parallel flat sides and one perpendicular flat side. The direction of the entering incident laser beam was perpendicular to the two parallel sides. The samples were positioned such that the Raman spectra were measured perpendicular to the third flat side. For the measurement of Raman spectra of hygroscopic glasses, the glasses were placed inside a cell that was filled with an index matching liquid (IML) prepared from a mixture of IR spectral grade paraffin oil and monobromonapthalene which had the same index of refraction as the glass. The IML was used to prevent attack of the glass by H_2O and CO_2 from the atmosphere. The spectra were measured in the Stokes region of the blue laser line (4880 A). The instrument was calibrated using indene with a precision of $\pm$ 3cm^{-1}.

RESULTS AND DISCUSSION

Most of the Raman and infrared spectra for the investigated potassium borogermanate glasses appear to involve the superimposition of bands found in the spectra of alkali borate and germanate glasses. The major difference between the spectra of the binary K_2O-containing systems and the ternary system is that the relative intensities of bands belonging to related boron-oxygen and germanium-oxygen groups are not the same for glass compositions having the same K_2O/B_2O_3 or K_2O/GeO_2 molar ratios. This result indicates that even though the major structural groups of the binary systems are also present for the ternary system, the relative concentrations of these groups are different for the latter system. However, band assignments can be made for the structural groups of the ternary system based on the assignments made earlier for related borate, germanate and borosilicate glasses and crystals. Table I lists the band locations for several pertinent Raman modes of various borate and germanate groups.

Potassium Borogermanate Glasses Containing 85% B_2O_3

The Raman spectra of $XK_2O \cdot (15-X)GeO_2 \cdot 85 B_2O_3$ glasses are illustrated in Figure 1. The strongest band for the 15 $GeO_2 \cdot 85 B_2O_3$ glass appears at 808cm^{-1} where a band is usually found for borate glasses and crystals containing boroxol rings. This band can be assigned to the breathing vibration of the boroxol ring possessing A_1'-symmetry[3]. A broad band is centered at 456cm^{-1}. A ring angle bending mode with E' - symmetry appears at 470cm^{-1} for vitreous B_2O_3. The shift of this band to lower wavenumbers occurring upon replacement of K_2O by

Table I. Useful Band Locations of Raman Modes of Various Borate Groups and Germate Units

Band Location (cm^{-1})	Origin
603	GeO$_6$ octahedral units
635	Metaborate groups
772	Six-membered borate ring containing one or two BO$_4$-units
780	Metagermanate groups
808	Six-membered borate ring
870	Digermanate groups

GeO$_2$ can be possibly explained by the addition of GeO$_4$-units into the glass network. Vitreous GeO$_2$ also shows a strong Raman band at 412cm^{-1} which has been assigned to a rocking mode with A_1' - symmetry[6]. The band with medium intensity at 720cm^{-1} can be assigned to the out-of-plane ring mode with A_2'' - symmetry.

The replacement of GeO$_2$ by K$_2$O in this series leads to a decrease in intensity of the band at 808cm^{-1} with a simultaneous increase for a new band at 772cm^{-1}. Such changes were observed earlier for alkali borate and borosilicate glasses which are rich in B$_2$O$_3$[3]. The similarity in the relative intensities of the two above-mentioned bands for the borogermanate glasses to those for binary K$_2$O-B$_2$O$_3$ glasses with equivalent K$_2$O concentrations suggests that K$_2$O has been mainly used in this series for changing the coordination of boron atoms from three to four. Similar to binary K$_2$O-B$_2$O$_3$ glasses with K/B-ratios less than ¼, K$_2$O has mainly converted boroxol rings to pentaborate and triborate groups.

Potassium Borogermanate Glasses Containing 65% B$_2$O$_3$

The Raman spectra of the XK$_2$O·(35-X) GeO$_2$· 65 B$_2$O$_3$ glasses are illustrated in Figures 2A and 2B. The strongest Raman bands for 35 GeO$_2$·65 B$_2$O$_3$ appear at 808 and 465 cm^{-1}, corresponding to the bands mentioned earlier for 15 GeO$_2$·85 B$_2$O$_3$. Again, replacement of GeO$_2$ by K$_2$O leads to a decreased intensity of the band at 808cm^{-1} with a simultaneous increase of the band at 772cm^{-1}. At 20% K$_2$O (K/B - ratio = $\frac{1}{3.25}$), the band at 808cm^{-1} completely disappears, indicating the elimination of boroxol rings in the glass network. For the binary K$_2$O·B$_2$O$_3$ system, this band disappears completely at lower K/B - ratios (¼). Thus, we can conclude indirectly that

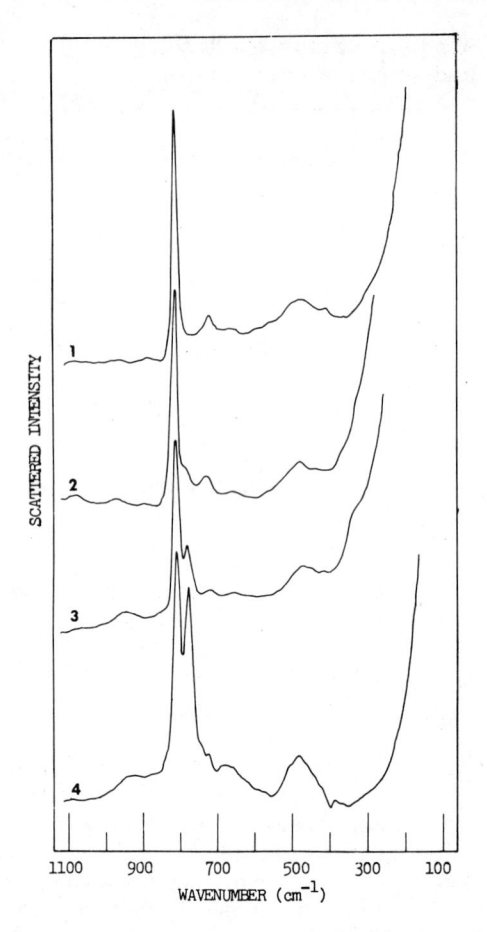

Fig. 1. The Raman spectra of $xK_2O \cdot (15-x)$ $GeO_2 \cdot 85$ B_2O_3 glasses.
1) $x = 0$; 2) $x = 5$; 3) $x = 10$; 4) $x = 15$.

although K_2O is converting boroxol rings to triborate and pentaborate groups at these concentrations, it is also changing the coordination of germanium atoms from four to six. The ν_s for GeO_6-groups is not observed in the 603-645cm^{-1} region due to overlapping of this band with the broad band centered at 495cm^{-1}. At 25 and 30% K_2O, a new band appears at ca 870cm^{-1} which is characteristic for digermanate groups with one non-bridging oxygen atom[5]. Apparently at these higher K/B ratios, alkali oxides are preferentially used in the formation of non-bridging oxygen atoms on GeO_4-tetrahedra rather than in the formation of diborate groups.

The Raman spectral results for the two above-mentioned glass series are consistent with structural predictions made earlier by Riebling et al[4] with partial molar volume data for sodium borogermanate glasses containing 65% B_2O_3. K_2O is mainly used to convert

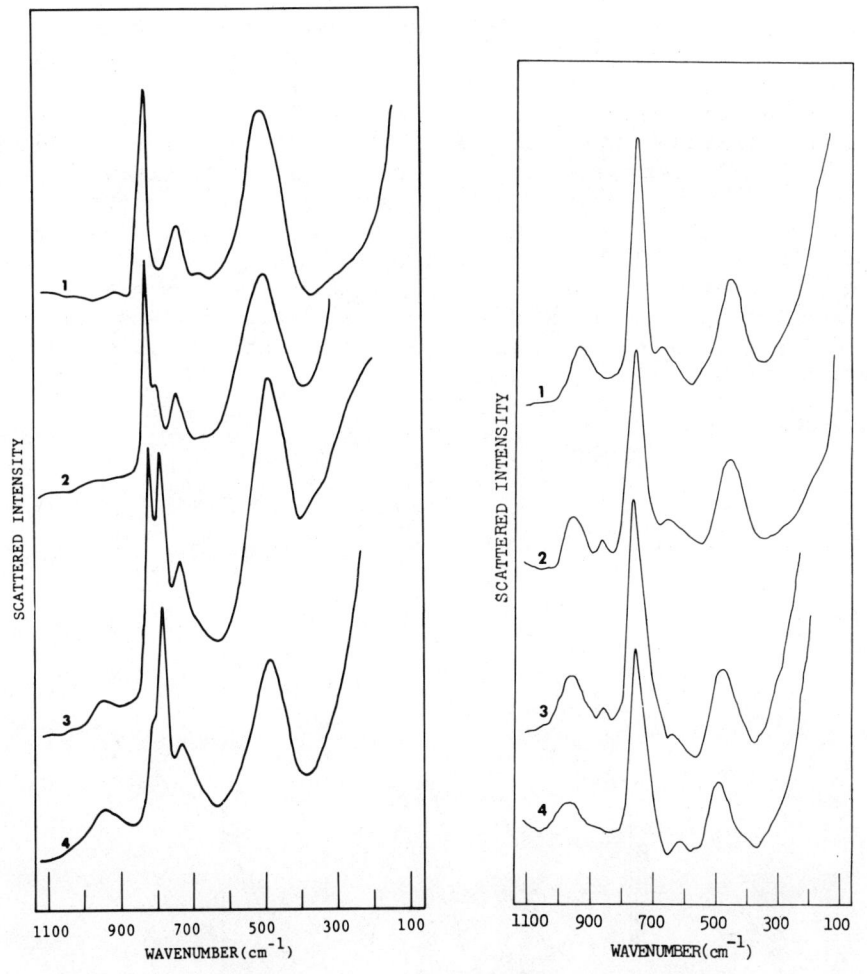

Fig. 2A. The Raman spectra of $xK_2O \cdot (35-x) GeO_2 \cdot 65B_2O_3$ glasses.
1) x = 0; 2) x = 5; 3) x = 10; 4) x = 15.
Fig. 2B. The Raman spectra of $xK_2O \cdot (35-x) GeO_2 \cdot 65B_2O_3$ glasses.
1) x = 20; 2) x = 25; 3) x = 30; 4) x = 35.

the coordination of boron from three to four. However, K_2O is also being used to convert GeO_4-units to GeO_6-units along with creating non-bridging oxygen atoms on GeO_4-units at higher K_2O concentrations. Infrared data is also consistent with this structural interpretation. An infrared band appears in the 900-1100 cm^{-1} region with K_2O addition, indicating the growth of BO_4-units.

Potassium Borogermanate Glasses Containing 80% GeO_2

The Raman spectra of $xK_2O \cdot (20-x) B_2O_3 \cdot 80 GeO_2$ glasses are illustrated in Figure 3. The broad Raman band at 545 cm^{-1} for $20K_2O \cdot 80GeO_2$

can be assigned to ν_s (O-Ge-O)-mode and to deformation modes.[5] The strong band at ca 870cm^{-1} which appears at 10% K$_2$O indicates the presence of digermanate groups with one non-bridging oxygen atom.[6] A band also appears at 790cm^{-1} at 15% K$_2$O indicating the presence of metagermanate chains.[5,6] A rapid increase in intensities for the bands associated with digermanate and metagermanate groups is observed with substitution of K$_2$O for B$_2$O$_3$. In addition to these bands, a shoulder appears at ca 605cm^{-1} which can be assigned to the ν_s-mode of GeO$_6$-units.[5] Down to 10% K$_2$O, this band persists along with the digermanate band, indicating the presence also of GeO$_6$-units. A band appears at ca 772cm^{-1} for the glass containing 5% K$_2$O which can be assigned to boroxol groups containing BO$_4$-tetrahedra.

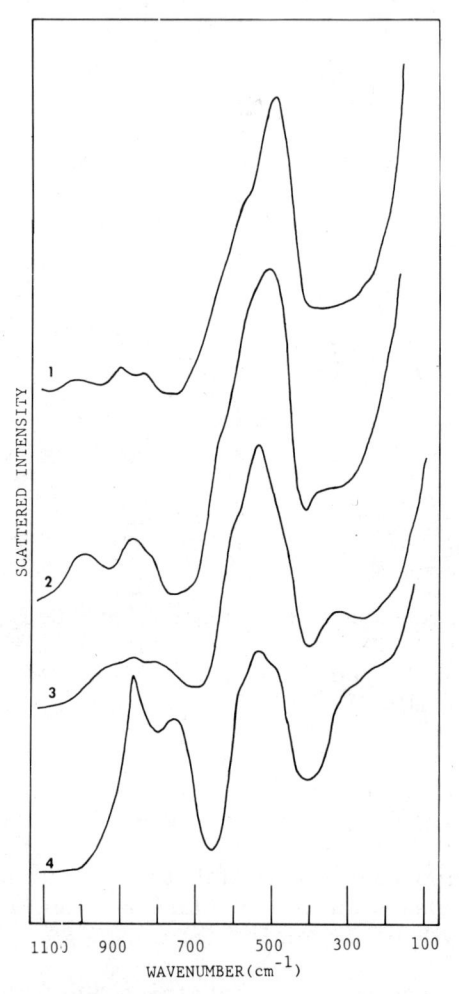

Fig. 3. The Raman spectra of xK$_2$O·20-x)B$_2$O$_3$·80GeO$_2$ glasses.
1) x = 5; 2) x = 10; 3) x = 15; 4) x = 20.

The infrared data is consistent with the Raman data. The asymmetric Ge-O-Ge stretching mode appears at 780cm^{-1} for 20K$_2$O· 80GeO$_2$. The shift of this infrared band to higher wavenumbers is consistent with a change in coordination of germanium.[7] Also, a band is noted at ca 1000cm^{-1} which indicates the presence of BO$_4$- units in the glass network.

Potassium Borogermanate Glasses Containing 65% GeO$_2$
___

The Raman spectra of xK$_2$O·(35-x)B$_2$O$_3$·65GeO$_2$ glasses are illustrated in Figures 4A and 4B. The spectra of 35B$_2$O$_3$·65GeO$_2$ is a composite of that of vitreous B$_2$O$_3$ and GeO$_2$. Again, the band at 808cm^{-1} decreases its intensity while the band at 772 cm^{-1} increases its intensity upon K$_2$O addition. At 10% K$_2$O, the former band disappears, indicating the disappearance of boroxol rings in the network. At this glass composition, a shoulder appears at 602cm^{-1} which can be assigned to ν_s (GeO$_6$). This band increases its intensity up to 25% K$_2$O. With further increase in alkali content, the band appears as a strong shoulder of the deformation mode at 540cm^{-1}, indicating conversion of 6-coordinated germanium to the 4-coordinated state. At 20% K$_2$O, a band appears at ca 870cm^{-1} which has been assigned earlier to digermanate groups which contain one non-bridging oxygen atom. The band at 770cm^{-1} appears as a shoulder on the strong digermanate band. Further replacement of B$_2$O$_3$ by K$_2$O leads to an increase in the intensity of the band at ca 870cm^{-1}. At 35% K$_2$O, a band appears at 780cm^{-1} which has been assigned to metagermanate chains. The borate-related bands in the 810-770cm^{-1} region are now overlapped by the metagermanate band. However, the infrared data indicates that boron atoms are 4-coordinated for these compositions.

Some interesting differences can be noted in the Raman spectra of xK$_2$O·(35-x)B$_2$O$_3$·65SiO$_2$[8] and xK$_2$O·(35-x)B$_2$O$_3$·65GeO$_2$ glasses. By comparing the relative intensities of the bands at 772 and 808cm^{-1} for these two glass series at 5%, one can conclude that the concentration of BO$_4$-tetrahedra is higher in the germanate system. At 10% K$_2$O (K/B - ratio = 1/2.5), the band at 808cm^{-1} disappears for the germanate system, indicating the absence of boroxol rings. However, this band is still present for the silicate system. At still higher K$_2$O concentrations, a band appears for the silicate series at 635cm^{-1} which is characteristic for metaborate chains. This band does not appear for the germanate series. Therefore, one can conclude that metaborate rings are comparatively unstable for the borogermanate system, and that non-bridging oxygen atoms appear at higher alkali oxide concentrations on digermanate and metagermanate groups rather than on metaborate groups. For the borosilicate system, non-bridging oxygen atoms are formed at higher alkali oxide concentrations simultaneously on both borate and silicate groups.

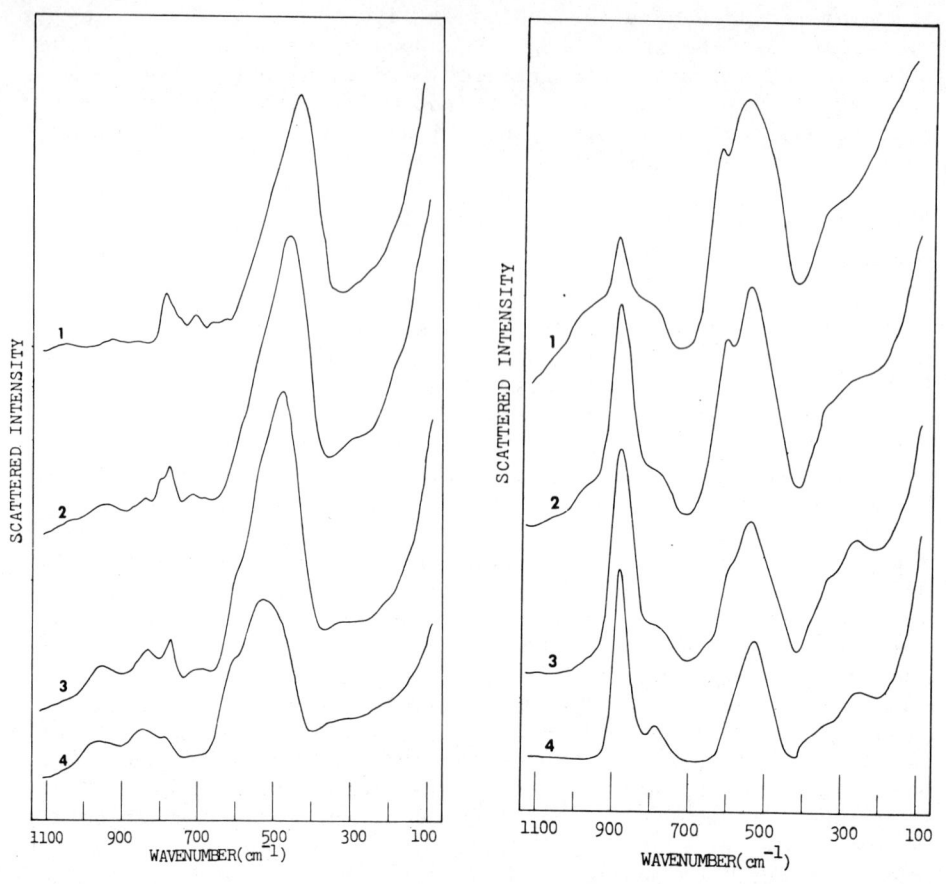

Fig. 4A. The Raman spectra of $xK_2O \cdot (35-x)B_2O_3 \cdot 65GeO_2$ glasses.
1) $x = 0$; 2) $x = 5$; 3) $x = 10$; 4) $x = 15$.
Fig. 4B. The Raman spectra of $xK_2O \cdot (35-x)B_2O_3 \cdot 65GeO_2$ glasses.
1) $x = 20$; 2) $x = 25$; 3) $x = 30$; 4) $x = 35$.

Structural interpretation of the infrared spectra is consistent with Raman data. At 20% K_2O, the Ge-O-Ge stretching mode at 810cm^{-1} splits into two bands. This splitting indicates the presence of non-bridging oxygen atoms in the GeO_4-groups. It continues up to 35% K_2O. The band in the 900-1000cm^{-1} persists up to 30% K_2O, indicating 4-coordinated boron atoms in the network.

Potassium Borogermanate Glasses Containing 25% K_2O

The Raman spectra of $25K_2O \cdot xB_2O_3 \cdot (75-x)GeO_2$ glasses are illustrated in Figures 5A, 5B and 5C. The Raman spectra of $25K_2O \cdot 75B_2O_3$

possess a strong band at ca 770cm^{-1} which is due to boroxol rings containing one or more BO$_4$-tetrahedra. Earlier studies have established that in the 20-35% K$_2$O region, binary K$_2$O-B$_2$O$_3$ glasses contain diborate groups along with triborate and pentaborate groups. This glass also possesses broad bands at 955 and 480cm^{-1}. Interesting spectral changes occur as B$_2$O$_3$ is replaced by GeO$_2$. A band at 690cm^{-1} becomes prominent for glasses containing 65% B$_2$O$_3$, and can be assigned to a wagging mode of the boroxol rings possessing E"-symmetry. At 60% B$_2$O$_3$, a shoulder appears at 873cm^{-1} which can be assigned to digermanate groups while at 55% B$_2$O$_3$, another band appears at 604cm^{-1} indicating GeO$_6$-units. In the case of glasses containing $\leqslant$ 50% B$_2$O$_3$(K/B - ratio $\geqslant$ ½), the absence of a band at 635cm^{-1} indicates the absence of metaborate groups in the network. For glasses in the binary K$_2$O-B$_2$O$_3$ system, for which the K/B - ratio is 1/2, borate is present as diborate groups possessing two BO$_4$-tetrahedra per ring. In the ternary system, we see that at this K/B - ratio, digermanate groups and GeO$_6$-octahedra are present, indicating that part of the alkali oxide has been utilized by the germanate groups. This result suggests that the predominating borate groups will be those which exist at a lower K/B - ratio (i.e. < 1/2) for the binary system. Thus, the major borate groups at this composition are triborate and pentaborates, and the concentration of borate groups containing two BO$_4$-tetrahedra (namely, diborate groups) is low. The above-mentioned trend continues up to 20% B$_2$O$_3$. Above that concentration, the borate band at 690cm^{-1} disappears. At 10% B$_2$O$_3$, the band at ca 770cm^{-1} appears as a shoulder of the strong digermanate band. At 5% and 0% B$_2$O$_3$, a band becomes prominent at 782cm^{-1} which can be assigned to metagermanate chains. Due to the overlapping of this stronger band at 782cm^{-1} with the borate band at ca 770cm^{-1}, no conclusion can be drawn from Raman spectra regarding the borates at this concentration. But on the basis of the infrared data it becomes clear that at lower B$_2$O$_3$-concentrations, boron atoms are mainly present in the 4-coordinated state. From these results, one can conclude that the metaborate and diborate groups occur at relatively small amounts as compared to the metagermanate and digermanate groups. The glass networks are mainly composed of pentaborate, triborate, digermanate groups, GeO$_6$-octahedra and metagermanate chains. The structural results reported by Riebling et al[4] for borogermanate glasses containing 23% Na$_2$O using partial molar volume data are consistent with the spectral results of this series. They predicted higher concentrations of GeO$_6$-octahedra along with BO$_4$-tetrahedra.

Potassium Borogermanate Glasses Containing 35% K$_2$O

The Raman spectra of 35 K$_2$O·xB$_2$O$_3$·(65-x)GeO$_2$ glasses are illustrated in Figures 6A and 6B. The spectrum of 35K$_2$O·65B$_2$O$_3$ possesses typical borate bands at 490cm^{-1} and 945cm^{-1} along with sharp bands at ca 770cm^{-1} and 635cm^{-1}. The band at ca 770cm^{-1} is assigned to

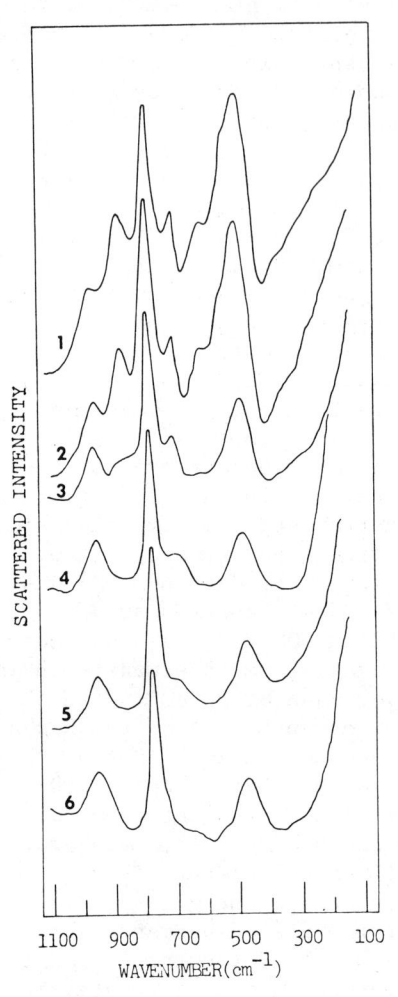

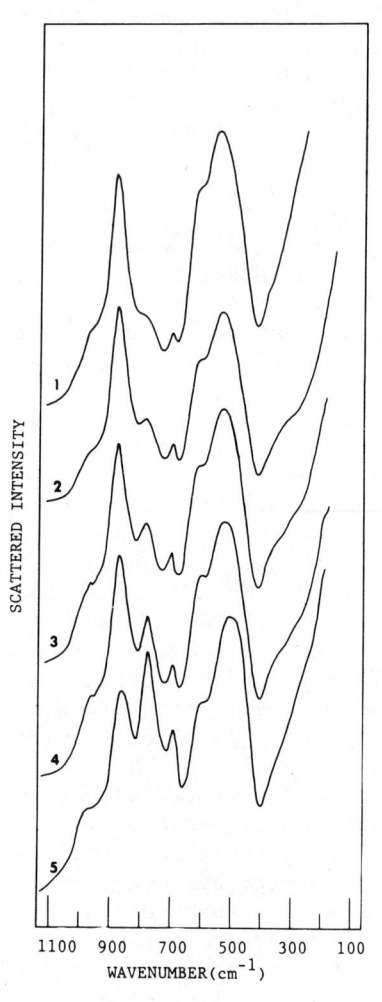

Fig. 5A. The Raman spectra of $25K_2O \cdot xB_2O_3 \cdot (75-x)GeO_2$ glasses.
1) x = 50; 2) x = 55; 3) x = 60; 4) x = 65;
5) x = 70; 6) x = 75.

Fig. 5B. The Raman spectra of $25K_2O \cdot xB_2O_3 \cdot (75-x)GeO_2$ glasses.
1) x = 25; 2) x = 30; 3) x = 35; 4) x = 40
5) x = 45.

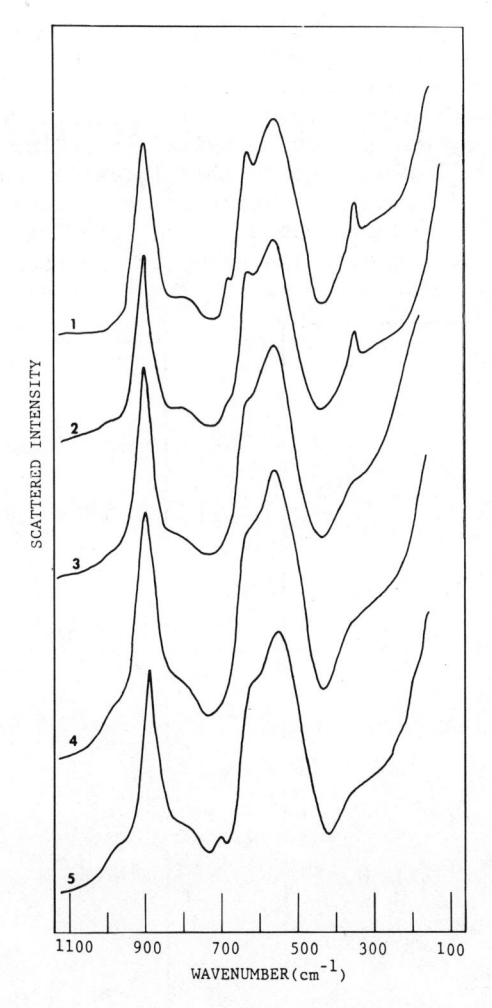

Fig. 5C. The Raman spectra of $25K_2O \cdot xB_2O_3 \cdot (75-x)GeO_2$ glasses.
1) x = 0; 2) x = 5; 3) x = 10; 4) x = 15; 5) x = 20.

boroxol groups containing one or two BO_4-tetrahedra, while the band at 635cm^{-1} has been assigned to the metaborate groups. Earlier investigators have already established that at 35% K_2O for the potassium borate system[9], most of the boron oxide is present in the glass network as diborate groups. This possibility is indicated here by a strong band at ca 770cm^{-1}. With the replacement of B_2O_3 by GeO_2, a strong band appears at 870cm^{-1} which can be ascribed to digermanate groups possessing non-bridging oxygen atoms. Also, the ratio of the relative intensities of the bands at ca 635cm^{-1} and 770cm^{-1} decreases. At 45% B_2O_3(the K/B - ratio=1/1.3) the band at ca 635cm^{-1} disappears completely, indicating the elimination of metaborate groups in the network. In the binary alkali borate system at a K/B-ratio of 1/1.3, most of the boron oxide will be in the metaborate state[8]. This result indicates the competition between metaborate and germanium oxide groups for K_2O. Because of this competition, boron oxide in this investigated system are mainly present as diborate, triborate and pentaborate groups. Formation of metagermanate cannot be detected at low germanate concentrations due to the overlapping of its characteristic band with the strong borate band at 770cm^{-1}. The replacement of B_2O_3 by GeO_2

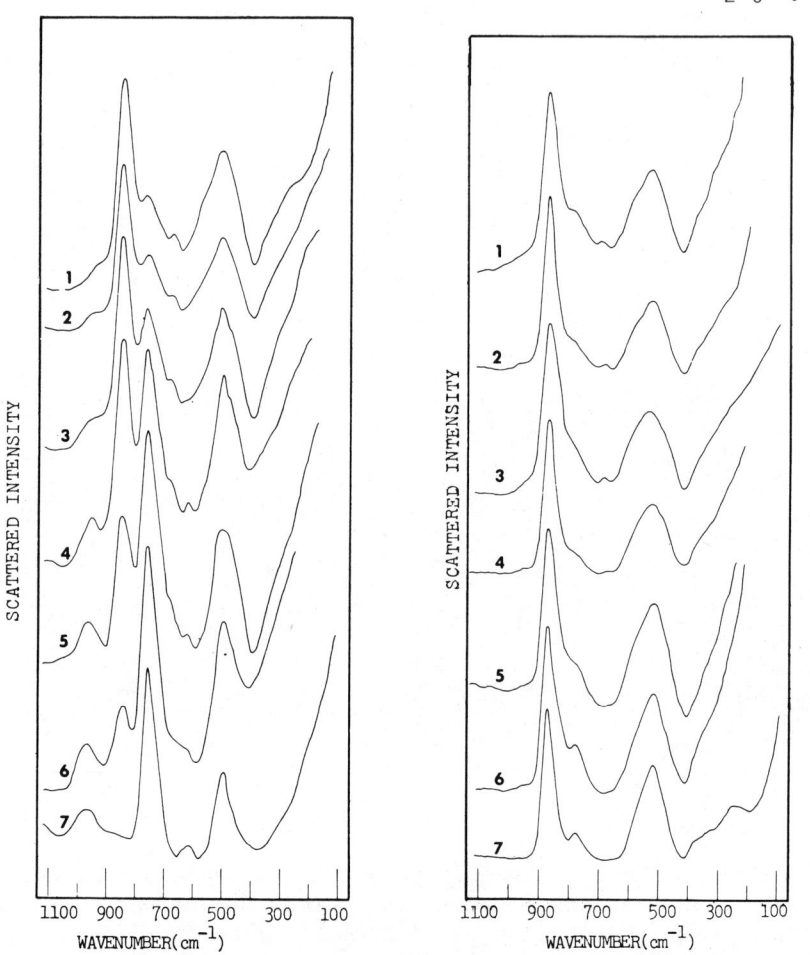

Fig. 6A. The Raman spectra of $35K_2O \cdot xB_2O_3 \cdot (65-x)GeO_2$ glasses.
1) x = 35; 2) x = 40; 3) x = 45; 4) x = 50;
5) x = 65; 6) x = 60; 7) x = 65.

Fig. 6B. The Raman spectra of $35K_2O \cdot xB_2O_3 \cdot (65-x)GeO_2$ glasses.
1) x = 30; 2) x = 25; 3) x = 20; 4) x = 15;
5) x = 10; 6) x = 5; 7) x = 0.

leads to a decrease in the intensity of this band so that at 25% B_2O_3, it appears as a shoulder of the digermanate band at 870cm^{-1}. At 10% B_2O_3, a band appears at 780cm^{-1}, indicating the presence of metagermanate groups in the network. From the analysis of the infrared spectra, we can conclude that the boron atoms are mainly present in the 4-coordinated state. The absence of a Raman band in the 603cm^{-1} - 645cm^{-1} region indicates the absence of the GeO_6-octahedra for the whole series. It appears that the number of non-bridging oxygen atoms are reduced in the network for glasses in this system as compared to those of the binary K_2O-B_2O_3 and K_2O-GeO_2 glasses containing the same alkali level (namely, 35% K_2O). This conclusion can be drawn from the absence of bands for meta-germanate and metaborate groups in the moderate to high B_2O_3- and GeO_2- concentration levels. The structural predictions made by Riebling et al for glasses containing 33.5% Na_2O are consistent with our results. They predicted high BO_4-tetrahedra concentrations along with high GeO_4-tetrahedra concentrations in the entire range.

CONCLUSIONS

Useful structural information can be obtained for glass compositions in the potassium borogermanate system using Raman spectroscopy. The results are consistent with those predicted earlier from partial molar volume data. The structural conclusion for various compositions are listed in Table II.

REFERENCES

1. R. D. Mauer, Glass Fibers for Optical Communications, Proc. Inst. Electr. Eng. Part B, 61:452 (1973).
2. P. W. Black, J. Irven, K. Byron, I. S. Few, and R. Worthington, Measurements on Waveguide Properties of GeO_2 SiO_2 Cored Optical Fibers, Electron Lett., 10:239 (1974).
3. W. L. Konijnendijk and J. M. Stevels, Structure of Borate and Borosilicate Glasses by Raman Spectroscopy, in: "Borate Glasses: Structure, Properties, Applications," edited by L. D. Pye, V. D. Fréchette, and N. J. Kreidl, Plenum Press, New York (1978).
4. E. F. Riebling, P. E. Baszyk, and D. W. Smith, Structure of Glasses and Melts in the $Na_2O \cdot GeO_2 \cdot B_2O_3$ System, J. Amer. Ceram. Soc., 50:641 (1947).
5. T. Furukawa and W. B. White, Raman Spectroscopic Investigation of the Structure and Crystallization of Binary Alkali Germanate Glasses, J. Mat. Sci., 15:1648 (1980).

Table II. Observed Major Borate and Germanate Groups Present in Potassium Borogermanate Glasses

Glasses		Borate Groups	Germanate Groups
85% B_2O_3		Pentaborate, Triborate and Boroxol groups.	Vitreous GeO_2 groups with 4 bridging oxygen atoms.
65% B_2O_3	K/B < 1/3.25	Pentaborate, Triborate and Boroxol groups.	Octahedral and vitreous GeO_2 groups.
	K/B > 1/3.25	Pentaborate, Triborate and Boroxol groups.	Digermanate, Octahedral and Tetrahedral groups.
80% GeO_2		Boroxol, Pentaborate and Triborate groups.	Octahedral and Tetrahedral groups, Metagermanate and Digermanate groups.
65% GeO_2	K/B < 1/2.5	Boroxol, Pentaborate and Triborate groups.	Tetrahedral and Octahedral groups.
	K/B > 1/2.5	Pentaborate and Triborate groups.	Octahedral, Digermanate and Metagermanate groups.
25% K_2O	K/B < 1/2	Mainly Pentaborate and Triborate groups along with a small amount of Diborate groups.	Tetrahedral and Octahedral groups.
	K/B > 1/2	Pentaborate and Triborate groups with a small fraction of Diborate groups.	Octahedral, Digermanate and Metagermanate groups.
35% K_2O	K/B < 1/1.3	Metaborate and Diborate groups.	Digermanate groups.
	K/B > 1/1.3	Mainly Diborate groups along with Triborate and Pentaborate groups.	Digermanate and Metagermanate groups.

6. H. Verweij and J. H. J. M. Buster, The Structure of Lithium, Sodium and Potassium Germanate Glasses Studied by Raman Scattering, J. Non-Cryst. Solids, 34:81 (1979).
7. M. K. Murthy and E. M. Kirby, Infrared Spectra of Alkali Germanate Glasses, Phys. Chem. Glasses, 5:144 (1964).
8. W. L. Konijnendijk, The Structure of Borosilicate Glasses, Philips Res. Rep. Suppl., No. 1: 135 (1975).
9. T. W. Bril, Raman Spectroscopy of Crystalline and Vitreous Borates, Philips Res. Rep. Suppl., No. 2: 79 (1976).

CHARACTERIZATION OF THE STRUCTURE AND NONSTOICHIOMETRY OF CaO-NiO SOLID SOLUTIONS

B. C. Cornilsen, E. F. Funkenbusch, C. P. Clarke,
P. Singh and V. Lorprayoon

Department of Chemistry and Chemical Engineering
Michigan Technological University
Houghton, Michigan 49931

INTRODUCTION

To understand structure-property relationships it is necessary to characterize the cation and anion nonstoichiometry and the structure of a system, for example, disorder in an antiferromagnetic material. For a ternary such as Ca:NiO the cation composition, i.e. Ca/Ni ratio, and oxygen nonstoichiometry must be defined. The presence of undesirable second phases should not be ignored; knowledge of such is critical. These phase equilibria are controlled by the temperature, cation composition, and oxygen partial pressure.

The latter variables define the point defect structure as well. High oxygen partial pressures introduce nickel vacancies, V_{Ni}, in metal deficit NiO_{1+x}. Since NiO is antiferromagnetic, the diamagnetic calcium ions and nickel vacancies act as magnetic impurities, reducing the antiferromagnetic ordering of the oxide. Disorder of antiferromagnetic NiO introduced by the latter or by higher temperatures can also be characterized. Dopants or impurities, including isovalent calcium, influence the point defect equilibria as well as the order. Our purpose has been to thoroughly define the structure and nonstoichiometry.

Raman spectroscopy, differential scanning calorimetry (DSC), X-ray fluorescence (XRF), and DC electrical conductivity have been used to characterize the above variables. In particular we have found that Raman spectra can provide extensive structural information and be used to monitor cation ratio (Ca/Ni), diamagnetic cation-induced disorder in NiO, second phase formation and oxygen nonstoichiometry. The latter was first suggested by the work of Dietz et al.[1] They observed a "defect-induced" band which appeared to vary in intensity for two samples of NiO with different, but undefined point defect contents. They did not attempt to quantify this observation.

EXPERIMENTAL

Sample preparation using the liquid mix technique (LMT) has been previously described.[2] This technique allows precise control of cation composition and the cations are homogeneously mixed at the atomic level. The cation composition was confirmed using XRF, and followed by the drop in Néel temperature, T_N, as measured with DSC.[2,3] The powders were hydrostatically pressed at 240 MPa and sintered into rods at 1550°C in air. Small pellets, 2 mm x 2 mm x 0.5 mm, cut from these rods have been equilibrated in controlled oxygen atmospheres at appropriate temperatures to define the oxygen nonstoichiometry and then quenched to room temperature in the same gas atmosphere. Each pellet was polished on a 15 μm diamond wheel. An oxygen partial pressure range between 1×10^{-7} atm and 1 atm, and temperatures between 1100 and 1400°C define oxygen excess nonstoichiometries, x, between 1×10^{-6} and 4×10^{-4} (for NiO_{1+x}).[4] CO_2/CO and O_2/Ar mixtures were used. NiO has a very low nonstoichiometry range, as compared with CoO, MnO, or FeO.[5] The nonstoichiometry of this polycrystalline NiO has been characterized using electrical conductivity measurements (DC). A four-probe method was used following standard procedures.[6]

Multiple scan Raman spectra were collected using a signal averager* and a four slit, double monochromator**, as described previously.[7] 10 to 15 multiple scans were collected with 5 cm^{-1} slits. Spectra were scanned of samples held at 12K. A closed cycle, liquid helium cryostat was used.[†] NiO is a poor scatterer because it strongly absorbs in the visible region of the spectrum. Signal averaging permits collection of quantifiable data despite this fact. Low temperature spectra display sharper bands and good resolution is attainable at low wavenumbers approaching the laser line.

---

* Model 570A, Tracor Northern, Inc., 2551 W. Beltline Highway, Middleton, WI 53562.
**Ramanor HG.2S, Instruments S. A., Inc., 173 Essex Ave., Metuchen, NJ 08840
† Model LTS-21-D70C refrigerator system and Model 21 cold head by Lake Shore Cryotronics, P. O. Box 29876, Columbus, Ohio 43229.

CaO-NiO SOLID SOLUTIONS 241

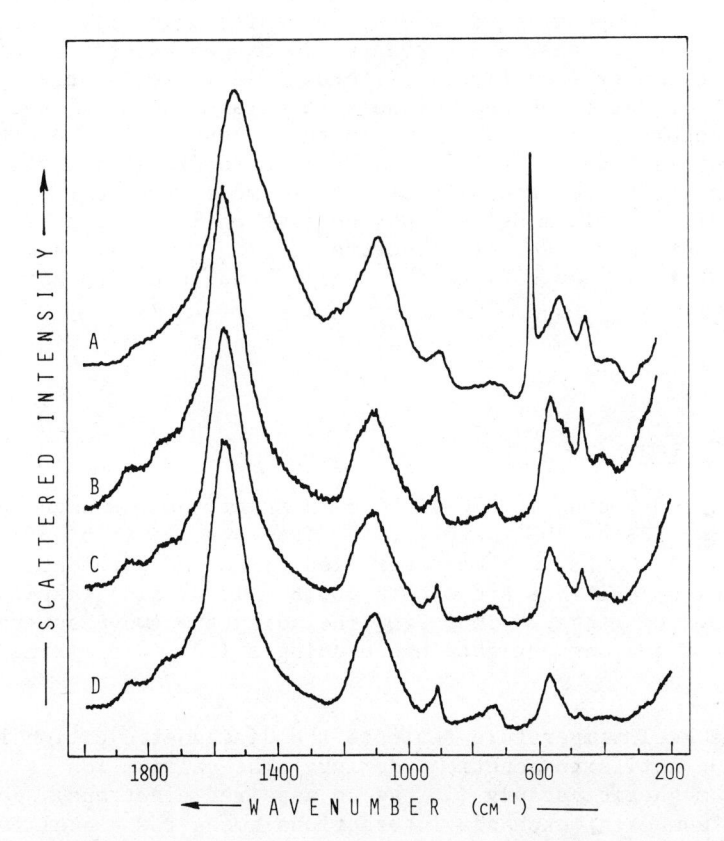

Fig. 1. Raman spectra of 2.5 mole % CaO-doped NiO (A) and NiO_{1+x} (B-D), scanned at 12 K. Oxygen partial pressures and temperatures at quench were (A) 1 atm., 1300°C; (B) 1 atm., 1400°C; (C) 0.21 atm., 1300°C; and (D) 1.12 x 10^{-7} atm., 1200°C.

RESULTS

The Raman spectra of stoichiometric NiO, defective NiO (NiO_{1+x}), and calcium-doped NiO are depicted in Figure 1. The spectra are sensitive to each of these composition variables. A two-magnon band at 1554 cm^{-1} shifts and broadens with dopant addition (Fig. 1a and b). Calcium doping induces a band at 610 cm^{-1} with a composition-dependent intensity (Fig 1a). The spectra are sensitive to the subtle changes in oxygen excess nonstoichiometry (see Figure 1b through 1d), represented by x. The most dramatic and quantifiable spectral change related to the nonstoichiometry is an increase in the intensity of the 560 cm^{-1}, longitudinal optical (LO) band. This intensity is quantified by ratioing it to the intensity of the 2LO mode (overtone of the longitudinal optic mode) at 1099 cm^{-1} which is constant and independent of the defect structure ($I_{LO}/I_{2LO} = R_I$). Also a more subtle shift in the 1554 cm^{-1} 2-magnon band intensity has been observed.

DISCUSSION

Cation Content

Calcium doping in NiO-rich solid solutions has been characterized using XRF and DSC methods.[2,3] The upper limit of precision attained in the LMT has been verified by XRF measurements (Figure 2). The compositions are within ±0.38 mole % (±1σ) over the entire system range, 0-100% CaO, including the two-phase region. Such precision was not obtainable using solid state mixing techniques.

The Néel temperature reflects the diamagnetic cation influence upon antiferromagnetic ordering. The calcium ion is effectively a magnetic vacancy (it has no unpaired electrons), decreasing the number of exchange interactions among the nickel ions. T_N is lowered because less thermal energy is required to totally disrupt the antiferromagnetic order and form the cubic phase. A linear dependence of T_N upon calcium concentration was observed in agreement with Ising model theory.[3,7] The phase transition enthalpy also decreased with calcium addition.[3]

Since Raman spectroscopy is sensitive to the calcium content as well, it can be used to characterize this content. Two-magnon Raman scattering (an overtone of a magnon which is a transition between the two spin states) is influenced by the dopant because the antiferromagnetic ordering is reduced.[7] A linear wavenumber dependence was observed, again in agreement with Ising model predictions. In addition the "calcium-induced" band (at 610 cm^{-1}) intensity is proportional to calcium concentration.

The Néel temperature, calcium-induced band intensity, or 2-magnon wavenumber can be used to monitor the calcium concentration in calcia-doped NiO, and used to define solvus limits as

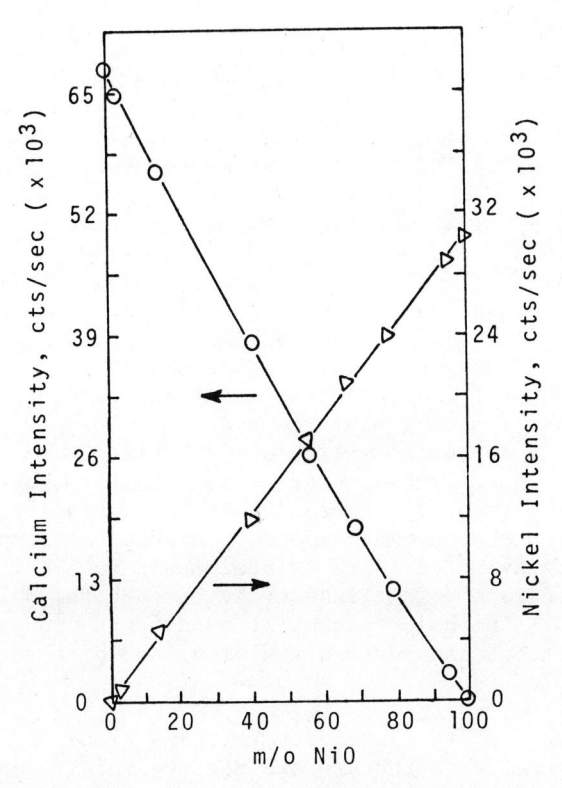

Fig. 2. X-ray fluorescence measurement of the calcium and nickel contents across the CaO-NiO system. The nonlinear term for each cation is proportional to the intensity of the other cation.

depicted in Figure 3. When a calcia-rich second phase precipitates, the calcium composition of the NiO-rich phase is fixed. The variable used to monitor the calcium composition of the nickel-rich phase, T_N, $\tilde{\nu}_{2M}$ or I_{Ca}/I_{2LO}, therefore becomes constant. Calcia-rich second phase precipitation can be monitored effectively with this method, i.e. by the indirect measurement of the loss of calcium in the NiO-rich phase. As little as 0.5 mole % CaO loss can be seen using this Raman technique.

Oxygen Nonstoichiometry

Samples with controlled oxygen nonstoichiometries have been prepared as previously described to produce oxygen excesses (x) between 1.5×10^{-5} and 4×10^{-4} for NiO_{1+x}. Figure 4 gives the results as a function of the p_{O_2} and temperature. This plot of $\log \Delta R_I$ vs $\log p_{O_2}$ demonstrates that at each temperature the ratio is proportional to the partial pressure of oxygen and therefore related to the nonstoichiometry or nickel vacancy content. The origin of this intensity change is point defect dependent. The slope of a plot of $\log \Delta R_I$ vs 1/T gives the heat of formation of a cation vacancy.[4,6] This treatment is analogous to the well known $\log \sigma$ vs 1/T plots. The slope at 1 atm oxygen partial pressure gives a formation energy of 27 kcal/mole in good agreement with 24 kcal/mole obtained from a thermogravimetric evaluation.[6] To quantify these changes the intensity ratio can be compared with nonstoichiometry data or electrical conductivity results. Thermogravimetric data is preferable to the latter because x is measured directly, not indirectly through the hole concentrations. The hole concentration depends upon the degree of vacancy ionization, and correlation of R_I and σ is therefore more complex.

To test whether the intensity ratio is proportional to the absolute point defect structure and not separately dependent upon temperature it is necessary to test correlation of R_I data with oxygen nonstoichiometry in NiO_{1+x}. The latter has been measured thermogravimetrically by Osburn and Vest.[4] Correlation with this literature data demonstrates that R_I and x are directly related (Figure 5). The error bar in Figure 5 depicts the reported uncertainty in the measurement of x. It should be emphasized that the maximum oxygen excess nonstoichiometry represented here is only 400 ppm. The slope of this curve represents the sensitivity of the Raman scattering to the metal vacancy concentration; a 1% change in relative intensity represents a 10 ppm change in excess oxygen. The precision of the Raman ratio is estimated to be within ±1% (±0.01 R_I). Therefore, the Raman technique may be more precise than thermogravimetric methods which are affected by buoyancy corrections, vaporization of samples at high temperatures, etc.[4]

CaO-NiO SOLID SOLUTIONS

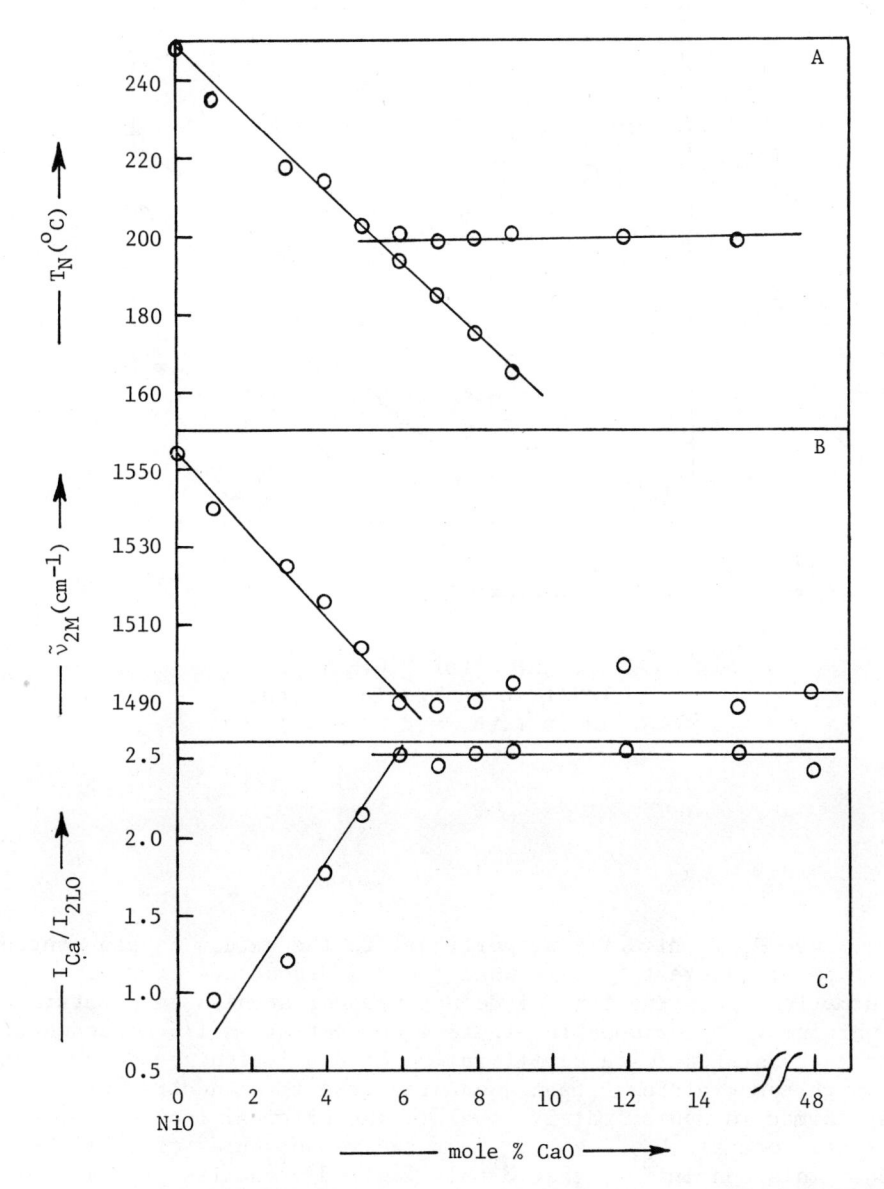

Fig. 3. The variation of Néel temperature (A), 2-magon wavenumber (B), and the intensity of the calcium-induced band (C) with respect to calcia content in Ca:NiO. Two-phase samples were quenched from 1300°C.

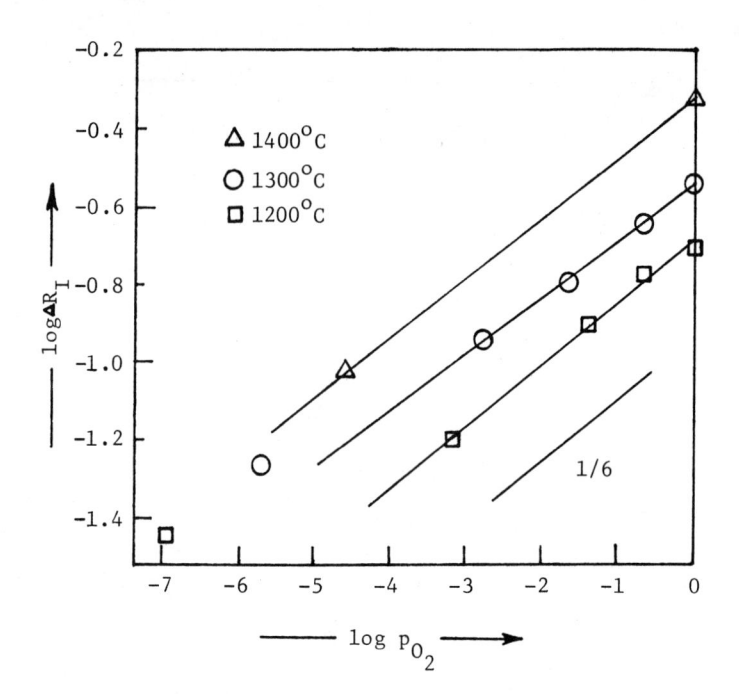

Fig. 4. Isotherms for NiO. ΔR_I is the relative change in intensity ratio ($\Delta R_I = R_I - 0.46$).

Since R_I is directly proportional to the excess oxygen content, x, it is of interest to aske what the P_{O_2} dependence of R_I is. Information as to the type of defect present at high temperature can be obtained. The slopes in Figure 4 demonstrate a 1/6 dependence. This is consistent with equilibration of doubly ionized defects at the high temperatures. Upon quenching, the total neutral vacancy content formed is approximately equal to the original ionized vacancy content, because $[V_{Ni}"] \gg [V_{Ni}]$ at the temperatures of equilibration. This conclusion, that doubly ionized vacancies predominate at these high temperatures (1200-1400°C), is in agreement with our electrical conductivity results and those in the literature.[4,6] The fact that Figure 5 does not display any change in slope suggests that the same electrical defect, $V_{Ni}"$, is present over this entire range of nonstoichiometry.

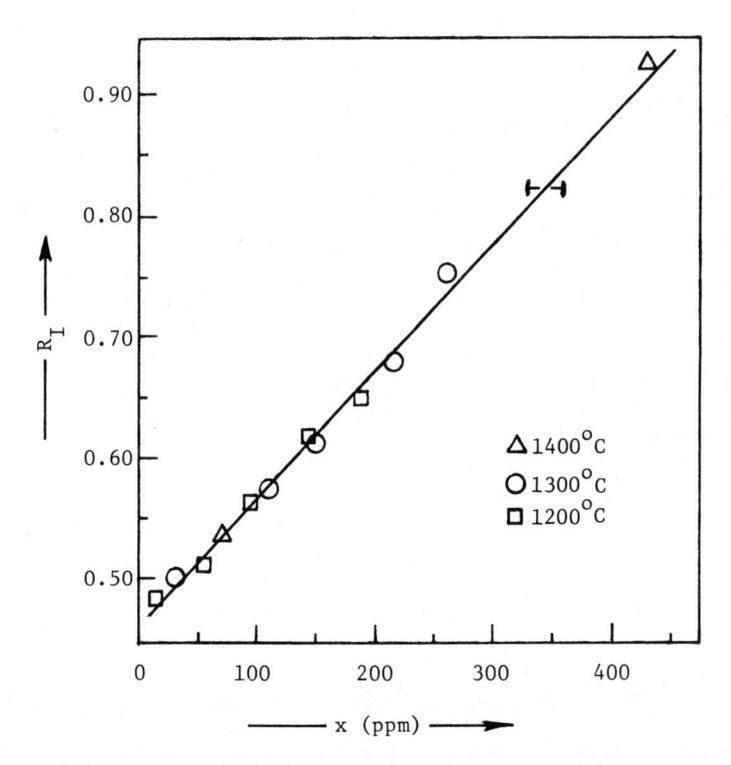

Fig. 5. Relation of the "defect-induced" Raman band intensity, R_I, and the nonstoichiometry, x in NiO_{1+x} (x values are taken from the thermogravimetric analysis of Osburn and Vest[4]).

CONCLUSIONS

We have shown that the point defect sensitive mode in the Raman spectrum of NiO is proportional to oxygen nonstoichiometry, x, over a wide range, to $x = 4 \times 10^{-4}$. Therefore this method can be used to monitor the oxygen nonstoichiometry in these carefully prepared samples. Samples and spectra must be thoroughly characterized so that some other variable does not lead to spurious results. The potential of Raman spectroscopy to measure the insitu point defect content of a given sample has extensive applications. This is not presently possible with electrical techniques. Oxygen nonstoichiometry is in general not monitored or measured directly, i.e. electrical techniques monitor charge carriers related to the defects, and most importantly electrical techniques are used to monitor the properties under controlled conditions of temperature and p_{O_2} only. A direct measure of x at room temperature (or even high temperature for that matter) is not possible. Only diffraction techniques, X-ray, neutron and electron, allow the measurement of point defect structures at the present time.

The Raman spectra of calcium doped NiO provide subtle measures of the calcium content. Calcia-rich second phase precipitation, has been monitored. Since it is possible to measure the oxygen nonstoichiometry and the concentration of cation dopant in NiO using Raman spectroscopy, it should be possible to measure the concentration of other cations including aliovalent species. This sets the background for the use of Raman spectroscopy to monitor the insitu point defect chemistry. Most importantly, very low nonstoichiometries, below 450 ppm in the present example, can be measured with high sensitivity and precision. This ability, together with that of characterizing the antiferromagnetic order is needed for structure - property studies.

ACKNOWLEDGEMENT: This work was supported by the National Science Foundation under grant number DMR78-05741.

REFERENCES

1. R. E. Dietz, G. I. Parisot, and A. E. Meixner, Phys. Rev. B, 4 2302-10 (1971).

2. E. F. Funkenbusch, P. Singh, and B. C. Cornilsen, in Processing of Metal and Ceramic Powders; R. M. German and K. W. Lay, Ed.; The Metallurgical Society of AIME: PA, 1982, pp. 65-72.

3. V. Lorprayoon, W. M. Lee, and B. C. Cornilsen in Proc. Eleventh North Amer. Thermal Analysis Soc. Conf.; J. P. Schelz, Ed.; NATAS, 1981; pp. 83-8.

4. C. M. Osburn and R. W. Vest, J. Phys. Chem. Solids, 32 1331-42 (1971).

5. R. W. Vest, "Defect Structure and Electronic Properties of Ceramics," in Physics of Electronic Ceramics, Part A; L. L. Hench and D. B. Dove, Ed.; Marcel Dekker, N.Y.; 1971.

6. S. P. Mitoff, J. Chem. Phys., 35 882-9 (1961).

7. E. F. Funkenbusch and B. C. Cornilsen, Solid State Commun., 40 707-710 (1981).

Vibrational Spectroscopies of Molecular Monolayers in Thin Film Geometries

J.R. Kirtley, J.C. Tsang, Ph. Avouris, and Y. Thefaine

I.B.M. Thomas J. Watson Research Center
Yorktown Heights, New York 10598

Molecular monolayers can be chemisorbed on the surface of oxidized evaporated thin metal films. The vibrational spectra of these molecular monolayers can be studied by inelastic electron tunneling spectroscopy (IETS) and, under certain circumstances, surface enhanced (SER) Raman spectroscopy. Tunnel junctions, formed by evaporating a second metal electrode on top of the molecular monolayer, provide a portable and durable model system for the study of, for example, chemisorption, catalysis, and lubrication, as well as an aid in the study of the tunneling and Raman scattering processes themselves. Advances in Raman instrumentation have made it possible to detect unenhanced molecular monolayers- avoiding some of the inherent disadvantages of tunneling and surface enhanced Raman spectroscopies.

INELASTIC ELECTRON TUNNELING SPECTROSCOPY

Inelastic electron tunneling spectroscopy has been reviewed in detail elsewhere[1-6] We will merely outline the experimental procedures, characteristics, and a few applications here. Junctions to be used for inelastic tunneling spectroscopy are prepared as follows: (Fig. 1). Thin metal films (approximately 100 nm thick and 1 mm wide) are evaporated through mechanical masks onto suitable insulating substrates (usually glass). The films are oxidized by exposure to air or an oxygen glow discharge to an oxide thickness of a few nanometers. The oxide is then coated with the molecular monolayer of interest. The molecular layer is either formed by exposing the films to a vapor, or by spinning on a solution (in a photoresist type spinner) of the molecular species of interest. In many cases the chemisorbed molecules are sufficiently tightly bound that they remain at the oxide surface, while the physisorbed molecules sublimate, during subsequent pumpdown and evaporation of the second metal electrode. The elimination of unwanted physisorbed molecules can be assisted by washing the sample with a non-polar solvent. Finally, a second metal film is evaporated through a second set of mechanical masks to form a series of crossed stripe metal-insulator- metal tunnel junctions.

Electrical contacts are made to the metal films, the samples are cooled to 4° K or less, and sensitive measurements are made of the current-voltage characteristics of the junctions in a 4-terminal geometry. When the insulating region between two metals is sufficiently thin, charge can transfer from one to the other by tunneling. If no external voltage is applied the two metals will reach equilibrium with their Fermi levels at the same energy, and the tunneling current in one direction

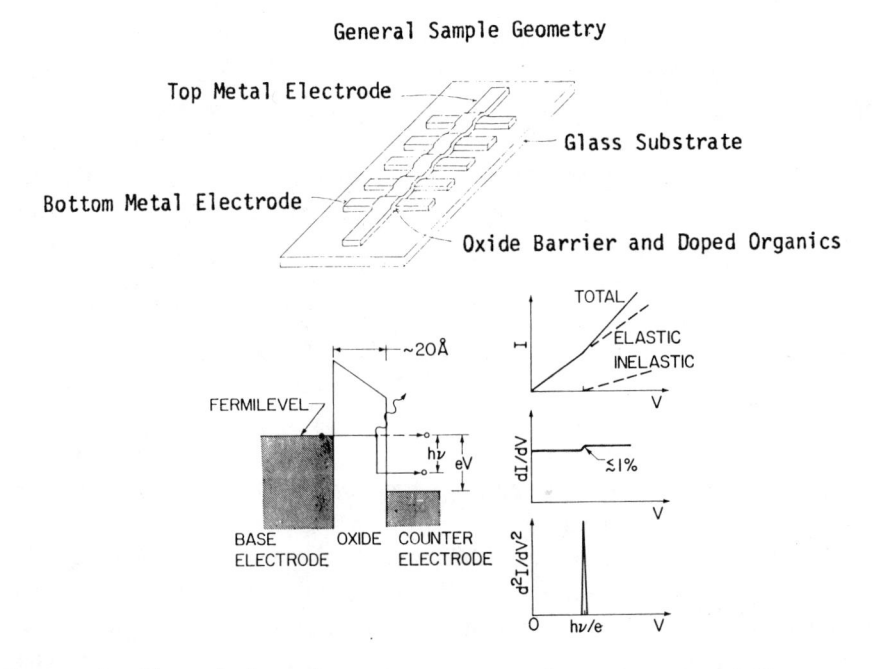

Figure 1. Tunnel junction geometry and energy band diagrams.

will be just canceled by the current in the opposite direction. If an external bias is applied, a net current flows (Fig. 1). The electrons will predominantly tunnel elastically: without losing energy in the process. But it is also possible for them to lose energy to an excitation in the barrier region. This can only happen if the bias voltage is larger than the particular energy loss, since the tunneling electron cannot enter states below the Fermi surface. Therefore the total conduction through the junction increases at the critical bias given by $eV = \hbar\omega$, where ω is the frequency associated with the excitation in the barrier. We will be primarily interested in molecular vibrations: other types of excitation have also been observed[7]. The inelastic tunneling process produces a step in the total conductance through the junction, or a peak in the second derivative of the current-voltage characteristic.

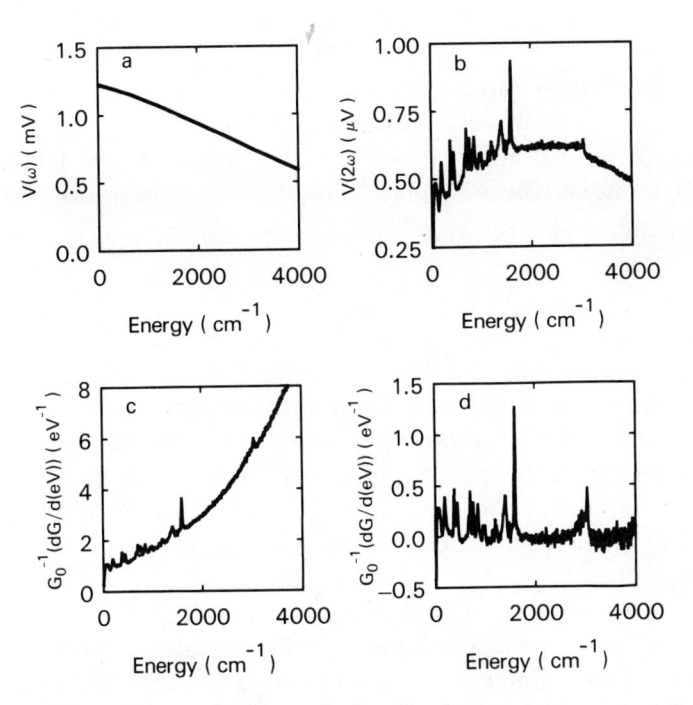

Figure 2. First (a) and second (b-d) derivative traces for an Al-Al$_2$O$_3$- 4-pyridine carboxylic acid-Pb tunnel junction.

The extra conductance due to the opening of an inelastic tunneling channel is typically quite small (from .01-1% of the total junction conductance). In Fig. 2 we show V(ω), the voltage due to a constant modulation current, across an Al-Al$_2$O$_3$-Pb junction doped with a molecular monolayer of 4-pydridine carboxylic acid at the oxide-Pb interface. The sample was run at 1.4 K with the current modulation chosen to give about 1 mV of voltage modulation. The signal in Fig. 2a corresponds to the dynamic resistance of the junction. It decreases (the conductance increases) as the junction bias is increased. Part of the change in conductance is due to the opening of inelastic tunneling channels, but most of it is due to changes in the shape of the barrier potential with voltage. V(2ω), the voltage at twice the modulation frequency, is related to the second derivative of the current- voltage characteristic by: $V(2\omega) = -(d^2V/dI^2)I_\omega^2/4$, where I(ω) is the (constant) current modulation across the junction. V(2ω) shows sharp peaks (Fig. 2b) corresponding to the vibrational spectrum of the molecular monolayer of 4-pyridine carboxylic acid. A quantity that is more closely related to the vibrational oscillator strengths than d^2V/dI^2, the change in the dynamic resistance with voltage, is d^2I/dV^2, the change in conductance with voltage. The normalized quantity $G_0^{-1}dG/d(eV) = -4V^{ref}(\omega)V(2\omega)/(V(\omega))^3$ is plotted in Fig. 2c. Here G is the voltage dependent differential conductance, and G^0 is the differential conductance at some reference voltage (taken to be 200 mV in this case). The peaks for this sample are superimposed on a large background which is due to the change in the barrier penetration probability with applied bias. This background has been subtracted out numerically in Fig. 2d.

The vibrational spectrum displayed in Fig. 2d is comparable to infrared and Raman spectra. The correspondence in energy is exact, in the sense that the peaks appear at voltages corresponding to vibrational frequencies through the relation eV = ℏω. Electron tunneling spectra contain potentially more information than either IR or Raman spectroscopy, since the tunneling electrons interact with both infrared and Raman active vibrational modes[8]. There may also be orientation information contained in the tunneling intensities[9]. A "selection preference " occurs because of image effects in the two metal electrodes, which favor vibrational modes with net dipole moments normal to the interface for molecules close to the interface, but favor vibrations parallel to the interface, for molecules deep within the tunneling barrier[10,11]. Several papers have tried to infer molecular orientations from IETS intensities[11,12].

The resolution available to IETS is limited by the temperature at which the spectrum is taken, and by the modulation voltage used. The contribution to the observed peak widths due to instrumental broadening is given roughly by:

$$\delta V = ((1.73eV^{rms}(\omega))^2 + (5.4k_bT)^2)^{1/2} \quad (1)$$

where $V^{rms}(\omega)$ is the root-mean-square modulation voltage across the junction, k_b is Boltzmann's constant, and T is the temperature at which the sample is run. The thermal contribution to the instrumental linewidth is essentially eliminated, and there are slight line shifts[13], if one or both electrodes are superconducting, but superconductivity is not required for inelastic electron tunneling. The signal observed in IETS is proportional to the square of the modulation voltage. The instrumental resolution is proportional to the modulation voltage. So, as in all spectroscopies, there is a tradeoff between resolution and signal. A common compromise is to run with modulation voltages of 1 mV, temperatures of 4.2K, and sweep times from 0-4000 cm^{-1} (0-0.5 volts) of 1 hour. Under these conditions, and with a supercoducting Pb electrode, the instrumental linewidth is about 12 cm^{-1} (1.5 meV).

The vibrational mode frequencies observed using IETS are often very different from their solution values. For example, in the spectrum of Fig. 2 the peak at approximately 1680 cm^{-1} due to a C=O stretching vibration is missing, replaced by the symmetric (1380 cm^{-1}) and antisymmetric (1550 cm^{-1}) COO^{-1} stretching mode vibrations: the 4-pyridine carboxylic acid undergoes an acid-base reaction in bonding to the aluminum oxide as a salt.

Fortunately, the presence of the counter metal electrode does not in general strongly perturb the vibrational mode frequencies[13]. For example, in the case of benzoic acid in an Al-Al$_2$O$_3$-Pb junction, the shifts from the no top electrode case (measured using large surface area alumina samples) were less than 1% for the C-H stretching modes of the benzoate ions and less than 0.2% for the ring deformation modes. In some instances the shifts due to the counter electrode can be important, especially for molecules with large dipole moments: shifts of 4% for hydroxyl ions on alumina under a silver electrode[13], and shifts of 5% for CO stretching modes under Au[14], as well as modifications of the spectral lineshapes for CN stretching modes[14], have been reported.

Inelastic tunneling spectroscopy has been used for a number of different applications, ranging from studies of chemisorption[15] and catalysis[16] to electron beam damage[17,18] and lubrication[19]. In one set of experiments McBride and Hall[20,21] doped a set of oxidized aluminum films with muconic acid (trans-trans- 1,3 butadiene dicarboxylic acid, HOOC-CH=CH-CH=CH-COOH) using the liquid doping technique. The tunneling spectra of the unheated samples showed that both ends of the muconic acid chemisorbed ionically to the oxide. Some samples

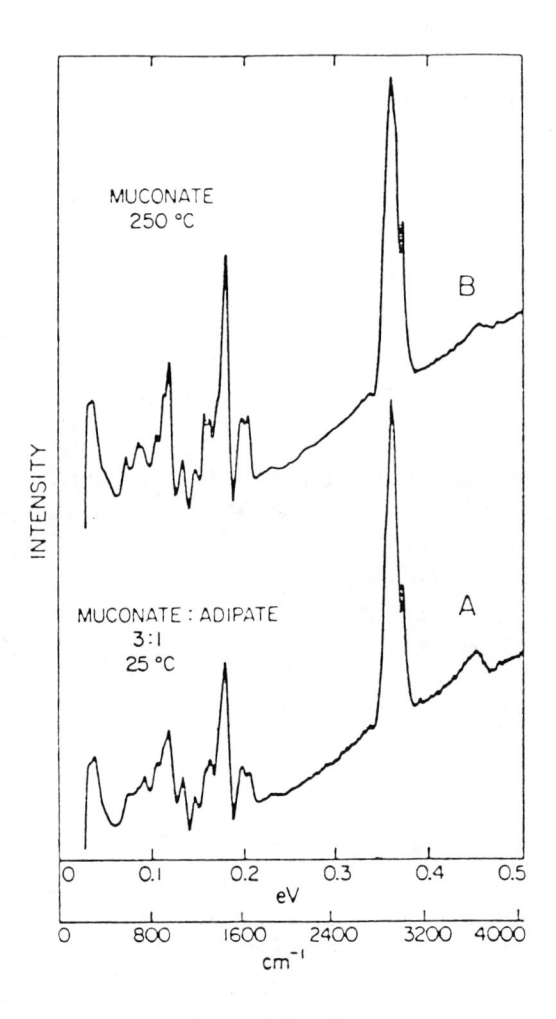

Figure 3. Tunneling spectra of muconic acid, which show that it reduces to the adipate when heated on alumina.

were returned to the vacuum system, and heated in the presence of water vapor to up to 400° by passing current through a heater strip evaporated on the back of the glass slide[22]. The films were then allowed to cool and the junctions completed by evaporation of a Pb counter electrode. Comparison of the tunneling spectra of the heated junctions with those of adipic acid, the unsaturated counterpart to muconic acid (see Fig. 3), as well as studies of the temperature dependence of the growth of the C-D vibrational intensity when using deuterated water, indicated that the surface species underwent a hydrogenation reaction using hydrogen (or deuterium) from the surface hydroxyl species, that the number of hydroxyl ions available to the reaction depended on the temperature of the substrate, and that the reaction stopped when the available hydrogen was used up. These results emphasized the importance of surface hydroxyl groups in this catalytic reaction on alumina.

Hansma, Kaska, and Laine[23] showed that dispersed metal catalysts could be modeled with tunnel junctions by evaporating very thin metal films (approximately 0.4 mass nm) of transition metals onto aluminum oxide. The metals formed highly dispersed 3 nm flattened balls on the surface of the oxide. Kroeker, Kaska, and Hansma[16] exposed highly dispersed Rh metal particles to a saturation coverage of CO, and then completed the junctions with a Pb counter-electrode. The junctions were then transferred to a high pressure cell and exposed to 10^7 Pa of hydrogen gas at temperatures up to 440 K. The Pb electrodes were highly permeable to hydrogen, but not to residual impurities in the pressure cell, allowing for a relatively clean oxide-Pb interface under extreme conditions.

The tunneling spectra of the junctions before exposure to hydrogen (Fig. 4) (including isotopic substitutions) showed the presence of at least three different CO species on the surface; two different linear-bonded species, and at least one bridge bonded species.

The tunneling spectra of the junctions after exposure to hydrogen (also in Fig. 4) showed two different species, one a formate like ion, and the other an ethylidene ($CHCH_3$) species. The formate ion is not thought to be an active intermediate in hydrocarbon synthesis, but the ethylidene species may well be a catalytic intermediate. Kroeker, Kaska, and Hansma[16] suggested a reaction pathway for the hydrogenation of CO on a supported rhodium catalyst with the formation of ethylidene as an intermediate.

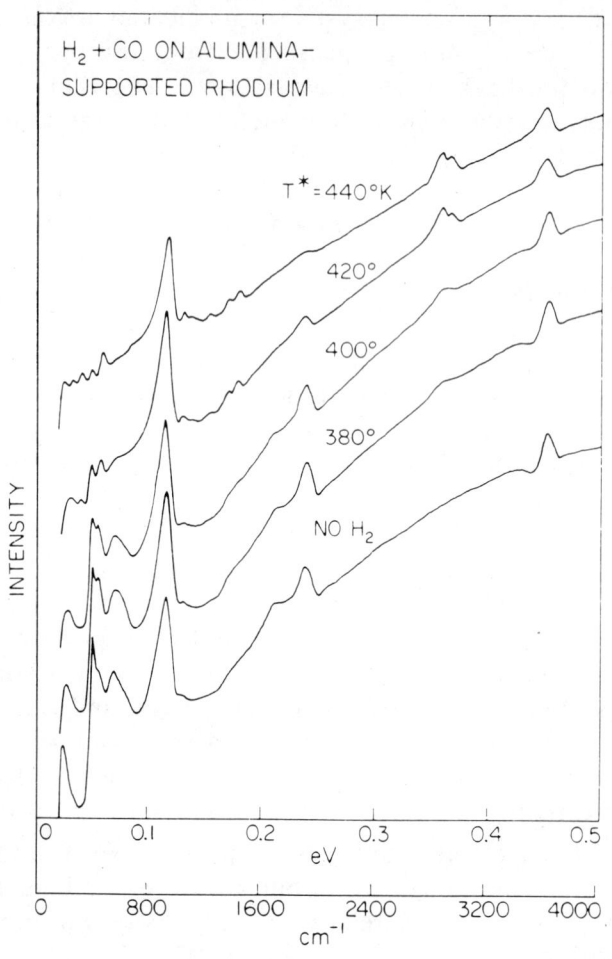

Figure 4. Tunneling spectra of CO on alumina supported rhodium, which show that it reduces when heated in the presence of hydrogen to an ethylidene species.

Inelastic electron tunneling spectroscopy has been shown to be a valuable technique for studying the vibrational spectra of model systems. However, there are difficulties: spectra must be taken at low temperatures, with the interface of interest coated with a metal counter-electrode. It would be valuable to have a surface vibrational technique which did not have these limitations. Raman spectroscopy may provide such a technique, if problems of sensitivity can be overcome.

SURFACE RAMAN SCATTERING

Raman scattering as a probe of surface structure has the advantages that it can be done at any temperature, with high intrinsic resolution, and without the need for ultra-high vacuum conditions. Unfortunately, typical molecular Raman scattering cross sections are so small that it is extremely difficult to detect molecular monolayers on smooth substrates with conventional techniques. It has been recognized for several years that some molecules have enormous Raman scattering cross sections when close to certain metal surfaces[24,25]. These large cross sections have made Raman scattering experiments from molecular monolayers on relatively small surface area samples possible[26].

The surface enhanced Raman (SER) effect was first observed in an electrochemical cell environment[24,25]. Pyridine molecules in water solution showed very large Raman scattering signals (10^6 times larger than in bulk solution) when adsorbed on a silver electrode that had been electrochemically cycled. It was not clear from these experiments what roles such factors as surface roughness had in the enhancement.

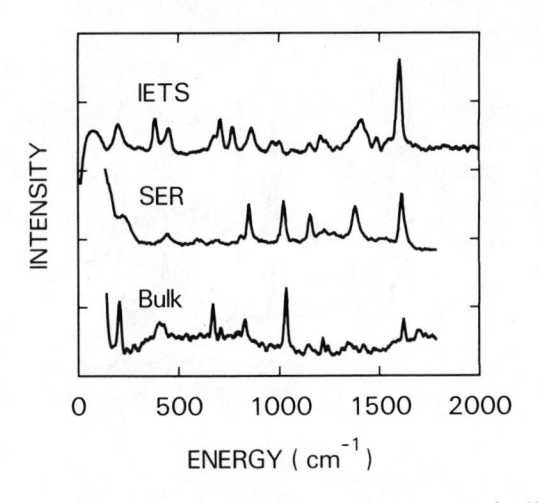

Figure 5. Tunneling, surface enhanced Raman, and bulk spectra of 4-pyridine carboxylic acid.

We were able to observe Raman scattering from molecular monolayers at the oxide-silver interfaces of Al-Al$_2$O$_3$-Ag tunnel junctions[27]. Fig. 5 shows a comparison of the tunneling and Raman spectra from 4-pyridine carboxylic acid in such junctions, along with the Raman spectrum of a bulk solution. It is clear from this figure that inelastic electron tunneling is sensitive to more of the vibrational modes than Raman scattering, and also that the intensity pattern of the surface enhanced Raman spectrum does not closely resemble that of the molecules in solution. Analysis of the tunneling spectra showed that there was indeed only a molecular monolayer present in these samples. These were the first observations of the surface enhanced Raman effect in a non-electrochemical cell environment: they showed that SER scattering was a general phenomenon.

The tunneling junction geometry allowed us to vary interface roughness independently of other variables. This was done by evaporating the junction structures on substrates coated with thin films of CaF$_2$[28]. The CaF$_2$ films formed small crystallites: the roughness of the substrate could be continuously varied by evaporating different thickness films of CaF$_2$. Fig. 6 shows the relative Raman scattering intensities from 4-pyridine carboxylic acid doped Al-Al$_2$O$_3$-Ag junctions on

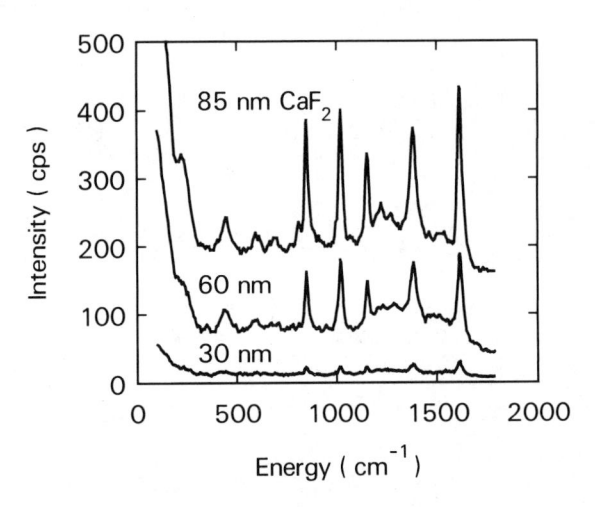

Figure 6. Surface enhanced Raman spectra of Al-Al$_2$O$_3$-4-pyridine carboxylic acid-Ag junctions on varying thicknesses of CaF$_2$.

30, 60, and 85 nm of CaF_2. The observed Raman scattering signal increased and then saturated at a CaF_2 thickness of about 120 nm, in a fashion similar to that observed for coupling light into surface electromagnetic resonances in optical reflectance measurements[29] and also similar to that observed for coupling light out of surface electromagnetic resonances in light emitting tunnel junction experiments[30]. Since these resonances (surface plasmon polaritons) propagate along the metal film surfaces at speeds slower than light in vacuum[31], they cannot couple to external radiation without the influence of surface roughness. These experiments indicated that coupling of external radiation to surface electromagnetic resonances through surface roughness played a strong role in the surface enhanced Raman effect.

It is possible to isolate the role of a single Fourier component of roughness in coupling of external radiation to surface resonances by fabricating the metal films on gratings[32]. In Fig. 7 we show the result of one such experiment, in which the surface enhanced Raman scatter-

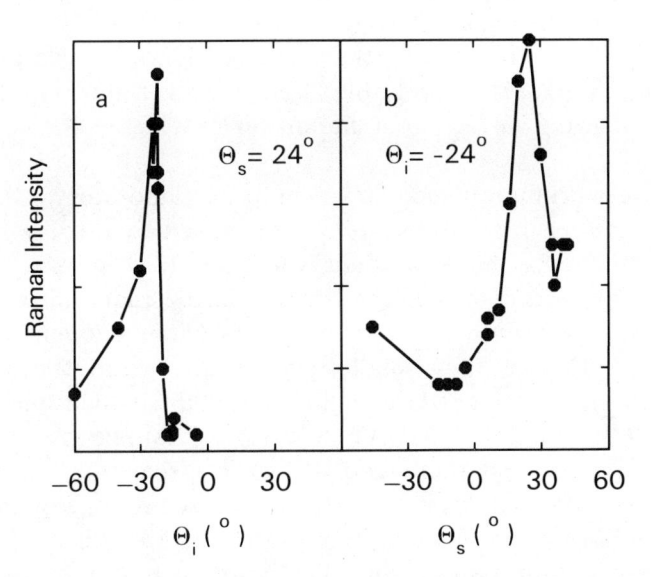

Figure 7. Surface enhanced Raman signal from $Al-Al_2O_3$-4- nitrobenzoic acid-Ag junction on a grating for: (a) fixed collection angle as a function of incident laser angle, and (b) fixed incident laser angle as a function of collection angle.

ing signal from a 4-nitrobenzoic acid doped Al-Al$_2$O$_3$-Ag junction on an 800 nm periodicity grating substrate is plotted. In Fig. 7a the Stokes shifted radiation is collected over a narrow angular aperture at 24° from the surface normal, while the incident laser angle is varied. Surface electromagnetic resonances are resonantly excited by the incident laser beam at an incident angle of -24°: the Stokes shifted intensity undergoes a strong resonance at that angle. Similarly, if the incident beam is held at the resonance angle, the Stokes shifted photons undergo an enhancement at the angle where surface electromagnetic resonances can resonantly couple out to external radiation through the surface roughness. These experiments were the first to definitively demonstrate the role of surface roughness coupling of external radiation to surface electromagnetic resonances in the surface enhanced Raman effect.

Detailed comparison of our experiments with theoretical estimates of Raman scattering cross sections on grating surfaces[33] have indicated that, although surface resonances play an important role in the surface Raman effect, they cannot account for the total enhancements observed. Nevertheless, it is clear that very large enhancements only occur under conditions in which strong resonances can occur: on noble metals, which have large negative dielectric constants with small losses in the optical region, and with rough surfaces. These conditions can be met in a variety of different ways, but it would also be useful to have a Raman scattering tool that will work for the more general (non-enhanced) low surface area monolayer case.

We are working on such a system in our laboratory[34,35] (see Fig. 8). Light scattered from the sample is dispersed by a triple monochromater, which blocks the laser light with a 0.22M subtractive double monochromater, and disperses the Stokes shifted radiation with a 0.5M spectrograph. The triple monochromater provides enough rejection to allow the detection of very small signal levels, without swamping the detector with stray incident laser light. The light is then imaged on the photocathode of a PAR 1254 SIT detector. Photoelectrons are stored on a silicon target, and then read off with a scanning electron beam into a multichannel analyser. Typically two scans, one with and one without illumination of the sample of interest, are run and subtracted numerically to obtain a spectrum. The detector is run at low temperatures (-40 C) to reduce the dark count levels and system noise.

This system has a major advantage over a standard Raman scattering setup. The optical multichannel detector allows detection of

VIBRATIONAL SPECTROSCOPIES OF MOLECULAR MONOLAYERS 261

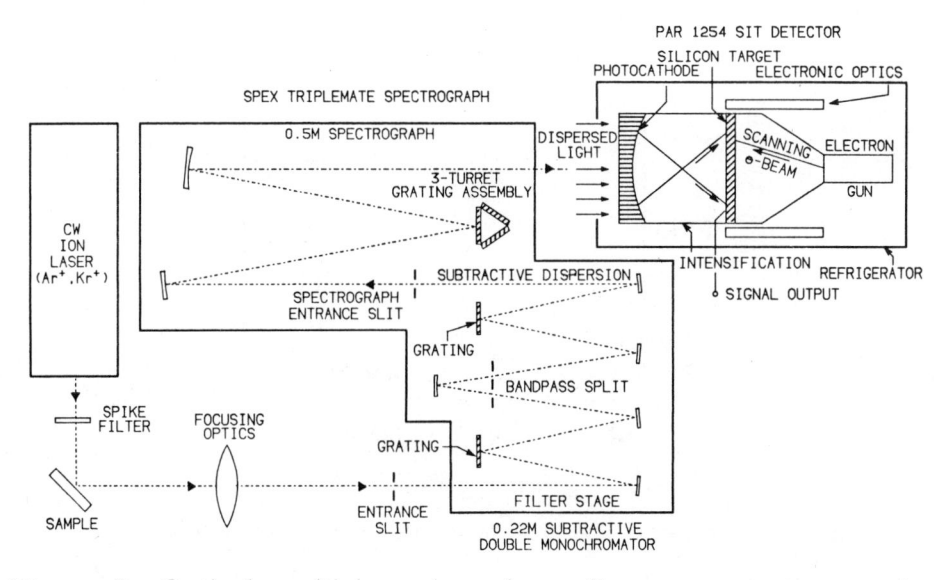

Figure 8. Optical multichannel analyser Raman spectroscopy system.

all of the Stokes shifted radiation in parallel, significantly speeding data acquisition rates. Nevertheless, the signal rates are quite small: scan times for our monolayer detection experiments of an hour are typical.

Fig. 9 demonstrates the power of this instrument. This is the Raman spectrum from a 3 mg/ml ethanol solution of 4-nitrobenzoic acid, with the spectrum of pure ethanol subtracted out. The solution concentration and scattering volumes were such that about 10^{14} molecules were being exposed to the incident laser radiation. Since typical surface concentrations are of order 10^{14} molecules/cm^2, we are approaching the sensitivity required for unenhanced Raman spectroscopy.

Fig. 10 shows the Raman spectra obtained from 4-nitrobenzoic acid on an oxidized aluminum film on an 800 nm periodicity, holographically produced grating. The spectrum of Fig. 10a, which resulted when a thick layer of the acid was spun on from a 3 mg/ml solution in ethanol, looked very much like the bulk spectrum of Fig. 9. The spectrum of Fig. 10b, which resulted from spinning on a more dilute 0.5 mg/ml solution, had the additional spectral features at 1354 and 1460 cm^{-1} seen in tunnel junction spectra and associated with the first monolayer on the surface[35]. The spectra of Fig. 10a and 10b were both taken

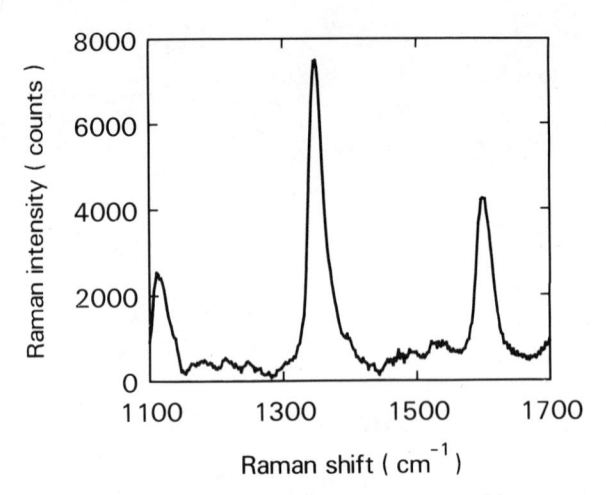

Figure 9. Solution Raman spectrum of 10^{14} molecules of 4-nitrobenzoic acid.

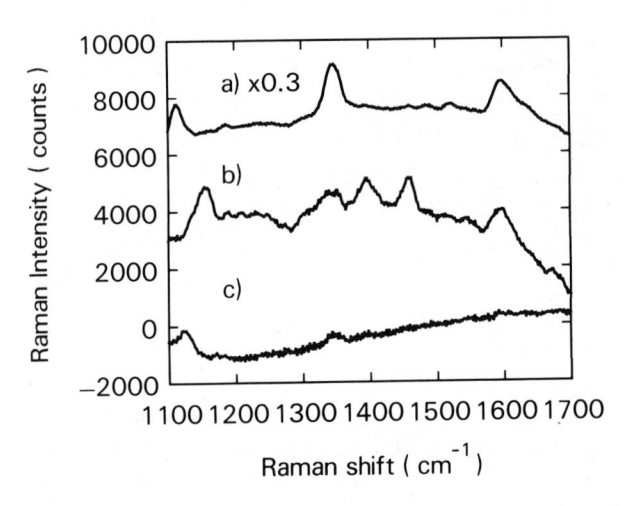

Figure 10. Raman spectra of 4-nitrobenzoic acid on an oxidized Al grating; (a) thick layer from 3 mg/ml solution, (b) monolayer from 0.5 mg/ml solution at surface plasmon resonance, and (c) monolayer off resonance.

VIBRATIONAL SPECTROSCOPIES OF MOLECULAR MONOLAYERS 263

with the incident laser beam at the appropriate angle for exciting surface electromagnetic resonances through the grating roughness. When the laser beam was moved off resonance for the molecular layer of Fig. 10b, the observed spectrum became about 20 times weaker (Fig. 10c): The field enhancement from the grating (which is at least 10 times smaller for Al than for Ag films) was still necessary to obtain good signal to noise ratios from molecular monolayers. These samples were run in air, so that lower incident laser power levels (50 mwatts) were required to maintain sample integrity than in a UHV environment[34], but with this system, studies of molecular monolayers on aluminum oxide, a catalyst of practical importance, become possible with the grating geometry.

This system can also be used to study bulk solid thin films, even for systems which show strong elastic scattering. We show in Fig. 11 the Raman spectrum of a 2μ layer of silicon rich (about 10% excess silicon) silicon dioxide on a saphire substrate[36]. This material consists of small islands of silicon included in a matrix of stoichiometric silicon oxide. The small particles scatter light strongly, making conventional Raman scattering techniques difficult. The Si rich silicon dioxide is

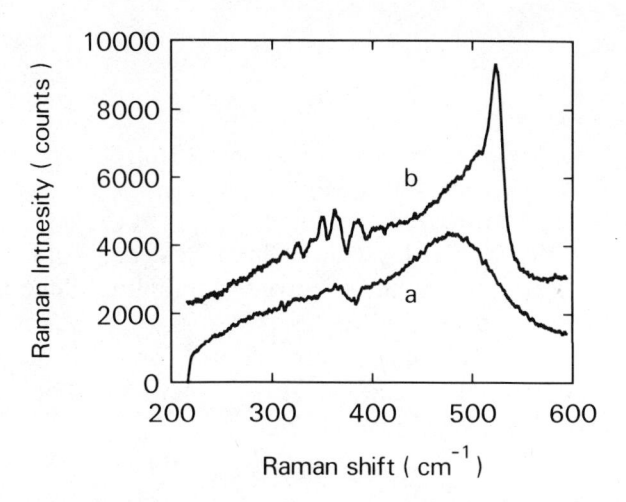

Figure 11. Raman spectra of unannealed (a) and annealed (b) silicon rich SiO_2 films.

used as the injector in Si-Si rich SiO_2-metal charge injection structures. The Si rich SiO_2 injects high current densities into the conduction band of the stoichiometric silicon oxide, allowing the use of these structures for programmable read-only memory cells[37,38]. The injection mechanism has been postulated to be tunneling from silicon island to silicon island through the silicon rich oxide, followed by Fowler-Nordheim tunneling into the conduction band of the stoichiometric silicon dioxide. The Raman spectrum in Fig. 11a shows the broad band peaking at about 480 cm^{-1} characteristic or either amorphous silicon or small Si crystallites. The oscillations between 300 and 400 cm^{-1} are due to incomplete subtraction of the saphire phonons from a control substrate. The Raman spectrum of Fig. 11b is from a sample annealed to 1100 C. The sharp band at 525 cm^{-1} is characteristic of crystalline silicon: the silicon islands either become larger or more crystalline when annealed. These experiments demonstrated the presence of amorphous silicon in a matrix of SiO_2 in these films. By studying the temperature dependence of the Raman scattering bands, Hartstein et. al.[36] were able to show that the Si islands annealed at temperatures considerably higher than for bulk amorphous silicon.

Surface vibrational spectroscopies have already demonstrated great power in dealing with interactions at interfaces. As instrumentation improves, they will be applied to a wide variety of systems.

We would like to thank A.M. Torrreson and M.T. Prikas for technical assistance in these experiments.

References

1. R.G. Keil, K.P. Roenker, and T.P. Graham, Appl. Spect. **30**,1(1976).
2. P.K. Hansma, Phys. Rep. **30C**,146(1977).
3. W.H. Weinberg, Ann. Rev. Phys. Chem. **29**,115(1978).
4. P.K. Hansma and J.R. Kirtley, Acc. Chem Res. **11**,440(1978).
5. T. Wolfram, ed. " Inelastic Electron Tunneling Spectroscopy ", Springer- Verlag, Berlin. 1978.
6. J.R. Kirtley, in " Vibrational Spectroscopies for Adsorbed Species ", ed. Alexis T. Bell and Michael L. Hair, American Chemical Society, Washington, D.C., 1980, pp. 217-245.
7. S. de Cheveigne, J. Klein, A. Leger, M. Belin, and D. Defourneau, Phys. Rev. **B15**,750(1977).
8. M.G. Simonsen, R.V. Coleman, and P.K. Hansma, J. Chem. Phys. **61**,3789 (1974).

9. D.J. Scalapino and S.M. Marcus, Phys. Rev. Lett. **18**,459(1967).
10. J.R. Kirtley, D.J. Scalapino, and P.K. Hansma, Phys. Rev. **B14**,3177 (1980).
11. J.R. Kirtley and J.T. Hall, Phys. Rev. **B22**,848(1980).
12. L.M. Godwin, H.W. White, and R. Ellialtioglu, Phys. Rev. **B23**, 5688(1981).
13. J.R. Kirtley and P.K. Hansma, Phys. Rev. **B13**,2910(1976).
14. A. Bayman, P.K. Hansma, and W.C. Kaska, Phys. Rev. **B25**,2449(1981).
15. R.M. Kroeker, W.C. Kaska, and P.K. Hansma, J. Chem. Phys. **74**,1(1981).
16. R.M. Kroeker, W.C. Kaska, and P.K. Hansma, J. Catal. **61**,87(1980).
17. M. Parikh and P.K. Hansma, Science **188**,1304(1975).
18. M. Parikh, J.T. Hall, and P.K. Hansma, Phys. Rev. **A14**,1437(1976).
19. A. Bayman and P.K. Hansma, Nature **285**,97(1980).
20. D.E. McBride and J.T. Hall, J. Catalysis **58**,320(1979).
21. D.E. McBride and J.T. Hall, J. Chem. Phys. **74**,4164(1981).
22. W.M. Bowser and W.H. Weinberg, Rev. Sci. Inst. **47**,583(1976).
23. P.K. Hansma, W.C. Kaska, and R.M. Laine, J. Am. Chem. Soc. **98**,6064(1976).
24. M.P. Fleischman, P.J. Hendra, and J. Mcquillen, J. Chem. Soc. Chem. Commun. **3**,80(1973).
25. D.L. Jeanmaire and R.P. van Duyne, J. Electroanal. Chem. **82**,329(1977).
26. a recent review of this field is included in: " Surface Enhanced Raman Scattering ", eds. R.K. Chang and T.E. Furtak, Plenum, New York, 1982.
27. J.C. Tsang and J.R. Kirtley, Solid State Commun. **30**,617(1979).
28. J.C. Tsang, J.R. Kirtley, and J.A. Bradley, Phys. Rev. Lett. **43**, 772(1979).
29. J.G. Endriz and W.E. Spicer, Phys. Rev. **B4**,4144(1971).
30. S.L. McCarthy and John Lambe, Appl. Phys. Lett. **30**,427(1977).
31. E.N. Economou, Phys. Rev. **182**,539(1969).
32. J.C. Tsang, J.R. Kirtley, and T.N. Theis, Solid State Comm. **35**,667 (1980).
33. S.S. Jha, J.R. Kirtley, and J.C. Tsang, Phys. Rev. **B22**,3973(1980).
34. detection of molecular monolayers using an optical multichannel analyser system has previously been reported by: A. Campion, J.K. Brown, and V.M. Grizzle, Surf. Sci. **115**,L153(1982).
35. J.C. Tsang, Ph. Avouris, and J.R. Kirtley, to be published.

36. A. Hartstein, J.C. Tsang, D.J. DiMaria, and D.W. Dong, Appl. Phys. Lett. **36**,836(1980).
37. D.J. DiMaria, R. Ghez, and D.W. Dong, J. Appl. Phys. **51**,4830(1980).
38. D.J. DiMaria, K.M. DeMeyer, C.M. Serrano, and D.W. Dong, J. Appl. Phys. **52**,4825(1981).

CHARACTERIZATION OF RARE EARTH SULFIDES

C. Lowe-Ma

Research Department (Code 3854)
Naval Weapons Center
China Lake, Calif.

ABSTRACT

Rare earth sesquisulfides have been known for many years. Very early it was recognized that these materials were refractories. Ceramics fabricated from ternary rare earth sulfides, such as $CaLa_2S_4$, have potential usefulness as sensor windows. Adequate characterization of these materials is crucial to obtaining optical quality ceramics. A variety of techniques have been used to characterize $CaLa_2S_4$. The results from these studies are yielding a better understanding of the effects of nonstoichiometry and moderately low-level impurities on the optical properties of $CaLa_2S_4$.

INTRODUCTION

There is increased interest in the optical community in improved optical-quality ceramic materials with good thermal-mechanical properties and high transmittance in the 8-12 micron wavelength region.[1,2] The rare earth ternary sulfides, MLn_2S_4, with the Th_3P_4 structure show promise of being optical-quality ceramics with adequate thermal-mechanical properties. The compounds are cubic, and, hence, in principle would be free of birefringence scattering. The materials are refractory with melting points in the range of 1800-1900°C. They have low fundamental vibrational frequencies. The materials are lightly colored, suggesting large band gaps.[3,4] Of the many possible ternary sulfides known, only $CaLa_2S_4$ has been investigated in detail by us.

The Naval Weapons Center (NWC) is involved in a collaborative research effort with Pennsylvania State University, Coors Porcelain

Company, and Raytheon to better understand $CaLa_2S_4$. Pennsylvania State, Coors, and Raytheon have kindly supplied $CaLa_2S_4$ samples to NWC for evaluation.

The primary emphasis of my research during the past year has been on the characterization of phase purity (stoichiometry) and the identification of minor impurities with the aim of improving the optical transmission of $CaLa_2S_4$. This has required the use of a number of characterization techniques, as no one technique has provided sufficient information about the material.

BACKGROUND

Zachariasen originally characterized some of the rare earth sesquisulfides as having the Th_3P_4 structure.[5] It was Zachariasen who proposed that materials such as La_2S_3 could be described as $La_{2.67}S_4$ with the sulfur fully occupying the sixteen sites of point set c (site symmetry 3) in space group $I\bar{4}3d$ and with 10.67 lanthanum atoms randomly distributed over the twelve sites of point set a (site symmetry $\bar{4}$). The sesquisulfide La_2S_3 (or $La_{2.67}S_4$) can, therefore, be described as a defect structure. Much of the early synthesis work on the rare earth sesquisulfides and ternary sulfides was done by Flahaut and coworkers.[6,7] Flahaut, et al., have pointed out that there is a continuous, homogeneous, range of stoichiometries between La_2S_3 and La_3S_4. There is also a continuous, homogeneous range of stoichiometries for the ternary $Ca_{1-x}La_2S_{4-x}$, for x = 1→0. In La_3S_4 or $CaLa_2S_4$ the lanthanum or calcium and lanthanum atoms are distributed over point set a fully occupying that site. La_3S_4 is reported to be a bluish-black degenerate semiconductor while La_2S_3 is a yellow-colored insulator and is transparent in the visible region as a single crystal.[8-10] $CaLa_2S_4$ is a light-yellow powder with a band gap of 2.70 eV.[11]

The Th_3P_4 type structure for lanthanum sulfide is only one of several known phases.[12,13] The Th_3P_4 structure may be stabilized by the presence of divalent cations.[13] In principle, then, the material of interest is stoichiometric $CaLa_2S_4$. The binary La_2S_3, although an insulator, should be avoided because it reacts with oxygen.[5,14] And lanthanum-rich or sulfur-deficient stoichiometries must also be avoided because of the electronic absorptions due to low band-gaps in these materials. A key question, which has yet to be answered, is what stoichiometry at or near $CaLa_2S_4$ can be made that has high (optimal) infrared transmittance?

SYNTHETIC PROCEDURES

Early in the course of this work it became clear that the reported literature syntheses usually gave a "$CaLa_2S_4$" product contaminated by other phases. These early syntheses relied on solid state diffusion, so complete reaction was difficult to achieve. Recently, two different synthetic methods have been developed that yield good $CaLa_2S_4$ powders. One method, developed at Pennsylvania State University, is based on the evaporative decomposition of solutions technique (EDS) which is used to prepare intimate mixtures of calcium and lanthanum oxides in micron-sized particles. In this technique stoichiometric amounts of calcium and lanthanum nitrates are dissolved in aqueous solution. This solution is then aspirated into a hot, vertical furnace. The nitrates undergo flash evaporation/decomposition. The very fine, intimately mixed oxides are collected at the bottom of the furnace using an electrostatic precipitator. This oxide mixture is then fired in a flat alumina boat in flowing H_2S at a temperature higher than 950°C for times up to several days.[15] A second method, developed at Raytheon, also involves dissolving stoichiometric amounts of calcium and lanthanum nitrates in aqueous solution. Ammonium carbonate is then added to the solution causing a white, fluffy, lanthanum-calcium carbonate powder to precipitate. Particle sizes from this method are reported to be submicron. After drying, this powder is fired in a high-purity alumina boat in flowing H_2S for 1-3 days at 950-1100°C.[16,17] Both methods give fine-grained, phase-pure, "$CaLa_2S_4$" powders with particle sizes of 3-10 microns. However, even after obtaining nominally phase-pure "$CaLa_2S_4$," the material may still not have the desired stoichiometry due to the possible homogeneity range.

POWDER CHARACTERIZATION

Several characterization techniques have been used to better understand the effects of stoichiometry and impurities on the optical properties of $CaLa_2S_4$.

Ideally, elemental analyses accurate to at least three significant figures on the phase-pure samples would indicate deviations off stoichiometry. These deviations could then be correlated with other physical properties such as lattice parameters, conductivity, and infrared transmission. However, until recently there were not enough phase-pure samples available for sufficient sampling. Early "$CaLa_2S_4$" powders were invariably contaminated by other phases and compounds.[18] In an effort to try and understand what methods would give reliable data about stoichiometry, samples for elemental analyses were analyzed at NWC and also sent to contract laboratories. Duplicate elemental analyses on available powders were run as "blind" samples; the analyzing laboratory was not told that some of the samples were duplicate. Overall, elemental analyses for calcium gravimetrically or by atomic absorption spectroscopy appear to be reasonably reliable. The results for the sulfur and lanthanum analyses are more variable. All four powders analyzed seem to be lanthanum-rich. Two powders prepared by the EDS technique appear to be calcium deficient.

Reports by Flahaut, et al., and Toide, et al., suggest that there may be small, but noticeable, lattice constant variations with deviations in stoichiometry.[19,20] The reported literature values for the (supposedly) end-members (Table 1) show some trends, but are not sufficiently reliable to be used to determine stoichiometry with confidence. These lattice constants were often reported with little or no characterization data, so it is not known what range of stoichiometries the lattice parameters represent. Although Toide, et al.,[20] report a linear Vegard's law plot for the solid solution $La_2S_3+CaS \rightarrow CaLa_2S_4$, a plot of stoichiometry versus lattice parameters obtained at NWC indicate that the curve is regular but probably not linear. Further work in this area is needed because the absolute values of the lattice parameters obtained at NWC and another laboratory on the same $CaLa_2S_4$ powders do not agree. However, the diffraction patterns have not yet been run with internal standards. Some data suggest that the lattice parameters obtained for ceramic samples may not lie on the lattice parameter versus stoichiometry curve.[21] This needs further clarification because placement of the ceramic in the diffractometer will drastically affect the apparent lattice parameter.

To check for cation impurities, qualitative energy-dispersive X-ray analyses (EDX) and semi-quantitative emission spectrographic analyses (EmS) were carried out on the powders and two ceramics. The only elements observed by EDX were Ca, La, and S. However, as indicated in Table 2, low-level cation impurities were detected by EmS. Further work needs to be done to determine if these impurities produce adverse effects on the optical properties.

Table 1. Literature Values of Lattice Constants for La_2S_3, La_3S_4, $CaLa_2S_4$.

Compound	a, Å	Reference
La_2S_3*	8.725 (1)	9, 10
La_2S_3	8.706 (1)	5
La_2S_3	8.731	6, 14
"La_2S_3" ($LaS_{1.44}$)	8.718 (1)	22
La_3S_4	8.730	6, 14
La_3S_4	8.720 (2)	23
"La_3S_4" ($LaS_{1.38}$)	8.723 (1)	22
$CaLa_2S_4$	8.687 (4)	6, 19
$CaLa_2S_4$	8.686	20
$CaLa_2S_4$	8.687	3
$CaLa_2S_4$	8.683 (3)	24

* Single-crystal value.

Table 2. Semi-Quantitative Spectrographic Analysis of Three $CaLa_2S_4$ powders.

Concentration Range	Powder #69 Sample 1	Sample 2	Powder EDS-411B	Powder EDS-411D
Principal (100-10%)	Ca, La	Ca, La	Ca, La	Ca, La
Major (10-1%)	--	--	Sr	Sr
Strong (1-.1%)	--	--	--	--
Medium (.1-.01%)	Si, Al, Cu, As*	Si, Al, Cu, As*	Mg, Si, Fe, Al, As*	Mg, Si, Fe, Al, As*
Weak (.01-.001)	Fe, Cr, Mg, Ni, Na*	Fe, Mg, Ni, Na*	Ni, Cu, Na*	Mn, Ni, Cu, Na*
Trace (.001-.0001%)	B	--	Mn, Ag	Ag

* Not positively identified.

Infrared diffuse reflectance with a Fourier transform infrared (FTIR) spectrometer has been a very sensitive technique for observing low-level anion impurities in the $CaLa_2S_4$ powders. Recent $CaLa_2S_4$ powders have all appeared to be phase-pure by X-ray diffraction. Yet, as shown in Figure 1, the observed IR diffuse reflectance spectra clearly indicate the presence of impurities. Without taking account of particle size effect, etc., a quick experiment based on doping $CaSO_4$ in KBr indicates that the level of impurities in these powders is about 0.01% (plus or minus an order of magnitude). The impurity absorptions have tentatively been assigned to water, carbonate or nitrate anions, sulfate, and sulfite or pyrosulfite anions. Based on the successful use of FTIR diffuse reflectance to detect anion impurities at $\sim$1-.001% level and tentatively identify them in $CaLa_2S_4$, this technique may have general applicability for the detection of moderately low-level absorbing species in a nonabsorbing (transparent) medium. This is especially useful if the impurities are below the detection limit by X-ray diffraction.

CERAMIC FABRICATION

Visually, ceramics fabricated from powders of $CaLa_2S_4$ vary from opaque to transparent and vary in color from black to dark yellow. Many of the early ceramics were opaque in the infrared region. There are four procedures currently being used to fabricate the $CaLa_2S_4$ ceramics. The procedures are: (1) hot press and then hot isostatic press (HIP); (2) sinter in H_2S and then hot press; (3) sinter in H_2S and then HIP; (4) hot press only. These procedures have resulted in

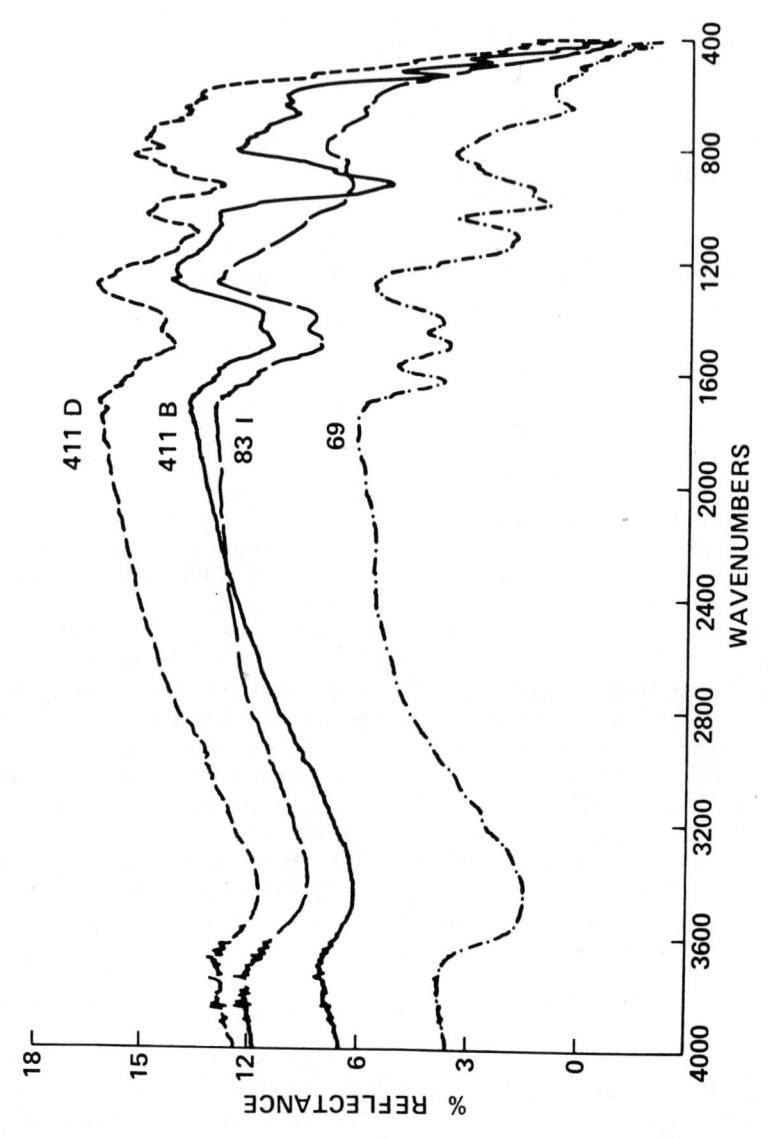

Figure 1. Infrared Diffuse Reflectance Spectra of CaLa$_2$S$_4$ Powders.

ceramics of variable optical quality. The best $CaLa_2S_4$ ceramics, to date, were fabricated using a procedure of sintering in H_2S and then HIP in argon. The best ceramic piece that we have examined is visually dark yellow, transparent, and transmits about 65-70% at 12 microns for a slab 0.0305 inches thick. (All quoted transmissions have not been corrected for reflection losses.) However, even this ceramic has some obvious flaws or second-phase precipitates visible with low-power magnification. Attempts to identify these "inclusions" are in progress. Under higher magnification other small "imperfections" can be observed.

CERAMIC CHARACTERIZATION

Infrared transmission curves for ceramic #83I and ceramics HP #119 and HP #34 are shown in Figure 2. These are "raw" transmission curves obtained on an FTIR spectrometer. Reflections losses and beam defocusing effects have not been taken into account. Ceramic #83I is optically superior to the other two and to previous ceramics.

During the past year and a half there have been improvements in all of the fabricated ceramics. Earlier ceramics often had transmittances of 0-5% in the 8-15 micron region.[18] The "raw" transmission curve for #83I has losses due to about 1-2% scattering. The absorption coefficient is about 2 cm^{-1} for #83I. This is about an order of magnitude better than the measured absorption coefficients for samples prepared a year ago. With continued feedback from our characterization studies, the optical-quality of the ceramics should continue to improve.

For samples fabricated solely by hot pressing, the quality of the ceramics may be limited by the quality of the starting powders. Spectra obtained for the starting powder (curve A), fabricated ceramic (curve B), and ground ceramic (curve C) are shown in Figure 3. It is clear that impurities in the starting powder are carried over into the ceramic. It is possible that some of these impurities may be due to hydrolysis/oxidation of the surfaces of the very finely divided powders. Optically, the best ceramics have been obtained from samples that have been sintered to better than 90% theoretical density in flowing H_2S. However, the hot pressed only samples do not pass through a "resulfurizing-cleaning" step prior to fabrication. Because hot pressing gives ceramics with smaller grain sizes than ceramics that have been sintered and "resulfurizing" may be necessary to remove anion impurities on the particle surfaces prior to densification, attempts will be made in the next few months to "hot press in H_2S."[25]

Many Auger Electron Spectroscopy (AES) and X-ray photoelectron spectroscopy (XPS) spectra have been obtained at NWC for 16 powder and ceramic samples of $CaLa_2S_4$. Duplicate spectra have been obtained both before and after polishing and different cleaning procedures. The

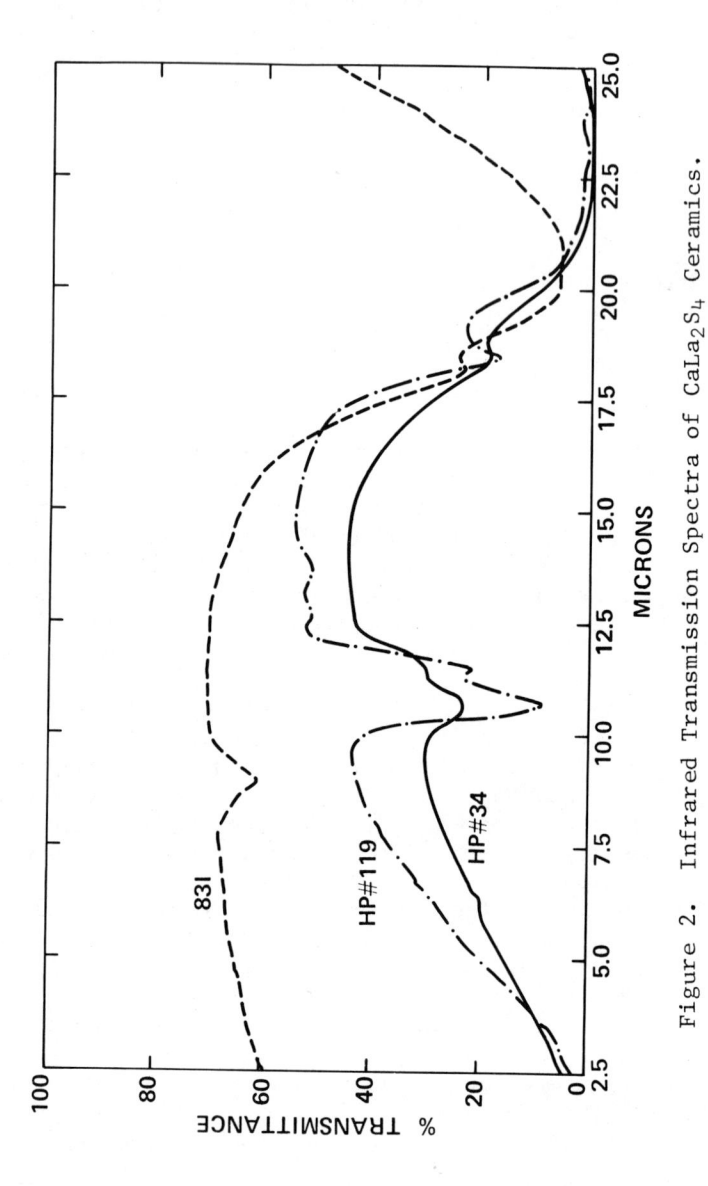

Figure 2. Infrared Transmission Spectra of $CaLa_2S_4$ Ceramics.

Figure 3. CaLa$_2$S$_4$ - Before and After Fabrication.

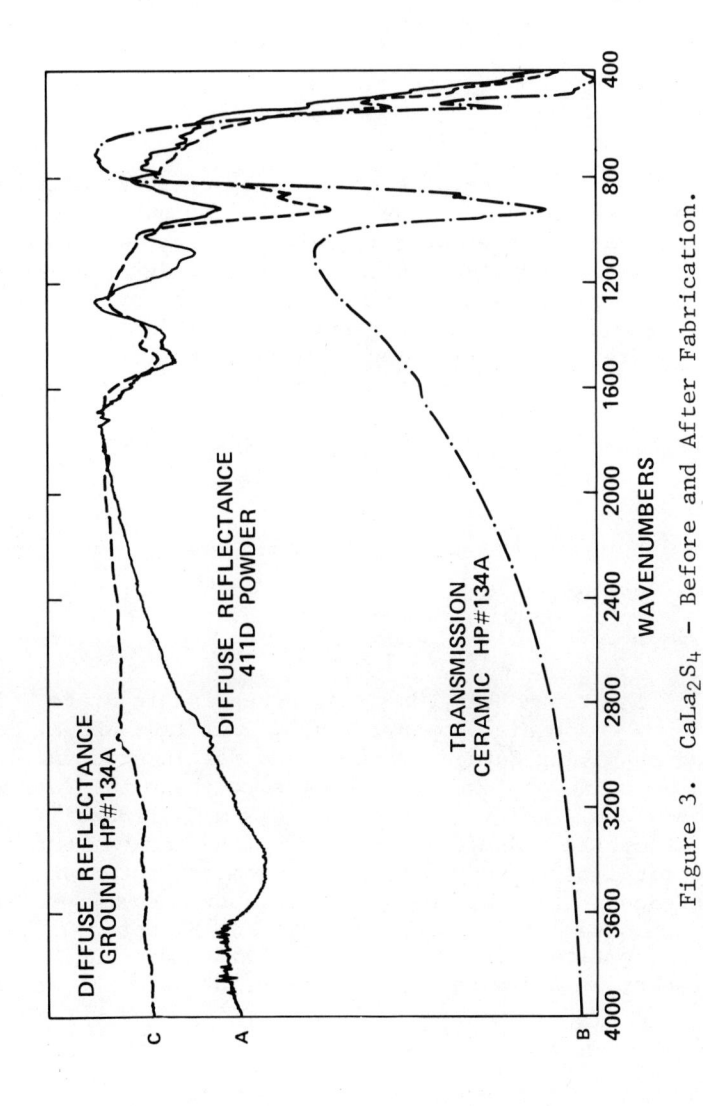

results are currently too ambiguous for discussion. Thus far we have been unable to correlate apparent oxygen and carbon impurities with the processing or handling and cleaning procedures. Unfortunately, in Auger the carbon and calcium peaks overlap causing ambiguities in the interpretation of the peaks in this region. Further work in this area is planned.

SUMMARY

Over the course of the present research efforts on ternary sulfides, there has been steady improvement in the purity of the starting $CaLa_2S_4$ powders and in the optical quality of the fabricated $CaLa_2S_4$ ceramics. This has been due, in large part, to information provided to the synthetic chemists and ceramicists from a variety of characterization techniques. As is apparent from the data that has been presented, no one characterization technique provides sufficient information about a material under development. It was only by using a combination of techniques that we have begun to understand how to improve $CaLa_2S_4$ and related ternary sulfides.

ACKNOWLEDGMENT

I would like to thank W. White at Pennsylvania State University, D. Roy at Coors Porcelain Company, and R. Gentilman and K. Saunders at Raytheon for supplying $CaLa_2S_4$ samples and for interacting constructively with us at NWC. I would also like to thank the following people for their invaluable help: M. Nadler (FTIR measurements); T. Atienza-Moore, H. Watson (XRF, SEM/EDX); A. Green (AES/ESCA measurements); M. Hills (optical microscopy); R. Muro (atomic absorption analyses); R. Kubin, R. Schwartz (DTA, TGA, optical fluorescence); P. Archibald (scattering measurements); and H. Bennett, J. Bennett, D. Burge (discussions about optics). R. McIntire, R. Smith, and the NWC Photography Laboratory helped with the manuscript preparation and graphic arts.

REFERENCES

1. S. Musikant, R. A. Tanzilli, R. J. Charles, G. A. Slack, W. White, R. M. Cannon, "Advanced Optical Ceramics, Phase 'O'-Final Report," Office of Naval Research, Contract No. N00014-77-C-0649, Report No. DIN: 78SDR2195, 15 May 1978.
2. S. Musikant, R. M. Cannon, Jr., C. E. Dulka, A. Gatti, J. J. Gebhardt, W. A. Harrison, I. C. Huseby, S. Prochazka, H. W. Rauch, Sr., G. A. Slack, R. A. Tanzilli, W. B. White, "Advanced Optical Ceramics, Phase I," Annual Report for 1 June 1978-31 May 1979, Office of Naval Research, Contract No. N00014-78-C-0466, Report No. DIN: 79SDR2297, 31 August 1979.
3. P. L. Provenzano and W. B. White, Synthesis and Crystal Chemistry of Sulfides and Tellurides with the Th_3P_4 Structure, Report No. AFCRL-TR-74-0560, NTIS No. AD-A008-490, August 1974.
4. P. L. Provenzano, S. I. Boldish, and W. B. White, Vibrational Spectra of Ternary Sulfides with the Th_3P_4 Structure, Mat. Res. Bull. 12:939 (1977).
5. W. H. Zachariasen, Crystal Chemical Studies of the 5f-Series of Elements. VI. The Ce_2S_3-Ce_3S_4 Type of Structure, Acta Crystallogr., 2:57 (1949).
6. J. Flahaut, M. Guittard, M. Patrie, M. P. Pardo, S. M. Golabi, and L. Domange, Phases Cubiques Type Th_3P_4 dans les Sulfures, les Séléniures et les Tellurures L_2X_3 et L_3X_4 des Terres Rares, et dans leurs Combinaisons ML_2X_4 avec les Sulfures et Séléniures MX de Calcium, Strontium, et Baryum. Formation et Propriétés Crystallines, Acta Crystallogr. 19:14 (1965).
7. J. Flahaut, "Sulfides, Selenides, and Tellurides," in: Handbook of the Physics and Chemistry of Rare Earths, Chap. 31, Vol. IV, K. A. Gschneidner, Jr. and L. Eyring, eds., North Holland Pub. Co., Amsterdam (1979).
8. F. K. Volynets, G. N. Dranova, N. V. Vekshina, and I. A. Mironov, Stoichiometry and Optical Properties of Hot-Pressed Lanthanum Sulfides, Izv. Akad. Nauk. SSSR, Neorg. Mat., 13(3):526 (1977).
9. A. A. Kamarzin, V. V. Sokolov, K. E. Mironov, Yu. N. Malovitsky, and I. G. Vasil'yeva, Physical-Chemical Properties of Stoichiometric Lanthanum Sesquisulfide Monocrystals, Mat. Res. Bull., 11(6):695 (1976).
10. A. A. Kamarzin, K. E. Mironov, V. V. Sokolov, Yu. N. Malovitsky, and I. G. Vasil'yeva, Growth and Properties of Lantanum and Rare-Earth Metal Sesquisulfide Crystals, J. Cryst. Growth, 52:619 (1981).

11. W. B. White, D. Chess, C. Chess, and J. V. Biggers, CaLa$_2$S$_4$: A Ceramic Window Material for the 8-14 μm Region, in "Emerging Optical Materials," Proc. of SPIE, 297:38 (1981).
12. JCPDS Powder Diffraction File.
13. A. W. Sleight and C. T. Prewitt, Crystal Chemistry of Rare Earth Sesquisulfides, Inorg. Chem., 7:2282 (1968).
14. M. Picon, L. Domange, J. Flahaut, M. Guittard, and M. Patrie, No. 41-Les Sulfures Me$_2$S$_3$ et Me$_3$S$_4$ des Éléments des Terres Rares, Bull. Soc. Chim, Fr., 221 (1960).
15. W. B. White, J. V. Biggers, D. Chess, D. Minser, C. A. Chess, and E. P. Plesko, Annual Report for 1 May 1980-31 October 1981, Office of Naval Research, Contract No. N00014-80-C-0526, October 1981.
16. R. W. Tustison, Final Technical Report for 1 June 1980-31 May 1981, Office of Naval Research, Contract No. N00014-80-C-0430, May 1981.
17. R. Gentilman and K. Saunders, private communication.
18. H. E. Bennett, D. K. Burge, C. K. Lowe-Ma, and J. M. Bennett, "Optical Properties of Advanced Missile Dome Materials," Naval Weapons Center, TP 6284, July 1981.
19. J. Flahaut, L. Domange, and M. Patrie, No. 340-Combinaisons Formées par les Sulfures des Éléments der Groupe des Terres Rares. VI.-Étude Cristallographique des Phases Ayant le Type Structural du Phosphure de Thorium Th$_3$P$_4$, Bull. Soc. Chim. Fr. 2048 (1962).
20. T. Toide, T. Utsunomiya, Y. Hoshino, and M. Sato, Preparation and Structures of Alkaline Earth-Lanthanum Double Sulfides using Carbon Disulfide as a Sulfurizing Agent, Bull. Tokyo Inst. Technol., 126:35 (1975).
21. W. B. White, private communication.
22. A. A. Grizik, A. A. Eliseev, V. A. Tolstova, and G. P. Borodulenko, Homogeneity Range of Rare-Earth Element Sulfides with a Structure of Th$_3$P$_4$ Type, Russ. J. Inorg. Chem. 23(3):330 (1978).

23. P. D. Dernier, E. Bucher, and L. D. Longinotti, Temperature Induced Symmetry Transformation in the Th_3P_4 Type Compounds La_3S_4, La_3Se_4, Pr_3S_4 and Pr_3Se_4, J. Solid State Chem., 15:203 (1975).
24. JCPDS Powder Diffraction File #29-339.
25. D. Roy, private communication.

APPLICATIONS OF ANALYTICAL MICROSCOPY TO

CERMAIC RESEARCH

>Paul F. Johnson
>
>New York State College of Ceramics
>Alfred University
>Alfred, NY 14802

INTRODUCTION

The objective of this paper is twofold. First, a definition of Analytical Electron Microscopy (AEM) will be developed through consideration of the development of instrumentation and techniques. Second, the usefulness of AEM for the characterization of ceramic materials will be demonstrated through discussion of specific applications. Details of the design and use of instrumentation, specific sample preparation requirements and interpretation of data obtained cannot be presented in a short review paper. However, the references are intended to point the reader to pertinent literature in each of these critical areas. While the recent literature available describing AEM is substantial, it is also specialized. Textbooks have been written covering topics presented in this review in a sentence. So, with the understanding that the field of AEM is ill-defined, broad and deep, let us proceed.

One of the more annoying problems facing the investigator concerned with the characterization of ceramic materials is the need to keep up with new (and changing) acronyms. This is particularly true of the area of AEM. Perhaps this reliance upon acronyms reflects the expense of AEM with the requirement for significant government funding. It is difficult to imagine preparing a manuscript describing electron optical systems without involving an abundance of acronyms. An attempt will be made here to take advantage of the alphabet soup associated with AEM through a systematic description of the various configurations of electron optical systems. There seems to be two good reasons for this approach; first, understanding the individual instruments and techniques which together make up AEM provides an understanding (and indeed a definition) of AEM. Second, the recent literature concerning characterization of ceramic materials is full of AEM acronyms. Successful interpretation of this literature depends upon understanding the techniques reported. Table I lists the acronyms used in this paper.

In considering virtually any method of instrumental characterization, the notion of the generalized analytical instrument is often useful. Figure 1 shows the three concepts involved in such an instrument; a source of energy, the interaction of energy and sample, and a detector system. In the case of electron optical systems, the energy incident upon the sample consists of a beam of high energy electrons. Three types of electron sources are in general use; the thermionic source,[1] the field emission source[2] and the LaB_6 source.[3] The interaction of incident electrons with the sample results in three categories of events; modification of the incident energy, generation of new energy species and modification of the sample itself. Electron beam − sample interactions give rise to the generation of number of detectible species as shown in Fig. 2. Table II lists the signals shown in Fig. 2, and indicates typical detector systems and resolution.

Virtually all of the detectable species shown in Fig. 2 are generated within the sample effectively simultaneously. No single detector system has yet been developed which can detect and sort the various species generated. Therefore a number of detector systems (each described by an acronym) must be available to the investigator to take advantage of the full capabilities of the electron optical analytical method. An electron optical instrument and associated detector systems define the Analytical Electron Microscope.

Historical Development of AEM

The history of electron optical techniques began in 1926 with the calculation of the trajectories of electrons in a cathode ray tube.[4] The first commercially available TEMs (early 1940's) consisted of a source of high energy electrons, a series of illumination lenses located between the source and the sample, series of projection lenses below the sample. A photographic plate served as the detector. Early TEMs provided two modes of characterization. Examination of Bragg diffracted incident electrons yielded information concerning the atomic and defect structure of the specimen, and, in another mode, mass, diffraction and/or phase contrast resulted in morphological information (i.e., the spatial relationships which exist between contrast features in a specimen). Chemical composition could often be deduced from structural and morphological analysis, but was not directly measurable.

Table I. A summary list of acronyms commonly used in AEM. Note that authors may intend either microscope or microscopy is referenced by the "M" in the instrumentation section below.

A. Signals

Acronym	Name	Characteristic Information
SE	Secondary Electron	Surface Topography
BSE	Backscattered Electron	Relative Atomic Number
AE	Auger Electron	Surface Composition
λ_C	Characteristic X-ray	Elemental Composition
λ_{CL}	Cathodeoluminescence	Molecular Bonding
E_D	Diffracted Electron	Crystal Structure
E_L	Low Loss Electron	Elemental Composition
E_{SC}	Specimen Current	Absorbed Electrons
EBIC	Electron Beam Induced Current	Electron-Hole Diffusion
E_S	Elastically Scattered Electron	Mass Contrast, etc.

B. Instrumentation

Acronym	Name
AEM	Analytical Electron Microscope
ASEM	Analytical Scanning Electron Microscope
ASTEM	Analytical Scanning Transmission Electron Microscope
EMP	Electron Microprobe
AES	Auger Electron Spectroscopy
EDS	Energy Dispersive Analysis
WDS	Wavelength Dispersive Analysis
EELS	Electron Energy Loss Spectroscopy

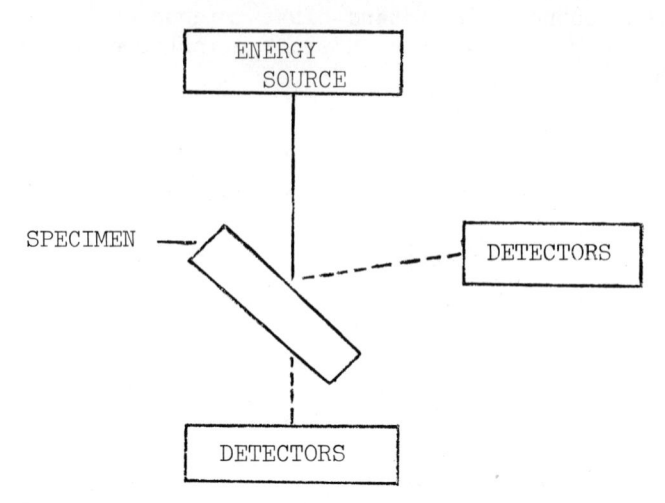

Fig. 1. The generalized analytical instrument energy interactions within the sample give rise to a number of detectible signals. Both reflected and transmitted detectors are shown.

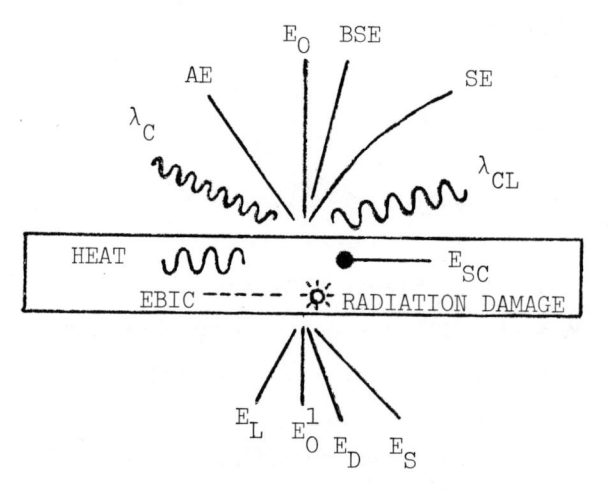

Fig. 2. Signals produced by interaction of high energy electrons with a ceramic material. Note that all signals shown at the top surface are also generated at the bottom surface of thin specimens.

Table II. AEM summarized; the signals, detectors, measured characteristics minimum spatial resolution and references to review articles. Refer to Fig. 2 for description of signals.

Detectible Signal	Analytical Instrument	Standard Detector	Typical Resolution	Characteristic Determined
SE	SEM STEM	Photomultiplier (Everhardt-Thornley)	6nm 2nm	Surface Topography
BSE	SEM STEM	Biased diode	1μm 10nm	Relative Atomic Number
AE	ASEM (AES)	Electron energy spectrometer	20nm	Surface composition
λ_C	ASEM EMP ASTEM	EDS/WDS WDS EDS	1μm 1μm 20nm	Elemental Composition
λ_{CL}	ASEM ASTEM	Monochromator with photo-multiplier		Molecular bonding
E_D	CTEM STEM ASTEM	Film or plate Diode Diode	1μm 10nm 10nm	Structure-atomic and defect
E_L	ASTEM	Electron energy spectrometer	20nm	Elemental composition, chemical binding energy molecular bonding, etc.
E_I, E_D	CTEM	Photofilm or plate	2nm	Morphology
SC	ASEM ASTEM	Electrometer	1μm 2μm	
EBIC	ASEM	Photofilm	1mm	Diffusion coefficients

Electron optical analysis of thick samples was made possible by the development of the scanning electron microscope (SEM).[5] In its earliest form, this instrument provided a means of recording the number of secondary electrons generated in the near surface of a thick specimen and detected nearby. The resulting image contrast provided information concerning the topography of the sample surface. The advantage of secondary electron analysis of the surface topography of thick specimens lay in the high spatial resolution available and the interpretability of the image. Even in the earliest commercial instruments, the electron probe size (and therefore, approximately, the spatial resolution) was 15-20nm.

The use of an electron optical system to measure directly the elemental composition of a specimen was first described by Castaing.[6] He made use of two major advantages of the SEM in developing the electron microprobe (EMP); the focused electron probe and the relatively large distance possible between the final lense and the specimen. He used a wavelength dispersive spectrometer (WDS)[7] to quantitatively measure the wavelength (and so the energy) of characteristic X-rays generated within the sample. Analysis of the characteristic wavelength spectrum provided a quantitative description of the elemental composition of a material composed of elements heavier than (approximately) Na. Though EMP was developed separately from SEM, it was obvious that the two instruments could easily be combined. Today the result of combining the SEM and EMP into a single instrument results in the analytical electron microscope, or more specifically, the analytical scanning electron microscope (ASEM).

The development of the Lithium drifted Silicon, Si(Li), detector[8] greatly enhanced the usefulness of the ASEM for the analysis of elemental composition of a material. The energy dispersive spectrometer (EDS) detector system based upon the Si(Li) detector provided a means of rapidly ($\sim$1 minute) qualitative determination of the elemental composition of materials. As was the case for the WDS detector system, detection of light elements of particular interest to the ceramist proved difficult.

The electron probe forming and scanning capabilities which resulted in the ASEM and EMP were applied to the TEM and resulted in the scanning transmission electron microscope (STEM).[9] The addition of this acronym to the literature resulted in renaming the TEM to CTEM, the "C" standing for "conventional". Even without adding detectors for composition analysis advantages were realized, particularly in the area of electron diffraction analysis of the structure of materials.[10,11] Electron probe sizes as small as 10.0nm provided sufficient diffracted electron signal for structure determinations. Thus, areas of the electron-thin samples as small as 10.0nm in diameter could be structurally analyzed.

New techniques for structure analysis were rapidly developed. The convergent beam electron diffraction (CBD) technique discovered by Kossell and Mollenstedt[12] and recently described by Steed[13] was made available on commercial instrumentation. The CBD technique nicely compliments traditional diffraction and can be particularly useful in measuring strain. High resolution electron microscopy (HREM)[14,15] is particularly useful for analysis of grain boundaries and polytypes. The ultimate in HREM is now commercially available; the imaging of a single atom.[16]

Developments in composition analysis of electron-thin specimens using the STEM paralleled the development of structure analysis techniques. The ability to form extremely small focussed electron probes on the surface of thin specimens, coupled with the fact that the electron beam does not scatter significantly in thin sections resulted in the ability to analyze characteristic X-rays which escape from a very small volume of the specimen. The activated (and for thin sections, the escape) volume of a specimen thinned to less than 100nm can be considered to be a cylinder as small as 20.0nm in diameter.[17] Thus, high resolution elemental composition analysis was possible. Given the sensitivity of EDS analysis under ideal conditions, the presence of just a few atoms within the activated volume could be detected.[18]

Of significant importance to the ceramist was the development of Electron Energy Loss Spectroscopy (EELS). The EELS detector[19] provided a means of measuring the energy loss of incident electrons due to single scattering events within the sample. Analysis of the resulting EELS Spectrum provided elemental composition determination with roughly the same spatial resolution as EDS but with the very real advantage of detection of the light elements carbon, nitrogen and oxygen. Detector sensitivity for light elements is particularly important to the ceramist since up to 75-80% (molar basis) of ceramic materials cannot be routinely detected by either EDS or WDS detector systems.

An additional advantage of EELS arises from detailed analysis of the fine structure of the energy spectrum. Not only is elemental analysis of a ceramic possible, but also molecular energy levels and chemical bonding states can be determined.[20] In addition, details of the electronic structure of atoms and, measurement of the local dielectric constant may also be measured.

The ability of the Analytical Scanning Transmission Microscope (ASTEM) to determine the composition, structure and morphology of a ceramic material with very high resolution makes it a significant and powerful analytical method. Only the light microscope can provide as complete a characterization of a single specimen and even the light microscope cannot provide the resolution of the ASTEM.

Limiting Factors in AEM

The development of AEM so briefly reviewed above was not without problems. Early use of EDS and WDS systems resulted in the reporting of elements described as being in the sample which in reality were systematic spectral contaminants, generated from sources within the specimen chamber of the microscope. Redesign of detector mount and the use of low atomic number collimators solved this problem.[21,22] As resolution was increased through lense and source improvements, radiation damage[23] and sample contamination[24] were recognized as significant problems. Radiation damage continues to be a significant problem and a topic of current research. Specimen contamination has been limited through the use of improved vacuum systems and improved sample preparation procedures.[25] Data interpretation for advanced techniques placed stringent requirments upon sample preparation techniques. Improved sample preparation procedures are constantly being reported.[26] Even defining and measuring the resolution of AEM instrumentation proved difficult.[27]

In most cases, data interpretation and reduction lagged well behind instrumental developments; those who purchased early versions of specific AEM hardware configurations found they could not make full use of the data made available to them because quantitative data reduction techniques were not available. Even when available, data reduction for composition analysis required access to a computer. Computerized methods for data reduction continues as a source of significant new development.

As these problems were solved, new hardware was developed to implement the solutions. Thus, the "generations" associated with AEM hardware were almost as short as those associated with computers; purchasers of AEM instrumentation in the mid-1970s found their equipment outdated in 3-5 years.

Interpretation of AEM Data

A number of excellent texts describe the theory fundamental to the interpretation of AEM data.[28-32] Clearly, the subject is too detailed to be covered in a brief review. However, a few specific methods of data presentation and interpretation deserve mention.

Perhaps the most intuitively interpretable and deceptively simple data made available through AEM are the topographical maps of surfaces provided by the SE mode of ASEM analysis. Intuition is surprising here; we have no direct experience with electrons upon which to base our interpretation. However, interpretation of the contrast of SE images in terms of a corresponding image produced by visible light proves successful (assuming that the photograph is properly oriented). The straight-forward interpretability of SE images often leads the investigator to ignore techniques for strengthening arguments based upon them. Too often SE image interpretation is limited to a qualitative comparison of features. For example, the microstructure of ceramic specimens prepared using different fabrication parameters are presented as a series of micrographs which the reader is asked to qualitatively interpret. The author, having spent many hours viewing the microstructures, makes a selection of micrographs which support his arguments. Certainly his arguments would be greatly strengthened if the techniques of quantitative microscopy (QM) and stereology were applied and presented. Statistical data can be developed (usually at the expense of substantial effort) which provide quantitative support for reported conclusions.

The techniques of QM are comprehensively presented by DeHoff and Rhines[33] and Underwood.[34] A number of commercial systems* are available for the quantitative analysis of both AEM and LM images. Public domain software packages are also available; for example, in reports by White, et.al.[35] Recent commercial development has resulted in combining EDS/WDS/EELS data acquisition and processing equipment with image analysis capabilities.[36,37]

Another application of the computer to data interpretation of AEM images is computer generated defect and structure imaging.[38-40] Calculated images can be considered "first principle images" in the sense that they represent images expected based upon application of physical theory. Techniques described in the above cited papers have resulted in the production of an atlas of TEM images.[41]

Perhaps the most widely used programs for computer data acquisition and reduction in AEM are in the area of EDS/WDS/EELS composition analysis. This area also represents one of the most important topics of continuing research. The modern student of AEM composition analysis no doubt would be lost without computer programs for data reduction such as MAGIC,[42] FRAME,[43] CORDATA[44] or Bence-Albee[45] (or commercial variations of these programs). The student is advised to try the hand calculations developed and

---

* Ortec Inc., Oak Ridge, TN 37830.
  Princeton Gamma-Tech, Princeton, NJ 08540.
  Tracor-Northern, Middleton, WS 53562.

described by Birks.[46] Two advantages are realized through application of Birks' methods. First, the reason a computer is useful is demonstrated because a systematic application of Birks' method is both hard work and time consuming. Second, and certainly more important, an appreciation and understanding of corrections for atomic number, absorption and fluorescence errors (commonly referred to as ZAF corrections) is developed. This is important because all empirical computerized methods of data reduction are to some extent variations of Birks' methods.[47]

The quantitative assessment of data made available through AEM remains an area of current research. Both AEM hardware and software improvements are constantly increasing the accuracy and precision of the description of the structure, composition and morphology of ceramic materials. The next section will review the use of both AEM hardware and software to applications in ceramics.

Applications of AEM for Analysis of Ceramics Materials

The 9 papers immediately following this review represent excellent examples of the application of the techniques of AEM for analysis of ceramic materials. The most general statement concerning application of AEM which can be made is that the technique can be used to quantitatively describe the structure, composition and morphology of materials, often using a single sample. Table III recasts Table II to describe the characteristics which can be quantitatively determined using AEM.

Table III. Examples of characteristics which can be determined using AEM.

Characteristic	Technique	Resolution
Composition	ASEM/EDS	1µm
	ASEM/WDS	1µm
	EMP	1µm
	ASTEM/EDS	20nm
	ASEM/WDS	40nm
	ASEM/EELS	20nm
	ASEM/AES	20nm
Structure	CTEM	0.1µm
	(A)STEM	10nm
Morphology	(A)SEM	6nm
	CTEM	2nm
	(A)STEM	2nm

The ceramic literature is rich concerning the use of AEM for qualitative description of the character of ceramic materials. A few specific examples are cited here. Votava and Reed described the structure of plaster slipcasting molds using the ASEM.[48] Votava discussed techniques which were used successfully to describe the effects of binders upon green ceramic compacts.[49] Johnson and Hench used ASTEM to describe anodized films formed on sintered tantalum anode capacitors.[50] Hench and Jenkins used CTEM to quantitatively describe agglomeration of ceramic powders.[51] Clarke describes the use of HREM for boundary analysis in SiN_3.[52]

Rather than continue to catalog applications papers, thus requiring selection of examples from among the hundreds available, perhaps it would be more useful to cite sources in which the interested reader would be certain to find papers describing and reviewing the applications of AEM. The May-June 1979 issue of the Journal of the American Ceramic Society was devoted to applications of AEM. Seimens,[53] Jeol[54] and Philips[55] publish widely available periodicals which contain interesting review and application papers. Scanning[56] is a recent journal for AEM. Other journals include Ultramicroscopy[57] and The Journal of Microscopy.[58] The Electron Microscopy Society of America (EMSA), the Microbeam Analysis Society (MAS) and SEM, Inc. publish the proceedings of their annual meetings. Finally, the biological literature represents a good source of information and applications of AEM.

The newsletters published by EMSA and MAS are both excellent sources of information. They contain occasional bibliographies of recent work, interesting review and applications papers and news. Both newsletters are a privilege of membership in these societies.

References

1. Busch, H. "Calculation of the Trajectory of Cathode Rays in Electromagnetic Fields of Axial Symmetry", Ann. d. Physik 81 974-93 (1926).

2. Gomer, R. Field Emission and Field Ionization, Harvard Univ. Press, Cambridge, Mass. (1961).

3. Broers, A.N. "A New High Resolution Reflecting Electron Microscope", Rev. Sci. Instrum. 40 [8] 1040-1045 (1968).

4. Davisson, C.J. and L.H. Germer, Diffraction of Electrons by a Crystal of Nickel", Phys. Rev. 30 [6] 705 (1027).

5. Zworykin, V.K., J. Hillier and R.L. Snyder, "Scanning Electron Microscope" ASTM Bulletin 117, American Society for Testing Materials, Phila. Pa. (1942).

6. Castaing, R. "Applications of Electron Probes to Local Chemical and Crystallographic Analysis", Ph.D. Thesis, University of Paris, ONERA pub. #55 (1952).

7. Friedman, H. and L.S. Birks, "Geiger Counter Spectrometer for X-Ray Fluorescence Analysis". Rev. Sci. Instrum. 19 [5] 323-330 (1948).

8. Bailey, N.A., J. Grainger and J.W. Mayer "Capabilities of Lithium Drifted p-i-n Junction Detectors When Used for Gamma-Ray Spectroscopy", Rev. Sci. Instrum. 32 [7] 865 (1961).

9. Hillier, J. and R.F. Baker, "Microanalysis by Means of Electrons", J. Appl. Phys. 15 [9] 663 (1944).

10. Geiss, R.H., "Electron Diffractions from Areas Less than Diameter", Appl. Phys. Lett. 27 [4] 174 (1975).

11. Warren, J.B. "Microdiffraction" in *Introduction to Analytical Electron Microscopy* eds. J.J. Hren, J.I. Goldstein and D.C. Joy. Plenum Press, NY. 369-384 (1979).

12. Kossell, W. and G. Mollenstedt "Electron Interference in Convergent Rays", Ann. Phys. 36 [1] 113-40 (1939).

13. Steeds J.W., "Convergent Beam Diffraction" in *Introduction to Analytical Electron Microscopy*, eds. J.J. Hren, J.I. Goldstein and D.C. Joy. Plenum Press, NY. 387-423 (1979).

14. Clarke, D.R., "High Resolution Techniques and Applications to Nonoxide Ceramics", J. Am. Ceram. Soc. 62 [5,6] 236-46 (1979).

15. Clarke, D.R. and G. Thomas, "Grain Boundary Phases in Hot Pressed MgO Fluxed Silicon Nitride", J. Am. Ceram. Soc. 60 [11-12] 491-95 (1977).

16. Isaacson, M., M. Ohtsuki and M. Utlaut, "Electron Microscopy of Single Atoms" in *Introduction to Analytical Electron Microscopy*, eds. J.J. Hren, J.I. Goldstein and D.C. Joy. Plenum Press NY. 343-368 (1979).

17. Goldstein, J.I., J.L. Costley, G.W. Lorimer and S.J.B. Reed, "Quantitative X-Ray Analysis in the Electron Microscope" in Scanning Electron Microscopy - 1977, 315-242, ed. O. Johari, ITRII/SEM/(1977).

18. Smolyo, A.P., H. Shuman and A.V. Smolyo, "Quantation, Minimumal Detection Limits and Applications of Biological

Electron Probe Analysis", p. 115 in Analytical Electron Microscopy - Report of a Specialist Workshop, Cornell University (1976).

19. Isaacson, M. and D. Johnson, "The Microanalysis of Light Elements Using Transmitted Energy Loss Electrons", Ultramicroscopy 1 [1] 33-52 (1975).

20. Joy, D.C. "The Basic Principles of Electric Energy Loss Spectroscopy" in Introduction to Analytical Electron Microscopy, eds. J.J. Hren, J.I. Goldstein and D.C. Joy. Plenum Press, NY. 223-244 (1979).

21. Hren, J.J., P.S. Ong, P.F. Johnson and E.J. Jenkins, "Modification of an Philips EM 301 for Optimum EDS Analysis" in Proceedings of the 34th Annual EMSA Meeting, 418, G.W. Bailey, ed. Claitor's Publishing, Baton Rouge, La. (1976).

22. Bently, J., N.J. Zaluzec, E.A. Kenik and R.W. Carpenter, "Optimization of an Analytical Electron Microscope for X-ray Microanalysis: Instrumental Problems in SEM 1979, ed. O. Johari, ITRII/SEM/1979, Washington DC, Chicago Press (1979).

23. Barnard, R.S., G. Das, B.J. Pletka and T.E. Mitchell, "Electron Irradiation Damage in MgO and Al_2O_3", Bull. Am. Ceram. Soc. 54 [4] 440 (1975).

24. Hren J.J., "Barriers to AEM: Contamination and Etching" in Introduction to Analytical Electron Microscopy eds. J.J. Hren, J.I. Goldstein and D.C. Joy. Plenum Press NY. 481-505 (1979).

25. Hren J.J., E.J. Jenkins and E. Aigeltinger, "Anti-Contamination Device for STEM" in Proceedings of the 35th Annual EMSA Meeting, 66, W. Bailey, ed. Claitors Publishing, Baton Rouge, La. (1977).

26. Heuer, A.H., R.F. Firestone, J.D.Snow, H.W. Green, R.G. Howe and J.M. Christie, "An Improved Ion Thinning Apparatus", Rev. Sci. Instrum. 42 [8] 1177-84 (1971).

27. Newbury, D.E. and R.L. Myklebust "Calculation of Electron Beam Spreading in Composite Thin Foil Targets". Microbeam Analysis (1980) 173.

28. Edington, J.W. Practical Electron Microscopy In Materials Science, V1-V5, Phillip's Technical Library, Eindhoven, Holland (1974).

29. Murr, L.E., <u>Electron Optical Applications in Materials Science</u>. McGraw-Hill Book Co., NY (1970).

30. Goldstein, J.I., D.E. Newbury, P. Echlin, D.C. Joy, C. Fiori and E. Lifshin, <u>Scanning Electron Microscopy and X-Ray Microanalysis</u>, Plenum Press, NY (1981).

31. Hirsch, P.B., A. Howie, R.B. Nicholson, D.W. Pashley and M.J. Whelan. <u>Electron Microscopy of Thin Crystals</u>. Butterworths, London (1965).

32. Hren, J.J., J.I. Goldstein and D.C. Joy (eds). <u>Introduction to Analytical Electron Microscopy</u>, Plenum Press, NY (1978).

33. DeHoff, R.T. and F.H. Rhines <u>Quantitative Microscopy</u>, McGraw-Hill Book Company, NY (1968).

34. Underwood, E.E. <u>Quantitative Stereology</u>, Addison Wesley Pub. Co., Reading, Mass. (1970).

35. White, E.W., W.R. Buessem, H.A. McKinstry and G.G. Johnson, Jr. "Quantitative SEM Characterization Methods for Ceramic Materials in All Process Stages and Applications to High Strength Aluminas and Silicon Nitride", Technical Report to ONR, Contract Number #N00014-67-A-0385-0019 (1975).

36. McCarthy, J.J., G.S. Fritz and R.J. Lee "Acquisition Storage and Display of Video and X-Ray Images" in <u>Microbeam Analysis</u>, R.H. Geiss, ed. 30-34. San Francisco Press (1981).

37. Lebiedzik, J., R. Edwards and B. Phillips, "Use of Microphotography Capability in the SEM for Analyzing Fracture Surfaces. 61-65 In SEM/1979, ed. Ohm Johari ITRII/SEM/(1979).

38. Sykes, L.J., W.D. Cooper and J.J. Hren. "Computer Simulated TEM Images" J. Nucl. Mat. (in press).

39. Humble, P. "Analysis of Defects Using Computer Simulated Images" <u>Introduction to Analytical Electron Microscopy</u>, eds. J.J. Hren, J.I. Goldstein and D.C. Joy. Plenum Press, NY. 551-574 (1979).

40. Head, A.K., P. Humble, L.M. Clarebourgh, A.J. Morton and C.J. Forward, <u>Computer Electron Micrographs and Defect Identificaton</u>. North Holland, Amsterdam (1973).

41. Sykes, L.J., W.D. Cooper and J.J. Hren. "Catalog of Computer simulated TEM Images of Small FCC and BCC Dislocation Loops". Oak Ridge Nation Laboratory Report, ORNL/TM-7619, Feb 1981. Available through NTIS.

42. Colby, J.W. "Quantitative Microprobe Analysis of Thin Insulating Films", p. 287-305 in Advances in X-ray Analysis. J.B. Newkirk, G.R. Mallett and H.G. Pfeiffer, eds. Vol. 11, Plenum Press, NY (1968).

43. Henoc, J., K.F.J. Heinrich and R.L. Myklebust, NBS Technical Note 769, U.S. Government Printing Office, Washington, DC (1973).

44. K.F.J. Heinrich, R.L. Myklebust, H. Yakowitz and S.D. Rasberry, "A Simple Correction Procedure for Quantitative Electron Probe Microanalysis". NBS Technical Note 719, U.S. Government Printing Office, Washington, D.C. (1972).

45. Bence, A.E. and A. Albee "Empirical Correction Factors for the Electron Microanalysis of Silicates and Oxides", J. Geol. 76 382 (1968).

46. Birks, L.S., Electron Probe Microanalysis, 2nd edition. Wiley-Interscience NY (1970).

47. Goldstein, J.I., D.E. Newbury, P. Echlin, D.C. Joy, C. Fiori and E. Lifshin. Scanning Electron Microscopy and X-ray Microanalysis. 305-391 Plenum Press, N.Y. (1982).

48. Votava, W.E., L.M. Holleran and J.S. Reed, "Microstructures and Properties of Permeable Mold Materials", Bull. Am. Ceram. Soc. 58 [2] 194-7.

49. Votava, W.E. "ASEM Analysis of Unfired Ceramic Compacts", Microbeam Analysis, R.H. Geiss, ed. 283-286 (1981).

50. Johnson, P.F. and L.L. Hench. "Characterization of Ceramic Microprocessing" in Processing of Crystalline Ceramics 87-98. Ed. H. Palmour, R.F. Davis and T.M. Hare. Plenum Press (1978).

51. Hench, L.L. and E.J. Jenkins, "Characterization of Agglomerates with Transmission Electron Microscopy" in Ceramic Processing Before Firing, ed. G.Y. Onoda and L.L. Hench, 75-84, John Wiley and Sons, NY (1978).

52. Clarke, D.R., "On the Detection of Thin Intergranular Films by Electron Microscopy", Ultramicroscopy, 4 [1] 33-44 (1979).

53. Seimens Review. Siemens Arliengesellschaft, Berlin and Munich. D8520 Erlangen 2, Postfach 3240, Fedral the Republic of Germany.

54. JEOL News. JEOL (U.S.A.) Inc. 11 Dearborn Rd., Peabody, Mass. 01960 (1963-present).

55. Norelco Reporter. Philips Electronic Instrument Co. 85 McKee Dr., Mahwah, NJ (1953 to present).

56. Scanning. Gehard Wetzstrock Publishing House Inc., N.Y. (1978 to present).

57. Ultra-Microscopy. North Holland Publishing Co., P.O. Box 211, 1000 AE Amsterdam, The Netherlands (1975 to present).

58. Journal of Microscopy. The Royal Microscopical Society. 37/38 St. Clements, Oxford XX41AJ, Great Britain (1969 to present under present title).

ANALYSIS OF SECOND-PHASE PARTICLES IN Al_2O_3

K.J. Morrissey and C.B. Carter

Department of Materials Science and Engineering
Bard Hall
Cornell University, Ithaca, NY 14853

INTRODUCTION

The preparation of alumina compacts usually involves adding small amounts of other oxides such as MgO, NiO, ZrO_2, etc. and it is well-known that such ceramic materials almost invariably also contain significant quantities of impurity elements such as Na, K, and Ca. A considerable amount of research effort has been devoted to locating both the impurities and the additives since these are often present in concentrations exceeding the solubility limit. An example of such research relates to the question of whether elements such as Ca and Mg segregate to grain boundaries or are localized at second-phase particles; at present it appears that the Ca is found uniformly at grain boundaries whereas the Mg is found predominently in spinel particles[1-4]. Recent studies by Blanc et al.[5] have shown that the Mg can also be found in combination with K in K-β'''-Al_2O_3.

The present study is concerned with identifying second-phase particles in Al_2O_3 compacts and, in particular, with determining the mechanism by which they form. As a result of such observations it becomes possible, in some cases, to infer the location of the impurity ions during the sintering process. Although not a primary intention of this paper, it will also become apparent how the second-phase particles can be expected to influence the properties of the compact. The present paper is intended to illustrate the advantages in combining different electron microscope techniques.

EXPERIMENTAL

The observations discussed in this paper were mainly made on one commercial alumina which had been prepared by hot pressing. The powder initially contained 0.25% MgO to act both as a sintering aid and to limit grain growth giving a mean grain size of 1-2μm in the near-theoretically dense product. The starting material also has a small amount of intentionally added Ni and the compact may contain as much as 200ppm of dissolved carbon from the graphite dies used in the hot-pressing. The material was prepared for examination in the transmission electron microscope (TEM) by ultrasonically cutting 3mm discs from thin slices and ion thinning these after mechanically polishing to a thickness of 20 to 50μm.

The high resolution TEM observations were made using a Siemens 102 TEM operating at 125kV. The compositional analysis was carried out in a JEM200CX scanning transmission electron microscope (STEM) fitted with a Tracor Northern energy dispersive x-ray spectroscopy (EDS) system incorporating a high take-off angle (70°) detector and operating at 200kV.

RESULTS AND DISCUSSION

The grain morphology and the geometric features of the two common second-phase particles are illustrated in fig. 1 which is a low magnification bright-field TEM image. The shape of the

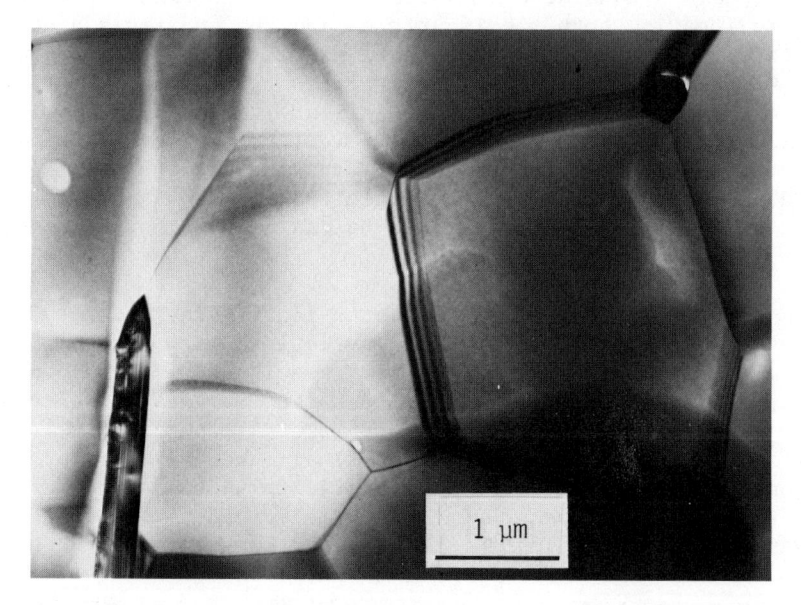

Fig. 1 Grain morphology and second-phase particles in a commercial Al_2O_3.

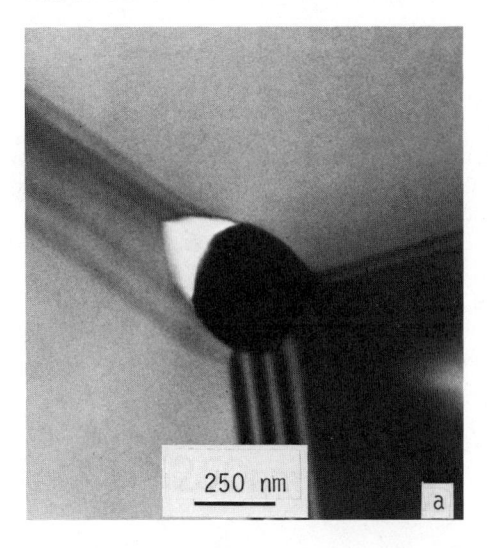

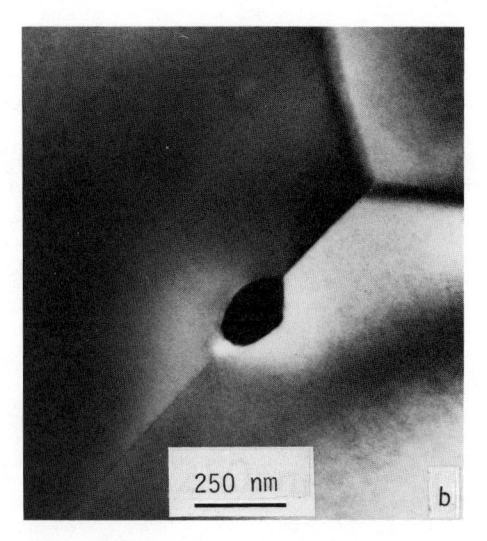

Fig. 2. Second-phase particles. (a) at a triple point, (b) in a grain boundary. Note the pore present in (a).

particles is typical of those found in this study. In amplitude contrast images (no objective aperture) the small particles always appear dark indicating that the average atomic number is greater in the particle. Such particles were invariably equiaxed with a diameter of ∼0.5μm. They were found both at triple points or on grain boundaries as shown in fig. 2. A common feature of these particles, particularly when at triple points, are the associated pores (see fig. 2a); such voids would naturally be expected to be detrimental to the mechanical strength of this material (see eg 7).

The EDS spectrum shown in fig. 3a is taken from a particle similar to those shown in fig. 2. The main constituent is Ni with various other elements, in particular Fe and Ga, also present. Since these particles are usually embedded in an Al_2O_3 matrix, the Al contribution to the spectrum could be caused by this material. The spectrum shown in fig. 3 was, therefore, obtained from a particle which protruded from the edge of the foil. Such particles were not common since they would presumably tend to drop out from the edge of a thin foil due to the associated pore. The Al peak is now considerably reduced and it is, therefore, concluded that this type of particle primarily contains Ni or NiO with considerable amounts of impurities. The observation implies that the particle is not Ni-Al spinel; other observations using, for example, electron energy loss spectroscopy (EELS) are necessary to distinguish unambiguously between Ni and NiO.

The elongated particle shown in fig. 1 is similar to those particles observed by Tighe[1] and Blanc et al.[5]. Tighe found that such grains were associated with extensive twinning. Blanc et al. found that Mg-Al spinel occurred as pseudolenticular inclusion and that platelets of K-β'''-Al$_2$O$_3$[6] grew in the spinel grains. Examples of spinel grains presented by Blanc et al. were ⪞1.5μm diameter and were identified as spinel by electron diffraction and semiquantitative x-ray microprobe analysis. The β'''-Al$_2$O$_3$ was identified in the study of Blanc et al., by electron diffraction and the presence of K was determined by microprobe analysis on areas that had previously been examined by TEM; Na was not detected in their platelets.

In the hot-pressed material investigated in the present study, the spinel and the β'''-Al$_2$O$_3$ were always found together. This

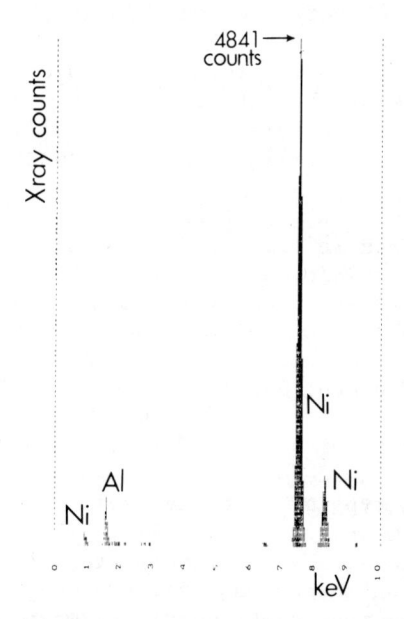

Fig. 3. Energy Dispersive x-ray Spectrum taken from a particle at the edge of the thin foil. The Al peak may arise from overlap with the matrix material.

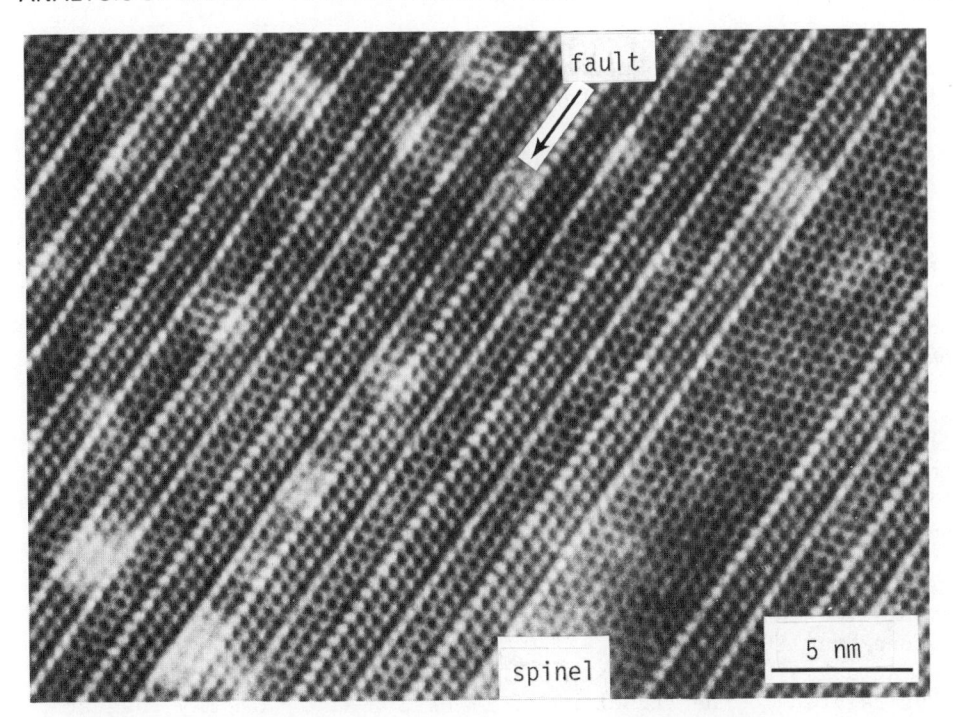

ANALYSIS OF SECOND-PHASE PARTICLES IN Al₂O₃   301

Fig. 4a. High-resolution image of β'''-Al$_2$O$_3$. The rows of white spots define the twin planes. A region of spinel and a stacking fault are also indicated.

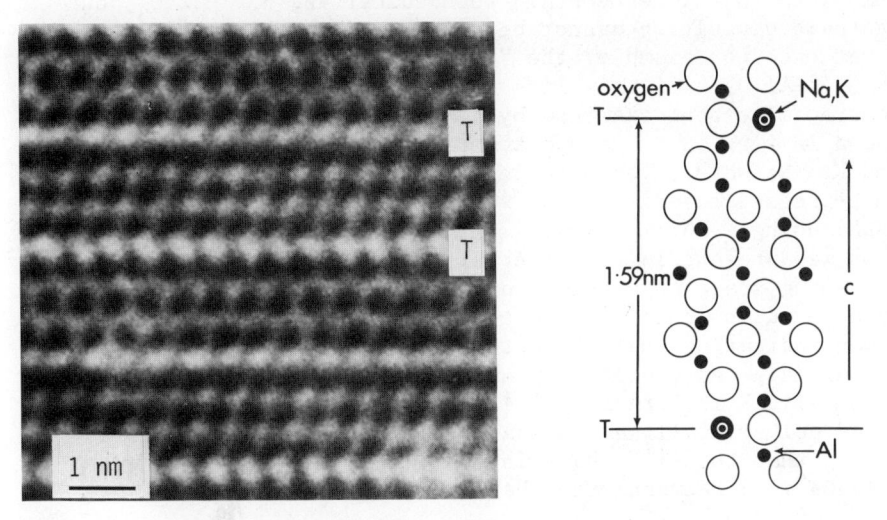

Fig. 4b. A comparison of the image with the β'''-Al$_2$O$_3$ structure. The structure diagram is that proposed in ref 6. T indicates the twin plane in each case.

conclusion was deduced from high resolution TEM images. An example of a high-resolution structure images is shown in fig. 4. The features of the image have been interpreted by comparison with the computed images of Bovin[7] and Matsui[8] for β'''-Al$_2$O$_3$[9]. The spinel blocks appear as rows of black or white spots depending on the conditions of defocus, they are black spots in this example, and the twin plane is defined in the image by the row of brighter white spots. The black spots in fig. 4a actually correspond to the projected positions of a pair of Mg ions in the spinel blocks. The spacing of the spinel {111} planes is 0.46nm and the point-to-point resolution of the Siemens 102 TEM is ∼0.35nm. The distance between the twin planes is 1.6nm as expected for β'''-Al$_2$O$_3$. It should be noted that the point-to-point resolution here is not sufficient to resolve individual atoms (the spacing between the 111 planes defined by the spots is 0.46nm while the d-spacing for the oxygen 111 planes is 0.23nm), however, there is a direct relation between the projected crystal structure and the high resolution TEM image.

The relationship between the β'''-Al$_2$O$_3$ and spinel is clearly emphasized in fig. 4 by the broad strip of spinel material. A second "fault" in the β'''-Al$_2$O$_3$ stacking is indicated by arrows and actually contains two spinel blocks rather than three. It is thus related to the structure of β- or β''-Al$_2$O$_3$ and may suggest a local deviation in Mg concentration. Such stacking errors can be introduced during the growth of β'''-Al$_2$O$_3$.

From this and other structural images, it is thus found that the elongated particles contain both spinel and β'''-Al$_2$O$_3$, but the actual composition cannot be determined unambiguously from such observations: for example, the "ideal" c-spacings of Na-β'''-Al$_2$O$_3$ and K-β'''-Al$_2$O$_3$ differ by less than ∼1%. These particles have, therefore, also been examined by EDS and a representative spectrum is shown in fig. 5a. The spectrum shows clearly the presence of Mg and K. No Na has yet been detected in any of the particles in the hot-pressed material; this result was confirmed by both EDS and EELS using a Philips EM400T. The possibility still exists that Na was present in the material, but was removed during the ion thinning stage of specimen preparation. Small quantities of Na were found in other, apparently amorphous, regions of the specimen indicating that the detectors were sufficiently sensitive to the anticipated concentrations. Na has now been detected together with K in larger β'''-Al$_2$O$_3$ particles in an extruded and sintered polycrystalline alumina. It is, therefore, presently concluded that the β''' phase in the hot-pressed material was K-β'''-Al$_2$O$_3$ in agreement with the results of Blanc et al.[6].

An interesting additional feature of the EDS spectrum shown in fig. 5a is the presence of Ar in quantities equal to, or in other examples, exceeding the concentration of potassium.

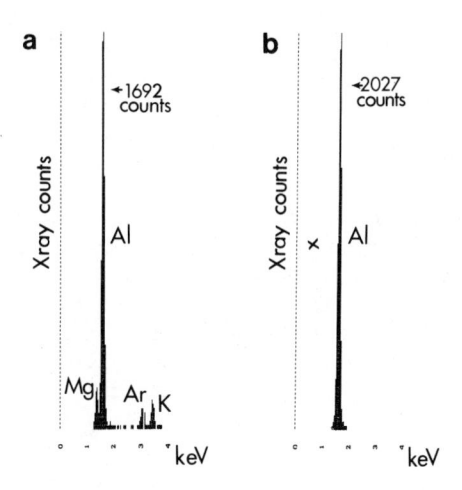

Fig. 5. Energy dispersive x-ray spectra. (a) from a β'''-Al_2O_3 particle, (b) from the Al_2O_3 matrix.

The argon K peak in an EDS spectrum coincides almost exactly with the Aluminum K-sum peak, but the count rate was in all cases well below that necessary for the Aluminum K-sum contribution to be significant; this conclusion was confirmed by noting that for similar count rates with the beam focused on the Al_2O_3 matrix no Argon was detected (cf fig. 5b). Argon can be introduced by Ar+ ion bombardment during ion-thinning, but such peaks are usually only just detectable above background. No argon was detected in either the Al_2O_3 matrix or separate samples of Mg-Al spinel. It is proposed that the argon can sit in the twin planes in the β'''-Al_2O_3 in a similar position to that occupied by the K ions. It is a little larger and would also be expected to loose its charge. Argon can, therefore, apparently be introduced during ion thinning since it has also been detected in material which had been sintered in air. There also exists the possibility that Ar could be trapped in β'''-Al_2O_3 particles during hot-pressing in an inert atmosphere.

The shape of the elongated particles is directly due to the crystal structure of the $\beta'''-Al_2O_3$. The longer axis is always parallel to both the basal plane of the $\beta'''-Al_2O_3$ and a {111} plane of the spinel. The spinel particle grows first; from subsequent observations this growth is thought to nucleate at triple points. The spinel grows both along the grain boundaries and normal to them as is evidenced by the triangular regions in fig. 6. When the spinel 'triangles' have grown to a critical size they then grow preferentially parallel to the {111} plane as illustrated schematically in fig. 7.

The $\beta'''-Al_2O_3$ grows into the spinel by a mechanism which is effectively chemical twinning[9] or, in this case, chemical microtwinning. The growth of the $\beta'''-Al_2O_3$ into spinel is illustrated in fig. 8 where the sample is actually a hot-pressed polycrystalline spinel compact and the EDS spectrum of the $\beta'''-Al_2O_3$ showed both K and Ar to be present. Two new twin planes are thus introduced simultaneously and the thickness of the $\beta'''-Al_2O_3$ particle increases by 3.2nm. A similar growth mechanism has been found for the elongated particles in Al_2O_3.

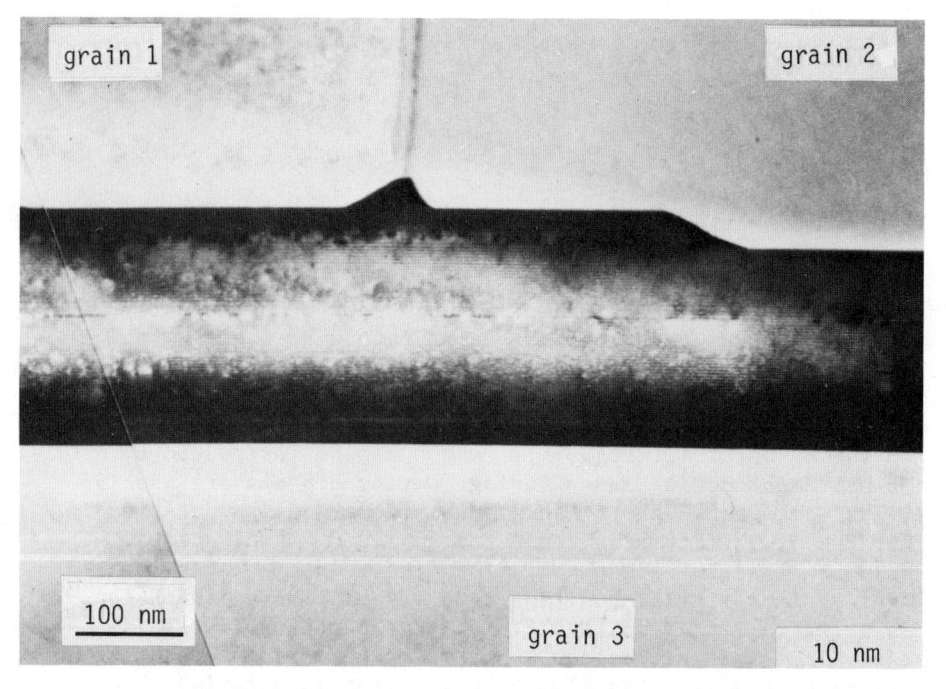

Fig. 6. A particle of $\beta'''-Al_2O_3$/spinel which has grown along the interface between grains 1 and 3, and 2 and 3. The spinel is also growing along the 1/2 grain boundary.

ANALYSIS OF SECOND-PHASE PARTICLES IN Al$_2$O$_3$

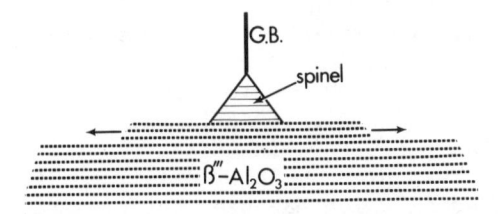

Fig. 7. Schematic diagram to show the mechanism for the growth of the β'''-Al$_2$O$_3$. G.B. is a grain boundary in the α-Al$_2$O$_3$ and the arrows indicate the motion of the interface.

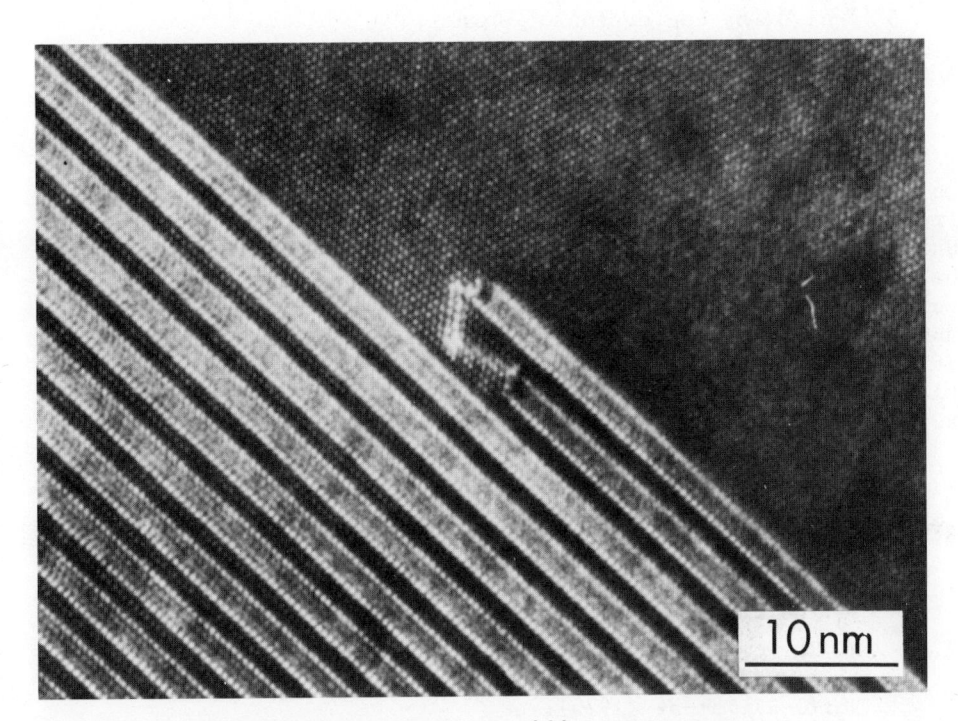

Fig. 8. High-resolution image of β'''-Al$_2$O$_3$ growing into a spinel grain by the process of chemical micro-twinning. Two microtwins are advancing in this example.

CONCLUSIONS

In a particular commercial Al_2O_3, two different types of second-phase particles have been at least partly identified using EDS and in one case by combining EDS with high-resolution structure imaging the crystal structure and interrelation of two phases has been resolved. The remaining questions to be answered include, is the Ni present as Ni, NiO or possibly in some cases as Ni-Al spinel; to resolve this ambiguity EELS will be used in further studies. Argon has been shown in this study to be trapped, presumably in the twin planes, in $\beta'''-Al_2O_3$. The argon has been found both in material which had been hot-pressed in an inert atmosphere, and in material which had been sintered in air. It is, therefore, concluded that the argon can be introduced in high concentrations during ion thinning, although this does not occur in the spinel or Al_2O_3; the argon may also be trapped during hot-pressing. Controlled experiments are in progress to further test these conclusions on materials with a better-known history.

Finally, the mechanism of growth of elongated grains of $\beta'''-Al_2O_3$ has been identified. It consists of a chemical microtwinning process whereby complete cells (c=3.2nm) are added to the $\beta'''-Al_2O_3$ by conversion of spinel.

ACKNOWLEDGMENTS

The authors would like to thank Professor Dr. Hermann Schmalzried for many stimulating discussions on the subject of solid-state reactions in ceramic oxides. Thanks are also due to Professor D.B. Williams and Graham Cliff for the use of Lehigh Universities Philips EM400T to test the Na detectability and for helpful discussions. Dr. David Joy provided helpful insight on sum peaks in EDS. The assistance of John Hunt in the EDS analysis and Ray Coles in maintaining the microscopes is also acknowledged. This research is part of a project being carried out in collaboration with Professor Schmalzried and is supported by the National Science Foundation under grant number DMR-8102994.

REFERENCES

1. N. J. Tighe, Microstructure of Fine-Grain Ceramics, in: "Ultrafine-Grain Ceramics".
2. W. H. Rhodes, D. J. Sellers, and T. Vasilos, Hot-Working of Aluminum Oxide: II, Optical Properties, J. Am. Ceram. Soc. 58:(1-2), 31-34 (1975).
3. W. C. Johnson and D. F. Stein, Additive and Impurity Distributions at Grain Boundaries in Sintered Alumina, J. Am. Ceram. Soc. 58:(11-12), 485-8 (1975).

4. W. C. Johnson and R. L. Coble, A Test of the Second-Phase and Impurity-Segregation Models for MgO-Enhanced Densification of Sintered Alumina, J. Am. Ceram. Soc. 61:(3-4), 110-114 (1978).
5. M. Blanc, A. Mocellin, and J. L. Strudel, Observation of Potassium β'''-Alumina in Sintered Alumina, J. Am. Ceram. Soc. 60:(9-10), 403-409 (1977).
6. M. Bettman and L. L. Turner, On the Structure of $Na_2O \cdot 4MgO \cdot 15Al_2O_3$, a Variant of β-Alumina, Inorg. Chem. 10:(7), 1442-1446 (1971).
7. J. O. Bovin, High Resolution Electron Microscopy Images of Defects in Mg- and Li-stabilized β''-aluminas, Acta Cryst. A35:572-80 (1979).
8. Y. Matsui and S. Horiuchi, Irradiation-Induced Defects in β''-Alumina Examined by IMV High-Resolution Electron Microscopy, Acta Cryst. A37:51-61 (1981).
9. S. Andersson and B. G. Hyde, Twinning on the Unit Cell Level as a Structure-Building Operation in the Solid State, J. Solid State Chem. 9:92-101 (1974).

MICROSTRUCTURAL CHARACTERIZATION OF ABNORMAL GRAIN GROWTH

DEVELOPMENT IN Al_2O_3

M. P. Harmer, S. J. Bennison and C. Narayan

Materials Research Center
Lehigh University
Bethlehem, PA 18015

ABSTRACT

Two major factors contribute to the difficulty of understanding the abnormal grain growth process in Al_2O_3. First, there are many possible causes for abnormal grain growth and second, there is a lack of detailed microstructural evidence. In the present work these problems have been minimized by studying the sequence of microstructural changes leading to the onset of abnormal grain growth in a very pure, monosized, Al_2O_3 powder. The data are analyzed in terms of a model based on inhomogeneous densification and solute breakaway.

1. INTRODUCTION

It is now well recognized that abnormal grain growth, i.e. the rapid growth of a few selected grains to sizes much larger than those of the average grain population, is an undesirable feature of sintering. One of the most convenient, effective and inexpensive ways of controlling (suppressing) abnormal grain growth is to make use of a sintering additive (solid solution if possible). While this method can work very well, as in the case of Al_2O_3 doped with MgO[1], the selection of a suitable host:additive combination remains largely empirical. This is in part due to an incomplete understanding of dopant effects in general, and in part due to the complex nature of the abnormal grain growth process itself. The present work deals with the latter issue, namely, obtaining a better understanding of the mechanisms responsible for abnormal grain growth development in Al_2O_3.

Two major factors contribute to the difficulty of understanding abnormal grain growth. These are:

1.1 The Many Possible Causes for Abnormal Grain Growth, viz.,

   a. *The Existence of a Wide Initial Particle Size Distribution.* Hillert[2], analyzed the conditions for obtaining stable normal grain growth, for the case of a fully dense, single phase material. The proposed criterion for stability is that the size of the largest grain must be less than twice the mean grain size. To maintain the system within the stable Hillert criterion, therefore, one should aim at keeping the original particle size distribution within as narrow a size range as possible. Furthermore, the conditions for preserving pore-boundary attachment and solute-boundary attachment during sintering are much favored by maintaining a very narrow grain size distribution[3].

   b. *The Occurrence of Breakaway Phenomena.* Grain boundary mobility is greatly influenced by the presence of attached pores, grain boundary phases, and solutes segregated to, or away from, the grain boundary region[4]; the result, in most cases, is a substantial slowing down of grain boundary movement. When local conditions favor breakaway of the boundary from the attached species, or when transitions amongst the various controlling regimes occur, the ensuing abrupt increase in boundary velocity will either (i) cause abnormal grain growth to develop if breakaway occurs inhomogeneously in the sample (the more likely case) or (ii) cause the normal grain growth process to speed up (without going abnormal) provided that breakaway occurs uniformly and homogeneously throughout the sample. In both cases, a detailed understanding of the interactions between moving boundaries and pores, solutes and second phases is much needed.

   c. *Inhomogeneous Densification.* Inhomogeneous densification results from density fluctuations and compositional variation within the unfired compact. Some common causes of microinhomogeneity are particle agglomeration and poor packing, the existence of pressure gradients during die pressing and nonuniform dopant distribution. Since inhomogeneous densification also implies inhomogeneous grain growth—both processes are mutually dependent[3], a likely consequence of inhomogeneous densification is a broadening of the original grain size distribution, and ultimately, abnormal grain growth.

   d. *Intrinsic Boundary Properties.* There is some evidence, both experimental[5] and theoretical[6], to suggest that grain boundary mobility is a function of grain orientation. The resulting anisotropy of grain boundary mobility provides an added incentive for nonuniform, and hence abnormal, grain growth. Recent computer simulation studies of grain growth, in which an orientation dependence was introduced into the grain boundary mobility, support this notion[7]. Certain dopants may be effective in preventing abnormal grain growth, therefore, by interacting with grain boundaries to minimize anisotropy effects (by preferred segregation to special high mobility boundaries for example).

Gleiter[6] has analyzed the effect of grain boundary structure on boundary migration. He argued that the boundary mobility is a function of the atomic step density at the boundary. Grain boundaries parallel to low index planes for example, are expected to have lower step densities, and hence lower mobilities, than other special boundaries, such as high angle boundaries with high step densities.

1.2 The Lack of Detailed Microstructural Evidence

Many of the diagnostic analyses of abnormal grain growth are based on the appearance of the final microstructure. Since each of the different mechanisms for abnormal grain growth described above can give rise to a similar final structure, the usefulness of this approach is very limited. Furthermore, much of the available microstructural evidence dates back to early work of the sixties and early seventies[8,9] when sample purity and powder preparation methods were somewhat inferior to present-day standards.

The aim of the present work is to update the existing microstructural evidence for abnormal grain growth development in Al_2O_3. Furthermore, by confining the study to a single phase, monosized powder, it is hoped that a much clearer interpretation of the mechanisms responsible for abnormal grain growth (recall section 1.1) can be made.

2. EXPERIMENTAL

Specimens were prepared from a high purity (99.99) Al_2O_3* powder of uniform particle shape and narrow particle size distribution. Pressed pellets (60% green density) containing carbowax as a lubricant were heated (100°C hr^{-1}) to 800°C and soaked at this temperature for 20 hours to remove the carbowax. Microstructural examination of the green compact revealed no evidence of extensive agglomeration or density fluctuation within the samples (see figure 1).

Sintering was performed inside of a high purity alumina work tube in a resistance heated furnace. All firings were done in air at 1630°C. Final bulk densities were measured to 0.01 Mgm^{-3} using the Archimedes method with toluene as the immersion medium.

---
*Atomergic Chemicals Ltd., 1.0±0.5 micron particle size.

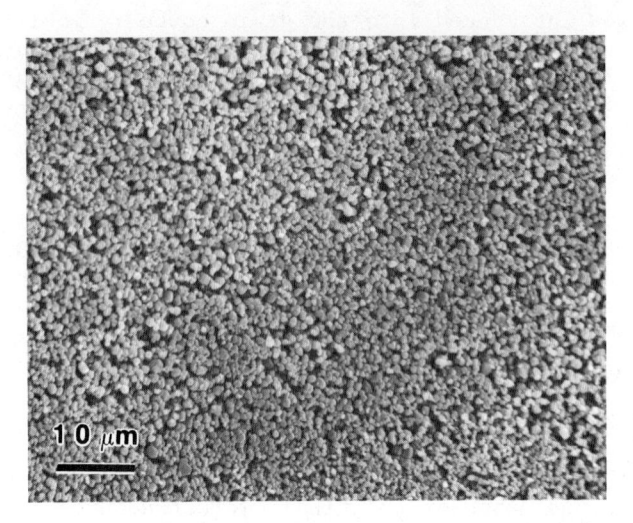

Fig. 1. Scanning electron micrograph showing uniform density within the as pressed green compact.

Polished sections for SEM examination were prepared by light mechanical grinding followed by diamond polishing using 15 micron and 0.25 micron pastes respectively. Grain boundaries were revealed by thermal etching at 1400°C for 30 minutes. Grain sizes were measured using the linear intercept method[10].

3. RESULTS

The dependence of the density on the firing time of samples is shown in figure 2. From this it can be seen that a limiting final density of ∼95% of theoretical is reached after approximately 60 minutes of firing time.

Early stages of microstructure development are shown in figures 3 and 4. The low magnification picture (figure 3) shows evidence of widespread inhomogeneous densification with the formation of many micro-dense island structures. One of these micro-dense regions is imaged at higher magnification in figure 4. From this it can be seen that the grain structure is uniform (∼1 μm mean grain size) and continuous across the island/matrix interface. Note the high density (almost theoretical) within the island structure in contrast to the low density of the surrounding porous matrix. Also note that all pores are intergranularly situated at this point.

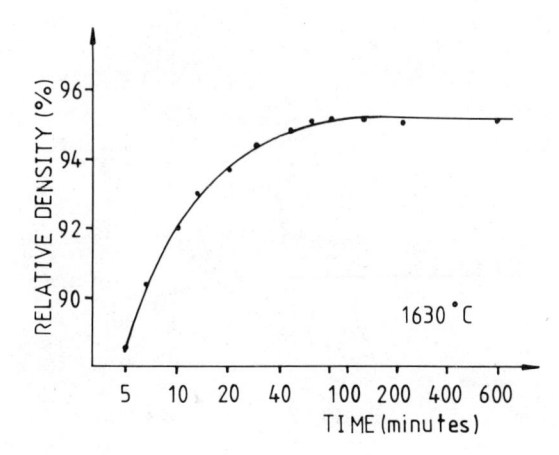

Fig. 2. Bulk density data for pure Al_2O_3, sintered in air (1630°C).

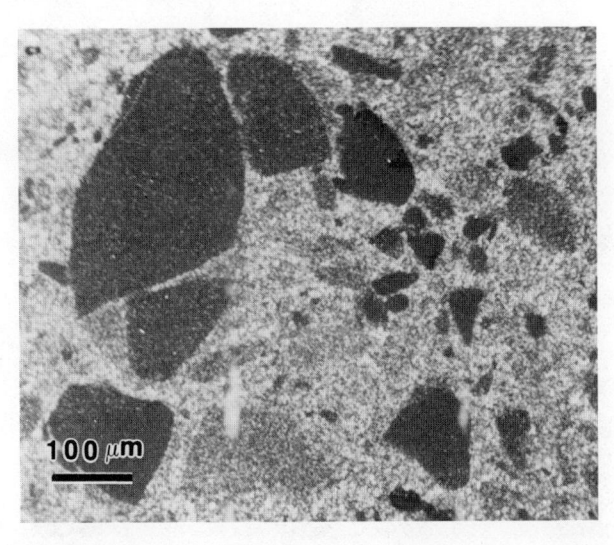

Fig. 3. SEM micrograph showing the formation of highly dense island structures (dark regions) due to inhomogeneous densification. (Bulk density, 88.5% theoretical).

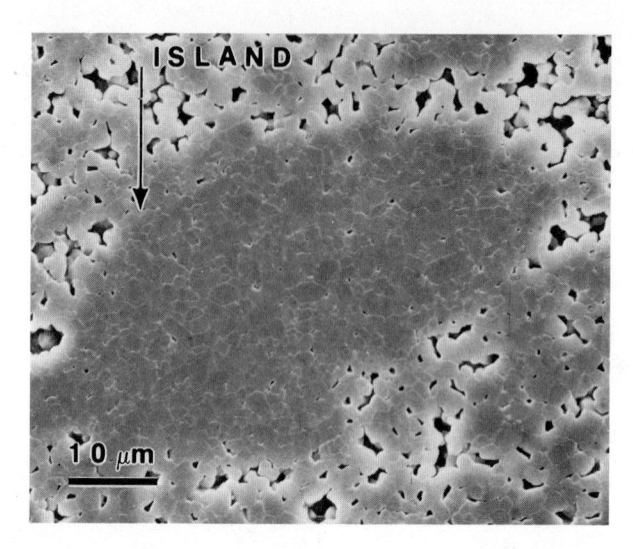

Fig. 4. SEM micrograph showing a micro-dense island. The grain size is uniform and continuous across the matrix/island interface (Bulk density, 88.5% theoretical).

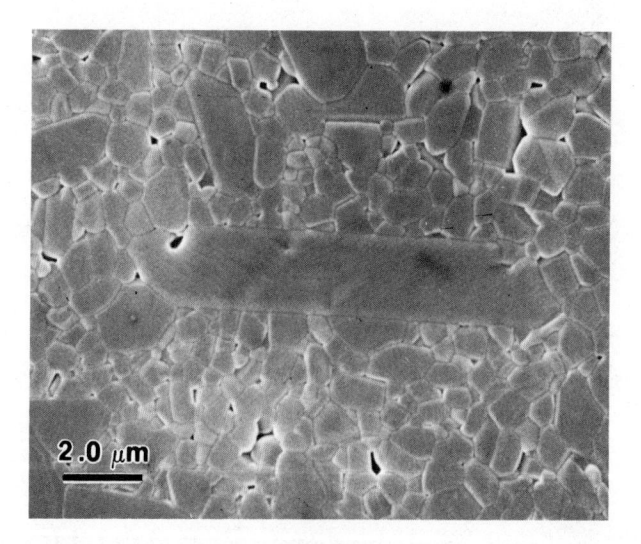

Fig. 5. SEM micrograph showing initiation of abnormal grain growth within a micro-dense island. Note the anisotropic growth pattern (Bulk density, 90.5% of theoretical).

Fig. 6. SEM micrograph showing impinging abnormal grains confined to micro-dense islands (Bulk density, 92% of theoretical).

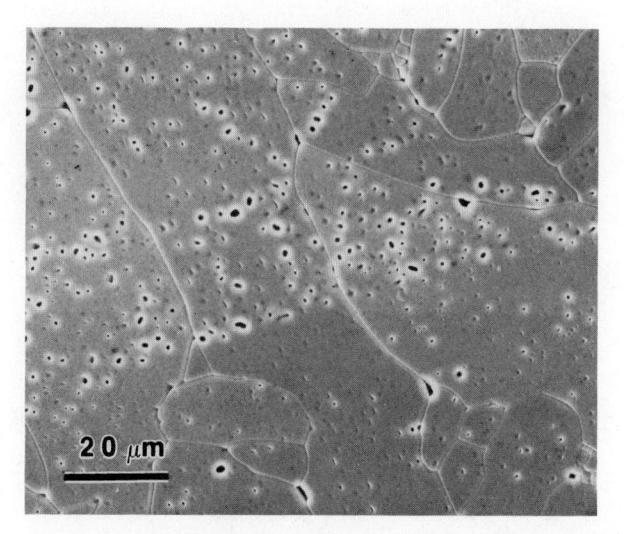

Fig. 7. SEM micrograph showing breakthrough of abnormal grains into surrounding porous matrix (top left of picture). The original island/matrix interface is clearly visible (Bulk density, 94% of theoretical).

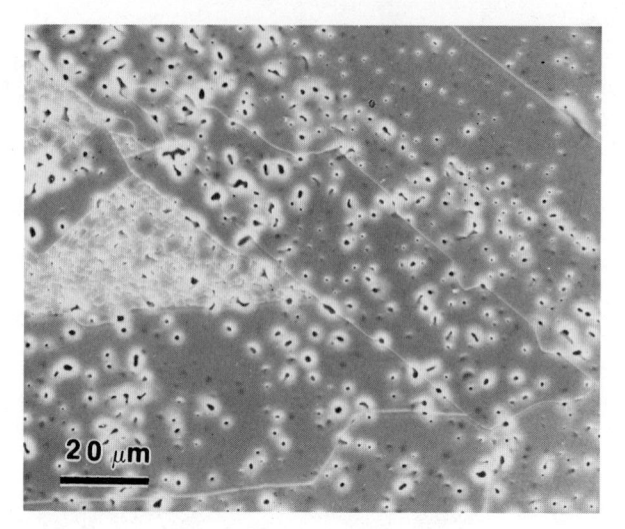

Fig. 8. SEM micrograph showing final sintered structure. Note the elongated grains and the presence of boundary cusps. (Bulk density, 95.5% of theoretical).

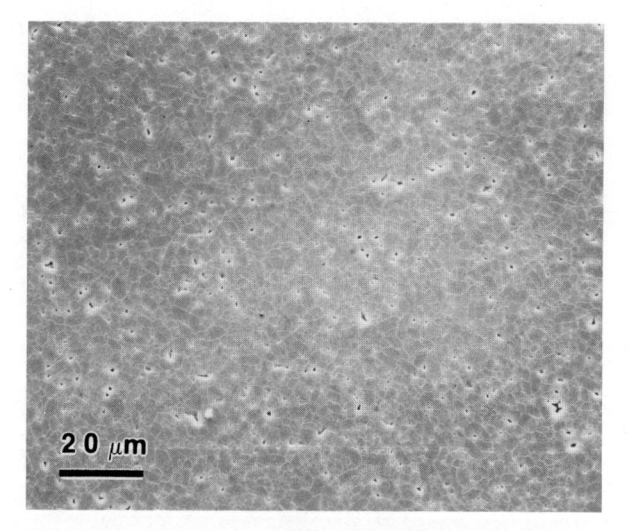

Fig. 9. SEM micrograph showing final sintered structure of MgO doped sample. Note the homogeneous microstructure (Bulk density, 98% of theoretical).

Abnormal grain growth commences at a slightly later stage of microstructural evolution (90% of theoretical density). In all cases, initiation occurs within the highly dense island structures (see figure 5). The abnormal grains grow in a very anisotropic fashion and develop a tabular habit (as shown). Abnormal grain growth continues to occur until the micro-dense islands become completely engulfed with impinging abnormal grains (see figure 6).

The last significant microstructural change occurs when the abnormal grains, contained within the micro-dense islands, can no longer restrain from breaking into the surrounding porous matrix (see figure 7). This results in the final structure shown in figure 8. (Pores trapped within the large, rapidly growing, abnormal grains.)

For comparison, the final microstructure of a MgO doped sample is shown in figure 9. The interesting feature to note here is the apparent ability of MgO to suppress inhomogeneous densification, a desirable feature which prevents the material from developing abnormal grain growth prematurely.

4. DISCUSSION

The results show that, in the undoped Al_2O_3, abnormal grain growth initiates within highly dense regions, the result of inhomogeneous densification. The results also indicate that MgO, when used as additive, suppresses inhomogeneous densification preventing abnormal grain growth from occurring prematurely.

Before speculating on the origin of the micro-inhomogeneity, it is worthwhile to consider some of the consequences of physical and chemical inhomogeneity on microstructure development using the available models. The approach used here is the one proposed by Cannon and Yan[7], and adapted more recently by Brook[11] to treat additive effects in Al_2O_3. In this approach, a simultaneous model of densification and grain growth is used to predict how the density and grain size of a microstructure develop during processing. The results are plotted in terms of density-grain size trajectories obtained under different sets of processing conditions such as of initial grain size (g_o), density (δ_o) and dopant content.

Several such plots are shown in figure 10 for pure and doped aluminas with different starting densities (0.75 and 0.90). The plots have been computed using Brook's data assuming a mechanism of lattice diffusion controlled densification, and surface diffusion controlled pore drag. The doped data represent MgO-saturated material where MgO has the effect of raising the lattice diffusion coefficient three times[12] and lowering the surface diffusion coefficient ten times[13].

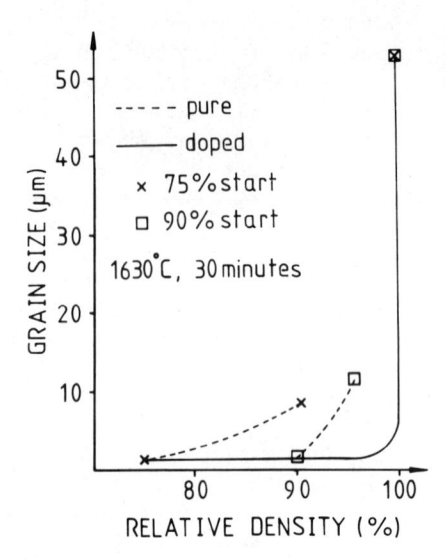

Fig. 10. The influence of starting density and doping on microstructure development in Al$_2$O$_3$. For data used in the Figure see ref. 11.

The data (figure 10) while only approximate in terms of the absolute values involved, demonstrate two important features. Firstly, chemical inhomogeneity (considered in terms of a mixture of doped and undoped material) is seen as a major potential contributory factor in the development of nonuniform microstructures; physical inhomogeneity is by comparison, shown to be much less significant. Secondly, doping can be seen as an effective means of minimizing the importance of physical inhomogeneity (i.e. the starting density has little effect on the end point density and grain size for a given firing schedule).

These two factors can help explain the effectiveness of MgO, as sintering additive, in the present work; undoped Al$_2$O$_3$ densifies inhomogeneously due to chemical heterogeneity (nonuniform background impurity distribution). MgO, when added to saturation, overrides any background effects and restores chemical homogeneity through its ability to distribute uniformly by vapor phase transport[14]. Consequently, inhomogeneous densification is suppressed and, as a result, premature abnormal grain growth is avoided*.

It remains to explain the mechanism of abnormal grain growth initiation within the highly dense island structures of the undoped

---

*See paper by Brook[11] for further discussion on the prevention of abnormal grain growth in MgO doped Al$_2$O$_3$.

specimens. The most plausible mechanism in our case (since we can dismiss the classical explanation based on poure-boundary separation) would be one such as that proposed by Cannon, et al.[15], based on solute breakaway, or abrupt changes in boundary velocity resulting from transitions between the low velocity and high velocity solute drag regimes. We suggest that anisotropy is also important, but more as a means of triggering the solute breakaway events, rather than providing an independent driving force for the abnormal grain growth process itself. The effect of MgO as additive, on the various breakaway phenomena, remains to be investigated.

5. CONCLUSIONS

Abnormal grain growth has been difficult to study in the past because of the many possible mechanisms involved (and their close interrelationships), and the general lack of detailed microstructural evidence available. The present work was an attempt to minimize some of these problems by confining the study to a very pure, single phase, monosized Al_2O_3 powder.

It was observed that abnormal grain growth initiated from within highly dense regions, the result of inhomogeneous densification. A mechanism based on sudden changes in boundary movement caused by solute breakaway, triggered by a strong tendency for anisotropic growth, was proposed.

The effectiveness of MgO as sintering additive in Al_2O_3 was explained in terms of its ability to suppress inhomogeneous densification. Further studies using different types of Al_2O_3 powders (pure and doped) are recommended.

REFERENCES

1. M. Harmer, E. W. Roberts and R. J. Brook, Trans. J. Brit. Ceram. Soc. 78, 22-25 (1979).
2. M. Hillert, Acta. Met. 13, 227 (1965).
3. K. Uematsu, R. M. Cannon, R. D. Bagley, M. F. Yan, U. Chowdhryand, H. K. Bowen, p. 190 in Proc. of Int. Symp. of Factors in Densification and Sintering of Oxide and Non-Oxide Ceramics, Edited by Shigayuki Somiya and Shinroku Salko, Tokyo, Japan (1978).
4. R. J. Brook, in "Treatise on Mat. Sci. and Tech." vol. 9, pp. 331-64, Edited by F. F. Y. Wang, Academic Press, New York (1976).
5. J. W. Rutter and K. T. Aust, Acta. Met. 13, 181-36 (1965).
6. H. Gleiter, Acta. Met. 17, 853 (1969).
7. M. F. Yan, Mat. Sci. and Eng. 48, 53 (1981).

8. R. Rossi, J. Burke, J. Amer. Ceram. Soc. 56, 654 (1973).
9. D. W. Budworth, Mineral Mag. 37, 833 (1970).
10. R. L. Fullman, Trans. AIME, 197, 447-52 (1953).
11. R. J. Brook, Proc. Brit. Ceram. Soc. 32, 7 (1982).
12. M. P. Harmer and R. J. Brook, J. Mater. Sci. 15, 3017 (1980).
13. C. Monty and L. E. Duigou, J. Symp. Chem. Met. at High T., Harwell, U.K., 7-10 September 1981, to be published High Temp.-High Press.
14. M. O. Warman and D. W. Budworth, Trans. J. Brit. Ceram. Soc., 66, 253 (1967).
15. a. A. M. Glasser, H. K. Bowen and R. M. Cannon, Mat. Sci. Res. 14, 227 (1981).
    b. M. F. Yan, R. M. Cannon and H. K. Bowen, pp. 276-307 in Ceramic Microstructures '76, Edited by R. M. Fulrath and J. A. Pask, Westview Press, Boulder, CO (1977).

ACKNOWLEDGMENTS

The authors are grateful to the National Science Foundation for financial support of this work through the Ceramics Division under Grant No. DMR-8116865. Discussions and experimental assistance from Mr. M. Lynch are also gratefully acknowledged.

PROPERTIES AND CHARACTERIZATION OF SURFACE OXIDES ON

ALUMINUM ALLOYS

        Karl Wefers

        Alcoa Laboratories
        Alcoa Center, PA 15069

ABSTRACT

    The technical properties of an aluminum alloy surface are strongly affected by the structure, morphology, and chemical composition of the oxide film.

    Using the thermal oxidation in air of three low magnesium alloys as examples, the complex growth mechanisms and compositions of surface films are illustrated. Continuous uniform surface layers only grow at temperatures well below 400°C. At high temperatures, oxides of locally varying thickness and non-uniform chemical composition develop due to preferential diffusion of Mg, recrystallization of the initially non-crystalline film and the presence of large intermetallic constituents in the metal surface.

    Because of their complex structure, these surface layers cannot be adequately characterized by a single analytical method. To fully describe the surface properties of an alloy, a number of parameters must be determined. While morphological and structural features can be analyzed by scanning and transmission electron microscopy, modern electron and ion spectroscopic techniques yield information about film thickness, elemental composition as well as lateral and vertical distribution of elements.

    Full text in ALUMINIUM, volume 57 (1981), No. 11, pp. 722/726.

THE CHARACTERIZATION OF MICROCRACKS IN BRITTLE SOLIDS

D. R. Clarke and D. J. Green

Structural Ceramics Group
Rockwell International Science Center
Thousand Oaks, CA 91360

ABSTRACT

The detection of microcracks in ceramics and other brittle materials is a key step in determining their influence on physical properties of this class of material. Various detection techniques are reviewed and discussed. In addition, new techniques involving scanning electron microscopy and porosimetry are introduced and illustrated with observations on the Al_2O_3-ZrO_2 system.

INTRODUCTION

The presence of microcracks is widely recognized to affect such diverse material properties as mechanical strength, thermal conductivity, acoustic transparency and optical transmission. It is also established that microcracking can commonly occur in ceramics, minerals and brittle polymers as a result of a variety of processes including anisotropic expansion, differential thermal expansion, phase transformations and irradiation-induced, differential expansion (swelling). Indeed, the acceptance of microcracking as a widely occurring phenomenon, has often led to experimental results being attributed to the presence or formation of microcracks, even in the absence of any direct observation of the microcracks themselves. Despite their importance, little attention has been paid to the generic problem of detecting microcracks in ceramics, especially those having sub-micron openings.

The direct observation of microcracks is important for several reasons. For example, although indirect techniques, such as elastic modulus measurements, can be used in a qualitative sense to detect microcracks, this approach indicates nothing about their spatial distribution or size. Moreover, although quantitative relationships between microcrack density and particular properties have been developed, these relationships have not been verified. The direct observation of microcracks would allow such work to be accomplished. On this basis, the techniques could then be applied to a number of outstanding problems concerning the role of microcracks, for example, measuring the size of the microcrack zone at a crack tip, detecting stress-induced microcracks and their contribution to toughening, determining their effect on leaching of ceramic nuclear waste forms, and assessing their influence on thermal shock resistance.

In this brief contribution, techniques for detecting and observing microcracks are described. These include direct observation of cracks in the transmission electron microscope and the use of decoration techniques for observing cracks in the scanning electron microscope. For this work, microcracks are defined as cracks whose size is smaller than or of the same order as that of the microstructure, e.g., grain size. The examples used to illustrate the techniques reviewed here are drawn from studies of zirconia based ceramics. When the tetragonal form of zirconia transforms to the monoclinic polymorph the transformation is accompanied by a 6-8% volume increase, and in certain circumstances the volume change is accommodated by the formation of microcracks in the ceramic.

INDIRECT TECHNIQUES FOR MICROCRACK DETECTION

As indicated microcracks influence various material properties. The inverse of this statement; that measurement of a particular property indicates the presence of microcracks is, however, not necessarily true because there may be other factors that are influencing that property. Notwithstanding this difficulty, such indirect techniques have often been employed, and these will be reviewed in this section. In addition, a new technique for indirect microcrack detection using porosimetry will be described.

Elastic Modulus Measurements

The most widely employed method for determining whether a ceramic has become microcracked is to measure changes in its elastic modulus. The method typically consists of measuring the resonance frequency of a vibrating beam or sphere and calculating the elastic modulus from this value. In the absence of any other microstructural changes the change in modulus in generally accepted as being a measure of the microcrack density. This technique is strictly appropriate only for samples of the simplest of geometries, hence, it is ill-suited for quality control purposes on ceramic components. In this case, it could be more suitable to measure the elastic constants using sonic velocity techniques of, for instance, surface waves.

Strictly, the presence of microcracks will affect not only the Young's modulus (E), but also the shear (μ) and bulk moduli (K). In the simplest, extreme case, i.e., assuming that the microcracks act as spherical pores, the effective shear and bulk moduli will be related to the volume fraction v_f of microcracks by the well established equations for porosity, namely[1]

$$\mu_{mc} = \mu(1-2v_f)$$

and

$$K_{mc} = K(1-3v_f) \tag{1}$$

In a more accurate description, if the microcracks are modelled as being "penny-shaped" discs then their effect on the elastic moduli can be calculated using the methodology introduced by Eshelby.[1] (The problem corresponds to inhomogeneity problem posed and solved by Eshelby.) This analysis was originally performed by Bristow[2] and leads to expressions for the modulus changes produced by N randomly oriented, penny-shaped cracks per unit volume (dilute concentrations of cracks) as follows,

$$-\Delta E/E = \frac{16(1-\nu^2))(10-3\nu)\, a^3 N}{45\,(2-\nu)}$$

$$-\Delta\mu/\mu = \frac{32(1-\nu^2)\, a^3 N}{45(2-\nu)}$$

and

$$-\Delta K/K = \frac{16(1-\nu^2)\, a^3 N}{9(1-2\nu)}$$

where a is the crack radius, ν is Poisson's ratio of the uncracked material and ΔE, ΔK and $\Delta\mu$ are the changes in the moduli. The analysis pertains to isotropic materials and to cases where the cracks are sufficiently well separated, that interaction between them is negligible.

Other treatments relating moduli changes to microcrack densities have also been developed. That of Salganik[3] (quoted by Hasselman and Singh[4]) leads to the same relations as presented above. A more general analysis, incorporating the effects of elastic interaction between the cracks, has been given by Budiansky and O'Connell.[5] Their equations for a random array of circular cracks are identical with these of Bristow[2] except that the Poisson ratio used is the effective Poisson ratio of the cracked body rather than that of the uncracked material.

While the elastic modulus measurements can strictly only be used as an indication of microcracking and for comparison purposes, valuable information has nevertheless been obtained. for instance, on the basis of elastic modulus changes, it is now recognized that for single phase ceramic materials possessing thermal expansion anisotropy there exists a critical grain size above which microcracks will form.[6,7] Likewise, there is also deduced to be a critical size for second phase particles in two-phase ceramics[8] at which microcracks will form in response to differential thermal contraction. As an example, Fig. 1 indicates the change in Young's modulus of a series of two phase Al_2O_3/ZrO_2 composites in which different proportions of the ZrO_2

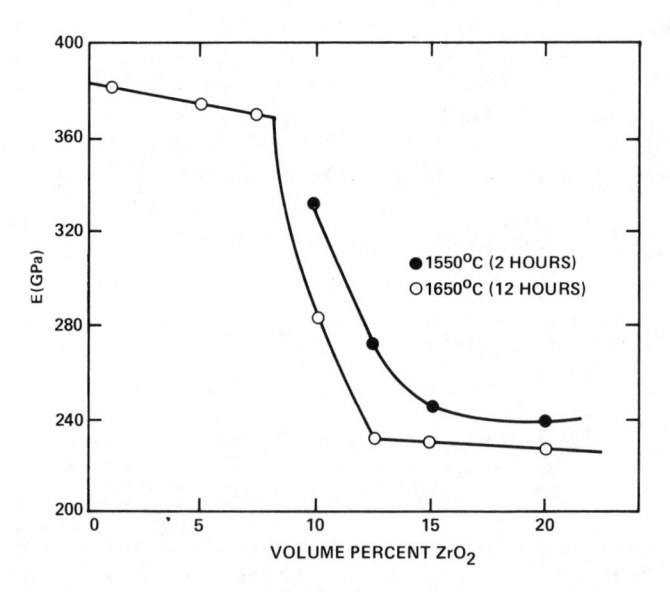

Fig. 1 Change in the elastic modulus of a series of Al_2O_3/ZrO_2 composites on annealing to transform the tetragonal zirconia phase to monoclinic.

content have been allowed to transform to the monoclinic polymorph thus introducing microcracks by controlled heat treatments.[9]

Thermal Diffusivity Measurements

Microcracks act as barriers to heat flow in solid material and as such their presence is expected to reduce thermal diffusivity. The relationship between thermal conductivity and microcrack density has been analyzed theoretically[10-11] and changes in thermal diffusivity have been used to monitor microcracking phenomena in various ceramic systems.[12-15]

Porosimeter Measurements

In fine-grained ceramics, microcrack openings are expected to be very small and hence the microcracks themselves can act like capillaries. Techniques have been developed for the measurement of ultrafine porosity on the basis of capillary action including the penetration under pressure of a liquid, such as mercury, into samples. It is clear therefore, that such a technique would also be applicable to microcracked materials, even where their crack openings are particularly small, i.e., less than 10 nm. In order to explore this concept two Al_2O_3/10 vol.% ZrO_2 materials were chosen that were known to have differences in their elastic properties from elastic modulus measurements. These two materials had been heat-treated differently so that the percentage of ZrO_2 that had been transformed to monoclinic was different, leading to materials with different microcrack densities. This idea was substantiated by differences in elastic modulus measurements and hence the samples were chosen to be analyzed using a porosimeter.* The results are shown in Fig. 2a and b. It can be seen that the composite in which the ZrO_2 is almost completely transformed to monoclinic has significant intrusion in the range of equivalent pore diameters from 10 to 60 nm. (If the assumption is made that the microcracks extend across the grain facet and that the grain size is ~0.5 μm, then the corresponding crack openings lie in the range 0.1-1.0 nm.) In comparison, the other sample, which contains mainly tetragonal ZrO_2 has relatively minor intrusion of the mercury. This difference, attributed to microcracking, is further substantiated by the appearance of the two specimens after intrusion. As shown in Fig. 3, the specimen with the significant intrusion has retained some of the mercury, giving it a uniform dark appearance, whereas the other specimen is partially blackened and only along big scratches.

---

*AutoPore 9200, Micromeritics, Norcross, Georgia, 30093.

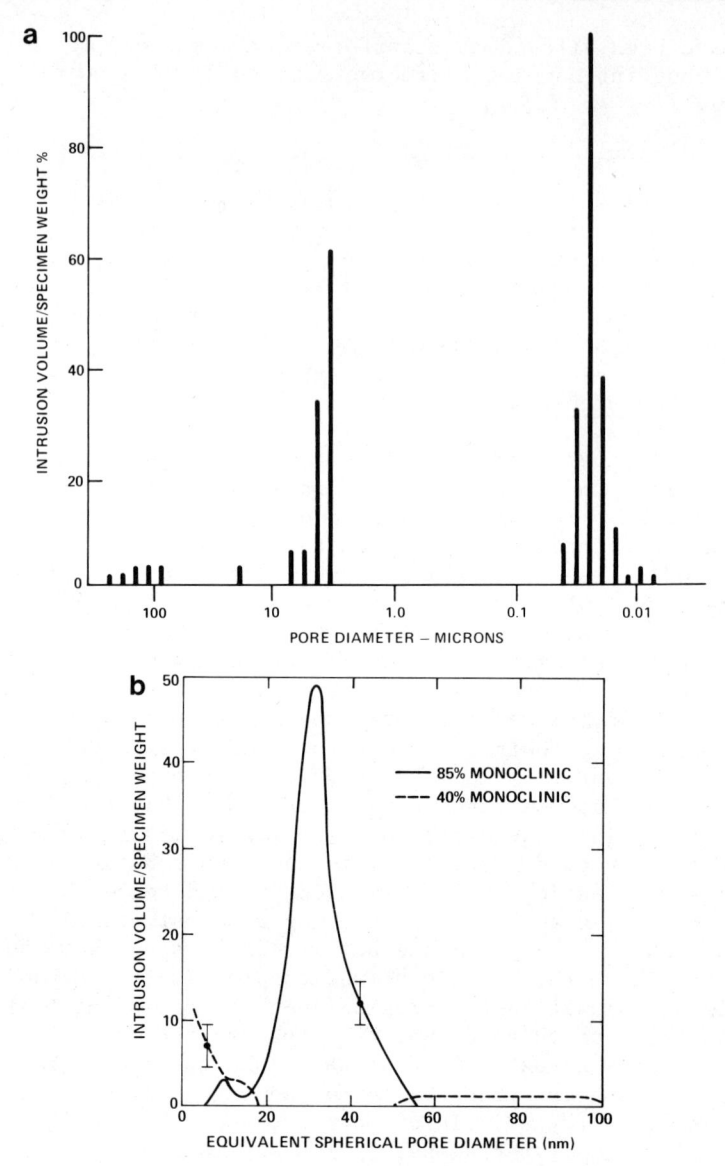

Fig. 2 (a) Pore size distribution for an $Al_2O_3/10$ v/o ZrO_2 material after the zirconia has transformed to the monoclinic state. Two distinct distributions are evident, one centered at ~4.0 μm, characteristic of remnent porosity, and the other centered at an equivalent pore diameter of ~40 nm. The latter is attributed to the presence of microcracks. (b) A comparison of the pore size distribution, in the size range 0-100 nm, of the two $Al_2O_3/10$ v/o ZrO_2 samples heat treated to produce differing proportions of the monoclinic phase.

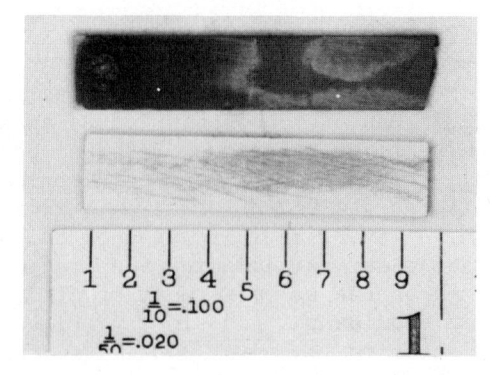

Fig. 3 Appearance of the two Al$_2$O$_3$/ZrO$_2$ samples after the porosimetry measurements reproduced in Fig. 2(b). The sample exhibiting significant intrusion (80% monoclinic) has a uniform dark appearance (due to retained mercury) whereas the other sample is only partially blackened (at surface scratches).

Other Indirect Techniques

The recent dedication of National User Facilities for Neutron Physics (at NBS and Brookhaven National Laboratory) has prompted consideration of Small Angle Neutron Scattering (SANS) for the detection of microcracks. SANS has already proven to be a viable, nondestructive technique for evaluating the size distribution of voids in creep-deformed metals and in reaction-bonded silicon nitride, and in work soon to be published, Case[16] has confirmed that a high density of microcracks is detectable by neutron scattering. By comparing the intensity of scattering from samples of yttrium chromite in a microcracked condition with that from annealed samples, Case has demonstrated that microcracks can act to scatter neutrons and has deduced the size distribution and median size of the microcracks in the material.[16] This work represents the first step in providing the methodology required to reduce the technique to one capable of providing a quantitative description of microcrack density and shape in materials.

Acoustic emission occurring as a result of differential thermal expansion or a diffusionless phase transformation has been used by a number of investigators to monitor microcracking but the technique cannot as yet be used to characterize microcracking since it has yet to be demonstrated that microcracking gives rise to an unambiguous and recognizable acoustic signature.

DIRECT TECHNIQUES FOR MICROCRACK DETECTION

In coarse-grained materials such as rocks or glassy materials, microcracks can often be observed using optical microscopy. Indeed, various techniques for such observations have been reviewed.[17] In many fine-grained ceramics, direct observation necessitates the use of electron microscopy. Prior to such observation, however, an indication of the presence of such microcracks can be determined using a dye penetrant. For example, Fig. 4 compares the two specimens of Al_2O_3/10 vol% ZrO_2 used for the elastic modulus and porosimetry (Fig. 2) measurements. It is clear that the material containing the higher fraction of monoclinic ZrO_2 shows significant absorption compared to the other specimen. In these materials, where the grain sizes are ~0.5-3 µm, spatial resolution of the dye-penetrated microcracks using optical microscopy was not successful.

Scanning Electron Microscopy

In many respects, the scanning microscope would seem to be the most appropriate imaging tool for observing microcracks because of its high resolution, large depth of field and its usefulness in microstructural characterization. The simplest

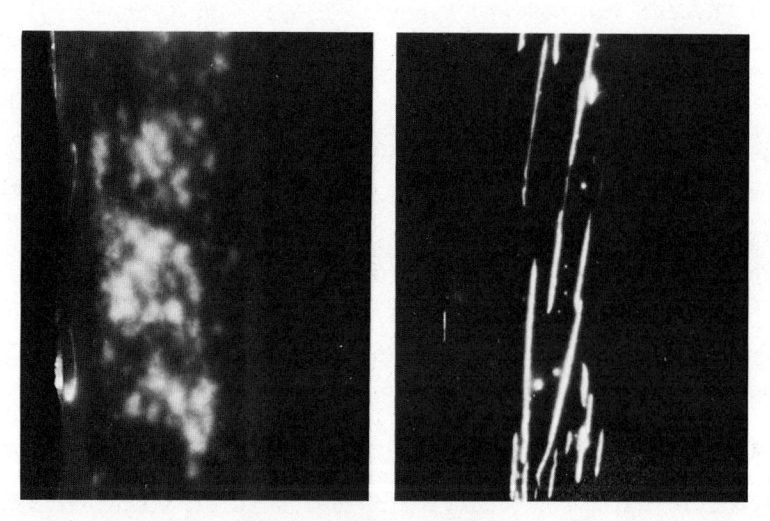

Fig. 4 Appearance of the two samples shown in Fig. 3 on dye penetration. The material containing the higher fractions of monoclinic ZrO_2 shows significantly higher, blotchy absorption of the fluorescent dye than the other sample where only large surface scratches have absorbed the dye.

method for observing microcracks in the scanning electron microscope is that of direct examination in the secondary electron imaging mode, such as illustrated by the micrograph of Fig. 5. Although relatively straightforward requiring only standard sample preparation the method is complicated by artifacts primarily because in this mode the cracks appear dark and it can be difficult to distinguish them. It can be difficult to recognize any cracks present in the presence of similar features, such as precipitates, inclusions or sharp edges. In consequence, it is an observational method best suited to observing relatively wide (>1 µm) cracks and cracks on highly polished samples where there is little other topographic contrast.

In view of these shortcomings, a new crack decoration procedure has been developed at the Science Center. The essence of the technique is to decorate cracks with a stain that appears with high contrast in one of the imaging modes of the scanning microscope. In the backscatter imaging mode, for instance, the

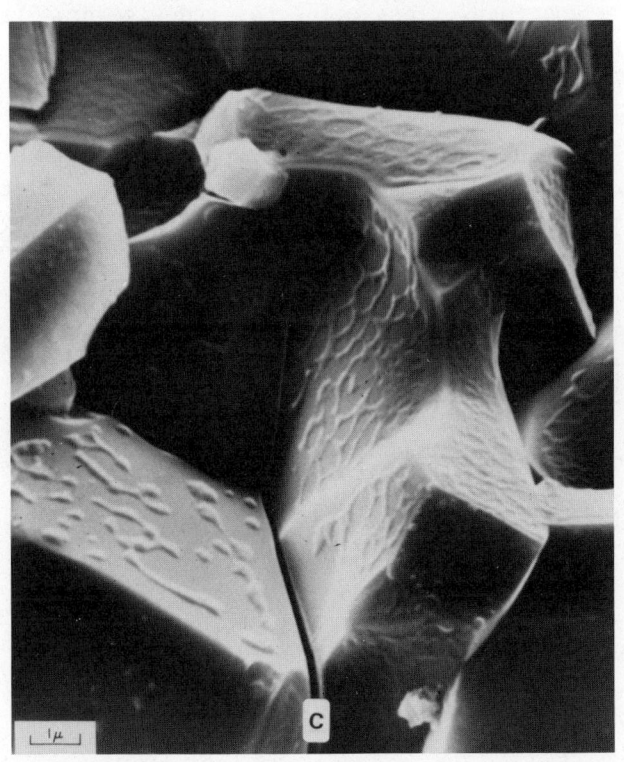

Fig. 5 Scanning electron micrograph of a debased alumina ceramic containing a microcrack along the facet, C.

contrast of a region is related to its average atomic number as shown by the curve of Fig. 6.[18] Thus, by infiltrating a heavy metal dye, such as silver nitrate, into the microcracks of a low atomic weight material, such as $Al_2O_3-ZrO_2$ or $MgTi_2O_5$, an image is in principle obtainable in which the microcracks appear as bright lines against a darker background. Such an effect is shown in the backscatter electron image of Fig. 7 for a micro cracked $Alf2_2O_3/ZrO_2$ sample. Successful implementation of the technique requires that the heavy element dye wets the microcracks so as to both enter them and remain inside; this has proven to be the most difficult step in the decoration.

The actual procedure we have utilized for crack staining may be summarized as follows:

(i)   Soak specimen in aqueous $AgNO_3$ solution (0.1 M) for several hours

(ii) Remove specimen, rinse off $AgNO_3$ and wipe dry

(iii) Allow specimen to dry prior to observation in SEM

Although staining for examination in the backscatter imaging mode is to be preferred because of the relatively high spatial resolution (<100Å) afforded by the backscatter electron mode, other contrast mechanisms can be employed. One alternative is to decorate the microcracks with a stain that is cathodoluminescent (emits optical photons when excited with an electron beam). The cathodoluminescent imaging mode has an intrinsically inferior resolution (~1000Å) to that of the more

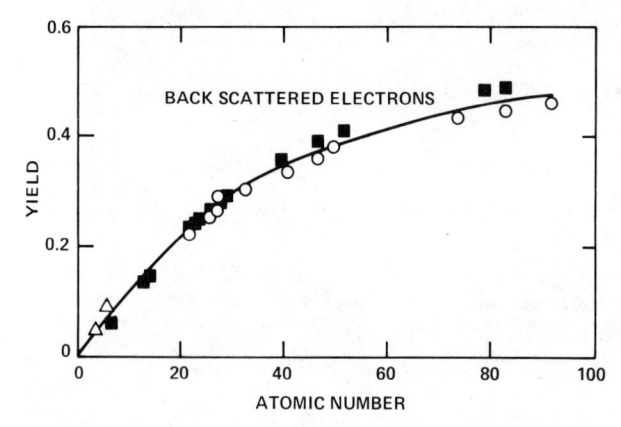

Fig. 6 Variation of primary backscatter electron yield with atomic number at 30 kV. The experimental points are from the work of Wittry (●) and Weinryb (o).

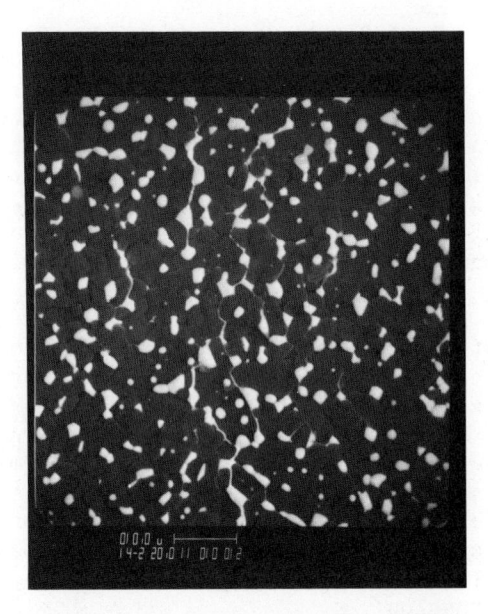

Fig. 7 Backscatter electron image in the scanning electron microscope of an Al_2O_3-ZrO_2 sample after infiltration with a silver nitrate solution. The microcracks are decorated with the silver salt and appear brightly delineated against the sample due to its higher effective atomic number. (The ZrO_2 grains also appear bright but have the expected grain shape.)

commonly used secondary electron or backscatter electron methods, but it might nevertheless find application in cases such as those where the microcracked material has a high atomic number.

Transmission Electron Microscopy

Microcracked materials are normally so weak and fragile that sample preparation difficulties generally preclude their examination by transmission electron microscopy. However, when samples can be prepared, the transmission microscope can be used to extend the observational range to microcracks having crack opening displacements down to ~1 nm. In order to achieve this resolution, it is necessary that the microscope be operated in the defocus imaging mode.[19,20] The technique, also known as Fresnel imaging, has recently been applied to the detection of very thin (~1-5 nm) intergranular phases in ceramics, to complement other observational techniques (diffuse dark field and lattice fringe imaging) developed for the same purpose.

An example of microcracks observed using the defocus technique is reproduced in Fig. 8 showing both an overfocus and underfocus image of a microfractured region in a single phase ZrO_2-Y_2O_3 ceramic. The out-of-focus contrast is particularly clear at microcrack, A and at the wider microcrack, B. At the exact focus condition of the microscope, the microcracks are barely visible except for particularly large cracks which can be seen through. If no microcracks were present, the defocus images of the region would not differ from an in-focus image except for being slightly out of focus.

The qualitative features of the observed defocusing contrast can be understood simply by analogy with, and extension of, Fresnel diffraction from a hole or edge. In passing through a thin specimen, the incident electron wave undergoes a phase shift ϕ, relative to a wave in the vacuum of the microscope, that is proportional to the mean inner potential, ΔV, within the specimen. (The electron enters a region of different refractive index in the specimen to that of vacuum.) The phase shift is also proportional to the thickness, t, of the specimen and is given by the expression

$$\phi = \frac{\pi}{\lambda E} \cdot t \cdot \Delta V$$

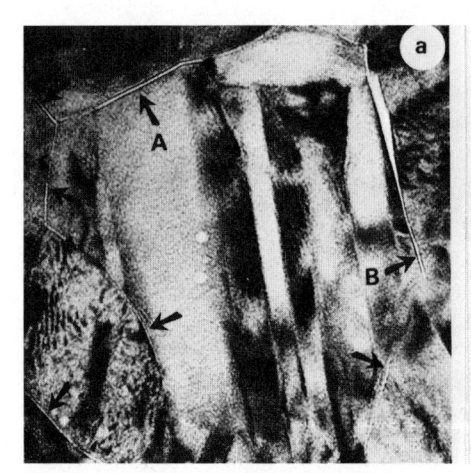

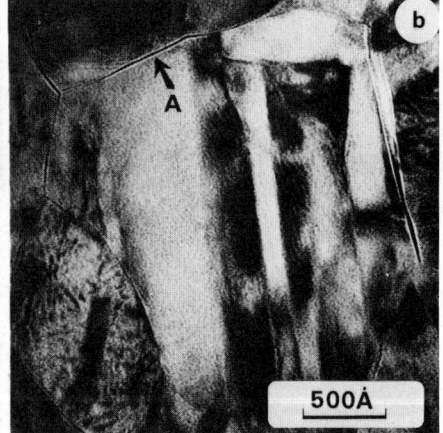

Fig. 8 (a) Underfocused and (b) overfocused transmission electron micrographs of a region containing microcracks (arrowed) in a sintered and fractured ZrO_2 -2.5 m/o Y_2O_3 alloy. The symmetrical double-fringe contrast, such as at A, is characteristic of the presence of a microcrack.

where λ is the electron wavelength, E is the accelerating voltage in electron volts. On defocusing the microscope, this phase shift is manifest at the edge of a sample by the formation of Fresnel fringes (Becke lines); a bright line outside of the sample edge on under-focusing and a dark line on over-focusing. In the case of a microcrack in a grain boundary oriented parallel to the electron beam, the electron wave suffers a phase shift in passing through the grains at either side relative to passing through the microcracked region. The analogy then is of two closely spaced edges, and a double set of Fresnel fringes is formed on defocusing. By direct analogy with Fresnel diffraction from an edge the following contrast features are expected when imaging edge-on a microcrack. In the focused micrograph, little or no contrast will be observed; in an over-focused micrograph ($\Delta f > 0$) the microcrack phase will be a dark band with a sequence of parallel bright and dark Fresnel fringes, whose intensity falls rapidly with distance from the center; on under-focusing ($\Delta f < 0$) the contrast will reverse to give a bright band at the film on microcrack with alternate dark and bright Fresnel fringes on either side. Similarly, the spacing of the first, strongest fringes increases with increasing defocus distance and also converge to the actual width of the microcrack or film at the exact focus condition is approached.

CONCLUSIONS

Various techniques for the indirect detection of microcracks, such as elastic modulus and thermal conductivity measurements are commonly used for ceramic materials. In addition, it is shown in this contribution, that porosimetry could also be used to characterize microcracks. These techniques are clearly a useful approach but the results can be ambiguous and do not indicate the size or distribution of the microcracks. There is, therefore, an important need to develop direct techniques of microcrack detection and it is shown here that heavy-element staining and cathodoluminescent dyes have great potential for such observations. These direct techniques will invariably give information that is not available using indirect techniques, thus complementing them allowing these latter techniques to be used in a more quantitative fashion than has hitherto been possible.

ACKNOWLEDGEMENT

The authors are indebted to the National Science Foundation under Grant No. DMR-8007445 for support of this work.

REFERENCES

1. J. D. Eshelby, "The Determination of the Elastic Field of an Ellipsoidal Inclusion and Related Problems," Proc. Roy. Soc. A241, 376-396 (1957).
2. J. R. Bristow, "Microcracks, and the Static and Dynamic Elastic Constants of Annealed and Heavily Cold-worked Metals," Brit. J. of Applied Phys. 11, 81-85 (1960).
3. R. L. Salganik, "Mechanics of Bodies with Many Cracks," Izv. Akad. Nauk. SSSR. Mekh. Iverd. Tela 8[4] 149-58 (1958).
4. D. P. Hasselman and J. P. Singh, "Analysis of Thermal Stress Resistance of Microcracked Brittle Ceramics," Bull. Am. Ceram. Soc. 58[9], 856-60 (1979).
5. B. Budiansky and R. J. O'Connell, "Elastic Modulus of a Cracked Solid," Int. J. Solid Struct. 12[1] 81-97 (1976).
6. J. A. Kuszyk and R. C. Bradt, "Influence of Grain Size of Effect of Thermal Expansion Anisotropy in $MgTi_2O_5$," J. Am. Ceram. Soc. 56[8], 420-23 (1970).
7. W. R. Manning, O. Hunter, Jr., F. W. Calderwood and D. W. Stacy, "Thermal Expansion of Nb_2O_5," J. Am. Ceram. Soc. 55[7], 342-47 (1972).
8. D. B. Binns, "Some Physical Properties of Two-phase Crystal-glass Solids," pp. 315-34 in Science of Ceramics. Edited by G. H. Stewart, Academic Press, New York, 1962.
9. D. J. Green, "Critical Microstructures for Microcracking in Al_2O_3-ZrO_2 Composites," J. Am. Ceram. Soc., to be published (1983).
10. J. R. Willis, "Bounds and Self-Consistent Estimates for the Overall Properties of Anisotropic Composites," J. Mech. Phys. Solids 25, 185-202 (1977).
11. D. P. H. Hasselman, "Effects of Cracks on Thermal Conductivity," J. Compos. Mater. 12[10] 403-407 (1978).
12. H. J. Siebeneck, D. P. H. Hasselman, J. J. Cleveland and R. C. Bradt, "Effect of Microcracking on the Thermal Diffusivity of Fe_2TiO_5," J. Am. Ceram. Soc. 59[5-6] 241-44 (1976).
13. H. J. Siebeneck, J. J. Cleveland, D. P. H. Hasselman and R. C. Bradt, "Effect of Grain Size and Microcracking on the Thermal Diffusivity of $MgTi_2O_5$," J. Am. Ceram. Soc. 60[7-8], 336-38 (1977).
14. W. R. Manning, G. E. Youngblood and D. P. H. Hasselman, "Effects of Microcracking on the Thermal Diffusivity of Polycrystalline Aluminum Niobate," J. Am. Ceram. Soc. 60[9-10], 469-70 (1977).
15. D. Greve, N. Claussen, D. P. H. Hasselman and G. E. Youngblood, "Thermal Diffusivity/Conductivity of Alumina with a Zirconia Dispersed Phase," Bull. Am. Ceram. Soc. 56[5], 54-15 (1977).

16. E. D. Case, unpublished work.
17. G. Simmons and D. Richter, "Microcracks in Rocks,"
18. See for instance D. R. Clarke, "High Spatial Resolution Analysis of Grain Boundaries: Techniques and Applications," in Adv. Ceram. $\underline{1}$, 67-90 (1981).
19. D. R. Clarke, "Observation of Microcracks and Thin Intergranular Films in Ceramics by Transmission Electron Microscopy," J. Am. Ceram. Soc. $\underline{63}$[1-2], 104-106 (1980).
20. M. Ruhle, C. Springer, N. Claussen and H. Strunk, "TEM Studies of $Al_2O_3-ZrO_2$ Composites," Proc. fifth Int. Conf. High Voltage Electron Microscopy, Kyoto, 1977.

X-RAY ENERGY DISPERSIVE SPECTROSCOPY OF

INTERGRANULAR PHASES IN β11 β' SIALONS

T. R. Dinger and G. Thomas

Department of Materials Science and Mineral Engineering
and the Materials and Molecular Research Division
Lawrence Berkeley Laboratory, University of California
Berkeley, California   94720

ABSTRACT

X-ray energy dispersive spectroscopy in conjunction with scanning transmission electron microscopy was used to qualitatively and semi-quantitatively characterize retained noncrystalline phases in several compositions of β11 β' sialons. Diffuse dark field imaging in conventional transmission mode was used to identify noncrystalline phases prior to x-ray microanalysis. Results show that partitioning of Al and impurity atoms does not occur for these compositions except in isolated cases. In addition, Fe impurities appear to be readily incorporated into the matrix phase.

INTRODUCTION

The use of silicon nitride ceramics for structural components of high temperature gas turbine engines and the problems associated with the fabrication of these ceramics are well documented. Alloying of Si_3N_4 with Al_2O_3 was first proposed in 1972 by Jack[1] in England and Oyama in Japan with the simultaneous discovery of the β' solid solution field in the Si_3N_4-SiO_2-Al_2O_3-AlN system. With this discovery came the realization that single-phase silicon nitride ceramics could theoretically be fabricated by incorporating the transient liquid phase necessary for densification into the crystalline matrix phase.

Numerous attempts at fabrication have resulted in alloys with properties comparable to those of hot-pressed silicon nitride. However, these alloys still retain various amounts of intergranular phase which degrades the high temperature properties of the ceramic. Retention of this noncrystalline intergranular phase has been attributed to volatilization of N_2 and SiO [3] during fabrication resulting in compositions lying outside the single-phase field, to segregation of impurities in the grain boundary phase [4,5], or to thermodynamic and microstructural restrictions.[6]

The purpose of the work presented in this paper was to determine whether segregation of impurities or Al partitioning was occurring, and whether these mechanisms could then be interpreted as a cause for the retention of noncrystalline intergranular phases in low Al β' solid solution alloys.

Because of the very small scale of the intergranular phase pockets (often smaller than 50 nm), analysis using transmission electron microscopy (TEM) and scanning transmission electron microscopy (STEM) in conjunction with x-ray energy dispersive spectroscopy was required to obtain the necessary spatial resolution.

EXPERIMENTAL PROCEDURE

Alloying of Si_3N_4 with Al_2O_3 is a process that is filled with compromise. Large additions of alumina result in readily sinterable ceramics with excellent oxidation resistance[1,7] but with poor thermal shock resistance[8-10]. With this in mind, compositions with low Al content corresponding to β11 (11 eq.% Al^{+3}) were chosen lying close to the β' solid solution field in the β'ss-Si_2ON_2-X_1 compatibility triangle. See Fig. 1. These compositions were chosen to suppress the formation of Si_2ON_2 and the mullite-like X_1 phase, therefore producing an essentially two-phase material containing β' solid solution and retained noncrystalline grain boundary phase.

The results of an emmission spectrographic analysis of the Si_3N_4, Al_2O_3, and AlN starting powders appear in Table I. The powders were weighed, milled, and mixed in an attrition mill with Al_2O_3 balls for 12 hours in n-hexane. The wear of the balls was taken into account in the final composition. The mixed powders were then dried, sieved, and hot-pressed in BN-coated graphite dies under 35 MPa pressure at 1770°C or 1780°C for 45 or 60 minutes. After firing, the surface layer was removed by mechanical grinding, and bulk density measurements were made by the Archimedes method. Phase analyses were made by x-ray diffraction using a Philips goniometer with Cu Kα radiation and a LiF single crystal monochromator. The compositions of the specimens, hot-pressing conditions, and results of the analyses are summarized in Table II. Since composition A did not fully densify, further microstructural analysis was not deemed necessary.

In order to prepare electron transparent thin sections, hot-pressed discs were first thinly sectioned on a diamond wafering saw

X-RAY ENERGY DISPERSIVE SPECTROSCOPY

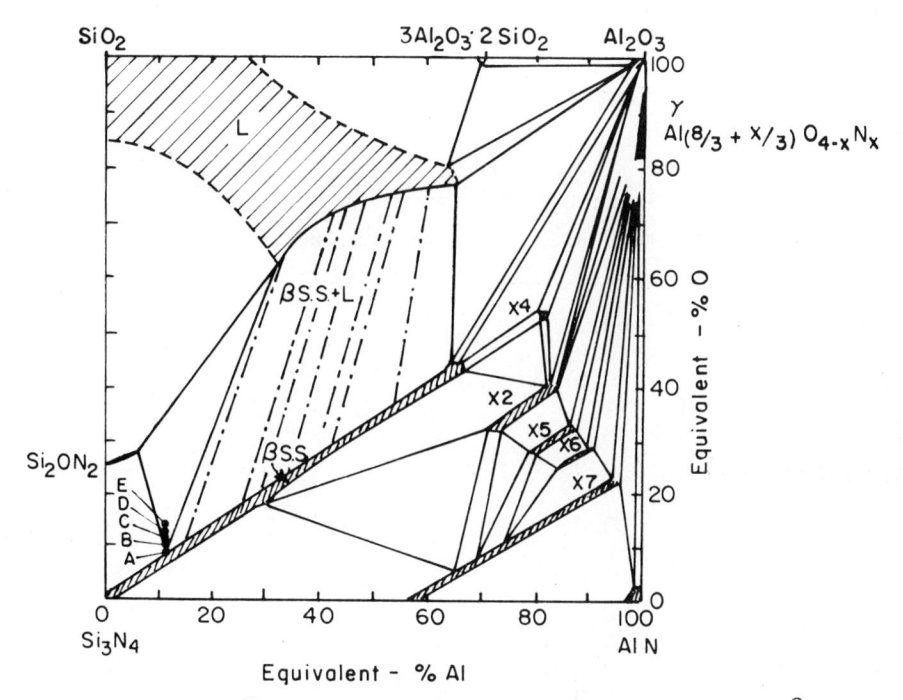

Fig. 1. Isothermal section of the Si-Al-O-N system at 1750°C after Naik et al.[11] Compositions B, C, D, and E were used in the investigation.

before being mechanically ground with SiC paper to thicknesses of approximately 50 μm. After attachment to 3 mm copper grids with a Ag-based cement, sections were further thinned by ion milling with 6 KV argon ions at a 20° angle of incidence. A thin layer of carbon was then evaporated onto the surface of the sample to help alleviate the problems associated with surface charging.

The electron microscopy of compositions B through E was done in a Philips EM400 transmission electron microscope equipped with scanning coils and a Kevex micro-x 7000 analytical spectrometer. Specimens were initially examined in conventional TEM mode using the technique outlined by Krivanek et al.[13] to identify noncrystalline intergranular phases. In this technique, only those electrons which have undergone diffuse scattering by the noncrystalline phase are allowed to form the image. Thus, the noncrystalline regions appear light on a dark background in dark field.

After the initial identification of intergranular phase pockets of sufficient size for x-ray microanalysis, the microscope was switched to scanning mode using a nominal probe size of 10 nm. All microscopy was done using electrons of 100 KeV incident energy.

Table I

Chemical Composition of Starting Powders*

Element	α-Si$_3$N$_4$ Starck, Berlin HCST 3510	AlN Starck, Berlin HCST 530	Al$_2$O$_3$ Alcoa A 16
Si$_{total}$	60.1	-	-
Al	0.06	65.03	52.9
O	2.0	3.12	47.4
N	37.1	31.85	-
Fe	0.01	-	-
K	0.001	-	-
C	0.4	-	-
Ca	0.03	-	-
Ti	0.002	-	-
Li	0.0005	-	-
Si$_{free}$	0.5	-	-
Mg	0.003	-	0.6
Na	0.003	-	-
W	0.01	-	-

* Emission spectroscopy. Data provided by Dr. J. Weiss and Dr. P. Greil, Max-Planck-Institut für Metallforschung, Stuttgart, West Germany.[12]

Table II

Chemical Compositions, Hot-Pressing Conditions, Bulk Density, and Phase Compositions of Sialons Examined **

	Eq. % Si	Al	O	N	Temp. (°C)	Time (min.)	Density (g/cm^3)	Phases (XRD)
A	89.4	10.6	7.5	92.5	1780	60	2.67	β'ss, Si$_2$ON$_2$
B	89.4	10.6	9.2	90.8	1770	45	3.150	β'ss, Si$_2$ON$_2$
C	89.4	10.6	10.9	89.1	1770	45	3.136	β'ss, Si$_2$ON$_2$
D	89.4	10.6	12.6	87.4	1770	45	3.116	β'ss, Si$_2$ON$_2$, X$_1$
E	89.4	10.6	14.0	86.0	1770	45	3.103	β'ss, Si$_2$ON$_2$, X$_1$

** Data provided by Dr. J. Weiss and Dr. P. Greil, Max-Planck-Institut für Metallforschung, Stuttgart, West Germany.[12]

X-RAY ENERGY DISPERSIVE SPECTROSCOPY

X-ray spectra[14] were obtained in the energy range of 0 to 10.24 KeV with a channel width of 10 eV. Spectra were acquired by count integration on the Si Kα peak. These integrations were for 10K, 20K, 50K, or 100K counts, with normal acquisition times being 300 to 350 seconds per 1024 channel spectrum depending on the thickness of the specimen. This technique permitted analyses for all elements of interest except nitrogen and oxygen.

DISCUSSION OF RESULTS

The initial examination of the series in conventional transmission mode resulted in the bright field/dark field micrograph pairs of Fig. 2 through Fig. 5, representing compositions B through E, respectively. As would be expected from the composition data and Fig. 1, the amount of noncrystalline phase and the size of noncrystalline phase pockets increases in the series moving from composition B to composition E. These grain boundary phases appear in light contrast in the dark field micrographs with the larger pockets being labeled (g).

The results of nearly 80 comparative spectra are represented, for the most part, by Fig. 6 through Fig. 10. As can easily be seen from Fig. 6, very little, if any, distinction can be made between the spectra corresponding to the noncrystalline intergranular phase (solid line) and to the crystalline phase (dots). This tendancy was strong in all four compositions examined indicating that little or no partitioning of Al had occurred. The Cr and Cu peaks apparent in all spectra are systems background and will not be treated further.[15] Note that the Fe impurity peak appears with equal intensity in both the crystalline matrix phase and the noncrystalline intergranular phase . The equivalent Si/Al ratios hint at different O/N ratios between phases as a reason for retention of the noncrystalline grain boundary phase.

The comparative spectra of Fig. 7 illustrate one of the rare instances where Ca impurities were found to be segregated in the grain boundary phase (dots).[16,17] Even in this case, the Ca peak is barely significant above background, but is definitely of higher intensity than that of the crystalline matrix phase (solid line). This effect was not apparent in the remainder of the sample or in other compositions of the series.

The spectra of Fig. 8 illustrate one of several instances in which the noncrystalline phase was found to be rich in Si. One possible explanation for this is that the crystalline phase probed corresponded to an X_1 phase grain exhibiting a much higher Al content. This does not seem probable, however, due to the distinctive elongated and often twinned morphology[18] of the X_1 phase which was not observed. A simpler and more acceptable explanation is that a local

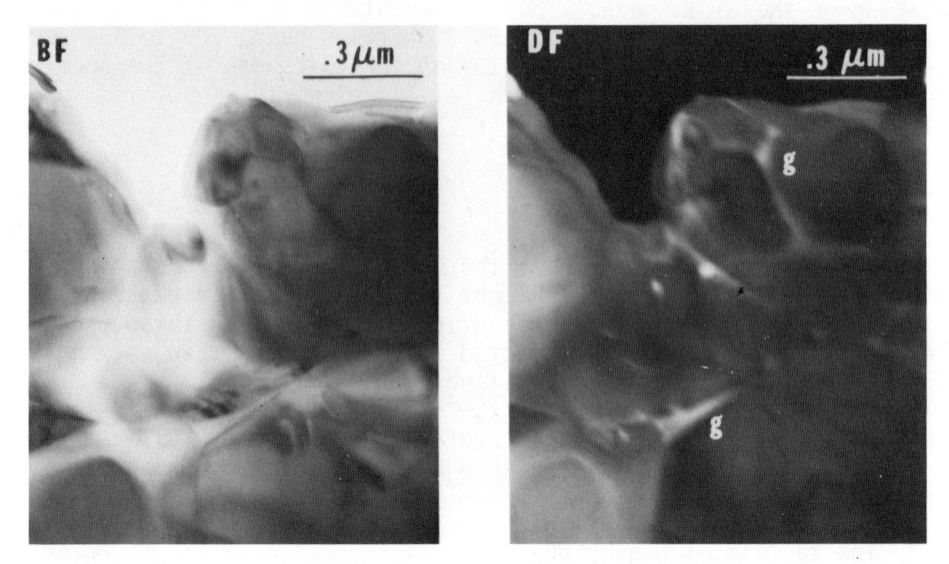

Fig. 2. Bright field/dark field pair of composition B. Note noncrystalline phases which appear in light contrast in dark field and are labeled (g).

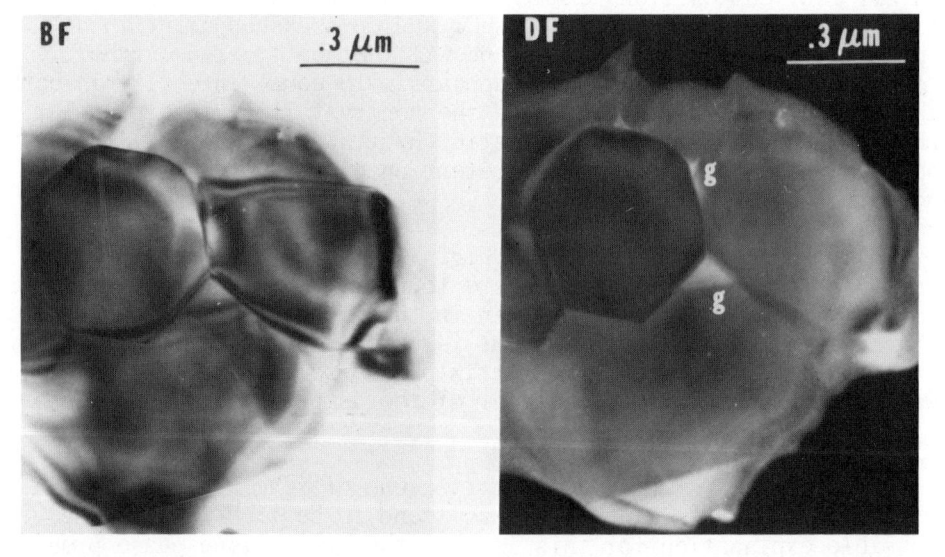

Fig. 3. Bright field/dark field pair of composition C. Noncrystalline grain boundary phases are denoted by (g).

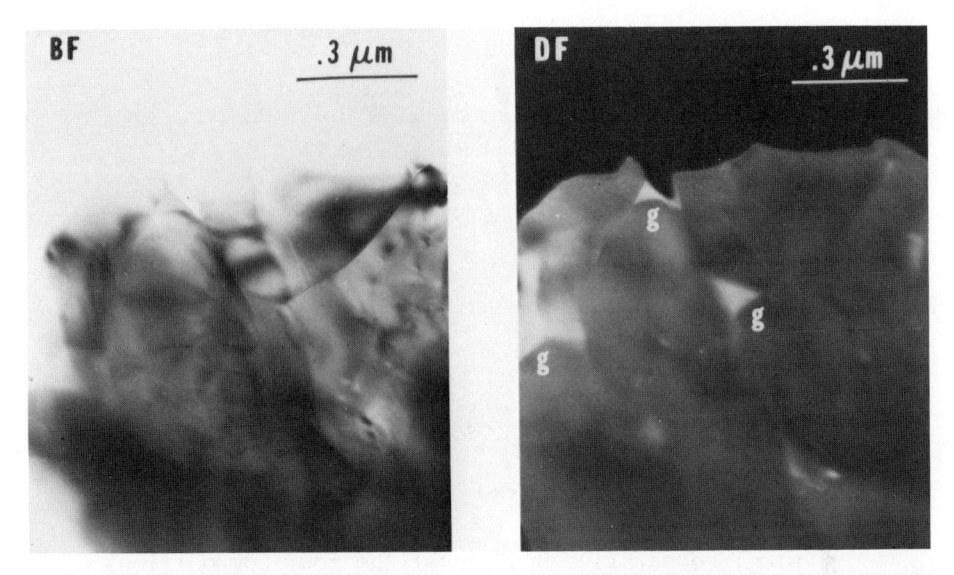

Fig. 4. Bright field/dark field pair of composition D. Noncrystalline phases are denoted by (g).

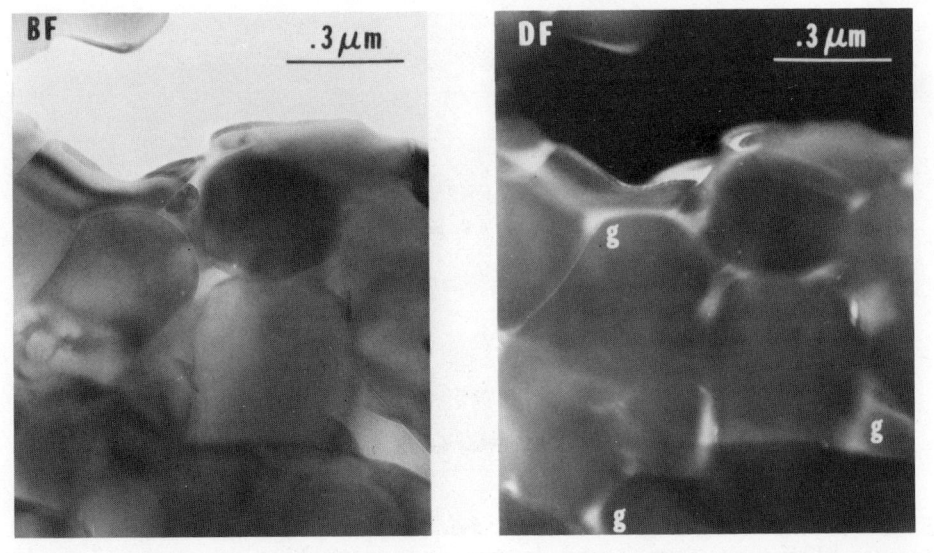

Fig. 5. Bright field/dark field pair of composition E. Noncrystalline phases are denoted by (g).

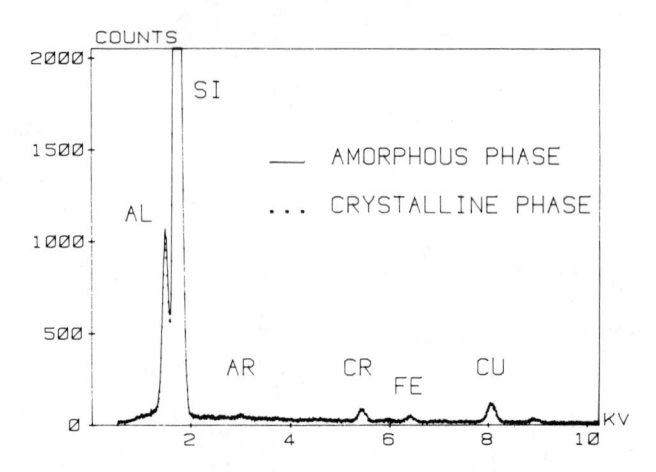

Fig. 6. Typical comparative spectra of the four compositions analyzed. The intergranular phase is denoted by (—) while the crystalline phase is denoted by (...).

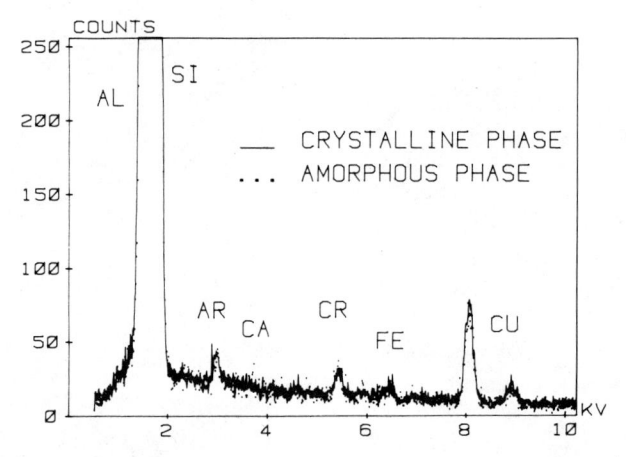

Fig. 7. Nontypical comparative spectra for composition D illustrating a very slight Ca enrichment in the noncrystalline intergranular phase.

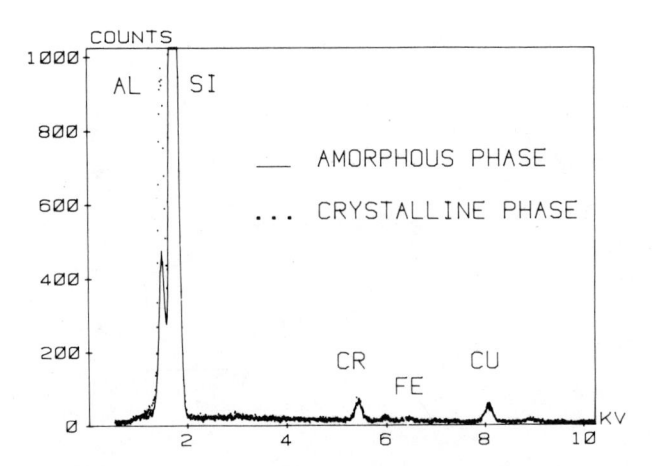

Fig. 8. Nontypical comparative spectra for composition E illustrating an apparent enrichment of the noncrystalline phase (—) with Si.

inhomogeneity has resulted in the noncrystalline phase being enriched with Si.

The STEM micrograph of Fig. 9 illustrates Fe-rich inclusions (arrowed and in dark contrast) which have perhaps provided the most interesting results of the investigation. The comparative spectra of Fig. 10 arise from the Fe-rich inclusion (arrow), the adjacent noncrystalline intergranular phase (—), and the adjacent crystalline matrix phase (..). Jack[19] has indicated that incorporation of impurities such as Ca and Mg into the sialon phase could occur in much the same way as the substitution of Al as long as charge neutrality is maintained. The homogeneous distribution of Fe in both the matrix and intergranular phases may indicate that this argument should be extended to include Fe impurities as well. Fe-rich inclusions of this type were found only in composition E, but in each case, the spectra were similar to those of Fig. 9.

CONCLUSIONS

The four $\beta 11$ β' sialon compositions examined by transmission electron microscopy, scanning transmission electron microscopy, and x-ray energy dispersive spectroscopy were found to contain various amounts of noncrystalline intergranular phase depending on the composition used. The amount of intergranular phase present increased, as expected, with increasing oxygen/nitrogen ratio. X-ray energy dispersive spectroscopy indicated that all compositions were homogeneous with

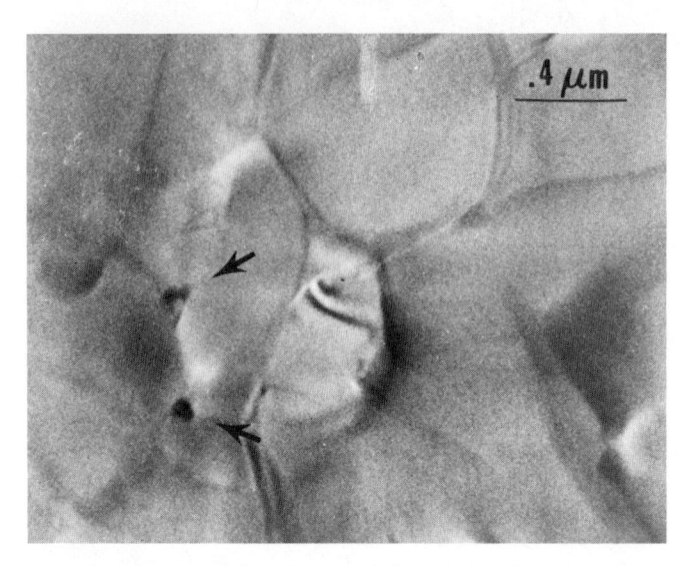

Fig. 9. STEM micrograph of Fe-rich inclusions and surrounding microstructure in composition E.

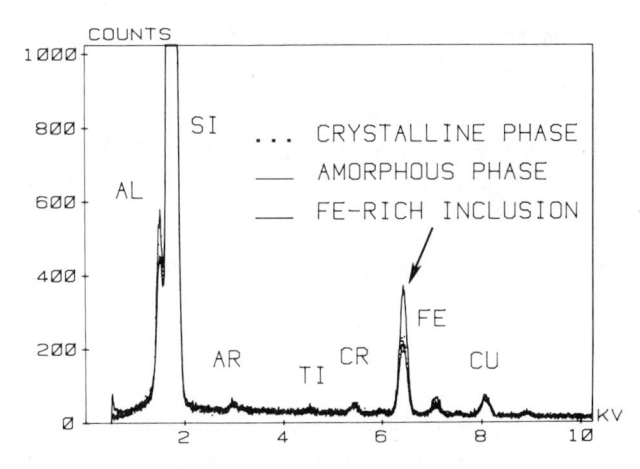

Fig. 10. Comparative spectra for Fe-rich inclusion (arrow) and surrounding crystalline (—) and noncrystalline (..) phases in comp. E.

respect to silicon/aluminum ratios between crystalline matrix and noncrystalline grain boundary phases with few exceptions noted. Impurity distributions were found to be equally homogeneous with the exception of Ca which tended to segregate in noncrystalline grain boundary phases. This is in contrast to earlier investigations which have found the intergranular phase to be a sink for several types of impurity atoms. In addition, localized Fe impurities have been found to be accomodated by both the matrix and intergranular phases in a homogeneous manner. This lends support to Jack's theory that impurity phases may be incorporated into the matrix in these solid solution materials.

ACKNOWLEDGEMENTS

This work was supported by the National Science Foundation under grant number DMR-77-24022 with facilities and support staff provided by the Director, Office of Energy Research, Office of Basic Energy Sciences, Division of Materials Sciences of the U. S. Department of Energy under contract number DE-AC03-76SF00098. Thanks go to Dr. J. Weiss and Dr. P. Greil of the Max-Planck-Institut für Metallforschung Institut für Werkstoffwissenschaften, for providing the sample material and supporting analysis data.

REFERENCES

1. K. H. Jack and W. I. Wilson, Ceramics Based on the Si-Al-O-N and Related Systems, Nature (London), Phys. Sci. 238:28-29(1972).

2. Y. Oyama and O. Kamagaito, Hot-Pressing of Si_3N_4-Al_2O_3, Yogyo Kyokai Shi 80:327-336(1972).

3. K. H. Jack, The Relationship of Phase Diagrams to Research and Development of Sialons, in: "Phase Diagrams: Materials Science and Technology, Volume 6-V: Crystal Chemistry, Stoichiometry, Spinoidal Decomposition, Properties of Inorganic Phases," A. M. Alper, ed., Academic Press, San Francisco(1978).

4. D. R. Clarke, The Microstructure of a 12H Mg-Si-Al-O-N Polytype Alloy: Intergranular Phases and Compositional Variations, J. Am. Cer. Soc. 63:208-214(1980).

5. O. L. Krivanek, T. M. Shaw, and G. Thomas, The Microstructure and Distribution of Impurities in Hot-Pressed and Sintered Silicon Nitrides, J. Am. Cer. Soc. 62:585-590(1979).

6. R. Raj and F. F. Lange, Crystallization of Small Quantities of Glass (or a Liquid) Segregated in Grain Boundaries, Acta Met. 29:1993-2000(1981).

7. S. C. Singhal and F. F. Lange, Oxidation Behavior of Sialons, J. Am. Cer. Soc. 60:190-191(1977).

8. F. F. Lange, H. J. Siebeneck, and D. P. H. Hasselman, Thermal Diffusivity of Four Si-Al-O-N Compositions, J. Am. Cer. Soc. 59: 454-455(1976).

9. M. Kariyama, Y. Inomata, T. Kujima, and Y. Hasegawa, Thermal Conductivity of Hot-Pressed Si_3N_4 by the Laser Flash Method, Bull. Am. Cer. Soc. 57:1119-1122(1978).

10. F. F. Lange, Silicon Nitride Alloy Systems: Fabrication, Microstructure and Properties, Rockwell International Science Center, Report No. J1376A/SN.

11. I. K. Naik, L. J. Gauckler, and T. Y. Tien, Solid-Liquid Equilibria in the System Si_3N_4-AlN-SiO_2-Al_2O_3, J. Am. Cer. Soc. 61: 332-335(1978).

12. P. Greil and J. Weiss, Evaluation of the Microstructure of β Solid Solution Materials with 11 Eq. % Al^{+3} Containing Different Amounts of Amorphous Grain Boundary Phase, to be published.

13. O. L. Krivanek, T. M. Shaw, and G. Thomas, Imaging of thin intergranular phases by high resolution electron microscopy, J. Appl. Phys. 50:4223-4227(1979).

14. See for instance, "Introduction to Analytical Electron Microscopy," J. J. Hren, J. I. Goldstein, and D. C. Joy, eds., Plenum Publishing Corporation, New York(1979).

15. J. Bentley, Systems Background in X-ray Microanalysis, in: "Analytical Electron Microscopy-1981," R. H. Geiss, ed., San Francisco(1981).

16. M. H. Lewis, B. D. Powell, P. Drew, R. J. Lumby, B. North, and A. J. Taylor, The formation of single-phase Si-Al-O-N ceramics, J. Mat. Sci. 12:61-74(1977).

17. D. R. Clarke, N. J. Zaluzec, and R. W. Carpenter, The Intergranular Phase in Hot-Pressed Silicon Nitride: I, Elemental Composition, J. Am. Cer. Soc. 64:601-607(1981).

18. A. Zangvil, The structure of the X phase in the Si-Al-O-N alloys, J. Mat. Sci. 13:1370-1374(1978).

19. K. H. Jack, The Crystal Chemistry of the Sialons and Related Nitrogen Ceramics, in: "Nitrogen Ceramics," F. L. Riley, ed., Noordhoff International Publishing, Leyden, The Netherlands (1977).

EM STUDY OF THE STRUCTURE AND COMPOSITION OF GRAIN BOUNDARIES IN $(Mn,Zn)Fe_2O_4$

I-Nan Lin, R. K. Mishra and G. Thomas

Department of Materials Science and Mineral Engineering
and the Materials and Molecular Research Division
University of California
Berkeley, CA 94720

ABSTRACT

Electron diffraction and microscopy studies supplemented by electron spectroscopic techniques such as Auger electron spectroscopy and energy dispersive x-ray spectroscopy were used to characterize the nature of grain boundary segregation in commercial grade $(Mn,Zn)Fe_2O_4$ samples containing small quantities of CaO. Chemical analyses by AES and EDAX show an enrichment of Ca near the grain boundary region. Convergent beam electron diffraction experiments show that the crystal symmetry of the spinel structure is distorted in the vicinity of the grain boundary. In-situ heating experiments in HVEM show the existence of a disordered phase at the sintering temperature. Lorentz microscopy in TEM shows the interaction of magnetic domain wall motion with grain boundaries. These chemical and structural features are correlated with electrical resistivity and magnetic permeability of the ferrites.

INTRODUCTION

The presence of any second phases at grain boundaries in polycrystalline ceramics has been of great interest for their effects on the mechanical or electronic properties. In the case of high temperature structural ceramics, such as Si_3N_4, the amorphous boundary phase is responsible for low temperature creep.[1] In the case of electronic materials, such as PZT, ZnO varistors and soft ferrites, the presence of a thin grain boundary layer drastically affects the electrical and magnetic properties.[2] The formation of the second phase at grain boundaries in ceramic materials is very common and a complete characterisation of these second phases can be done only by modern techniques, such as transmission electron microscopy (TEM), analytical electron microscopy (AEM) and Auger electron spectroscopy (AES). In the present work, the physical and chemical characterization of an amorphous grain boundary phase in $(Mn,Zn)Fe_2O_4$ using the above-mentioned techniques and the effects of this phase on the electrical and magnetic properties of the material will be discussed.

The $(Mn,Zn)Fe_2O_4$ is a soft ferrite with high initial permeability (μ_i). Small amounts of CaO are added in commercial $(MnZn)Fe_2O_4$ to increase the electrical resistivity through the formation of an insulating layer along the grain boundaries.[3] The existence of a thin intergranular amormphous phase has been observed by Mishra et al.[4] using lattice fringe image microscopy[5] and also by Lin et al.[6] using high resolution dark field microscopy.[7] It has been shown that the grain boundary phase is enriched in Ca. A complete characterization of the grain boundary and its effects upon the magnetic properties and sintering mechanism has not been ascertained, but will be discussed in this work.

EXPERIMENTAL

Sintered specimens of MnZn ferrite with the nominal composition of $MnO:ZnO:Fe_2O_3$ = 26.9:19.8:53.3 (mole %) and with the major additive of CaO (2543 ppm) were supplied by TDK Electronics Co. of Japan. A 1mm x 5mm x 10mm slice of sintered material was fractured in-situ in a Physical Electronics 590 scanning Auger electron microscopy[8,9] and the chemical composition was determined from Auger electron spectra. Electron transparent thin foils were prepared from the bulk sample by an ion milling technique. The magnetic domain wall structure and its interaction with grain boundaries were studied in a Philips EM301 microscope using Lorentz microscopy (LM).[10,11] The symmetry of the crystal structure was studied in a Philips EM400 microscope using convergent beam electron diffraction (CBD).[12] The behavior of Ca-doped grain boundaries at high temperature was studied by heating a thin foil in the hot stage of the Osaka University 3MeV high voltage electron microscope (HVEM) to a temperature of 1400°C.

RESULTS

A typical AES spectrum from a fractured surface of MnZn ferrite (Fig. 1b) clearly shows the existence of Ca. Careful experiments show that the Ca signal comes from those regions of the surface that are intergranular fracture. The Ca-signal completely disappears after the fractured surface is sputtered in-situ by argon ions for ʊ six minutes. Since the escape depth of the Auger electrons is less than 20Å and the sputtering rate is only about 10Å/min., the thickness of the Ca containing phase is about 60Å. Besides the Ca signal indicated in Fig. 1b, the spectra from intergranular fracture surfaces also have Fe and Mn signals. This implies that the Ca-containing phase is not a pure CaO phase as proposed by other authors.[3,13,14] Although the chemical composition is very difficult to determine quantitatively, it can be concluded that the Ca-containing phase is an intermediate compound of CaO and MnZn ferrite. The existence of the intermediate compound has also been confirmed by Lin et al. by SEM hot stage experiments. The Ca signal is observed only when the electron probe, 500A in diameter, is placed at intergranular fracture surfaces. No Ca peak is observed when the electron probe is placed at transgranular fracture surfaces (Fig. 1a). This indicates that the Ca-containing layer only forms a thin layer along grain boundaries. This grain boundary phase exists as an amorphous phase,[4,6] as has been shown elsewhere. It occurs due to the difficulty[15,16] of crystallization of thin layers of liquid trapped between grains.

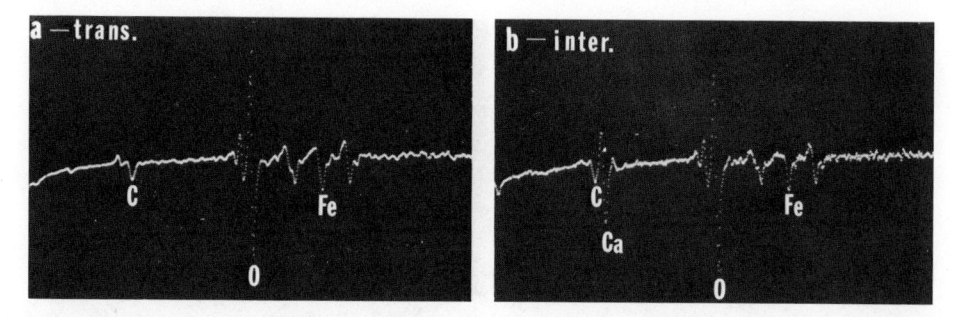

Fig. 1. Auger electron spectrum of in-situ fracture surface: (a) transgranular and (b) intergranular fracture surface. The Ca signal is observed only at intergranular fracture surface.

The interaction of grain boundaries and magnetic domain walls is shown in Fig. 2. When the electron microscope is operated in the Lorentz mode, the images of the magnetic domain walls appear as either bright or dark fringes, depending on the relative orientation of the magnetization vector in the two adjacent magnetic domains. The domain walls reverse the contrast between overfocus and underfocus. The technique can be used to differentiate between the domain walls and other linear defects such as grain boundaries and dislocations.

The domain configurations in Fig. 2 show the sequence of domain wall positions with increasing applied magnetic field. The domain walls which appear as bright fringes in the overfocused situation, labelled O, are plotted as solid lines and the ones which appear as bright fringes in the underfocused situation, labelled U, are plotted as dashed lines in the central sketch. The grain boundaries, on the other hand, are plotted as dotted lines. As the applied magnetic field increases, the domains B and C will grow; the solid line will move to the right and the dashed line will move to the left. The domain A, which is bounded partly by grain boundaries, is not able to change as the domain wall 1-1 is unable to move toward the grain boundary under a moderate applied magnetic field (Figs.

2a and 2b). Only after the applied magnetic field is raised appreciably will the domain A change its magnetization vector, and the domain wall 1-1 will jump to the grain boundary abruptly. This domain wall will not be able to move across the grain boundary even when the magnetic field is increased further (Fig. 2c).

While the Ca-segregated grain boundaries will stop the domain wall motion completely, it has been observed that the domain wall can move across the grain boundaries when they are free of segregation[17]. Lorentz microscopy techniques in TEM thus provide a promising approach for in-situ

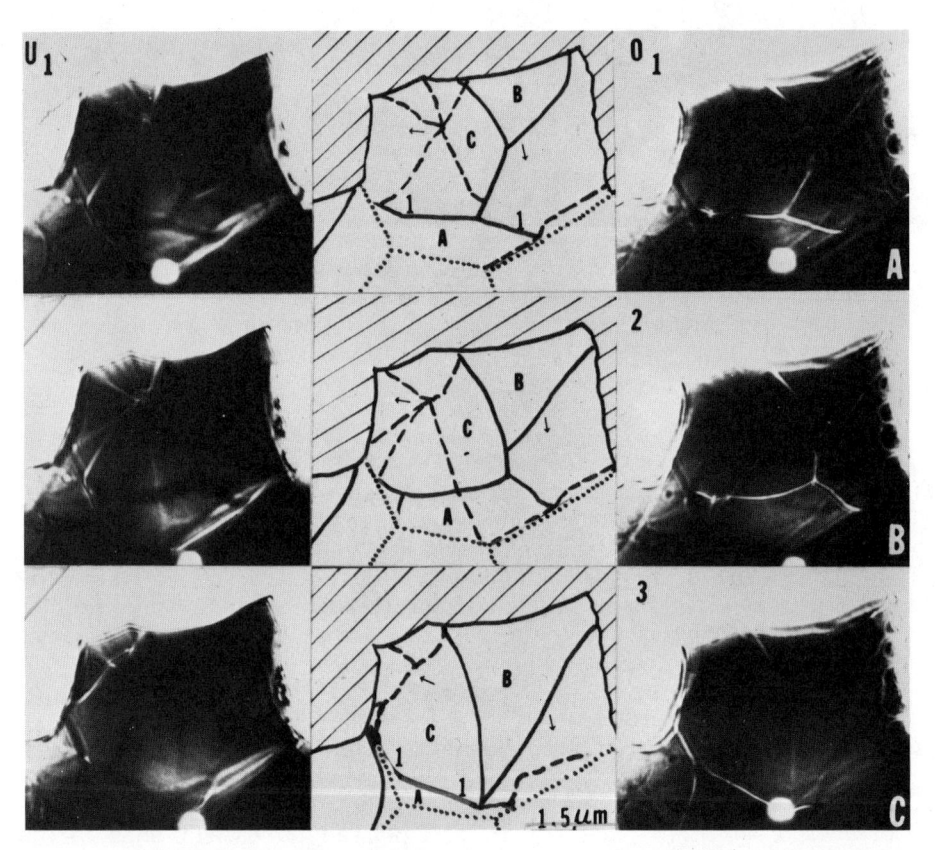

Fig. 2. Interaction of grain boundaries with magnetic domain wall motion: domain configurations under a situation where a (a) zero, (b) moderate and (c) high magnetic field is applied. (i) the domains B and C grow as applied magnetic field is increasing; (ii) domain A will not grow under moderate applied magnetic field and it changes its magnetization abruptly under higher applied magnetic field.

EM STUDY OF GRAIN BOUNDARIES IN (Mn,Zn)Fe$_2$O$_4$

study of the interaction of the microstructure with domain wall motion, an experiment that is not possible by other techniques.

The nature of the grain boundary region was further examined by convergent beam electron diffraction (CED). In this technique, a convergent electron beam is used to form the diffraction pattern instead of a defocussed parallel beam. The CBD pattern typically consists of disks, each of which corresponds to a Bragg diffraction spot in conventional diffraction mode. The sharp line pattern in the central disk, called high order Laue zone (HOLZ) lines, and the intensity distribution in the surrounding disks are sensitive to the changes in both crystal symmetry and lattice parameter.[13]

The CBD patterns of the Spinel ferrite (Mn,Zn)F$_2$O$_4$ with the electron beam incident along the [001] zone axis are shown in Fig. 3. When the electron probe, 0.2μ in diameter, is put in the interior of the grain, both the Holz lines and the intensity distribution show 4mm crystallographic cymmetry for Fd3m symmetry of the spinel structure (Fig. 3a). If the probe is placed in the region near grain boundaries, however, the CBD pattern no longer possesses such a high symmetry; the HOLZ lines have only 2mm symmetry and the intensity distribution is only of m symmetry (Fig. 3b). The lowering of the crystallographic symmetry in the region near the grain boundary is ascribed to the dissolution of a small amount of Ca ions which are much larger than the other cations in (Mn,Zn)Fe$_2$O$_4$.

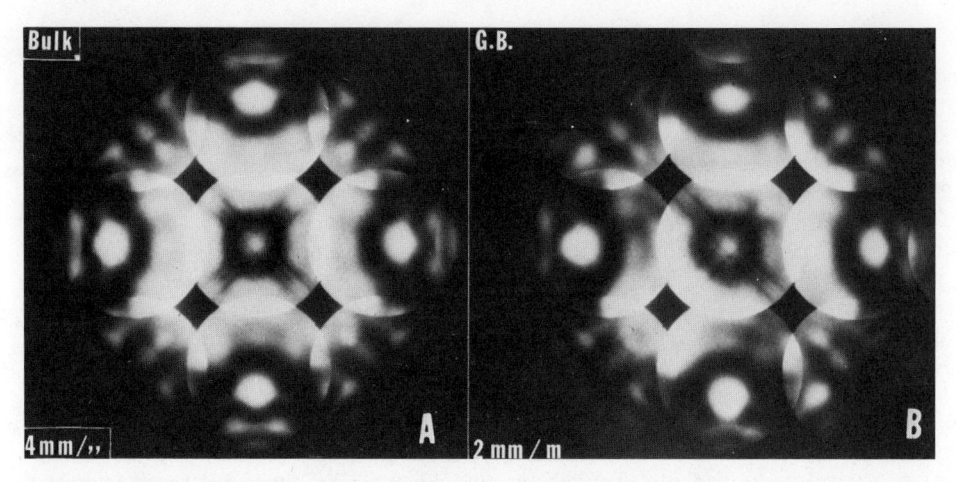

Fig. 3. CBD patterns of (Mn,Zn)Fe$_2$O$_4$ with electron beams incident along [001] zone axis and electron probe placed at (a) interior of the grain and (b) region near grain boundary respectively. The symmetry of the Holz lines in the central disk and the intensity distribution of the surrounding disks have lower symmetry in the region near grain boundaries.

The segregation of Ca at grain boundaries will affect the sintering mechanism of $(Mn,Zn)Fe_2O_4$. In order to study the behavior of the Ca-segregated grain boundaries at the sintering temperature, a thin foil was heated in-situ in a hot stage using a high voltage electron microscope (HVEM). The results are shown in Fig. 4.

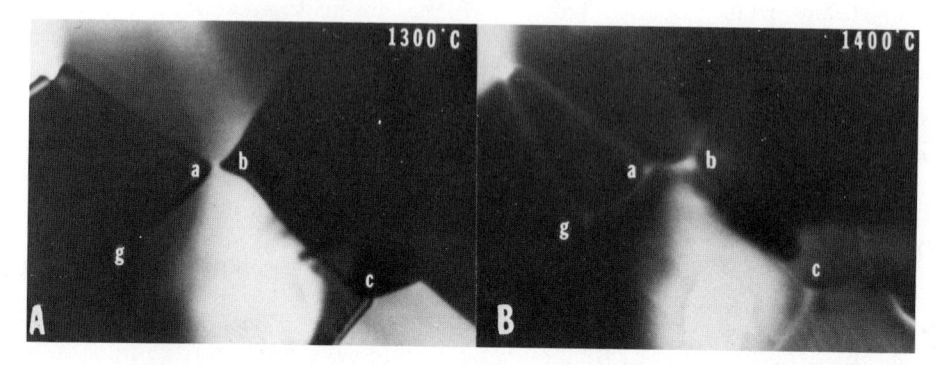

Fig. 4. The grain boundary structure of $(Mn,Zn)Fe_2O_4$ at (a) 1300°C and (b) 1400°C respectively. (i) the grain boundary images become diffuse at higher temperature, (ii) grain boundary migration occurs at 3-grain junctions a and b.

For temperatures below 1300°C, there is no noticeable reaction observed. However, when the temperature is raised to 1400°C, there are two important reactions observed near the grain boundaries: i) The image of grain boundaries changes from a sharp line (or fringes) to a diffuse line. The existence of a disordered phase at the grain boundaries at 1400°C is inferred from this observation. (ii) The positions of the three grain junctions are shifted due to grain boundary migration. The migration is more prominent in the junctions where the angle between grain boundaries originally deviated considerably from 120°C (e.g. junction a, b in Fig. 4).

DISCUSSION

The addition of small amounts of Ca is known to lead to the formation of a secondary phase along grain boundaries. It is concluded from the AES analysis that this grain boundary phase is an intermediate compound of CaO and MnZn-ferrite, and previous studies suggest that it exists in an amorphous form. In addition, the region near the grain boundaries must possess higher magnetic anisotropic energy and magneto-striction energy such that the magnetization vectors can change the directions only under higher applied magnetic fields, as observed in Lorentz microscopy experiments. The increase in the magnetic energy of this region is ascribed to the lowering of crystal symmetry or localized strain, as confirmed by CBD technique. The effect of Ca-addition on the magnetic properties is to modify the grain boundary structure and its interaction with magnetic domain walls.

In the HVEM experiments the tendency of the grain boundaries to make equal angles with each other implies that the grain boundary surface energy is isotropic at high temperatures. This also supports the existence of either a disordered phase or a liquid phase at the grain boundaries, as the solid-liquid interfacial energy is known to be less anisotropic than the solid-solid interfacial energy. Furthermore, the existence of a eutectic liquid between CaO and the CaO-ferrite intermediate compound has been observed by in-situ heating of this material in SEM.[6] In summary, these results, namely the existence of eutectic liquid, diffuse grain boundary image and grain boundary migration, strongly imply that the addition of CaO leads to the formation of a liquid phase at sintering temperatures and such a phase remains as an intergranular phase after the sample has been cooled.

CONCLUSION

The grain boundary phase in MnZn-ferrite plays an important role in determining the electrical and magnetic properties. Addition of a small amont of Ca in these materials increases the electrical resistivity, as well as the coercivity of the materials, but lowers the initial permeability. All these effects can be ascribed to the segregation of Ca at grain boundaries in the form of an amorphous intermediate phase.

The interaction of the grain boundaries with magnetic domain wall motion studied by Lorentz microscopy and CBD technique in TEM provides an understanding of the effect of microstructure on the magnetic properties. In-situ heating in HVEM results in an understanding of the behavior of grain boundary phase at sintering temperatures. Finally, these TEM techniques in conjunction with analytical techniques such as AES and x-ray analysis in STEM, provide a promising approach towards understanding the effect of grain boundaries upon the behavior of ceramic materials.

ACKNOWLEDGEMENTS

This work was supported by the Director, Office of Energy Research, Office of Basic Energy Sciences, Materials Sciences Division of the U. S. Department of Energy under Contract No. DE-AC03-76SF00098. Special thanks from Lin to INER (R.O.C.) for financial assistance. Special thanks are due to TDK Electronics Co. of Japan for supplying the MnZn ferrite samples and Professor H. Fujita for arranging the hot stage HVEM experiments. Drs. H. Mori and M. Komatsu performed the in-situ HVEM experiments on the 3MeV microscope at Osaka University.

REFERENCES

1. R. C. Bradt and R. E. Tressler, "Deformation of Ceramic Materials", Plenum, New York (1975).
2. L. M. Levinson, "Grain Boundary Phenomena in Electronic Ceramics; Advances in Ceramics, Vol. 1", The American Ceramic Society, Inc., Columbus, Ohio (1980).
3. The Akashi, Tran. Japan. Inst. Metals 2:171 (1961).
4. R. K. Mishra, E. K. Goo and G. Thomas, Amorphous Grain Boundary Phase in Ferrimagnetic $(Mn,Zn)Fe_2O_4$ and Ferroelectric PZT Ceramics in: "Surface and Interface in Ceramic and Ceramic-Metal Systems", J. Pask and A. G. Evans, eds., Plenum Press, New York (1981).
5. J. Spence, "Experimental High-Resolution Electron Microscopy", Clarendon Press, Oxford (1981).
6. I. N. Lin, R. K. Mishra and G. Thomas, "Ca-segregation in $(Mn,Zn)-Fe_2O_4$, in press, IEEE Magnetics, 1982.
7. O. L. Krivanek, T. M. Shaw and G. Thomas, J. Appl. Physics 50:4423 (1979).
8. L. E. Davis et al. "Handbook of Auger Electron Spectroscopy, 2nd ed.", Physical Electronics Industries, Inc. (1972).
9. P. M. Hall and J. M. Morabito, Compositional Depth Profiling by Auger Electron Spectroscopy, CRC Critical Review in Solid State and Materials Sciences, 53 (1978).
10. G. Thomas and M. J. Goringe, "Transmission Electron Microscopy of Materials", John Wiley and Sons, N. Y. (1979).
11. P. B. Hirsch et al., "Electron Microscopy of Thin Crystals", Butterworths, London 388, (1971).
12. J. W. Steeds, Convergent Beam Electron Diffraction, Ch. 15 in: "Introduction to Analytical Electron Microscopy", J. J. Hren, J. I. Goldstein and D. C. Joy, eds., Plenum, New York (1979).
13. M. Paulus, Properties of Grain Boundaries in Spinel Ferrites, Ch. 3 in: "Materials Science Research, vol. 3" (1966).
14. P. F. Bongers, et al., Defects, Grain Boundary Segregation and Second Phases of Ferrites in Relation to Magnetic Properties, p. 265 in: "Ferrites: Proceedings of the ICF 3", H. Watanake, S. Iida and M. Sugimoto, eds. (1980).
15. E. K. W. Goo, R. K. Mishra and G. Thomas, Transmission Electron Microscopy of $Pb(Zr_{0.52}Ti_{0.48})O_3$, J. Amer. Ceramic Soc. 64:517 (1981).
16. R. Raj, Morphology and Stability of the Glass Phase in Glass-Ceramic Systems, J. Amer. Ceramic Soc. 64:245 (1981).
17. I. N. Lin, Microstructure and Magnetic Domain Wall Motion, P324 in: "Proceedings of EMSA", Claitors Publishing House, New Orleans (1981).

CROSS-SECTIONAL TRANSMISSION ELECTRON

MICROSCOPY OF SEMICONDUCTORS

D. K. Sadana

Department of Materials Science and Mineral Engineering
and, Lawrence Berkeley Laboratory
University of California, Berkeley, CA 94720

ABSTRACT

A method to prepare cross-sectional (X) semiconductor specimens for transmission electron microscopy (TEM) has been described. The power and utility of XTEM has been demonstrated. It has been shown that accuracy and interpretation of indirect structural-defects profiling techniques, namely, MeV He$^+$ channeling and secondary ion mass spectrometry (SIMS) can be greatly enhanced by comparing their results with those obtained by XTEM from the same set of samples.

INTRODUCTION

For a 100 kev microscope, the maximum thickness of silicon that can be imaged under the bright-field condition is ~1 micron. Therefore, good quality specimen preparation for the TEM studies is one of the essential requirements. The specimens can be prepared in three ways so that either top or edge-on (90° cross-section) or low-angle (1-3°) beveled view of the specimen can be seen. For the top view and low angle beveled view specimens, the thinning is performed by a chemical jet that ejects an HF:HNO$_3$ solution or by low energy (6-10 kev) ion milling. However, for cross-section type specimens, especially from the ion implanted samples where the surface layers of thicknesses of only a few 1000 Å are to be viewed, the preparation method involves metchanical thinning followed by ion beam milling.[1] It is nearly impossible to prepare such specimens by chemical thinning.

The cross-sectional TEM (XTEM) is a very powerful method because the buried damage regions and the defects at the sub-surface interfaces can be directly viewed with very high resolution.

PREPARATION TECHNIQUE FOR CROSS-SECTION SPECIMENS

The first stage for specimen preparation is to mechanically prepare a cross-section specimen ~25 µm thick. This is schematically illustrated in Fig. 1a. Two pieces of specimen of dimensions 5 mm x 1 mm are glued together face-to-face with contact adhesive, and then mounted on a glass disc with wax with a piece of silicon slice ~6 mm x 9mm on either side for support. Then, the surface is polished flat with 240 grit SiC paper followed by another 600 grit SiC paper and 6 µm diamond paste. With some of the III-V compounds, such as InAs, a final 1 µm polish is given to improve the surface finish. Then the specimens are turned over, remounted with wax, and the polishing sequence repeated to give a final specimen less than 25 µm thick.

Subsequent thinning is done by a 6 kV, 50 µA Ar^+ ion-beam. The specimen is mounted with the mechanically polished surface making an angle of ~20° with the ion-beam. It is thinned from one side only at a time, and the specimen is _not_ rotated. The specimen is orientated with the edge of interest furthest from the ion-beam. After ~1 hr it is turned over and then the other side is thinned until a semi-circular shaped hole occurs at the edge. This procedure preserves the specimen edge without the need for an added protective coating, but occasionally gives rise to some readily recognizable ion beam thinning surface structure. On the other hand, if a protective coating is used, the specimen can be rotated during ion-beam thinning, and such surface structure is markedly reduced.

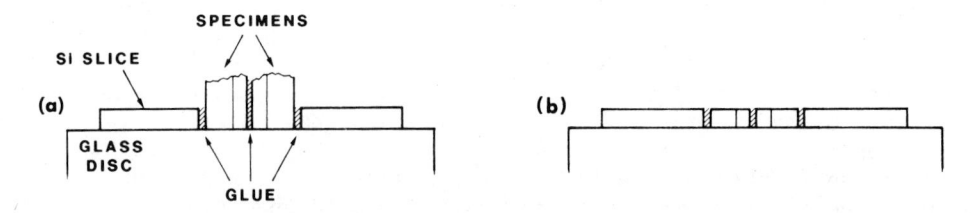

Fig. 1. A schematic diagram illustrating mechanical polishing: (a) initial mounting of the specimens, (b) after the first side has been polished flat.

Figure 2 demonstrates the power and utility of the XTEM. In this case, arsenic was implanted at 11 MeV into (100) Si at room temperature and the micrographs were recorded for bright-field conditions. The dark band in Figure 2 a is the image of the amorphous region which is present at a mean depth of ~4 microns below the surface. This damage is four times deeper than can be imaged by a conventional 100 keV TEM. The annealing behavior of the damage in Fig. 2a is shown in Fig. 2b. The regrowth features indicate that the solid phase recrystallization proceeds from both upper and lower interfaces. Dislocations remaining after the recrystallization are visible throughout the previously amorphous and partially amorphous regions.

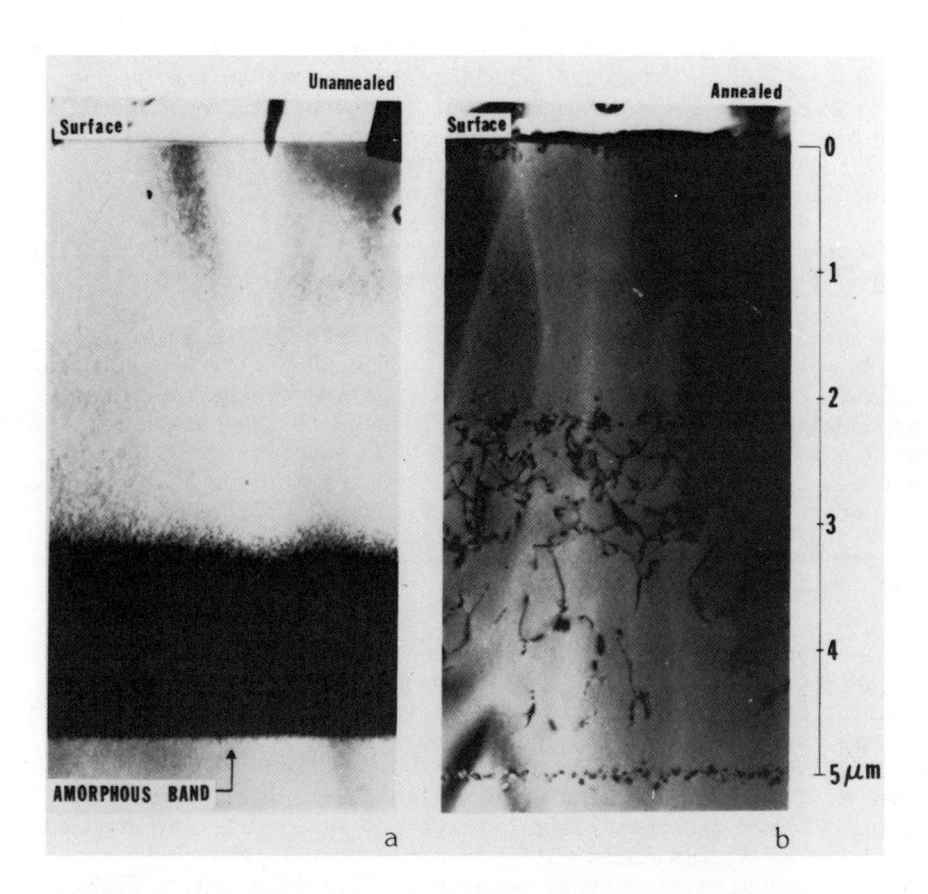

Fig. 2. Damage distribution obtained by XTEM from an 11 MeV As implanted (100) Si specimen (RT implant), (a) before annealing (b) after two step annealing; 550°C/16 hrs + 945°C/15' in N_2 atmosphere.

LIMITATIONS

Although room temperature ion beam thinning is used routinely for preparing various types of semiconductor, it has been reported that for some III-V compounds, in particular indium phosphide, cooling the specimens is essential during the thinning.[1]

STRUCTURAL DEFECTS PROFILES FROM XTEM AND OTHER TECHNIQUES

The XTEM when combined with indirect structural defects profiling techniques, such as Rutherford backscattering (RBS) in a channeling orientation can greatly enhance the capability of RBS/channeling from merely giving the extent of disorder to characterizing the nature of defects present in the material.[2] This is demonstrated in Fig. 3 where a network of stacking faults appears as a gradually rising dechanneling slope in the channeling spectrum of a 1.6 MeV He$^+$ beam. The dechanneling from a band of defect clusters (~50 Å across) on the other hand has been shown to appear as a broad hump in the channeling spectrum. The dislocation loops produced the least dechanneling.[3]

Comparisons of cross-section TEM images of single and multiple layers of damaged Si with channeled RBS spectrum has shown that there is a good qualitative agreement between the damage results from the two techniques. The discrete damage layers as observed by TEM appear as the discrete peaks in the channeled RBS spectra. The mean depths of the damage layers from the two methods also agreed with each other, however, the widths of the damage layers as calculated from the RBS/channeled consistently gave higher values. These comparisons have resulted in improved procedures for determination of damage widths by channeled RBS.[3]

b) TEM/SIMS

Another correlation study is now underway which is expected to enhance the utility of the SIMS technique from its present impurity profiling capability to probing the structural defects.[4] In this case, a fast diffusing impurity which does not react chemically with the semiconductor to be profiled and also is least susceptible to oxidation is used. A good example is Ag in Si. Figure 4 shows the comparison between the Ag profile obtained by SIMS from an Ag implanted (100) Si after annealing at 550°C and the corresponding structural defects profile obtained by XTEM. The conditions for Ag implantation were as follows: 300 keV, RT, 10^{15} cm^{-2}. Before the annealing; the Ag distribution was gaussian and the corresponding XTEM micrograph (not included), showed an amorphous layer extending from the surface to a depth

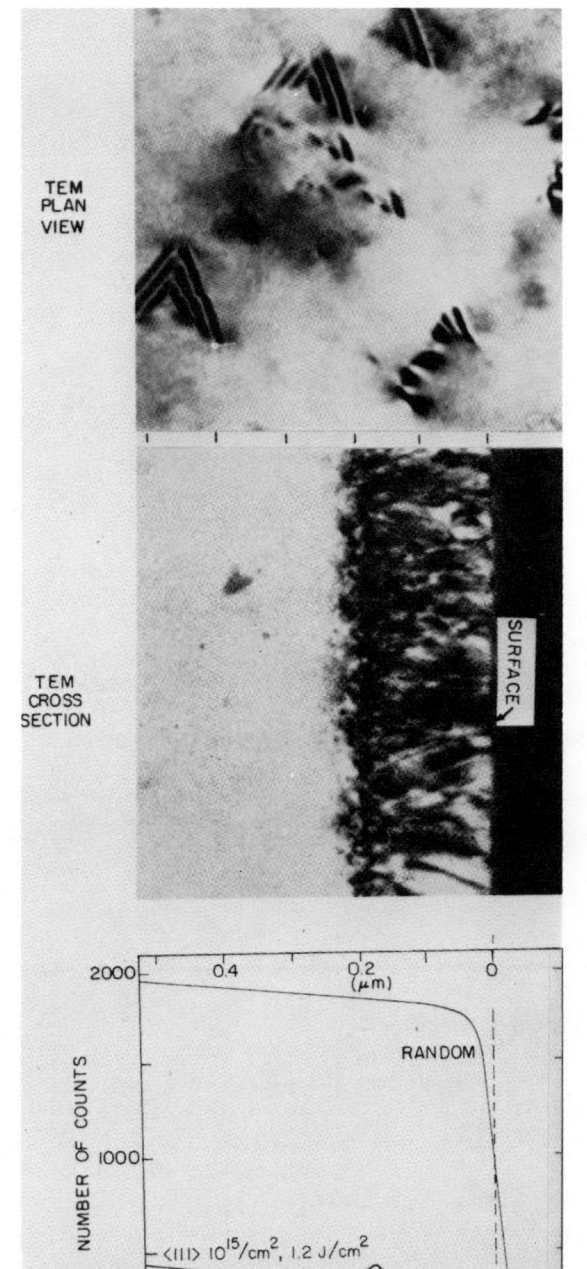

CROSS-SECTINAL TRANSMISSION ELECTRON MICROSCOPY 363

Fig. 3. 1.6 MeV He+ dechanneling by stacking faults: (a) Top view bright-field TEM micrograph showing stacking faults, (b) depth distribution of stacking faults in the specimen (a), and (c) channeling spectrum from the same specimen. Experimental conditions; P+ → (111) Si, $10^{15} cm^{-2}$, 120 keV, Q switched ruby laser annealed at 1.2J cm^{-2}.

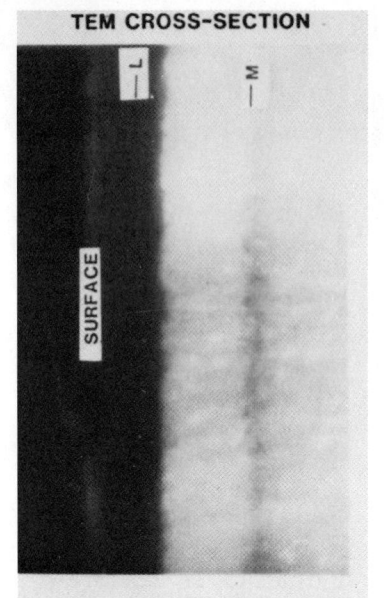

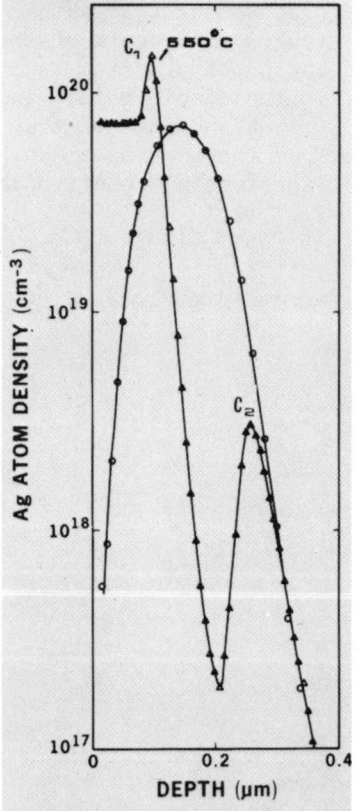

Fig. 4. Ag segregation to defects rich regions in (100) Si: (a) brightfield XTEM micrograph, (b) Ag depth distribution from SIMS. Experimental conditons: Ag → (100) Si, 10^{15}cm^{-2}, RT; annealed at 550°C for 15 minutes.

of 2200 Å. However, after annealing at 550°C, the solid phase epitaxial recrystallization occurred at the amorphous/crystalline interface and the interface advanced toward the surface. In addition, a band of fine clusters formed just below the original amorphous/crystalline interface at a mean depth of 2400 Å. The corresponding Ag profile shows that Ag tends to remain in th damaged region only (Fig. 4). The flat Ag distribution extending from the surface to 900 Å correlates precisely with the width of the remaining amorphous region. A small peak in Ag distribution occurs where the band of fine defects is observed. Further experiments on Ag redistribution in the Si samples with other types of defects, such as dislocation loops, line dislocations, polycrystalline region, etc., indicate that Ag easily segregates locally to the regions containing defects[4] and therefore can be used as a tracer of the microdefects in Si.

SUMMARY

1. The preparation method for XTEM specimens has been described.

2. The power and utility of the XTEM has been demonstrated.

3. It has been shown that accuracy and interpretation of the indirect structural defects profiling techniques namely, RBS in a channeling oreintation and SIMS can be greatly enhanced by comparing their results with those obtained by XTEM from the same samples.

ACKNOWLEDGEMENTS

The authors would like to thank Peter Byrne of the Electrical Engineering Department of U.C. Berkeley, and Robert Wilson of Hughes Research Labs, Malibu, CA, for providing the specimen of Fig. 2 and SIMS data of Fig. 4, respectively. The financial support for this work came through the Director, Office of Energy Research, Office of Basic Energy Sciences, Materials Research, Division of the U.S. Department of Energy under Contract No. DE AC03-76SF-00098.

REFERENCES

1. J. Fletcher, J. M. Titchmarsh and G. R. Booker, Inst. Phys. Conf. Series (London) 52, 153 (1980).
2. D. K. Sadana, M. Strathman, J. Washburn and G. R. Booker, J. App. Phys. 51, 5718 (1980).
3. D. K. Sadana and J. Washburn, Phys. Rev. B 24 3626 (1981).
4. R. G. Wilson and D. K. Sadana (unpublished).

MICROSTRUCTURAL CHARACTERIZATION OF NUCLEAR WASTE CERAMICS

F.J. Ryerson

Lawrence Livermore National Laboratory
University of California
Livermore, Ca. 94550

D. R. Clarke

Rockwell Science Center
P.O. Box 1085
Thousand Oaks, Ca. 91360

ABSTRACT

Characterization of nuclear waste ceramics requires techniques possessing high spatial and x-ray resolution. XRD, SEM, EMPA, TEM and AEM techniques are applied to ceramic formulations designed to immobilize both commercial and defense-related reactor wastes. These materials are used to address the strengths and limitations of the techniques above. An iterative approach combining all these techniques is suggested.

INTRODUCTION

The continuing development and improvement of scanning electron microscopy (SEM), electron probe microanalysis (EMPA), transmission electron microscopy (TEM) and analytical electron microscopy (AEM) have fostered dramatic improvements in chemical analysis and microstructural characterization [1]. The importance of microstructural characterization to better understanding polyphase crystalline materials has now been demonstrated for a number of materials. Principal concerns of such investigations include the presence, distribution and composition of intergranular amorphous films, chemical inhomogeneities within individual phases, and chemical segregation due to the formation of discrete precipitates or crystallographically coherent exsolution lamallae.

In many ceramic materials the formation of a high temperature liquid phase is required in order to aid densification. This is prevalent in the fabrication of covalent solids such as nitrogen or carbide ceramics. After densification, these high temperature liquids may remain within the material as a combination of intergranular glass and crystals. These secondary phases exhibit a major effect upon the properties of the final material. Typical examples of such properties include the high-temperature mechanical properties of Mg, Y, Al, and Sc-fluxed Si_3N_4 alloys [2-7] and the anomalously high dielectric constants of $NaNbO_3$ fluxed $BaTiO_3$ ceramics [8]. In some instances the intergranular phases reflect the chemistry of impurities in starting materials as well as that of additives. A case in point is the presence of chlorine within the intergranular glass in Si_3N_4 alloys [6,7].

In geologic material, the presence of molten intergranular phases has been postulated as the cause of anomalously high electrical conductivity and seismic attenuation within regions of the earth's mantle [9]. A number of experiments have been designed to test these hypotheses. Samples quenched from high temperatures (greater than 1200°C) and pressure (greater than 20kb) are characterized by TEM analysis in order to determine the presence or absence of resulting intergranular glass phases [10,11]. In another application, TEM analysis is employed to determine the coarsening of exsolution lamallae in important rock-forming minerals, i.e., pyroxenes and feldspars [12]. Time-temperature histories for samples containing these phases can be elucidated on the basis of such data.

The characterization of ceramic materials for the long-term immobilization of high-level nuclear waste presents a number of special problems. Firstly, nuclear waste from both commercial reactors and as by-products of nuclear weapons production are chemically complex, representing a large portion of the naturally occurring elements as well as the actinides. The utilization of ceramics for immobilization of these wastes is predicated upon the incorporation of radionuclides as essential structural constituents within a limited number of crystalline phases. Formation of the desired assemblages is accomplished by blending of a specified suite of additive components with the waste stream. This allows the waste composition to be altered such that it enters the desired phase volume. Armed with the necessary data regarding radionuclide substitution mechanisms and desired waste loading, simple algorithms can be constructed for the formulation of optimum additive proportions [13].

In characterizing the densified product, the initial concern is simply that of establishing whether or not the proposed assemblage was indeed produced. In large part, powder x-ray diffraction can answer this question. However, a number of major concerns cannot be satisfied by this method alone. These include determinations of the following.

1. Quantitative chemical analysis of the phases formed, and distribution of radionuclides among them.

2. Presence of radionuclide-rich phases in amounts below detectability by XRD or hidden below reflections of major phases.

3. Presence, composition and distribution of intergranular glass,

4. Overall textural and compositional homogeneity.

These features are crucial to satisfying the intended performance criteria of these materials, namely, long-term stability in a geologic repository. Hence, microstructural characterization and chemical analyses of individual phases is required. As these materials are typically fine-grained (less than 1µM), techniques with high spatial resolution are required.

In this paper, the characterization of two different types of titanate-based nuclear ceramics is described in light of the concerns expressed above.

The first material is designed to incorporate nuclear wastes produced as by-products of nuclear weapons production which are now in tank storage at the Savannah River Plant, Aiken, North Carolina [14]. The second material is designed to incorporate waste from commercial reactors. These are designated as DTC (defense titanate ceramic) and CTC (commercial titanate ceramic) [15]. These forms are contrasted to illustrate the implications of waste stream composition on additive formulations and sample microstructure.

TITANATE-BASED NUCLEAR WASTE CERAMICS

The use of titanate-based ceramics for nuclear waste disposal was introduced by Ringwood et.al. under the acronym, SYNROC (<u>syn</u>thetic <u>roc</u>k), and was originally intended for the immobilization of commercial reactor waste [16,17]. Perovskite ($CaTiO_3$), zirconolite ($CaZrTi_2O_7$) and "hollandite" ($BaAl_2Ti_6O_{16}$) are the principal phases of interest. It is assumed that the distribution of radionuclides in this assemblage is as follows: alkalis enriched in hollandite, alkaline earth, lanthanides and trivalent actinides enriched in perovskite, and tetravalent actinides and uranium enriched in zirconolite. In practice, Mo, Ru, Re, and Pd are reduced to the metallic state and reside in a metallic alloy. It is interesting to note that Dosch and coworkers, working independently, produced assemblages containing some of the same phases [18]. CTC has a composition and phase assemblage closely resembling SYNROC-C.

In applying the perovskite-zirconolite-hollandite formulation to defense wastes, the high concentration of transition metals, alumina, sodium and silica as inert processing contaminants must be considered [14]. In such compositions hollandite proved to be unstable requiring the addition of a new cesium host phase. Since silica and sodium were present in the waste stream, the alkalis were recast as an alkali aluminosilicates, nepheline ($NaAlSiO_4$) and pollucite ($CsAlSi_2O_6$). The transition metals form spinels and metal alloys. DTC has been formulated on this basis [13].

It should be noted that a large portion of the transition metals require reduction to the form spinels. Hence, proper redox control must also be verified during sample characterization.

SAMPLE PREPARATION

Both DTC and CTC starting powders were produced by spray drying slurries of additive components and aqueous solutions of waste stream constituents. The powders were fired in air at 600°C to devolatilize the samples. This was followed by a reducing calcine. DTC powders were reduced in mixtures of CO/CO_2 mixed to approximate the Ni-NiO buffer at 800°C. Samples were then uniaxially hot pressed in graphite dies at 1100°C and 1250°C for DTC and CTC, respectively. XRD of the powders prior to pressing indicates that they are partially crystalline, usually containing minor perovskite, ZrO_2 and spinels.

ANALYTICAL TECHNIQUES

As mentioned previously, the goals of wasteform characterization are to identify the phases within the form and to obtain qualitative and quantitative compositional data relating to the distribution of radionuclides. In order to meet these requirements, XRD, SEM, wavelength dispersive (WDS) EMPA, TEM and AEM techniques are employed. It is necessary to use a number of

NUCLEAR WASTE CERAMICS

analytical techniques in conjunction in order to adequately characterize these materials. No single technique has proven capable of meeting all these requirements. In some cases it has even been necessary to fabricate additional samples under a variety of conditions in order to resolve ambiguities in characterization. In this paper, DTC and CTC wasteforms are used to illustrate the strengths and limitations of the various techniques in addressing the problems above.

Since the techniques used in this study are will known, they are not described here in detail. One exception, however, is the observation of intergranular amorphous phases by diffuse dark field imaging of electrons diffracted from these phases [19]. This method is illustrated schematically in Figure 1. Electrons diffracted from an amorphous phase give rise to a series of concentric rings of decreasing intensity. These rings correspond to the peaks on the radial distribution function of the amorphous phase. In the case of silicate glasses, the most intense peak is produced at 3.4 A^{-1} corresponding to the Si-Si bond length [20]. Hence, silicate glasses may be imaged by placing the objective aperture at the corresponding distance in the back focal plane of the TEM. Discrete diffraction spots from adjacent crystalline material should, of course, be excluded. An example of this type of imaging from a "supercalcine ceramic" is illustrated in Figures 2 and 3 [21]. The diffuse

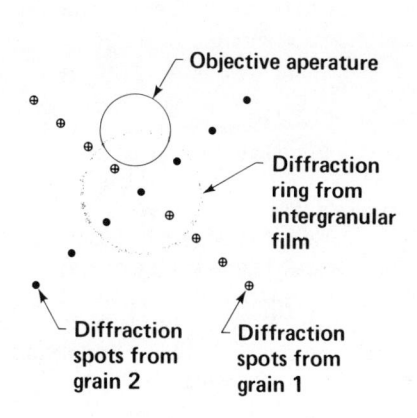

Figure 1. Schematic illustration of the positioning of the objective aperture in the back focal plane in order to obtain a diffuse dark field image of a noncrystalline phase.

Figure 2. Diffraction pattern recorded from an amorphous region in the "supercalcine" ceramic in Figure 3.

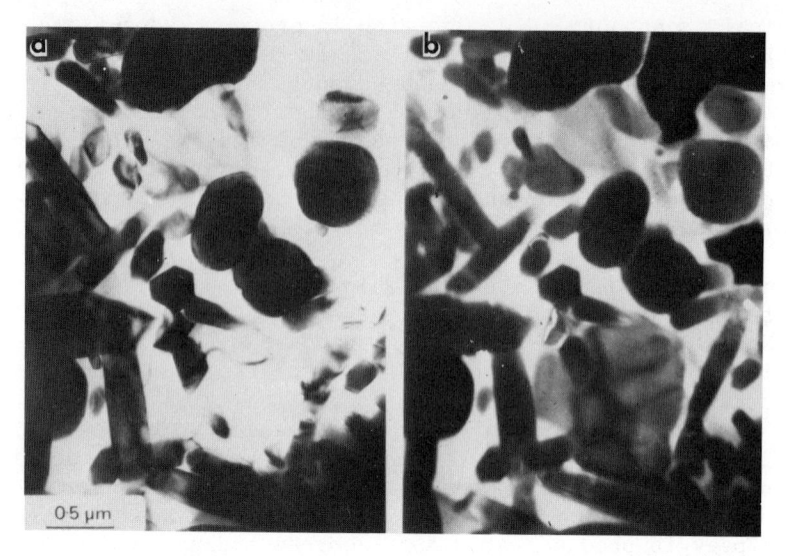

Figure 3. TEM images recorded from "supercalcine" ceramic (a) bright field image, (b) diffuse dark field image produced by imaging the diffraction ring in Figure 2.

ring in the diffraction pattern from an amorphous phase is imaged in dark field in order to "light up" the amorphous regions (Figure 3b). It should be noted that the contrast in bright field is not sufficient to distinguish the amorphous from crystalline phases simply on the basis of atomic number contrast (Figure 3a)

DEFENSE TITANATE CERAMIC

DTC can be used to illustrate the difficulty in obtaining quantitative chemical analyses in chemically complex, fine-grained (1µM) polyphase ceramics. Samples fabricated at 1100°C typically have an average grain size of 1um with some grains as large as 3µM (Figure 4). XRD establishes the presence of the predicted major phases, zirconolite, perovskite, nepheline, an ulvöspinel and a hercynitic spinel (Figure 5a), but does not allow chemical analysis of the observation of amorphous phases. Hence, microstructural techniques are required.

A backscattered electron image from a 1100°C run obtained with an accelerating voltage of 15 KV is capable of resolving individual phases (Figure 4). The corresponding K-alpha line x-ray images for Si, Fe, and Ca, obtained at the same accelerating voltage, also resolve the chemical segregation within the sample (Figure 4). However, throughout most of the sample, it is difficult to correlate regions of high iron concentration

with those of low calcium concentration, for instance. This
correlation is expected based upon the mineralogy observed in
XRD, and its absence indicates the near equality of x-ray excitation volume and grain size. The correlation can be observed
in a single 2 μM region, "A", in Figure 4. However, an EDS
spectra obtained from this region indicates the presence of Ca
in this region (Figure 6). Since calcium is not known to be
incorporated in spinel structures, the calcium signal must be
generated from adjacent grains. Hence, grains of 2 μM or less
in diameter cannot be analyzed by EMPA techniques at 15 KV.
Lower accelerating voltages would increase the spatial resolution, but also increase the detectability limits, or totally
preclude analysis, of some important radionuclides, i.e. the
lanthanides. Obviously, either the grain size of the sample, or
the spatial resolution of the analytical technique must be increased.

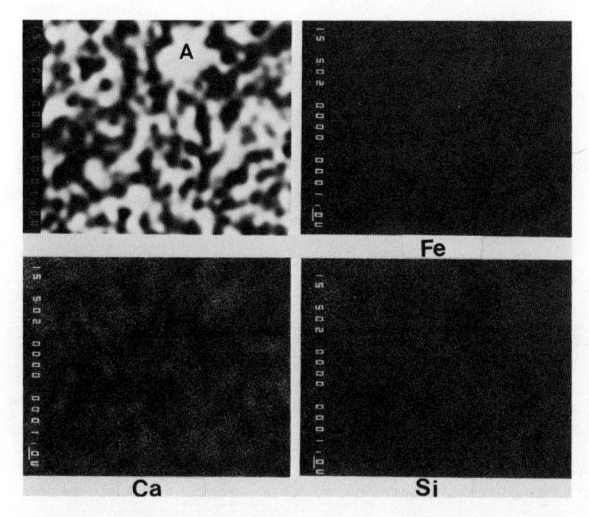

Figure 4. Backscattered electron and Fe$_K$, Ca$_K$, and Si$_K$
x-ray images form DTC pressed at 1100°C.

In an attempt to increase grain size, a sample of the identical composition was pressed at 1200°C. This produced a
sample with grains as large as 10μM in diameter (Figure 7).
X-ray images for a number of elements can now be used to discriminate the phases chemically. The lanthanides and strontium
images correlate with calcium, while uranium correlates with
zirconium. This indicates the immobilization of strontium and
the lanthanides in perovskite and uranium in zirconolite.
Phases of this grain size are also suitable for quantitative
analyses by EMPA. Typical analyses of major phases are given in
Table I.

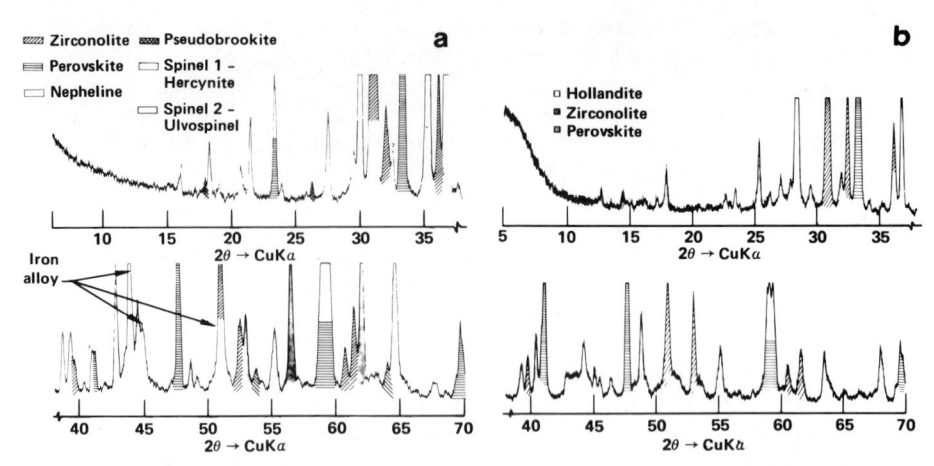

Figure 5 X-ray diffraction patterns, (a) DTC (LLLS04HP4A) and (b) CTC (S08HIP2A) (courtesy of P.E.D. Morgan)

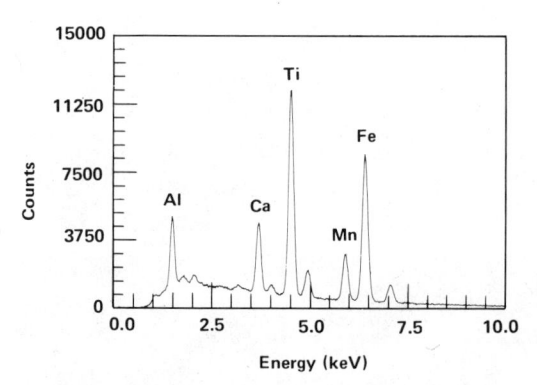

Figure 6. EDS of a high-Fe, low-Ca region (area "A"), Fig. 4.

TABLE I. Microprobe Analyses (wt%) of DTC Phases

	PER	ZIR	SP 1	SP 2	GLASS
SiO_2	0.13	1.44	-.-	-.-	34.49
Al_2O_3	0.72	3.19	11.81	51.42	30.43
TiO_2	49.77	23.43	25.27	4.21	5.82
ZrO_2	0.92	30.33	-.-	-.-	1.58
Na_2O	2.11	0.66	-.-	-.-	9.51
CaO	26.05	7.88	-.-	-.-	6.96
SrO	0.60	0.25	-.-	-.-	0.04
MnO	0.86	2.22	12.42	8.13	1.90
FeO	1.70	11.52	47.53	35.97	5.21
NiO	0.19	5.10	0.03	0.58	0.00
Ce_2O_3	5.14	0.66	-.-	-.-	0.04
Nd_2O_3	4.63	0.97	-.-	-.-	0.00
UO_2	5.96	11.92	-.-	-.-	0.00
Total	98.77	99.57	97.06	100.31	96.38

NUCLEAR WASTE CERAMICS 375

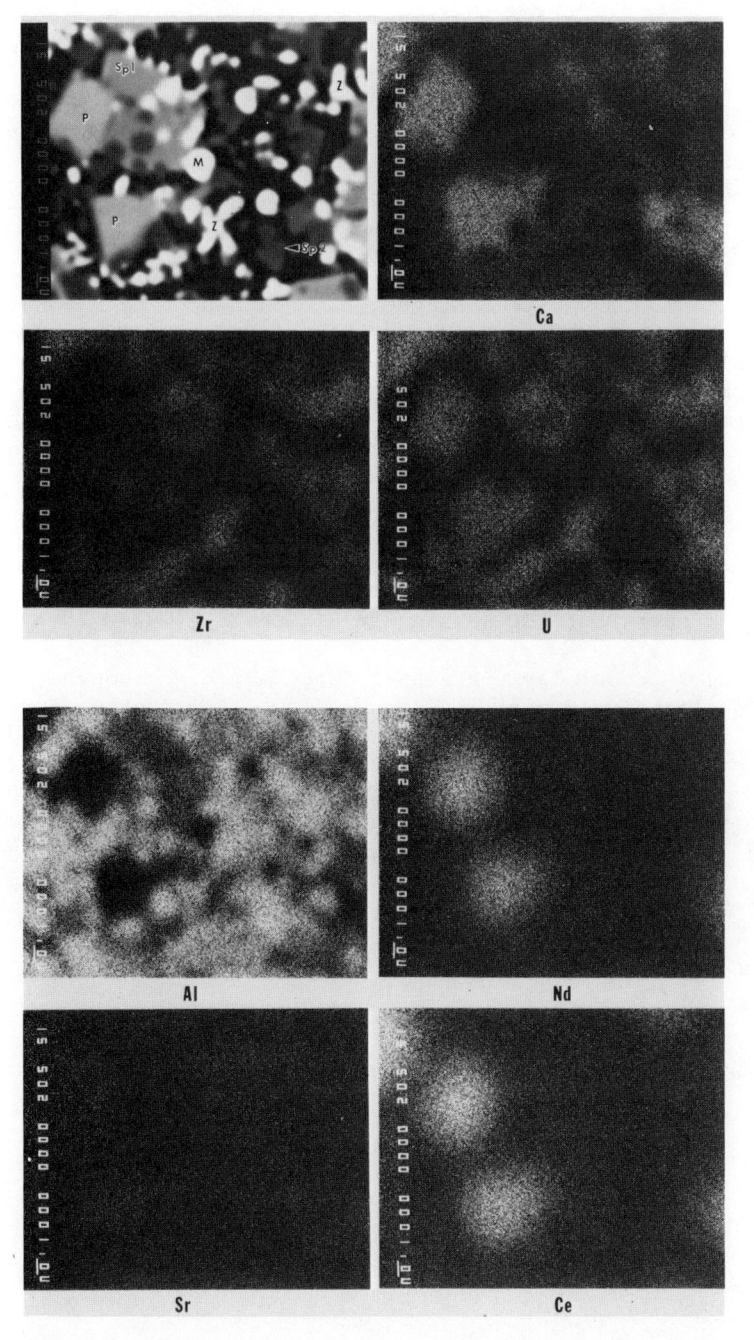

Figure 7 Backscattered electron and x-ray images for Ca_K, Al_K, Zr_L, Ce_L, Nd_L, Sr_L and U_M lines for DTC pressed at 1200°C.

Quantitative chemical analyses are extremely important for a number of applications. Of particular importance is their application to quantitative modeling of wasteform leaching. Ceramic wasteforms dissolve incongruently, certain phases (i.e. silicates) dissolving more readily than others. Hence, in order to predict elemental leach rates it is necessary to obtain quantitative analyses of these phases [22].

Quantitative analyses coupled with crystalline phases stoichiometries can also be used to determine the redox states of multivalent cations. For instance, DTC contains a high concentration of iron which may occur as either Fe^{+2} or Fe^{+3}, and resides almost totally in spinels. Hence, the redox state of iron in spinels is indicative of that in the entire sample [23]. The cation to oxygen ratio in spinel is 3 to 4, so that the microprobe analyses (all iron as FeO) can be recast on this basis to determine Fe^{+2}/Fe^{+3}. This amounts to assigning the cations to ulvöspinel (M_2TiO_4), hercynite (MAl_2O_4) and magnetite (Fe_3O_4) end members (M= a divalent cation, Mn, Ni, Fe^{+2}). Magnetite is the only endmember containing Fe^{+3} (FeO + $2Fe_2O_3$). The results of this type of recalculation are shown for a number of DTC samples in Figure 8.

The chemistry of large grains produced at high temperatures must be qualitatively verified for the lower temperature experiments by TEM and AEM analysis. Energy dispersive analyses obtained in the AEM for perovskite and zirconolite in 1100°C DTC samples confirm the major element and radionuclide distribution in these phases (Figure 9). Uranium is found in both phases while Nd can be distinguished in perovskite. Cerium, the other

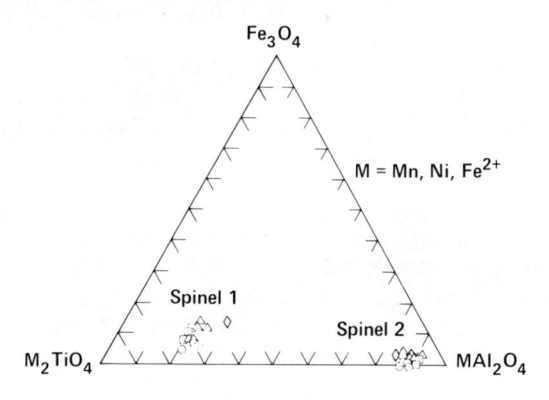

Figure 8. Microprobe analyses of spinels from DTC recast an mole% M_2TiO_4 (ulvospinel), MAl_2O_4 (hercynite) and Fe_3O_4 (magnetite) endmembers.

NUCLEAR WASTE CERAMICS

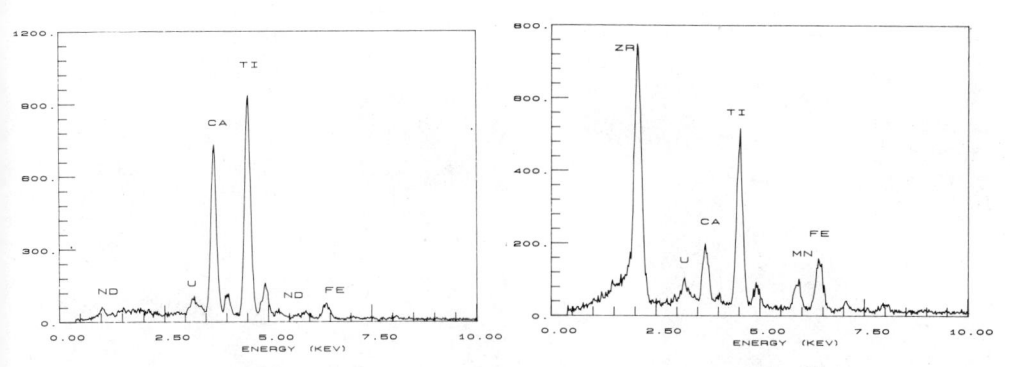

Figure 9. EDS results obtained from DTC in the AEM, (a) perovskite, (b) zirconolite

abundant lanthanide in these samples, is also present in perovskite, but its L-lines are convoluted with the Ti K-lines. The Ce K-lines can be distinguished at higher voltages (alpha=34.57 KV). Lower count rates are exhibited for lanthanide K-lines, however. In general, AEM allows sufficient overvoltage to excite all the necessary x-ray transitions. Hence, unconvoluted lines may be distinguished for most radionuclides. However, low count rates for some lines, coupled with small sample areas may yield unacceptably high detection limits.

The higher spatial resolution of TEM and AEM analyses allow better characterization of glassy and minor phases in DTC samples. UO_{2+x} was found included as small (~500Å) grains in a glassy matrix (Figure 10). The glass was distinguished by diffuse dark field imaging and undergoes radiation damage under the 100 KV electron beam. This radiation damage is accompanied by alkali loss during analysis. This is particularly troublesome with regard to cesium analysis in the glass phase. Cesium Klines (alpha=30.54 KV) must be used for analysis due to the convolution of its L-lines with the Ti K-lines. The reduced count rates of the K-lines relative to the L-lines, accompanied by the volatilization of alkalis can make even qualitative analysis of cesium distribution difficult. This glassy phase appears not only at triple junctions, but also along grain boundaries (Figure 10b). This most surely indicates a continuous glassy phase throughout the sample. Since silicate glasses in these samples are particularly susceptible to dissolution, data regarding its distribution is critical to understanding the sample's leaching behavior [22,23].

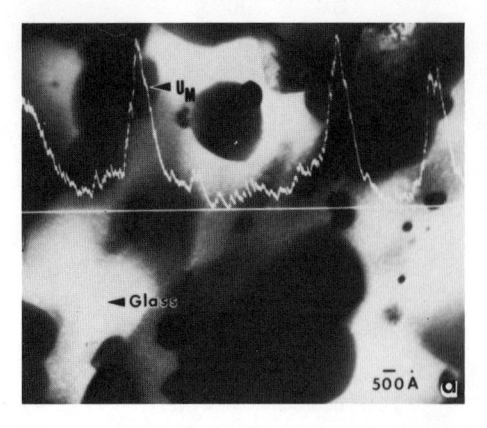

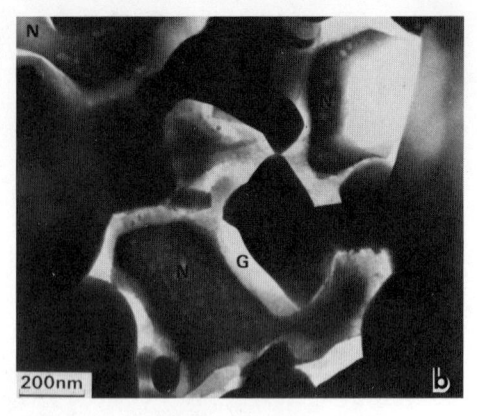

Figure 10. Bright field images of DTC thin foil. (a) U_M line profile indicating UO_{2+X} in the glassy matrix (courtesy of C. Echer), (b) glass at grain boundaries and triple junctions.

COMMERCIAL TITANATE CERAMICS (CTC)

Critical to the microstructural development of DTC was the presence of a glassy intergranular silicate phase. CTC contains no silica and, hence, should not include a continuous glassy network. Grain boundaries in CTC are devoid of glass as expected, but inspection of triple junctions by diffuse dark field imaging and convergent beam diffraction indicates the presence of a glassy phase (Figure 11). The glass phase is readily damaged under the electron beam, and EDS analysis indicates a phosphate rather than silicate chemistry (Figure 12). The glass is approximately 30wt% P_2O_5 (EMPA semiquantitative analysis), and the entire sample contains 0.15wt% P_2O_5. Assuming that the glass density is 50% that of the coexisting titanate crystals, the volume percentage of glass should be less than a percent of the entire sample Hence, it appears likely that the glass is not continuous throughout the wasteform, but instead, isolated at triple junctions. Further microstructural analysis coupled with quantitative leaching models are necessary to resolve this question.

The presence of five major phases are observed in backscattered electron and x-ray images of CTC (Figure 13). As predicted, perovskite, zirconolite and hollandite coexist with an alloy and magnetoplumbite (nominally $CaAl_{12}O_{19}$). Compositions determined by EMPA are given in Table II. It should be noted that only the 3 major phases were positively identified in XRD (Figure 5). Uranium is again distributed between perovskite and zirconolite; perovskite is the major host phase for the lanthanides. Cesium is incorporated in hollandite, and strontium is distributed between the magnetoplumbite and perovskite.

NUCLEAR WASTE CERAMICS

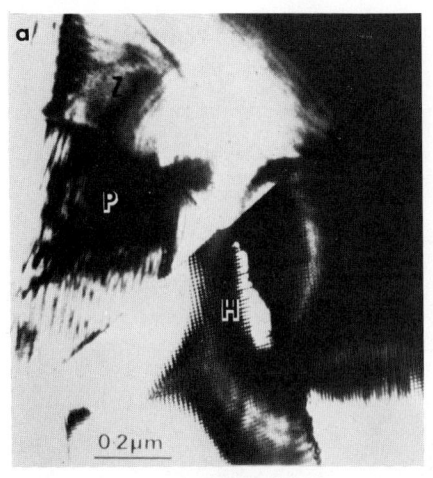

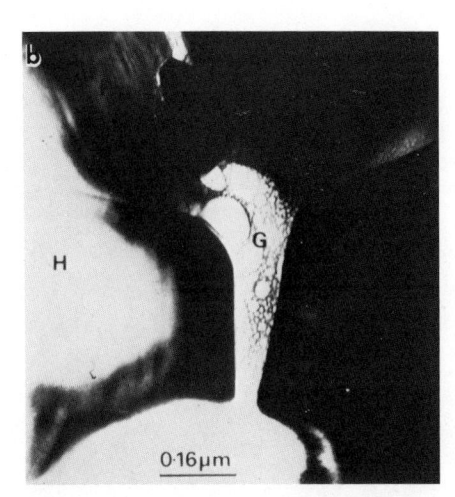

Figure 11. Bright field images of CTC thin foil. (a) Glass-free perovskite (twinned phase)-hollandite grain boundary, (b) phosphate glass at triple junction displaying radiation damage.

Hollandite chemistry can be used in much the same manner as that of the spinels in DTC in determine the redox state of Ti. The cation to oxygen ratio in hollandite is 9 to 16, requiring that the number of trivalent cations be equal to 2Ba + Cs. Hence, hollandite compositions can be resolved into the components $(Ba,Cs)Al_2O_4$, $(Ba,Cs)Ti_2O_4$ (trivalent Ti) and TiO_2 (tetravalent Ti). The resulting ternary can be contoured for $Ti^{3+}/Ti^{3+}+Ti^{4+}$, which increases for samples pressed in graphite compared to those sintered in air (Figure 14).

The analysis of hollandite also indicates the necessity of WDS analysis capabilities for these materials. The energy dispersive spectra obtained from hollandites by AEM are characterized by convoluted Ti_K, Ba_L, and Cs_L x-rays (Figure 15). Hence, quantitative analysis is dependent upon the accuracy of deconvolution algorithms. In WDS analysis the peaks are fully resolved, however.

The metallic grains in CTC are typically less than an micron in diameter requiring AEM analysis. EDS of these grains indicates that the noble metals are reduced to the metallic state and are immobilized in these grains. The reduction of Mo to the metallic state is of particular importance, as unreduced Mo may cause the formation of a water soluble cesium molybdate.

TABLE II. Microprobe Analysis (wt%) of CTC Phases

	Ho	Per	Zir	Mpb
TiO_2	68.08	57.32	45.92	10.12
ZrO_2	-	0.46	34.43	-
Al_2O_3	9.45	0.21	2.27	76.80
CaO	0.74	30.29	13.39	5.15
BaO	18.10	0.48	0.17	0.74
NiO	0.45	0.00	0.00	0.76
FeO	0.03	0.02	0.19	2.47
MnO	0.03	0.10	0.09	0.00
SrO	0.22	0.63	0.00	0.37
Ce_2O_3	-	3.22	0.35	0.13
Nd_2O_3	-	2.30	0.37	0.00
Eu_2O_3	-	0.10	0.07	0.00
Gd_2O_3	-	0.35	0.08	0.00
Na_2O	0.00	2.02	0.47	1.37
Cs_2O	1.12	0.06	0.06	0.00
UO_2	-	1.35	0.75	0.00
P_2O_5	-	0.00	0.00	0.00
Total	98.22	98.91	98.61	97.91

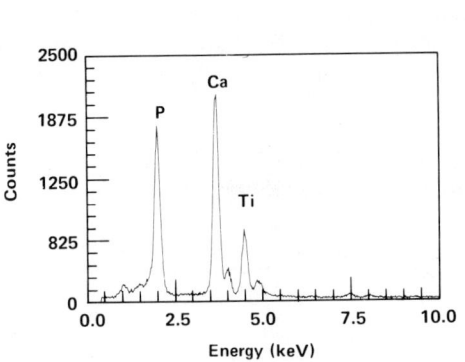

Figure 12.
EDS analysis of phosphate glass in CTC (Figure 11b).

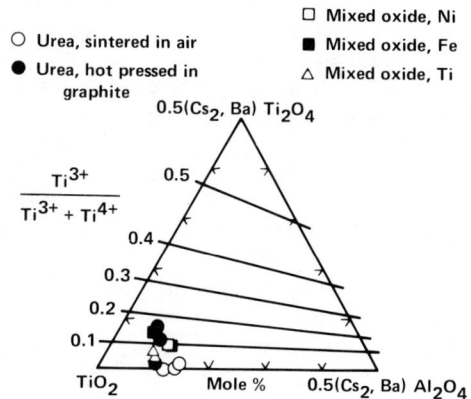

Figure 14.
Hollandite analyses (EMPA) from CTC recast as fictive endmember components and contoured of $Ti^{3+}/Ti^{3+}+Ti^{4+}$ (see text).

NUCLEAR WASTE CERAMICS 381

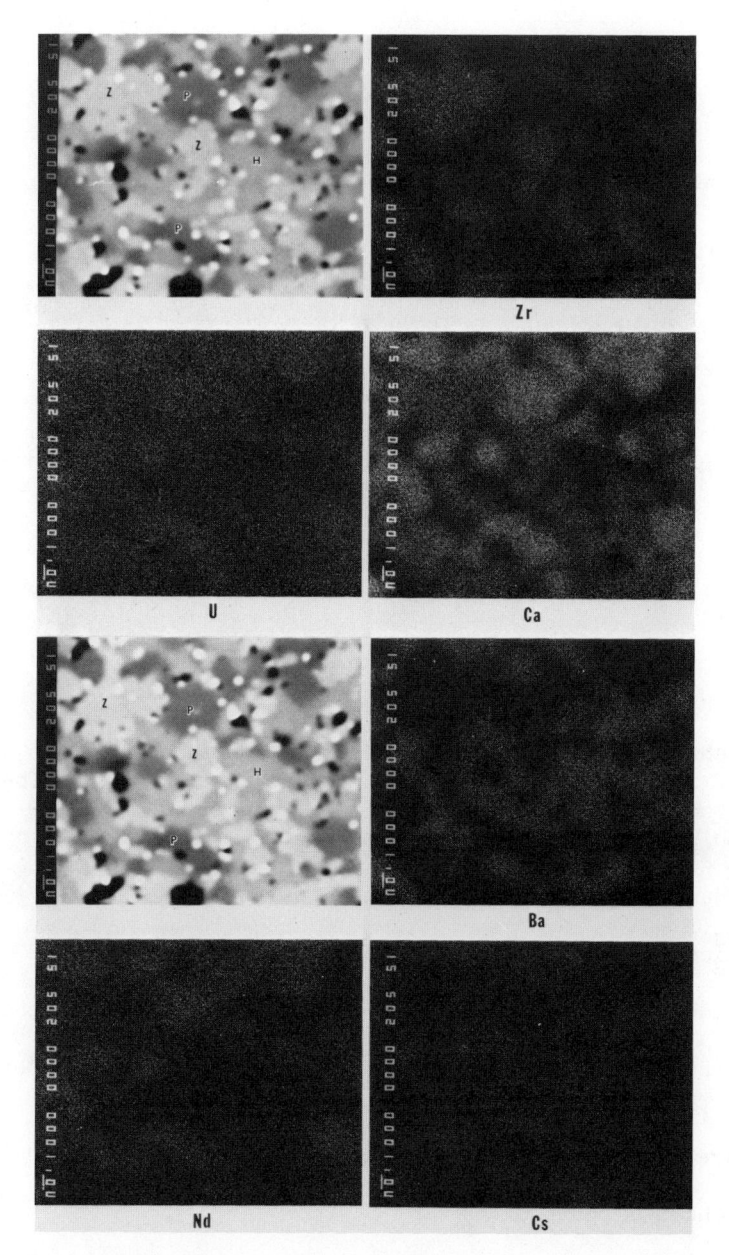

Figure 13. Backscattered electron image and x-ray images form CTC.

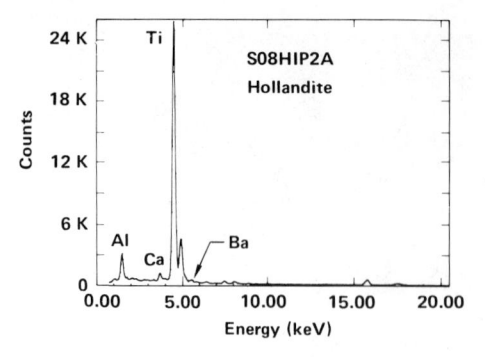

Figure 15.
EDS of hollandite from CTC obtained by AEM. Cs_L, Ba_L and Ti_K lines are convoluted.

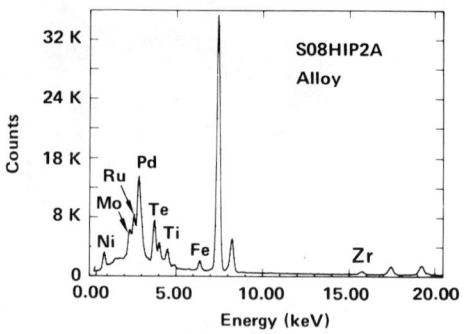

Figure 16.
EDS of metal grains in CTC obtained by AEM indicating immobilization and reduction of noble metals.

DISCUSSION

The characterization of nuclear waste ceramics imposes a number of stringent requirements upon the investigator. These are due to both the nature of the problem and the material itself. Solutions to the problem require,

1. quantitative analyses of all phases

2. identification of intergranular phases

3. identification of phases present in amounts less than 1%

4. the microstructural distribution of phases with poor leach resistance.

The nature of the materials provide difficulties due to their,

1. chemical complexity

2. fine-grained nature

Limitations in the various analytical techniques prevent any single technique from completely fulfilling our characterization needs. Hence, a number of techniques must be employed in an iterative approach.

XRD provides a rapid method of determining the phase assemblage. However, no data is obtained for phase chemistry and microstructural phase distribution. Minor phases can also be missed due to detection limits and/or overlapping reflections from major phases. EMPA and SEM techniques complement XRD by providing chemical and microstructural data. WDS analysis is particularly useful in complex chemical systems due to its excellent x-ray resolution (compared to EDS). However, the spatial resolution is not always sufficient in these techniques. As such, they must be complemented by TEM and AEM analysis with their superior spatial resolution. Unfortunately, quantitative AEM analysis is still in the developmental stage, and, should EDS techniques be employed, x-ray resolution is again a limitation. Another feature which must be addressed is the effect of radiation damage and volatile loss upon AEM results.

Characterization of nuclear waste ceramics requires the application of a number of techniques possessing both high spatial and x-ray resolution. Characterization at the "100% confidence level" is most likely impossible due to the inability of any single technique to meet all the analytical constraints. However, we propose that the iterative approaches outlined above provide credible solutions to a majority of the problems.

REFERENCES

1. A. H. Heuer and N. J. Tighe (ed.), Application of Electron Microscopy to Engineering Practice in Ceramics, J. Am. Ceram. Soc. 62: 225-305 (1979).
2. F. F. Lange, Silicon Nitride Polyphase Systems: Fabrication, Microstructure, and Properties, International Metals, Rev. 247 (1980).
3. D. R. Clarke, Densification of Silicon Nitride Alloys Using a Eutectic Liquid: An Experimental Test, in "Sintering - Processes," G. C. Kuczynski, ed., Plenum Press, New York, N.Y. (1980).
4. G. R. Terwilliger and F. F. Lange, Hot-Pressing Behavior of Si_3N_4, J. Am. Ceram. Soc. 57:25 (1974).
5. L. K. V. Lou, T. E. Mitchell, and A. H. Hener, "Impurity Phases in Hot Pressed Si_3N_4, J. Am. Ceram. Soc., 61:392 (1978).
6. D. R. Clarke, N. J. Zaluzec, and R. W. Carpenter, The Intergranular Phase in Hot-Pressed Silicon Nitride: I. Elemental Composition, J. Am. Ceram. Soc. 64:601-607 (1981).
7. D. R. Clarke, N. J. Zaluzec, and R. W. Carpenter, The Intergranular Phase in Hot-Pressed Silicon Nitride: II. Evidence for Phase Separation and Crystallization, J. Am. Ceram. Soc. 64:608-611 (1981).

8. L. E. Cross, and D. E. Payne, "The Role of Internal Boundaries upon the Dielectric Properties of Polycrystalline Ferroelectric Materials," to be published.
9. H. S. Waff, Theoretical Considerations of Electrical Conductivity in a Partially Molten Mantle and Implications for Geothermometry, Jour. Geophys. Res., 79: 4003-4010 (1974).
10. H. S. Waff and J. R. Balau, Equilibrium Fluid Distribution in a Ultramafic Partial Melt Under Hydrostatic Stress Conditions, Jour. Geophys. Res. 84:6109-6114 (1979).
11. J. R. Balau, H. S. Waff, and J. A. Tyburezy, Mechanical and Thermodynamic Constraints on Fluid Distribution in Partial Melts, Jour. Geophys. Res. 84:6102-6108 (1979).
12. R. A. Yund, A. C. McLaren, and B. E. Hobbs, Coarsening Kinetics of Exsolution Microstructure in Alkali Feldspar, Contrib. Mineral. and Petrol. 48:45-55 (1974).
13. F. J. Ryerson, K. Burr, and R. B. Rozsa, "Formulation of SYNROC D Additives for Savannah River Plant High-Level Radioactive Waste," Lawrence Livermore National Laboratory, Livermore, CA Rept. UCRL-53237 (1981).
14. J. Campbell, C. Hoenig, F. Bazan, F. Ryerson, M. Guinan, R. Van Konynenberg, and R. Rozsa, "Properties of SYNROC D Nuclear Waste Form: A State-of-the-Art Review", Lawrence Livermore National Laboratory, Livermore, CA Rept. UCRL-53240 (1982).
15. F. J. Ryerson, C. L. Hoenig, and F. Bazan, SYNROC C: Preparation and Phase Equilibria, in: "High Temperature Materials Chemistry," (ed., D. D. Cubicciotti and D. L. Hildenbrand). The Electrochemical Society, 144-161 (1982).
16. A. E. Ringwood, S. E. Kesson, N. G. Ware, W. Hibberson, and A. Major, "Immobilization of High-Level Nuclear Reactor Wastes in SYNROC," Nature, 278:219-223 (1979).
17. A. E. Ringwood, S. E. Kesson, N. G. Ware, W. Hibberson, and A. Major, "The SYNROC Process: A Geochemical Approach to Nuclear Waste Immobilization," Geochem. J., 13:141-165, (1979),
18. R. G. Dosch, A. W. Lynch, T. J. Headly, and P. F. Hlava, Titanate Waste Forms for High-Level Waste - An Evaluation of Materials and Process, in: "Scientific Basis for Nuclear Waste Management," J. G. Moore, ed., 3:123-130 (1981).
19. D. R. Clarke, Application of Electron Microscopy to the Processing of Ceramic Materials, Ultramicroscopy, 8:95-108 (1982).
20. M. Taylor, and G. E. Brown, Jr., Structure of Mineral Glass- I. The Feldspar Glasses $NaAlSi_3O_8$ $KAlSi_3O_8$, $CaAl_2Si_2O_8$, Geochim. Cosmochim. Acta, 43:61-75 (1979).
21. G. J. McCarthy, and D. R. Clarke, Microstructural Characterization and Crystal Chemistry of the Synthetic Minerals in Current Supercalcine Ceramics, to be published.

22. F. J. Ryerson, F. Bazan, and J. H. Campbell, Dissolution of a Nuclear Waste Ceramic: An Experimental and Modelling Study, J. Am. Ceram. Soc. (submitted).
23. F. J. Ryerson, SYNROC-D: Phase Chemistry and Microstructural Characterization, J. Am. Ceram. Soc., (submitted).

Work performed under the auspices of the U.S. Department of Energy by Lawrence Livermore National Laboratory under Contract W-7405-Eng-48.

SEMIAUTOMATIC IMAGE ANALYSIS FOR CERAMIC POWDER AND MICROSTRUCTURE CHARACTERIZATION

D. W. Readey, E. Bright, S. S. Campbell, J. H. Lee,
R. S. Pan, T. Quadir, and K. A. Williams

Department of Ceramic Engineering
The Ohio State University
Columbus, Ohio 43210

INTRODUCTION

Research on the science of ceramic powder processing and microstructure development requires procedures for particle size analysis and microstructure evaluation. Since the goal is to produce and utilize submicron, nonagglomerated powder, it becomes clear after surveying available particle size analysis techniques, that they all have deficiencies, particularly for small particles. Regardless of the technique, micrographs of powders are necessary to evaluate the state of agglomeration and shape anisotropy. Thus, image analysis is an appropriate technique. Completely automatic image analyzers are expensive and require a perhaps better than average quality micrograph, particularly with regard to gray scale, than might always be available. Therefore, semiautomatic image analysis is indicated. The technique can be used for microstructure analysis equally as well. The main advantages of semiautomatic image analysis for powder characterization and microstructure evaluation over other techniques are:

    cost
    multifunctional
        powder characterization
        microstructure evaluation
        other (curve digitizing, intergation, etc.)
    operator judgement can be exercised
    gray scale no problem
    ease of operation
    rugged

The last two features are particularly important for environments in which several operators use the equipment. Many particle size measurement techniques require a great deal of expertise and care in their use and are not well-suited to multi-operator use.

With the availability of microcomputers, one need only attach a graphics tablet, place a micrograph on the tablet and perform point, line, area, or other parameter analysis, assuming the software is available. Thus, for a modest hardware expenditure and perhaps twice as much for software (depending on the availability and cost of programers), a great deal of analysis can be performed. The results presented here were obtained with a commercial instrument* which conveniently provides statistics and quantitative stereology[1] software as part of the system. Our experience has been that with as few as 300 particle or grain size measurements data are quite reproducible. The time required for measurement of 300 particles is only on the order of thirty minutes. The main problems with image analysis of powders are proper dispersion and sampling shared by all powder analysis techniques. Proper dispersion is easily evaluated in micrographs and sampling can be facilitated by taking photographs at random. The purpose of this paper is to demonstrate the application of the technique to ceramic powder preparation, to microstructure development, and to the effect of powder morphology on dissolution or corrosion kinetics of oxides.

POWDER PREPARATION

Gas Phase Hydrolysis

Nickel oxide powders were prepared by gas phase hydrolysis of nickel chloride by the following reaction:

$$NiCl_{2(g)} + H_2O_{(g)} = NiO_{(s)} + 2HCL_{(g)}.$$

Figure 1 shows a typical powder. Although not perfectly spherical, the particle shape is generally isometric. More importantly, the particles are nonagglomerated individual crystallites. Figure 2 is a particle size distribution obtained from the image analyzer and compared to a log-normal distribution. Invariably these powders exhibited log-normal distributions. If the supersaturation is defined as:

$$\text{Supersaturation} = K_{eq} \left(\frac{P_{H_2O} \; P_{NiCl_2}}{P^2_{HCl}} \right)$$

---

*Zeiss Videoplan, Carl Zeiss, Inc., New York, N.Y.

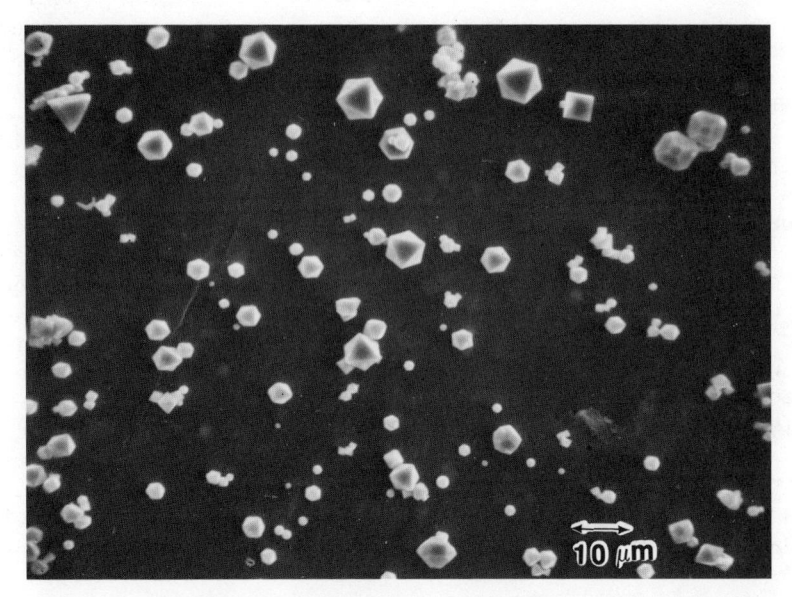

Fig. 1. NiO powder produced by vapor phase hydrolysis.

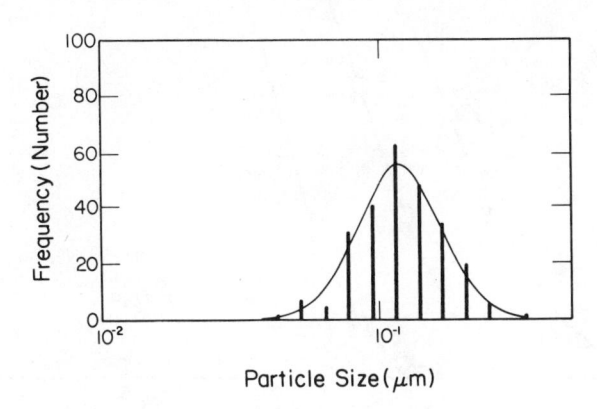

Fig. 2. Log-normal distribution of NiO powder produced by vapor phase hydrolysis.

then Figure 3 shows that the average particle size decreases and the distribution broadens as the supersaturation increases. This strongly suggests that the powder is homogeneously nucleated. As expected, Figure 4 shows that as the flow rate is decreased through the reaction zone the average particle size increases due to the longer growth time. In addition to demonstrating the quantitative nature of the results, these data also show the small differences in powder characteristics that can be detected with quantitative image analysis.

Aqueous Precipitation

Powders were also prepared by aqueous precipitation since this continues to be an important technique in preparing ceramic powder precursor materials. Specifically, nickel hydroxide is being homogeneously precipitated from nickel chloride solutions by decomposition of urea. In contrast to the vapor process, this technique results in small hydroxide agglomerates as seen in Figure 5. Each hydroxide crystallite has a planar habit reflecting the hydroxide crystal structure. However, the agglomerates themselves are quite small and suprisingly spherical. Interestingly, Figure 6 shows that the agglomerates follow rather

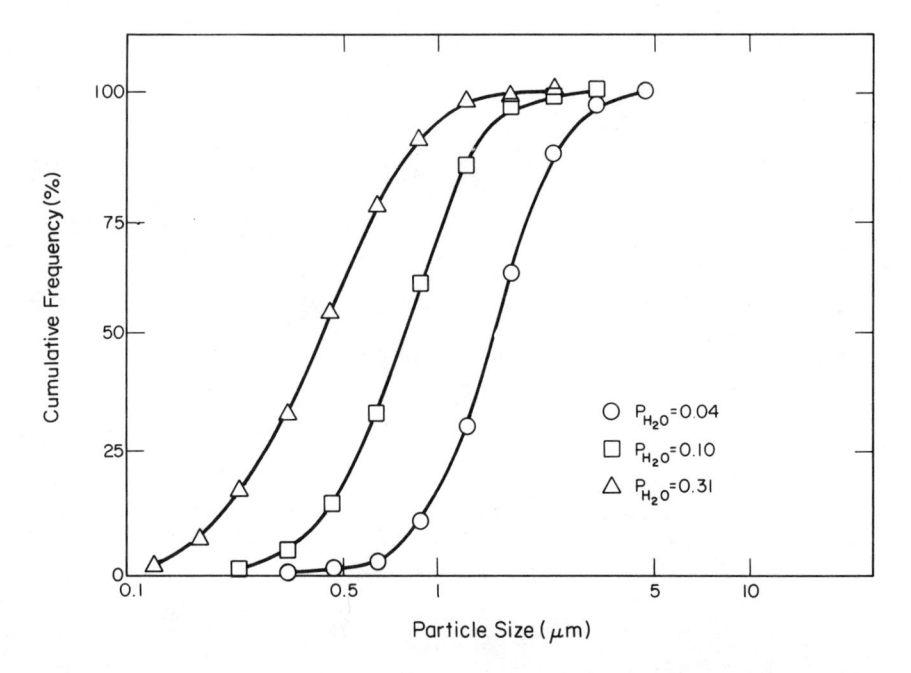

Fig. 3. The effect of supersaturation on NiO particle size distributions produced by hydrolysis, P_{NiCl_2} = 0.1 atm, T = 900°C.

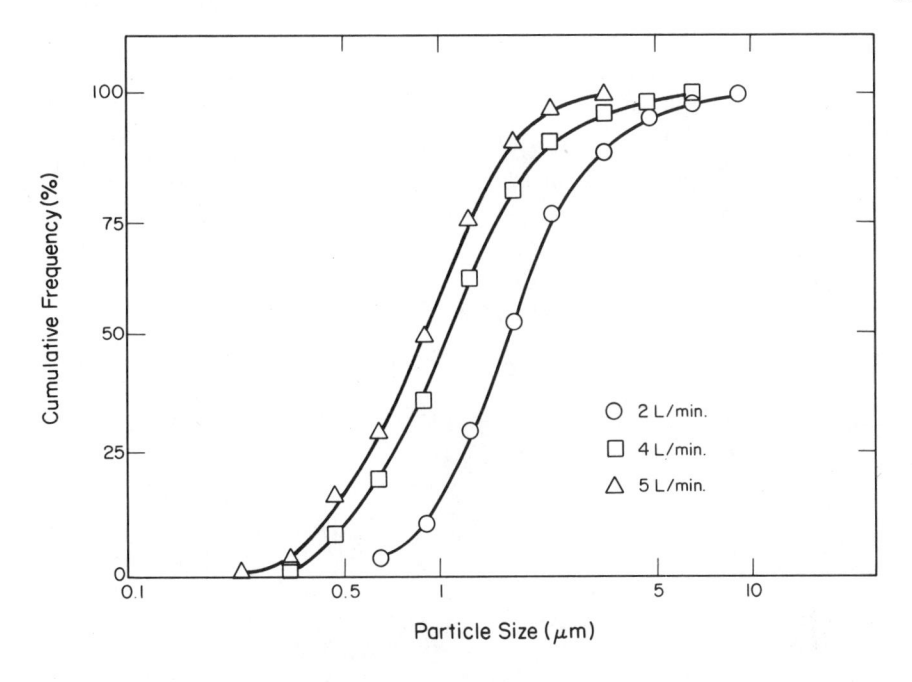

Fig. 4. The effect of flow (liters/min) rate on NiO particle size distributions

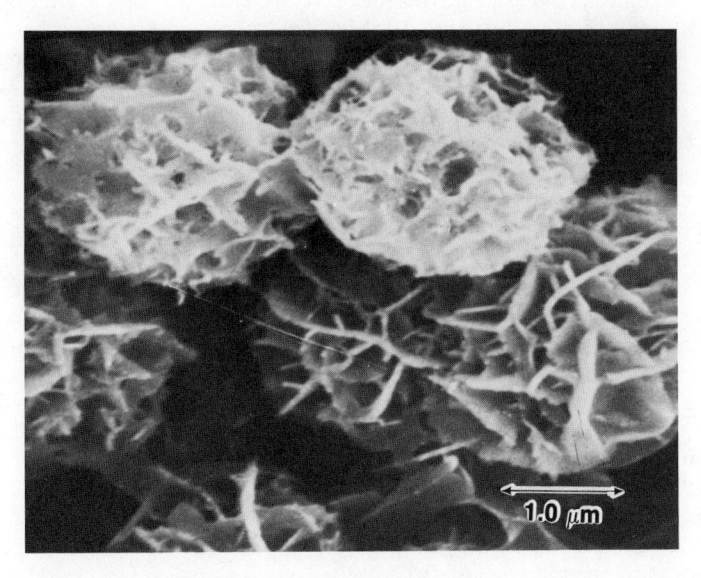

Fig. 5. Nickel hydroxide agglomerates produced by decomposition of urea.

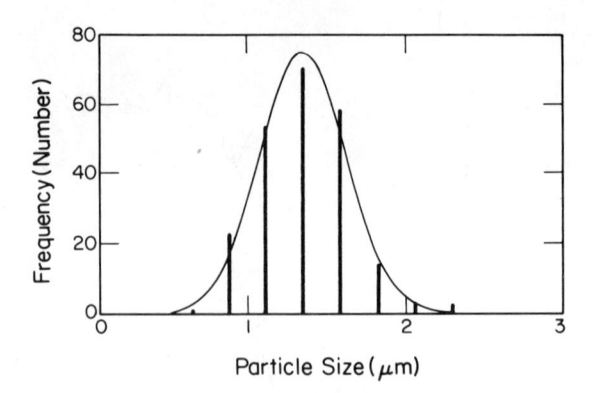

Fig. 6. Nickel hydroxide agglomerate size distribution from urea decomposition.

closely a normal distribution. The agglomerate morphology, primarily the hydroxide crystallite size, and the shape of the distribution depend strongly on the precipitation conditions. These results demonstrate the important advantage that image analysis has over other particle measurement techniques; namely, the ability to observe particle morphology as well as size parameters.

AQUEOUS DISSOLUTION KINETICS

Dissolution kinetics of oxide powders shows how microscopy and image analysis can be invaluable for the interpretation of data. The rate of dissolution of oxides in aqueous environments is not well understood[2] yet some interesting theory exists[3]. With recent interest in photoelectrolysis[4] the stability of oxides and other compound semiconductors in corrosive aqueous media is receiving more attention. Under normal circumstances, the rate of dissolution of most oxides is slow. Hence, to enhance observability, most researchers have used high specific surface area powders produced from some precursor salt such as sulfate[5]. With such powder, complex nonlinear dissolution kinetics are usually observed[6] as shown in Figure 7. Such nonlinear behavior has been ascribed to a number of phenomena including a nonuniform, nonequilibriam distribution of point defects in a powder particle. The data in Figure 7 demonstrate that the kinetics do depend strongly on the oxygen content of the atmosphere in which the nickel sulfate was decomposed. However, as Figure 8 demonstrates, the main effect of the calcining atmosphere is to change the crystallite size and agglomerate density[7]! In short, the nonlinear dissolution kinetics are an artifact caused by the presence of agglomerates. There are actually two effects of the oxygen pressure. One is indeed an effect on dissolution produced by point defects which is completely masked by the presence of agglomerates. The other

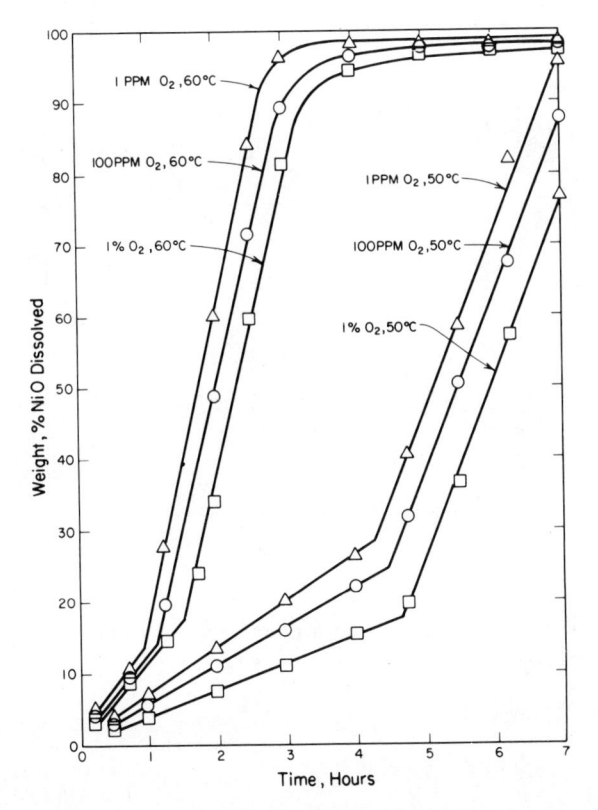

Fig. 7. The effect of the oxygen content of the calcining atmosphere on the dissolution of NiO agglomerates in HCl (pH = 1.0) at 50°C and 80°C.

is the effect of oxygen pressure on nonstoichiometry and mass transport during calcining which determines the agglomerate morphology. An argument can be made that particle growth and densification are controlled by nickel diffusion. As the oxygen pressure increases, the nickel vacancy concentration and diffusion coefficient increases[8]. Hence particle size and agglomerate density increase with oxygen pressure. Microscopy permits observation of what actually is dissolving and image analysis quantifies it.

The dissolution kinetics of TiO_2 reinforces this argument[9]. Figure 9 shows the largely nonagglomerated TiO_2 powder, and Figure 10 the dissolution data. Figure 10 clearly demonstrates that the absence of agglomerates gives less complex dissolution kinetics. The amount dissolved tends to be linear until solubility under the given conditions is approached. Amounts dissolved in excess of 100 percent were due to unstable analysis standards. Again in this case, microscopy and image analysis provides characterization in a particle size range unaccessible to other techniques.

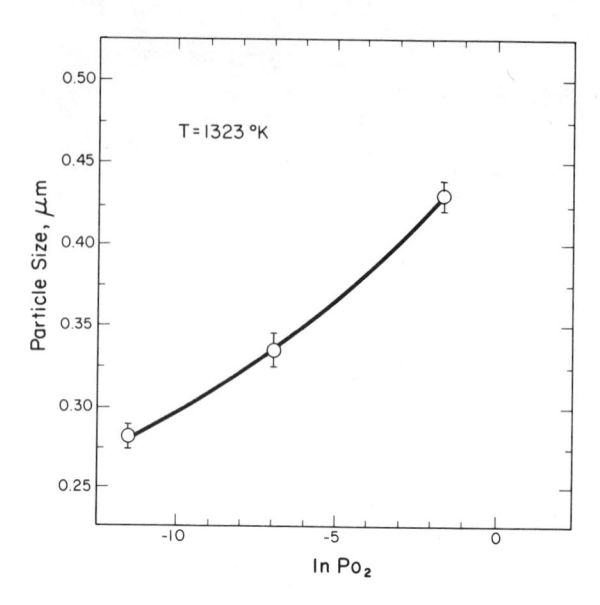

Fig. 8. The effect of oxygen content of the calcining atmosphere on agglomerate crystallite size.

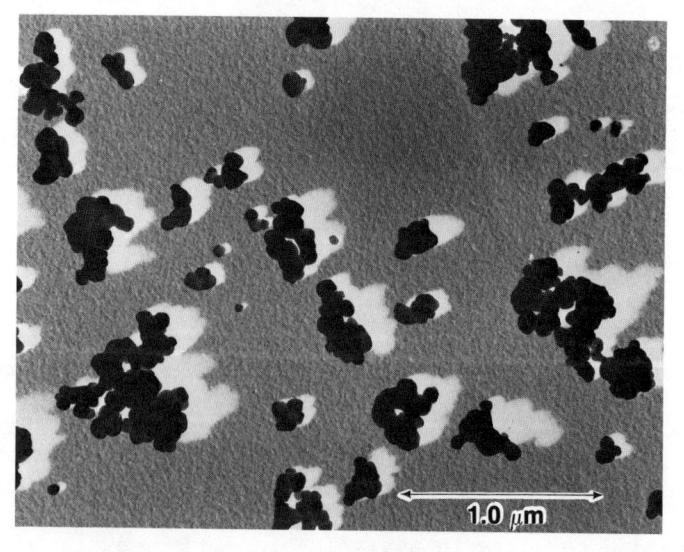

Fig. 9. Micrograph of largely nonagglomerated TiO_2 powder.

SEMIAUTOMATIC IMAGE ANALYSIS

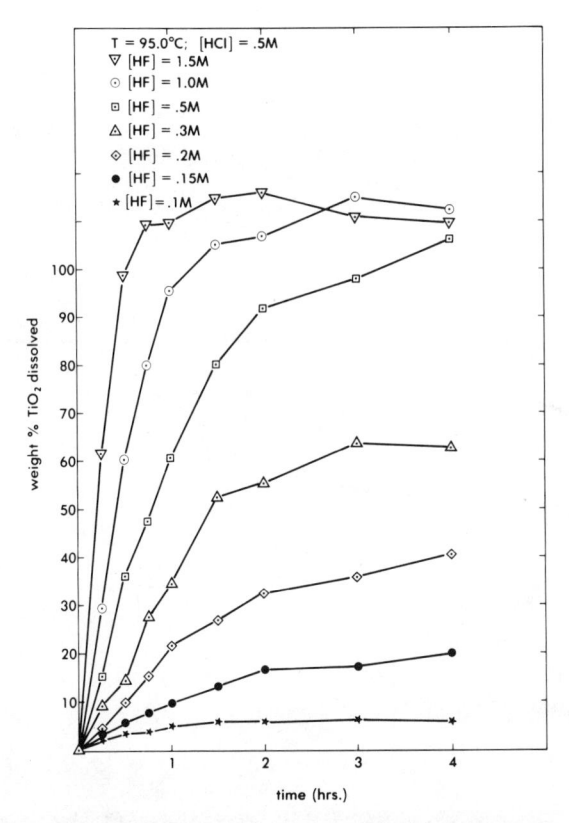

Fig. 10. Dissolution Kinetics of nonagglomerated TiO$_2$ powder.

MICROSTRUCTURE DEVELOPMENT

As a final example of the utility of image analysis, microstructure development in porous ceramic powder compacts with high rates of vapor transport is examined. Conventional sintering models suggest that under high rates of vapor transport, interparticle neck growth will be enhanced and the driving force for sintering will be decreased[10]. Vapor transport can be enhanced in the H$_2$ - ZnO and HCl - Fe$_2$O$_3$ systems according to the following reactions:

$$ZnO_{(s)} + H_{2(g)} = Zn_{(g)} + H_2O_{(g)}$$

$$Fe_2O_{3(s)} + 6HCl_{(g)} = Fe_2Cl_{6(g)} + 3H_2O_{(g)}$$

In both cases, enhanced vapor transport decreases the sintering or densification rate and can even stop it completely as shown in Figure 11. However, the decrease in shrinkage is not due to neck growth but to rapid particle growth as is amply demonstrated in Figure 12. The results for both systems are quite similar. Figure 13 shows that the average particle size, as determined

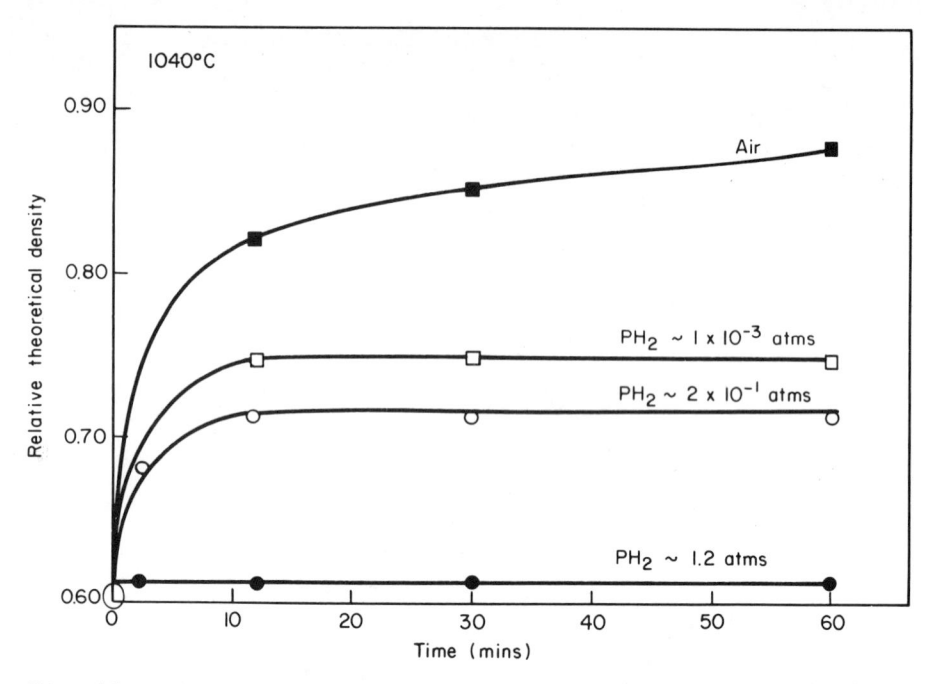

Fig. 11. Shrinkage of ZnO compacts fired in different hydrogen containing atmospheres.

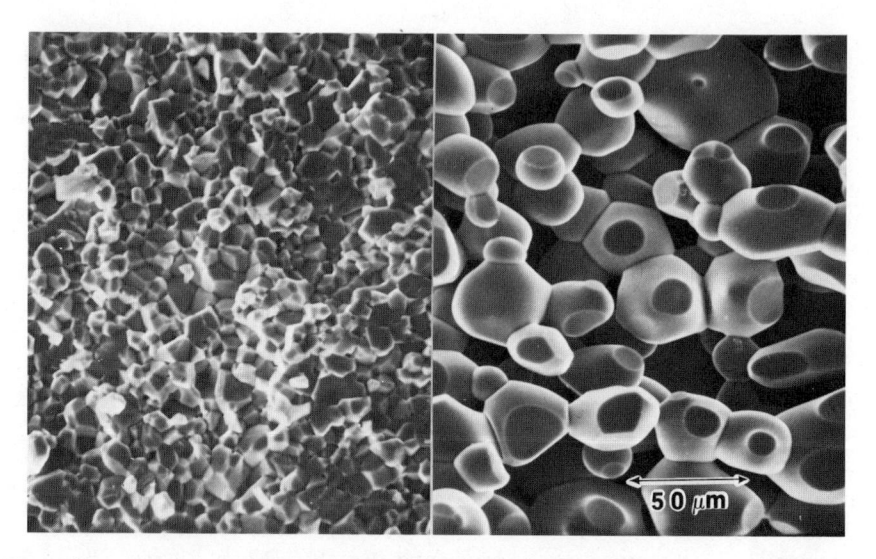

Fig. 12. Left: Fe_2O_3 sintered in air for 5 hrs. at 1200°C. Right: Same sintered in He + 0.1 atm. HCl. Both at the same magnification.

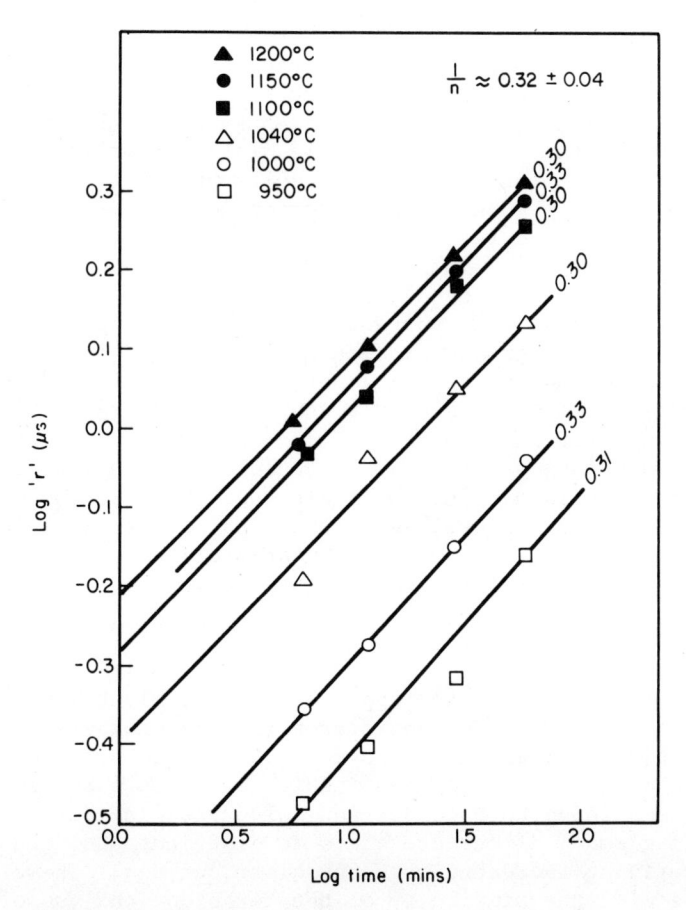

Fig. 13. ZnO average particle size as a function of time during sintering in hydrogen.

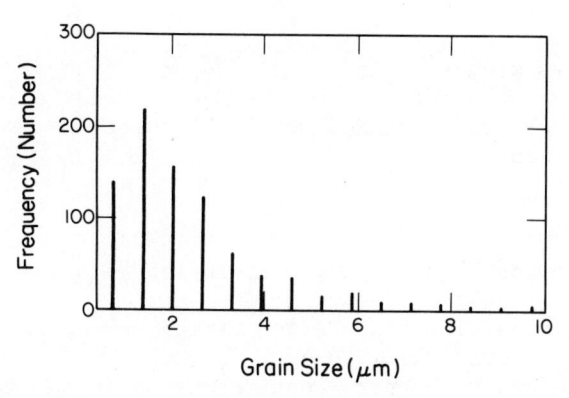

Fig. 14. Fe_2O_3 particle size distribution after sintering in 0.1 atm. HCl at 1050°C for 3 min.

by image analysis, grows as $t^{1/3}$ as predicted by classical Ostwald Ripening theory[11]. However, the particle growth or coarsening does not fit the theory in all respects. Figure 14 gives a typical particle size distribution clearly demonstrating skewness to large particles while the classical Ostwald distribution is skewed in the opposite direction. The same results were obtained directly from SEM fractographs and from plane sections of glass-impregnated samples and corrected for a plane section through a three dimensional particle distribution. The results deviate from the theory in the temperature dependence of the particle growth rate as well. Theory predicts that the apparent "activation energy" for the gas phase transport growth process should be the enthalpy for the reactions of Zn and Fe_2Cl_6 vapors given above[11]. However, the activation energy is closer to that for grain growth[12]. This is consistent with the Grescovich-Lay[13] mechanism of particle coarsening and suggests that grain boundary motion is the rate-controlling step in particle or grain coarsening in the early stages of sintering as well as in the final stage of isolated porosity. Therefore, control of boundary motion or grain growth is important during all stages of sintering.

CONCLUSIONS

Several examples of the application of semiautomatic image analysis to ceramic powder and microstructure characterization have been given. The primary purpose was to demonstrate the value of the technique: in being applicable to a wide particle size range; in examining both powder and polycrystaline bulk solids; in detecting small differences in size distributions; and in clearly defining the details of the distributions. These advantages plus the simplicity of the technique and use of operator judgement far outweigh the negligible time involved in obtaining the data.

AKNOWLEDGEMENT

The authors wish to acknowledge support of these several projects through contract No. N00014-80-C-0523 for the Office of Naval Research and Grant No. DMR-8112557 from the National Science Foundation.

REFERENCES

1. E. E. Underwood, "Quantitative Stereology", Addison-Wesley, Reading (1970).
2. D. A. Vermilyea, The Dissolution of Ionic Compounds, J. Electrochem. Soc. 113:1067 (1966).
3. N. Valverde and C. Wagner, Considerations on the Kinetics and the Mechanism of the Dissolution of Metal Oxides in Acidic Solution, Ber. Bunsen - Gesellschaft 80:330 (1976).

4. A. J. Bard, Photoelectrochemistry, Science 207:139 (1980).
5. H. Kametani and K. Azuma, Dissolution of Calcined Ferric Oxides, Trans. AIME 242:1025 (1968).
6. K. Nii, On the Dissolution Behavior of NiO, Corrosion Sci. 10: 571 (1970).
7. R. S. Pan, Oxygen Partial Pressure During Calcination and Dissolution Kinetics of Nickel Oxides in Hydrochloric Acid, M. S. Thesis, The Ohio State Univ., 1981.
8. Y. D. Tretyakov and R. A. Rapp, Nonstochiometries and Defect Structures in Pure Nickel Oxide and Lithium Ferrite, Trans. AIME 245:1235 (1969).
9. E. Bright, The Dissolution Kinetics of TiO_2 and $SrTiO_3$ in Hydrochloric-Hydrofluoric Acid Solutions, M.S. Thesis, The Ohio State Univ., 1982.
10. R. L. Coble, Initial Sintering of Alumina and Hematite, J. Am. Ceram. Soc. 41:55 (1958).
11. I. M. Lifshitz and V. V. Slyoz, The Kinetics of precipitation from Supersaturated Solid Solutions, J. Phys. Chem. Solids 19:35 (1961).
12. T. K. Gupta and R. L. Coble, Sintering of ZnO: I, Densification and Grain Growth, J. Am. Ceram. Soc.51:521 (1968).
13. C. Grescovich and K. W. Lay, Grain Growth in Very Porous Al_2O_3 Compacts, J. Am. Ceram. Soc. 55:142 (1972).

ACOUSTIC CHARACTERIZATION OF STRUCTURAL CERAMICS

B. T. Khuri-Yakub

E. L. Ginzton Laboratory
W. W. Hansen Laboratories of Physics
Stanford University
Stanford, California 94305

INTRODUCTION

The ceramics silicon nitride (Si_3N_4) and silicon carbide (SiC) are becoming important structural materials for applications such as turbine blades, ball bearings, heat exchanger tubing, and combustion engine piston caps. The importance of structural ceramics stems from their high strength at elevated temperatures (1000°C-1400°C), thus making turbine engines more efficient in terms of energy conservation and reduced pollutants in combustion products. Structural ceramics are also light weight (density ≈ 3.28×10^3 kg/m^3) and the raw materials (Si, C, N) are abundant, which reduces the cost of ceramic components over their metal and powder metal counterparts. The advantages of structural ceramics are somewhat offset by the fact that they are brittle; consequently very small defects can lead to a dramatic decrease in their strength.

The defects that lead to fracture in structural ceramics can be classified as either extrinsic or intrinsic. Intrinsic defects are induced during the manufacturing processes and are generally volumetric voids or inclusions. Extrinsic defects are generally surface breaking cracks that form due to machining, impact damage, or thermal cycling.

The fracture of ceramics usually occurs by direct extension of pre-existing defects (Evans and Landon, June 1976). The largest sharp crack (radius a_m) that can be tolerated in a component, that is expected to survive for a time t_i after inspection, is given by Evans and Landon, 1976):

$$a_m = \left(\frac{K_C}{\sigma_a Y}\right)^2 \left(\frac{2}{2 + K_C^{n-2} A t_i \sigma_a^2 Y^2 (n-2)}\right)^{2/(n-2)}$$

where K_{IC} is the stress intensity factor for mechanical extension of the flaw, σ_a is the pertinent level of applied tensile stress, Y is a parameter that depends on the flaw shape, and A and n are parameters that define the susceptibility of the material to slow crack growth. Defects in the range of 10-100 μm have to be detected, thus placing the first requirement on any nondestructive evaluation (NDE) technique.

The NDE problem of structural ceramics is further complicated by the fact that different modes of failure have been associated with the presence of different types of inclusions. The expected failure mode depends on the elastic modulus, fracture toughness, and thermal expansion coefficient of the inclusion as compared to the host material. Strength-size relations for various types of inclusions in silicon nitride (Evans, September 1979) shows that a WC inclusion has no effect on the strength of Si_3N_4, while a silicon inclusion decreases the strength of Si_3N_4 dramatically. It is thus imperative to determine the type and size of inclusions using NDE techniques.

Fracture due to surface cracks can be predicted by using standard fracture mechanics relations for mode I (Sih, 1973):

$$K = F(\theta)\, 2\sqrt{a/\pi}\, \sigma$$

where K is the stress intensity factor. The parameter $F(\theta)$ describes the variation of the stress intensity factor K around a semicircular crack periphery of radius a; σ is the applied stress. Fracture due to a semicircular crack in a ceramic occurs for a stress σ_p given by

$$\sigma_p = \frac{K_{IC}}{F(\theta)} \frac{1}{2} \sqrt{\frac{a}{\pi}}$$

For surface cracks in Si_3N_4, crack sizes from 50 to 100 μm in radius have to be detected.

Besides defect characterization, NDE techniques can be applied to characterize the host material. Material characterization is needed because it can be used to adjust manufacturing processes (hot pressing, sintering, injection molding). Once a material is produced without porosity or large grains, thus achieving full strength, NDE can be applied to find and characterize isolated defects.

HOST MATERIAL CHARACTERIZATION

Structural ceramics are typically manufactured by a variety of techniques, such as injection molding, uniaxial hot pressing, compression molding, slip casting, sintering, etc. Besides the presence of isolated defects, one reason for ceramic parts not attaining full strength is the presence of distributed porosity throughout the part leading to a decrease in strength. The decrease in density is also manifested by a decrease in the velocity and an increase in the propagation loss of acoustic waves propagating in ceramic parts. Acoustic velocity and propagation loss measurements should be used to characterize ceramics and correct the manufacturing process until the desired material strength is obtained (Chou et al., August 1980).

Knowledge of acoustic attenuation in ceramics is also needed to calibrate acoustic techniques used for defect detection and characterization. Traditionally, acoustic attenuation due to grain scattering has been treated in one of two extreme regimes: the Rayleigh regime where $\lambda > 10D$ (λ is the wavelength and D is the average grain diameter), and the stochastic regime where $\lambda < 0.1D$ (Papadakis, 1968). This traditional approach does not explain the behavior of attenuation in the intermediate frequency range when $\lambda \approx D$. Evans et al. (May 1978) have shown that, in this range, attenuation is dominated by large scatterers. Thus, the largest scatterer in a small volume is regarded as the main scatterer which is surrounded by a finer scale microstructure, considered the homogeneous host material. Figure 1 shows a comparison of predicted and measured attenuation of Si_3N_4 samples with MgO, ZnS, and porous large size scatterers. The agreement between theory and experiment is excellent, indicating the potential of attenuation measurements for characterizing microstructure. It is also seen from Fig. 1 that the propagation loss varies roughly as the square of the frequency and is about 10-30 dB/cm at 50 MHz, depending on the type of large grains or pores. Acoustic NDE techniques for this material are thus limited to operate at about 25 MHz. For ceramics of very high quality, NDE techniques at frequencies of the order of 400 MHz can be used; defects of the order of 25 μm can be detected.

INTRINSIC DEFECT DETECTION AND CHARACTERIZATION

The basic requirement for an NDE technique is to find defects in the size range of 10-100 μm and determine the size, shape, orientation, and type of material that constitutes the defects. Several NDE techniques have been developed or adapted to address the problem of defect characterization in structural ceramics.

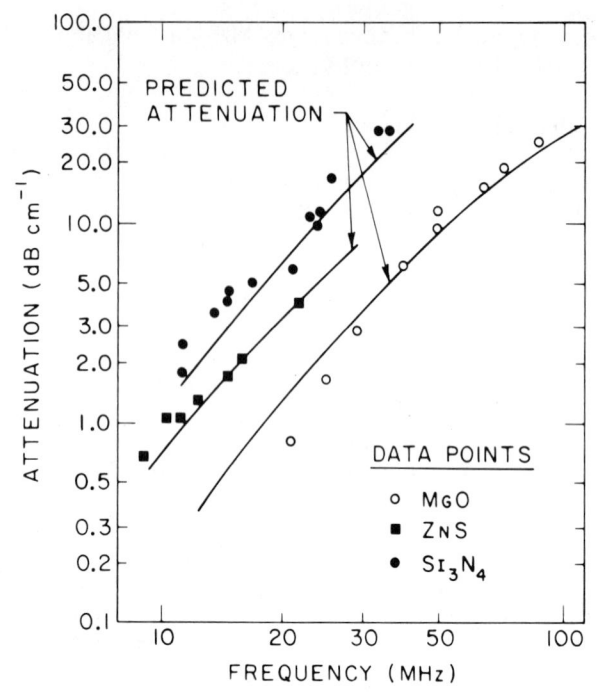

Fig. 1. A comparison of the predicted and measured attenuation for grains in MgO and ZnS, and pores in Si_3N_4.

The most promising NDE techniques that have been developed so far are: X-ray, microwave, and ultrasonic [acoustic microscopy, C-scan imaging, and A-scan (low-frequency, high-frequency)]. This paper will deal with ultrasonic techniques only.

Ultrasonic Techniques

An advantage of ultrasonic waves is that they propagate in solid materials with a velocity of the order of 10^4 m/sec. The relatively slow velocity of ultrasonic waves compared to microwaves results in a regime of operation where the wavelength is comparable to the size of the defects of interest, hence resulting in a large increase in sensitivity. Scattering of ultrasonic waves by defects is used for defect detection and characterization. Several types of ultrasonic techniques are used for the NDE of ceramics, namely pulse-echo, pitch-catch, back-reflection, and imaging. The following sections outline some of the more promising ultrasonic techniques.

Scanning Laser Acoustic Microscopy (SLAM). In SLAM, a sample is viewed by placing it on a stage where it is insonified by plane acoustic waves, typically at 100 MHz , and a laser beam is used to measure the surface displacement at the other side of the sample. Thus, the measured displacement represents an interference between the incident signal and the scattered signal by a defect that is present in the path of the acoustic beam (Kessler and Yuhas, April 1979). SLAM has been used to look at biological cells, defects in thin metal bellows, and defects in both Si_3N_4 and SiC. Complex shapes, such as turbine blades, can be handled by the SLAM.

A computer calculation to synthesize the signals that would be observed by SLAM from spherical defects has been developed by Chou et al. (1979). The character of the rings observed changes as the observation plane is moved away from the defect. For an observation plane tangent to the defect, a void shows no transmission (dark spot) as expected, while an inclusion shows a very strong transmission due to the focusing of the acoustic beam within the inclusion because its velocity is typically smaller than that of Si_3N_4. The overall size and distribution of the rings observed is in good agreement with the experimental results obtained with the SLAM. Different defects, with a distance to the edge of the sample of the order of 30a , are indistinguishable from each other by direct observation of a SLAM output. Thus, by itself, a SLAM provides only qualitative data on the presence of a flaw. It is possible, however, to use holographic reconstruction techniques to determine the type and size of the defect observed (Chou et al., 1979). Thus, with a reconstruction scheme, a SLAM is an excellent tool for the NDE of ceramics.

A-Scan (Low-Frequency, High-Frequency). Because the propagation loss of good quality ceramics is very low, about 3 dB/cm at 300 MHz , it is possible to use A-scan pulse reflection techniques at frequencies as high as 500 MHz . Defect characterization using A-scan systems is based on the difference in the backscattered power spectra of different types of defects in the Si_3N_4 or SiC host matrix (Chou et al., August 1980). High-frequency systems have been successful in finding defects in the size range of 25-100 μm , and in isolating voids and cracks from inclusions. The main difficulty in using high-frequency systems is that the samples under test have to be polished to allow for easy contacting with the buffer rod. Another difficulty is that theories assume that defects are spherical, while in practice defects have odd shapes and are typically made of several compounds.

Operating at low ka (ka < 0.5), two features about the defect are extracted from an ultrasonic measurement, namely A_2 and d , where A_2 is the low-frequency (long wavelength) scattering coefficient (Richardson, 1978) and d is the radius estimate derived from the Born inversion (weak perturbation

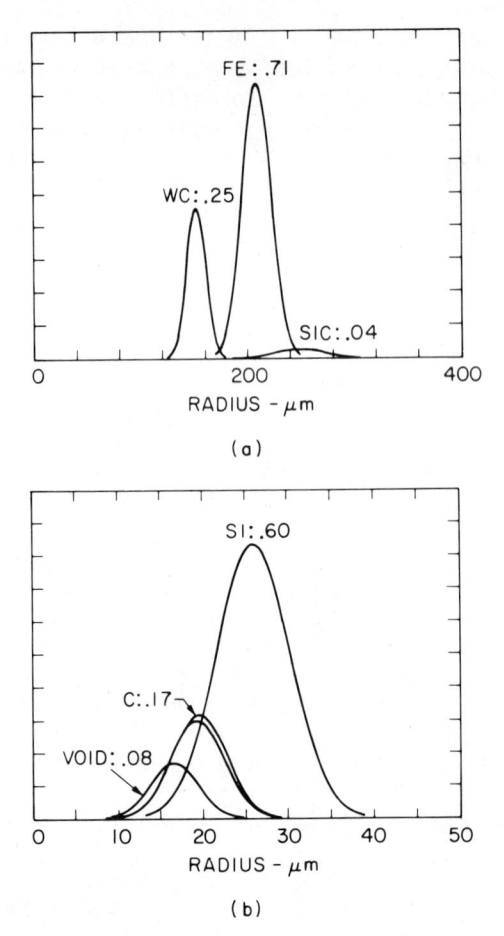

Fig. 2. (a) Joint probability of flaw type and radius for nominal 200 μm radius Fe inclusion; and (b) Joint probability of flaw type and radius for nominal 50 μm radius Si inclusion.

theory) (Rose et al., 1980). One or more pairs of A_2 and d (from interrogating the inclusion from one or more directions) are used as the input, along with error estimates, to a statistical inversion algorithm (Fertig and Richardson, July 1980). This algorithm estimates the size, shape, orientation, and material context of the flaw. Typical results of such an algorithm are shown in Fig. 2 (Ahlberg et al, 1979). The Fe inclusion of Fig. 2a is determined with very good accuracy, while the Si defect of Fig. 2b is determined with an error of 40%. The major difficulty with low ka (low-frequency) scattering measurements is that small defects can be easily missed (undetected), especially if the host material exhibits grain scattering from large grains or large pores.

C-Scan Imaging. C-scan is the most common method of implementing high-speed ultrasonic inspection of structural materials. Typically, a transducer in water operating in the pulse-echo mode is moved in a two-dimensional X-Y raster across the surface of the specimen under test. Echoes returning from the sample are range-gated to select only those echoes occurring between the front surface and back surface of the sample. The gated echoes are used to intensity-modulate a display which is scanned in the same raster pattern as the transducer. In ceramics, because the acoustic velocity is very large and small defects are of interest, frequencies of the order of 50 MHz and 100 MHz have to be used. Thus, severe requirements on location accuracy are placed on the mechanical scanning system. The transducer has to be broadband, efficient, and of small size so that sensitivity and location accuracy are maintained. Acoustic transducers operating at a frequency of 50 MHz, with a beamwidth of 1 mm, a compact impulse response, and a round-trip insertion loss of 12 dB for a C-scan system with a relocation accuracy of 8 μm, have been demonstrated (Khuri-Yakub et al., July 1980). C-scan systems are very useful for quick location of defects. Once defects are found, one of many other techniques, such as synthetic aperture imaging or A-scan methods, can be used to characterize the detected defects.

EXTRINSIC DEFECT DETECTION AND CHARACTERIZATION

Extrinsic defects, such as surface cracks, are introduced during machining, manufacturing, or are due to impact damage or thermal cycling. Surface cracks are responsible for the fracture of more than 60% of ceramic parts. Thus, even though more effort has been concentrated on the bulk defect problem, the surface crack problem is also important. Surface cracks with a radius in the 25-50 μm range must be detected and sized. The most promising NDE techniques that have been developed for this purpose are: X-ray (dye-enhanced radiography) and ultrasonic (1) photoacoustic spectroscopy, (2) scanning laser acoustic microscopy (SLAM), and (3) pulse echo (high-frequency, low-frequency). Again, we will only deal with the ultrasonic techniques.

Ultrasonic Techniques

Photo-Acoustic Spectroscopy. In this technique, a sample is placed inside a specially-designed cell containing a suitable gas and a sensitive microphone. The sample is then illuminated with a chopped laser beam, as shown in Fig. 3. Light absorbed by the sample is converted in part into heat. The resulting periodic heat flow from the sample to the surrounding gas creates pressure fluctuations which are detected by the microphone as a signal which is phase-coherent at the chopping frequency. The resulting

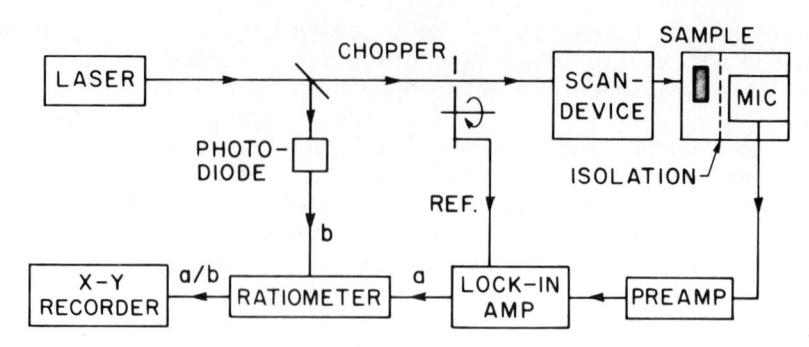

Fig. 3. Schematic diagram of the laser photoacoustic scanning system.

photo-acoustic signal is directly related to the amount of light absorbed by the sample. Any change in the amount of light absorption due to a surface crack or a subsurface flaw is indicative of the presence of that flaw. The laser beam is raster scanned across the surface of the sample, and the resulting signal is used to modulate the intensity of a recorded image, or the amplitude in a three-dimensional projection (Wang et al., May 1978). Surface cracks 100 μm in length have been detected.

Scanning Laser Acoustic Microscope (SLAM). The application of the SLAM to the characterization of surface flaws has been carried out by Kupperman et al (March 1979) and Kessler and Yuhas (April 1979). Figure 4, a result of a SLAM test, shows the difference between a shadowed region from a surface crack and a clear (unflawed) region. Notice that the wavefronts on the surface have a periodicity corresponding to the wavelength of the surface wave scattered by the crack. The presence of the crack is clearly evidenced in Fig. 4, and a rough estimate of its size is induced from the size of the wavefront at the crack location. For partially closed cracks, this technique will run into difficulties because the shadowing effect will be diminished.

Pulse-Echo Techniques (High-Frequency, Low-Frequency). With this technique, surface acoustic waves (SAW) are excited on nonpiezoelectric materials by a wedge which converts a longitudinal or shear wave into a SAW. The SAW propagates on the surface of the material under test, and upon incidence on a crack, a reflected wave is generated and is used to detect and characterize the crack. For maximum sensitivity, a SAW, with a wavelength of the same order of magnitude as the crack size, has

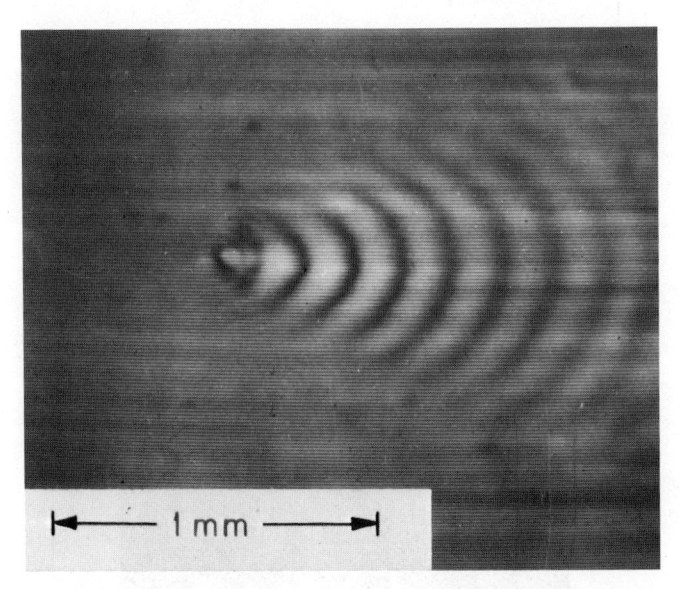

Fig. 4. Acoustic micrograph of a surface crack in hot-pressed Si_3N_4 (NC-132).

to be used. In the NDE of structural ceramics. this corresponds to operating at a frequency of 100 MHz, where the SAW wavelength is about 60 μm. At such high frequencies, conventional wedge transducers are not realizable, and a technique has been developed to excite high-frequency SAW on ceramics (Khuri-Wakub and Kino, May 1979). A schematic diagram of the transducer configuration used is shown in Fig. 5. A SAW is excited on piezoelectric material by an interdigital transducer. When it is incident on the water couplant, the SAW is converted into a longitudinal wave which is reconverted to a SAW on the ceramic. The conversion loss is only 3 dB. The technique has been very successful in detecting cracks with radii as small as 25 μm (Khuri-Yakub et al., January 1980). Because SAWs propagate on curved surfaces, the technique has been used to detect 50 μm cracks in the neck region of Si_3N_4 turbine blades. The main disadvantage in this technique is that the water coupling length has to be adjusted by observing it with an optical microscope with a magnification of 10x . The sizing of surface cracks at high ka can be done either by measuring the reflection coefficient versus frequency at normal incidence or by measuring the reflection coefficient at constant frequency versus incidence

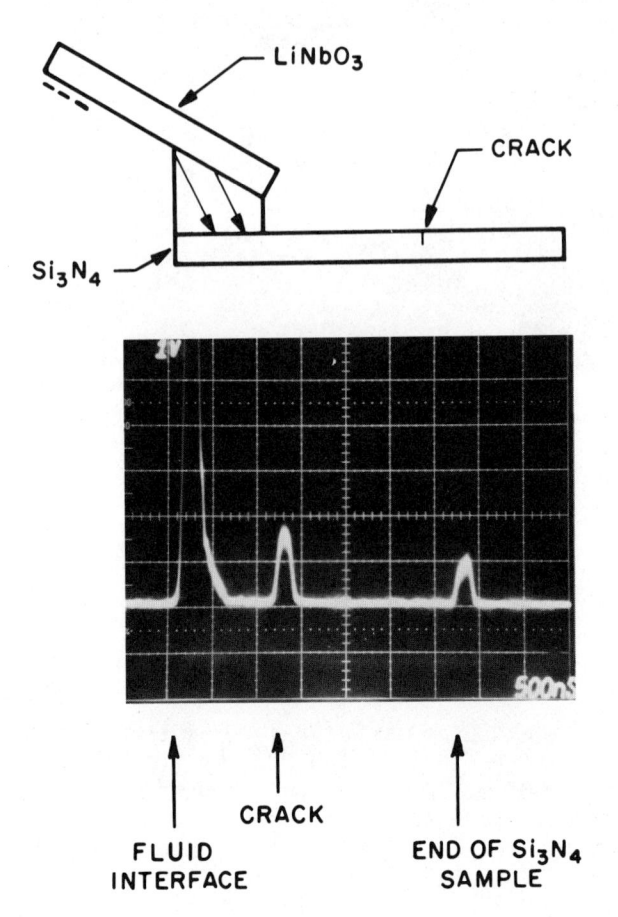

Fig. 5. A schematic diagram of the configuration used to excite high-frequency surface waves on ceramics.

angle (Domarkas et al., October 1978; Tittmann et al., July 1978).

An alternative technique is to use much lower frequencies (< 10 MHz), where the crack size is much smaller than the wavelength. One measurement of the reflection coefficient of a surface crack at a known location in the long wavelength regime is sufficient to determine the size of the crack and the maximum value of the normalized stress intensity factor (SIF) around its edge (Resch et al., February 1979). A low-frequency (~8 MHz) wedge transducer is used to excite and receive the SAW and to measure the SAW reflection coefficient from the crack. Given the fracture toughness of the material, the fracture stress of the part containing the crack can be predicted (Tien et al., July 1980; Khuri-Yakub et al., January 1980). The technique can be used to measure the size of cracks in the 50-250 μm range and to predict fracture stress with an accuracy of 10%.

CONCLUSIONS

At the present time, there is no one technique that can satisfy all the ceramic NDE requirements. We use one technique for locating defects and one or several other techniques for defect characterization.

ACKNOWLEDGMENT

This work was sponsored by the Ames Research Laboratory for the Defense Advanced Research Projects Agency and the Air Force Materials Laboratory under Contract No. SC-81-009 and the Office of Naval Research under Contract No. N00014-78-C-0283.

REFERENCES

Ahlberg, L. A., Elsley, R. K., Graham, L. J., and Richardson, M. P., "Measurement and Characterization of Inclusions in Ceramics," Proc. Ultrasonics Symp. (1979).
Chou, C. H, Khuri-Yakub, B. T., and Kino, G. S., "Transmission Imaging: Forward Scattering and Scatter Reconstruction," K. Wang (ed.) Acoustic Imaging, Vol. 9. Plenum Press, New York (1979).
Chou, C. H., Khuri-Yakub, B. T., Kino, G. S., and Evans, A. G., "Defect Characterization in the Short Wavelength Regime," JNDE (August 1980).
V. Domarkas, B. T. Khuri-Yakub, and G. S. Kino, "Length and Depth Resonances of Surface Cracks and Their Use for Crack Size Estimation," Appl. Phys. Lett., 33:557 (October 1978).
Evans, A. G., and Landon, T. G., "Structural Ceramics," B. Chalmers, J. W. Christian, and T. B. Massalski (Eds.) Progress in Materials Science, Vol. 19. Pergamon Press, Oxford Press, 171-241 (1976).
Evans, A. G., Tittmann, B. R., Ahlberg, L., Khuri-Yakub, B. T., and Kino, G. S., "Ultrasonic Attenuation in Ceramics," J. Appl. Phys. 49:2669 (May 1978).

Evans, A. G., "Structural and Microstructural Design in Brittle Materials," 3rd International Conf. of Materials (September 1979).

Fertig, K., and Richardson, J. M., "Unified Inversion Algorithms and Computer Simulation," Proc. DARPA/AFML Review of Progress in Quantitative NDE (July 1980).

Kessler, L. W., and Yuhas, D. E., "Acoustic Microscopy," Proc. IEEE. 67:526 (April 1979).

Khuri-Yakub, B. T., and Kino, G. S., A New Technique for Excitation of Surface and Shear Acoustic Waves for Nonpiezoelectric Materials," Appl. Phys. Lett. 32:513 (May 1979).

Khuri-Yakub, B. T., Kino, G. S., and Evans, A. G., "Acoustic Surface Wave Measurements of Surface Cracks in Ceramics," J. Am. Ceram. Soc. 63:65 (January 1980).

Khuri-Yakub, B. T., Chou, C. H., Liang, K., and Kino, G. S., "NDE for Bulk Defects in Ceramics," Proc. DARPA/AFML Review of Progress in Quantitative NDE (July 1980).

Kupperman, D. S., Yuhas, D., Sciammarella, C., Lapinski, N. P., Fiore, N., "NDE Techniques for SiC Heat-Exchanger Tubes," Argonne National Laboratory Report No. ANL-79-4 (March 1979).

Papadakis, E. P., "Ultrasonic Attenuation Caused by Scattering in Polycrystalline Media, W. P. Mason (Ed.) Physical Acoustics, Vol. 4B. Academic Press, New York, 269-329 (1968).

Resch, M. P., Khuri-Yakub, B. T., Kino, G. S., and Shyne, J. C., "The Acoustic Measurement of Stress Intensity Factors," Appl. Phys. Lett. 34:182 (February 1979).

Richardson, J. M., "The Inverse Problem in Elastic Wave Scattering at Long Wavelengths," Proc. Ultrasonic Symp. (1978).

Rose, J. H., Varadan, V. V., Varadan, V. J., Elsley, R. K., and Tittmann, B. R., "Testing the Inverse Born Procedure for Spherical Voids," Proc. Recent Development in Classical Wave Scattering (1980).

Sih, G. C., Handbook of Stress Intensity Factors, Lehigh University, Pennsylvania (1973).

Tien, J., Khuri-Yakub, B. T., Kino, G. S., Evans, A. G., and Marshall, D., "Surface Wave Measurements of Surface Cracks in Ceramics," Proc. DARPA/AFML Review of Progress in Quantitative NDE (July 1980).

Tittmann, B. R., Cohen-Tenoudji, F., deBilly, M., Jungman, A., and Quentin, G., "Simple Approach to Estimate the Size of Small Surface Cracks with the Use of Acoustic Surface Waves," Appl. Phys. Lett. 33:6 (July 1978).

Wang, Y. H., Thomas, R. L., Hawkins, G. F., "Surface and Subsurface Structure of Solids by Laser Photoacoustic Spectroscopy," Appl. Phys. Lett. 32:538 (May 1978).

CHARACTERIZATION OF CERAMICS BY ACOUSTIC MICROSCOPY

D.E. Yuhas and L.W. Kessler

Sonoscan, Inc.
530 East Green Street
Bensenville, IL 60106

INTRODUCTION

The production of reliable and long-lived structural ceramic components depends on the development of effective non-destructive evaluation techniques. The brittle failure of ceramics under stress from small defects make flaw detection highly important. Because ceramics such as silicon nitride and silicon carbide have critical flaw size, an order of magnitude smaller than metals, testing methods adequate for metals may not be appropriate for ceramic materials.

Acoustic microscopy refers to the production of magnified views of objects using high frequency sound waves. The transparency of many ceramics to acoustic waves makes it possible to probe the interior non-destructively. By virtue of their low acoustic attenuation and small grain size, high frequency sound waves can easily penetrate thick ceramic materials. Components as thick as a few centimeters have been evaluated non-destructively. Utilization of high frequencies > 100 MHz insures detection of small flaw sizes which are required to evaluate ceramic materials. Several important applications in the area of non-destructive testing and quality control have been demonstrated. Additional applications in the areas of failure analysis and materials characterization are being explored.

Magnified acoustic images of ceramics provides a convenient data format for conveying the large quantity of information encountered in testing applications. In addition, defect morphology, available only through images has been useful in sizing and characterizing flaws. Interpretation of acoustic images has arisen empirically by comparing images of defects with the results of destructive analysis. In recent years, interpretation and quantitative analysis of images has been aided by theoretical descriptions of sound scattering. In this investigation, we describe the imaging technique, image interpretation and applications of SLAM to ceramics. Characteristic images of cracks, porosity, inclusions and laminar flaws will be presented.

THE SCANNING LASER ACOUSTIC MICROSCOPE (SLAM)

The key attribute of the SLAM technology is the production of real-time, high resolution acoustic images. Both the speed and resolution arise from the use of a focused, scanning laser beam used as a detector of the sound waves. A description of the basic principles of operation have been presented previously[1-7]. For completeness, a brief summary is presented here.

Figure 1 is a micrograph of the SONOMICROSCOPE 100[8]. The instrument produces full grey scale acoustic images at frequencies from 30-500 MHz. Many of the results appearing in this paper are obtained at 100 MHz where the field of view is 2 x 3 mm, resolution is about ½ wavelength, or 30 microns in silicon nitride (shear waves) and images are produced in real-time. The transmitting transducer is a piezoelectric element. A focused laser scans the insonified zone and acts as the receiving transducer. By synchronizing the

Fig. 1. Commercially available Scanning Laser Acoustic Microscope (SLAM).

laser scan with a television monitor, real-time acoustic images are obtained.

As a by-product of the technique, an optical image of the scanned surface is produced on a separate TV monitor. This image is quite useful in documenting the location of flaws.

Figure 2 shows a sound cell in which the piezoelectric element is bonded to a fused quartz delay line. Alternately, the piezoelectric element can be coupled to the specimen using a fluid path. In both cases, images are obtained by inserting the sample between the piezoelectric transducer, the source, and the coverslip, the detector plane. A fluid couplant is required in both cases. The coverslip, which is a transparent plastic with a semireflective optical coating, provides the interface required to record the acoustic image. The optical image is obtained by using the portion of light transmitted through both the sample and the coverslip. Optical images can also be obtained using light reflected directly from the surface of optically opaque ceramic samples.

Two acoustic imaging modes are commonly employed in the SLAM technique, the acoustic amplitude mode and the acoustic interferogram. Figure 3 shows an amplitude image (a) and interferogram (b) obtained in the same silicon nitride test sample. The amplitude image shows the relative transmission level over the field of view. In the amplitude mode, bright areas correspond to zones in the sample with good acoustic transmission, whereas the darker areas are attenuating. The interferogram has a characteristic set of vertical fringes which are produced by electrical mixing of the detected acoustic signal with a phase reference. In principle the

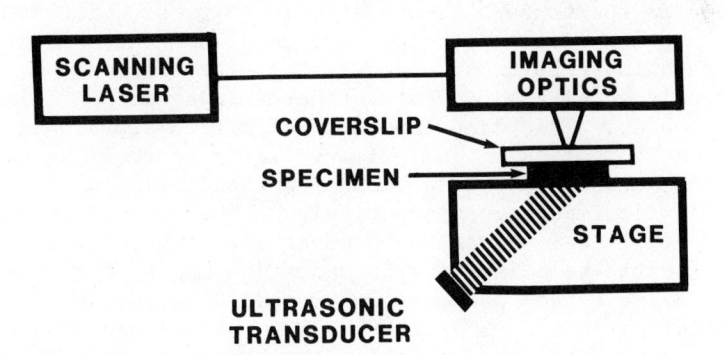

Fig. 2. Insonification geometry used for SLAM imaging with the "glass" sound cell. A fused quartz stage conveys sound from transducer to sample. For the liquid cell, a direct fluid contact between transducer and specimen is made.

interferogram is 85 microns. It can be determined from this figure that the silicon nitride sample shown is fairly clean and uniform in its elastic properties. The horizontal streaks in the acoustic images are due to surface texture caused by a grinding operation. Though visible, the grinding marks do not interfere with the visualization of the buried flaws. By way of contrast, other ultrasonic inspection techniques[10] require a high degree of surface flatness and polish. Therefore, poorly prepared samples such as this are ordinarily difficult to inspect.

FLAW IMAGE CHARACTERISTICS

The differentiation of various flaw types is made by comparing their characteristic signatures. In order to acquaint the reader with the various flaw characteristics, a series of micrographs illustrating a variety of flaw types is presented.

Influence of Porosity

Figure 4 shows an acoustic amplitude micrograph obtained on a reaction sintered silicon nitride turbine airfoil. This is a much more porous material than shown in Figure 3 as evidenced by the high attenuation regions (dark zones). Destructive analysis of these samples indicates the presence of pores ranging in size from 1 to 50 microns. The larger pores are directly resolvable acoustically while the smaller pores give rise to the texture seen in this micrograph. A qualitative correlation was found between the porosity distribution determined acoustically and that obtained destructively by sectioning. Quantitative measurements of average attenuation can be used to determine average porosity values while the images are useful in detecting local zones of increased porosity. Recent investigations on silicon carbide have shown substantial variations in the texture of acoustic micrographs which can be related to grain size and the percentage of unreacted material[3].

Ultrasonic techniques, and acoustic microscopy in particular, are extremely sensitive methods for determining mechanical continuity in samples. This makes the method particulary sensitive for detecting delaminations. The primary factor controlling the sound propagation across the boundary between two dissimilar materials is the acoustic impedance discontinuity. The specific acoustic impedance (Z) is defined as the product of the bulk density and the acoustic velocity C. The acoustic impedance for a delamination (e.g., air gap) is very low, Z_a= .0004 x 10^6 Ns/M^3, compared to a typical ceramic Z_c=36 x 10^6 Ns/M^3. This large discontinuity leads to almost total reflection at the boundary and a dark zone on the SLAM micrograph. Figure 5 shows an example of the type of sensitivity which might be expected for a delamination in a ceramic material. The figure shows a plot of percent transmission as a function of gap width for a delamination in barium titanate (a

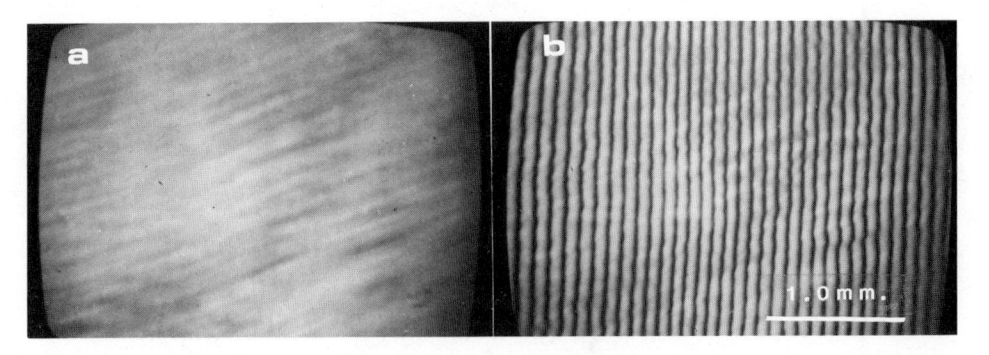

Fig. 3. Acoustic amplitude micrograph (a) and interferogram (b) obtained on a 0.12 inch thick silicon nitride tensile test bar. This fully dense material shows excellent acoustic transmission at 100 MHz.

interferogram is an acoustic hologram containing all three dimensions of information throughout the sample volume. In our employment of this mode, the hologram is not "reconstructed," rather graphical analysis of the fringes yield quantitative plastic property information. In particular, lateral shifts of the fringes indicate sonic velocity variations within the sample or thickness changes. In the interferogram (Fig. 2b), a lateral shift of one fringe corresponds to a 1.0 per cent velocity variation[9]. The field of view in this sample is 3 mm across; the spacing between the fringes in the

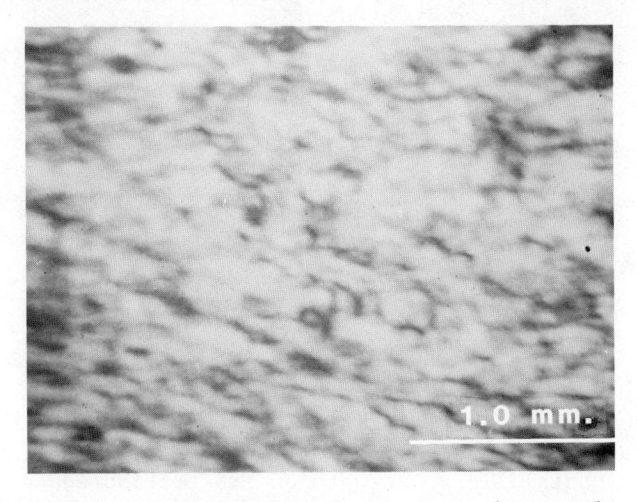

Fig. 4. 100 MHz acoustic amplitude micrograph revealing porosity in reaction sintered silicon nitride turbine airfoil.

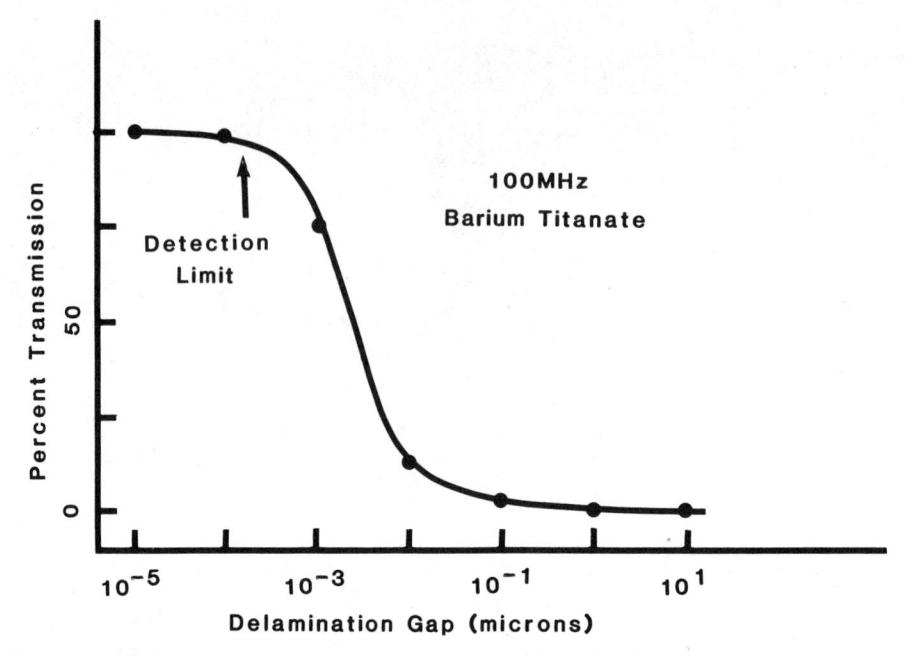

Fig. 5. Calculated acoustic transmission across a delamination as a function of gap size.

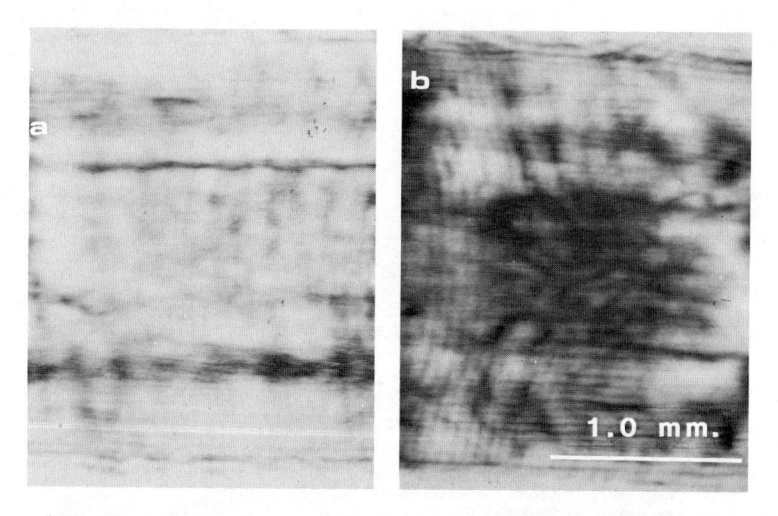

Fig. 6. 100 MHz acoustic amplitude micrographs comparing a "good" capacitor (a) with one containing a delamination (b). The horizontal linear structures are attributed to the internal metallization. Destructive analysis revealed a delamination gap of less than one micron.

typical ceramic capacitor material). This curve was calculated using simple impedance considerations and thin layer resonance formulation[11]. Local variations in acoustic transmission of a few percent are easily discernible in the micrographs, thus gaps less than fractions of microns can be detected at frequencies of 100 MHz even though the wavelength may exceed 60 microns.

Figure 6 shows a comparison between a good and delaminated ceramic capacitor. Both micrographs were obtained at a frequency of 100 MHz. The measured gap size in the delaminated chip was found to be on the order of 1 micron after destructive sectioning.

The physics of crack detection is identical to that for delaminations; the image morphology is quite different. The image contrast for cracks (whose size is much larger than a wavelength) is controlled by the impedance discontinuity at the gap. For example, the percent transmission across a crack gap is given by Figure 5. The image characteristics are quite different and are attributed primarily to geometrical effects.

The differences in the SLAM images are illustrated in Figure 7. For delaminations the plane of the gap is generally parallel to the detector plane, i.e. top surface of the sample. The shadow, if diffraction effects are ignored is the same size and shape as the delamination. The image looks the same as if we could see inside the material from the top and pick out the delamination. The delamination appears as we would expect it to look optically. The crack is quite different. Typically, cracks will be oriented such that its plane has a component perpendicular to the top surface.

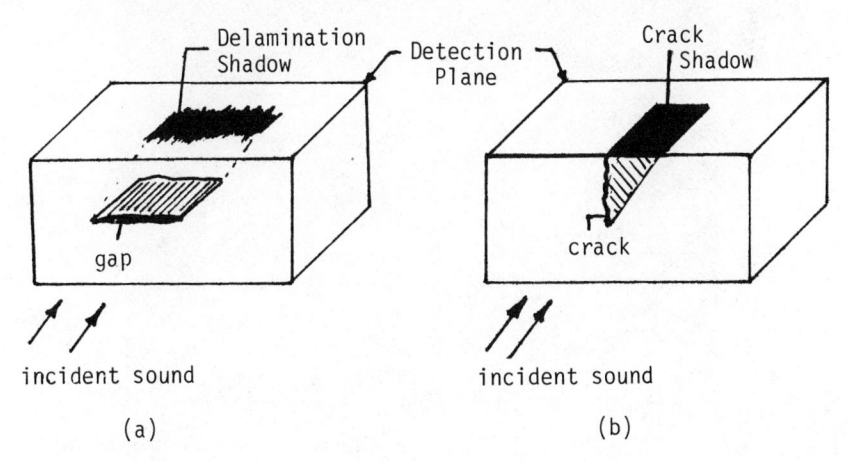

Fig.7. Schematics showing SLAM orthographic projection images of a delamination (a) and crack (b).

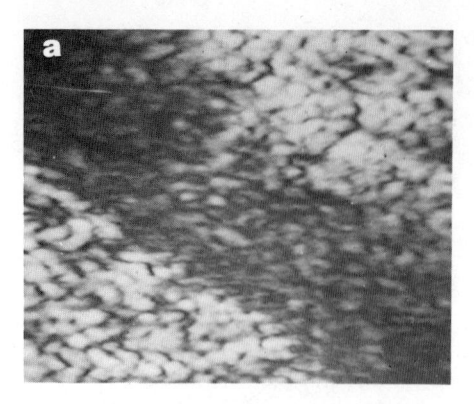

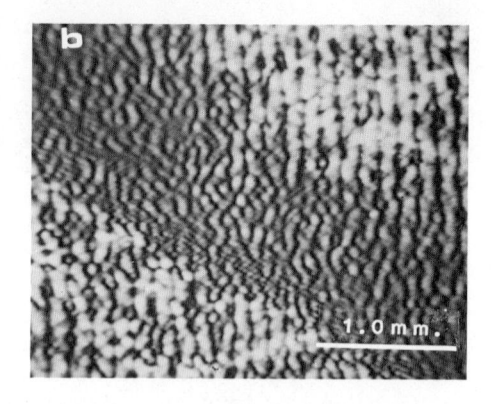

Fig. 8. Acoustic amplitude micrograph (a) and interferogram (b) showing SLAM crack shadow found in an alumina substrate. Measurement of the width of the shadow can be used to measure crack extension.

Thus, a crack extending beneath the surface casts a rather large shadow. The shadow size is of course dependent on the insonification direction and the depth of the crack beneath the surface. As the insonification direction is changed (e.g. the sample rotated on the stage), the characteristics of the image will change. This makes it easy to distinguish cracks from delaminations as well as permitting the measurement of crack extension. The large crack shadow

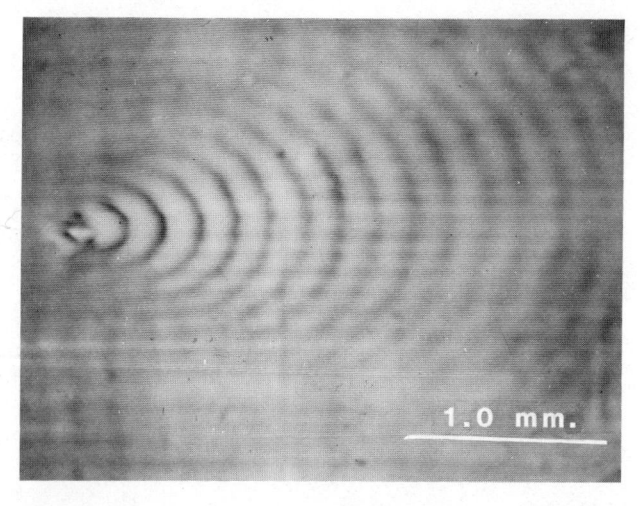

Fig. 9. Characteristic SLAM image showing mode conversion at the site of a flaw produced by Vicker's indentation (5Kg load).

makes cracks easily detectable even though the gap size may be less than a few microns. The crack images are not as intuitive as the delaminations.

Figure 8 shows acoustic micrographs of a fracture opening to the surface of an alumina sample. The micrographs are oriented such that the sound field propagates out of the plane of the paper at an angle of 45° from left to right across the field of view. Sound propagating through the sample is attenuated at the fracture interface resulting in a shadow to the right of the fracture. The primary feature that distinguishes a surface opening crack is the abrupt and sharply defined onset of the shadow region. This is seen as a sharp boundary separating the light and dark portions of the micrographs in Figure 8. The interferogram fringes are almost obliterated in this area, indicating almost total sound attenuation. By measuring the length of the shadow and using the known propagation angle, measurement of crack extension beneath the surface is made. In Figure 8, the propagation angle was 45°; the shadow is approximately 1 mm in length, thus the crack extends 1 millimeter below the surface.

The image features of small surface flaws and buried flaws within a ceramic host are substantially different from cracks and delaminations. The unique image features can be attributed to the propagation characteristics of elastic waves; specifically, mode conversion and diffraction. The former is purely an acoustic phenomenon; the latter occurs whenever the flaw size is comparable to acoustic wavelength.

Mode conversion refers to a conversion of the propagating acoustic wave from one form to another. This occurs whenever an elastic discontinuity is encountered. In many situations, its effects on SLAM image characteristics is negligible. However, for small surface flaws, such as cracks resulting from faulty machining or cracks induced by microhardness testers (Vicker's indentation), the mode conversion process dominates the image characteristics. In this particular case, a portion of the bulk wave energy (shear or compressional waves) impinging on the flaw is converted to a surface wave. The interference of the surface wave and the unaffected bulk wave gives rise to a characteristic cone-shape "ring pattern." The ring spacing is controlled by the propagation direction, bulk wave velocity, and surface wave velocity. The shape of the cone amplitude distribution is controlled by the flaw size, orientation, and angular response of the laser detector[2].

A typical surface flaw SLAM image is shown in Figure 9. The mode conversion phenomenon makes the detection of small surface flaws (flaws whose size are less than a wavelength) quite easy. Detailed analysis of the relationship between surface flaws and image characteristics has yet to be done.

The last major flaw type to be discussed is the detection and image characteristics of isolated subsurface flaws. Because the flaw sizes are often comparable to a wavelength and buried several wavelengths beneath the surface, the images are dominated by diffrac-

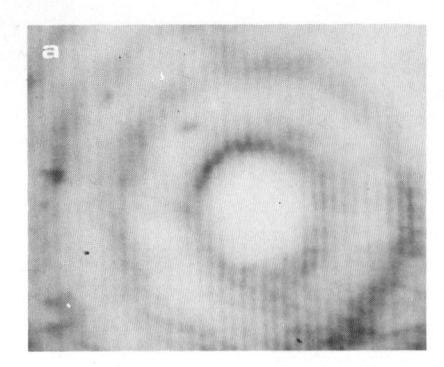

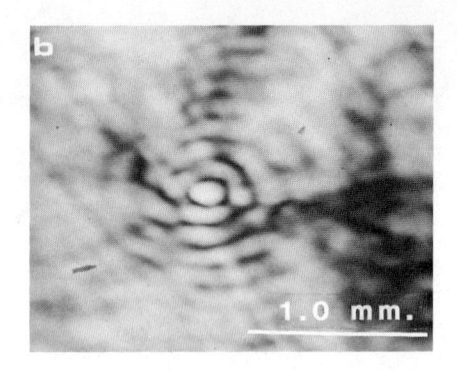

Fig. 10. 100 MHz acoustic micrographs showing diffraction ring patterns for (a) iron inclusion located 3.1 millimeters below the surface and (b) silicon inclusion located 0.9 mm beneath the surface of a silicon nitride disc.

tion phenomena. Figure 10 compares images of two isolated, buried flaws, (a) an iron inclusion located 3.2 mm beneath the surface, and (b), a silicon inclusion located 0.9 mm beneath the surface. In both cases, the flaw depth is measured by acoustic stereographic techniques[5].

The typical characteristics of SLAM images of inclusions is well illustrated in Figure 10. The primary features which distinguish this flaw type from other defects are the - 1) bright center with acoustic transmission greater than or equal to the background structure, 2) ring pattern due to diffraction of sound by the inclusion, and 3) the well-defined boundary of the flaw. These image characteristics seem to be prevalent for solid inclusions observed so far, and they are quite different from porosity variations and laminar flaws. The diameter of the first ring is 160 microns in Figure 10b, and the flaw is 900 microns below the surface. To obtain the relationship between SLAM image size and the actual flaw size, it is necessary to account for the effects of diffraction and beam spreading. This point has been investigated in greater detail by ourselves[14] as well as others[15]. The flaw images are presented to illustrate the unique morphology of solid inclusions.

Detailed quantitative analysis of SLAM images from isolated flaws requires theoretical descriptions of scattering. Direct scattering calculations based on the work of Ying and Truell[12,13] have been used to produce computer generated SLAM images[14]. Comparison of calculated and real SLAM images have shown good qualitative agreement. Less satisfactory quantitative agreement in image size and image contrast determination has been achieved[15]. More work is needed to incorporate shear wave scattering into the calculation and to account for roughness at the flaw boundaries.

The interpretation of SLAM images on buried flaws can be greatly simplified by producing flaw images which are focused. This can be accomplished by focusing the insonifying acoustic waves. This approach has the disadvantage that focused waves are apertured by critical angle reflection (analogous to total internal reflection observed optically). This limits both resolution and depth of penetration. An alternate approach involves computer holographic reconstruction of standard SLAM images. This method can provide good resolution while maintaining depth of penetration. Impressive inversion results have recently been obtained[14]. The results of Chovetal suggests that holographic inversion techniques will permit determination of size, type, and depth of flaws from standard SLAM images.

SUMMARY

Examination of a variety of different flaws in ceramics has led to the development of a series of unique image characteristics. These acoustic signatures can be used to differentiate various flaw types. Many of the interpretations rely on defect morphology. Thus, underscoring the utility of the imaging approach. This is particularly important, for example, in detecting solid inclusions in porous ceramics. Acoustic images also provide a handle on the determination of flaw sizes. The extension of cracks below the surface are directly obtainable from micrographs. Estimates of the size of solid inclusions are also obtainable. However, in this case, it may be necessary to correct recorded image size for beam spreading and diffraction effects. Current development work is concentrated on development of a better understanding of SLAM images as well as exploring inversion techniques such as holographic reconstruction. Successful implementation of inversion techniques will lead not only to simplified image interpretation but also to quantitative characterization data.

REFERENCES

1. D.S. Kupperman, C. Sciammarella, N.P. Lapinski, A. Sather, D. Yuhas, L. Kessler, and N.F. Fiore, "Preliminary Evaluation of Several NDE Techniques for Silicon Nitride Gas-Turbine Rotors," Argonne National Laboratory Report ANL-77-89, January 1978.
2. D.E. Yuhas, Characterization of Surface Flaws by Means of Acoustic Microscopy, First International Symposium on Ultrasonic Materials Characterization, June 7-9, 1978, National Bureau of Standards, Gaithersburg, MD., ed by: H. Berger, pp. 347-358 1980.
3. D.S. Kupperman, L. Pahis, D. Yuhas, and T. McGraw, Acoustic Microscopy for Structural Ceramics, Ceramics Bulletin, Vol. 59, No. 8, (1980).

4. D.E. Yuhas and L.W. Kessler, Scanning Laser Acoustic Microscope Applied to Failure Analysis, Proc. ATFA - 78 IEEE, Inc., New York, NY Catalog No. 78CH1407-6 REG6., pp. 25-29 (1978).
5. Acoustic Microscopy, SEM and Optical Microscopy: Correlative Investigations in Ceramics, Scanning Electron Microscopy 1979, SEM, Inc., AMF O'Hare, IL 60666, pp. 103-110 (1979).
6. D.E. Yuhas, T.E. McGraw, and L.W. Kessler, Scanning Laser Acoustic Microscope Visualization of Solid Inclusions in Silicon Nitride, Proc. ARPA/AFML Conf. on Quantitative NDE., LaJolla, CA, pp. 683-690 (1979).
7. L.W. Kessler and D.E. Yuhas, Acoustic Microscope 1979, Proc. IEEE, 67, (4) pp. 526-536 (1979).
8. Commercially available under trade name SONOMICROSCOPE^{TM} 100, Sonoscan, Inc., Bensenville, Illinois 60106.
9. S.A. Goss and W.D. O'Brien, Direct Ultrasonic Velocity Measurements of Mammalian Collagen Threads, J. Acoust. Soc. Amer. 65 (2) pp. 507-511 (1979).
10. G.S. Kino, Nondestructive Evaluation, Science, Vol. 26, pp. 173-180, Oct. (1979).
11. J. Krautkramer and H. Krautkramer, Ultrasonic Testing of Materials, Springer-Verlag, New York, p. 27 (1977).
12. C.F. Ying, R. Truell, Scattering of a Plane Longitudinal Wave by a Spherical Obstacle in an Isotropically Elastic Solid, J. Appl. Phys., 27, 1086 (1956).
13. G. Johnson, R. Truell, Numerical Computations of Elastic Scattering Cross Sections, J. Appl. Phy., 36, 3466 (1965).
14. C.H. Chou, B.T. Khuri-Yakub and G.S. Kino, Transmission Imaging; Forward Scattering and Scatter Reconstruction, Acoustical Imaging, Vol. 9, pp. 357-377, Plenum Press, New York, (1980).
15. D.E. Yuhas, M.G. Oravecz, and L.W. Kessler, Quantitative Flaw Characterization by Means of the Scanning Laser Acoustic Microscope (SLAM), In Rev. Progress in Quant. NDE ed. D.O. Thompson and D.E. Chimenti, Plen pp. 761-766.

DETERMINATION OF SLOW CRACK GROWTH USING AN AUTOMATED TEST TECHNIQUE

Helen H. Moeller and Robert L. Farmer

Babcock & Wilcox
Lynchburg, Virginia

INTRODUCTION

Recently, ceramic materials have received increased attention for use in structural applications with high operating temperatures. The design of ceramics for these critical applications depends heavily upon the assessment of the material's mechanical reliability. Traditionally, ceramic materials are tested for strength by obtaining rapid fracture under well defined stress states. In addition to flexural testing, tests such as compression, creep and fracture toughness are commonly performed. However, it is known that ceramics do experience delayed fracture.

Delayed fracture occurs when stresses lower than that required for rapid fracture are applied to a material with sufficient time for flaw growth and subcritical crack growth. While fracture will not immediately occur in these applications, eventually failure does occur.

With the increased emphasis of developing ceramics for structural applications, more attention is being given to characterizing ceramic materials for subcritical crack growth. This is accomplished by experimentally determining the relationship between crack velocity and stress intensity. There are two widely used testing methods which evaluate a single crack growth without interference from extraneous cracks. These methods are known as double cantilever beam and double torsion. After investigation into both of these techniques, it appeared that the double torsion method would be more convenient to use for both room temperature and high temperature testing.

Williams and Evans[1] developed the mathematical expression relating the double torsion specimen dimensions and load to stress intensity. There are three different test techniques which can be used to study crack propagation using the double torsion method. These techniques are categorized according to the method used to obtain crack growth rate information and described as constant displacement, constant displacement rate, and constant load. The constant displacement rate is a convenient method to use because during the testing crack length does not need to be measured, however lower crack velocities are generally obtained with the constant displacement and constant load techniques.

The double torsion technique provides the data for, and thus the capability of, predicting long-term service performance with short term testing. However, due to the labor-intensive nature of this testing to yield one complete set of data, the routine performance of this test is usually not cost effective.

A more economical test technique needed to be identified to characterize subcritical crack growth. The double torsion method still appeared to be the most convenient method to use. A reduction in labor was sought through the automation of the testing which included both the control of test parameters and post data analysis by a minicomputer.

In addition, it was anticipated that an automated test technique would result in a more efficient use of samples, thus reducing the cost of sample preparation. The software development for this testing is described further in the following sections.

EQUIPMENT

An automated, servo-hydraulic universal testing machine* was used for this test development. Accessory equipment included a 5 gallon per minute servovalve, extensometer (0.1 inch maximum range) and a 200 pound capacity load cell. All testing was controlled by the computer through the use of a digital-analog converter.

SPECIMEN PREPARATION AND APPARATUS

This study required the use of a material with a known tendency to subcritical crack growth. Freiman[2] has reported the relationship between crack velocity and stress intensity for

---

*Manufactured by MTS (Model 880.14 with PDP 11/04 computer)

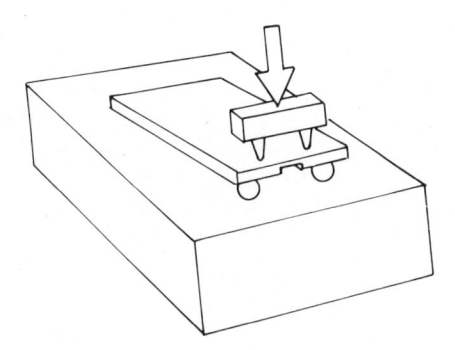

Figure 1. Schematic of Double Torsion Sample and Fixture.

alumina*. The data to determine this relationship was obtained using a constant load double torsion test. As a result of this published data, an alumina** material was chosen for this activity.

Specimens consisted of alumina plates (1.0"x0.125"x4.0"). A crack guide groove (0.075"x0.25"x4.0") was ground into the sample. The sample was notched and a crack induced into the sample by a thermal shock technique. The plate was then tested in point loading using a four point bending fixture at one end of the sample. Figure 1 shows a schematic of the double torsion sample and fixture. Figure 2 shows the double torsion apparatus located on the actuator of the testing machine.

DISCUSSION

Software Development and Testing

The first activity consisted of developing software to control the test parameters for the double torsion testing. The constant displacement method was used for this testing. Each cycle of the testing consisted of stressing (at a specified displacement rate) a sample containing a crack until the maximum load was determined. At that point the test was quickly terminated, allowing another cycle to be performed on the sample. Automation of this test required load and displacement data to be collected, analyzed and stored by the computer while stress was being applied to the sample. The data analysis during testing consisted of calculating the slope of the load versus displacement through a specified number of points (slope window). This data collection and analysis continued until the slope became zero. At that point, the test was automatically terminated, and the stress quickly relieved.

---
* AD85, alumina manufactured by Coors Porcelain Company
**AD85 alumina manufactured by Coors Porcelain Company

A test was considered successful if, the stress was relieved while the slope of the load versus displacement curve was approximately zero. The initiation of a response by the testing machine required the data acquisition rate and slope window to vary with ramp rate. Table I shows selected displacement rates, and the corresponding data collection rate and slope window for these displacement rates. The number of points in the slope window was approximately equal to the number of points collected for one second.

Table I

Test Parameters For
Automated Double Torsion Testing

Displacement Rate in/min$\times 10^{-1}$	Data Collection Rate Points/sec	Slope Window No. Points
0.02	3.5	3
0.1	17	17
0.6	102	102
1.0	170	171
4.0	682	682

Using the equipment, as previously described, it was found that this testing was limited to 8×10^{-4} in/min displacement rate, which resulted in crack velocities of 1×10^{-3} m/sec in the alumina specimens. Since lower crack velocities are desirable, alternate hardware with higher resolution was selected and installed. This equipment included a hardware segment generator (HSG), 1 gallon per minute servovalve and extensometer (.01 inch maximum range). The HSG was a digital to analog converter with increased sensitivity over the present digital to analog converter.

The software was modified to include the HSG and the testing was repeated as described above. Table II shows selected rates and the corresponding data collection rate and slope window. Thus far, testing has been conducted with ramp rates as low as 6×10^{-5} in/min. This ramp rate yielded a crack velocity of 7×10^{-5} m/sec in the alumina sample.

Table II

Test Parameters for
Automated Double Torsion Testing
Using a Hardware Segment Generator

Displacement Rate in/min×10^{-3}	Data Collection Rate Points/sec	Slope Window No. of Points
0.01	.02	3
0.06	.1	3
0.1	.2	3
0.4	.7	3
1.0	1.7	3

A crack velocity (V) was calculated for each successful test by considering the displacement rate, the applied load, and the appropriate specimen dimensions. The corresponding stress intensity factor (K) was also calculated using the appropriate specimen dimensions, the applied load and Poisson's ratio for the material. All post-data analysis was performed by the minicomputer.

DISCUSSION OF RESULTS

The crack velocity-stress intensity relationship generated from this series of tests is shown in Figure 3. All values were calculated from the preliminary data obtained using the automated double torsion test technique. This data was compared to the data reported by Freiman (shown in Figure 3).

The relationship between crack velocity and stress intensity is described by:

$$V = AK^n$$

where A and n are constants for a given material and test conditions. A material's susceptability to subcritical crack growth is inversely related to n. To determine n, the best fitting straight line of log V versus log K is derived using the least squares method. The slope of the straight line, n, is then determined.

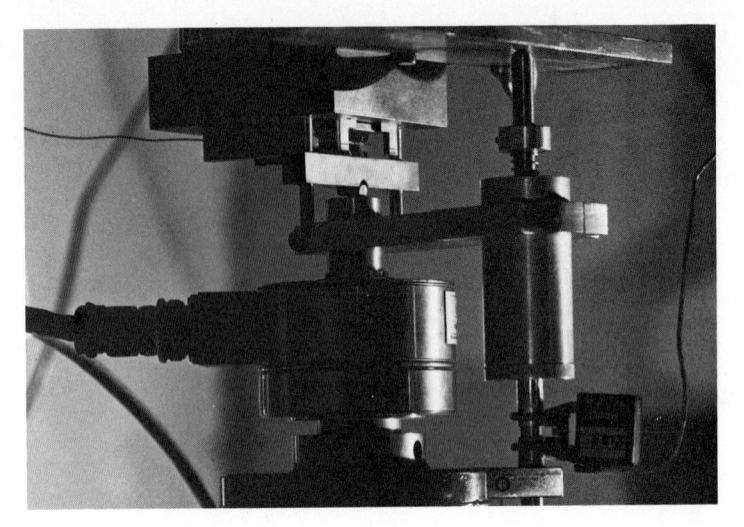

Figure 2. Double Torsion Apparatus.

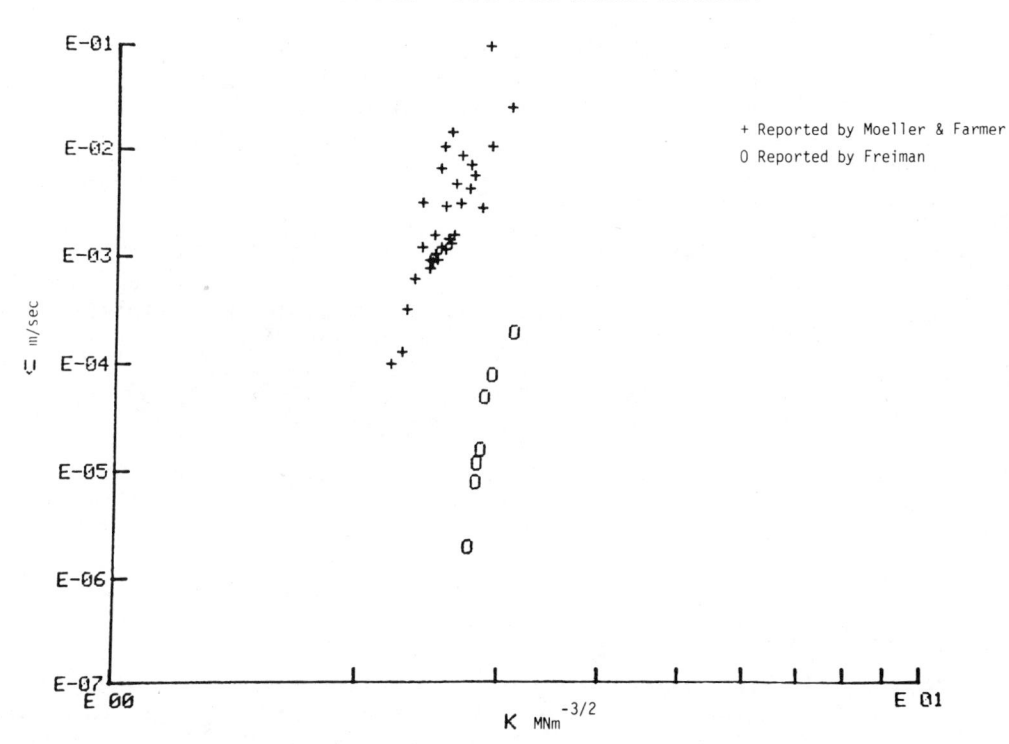

CRACK VELOCITY VERSUS STRESS INTENSITY FOR ALUMINA

+ Reported by Moeller & Farmer
O Reported by Freiman

$\dot{c}$ m/sec

K MNm$^{-3/2}$

"B&W D.T." COPYRIGHTED 1982 BY BABCOCK & WILCOX

Figure 3

Using corresponding values of crack velocities, n was determined for both sets of data. From the data reported by Freiman the slope was calculated to be 26 while the data generated in this study resulted in a calculated slope of 23. Based upon the agreement of the two slopes, it was concluded that the automated test technique developed in this study was valid. Therefore, this development of the automated test technique will be continued such that the current capabilities will be expanded to include data resulting from low crack velocities ($<10^{-5}$ m/sec).

Once the development work is complete, the automated technique will be used to characterize subcritical crack growth of materials such as silicon nitride and partially stabilized zirconia. It is not anticipated that the automated testing technique will be limited to room temperature testing but will be expanded to include elevated temperatures.

CONCLUSIONS

It has been shown that double torsion testing can be effectively automated. An automated testing system was used to control test parameters and post data analysis for double torsion testing. During the testing, the computer was continually analyzing load and displacement data by calculating the slope of the load versus displacement curve through a specified number of data points. When the computer detected the slope as being equal to zero, the test was automatically terminated. To successfully achieve the response from the testing machine at the appropriate time, the data collection rate and slope window varied with ramp rate. The data generated from this testing showed good agreement with published data.

REFERENCES

1. D. P. Williams and A. G. Evans, "A Simple Method for Studying Slow Crack Growth," Journal of Testing and Evaluation, Vol. 1, No. 4, July 1973.

2. S. W. Freiman, et al., "Slow Crack Growth in Polycrystalline Ceramics," Fracture Mechanics of Ceramics, Vol 2, Plenum Press, New York, 1974.

BIAXIAL COMPRESSION TESTING OF REFRACTORY CONCRETES

Albert H. Bremser* and Oral Buyukozturk**

*Babcock & Wilcox
 Lynchburg, Virginia
**Massachusetts Institute of Technology
 Cambridge, Massachusetts

INTRODUCTION

In the past, the design of large refractory lined process vessels, such as used in coal gasification, has been based on operating experience. During the 1970's, efforts were funded by the Department of Energy (DOE) to develop mathematical models that would predict the thermo-mechanical behavior of refractory concrete lined coal gasification process vessels.[1,2] The purpose of the mathematical modeling efforts was to provide a scientific basis for optimizing the design of refractory concrete lined process vessels. An optimum design is one where the tendency of the lining to crack is eliminated or minimized. The development of cracks in refractory linings is of concern since cracks allow hot and corrosive gases to reach the outer steel shell of the vessel.

During these modeling efforts, 2 and 3 dimensional models were generated which required multi-axial physical property data. Since little or no multi-axial physical property data existed for refractory concretes, data had to be estimated from uniaxial data. Thus, a joint program between the Massachusetts Institute of Technology (MIT) and Babcock & Wilcox (B&W) was initiated to generate biaxial compression data for two refractory concretes. These materials are currently used in process vessels and were previously used in an earlier study of gasifier lining design.[2]

The two types of refractory concretes tested were a dense 90 percent alumina castable, which is used for the working lining, and an insulating castable, which is located between the working lining and the shell.

CONSTRUCTION OF TEST EQUIPMENT

Ideally, in biaxial compression testing, specimen loading should be accomplished using four rams whose movements are synchronized such that the center of the sample remains stationary. This procedure minimizes end effects. However, such an approach was outside the scope of this property determination effort. Therefore, the approach used consisted of a system of three interconnected hydraulic cylinders. One cylinder is used to actuate the other two cylinders, which apply the load to the sample in the two orthogonal directions. The ratio of the orthogonal loads is set by the ratio of the effective areas of the two cylinders.

Since testing was to be accomplished at temperatures up to 2000°F, strain measurement devices could not be mounted directly on the sample. Also, flexure of the test frame had to be considered. Thus the approach was to cast pins in the sample and measure the relative displacement of the pins by the use of extension rods which extended outside the furnace. Since the tests were to be conducted in an isothermal state, the extension rods could be allowed to come to thermal equilibrium. Attachment of an extensometer directly to the extension rods eliminated effects of test frame deflection as the load was applied.

Figure 1 is a schematic of the overall test set-up which shows the frame, the hydraulic lines, the placement of the drive cylinder in the MTS frame, the hydraulic pump used to position the rams, and an auxiliary pressure gauge.

The load frame was constructed from an 8 inch by 8 inch I-beam having a web thickness of approximately 0.40 inches. The interior dimensions of the frame were 3 feet by 3 feet. The rams and hydraulic cylinders for applying the biaxial loading were located inside the frame, along with the furnace and strain measuring devices. The test frame was constructed such that the rams and cylinders were located 45 degrees to the horizontal. This was done so the system would be symmetrical with respect to the rams and especially the strain sensing devices.

The furnace used for heating the sample was a split box with an inside working area of 9 inches by 9 inches by 5 inches. Ceramic plate heaters, rated to 2200°F, covered the 9 by 9 inch faces of the furnace. The heaters were protected by perforated sheet metal.

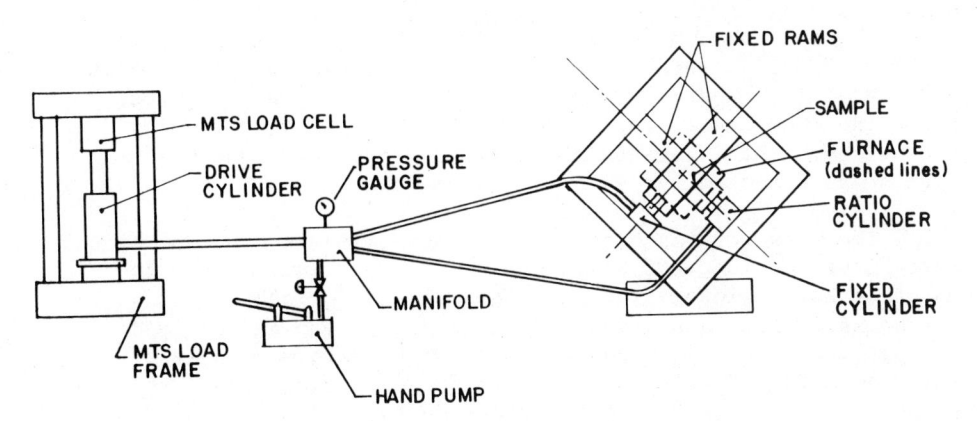

Figure 1. Schematic of Test System

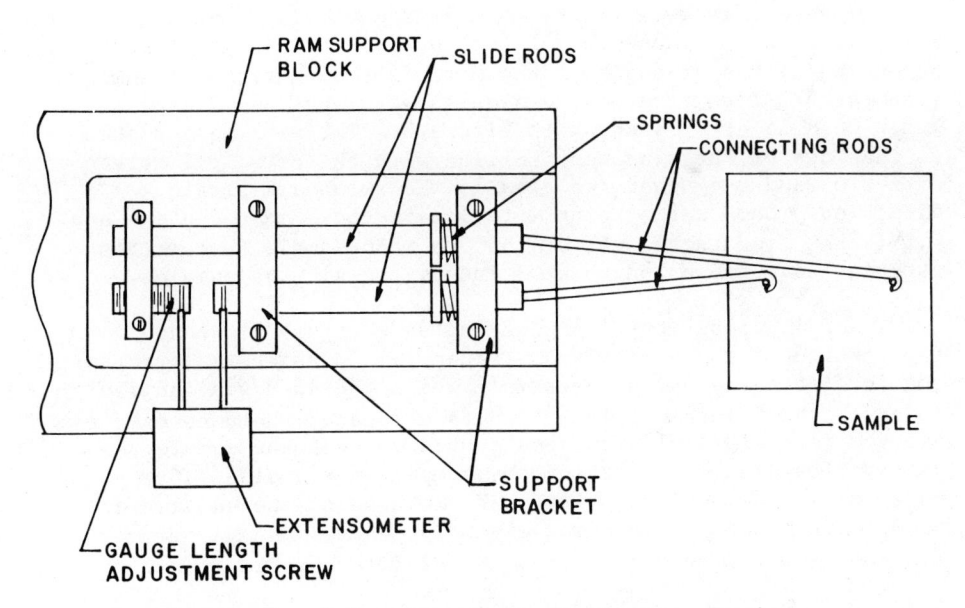

Figure 2. Schematic of Strain Sensing Device

The components of the strain measuring devices are shown in Figure 2. Each device consisted of two slide rods, made of centerless ground drill rod, which slide in reamed bronze brushings contained in two support brackets. The brackets are mounted in pockets which are milled in the steel ram support block. The slide rods are connected to the alumina pins, which are cast into the sample, by means of inconel connecting rods. The slide rods are spring loaded to insure a minimum of play in the system. The movements of the slide rods due to sample strain was detected by use of extensometers.

The rams were machined from high alumina brick. The brick were first cut longitudinally using a diamond cut-off saw to obtain a brick piece which was nominally 1-1/4 inches in thickness. Then these pieces were cut to length and the ends were ground with a diamond cup wheel on a surface grinder so that the ends were flat and parallel to each other and perpendicular to two adjacent reference faces of the ram. These reference faces were marked and were used for orientation purposes during ram installation and sample loading.

SAMPLE DESCRIPTION AND PREPARATION

The samples were in the form of square flat plates which were 5 inches x 5 inches and 1 inch or 3/4 inch in thickness, depending on the strength of the material being tested. Cast into the 5x5 inch face of the samples were four alumina pins which were located as shown in Figure 3. The pins were placed so that two of the pins were in line with the principal strain direction and the other two pins in the secondary strain direction. Each set of pins were located on 3 inch centers and 1.5 inches from the center of the 5x5 inch sample face. This established a gauge length of 3 inches for all the samples.

Acrylic plastic (Plexiglas) molds were used in the fabrication of the specimens. The acrylic plastic resulted in smooth flat surfaces on the sample where the load was applied. Holes in the 5 inch x 5 inch mold face located the pins cast in the samples. The mold was constructed so that the sample was cast on its side. By casting the sample on its side, parallelism, perpendicularity, and flatness of the sample faces to which the load was applied was maintained.

TEST PROCEDURE

Testing was conducted using an MTS 880.14 Automated Test System which is controlled with a Digital pdp 11/04 computer. During preliminary tests, it was found that performing the test

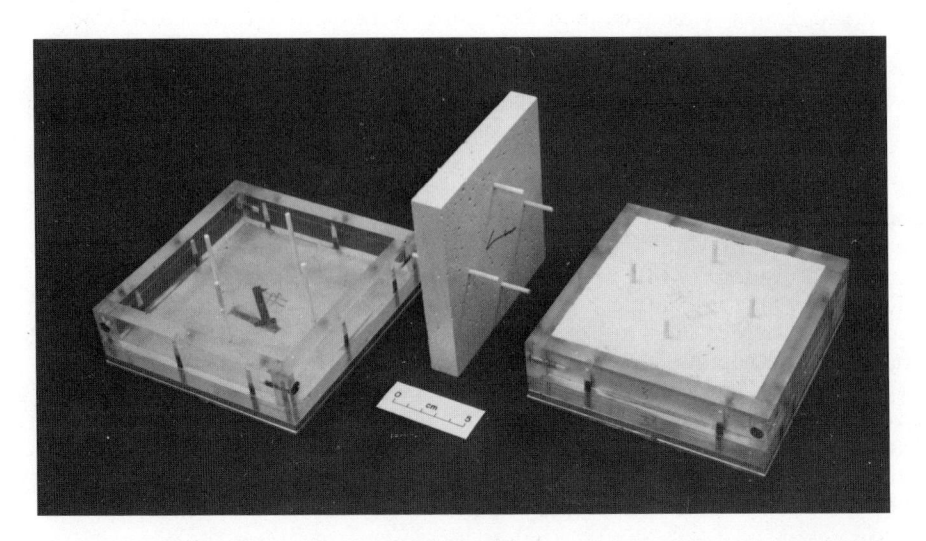

Figure 3. Specimen Mold and Test Specimens

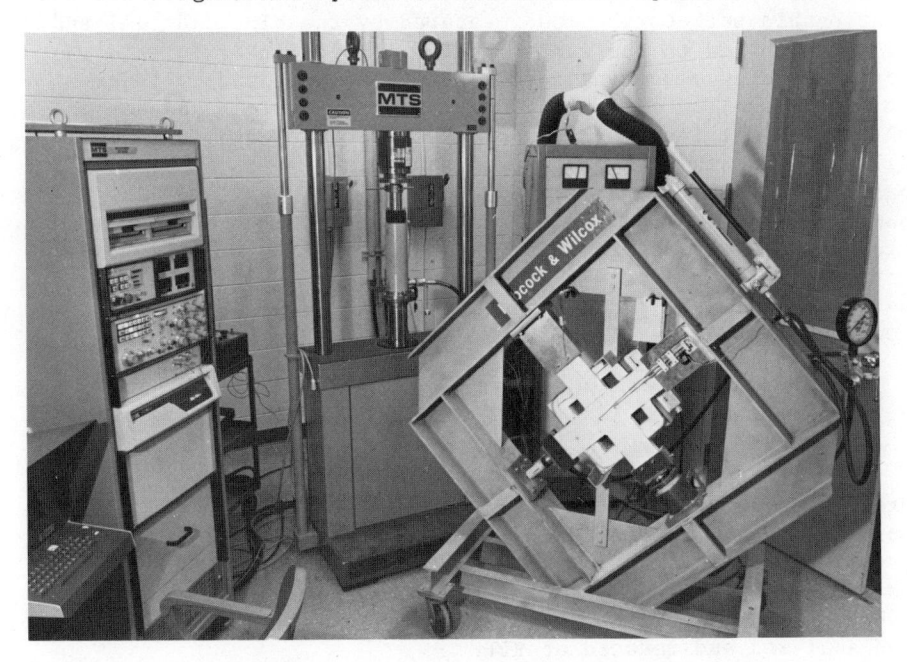

Figure 4. Overall Test Set-Up

where the strain measured in the sample is used to control the
loading was not possible. This was due to the "softness" of the
hydraulic lines which caused too long a response time to allow
the strain signal to be used as a control parameter. Thus the
MTS actuator stroke was used as the load control. This mode
provided a control somewhere between strain and load control.

A computer program was written to control the loading of
the sample, obtain the displacement and load data, calculate the
loading of the sample in the primary load direction, print out
the displacement and primary load data, and plot both the
load-displacement and stress-strain curves. The program had an
added feature that during the test, the data collection rate
could be changed once a predetermined load was applied to the
sample. The inputs to the program were sample identification,
sample dimensions, the MTS load cell to primary sample load
conversion, MTS actuator stroke movement (0.05 in/min for all
tests), and the MTS load value for rate of data acquisition
change. The computer program was also set-up so that the raw
data could be stored on a magnetic disc.

Load cylinder calibrations were performed to determine the
relationship between the load measured by the MTS load cell and
the load applied to the sample by the load cylinders. This was
accomplished by measuring force exerted by the load cylinder
(uniaxial case) or cylinders (biaxial case) through the use of a
20KIP load cell in the primary load direction and a 10KIP load
cell in the secondary load direction (15 and 50 percent biaxial
load conditions). In the case of 100 percent biaxial load
conditions the 20 KIP load cell was used for both cylinders.
Once the data was obtained, a linear regression analysis was
performed. In all cases the correlation was 0.999. The overall
test set-up used in performing the biaxial compression tests is
shown in Figure 4.

TEST RESULTS

Specimens of both dense and insulating refractory concrete
were tested at B&W under each of the four stress ratios
(uniaxial, 15.3% biaxial, 48.4% biaxial, and 100% biaxial), and
at the four prescribed temperature levels (RT, 1000°F, 1500°F,
and 2000°F for the dense material, and RT, 500°F, 1000°F, and
1500°F for the insulating material. The raw data obtained was
then analyzed and reduced at MIT.

The analysis of the data showed that the high temperature
ceramic paper, placed between the sample and the ram, was not
fully effective in eliminating frictional effects. In addition
the analysis indicated that some eccentricity of loading existed

which would lead erroneous valves of the initial Young's modulus. A correction procedure was established to account for both the frictional effects and the loading eccentricity.(3)

The results of the biaxial testing were as follows:

1. At a given biaxial condition the ultimate strength of the refractory concretes tested did not vary much with increases in temperature, but the peak strain increased considerably with temperature.

2. At all test temperatures the influence of the confining stress on the ultimate strength is significant, with the maximum strength occurring at about the 50 percent stress ratio.

3. The biaxiality effect was more prominent in the dense refractory concrete than in the insulating refractory concrete tested.

SUMMARY

Test equipment was designed and constructed to perform biaxial compression testing at temperatures up to 2000°F. A dense and an insulating refractory concrete were tested at various load ratios over a series of temperatures. Both types of refractory concretes exhibited a maximum in ultimate strength when the confining stress was approximately 50 percent of the primary stress.

ACKNOWLEDGEMENT

This work was performed on Subcontract No. 7862 between Massachusetts Insitute of Technology and Union Carbide Corporation issued under its principal contract (W-7405-ENG-26) with the U.S. Department of Energy.

REFERENCES

1. O. Buyukozturk and T. M. Tseng, Thermomechanical Behavior of Refractory Concrete Linings, J. Am. Ceram. Soc. 65:301(1982).

2. "Improvement of the Mechanical Reliability of Monolithic Refractory Linings for Coal Gasification Process Vessels," Compiled by R. A. Potter, Report No. LRC-5258, Babcock & Wilcox, Lynchburg, Va., Sept. 1981.

3. T. M. Tseng, "Thermomechanical Behavior of Refractory Concrete Lined Vessels," Ph.D. Thesis, Massachusetts Institute of Technology, Sept. 1982.

MECHANICAL TESTING OF GLASS HOLLOW MICROSPHERES

Paul W. Bratt, J.P. Cunnion and
Bruce D. Spivack

The PQ Corporation
Research and Development Center
Lafayette Hill, PA 19444

INTRODUCTION

Glass hollow microspheres are used in a number of applications that require their introduction into a matrix material through a variety of mixing operations. In order to survive this processing, the spheres must be able to withstand tremendous pressures. To characterize the strength of the microspheres as well as a comprehensive understanding of sphere mechanical properties, equipment was designed and constructed that could individually test spheres. By the use of Classical Buckling Theory for isostatic compression and by developing a theory for failure under uniaxial compression, sphere strength can accurately be determined.

EXPERIMENTAL

Interferometric Measurements

Wall thickness measurements were determined using a Leitz Linnik interference microscope. By placing the microsphere on a mirror that appears to be slightly non-parallel to the internal reference mirror, a parallel array of interference fringes is visible on the mirror and a series of concentric interference rings appear within the sphere. Due to the increased path length for light traveling through the microsphere, a reference fringe seen in the center of the sphere would also appear on the mirror (but displaced). The number of fringes between the center of the sphere and this reference fringe is a direct measure of sphere wall thickness. According to Weinstein[1]

$$t = \frac{x\lambda}{4(n'-1)}$$

where t is the average wall thickness, x is the number of fringes, λ is the wavelength of the light used and n' is the index of refraction (1.5 for glass). The sphere diameter is determined by scanning across the sample with a moving filar eyepiece that digitally displays position in tenths of microns.

Isostatic Compression

Buckling pressure measurements were made using an isostatic pressure chamber capable of pressures up to 40,000 psi. The test is performed by inserting a single microsphere into a specially designed holder that is placed into the pressure chamber and torqued into position. After aligning the holder, the sphere is illuminated through a 1" thick quartz window at the rear of the cell and observed with a stereomicroscope through the front window. The system is then pressurized hydraulically using a transparent mineral oil and a hand operated pump. This operation allows continual observation of the sphere during pressurization until failure occurs. The critical pressure is then recorded either directly on a dial gauge transducer or through a pressure transducer connected to a chart recorder.

Uniaxial Compression

Uniaxial compression measurements were made using a device similar to one developed at Los Alamos Scientific Laboratory[2] with the addition of a load independent, motorized actuator capable of constant speeds as slow as 1 micrometer/second. This positioner translates an X-Y-Z displacement stage supporting a Linear Voltage Differential Transducer (LVDT). The microsphere is placed on a flat tipped, conical rod and attached to the LVDT. Compression is achieved by advancing the stage against a stationary platen until failure occurs. The resulting force is output from the LVDT to a chart recorder and a force vs. time graph produced.

Computer code was written that allows the computer to digitize the graph directly by scanning with a fiber optic pen and store the coordinates in memory. A transformation is performed converting the coordinates into stress-strain data and then replotting the graph as stress vs. strain. A least squares linear regression of the initial straight line portion of the curve is performed and the Young's modulus of the sphere obtained. The sphere strength is then calculated as the stress on the sphere at failure.

DISCUSSION

Due to the number of parameters that influence the mechanical properties of a material, a single expression for "strength" cannot always be determined. Materials behave differently depending on their size and shape, the direction and magnitude of the force being applied, and the rate at which it is applied. Once these parameters are specified, quantitative measurements can be taken.

While in hollow sphere testing the general shape is defined, the wall thickness and diameter are not fixed. One of the most critical parameters in the hollow sphere structure is the uniformity of the sphere wall. An uncentered void cavity (a thin spot in the wall) will concentrate the stress on the sphere and cause premature failure. When observed in the interference microscope, these irregularities are easily seen as non-concentric, non-circular rings within the sphere. Only the most perfect spheres are used for further testing to avoid the pitfalls of these imperfections.

Under uniform isostatic compression, the stress (σ) on a hollow sphere is given by[3]:

$$\sigma = \frac{Pa}{4}$$

Where a is the aspect ratio (diameter/wall thickness) and P is the pressure per unit area. If the pressure is increased to some critical level, the spherical symmetry is perturbed and the sphere wall buckles catastrophically. This critical buckling pressure[3] (P) is defined as:

$$P = \frac{8E}{a^2 \sqrt{3(1-\nu^2)}}$$

where ν is Poisson's ratio (0.21 for glass) and E is Young's modulus. Combining these equations yields a true sphere strength

$$\sigma_{critical} = \frac{2E}{a\sqrt{3(1-\nu^2)}}$$

Under uniaxial compressive forces, the microsphere experiences a much different type of stress than in isostatic compression. Rather than a uniform force, the sphere "sees" a compressive force on the axis and a tensile force on the sphere surface at the equator. Unlike the case for solid or thick walled spheres[4], failure does not initiate from the point of contact of the force, but from the tensile force on the equator. When this force exceeds the tensile strength of the glass, fracture occurs and a circumferential crack forms around the equator splitting the sphere into two hemispherical caps.

Modeling this deformation mechanism similar to the ASTM standardized flexure test, the stress on the surface is

$$\sigma = \frac{3F(d/3)}{2(\pi d)t^2} = \frac{F}{2\pi t^2}$$

where F is the uniaxial force, d is the sphere diameter and t is the wall thickness. Under this mode of deformation the strain (ε) on the sphere is:

$$\varepsilon = \frac{6et}{(d/3)^2} = \frac{54et}{d^2}$$

where e is the deflection. Young's modulus can then be calculated from $E = \sigma/\varepsilon$.

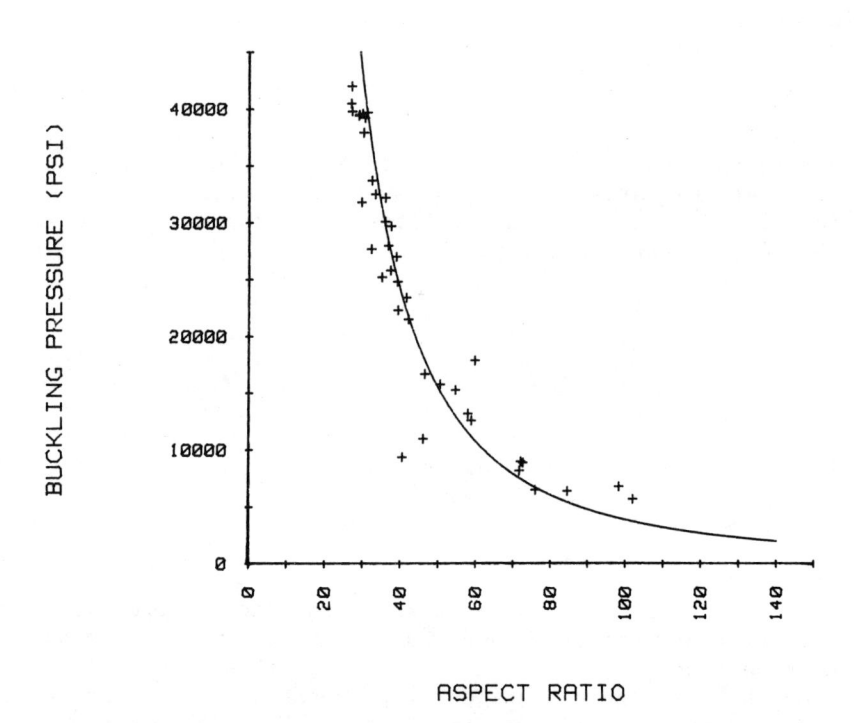

Fig. 1. Isostatic compression data for a typical hollow glass microsphere composition. The line is calculated from classical buckling theory using an average modulus of 8.2×10^6 psi.

RESULTS

Data generated in isostatic compression clearly show a good correlation with buckling pressure theory. Figure 1 shows a plot of buckling pressure vs. aspect ratio for a typical hollow glass sphere composition. The line drawn is a plot of the buckling equation using a calculated average modulus of 8.2×10^6 psi.

The average modulus, aspect ratio and buckling pressure for four different sphere compositions is shown in Table 1. Generally, it was found that the higher the modulus the better the sphere mechanical properties. However, in testing a variety of sphere compositions, an experimental geometrical influence on modulus was observed. If the average aspect ratio of the tested spheres is significantly higher, the apparent modulus is also higher. This effect can be seen in composition #4 and occurs even though the critical buckling pressure is much lower for these high aspect ratio spheres. This effect could be eliminated by specifying a standard geometry such as is done in ASTM mechanical testing. However, in cases where it is difficult or impossible to test only spheres of a particular average aspect ratio, a conversion factor could be used to normalize the modulii.

Uniaxial compression data also demonstrates a good correlation with the proposed theory. Unlike isostatic compression, the uniaxial force is strongly related to the thickness of the sphere wall. Figure 2 shows a plot of the critical breaking uniaxial force vs. the sphere wall thickness. The line drawn is based upon the wall flexure theory using the calculated average strength. Where an average of 24,000 psi is required to buckle a sphere under isostatic compression, only 8 grams of force is needed when the compression is applied uniaxially. Table 2 presents data obtained under uniaxial compression for the same compositions as in the

Table 1. Isostatic Compression Results

Composition	Average Modulus ($\times 10^6$)psi	Average Aspect Ratio	Average Pressure ($\times 10^4$)psi
1	8.2±2.0	46.2±19.9	2.4±1.2
2	7.6±1.4	45.5±10.4	1.8±0.7
3	4.0±1.2	46.5±19.5	1.3±0.9
4	11.3±4.1	134.2±29.1	0.4±0.2

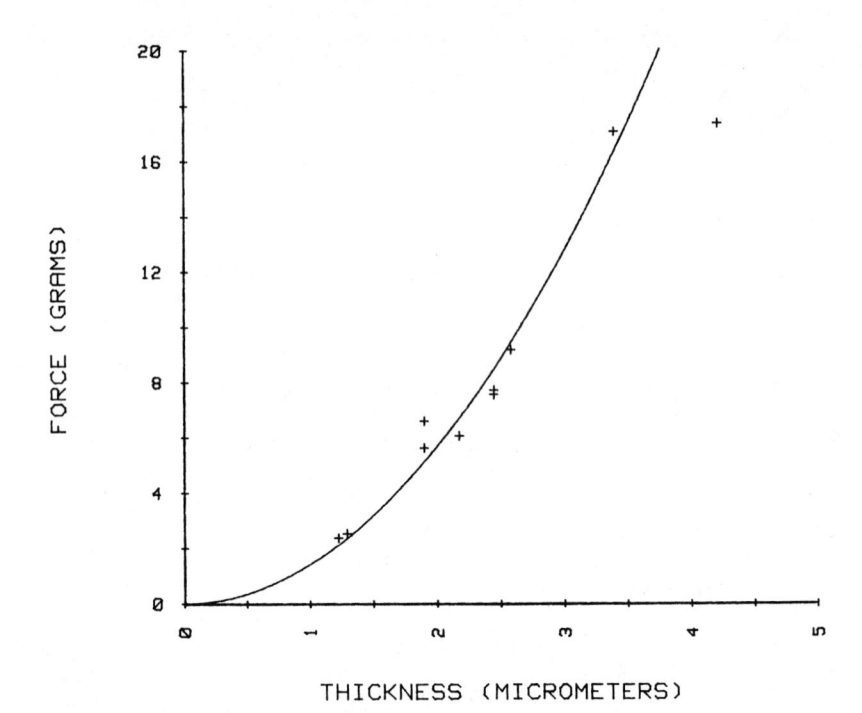

Fig. 2. Uniaxial Compression data showing the relationship between breaking force and wall thickness. The line is calculated from the wall flexure theory using 3.2×10^5 psi as the average strength.

Table 2. Uniaxial Compression Results

Composition	Average Strength $(\times 10^5)$psi	Average Thickness Microns	Average Force Grams
1	3.2±0.5	2.4±0.9	8.2±5.0
2	4.3±1.4	1.1±0.2	2.4±1.1
3	2.3±1.2	1.0±0.4	1.1±1.4
4	1.2±0.3	0.77±0.13	0.3±0.1

previous table. While composition 1 has a slightly higher average modulus under isostatic compression and a much higher uniaxial breaking force, the overall uniaxial strength is higher for composition 2 due to its much thinner walls.

CONCLUSIONS

Typically, a complex system of forces will be acting on a sphere during processing and in any end use application. While under isostatic compression, surface flaw propagation is inhibited allowing a good value of sphere modulus to be obtained. However, uniaxial compression appears to be more indicative of actual processing conditions and provides a better value of sphere strength. The combination of these tests should prove valuable in evaluating spheres in a variety of applications.

REFERENCES

1. B. W. Weinstein, White light interferometric measurement of the wall thickness of hollow glass microspheres, Journal of Applied Physics, 46:5305 (1975).
2. J. V. Milewski and R. G. Marsters, Tensile testing of glass microshells, J. Vac. Sci. Technol., 18:1279 (1981).
3. S. P. Tumoshenko and J. M. Gere, "Theory of Elastic Stability", McGraw-Hill, New York (1961).
4. A. H. Jones, R. A. Cutler, and S. R. Swanson, Lightweight proppants for deep gas well stimulation, DOE/BC/10038-19 (DE8200950) (1981).

THE RENAISSANCE OF X-RAY POWDER DIFFRACTION

Robert L. Snyder

New York State College of Ceramics
at Alfred University
Alfred, N.Y.

INTRODUCTION

Researchers have recognized the power of X-ray powder diffraction techniques since the first powder experiment in 1913. Early workers used powder diffraction to qualitatively distinguish phases and quantitatively analyze them[1]. However it was not until the 1938 paper of Hanawalt, Rinn and Frevel[2] that a general procedure for characterizing powders by diffraction began to develop. The principal impetus leading to the establishment of powder diffraction as one of the most important methods of materials characterization was the development of the powder diffractometer by William Parrish[3]. Applications of this technique expanded rapidly through to the 1960's. During the 1960's, however, a significant plateauing of developments occurred and has persisted to recent years.

Four reasons may be discerned as causing the decreased growth rate of powder diffraction techniques. These include the low average accuracy of the measurement of diffraction line positions and the generally accepted limit of phase detectability of one to two percent. The very large errors in diffraction line intensities have also been a principal hindrance. The last limitation of powder diffraction applications is a more subtle one involving the difficulty the noncrystallographer encounters in attempting to use the computer programs which embody some of the more powerful methods. Recent work has produced solutions to each of the above problems and has opened doors to a number of new and exciting applications of powder diffraction.

INACCURACY IN THE MEASUREMENT OF 2θ

The qualitative analysis of phases in an unknown depends both on the accuracy of the measurement of the unknown pattern and the accuracy of the reference patterns in the powder diffraction file (PDF). As the quality of both of these patterns increases the problem of pattern recognition becomes easier. The current PDF as collected and distributed by the JCPDS[4] contains about 38,000 patterns of highly variable quality. A recent study[5] isolated the approximately 3,000 cubic patterns from sets 1 to 24 of the PDF and from the known lattice parameters, computed the 2θ positions of each observed line. Twenty percent of the patterns had to be rejected from further analysis because they contained lines which could not be indexed within $0.5°$ 2θ Cu. The average $\Delta 2\theta$ ($|2\theta_{obs} - 2\theta_{calc}|$) for the 2,000 indexable cubic patterns is $0.091°$. However if only the patterns determined by the JCPDS sponsored associateship at the U.S. National Bureau of Standards are considered the average $\Delta 2\theta$ is $0.015°$; nearly an order of magnitude better than the average pattern extracted from the literature. A figure of merit, F_N[6], has been devised for rating the metric quality of powder patterns,

$$F_N = \frac{1}{|\overline{\Delta 2\theta}|} \frac{N_{obs}}{N_{poss}}$$

where N_{obs} is the number of observed lines and N_{poss} is the number of possible, space group allowed, lines in the range of N_{obs}. Figures 1 and 2 show the distribution of the F_N function for the indexable PDF cubic patterns and for the NBS cubic patterns. The principal reason for the dramatic difference between these two figures is that the NBS patterns have been corrected for instrumental aberations by applying an internal standard. Figure 3 shows the average F_N for each set of the PDF. Since a new set is released each year, Fig. 3 indicates that the average published cubic powder pattern was not significantly improving with time up through set 24, even though the NBS group has made no secret of how to obtain high quality patterns. It was this fact which gave the prime impetus to the development of digital automated procedures for collecting and treating powder diffraction data. The exacting and laborious procedures needed to produce high quality data can be performed in an insignificant amount of time by a well programmed computer.

COMPUTER AUTOMATED POWDER DIFFRACTION

The first automated powder diffractometer was developed by Rex[7] in the mid 1960's. However it took the microelectronics revolution of the 1970's to initiate the general techniques which have led to today's automated instrumentation. Although the principal thrust in the early 1970's was to develop the hardware interfaces needed to allow a computer to control a diffractometer, this work rapidly gave way to the much more serious problems of devising algorithms for the control of the instrument and the processing of the digital data[8].

RENAISSANCE OF X-RAY POWDER DIFFRACTION   451

QUALITY OF POWDER DIFFRACTION STANDARDS
PDF Cubic Data : AV. 2 THETA MERIT

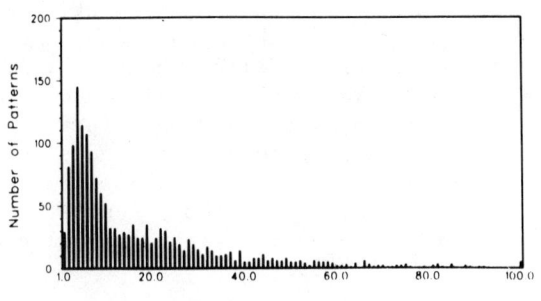

Figure 1. Distribution of the F_N merit function for the indexable cubic PDF pattern.

QUALITY OF POWDER DIFFRACTION STANDARDS
True NBS Cubic Data : AV. 2 THETA MERIT

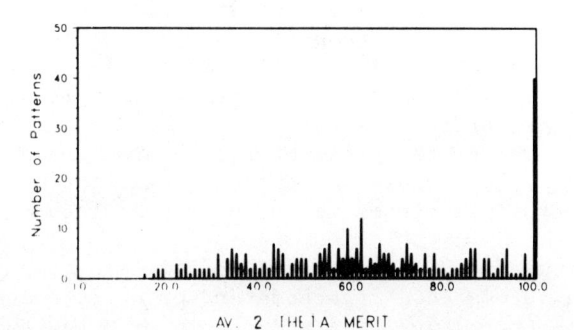

Figure 2. Distribution of the F_N merit function for the NBS cubic pattern.

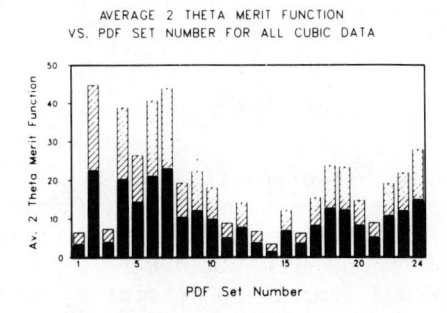

AVERAGE 2 THETA MERIT FUNCTION
VS. PDF SET NUMBER FOR ALL CUBIC DATA

Figure 3. Distribution of the average F_N merit function for the cubic patterns in each set of the PDF.

In retrospect the hardware aspects of automating a diffractometer look very simple although at that time they were formidable. There are four elements to the automation of a conventional diffractometer: 1. The replacement of the synchronous θ-2 motor with a stepping motor and its associated electronics; 2. The replacement of a conventional scaler/timer with one which can be remotely set and read; 3. The conversion of the various alarms, limit switches and shutter controls to computer readable signals and lastly; 4. The creation of a computer interface which will allow a computer to control items 1 through 3. These four items are easily obtained today by direct purchase of modules which often plug directly onto the lines of a modern minicomputer.

The major impact of computer automation on improving the accuracy of the measurement of diffraction angles has come from the algorithms which bring much more "intelligence" to the process than has been conventionally used in manual measurements. There are two areas here which need to be considered, the first is the algorithms which control the collection of data and the second those which reduce the data to d values and intensities. The first generation of control algorithms[8-10] were principally nonoptimizing, move and count methods. Recently a second generation algorithm has been developed[11-12] which brings full optimization to all aspects of data collection except for full pattern scans, to be used for qualitative phase analyses, which are still obtained by the nonoptimized move and count method. We should see the third generation of completely optimizing control algorithms in the very near future.

The techniques used to process the digitized step scan data produced by an automated diffractometer, or an automated film reading densitometer, have been and continue to be the principal area of research and development. The methods for locating peaks in a digitized pattern usually involve the following five steps: 1. Background determination, 2. Data Smoothing, 3. Spectral Stripping, 4. Peak Location and 5. Profile Fitting. A recent critical review of these procedures has been written[13] so we will confine ourselves here to a description of the Alfred University Data Reduction system[10] as a typical example of the approaches taken to the problem.

Background Determination

We perform the operation of differentiating peaks from background noise in two discrete steps[14]. The first is to linearize the pattern in order to remove the low angle curvature due to the divergence slit and the broad maximum resulting from amorphous scattering. The second step is to determine the threshold of statistically significant data. These procedures are illustrated in Figure 4. This pattern was obtained from a five milligram sample placed on a glass slide with a small depression etched in it to

Figure 4. Results of the linearization and threshold determination procedures. Upper trace is the raw data, lower trace is the linearized pattern and the lower smooth line is the determined threshold level. 3 mg sample on a glass slide.

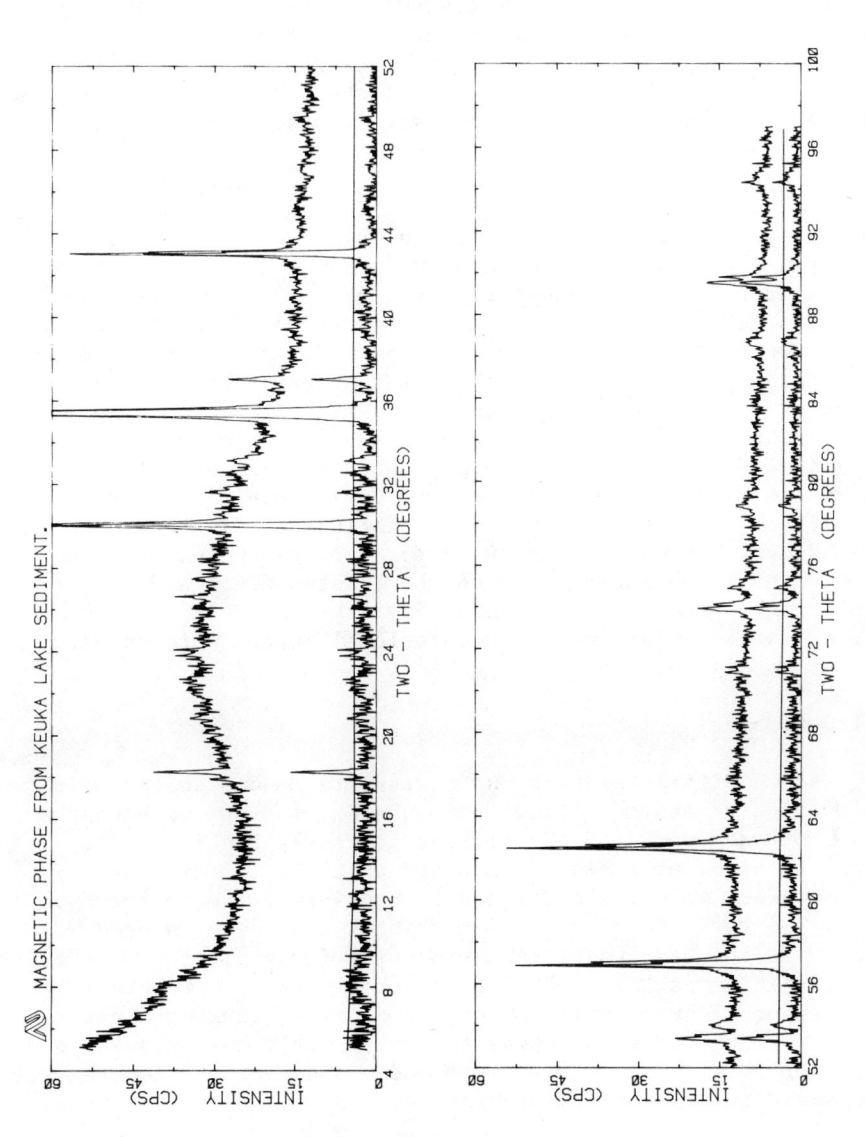

avoid sample displacement error. Due to the very small sample size the pattern was counted at 20 seconds per .04° step. The raw data shown in the upper curve exhibit both the effect of the amorphous sample holder and the increased low angle intensity due to the 1° divergence slit. The lower tracing shows the pattern after linearization. Note that all traces of the distortions have been removed. The smooth line on top of the background noise of the linearized pattern is the computer established threshold level. The peaks rising above this line are statistically significant.

Data Smoothing

Statistical fluctuations and the possible presence of noise spikes in the intensity measurements can lead to the detection of false peaks in the regions above the threshold. In order to avoid this a quadratic or cubic polynomial is fit to an odd number of raw data points using least squares regression. The point in the middle of the interval is replaced with the point computed from the interpolating polynomial. This process is illustrated in Figure 5. Subsequent smoothed data points are produced in a similar manner by selecting the odd number of raw data points to start at each successive raw data point in the peak or peak group. As this "digital filter" slides over the data, statistical fluctuations are greatly reduced. However, there is also a corresponding loss in peak resolution which increases with the 2θ step width and the number of points used in the filter. Savitzky and Golay[15] have produced an extremely efficient computational method for applying this filter to digital data.

Spectral Stripping

Most modern diffractometers use a graphite monochrometer which, due to its high mozaisity, allows both the $K_{\alpha 1}$ and $K_{\alpha 2}$ wavelengths through. For most work the $K_{\alpha 2}$ diffraction peaks can be readily recognized by a computer algorithm based on their location and height. However, when it is desired to completely remove the $K_{\alpha 2}$ peaks from the raw data a modified Rachinger procedure is used[16]. Figure 6 shows the results of this procedure for a portion of the quartz diffraction pattern. This procedure is quite effective. However, it does leave a small amount of Fourier "ringing" on the high angle tail of the high intensity peaks. This may be seen in the peak near 60° in Figure 6. Platbrood[17], has further generalized this procedure to apply to any radiation.

Peak Location

A second derivative method is used to locate peaks because the first derivative is insensitive to shoulders which indicate overlap. The Savitsky Golay polynomial smoothing procedure automatically produces a curve whose second derivative may be evaluated. The

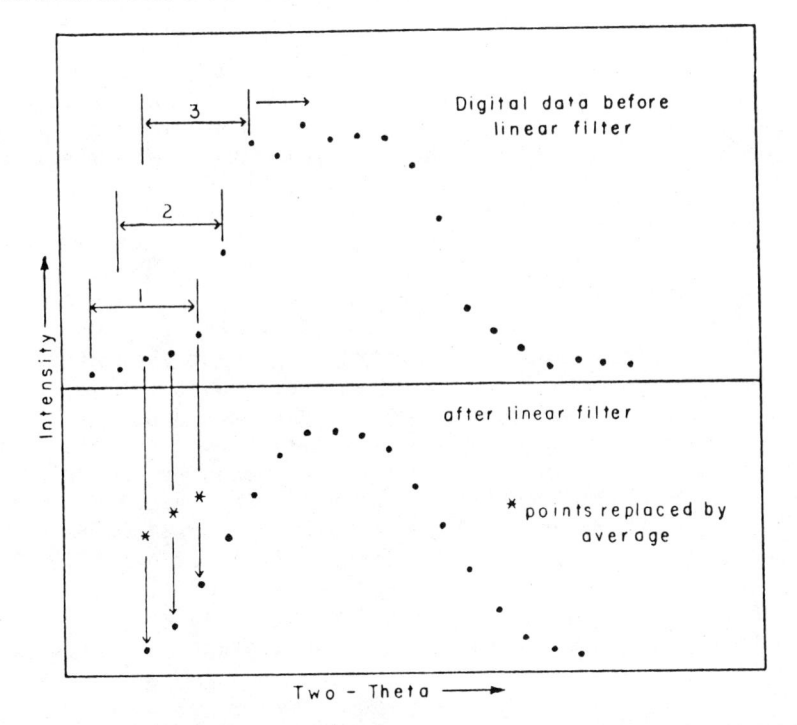

Figure 5. Effect of a linear digital filter on data with random noise.

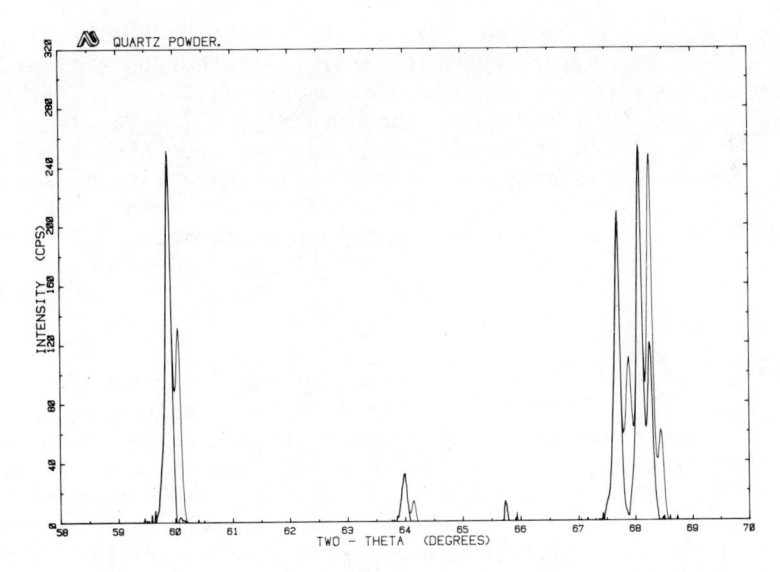

Figure 6. Results of the $K_{\alpha 2}$ elimination procedure. Outer tracing is the raw data. Inner curve is data after $K_{\alpha 2}$ elimination.

minima in the second derivative indicate peaks in the raw data. The
number of peaks found in a procedure will depend on the amount of
data smoothing and the signal to noise ratio, which is a simple
function of the count time. The number of false peaks found can be
minimized by correlating the smoothing parameters with the count
time.

The accuracy of the peak locations will depend in some degree on
each of the above four steps. However, to achieve absolute accuracy
it is essential that the data be corrected for the aberations
introduced by the instrumental measurement technique. One of the
most serious errors in diffractometry is the displacement of the
sample from the focusing circle. Displacements in the micron range
can cause peak shifts of hundreths of a degree. Since the X-ray
beam penetrates into the sample, there will always be some effective
sample displacement. This can only be corrected by mixing an
internal standard with accurately known lattice parameters, into an
unknown and then correcting all observed peak locations based on the
position of the internal standard lines. It is principally the use
of the internal standard calibration technique which has caused the
data produced by the National Bureau of Standards group to be a full
order of magnitude more accurate than the average published powder
pattern.

The use of computer automated procedures can affect accuracy in
two ways: the first is to remove the computational drudgery from
making internal standard corrections and the second is to allow the
routine application of an external standard calibration curve[10,13].
Our studies have shown that approximately half of the error in peak
locations is due to instrumental error. All instrumental error can
be removed by running a series of known standards and plotting $\Delta 2\theta$
(i.e., $2\theta_{obs} - 2\theta_{calc}$) vs. 2θ as shown in Figure 7. The least squares
fitting of a polynomial to this curve allows the polynomial
coefficients to be stored on a disk file. These may be used to
correct all patterns for instrumentally induced error. However, it
should be emphasized that the error in peak location can be reduced
by another factor of two, if an internal standard is used.

Profile Fitting

The last aspect of computer reduction of digitized data has by
far produced the most dramatic impact on the precision of peak
location. The fitting of analytical profiles to the observed data
using some type of optimization procedure (like Simplex or least
squares) has produced average $\Delta 2\theta$ values from $0.001°$[18] to $0.0001°$[19].
This subject of very active research has recently been reviewed[20].
The technique is illustrated in Figure 8 where a modified Lorentzian
function has been placed at each point where the second derivative
indicated the presence of a peak. The difference between the
observed data and the computed profile is then minimized using an

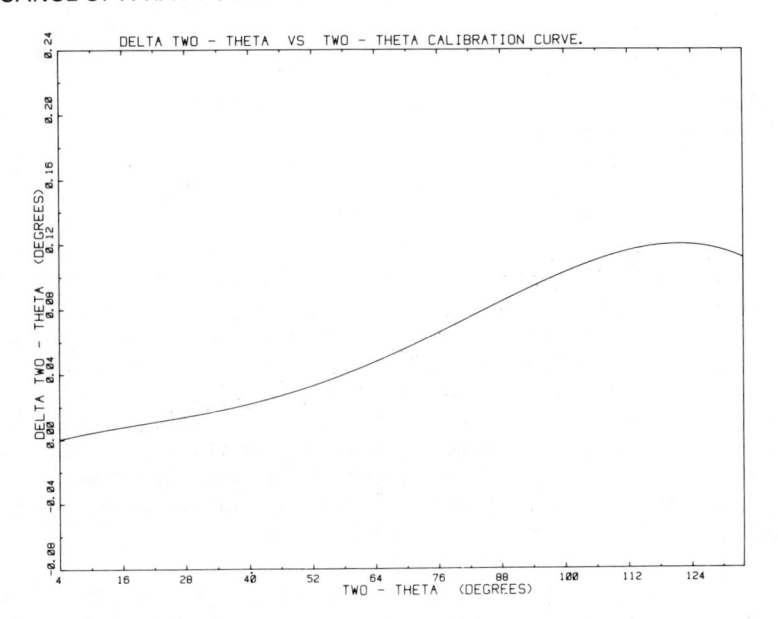

Figure 7. Calibration curve. Dots are the Δ2θ values for the standard materials. Solid curve is the least squares polynomial.

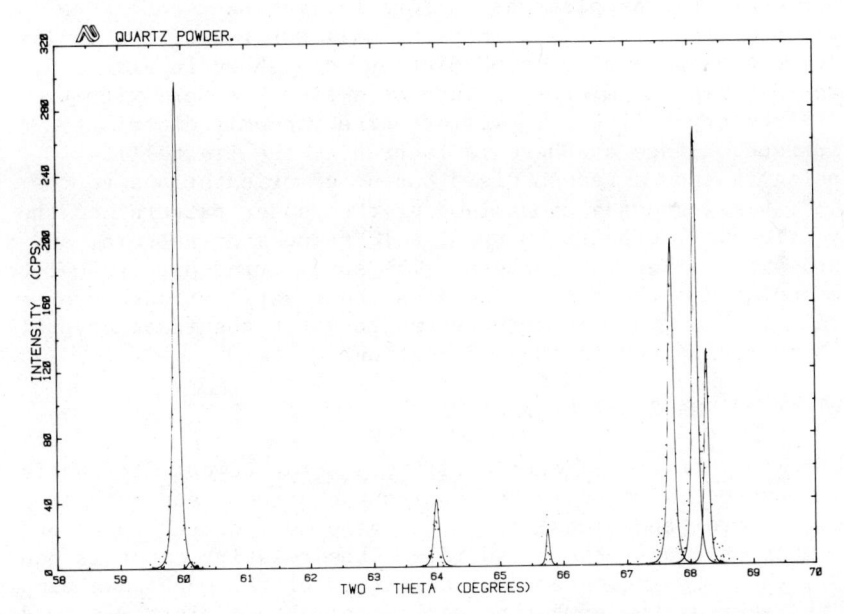

Figure 8. Linearized, threshold removed and $K_{\alpha 2}$ stripped quartz powder pattern (dots). Smooth curves are non-refined Lorentzian's located at the positions indicated by the peak finding procedure.

optimization technique. Profile fitting not only produces extremely high precision for peak locations, but it also dramatically improves their standard deviations, in that a large number of observations go into the location of each peak.

THE ACCURACY OF DIFFRACTION INTENSITIES

The second traditional limitation of X-ray powder diffraction is the large uncertainty in observed intensity values due primarily to the preferred orientation of crystallites in powders. This problem has strongly limited the application of powder diffraction techniques to qualitative, quantitative and structural analysis. This problem can be seen in the very first published powder patterns and remained in need of a general solution until 1979. The spray drying of powders[21-22] has been shown to remove preferred orientation and produce intensity values which are accurate to a few percent.

Spray drying is a technique in which a powder is suspended in a non-dissolving liquid. A small amount of an amorphous binder like polyvinyl alcohol is added to the suspension along with a defloc-culant to keep the particles in a fifty weight percent suspension. The slurry is then atomized through a spray nozzle into a heated chamber where the droplets dry before falling back onto a collector tray. The dried droplets which are small spheres with a typical average agglomerate size of 50 microns, are shown in Figure 9. Figure 10 shows the microstructure of a single sphere with the needle-like crystallites of wollastonite randomly distributed over its surface. Since a sphere is isotropic, the crystallites composing it show a random distribution of orientations to the X-ray beam. Table 1 shows a comparison of the powder patterns of the spray dried wollastonite to the ideal, calculated pattern, and to the non-spray dried PDF pattern. The spray dried pattern matches the calculated pattern as closely as the crystal structure of a real mineral of this type can be expected to match the ideal crystal structure model used in the calculation.

THE LIMIT OF PHASE DETECTABILIITY

The limit of phase detectability by X-ray powder diffraction is usually stated as being in the one to two percent concentration range. However the signal to noise ratio is a direct function of the number of counts collected (i.e., the relative error is equal to $\sqrt{I}/I$). Thus, to increase the sensitivity of the method we simply need to increase the number of counts collected. This may be done in three ways. The first is to increase the count time an automated instrument spends at each angular increment in a pattern. This is illustrated in Figure 11 which shows the powder pattern of a hot pressed sample of elemental Si counted for 0.1, 1.0 and 10.0 seconds

RENAISSANCE OF X-RAY POWDER DIFFRACTION

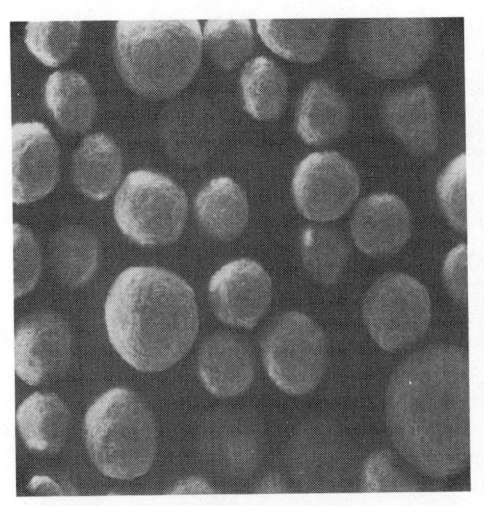

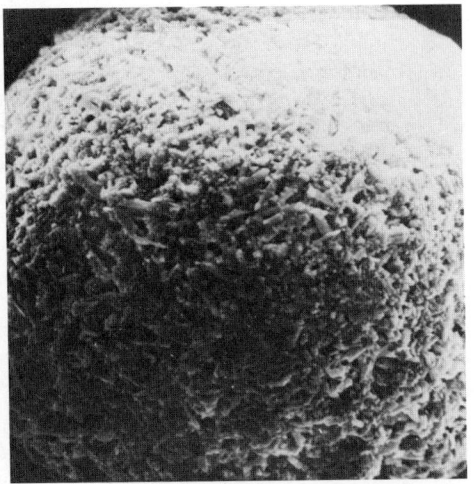

Figure 9. Spray dried Wollastonite agglomerates (200X).

Figure 10. Closeup of a single Wollastonite agglomerate (600X).

Table I. Powder Patterns for Wollastonite ($CaSiO_3$). Non-spray Dried (NSD), Calculated (CAL) and Spray Dried (SD).

d(Å)	H	K		NSD	CAL	SD	d(Å)	H	K		NSD	CAL	SD
7.687	1	0	0	25	13	9	2.928	-2	-1	1	0	8	5
7.037	0	0	1	0	1	0	2.801	2	2	1	5	4	5
5.459	-1	0	1	8	1	0	2.745	2	1	1	0	1	0
4.958	1	0	1	0	4	3	2.730	-2	0	2	0	6	3
4.704	1	1	0	0	2	0	2.722	1	1	2	30	4	5
4.679	-1	1	1	0	3	0	2.687	1	2	1	0	2	0
4.429	1	-1	1	0	2	0	2.613	-3	1	0	0	1	0
4.052	-1	-1	1	0	8	5	2.531	0	-2	2	0	3	5
3.843	2	0	0	85	22	19	2.477	1	-2	2	25	14	16
3.770	-2	1	0	0	5	0	2.381	3	-1	1	0	1	0
3.519	-2	0	1	75	29	29	2.356	-3	2	0	0	7	0
3.443	-2	1	1	0	7	5	2.352	2	2	0	0	9	0
3.323	-1	0	2	100	35	27	2.346	0	0	3	0	9	13
3.244	2	0	1	13	2	0	2.339	-2	2	2	30	7	0
3.206	-1	2	1	0	3	0	2.306	-1	0	3	50	16	12
3.183	0	-1	2	0	6	3							
3.107	-1	1	2	0	11	0							
2.981	-2	2	0	30	100	100							

Disagreement index $R = \Sigma |I_{obs} - I_{CAL}| I_{CAL} / \Sigma I^2_{CAL}$:

$R_{NSD} = 100.17\%$ $R_{SD} = 7.37\%$

per point. These patterns were passed through the linearization and threshold finding procedures previously described. The increase in the number of peaks breaking through the threshold line with increasing count time is due to the decreasing relative error. All of the peaks other than the three lines due to Si are below 1% relative intensity. These peaks have been identified as being due to minor impurity concentrations of SiC, Si_3N_4 and SiO_2. It should be noted that a scan with .03° steps and .1 sec. count times would have easily allowed the identification of the principal phase in less than four minutes!

Another way to increase the number of counts is to make use of the relatively new position sensitive detectors. These detectors, in scanning mode, greatly enhance the observed counting rate. Gobel[23] has used this type of detector to measure accurate diffraction patterns at rates from 60 to 200°/min. He has also applied them to the study of short time duration phenomena. This detector may enable routine diffraction analysis into the parts per million concentration range. We have recently shown[24] that one part per million of SiO_2 in Al_2O_3 could be observed at a >3σ significance by counting for 3,000 seconds per point with a conventional detector.

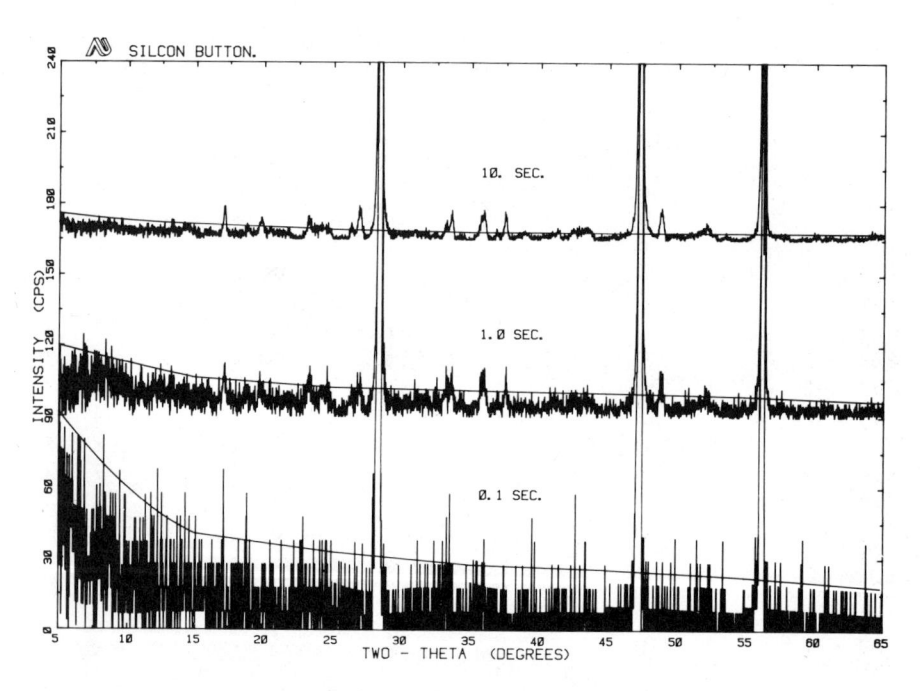

Figure 11. The powder pattern of an impure sample of silicon taken at 0.1 sec. count times per point (lower trace). 1.0 sec. (middle trace) and 10.0 sec. (top trace). Smooth line indicates the computer determined threshold level.

The second way to improve the signal to noise ratio in a powder pattern is to increase the intensity of the incident X-ray beam. This can be done by using a rotating anode generator or by using the new synchrotron X-ray souces. We should see new developments in this area in the near future as the new National Synchrotron Light Source at Brookhaven National Laboratory comes on line.

IMPROVED COMPUTER PROCEDURES

At the moment some of the most powerful research techniques involving X-ray powder diffraction are embodied in a growing number of computer programs. These include methods for calculating powder patterns from crystal structure information, indexing even low symmetry patterns, refining lattice parameters by least squares, computing figures of merit, evaluating the quality of powder patterns, carrying out crystallite size, stress and orientation analyses and automated phase identification. Two principal difficulties hinder the wide spread use of these techniques. The first is that the program often requires a rather high degree of computer literacy. The second is the requirement that the user be conversant with the somewhat complex notation of X-ray crystallography. The development of computer procedures which are more user friendly and oriented toward the nonexpert is extremely time consuming. However the growth of companies which manufacture and market automated diffractometers has stimulated progress in this area. We have already begun to see much more friendly software developed around these commercial systems. The use of the computer to translate the powerful tools available into procedures usable by the noncrystallographer and noncomputer expert is currently underway. The near future should bring these tools to a much wider user group.

APPLICATIONS OF COMPUTER AUTOMATED X-RAY DIFFRACTION

The introduction of computer control and analysis of X-ray powder diffraction, along with developments like spray drying, position sensitive detectors and synchrotron sources has eliminated each of the reasons that caused the plateauing of applications of this method in the 1960's. Today we are experiencing a rebirth in powder diffraction techniques which promise to lead us into new realms of materials characterization. Before ending this review of the causes of this renaissance I will survey a few of the most recent and exciting applications of powder diffraction techniques.

Automated Phase Identification

The first generation of computer search/match algorithms were forced to assume very wide error windows on both the d's and I's of a powder pattern. The new levels of accuracy obtainable with automated analysis of spray dried powders has permitted the development of a second generation algorithm[25] which assumes very accurate data. The general success of this approach is only hindered by the poor quality of some of the existing reference patterns. However the JCPDS has undertaken a significant effort to obtain highly accurate references for the most common phases. In addition the journals have adopted a set of standards[26] which should improve the quality of future published data. As the quality of the PDF improves we can expect to see rapid, routine, reliable computer identification of unknowns.

Quantitative Analysis

The effect of spray drying[22] and computer optimization of the data collection[11-12] has allowed the development of a sophisticated system for quantitative analysis. The NBS*QUANT82[27-28] programs carry out data analysis far more complex than manual methods would permit. They allow, for example, routine analysis using overlapped lines and the application of chemical constraints. The very high precisions obtained by the use of the AUTO control algorithm on spray dried samples and the use of NBS*QUANT82 to process the data were instrumental in recently exposing the seriousness of the effects of crystallite size and absorption coefficient on the accuracy of quantitative analysis[29]. Effects like these have previously been obscured by the much larger effects of preferred orientation and statistical errors in manual data.

Micro Phase Analysis

As previously described the improved signal to noise ratio now obtainable has allowed analysis of phases in the one part per million concentration range. We should soon see routine analysis work at extremely low concentrations even allowing analysis of crystalline grain boundry phases.

Analysis of Rapid Events

The use of position sensitive detectors has permitted the collection of entire powder patterns in times on the order of one minute. Gobel[23] has followed the kinetics of a rapid phase transition by multiple scans of a limited 2θ range. The extension of powder diffraction analysis to phenomena which occur in the few second time regime should produce a wealth of new information.

Crystal Structure Analysis

Every one of the developments reviewed in this paper has a bearing on this topic. The increasing levels of accuracy in both 2θ and intensity, the new developments in experimental procedures and the progress currently being made[30] on Reitveld refinement techniques all lead to the most exciting potential application of all, routine structure analysis from powders. Since one of the most serious impediments in the development of materials science is our inability to readily characterize the structures of real materials, which may never be obtained as single crystals, this application holds the promise of having the most far-reaching effects on both science and technology.

REFERENCES

1. L. Navias, "Quantitative Determination of the Development of Mullite in Fired Clays by an X-ray Method," J. Am. Cer. Soc. 8, 296 (1925).
2. J. D. Hanawalt, H. Rinn, L. K. Frevel, "Chemical Analysis by X-ray Diffraction," Ind. Eng. Chem. Ed., 10, 457 (1938).
3. W. Parrish and S. G. Gordon, "Precise Angular Control of Quartz Cutting by X-rays," Am. Mineral., 30, 326 (1945).
4. JCPDS - International Centre for Diffraction Data, 1601 Park Lane, Swarthmore, Pennsylvania 19081.
5. R. L. Snyder, Q. C. Johnson, E. Kahara, G. S. Smith and M. C. Nichols, "An Analysis of the Powder Diffraction File," Lawrence Livermore Laboratory (UCRL-52505), 61 pages (June 1978).
6. G. S. Smith and R. L. Snyder, "F_N: A Criterion for Rating Powder Diffraction Patterns and Evaluating the Reliability of Powder Pattern Indexing," J. Appl. Cryst., 12, 60-65 (1979).
7. R. W. Rex, "Numerical Control X-ray Powder Diffractometry," Adv. X-ray Anal., 10, 366-73 (1966).
8. C. L. Mallory and R. L. Snyder, "Evaluation of Powder Diffraction Peak Finding Algorithms," Adv. X-ray Anal., 22, 121-131 (1979).
9. R. P. Goehner and W. T. Hatfield, "A Microcomputer Controlled Diffractometer," Adv. X-ray Anal., 22, 165-167 (1979).
10. C. L. Mallory and R. L. Snyder, "The Alfred University X-ray Powder Diffraction Automation System" N.Y.S. College of Ceramics Technical Publication No. 144, 172 pages (1979).
11. R. L. Snyder, C. R. Hubbard and N. C. Panagiotopoulos, "A Second Generation Automated Powder Diffractometer Control System," Adv. X-ray Anal. 25, 245-260 (1982).
12. R. L. Snyder, C. R. Hubbard and N. C. Panagiotopoulos, "Auto: A Real Time Diffractometer Control System," National Bureau of Standards Publication NBSIR 81-2229, 102 pages (1981).
13. R. L. Snyder, "Accuracy in Angle and Intensity Measurements in X-ray Powder Diffraction," Adv. X-ray Anal., (in press 1983).
14. C. L. Mallory and R. L. Snyder, "Threshold Level Determination from Digital X-ray Powder Diffraction Patterns," in Accuracy in Powder Diffraction, National Bureau of Standards Special Publication 567, p. 93 (1980).

15. A. Savitzky and M. Golay, "Smoothing and Differentiation of Data by Simplified Least Squares Procedures," Anal. Chem, 36, [8] 1627-39 (1964).
16. J. Ladell, A. Zagofsky, and S. Pearlman, "CuK$_{\alpha2}$ Elimination Algorithm," J. Appl. Crystallogr., 8, 499-506 (1970).
17. G. Platbrood, J. M. Quitin and H. Barten, "Application of the Modified Snyder's Program for the Data Processing of An Automated X-ray Powder Diffractometer," Adv. X-ray Anal. 25, 261-272 (1982).
18. A. Brown and J. W. Edmonds, "The Fitting of Powder Diffraction Profiles to An Analytical Expression and the Influence of Line Broadening Factors," Adv. X-ray Anal., 23, 361-374 (1980).
19. W. Parrish and T. C. Huang, "The Accuracy of the Profile Fitting Method for X-ray Polycrystalline Diffraction," NBS Special Publication 567, "Accuracy in Powder Diffraction," 95-110 (1980).
20. S. A. Howard and R. L. Snyder, "An Evaluation of Some Profile Models and the Optimization Procedures used in Profile Fitting," Adv. X-ray Anal. (in press 1983).
21. S. T. Smith, R. L. Snyder and W. E. Brownell, "Minimization of Preferred Orientation in Powders by Spray Drying," Adv. X-ray Anal., 22, 77-88 (1979).
22. S. T. Smith, R. L. Snyder and W. E. Brownell, "Quantitative Phase Analysis of Devonian Shales by Computer Controlled X-ray Diffraction of Spray Dried Powders," Adv. X-ray Anal., 22, 181-192 (1979).
23. H. E. Gobel, "The Use and Accuracy of Continuously Scanning Position Sensitive Detector Data in X-ray Powder Diffraction," Adv. X-ray Anal., 24, 123-138 (1981).
24. M. Bliss, "The Application of X-ray Powder Diffraction to Micro-phase Analysis," N.Y.S. College of Ceramics, Senior Thesis (1981).
25. R. L. Snyder, "A Hanawalt Type Phase Identification Procedure for a Minicomputer," Adv. X-ray Anal. 24, 83-90 (1981).
26. L. D. Calvert, et al., "Standards for the Publication of Powder Patterns: The American Crystallographic Association Subcommittee Final Report," Accuracy in Powder Diffraction, NBS Special Publication 567, p. 513-536 (1980).
27. R. L. Snyder and C. R. Hubbard, "NBS*QUANT82: A System for Quantitative Analysis by Automated X-ray Powder Diffraction," NBS Special Publication, (in press 1982).
28. C. R. Hubbard, C. R. Robbins and R. L. Snyder, "XRD Quantitative Analysis Using the NBS*QUANT82 System" Adv. X-ray Anal. (in press 1983).
29. J. Cline and R. L. Snyder, "The Dramatic Effect of Crystallite Size on X-ray Intensities," Adv. X-ray Anal., (in press 1983).
30. Young, R. A. and Wiles, D.B., "Application of the Reitveld Method for Structure Refinement with Powder Diffraction Data," Adv. X-ray Anal., 24, 1-23 (1981).

CHARACTERIZATION OF IMPERFECTIONS IN PLASMA-SPRAYED TITANIA

C.C. Berndt, R. Korlipara, R.A. Zatorski and H. Herman,
A. Jonca, T. Templeton and R.K. MacCrone

Dept. of Materials Science & Engineering, SUNY at Stony Brook, N.Y. 11794; Dept. of Physics and Dept. of Materials Science, R.P.I., Troy, N.Y. 12180

INTRODUCTION

Plasma spraying is widely used for the fabrication of protective coatings. There are a number of features that make the process of great utility:

1. The coatings may be applied under a variety of conditions,
2. A large number of different materials may be sprayed,
3. A variety of substrates may be used.

It can be appreciated that since the particles are subjected to a very high temperature followed by rapid solidification at the substrate, various metastable phases may form. In addition, the atmosphere and environment in the plasma may cause severe deviations from stoichiometry. Finally, there exists the possibility that unusual reactions at the substrate surface may occur.

Plasma-sprayed oxides are frequently employed as protective coatings in a wide range of applications. In use, the coatings commonly undergo thermal cycling coupled with mechanical deformation. Survivability of the coating-substrate system will depend on the degree of phase stability of the coating and the mechanical and chemical compatibility between coating and substrate. Adhesion will of course be central to any consideration of utility, and characterization of the plasma-sprayed oxide will commonly include tensile adhesion strength measurements[1].

When the oxides are plasma sprayed they undergo rapid melting followed by ultra rapid solidification at the substrate. Of course, the first layer to solidify will be "liquid quenched" at rates of the order of $10^6 C/sec^2$. Subsequent layers will be quenched at somewhat slower rates due to the thermal insulative properties of the previously deposited oxide. Nevertheless, the coating having undergone rapid quenching will contain a high fraction of metastable phases. This has been observed for virtually all plasma-sprayed oxides which have been subject to phase analysis[2]. It is the high degree of metastability of the oxide phase which can, under thermal-mechanical cycling, give rise to phase transformations with attendant crack initiation from volume changes. Such a situation can be deleterious to the coating system performance.

Titania is commonly sprayed for wear applications. The as sprayed coating accepts further finishing to become mirror-like, having good friction and wear properties. Titania and titania-alumina alloys in composite, in calcined, and in fused powder form have been studied extensively from the engineering point of view. Some fundamental studies have been carried out as well[3]. Transmission electron microscopy, for example, has indicated the presence of Magneli phases, due, no doubt, to the occurrence of low oxygen partial pressures within the plasma flame as well as the rapid quenching[3].

Perhaps because of the obvious complexities of the process, and the difficulties in controlling the preparative and spraying variables, there has been limited fundamental research into the structure and properties of plasma-sprayed materials, particularly oxides. Therefore, despite the hazards of investigating so uncontrolled a situation, we have undertaken a magnetic and morphological study of plasma sprayed TiO_2.

The coatings were studied by examining the EPR spectra at various temperatures, the dc magnetization, their microstructure, and their chemistry using electron microprobe analysis.

EXPERIMENTAL METHODS

Plasma spraying

The titania powder was fused and mechanically reduced to proper size by the Muscle Shoals Mineral Co. The particle distribution was such that 100% passed through a sieve of 53 microns, 32% through a sieve of 30 microns and 12% through a sieve of 20 microns. The purity of the powder was determined by spark analysis, and was found to contain Fe an Al in the ppm range. The plasma gas consisted of a mixture of argon and hydrogen in a 6 : 1 ratio.

The substrate was low-carbon steel, which was cleaned and

polished prior to plasma spraying. Normally substrates are grit-blasted or abraded to develop a so-called "anchor pattern" and, thus, to enhance adhesive bonding. In the present experiments, however, it was desired to remove the coating from the substrates.

Magnetic Measurements

The EPR measurements were made using a spectrometer operating at 9.25 GHz. The temperature variation was effected by blowing liquid nitrogen or liquid helium through an insert dewar. The temperature could be controlled to within 1K.

The dc magnetization measurements were made in a vibrating specimen magnetometer. The sensitivity of the instrument is better than 5.0×10^{-5} gauss cm^3, and temperatures to 30K are possible.

EXPERIMENTAL RESULTS

Substrate-Sprayed Specimens

The room temperature EPR spectra of the material sprayed onto the steel substrate are shown in Fig. 1. As can be seen, a broad spectrum results, the magnetic field of maximum absorbtion H_c depending on the orientation of the magnetic field relative to the substrate-coating normal. When the magnetic field is normal to the plane of the coating, curve A, H_c is larger than when the field is parallel to the plane of the coating, curve B.

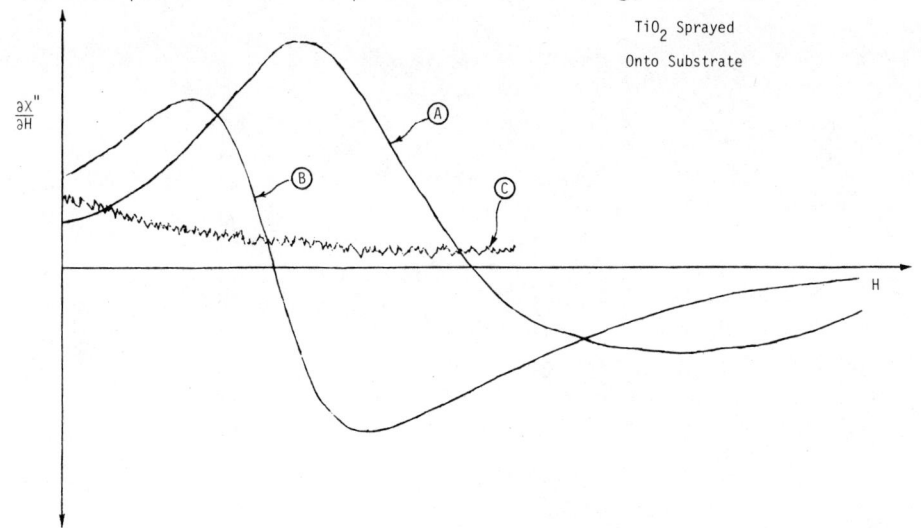

Fig. 1. EPR spectra at 300K of coating sprayed onto carbon steel substrate. A: field perpendicular, B: field parallel to the coating. C: see text.

The broadness of the resonances and their orientation strongly suggests that significant demagnetizing factors may be present in a ferro- or ferri-magnetic resonance due to structures at the iron-oxide coating interface. To test for this, 20% of material was removed from two specimens: in one the material was removed starting from the interface surface inwards; in the other, the material was removed from the final exposed surface inwards. The result was conclusive evidence that the resonance arises at the interface: the resonance disappears when material from the interface is removed, curve C, which was taken at much higher sensitivity.

The obvious candidate for the resonance is some compound of iron. Microprobe analysis was used to test for the presence of iron, which was detected.

The substrate-sprayed specimen was cooled down to liquid nitrogen temperatures and below, and the EPR spectra were obtained. The data are shown in Fig. 2. There are several features of note.

1. The broad ferromagnetic resonance has disappeared. This is characteristic of a highly conducting magnetic species which is metallic in nature and arises from the decreasing skin depth as the temperature and resistivity decrease. This behavior is certainly suggestive of the presence of metallic iron.

2. A new resonance has appeared with g=1.95. This resonance decreases in width and increases in strength as the temperature decreases. The resonance is due to Ti^{+3} ions.

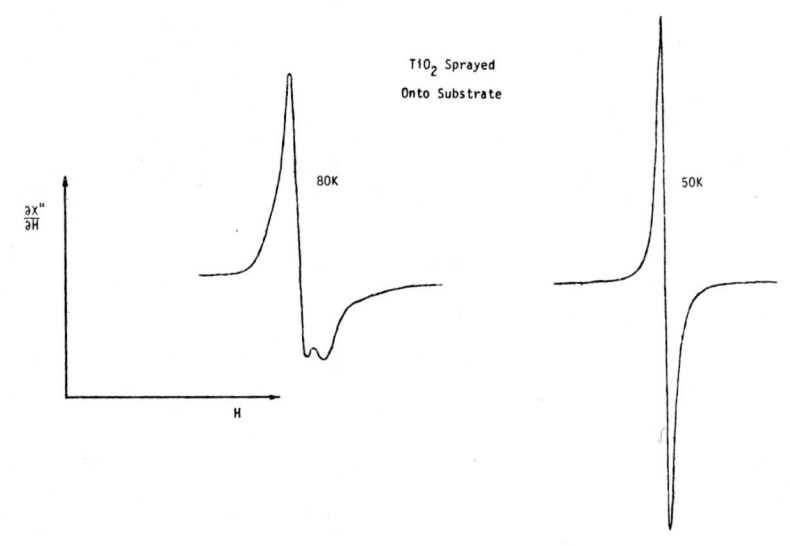

Fig. 2. Typical EPR spectra at temperatures below 80K of coating sprayed onto carbon steel substrate.

DISCUSSION

Adhesion

As mentioned in the introduction, the mechanical adhesion of plasma sprayed coatings onto the substrates is a technical issue of great importance. The question essentially is to what extent does melting of the substrate at the interface and subsequent fusing with the coating particles contribute to the adhesion. The alternative discussed mechanism of adhesion is one of purely mechanical nature, in which the molten particles flow around and into the surface asperities.

The ferromagnetic resonance results provide further evidence for "chemically-based" adhesion. The very strong resonance, and the microprobe results indicate that iron becomes incorporated into the interface in significant amounts. It is difficult to see how this could occur without some melting taking place. Indeed, the amount of iron present in the coating interface is so large that the inter-ion coupling precludes the observation of the characteristic and strong g=4.3 resonance of iron.

The decrease in the intensity of the resonance with decreasing temperature is in accord with the suggestion that pure iron is present, the decrease in intensity of the resonance resulting from the decreasing microwave skin depth with temperature as the resistivity decreases. However, the decrease of the resonance to essentially zero at 80K foreshadows a more complicated situation.

In principle, the shift in the resonance with orientation, Fig. 1, may be used to identify the magnetic species on the simplifying assumption that the shift in H is due to differences in demagnetizing factors alone (no strain effects or texture effects) of a pancake-like structure containing the magnetic phase. In this case the change in H_c is due to the differences in the demagnetizing factors, 4π,

$$\Delta H_c = 4\pi M_s$$

where M_s is the magnetization of the magnetic phase present.

From the data in Fig. 1, we find M_s to be 79.5 gauss. This value compared with 1700 gauss for metallic iron and 480 gauss for magnetite Fe_3O_4 strongly suggests that the magnetic species is not iron. This is not in agreement with the observation that the resonance disappears with decreasing temperatures.

dc magnetization measurements can be used to determine the volume of magnetic species present at the interface: at high fields and low temperatures the magnetic moment will saturate to the value M_sV, where V is the volume of the magnetic species present.

The magnetization of a coating has been measured to 10 kOe, with H parallel to the plane of the coating. The results are shown in Fig. 3, a measurement at room temperature.

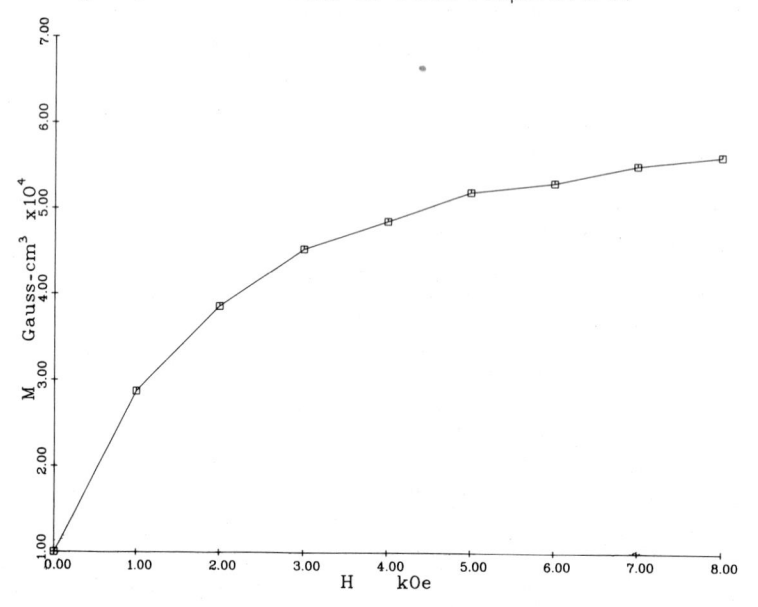

Fig. 3. Magnetization versus magnetic field (parallel to plane of coating) at room temperature.

The magnetization shows signs of saturating to the value $M=6.0\times10^{-4}$ gauss cm^3. On the assumption that

$$M = M_s V$$

we can estimate the volume of the magnetic material present, and hence the thickness of the magnetic layer knowing the area of the specimen measured. In this way we find a thickness estimate of 1.0×10^{-5} cm. Using M=480 gauss and 5.0×10^{-5} cm^3 using the measured value of 80 gauss. These values are reasonable. We also note that a significant amount of titanium may become incorporated in this ferrite. It is thus not clear why the large resonance decreases with temperature. Further study of the magnetization as a function of temperature is needed to answer this question.

The defect state

The EPR measurements also provide information about the defect state of the titania. The resonance at g=1.95, as is well known, is due to the Ti^{+3} ion.

Temperature dependency of the integrated intensity has been determined. The results are shown in Fig. 4, where the 1/T dependence indicates a simple Curie dependence.

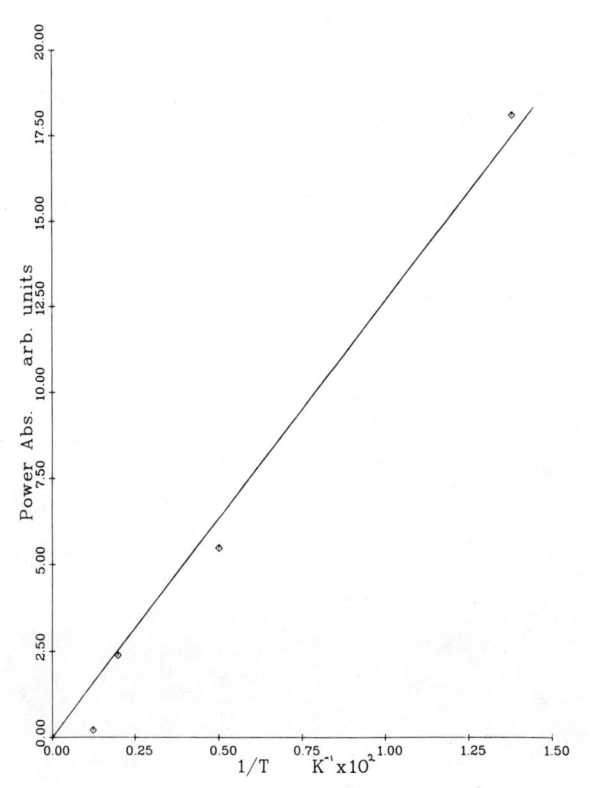

Fig. 4. Integrated intensity of the g=1.95 EPR line as a function of temperature.

This result implies in consequence that the number of Ti^{+3} ions is not a function of temperature, i.e., the electron traps (the Ti^{+4} ions) are completely filled at these temperatures or that all the available electrons are trapped.

CONCLUSIONS

Magnetic measurements have been shown in this preliminary and short investigation to be a powerful method of obtaining new information about plasma-sprayed oxides. Specifically, the investigation has shown that significant interaction in the form of substrate melting occurs on the impingement of the molten particles, lending support to the ideas of Dallaire[4]. Although the temperature dependence suggests metallic inclusions, the deduced magnetization is not in agreement with the magnitude appropriate for iron. Indeed, the magnetization is less than the likely ferritic phases. The estimated thickness of the magnetic phase is very reasonable.

As expected, the predominant ionic defect detected was the Ti^{+3} ion, in concentrations very similar to titania, reduced under moderate conditions. In principle, the Ti^{+3} ion concentration can be determined quantitatively and related to the spraying parameters.

ACKNOWLEDGEMENTS

One of us (R.K.M.) would like to acknowledge the support of DOE Contract No. DE-AC02-79ER10428. Partial support (A.J. and T.T.) under NSF grant DMR80-11439 is also gratefully acknowledged.

REFERENCES

1. D.A. Gerdeman and N.L. Hecht, "Arc Plasma Technology in Materials Science", Springer-Verlag, New York, 1972.

2. S. Safai and H. Herman, "Plasma-sprayed Materials" in Ultra-rapid Quenching of Liquid Alloys, Ed., H. Herman, vol 20, Treatise on Materials Science and Technology, Academic Press, New York 1981.

3. S. Safai, Ph.D. Thesis, SUNY at Stony Brook, New York 1979.

4. S. Dallaire, Int. Conf. on Metallurgical Coatings and Process Technology-1982, San Diego, April 4-9, 1982.

CHARACTERIZATION OF THE MECHANICAL PROPERTIES OF PLASMA-SPRAYED

COATINGS

N.R. Shankar, C.C. Berndt and H. Herman

Dept. of Materials Science & Engineering
State University of New York
Stony Brook, N.Y. 11794

ABSTRACT

The heterogeneous nature of plasma-sprayed coatings plays a major role in their mechanical properties. The fundamental deformation modes are usually classified as adhesive or cohesive. However, these failure mechanisms are poorly understood, especially when they are related to coating performance under service conditions.

Strain deformation of plasma-sprayed stainless steel and yttria-stabilized zirconia coatings, with and without a NiCrAlY bond coat, were measured and the acoustic emission (AE) generated during the test was simultaneously monitored. Comparison of the results with coatings heat treated for 10 hours at 1150°C and stainless steel coatings showed that AE correlates with the fracture mode. Heat treatment affected the bond strength and also the type of fracture. Coating and fracture characterization by SEM and EDA, coupled with AE results, show potential for studying coating adhesion and integrity.

INTRODUCTION

The failure mechanisms of plasma-sprayed ceramic coatings are not completely understood. The response of a coating system to mechanical and thermal stresses is influenced by a highly oriented microstructure and by the incorporation of an intermediate bond coat[1,2,3].

The durability of coatings depends both on coating properties and, more fundamentally, on their adhesion strength to the substrate. The adhesion of coatings is conventionally determined by the Tensile Adhesion Test, ASTM C633-69 (TAT)[4]. The end of a 25mm diameter rod is coated and then epoxied to a like rod (support bar), thus forming a tensile specimen. If the epoxy has a tensile strength greater than that of the adhesion strength, the latter can be determined by a tensile test.

The TAT results, due to both the nature of the test and the coating microstructure, exhibit wide scatter[5,6]. Also, coating failures often show both cohesive and adhesive fracture regions and, therefore, characterization of the bond strength by a single number is not realistic and, furthermore, does not lead to an understanding of coating failure mechanisms.

In this study, the TAT was coupled with the technique of Acoustic Emission (AE)[7] in order to study coating failure. AE, in the present context, refers to elastic waves arising due to the generation and propagation of cracks. Previous work done on ZrO_2-8 wt% Y_2O_3 plasma-sprayed coatings showed that AE can, in fact, differentiate between failure mechanisms and that cumulative AE counts can be correlated with bond strength[8].

This paper describes similar tests performed on ZrO_2-7 wt% Y_2O_3 coatings, with and without a NiCrAlY bond coat. Coatings were tested in as-sprayed and heat-treated conditions and contrasted with stainless steel coatings. Coatings were subjected to repeated, increasing loading until rupture occurred.

EXPERIMENTAL PROCEDURE

Yttria-stabilized zirconia (YSZ), with and without a NiCrAlY bond coat, and stainless steel were plasma sprayed onto ends of cylindrical steel substrates conforming to specifications of ASTM Tensile Adhesion Test C633-69. Prior to plasma spraying, the substrate surface was prepared by grit blasting. The average deposit thickness of the stainless steel metallic coating was 0.4mm and that of the ceramic coating was 0.5mm. Experimental details of plasma spraying and the heat treatment procedure have been described previously[8]. Following the heat treatment, the coatings were furnace cooled to room temperature.

YSZ powder was also plasma sprayed into water using the same spray parameters as that used to produce the coatings. This was done to characterize the effect of the plasma effluent on powder shape, distribution, and degree of melting. Particle size distribution of all the powders was determined by sieving in 7.5cm diameter sieves of woven copper.

The experimental arrangement of AE monitoring during TAT and specimen preparation technique have been described in a previous publication[8]. A special TAT jig has been constructed with a flat face on the support bar, which allows the attachment of the AE transducer. As the tensile load is applied, cohesive (through the coating) and/or adhesive (through the coating-substrate interface), failure occur(s). The propagating cracks give rise to acoustic signals which is detected and registered by the AE setup. The output from the AE transducer was amplified to 80 dB gain. The amplified voltage was then compared with an automatic threshold voltage (which compensates for background and system noise), and each crossing of this threshold voltage was counted by the signal processor as a ring-down count. Also monitored was the RMS voltage, which is related to the intensity of the generated AE, and the AE events, which represent each acoustic pulse. The force vs extension curve was plotted and some specimens were: (i) subjected to repeated loadings; and (ii) held at certain loads before being loaded to rupture.

The powders, sprayed surfaces and the fracture surfaces of the coatings were examined by scanning electron microscopy (SEM) and by energy dispersive analysis (EDA). Fracture surfaces of both the substrate and support bar pieces were also examined by optical stereomicroscopy to characterize the fracture on a macroscopic scale.

RESULTS

Powder and Coating Characterization

SEM's of the YSZ powder are shown in both the as-received and water-quenched conditions in Fig. 1 and their characteristics are listed in Table 1. Fig. 1d, for water-quenched YSZ, shows a large amount of partially molten particles and microcracking. Water quenching of YSZ did not significantly alter the particle size distribution. Both of the as-received metallic powders, on the other hand, exhibited much narrower size ranges.

The surfaces of the metallic coatings revealed more particle spreading than the case of the ceramic coatings; while the latter also exhibited more cracking. Surface cracks in the YSZ coatings increased in number and length with heat treatment; particularly for the duplex coating have a NiCrAlY bond coat. X-ray diffraction of the heat-treated YSZ coating revealed a slight increase in the tetragonal to cubic phase ratio. Heat treatment of the YSZ + NiCrAlY coating modified the coating color from yellow to gray indicating a slight change in the stoichiometry of the oxide. EDA analysis of the oxide coating surface showed the absence of Ni, Cr and Al; while the presence of Y was masked by the Zr peak.

Table I. Powder Characteristics

Powder	Composition wt %		Size Range μm	Production Method
Stainless Steel	Fe	67.4	−53+15	Atomized
	Cr	17.0		
	Ni	12.0		
	Mo	2.5		
	Si	1.0		
	C	0.1		
NiCrAlY	Ni	76.7	−74+30	Atomized
	Cr	17.2		
	Al	5.8		
	Y	0.3		
Yttria Stabilized Zirconia	Zr	92.4	−90+15	Fused
	Y	7.2		
	Fe_2O_3	0.1		
	Trace Oxides Rest			

Fig. 1. Morphology of the powders: (a) Stainless steel; (b) NiCrAlY; (c) Yttria stabilized zirconia; (d) YSZ in water quenched state showing partial melting and microcracking.

AE Monitoring of TAT

Tensile adhesion testing was done under a constant force rate of 500 N sec^{-1} with the testing machine in the operation mode of "Load Control". Thus, on unloading there was no crack closure and compression which would have occurred under "Stroke Control". It was possible, therefore, to use the permanent set extension as a measure of the irreversible deformation undergone by the coating.

Acoustic emission was monitored from the initiation of the load and during the subsequent load cycles. Threshold voltages and gains were adjusted so that there were no significant AE counts from a TAT specimen with no coating. Ring-down count rate, RMS voltage, and AE event output were measured during the test, with simultaneous recording of the force vs displacement curve.

The two types of AE spectra that were observed are shown in Fig. 2 and described below.

1) The count rate increased gradually to an approximately constant value prior to increasing at failure - i.e., the curves are skewed left (Fig. 2a(i)).
2) The count rate increased initially and then decreased to a lower level prior to increasing at failure - i.e., the curves are skewed right (Fig. 2b(i)).

The RMS voltage vs time curves generally exhibited corresponding features with respect to their skewness. There was also a variation in the fluctuation of the RMS voltage, and this is depicted in Fig. 2, where high and low voltage ranges are observed. The event output exhibited similar features to the RMS voltage with respect to their skewness, and, hence, further discussions will be limited to AE counts and RMS voltage. In all of the tests, RMS voltage and count rate increased significantly during the final failure of the coating.

The AE response during a loading cycle is described below. There was a net extension when the specimen was loaded to forces less than the fracture value, and AE signals were produced only during loading and not during unloading. Further loading cycles up to or less than the previous load did not produce further AE signals nor any further extension on unloading. Holding at a constant load produced additional AE counts for some time, and the corresponding RMS voltage was reduced to a low level. When the load level at which the specimen was previously held was exceeded there was an incubation period of a few seconds prior to the re-initiation of an AE signal.

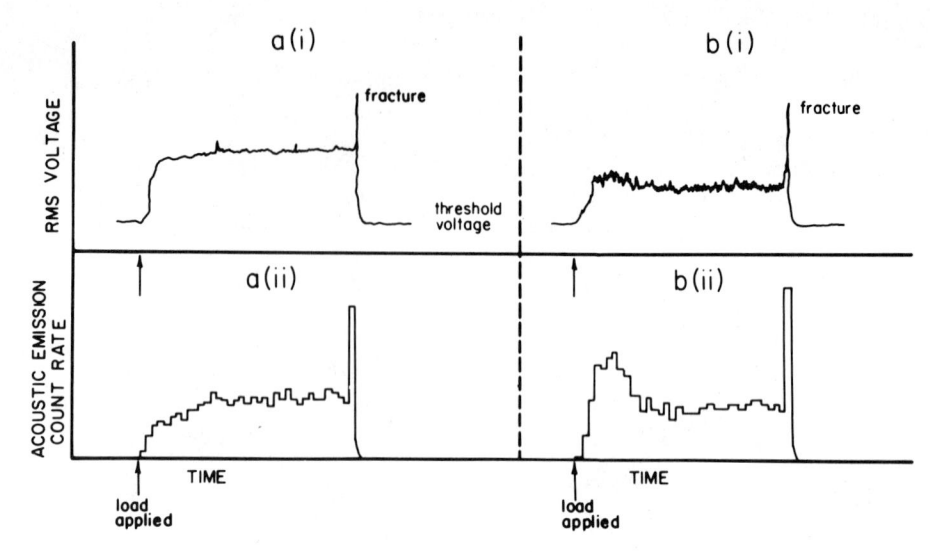

Fig. 2. Two types of AE spectra observed during the TAT: a(i) The AE RMS voltage increases gradually to an approximately constant value prior to increasing at fracture. a(ii) The corresponding AE count rate vs time curve. b(i) The AE RMS voltage increases initially and then decreases to a lower value prior to increasing at failure. Also the RMS voltage fluctuates more than in the case of a(i). b(ii) The AE count rate corresponding to curve b(i).

Table II is a summary of all the TAT tests. Specimens in each category are numbered according to their decreasing bond strengths, with Specimen 1 having the highest bond strength, etc. Likewise "a" has the highest bond strength among the heat-treated coatings.

Stainless Steel Coatings

The cumulative AE counts vs load for stainless steel coatings is shown in Fig. 3. The high slopes at failure are due to high AE counts generated at fracture. Specimens 1, 2 and 3 failed in a cohesive mode, and the AE count plot has a lower slope prior to failure than that during the center part of the test period. The forces up to which the various load cycles were carried out are indicated by the letter "U" (i.e., unloading), while "H" represents "holding" at a specific load. The generation of AE counts at hold, and the subsequent time lag of AE counts in the next loading, results in "steps" in the AE count vs force plot, as can be seen for the case of Specimen 3 and 4. The dashed lines at the 'steps', indicate interpolation of AE counts as if there were no

TABLE II. Summary of AE/TAT Data

Sample Number	Bond Strength Mpa	Net Extension mm	Type of Fracture*	Cumulative AE Counts	Average AE Counts During Test	Shape of Cumulative AE vs Force Plot**
As-Sprayed Stainless Steel Coatings						
1	65.9	0.200	C	88,400	1403	CC
2	59.2	0.025	C	37,550	647	CC
3	51.4	0.035	C	45,350	907	CC
4[+]	50.4	0.048	C	120,700	2514	CC
5	44.1	0.063	A	14,500	345	L
As-Sprayed YSZ Coatings						
1	34.8	0	A	7,450	226	L
2	30.9	0.094	A	20,680	689	L
3	24.4	0.073	A	7,670	319	L
4	20.8	0.065	A	8,650	433	L
5	18.2	0.071	A	8,900	494	L
Heat-Treated YSZ Coatings						
a	22.9	0.017	A	10,400	472	CV
b	20.8	0.080	A	12,120	606	CV
c	19.2	0.048	A	4,730	262	L
d	17.7	0.070	A	8,200	482	CV
e	15.6	0.005	A	7,200	480	CV
As-Sprayed YSZ + NiCrAlY Coatings						
1	54.8	0.103	M	43,450	945	L
2	36.3	0.040	M	37,700	1077	L
Heat-Treated YSZ + NiCrAlY Coatings						
a	65.4	0.092	C	51,575	819	CC
b	64.4	0.065	C	38,200	670	CC
c	48.6	0.140	M	18,300	425	L

* C - Cohesive; A - Adhesive; M - Mixed (i.e., C + A).
[+] Sample 4 was flawed.
** Shape is defined with respect to the abscissa as: CC - Concave; CV - Convex; L - Linear.

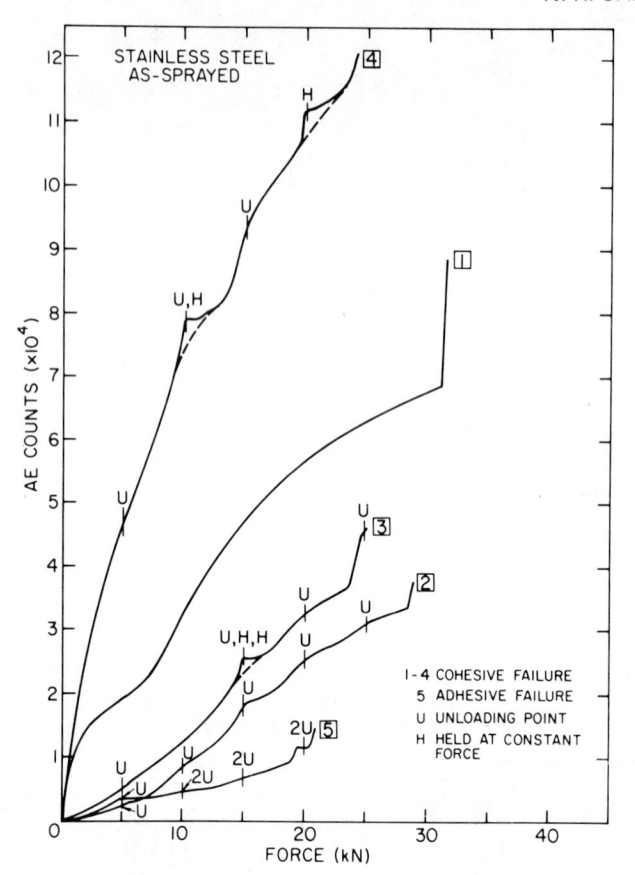

Fig. 3. Cumulative AE counts vs force for as-sprayed stainless steel coatings.

holding at a specific load. The smooth shape of these composite curves indicates that no additional AE was generated as a consequence of the intermittent nature of the test.

The occurrence of a pre-existing crack should be emphasized! (i.e., when a pre-existing crack is present, the AE count-force plot rises abruptly as shown by the curve for Specimen 4.) The RMS voltage showed a large fluctuation, and along with the AE count rate was skewed right. This sample failed at the lowest load among those specimens that failed cohesively and as well generated the maximum AE counts and had a very high average RMS voltage value.

Specimen 5 generated the least AE counts and failed adhesively at a much lower bond strength than the other stainless steel coatings. Also, the curve was not concave to the abscissa as were curves of Specimen 1, 2 and 3, but appeared linear.

Net extension of the coating was measured by the difference in the stroke positions between TAT specimens with and without a coating (shown in Table II). The extension (and as well the AE counts) was higher for those specimens which failed in the cohesive mode. Cohesive failure (Specimens 1 to 3) exhibited interlamellar tearing on both fracture surfaces of the TAT specimen, as shown in Fig. 4a. Fig. 4b indicates the ductile nature of the fracture.

The adhesive fracture surface, on the other hand, was smoother and devoid of any tearing. In some regions, the topology of the grit blasted substrate was revealed on the coating fracture surface.

YSZ Coatings

Figs. 5a and b show the plot of AE counts vs force for the yttria-stabilized coatings in the as-sprayed and heat-treated conditions. Heat treatment reduced the bond strength and the average count rate, during the TAT increased slightly (Table II). Also, the AE count-force curves were convex towards the abscissa and failure counts were higher than for the as-sprayed coatings. The RMS voltage fluctuations were larger for those coatings which yielded at higher counts.

As-sprayed YSZ coatings (Specimens 2 to 5) showed a greater net extension when they yielded more AE counts and higher fluctuations in RMS voltage. Specimen 1 exhibited a low RMS fluctuation and very low average AE count rate and this correlated with a negligible net extension. Heat-treated coatings showed similar correlation of net extension with RMS fluctuations, but no correlation with AE counts.

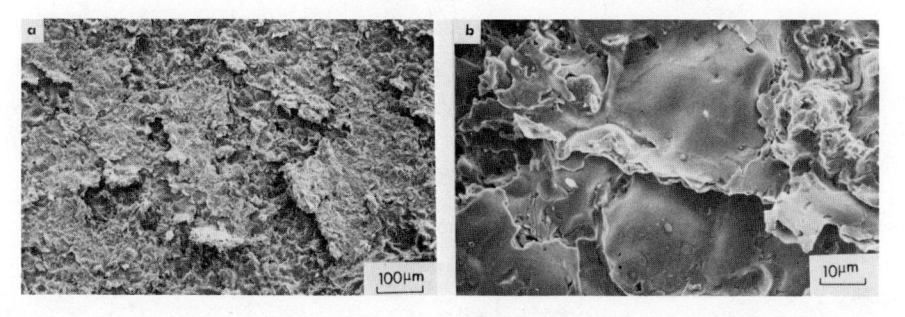

Fig. 4. Cohesive fracture of as-sprayed stainless steel coating: (a) Area revealing interlamellar tearing; (b) magnified image indicating ductile fracture.

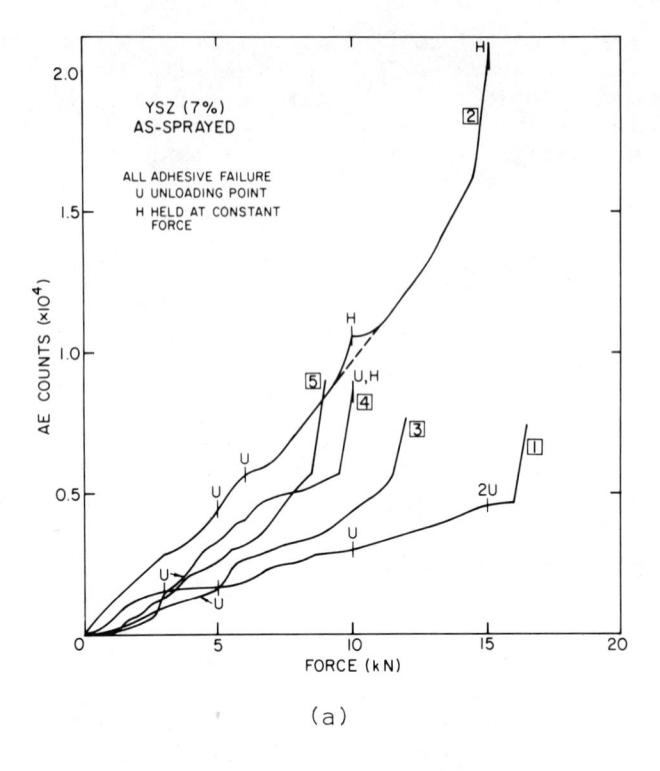

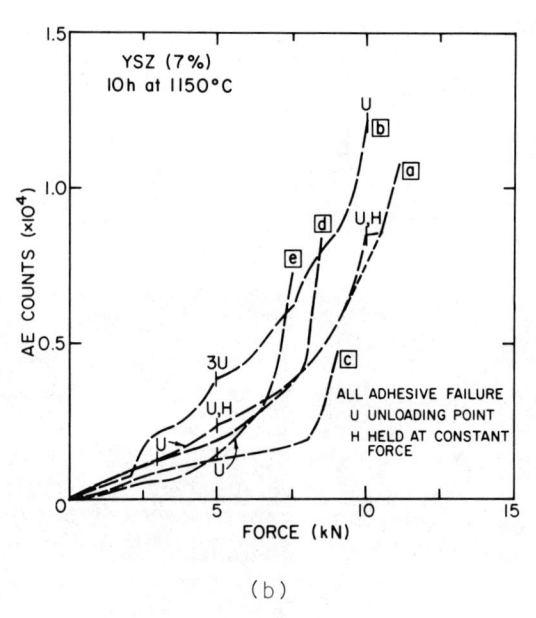

Fig. 5. Cumulative AE counts vs force curves for: (a) As-sprayed YSZ coating; (b) heat-treated YSZ coating.

Fracture surfaces of both the as-sprayed and heat-treated coatings were all adhesive in nature. Optical microscopic examination showed a uniform texture of the fracture surfaces with no tearing, as in the case of the stainless steel coatings which failed adhesively. The adhesive fracture surfaces (i.e., the substrate surface and the corresponding support bar-attached coating surface) of the as-sprayed and heat-treated specimens did not exhibit the same features. The common feature was that both series of specimens revealed sections of the ceramic coating on the substrate surface, as well as regions of the substrate. However, the mating fracture surfaces of the as-sprayed coating showed regions of the detached substrate, which appeared as well defined gray-metallic areas. The heat-treated YSZ fracture surface, on the other hand, did not exhibit these regions but showed signs of incipient melting and densification of the ceramic.

SEM fractography allowed the microscopic nature of the cracking process to be followed. The fracture surfaces of the coatings (as-sprayed and heat-treated) which remained on the support bar after a TAT reflected the topology of the grit-blasted substrate (Fig. 6a). The substrate fracture surface of the as-sprayed coating revealed areas of the grit-blasted substrate and regions where YSZ particles (or parts and groups of particles) were firmly attached to the substrate. The particle edges showed a brittle fracture morphology. The substrate fracture surface of the heat-treated samples, on the other hand, showed incipient melting and clusters of small spherical particles (Fig. 6b). EDA of these particles showed the presence of both Zr and Fe. Precise concentration ratios were difficult to estimate due to the small size of the particles and the low resolution of the electron microprobe analyzer.

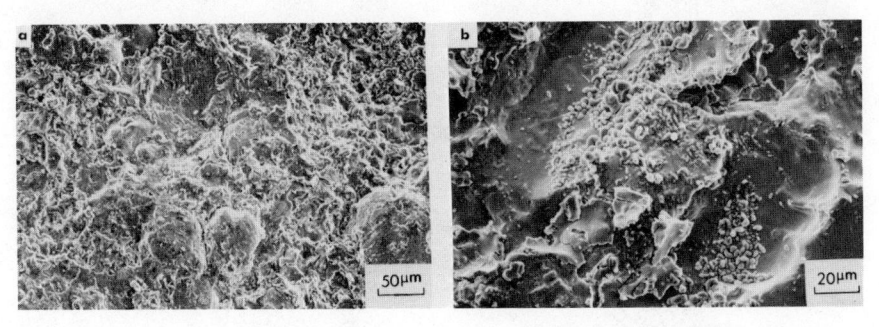

Fig. 6. (a) Adhesive fracture surface (typical of both the as-sprayed and heat-treated YSZ coatings) reflecting the topology of the grit blast substrate. (b) Substrate fracture surface of the heat-treated YSZ coatings revealing incipient melting and with clusters of small spherical particles on it.

YSZ + NiCrAlY Coatings

Yttria-stabilized zirconia, with the incorporation of a NiCrAlY bond coat, had higher bond strengths and yielded higher average AE counts, both in the as-sprayed and heat-treated conditions, as compared to the YSZ coatings (Table II and Fig. 7). The bond strength of the YSZ + NiCrAlY coatings increased after heat treatment. The AE counts vs force response was linear for the as-sprayed coatings (curves 1 and 2 of Fig. 7) and curved for the heat-treated coatings (curves a and b). Optical microscopic examination of the as-sprayed YSZ + NiCrAlY coatings indicated a "mixed" mode of failure, where fracture occurred mostly between the YSZ layer and the NiCrAlY bond coat and partly between the NiCrAlY layer and the substrate. Specimen "c", of the heat-treated coatings, unlike "a" and "b" specimens which fractured cohesively, failed at the substrate bond coat interface (adhesive failure) and also through the YSZ coating (cohesive failure).

Scanning electron microscopy of the as-sprayed coatings was used to characterize the morphology of adhesive fracture between YSZ and NiCrAlY. Fig. 8a shows the fracture appearance of the

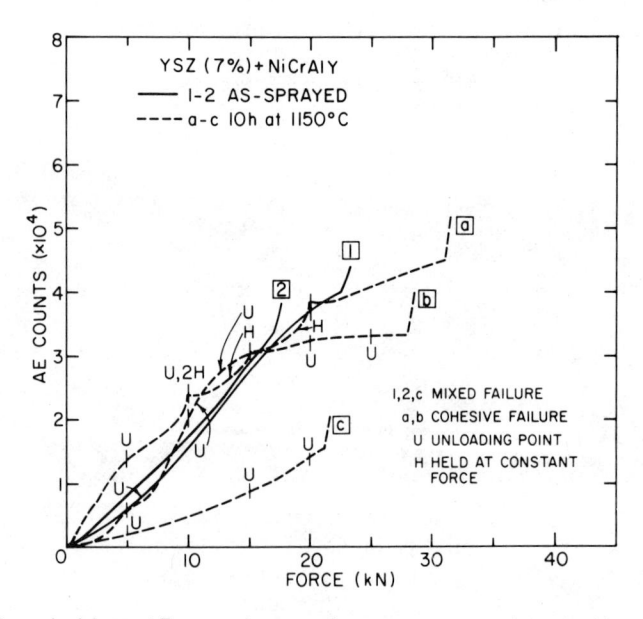

Fig. 7. Cumulative AE counts vs force curves for both the as-sprayed and heat-treated YSZ + NiCrAlY coatings.

coating which remains on the support bar after the TAT, and it can be seen that the topology compliments the coating surface structure. There was also more cracking than in the case of coatings without bond coat (Fig. 8a). Fig. 8b shows the NiCrAlY side (i.e., substrate side) of the adhesive fracture showing the crack free NiCrAlY matrix with fractured YSZ particles embedded in it. After heat treatment, the fracture changed to the cohesive mode, as shown in Fig. 8c. This fractograph of the coating on the support bar shows interlamellar tearing similar to that of stainless steel coatings. However, the fracture appearance of the ceramic appears brittle rather than ductile, as in the case of the metal coating. The substrate side for this cohesive fracture, Fig. 8d, shows brittle-fractured particles of YSZ.

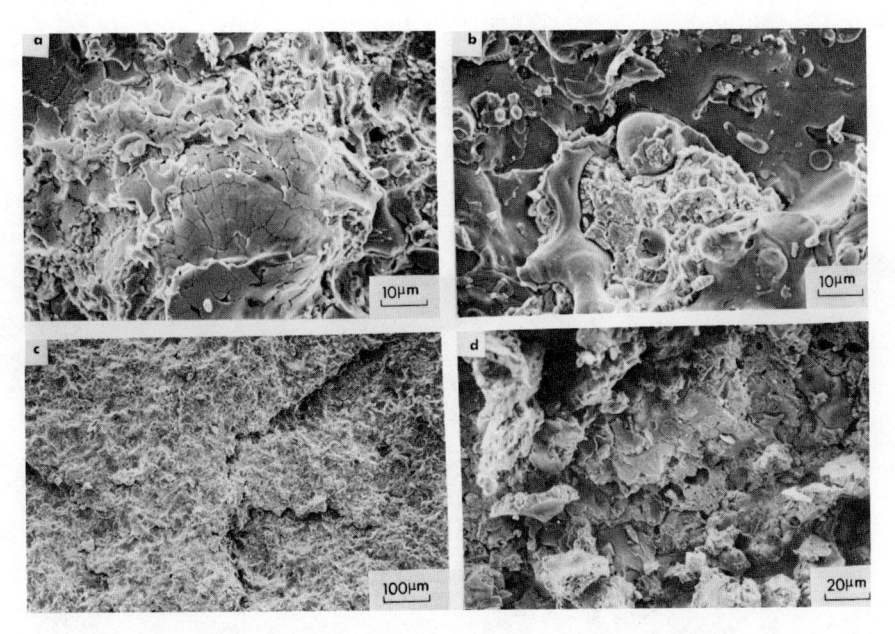

Fig. 8. (a) Adhesive fracture surface of YSZ + NiCrAlY coating showing topology complimentary to bond coat surface and showing a lot of cracking; (b) NiCrAlY side of the above fracture showing crack free NiCrAlY matrix with a brittle-fractured YSZ particle embedded in it. (c) Cohesive fracture of the YSZ + NiCrAlY coating showing lamellar tearing; (d) magnified image of a part of Fig. (c) showing the brittle nature of the fracture.

DISCUSSION

An important feature of these results is the correlation of the AE data with the mechanical behavior of the coatings. Fig. 9 shows that different fracture modes are characterized by their average AE count rate during TAT, with adhesive failure having a lower average AE count rate. On consideration of failure mechanisms, this behavior is to be expected, since coatings which fail in a cohesive manner exhibit a high degree of deformation in the form of interlamellar tearing, indicating greater cracking and, hence, higher average AE counts. In addition, metallic coatings, such as stainless steel and NiCrAlY bond coat, undergo plastic deformation and, thus, have a greater average AE count rate: See Table II.

The shape of cumulative AE vs force plot also corresponds to the mode of fracture. As can be seen from Table II, all cohesive fracture resulted in concave AE plots with respect to the absicca, indicating that more AE is generated during the beginning of the test than towards the end. This in turn indicates more cracking during the initial portion of the test: since AE was generated only on loading and when there was net extension. The cohesive

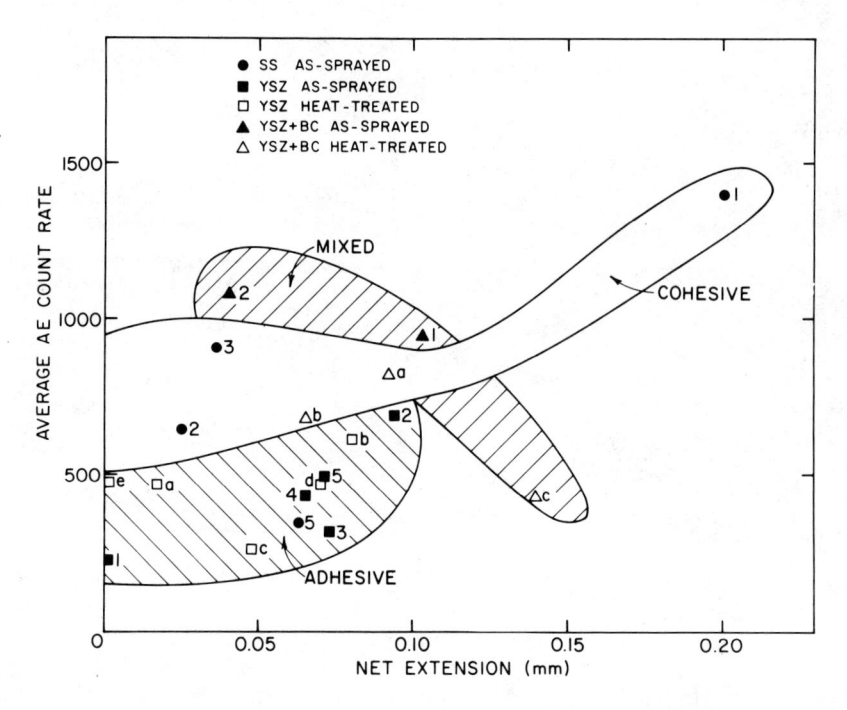

Fig. 9. Average AE count rate plotted vs net extension for the indicated coatings showing regions of different fracture modes.

fracture surfaces show a considerable interlamellar tearing, indicating crack generation at various nuclei. These initial cracks combine together towards the end of the TAT, resulting in a decrease of the AE count rate. The analysis is complicated due to cracks acting as acoustic attenuator[9,10].

Heat-treated YSZ coatings failed adhesively and had a lower bond strength than the as-sprayed coatings and incipient melting of the interface resulted in a smooth fracture surface. These coatings showed a convex AE plot indicating that more AE counts were generated towards the end of the TAT. Thus, most of the energy in the heat treated coating was released toward the end of the test as is also shown by their higher counts at failure than the as-sprayed YSZ coatings.

The as-sprayed YSZ coatings which failed in a adhesive manner showed considerable fractured YSZ particles embedded in the substrate. YSZ with NiCrAlY bond coat, both in the as-sprayed and heat-treated conditions, failed in a mixed mode and showed regions of fractured YSZ particles embedded in the NiCrAlY surface; Fig. 8(b). Thus, all coatings had regions of cohesive coating fracture and regions of adhesive interfacial delamination. This combination of fracture modes results in a linear AE plot and appears to be a linear combination of the concave plot of the purely cohesive mode and the convex plots of the purely adhesive mode. But the mixed mode fracture does not necessarily have an intermediate average AE rate, and reasons for this are unclear. It could be that the mixed mode, having a more tortuous fracture surface, resulted in greater cracking and hence higher average AE counts.

A plot of cumulative AE shows a parabolic correlation with the bond strength (Fig. 10). This is similar to the result of previous work[8] and shows that higher bond strength results in greater cracking and hence higher AE counts. But no correlation was found between net extension, bond strength and average AE count rate. This is surprising since an increase in net extension usually indicates an increase in the bond strength (or yield point) when the Young's modulus is assumed to be constant. This behavior may arise from the large statistical variation of bond strength in any group of tensile adhesion tests. It is likely that the net-extension depends on where the delamination occurs with respect to the interface. That is, failure which occurs closer to the coating-substrate interface will not experience as great a net extension as when failure requires bending of the individual lamellae.

CONCLUSIONS

Acoustic emission was monitored during Tensile Adhesion Testing of stainless steel, YSZ and YSZ + NiCrAlY plasma-sprayed coatings. The latter two coatings were tested both in the as-sprayed

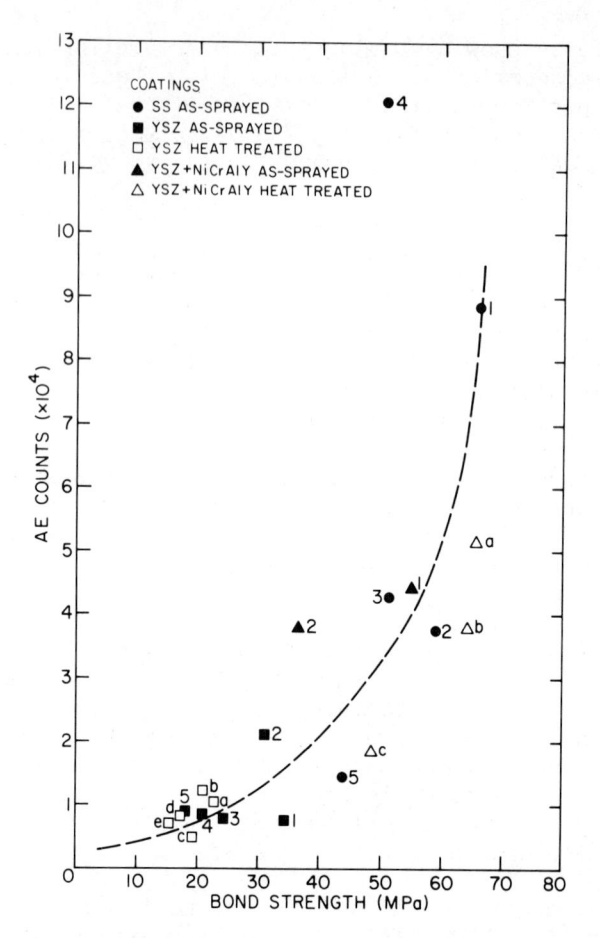

Fig. 10. Cumulative AE counts vs bond strength for all the coatings.

condition and after heat treatment of 10 hours at 1150°C. The AE response of these coatings correlated with their bond strength and fracture surface morphology. Heat treatment decreased the bond strength of YSZ coatings but increased it for the YSZ + NiCrAlY coatings. Also heat treatment altered the type of fracture and changed the characteristics of the generated AE. Extent of strain deformation and average AE count rate of the coatings depended on the type of fracture. Results show that AE can differentiate between different fracture modes and has potential for fundamental coating failure and adhesion studies.

REFERENCES

1. I.A. Fisher, "Variables Influencing the Characteristics of Plasma-Sprayed Coatings", Int. Metall. Revs., 17, 117-129 (1972).
2. S. Stecura, "Two-Layer Thermal-Barrier Systems for Ni-Al-Mo

Alloy and Effects of Alloy Thermal Expansion on System Life", Am. Ceram. Soc. Bull., 61 (2) 256-262 (1982).
3. R. J. Bratton, S. K. Lau and S. Y. Lee, "Evaluation of Present Thermal Barrier Coatings for Potential Service In Electric Utility Gas Turbines", NASA Contract Rep. NASA CR-165545 (1982).
4. ASTM C633-69, American Society for Testing and Materials Standard Titled "Standard Method of Test for Adhesion or Cohesive Strength of Flame-Sprayed Coatings".
5. F. J. Hermanek, "Determining the Adhesive/Cohesive Strength of Thin Thermally Sprayed Deposits", Weld. J. (Miami, Fla.) 57 (11) 31-35 (1978).
6. R. L. Apps, "The Influence of Surface Preparation on the Bond Strength of Flame-Sprayed Aluminum Coatings on Mild Steel", J. Vac. Sci. Technol., 11 (4) 741-746 (1974).
7. W. Swindlehurst, "Acoustic Emission - I: Introduction", Non. Destr. Test. 6 (3) 152-158 (1973).
8. N. R. Shankar, C. C. Berndt and H. Herman, "Failure and Acoustic Emission Response of Plasma-Sprayed ZrO_2-8 wt% Y_2O_3 coatings" 6th Annual Conference on Composites and Advanced Ceramic Materials, Cocoa Beach, Florida, Jan. 17-21 (1982).
9. H.N.G. Wadley, C. B. Scruby and J. H. Speake, "Acoustic Emission for Physical Examination of Metals", Int. Met. Revs., 25 (2) 41-64 (1980).
10. R. L. Cox, D. P. Almond and H. Reiter, "Ultrasonic Attenuation in Plasma Sprayed Coating Materials", Ultrasonics, 19 (1) 17-22 (1981).

ANALYSIS OF SILICON NITRIDE

Gary Czupryna and Samuel Natansohn

GTE Laboratories, Inc.
40 Sylvan Road
Waltham, MA   02254

INTRODUCTION

Ceramics made of silicon nitride are currently undergoing extensive evaluation as silicon nitride components in advanced heat engines as well as in varied vehicular applications. Impurities in the material have deleterious effects on the performance of silicon nitride ceramics, especially under conditions of high thermal and mechanical stress. In order to qualify the silicon nitride for structural ceramic applications it is necessary to obtain accurate chemical characterization of both the initial powder as well as the dense consolidated ceramic material.

Chemical analysis of silicon nitride, particularly of components which have been consolidated close to theoretical density, is complicated by the difficulty of dissolving such ceramic pieces. The prevalent techniques which are based on digestion with a mixture of nitric and hydrofluoric acids in a pressure vessel are lengthy and cumbersome. This report describes a viable method for dissolving both silicon nitride powders and ceramics by alkali fusion. The concentration of the impurities can then be determined in the resulting solution using a dc plasma atomic emission spectrometer.

EXPERIMENTAL

Apparatus

The instrument used for these impurity determinations was a Spectraspan IV (SpectraMetrics, Andover, MA) dc argon plasma atomic emission spectrometer. The Spectraspan IV utilizes a Czerny-Turner type of spectrometer. The spectrometer is equipped with an Echelle grating of 0.75 in focal length. The argon plasma is formed between two carbon anodes and a tungsten cathode in an inverted Y configuration[1]. The dissolved silicon nitride sample was fed into the instruments nebulizer at a rate of 3 ml/min. The argon pressure in the nebulizer was maintained at 50 psi and in the sleeves that house the electrodes at 30 psi. All determinations were made using 50 x 300 µm entrance slits and 25 x 300 µm exit slits.

Sample Preparation

About 200 mg of solid sintered Si_3N_4 pieces or about 500 mg Si_3N_4 powder were fused in a Pt crucible containing 12 g of an alkaline flux. The flux consisted of a mixture of 90 w/o Na_2CO_3 and 10 w/o of Na_2SO_4. The Pt crucible was heated to above 800°C with a Bunsen burner for 15 minutes (to completely fuse the charge). The fused plug was placed in a beaker containing 500 ml of a 1 M HNO_3 solution and stirred until completely dissolved. Stirring enhanced dissolution and helped prevent the formation of a solid residue. In order to provide a comparable matrix, standard and blank solutions were prepared to contain the same concentration of alkaline flux and nitric acid as the Si_3N_4 samples. Standard solutions were prepared by serial dilution of atomic absorption reference standards obtained from either Fisher Scientific Co. or Aldrich Chemical Co. Independent analysis corroborated the concentration values of the standard solutions.

Procedure

A single determination consisted of at least three measurements, each of which comprised two samplings of the analyte with a signal integration time of 10 seconds each. Several of these determinations, at least two of which were done on different days, were used to determine the concentration of the samples. The calibration of the spectrometer was checked periodically (after no more than three determinations) to compensate for potential instrumental drift and to improve analytical precision. Because of excessive background noise, it was found easier to peak the instrumental response at a particular wavelength by using a standard made just in deionized water rather than in the salt-containing solutions. After the instrument was peaked at the desired wavelength, the magnitude of the background and reference signal levels were obtained using the appropriate salt matrix solutions.

RESULTS AND DISCUSSION

The major problem in the analysis of silicon nitride is its dissolution in sufficient concentration so as to permit the determination of low levels of contaminents. Contrary to a published report,[2] it was found that the alkali fusion technique outlined above is equally applicable to whole pieces of sintered silicon nitride as it is to powders. It is easier, simpler and takes a considerably shorter time than the pressure vessel treatment. This method also has the advantage of bringing all the elements into solution, permitting direct determination of all components, silicon as well as trace elements. In the hydrofluoric-nitric acid treatment, silicon is volatilized as silicon tetrafluoride and must be determined by an alternate analytical method.

The use of dc plasma emission spectrometry for quantitative chemical analysis is affected by several factors. Among them are:

a. The total dissolved solids content. This is particularly important in solutions resulting from alkaline fusion which contain about 2.5 w/o of dissolved solids.[3] A substantial background signal is observed in such systems. Also evident is line broadening which makes it difficult to resolve wavelengths that are close together.[4] The background noise in this system was compensated for by the addition to both the reference standard and blank solutions of a requisite amount of sodium salts, commensurate with the flux composition. This permitted the accurate determination of the signal level due to the unknown and also negated the effect of any impurities present in the reagents used.

b. Spectral line overlap. Particular care needs to be taken to eliminate spectral interferences. The emission lines used in the quantitative determination of the various impurities were selected not only because of their relative intensity but also because of their isolation from wavelengths common to other elements in the system.

c. Physico-chemical effects. Components of the matrix may affect the ionization of the system and either enhance or suppress the analyte signal.[5] A case in point is silicon, the major component of the solutions resulting from fluxed silicon nitride samples. It is present in concentrations 100 to 1000 times greater than that of the impurity being analyzed.

The interference effect of silicon on the determination of magnesium in a fluxed silicon nitride sample is demonstrated by the data in Table I. They show that with increasing concentration of Si the signal intensity of a 1 µg/ml Mg solution is increasingly suppressed. These determinations were carried out in deionized water so that any observed interference could be directly attributed to silicon. The signal was monitored for each doped solution at three different wavelengths. The data in Table I is normalized with respect to the maximum signal obtained with a 1 µg/ml magnesium solution at each spectral line. When monitoring the 285.21 nm spectral line suppression was observed even at as low a Si/Mg ratio as 1:1. At 100:1 Si/Mg ratio, silicon suppresses the magnesium

Table I. Silicon Interference Effects on Mg Measurements

Sample	Signal at 279.55 nm	Signal at 285.21 nm	Signal at 383.82 nm
1 µg/ml Mg	100	100	100
1 µg/ml Mg + 1 µg/ml Si	99	80	99
1 µg/ml Mg + 10 µg/ml Si	58	51	65
1 µg/ml Mg + 100 µg/ml Si	47	41	53
1 µg/ml Mg + 200 µg/ml Si	45	31	49
1 µg/ml Mg + 1000 µg/ml Si	34	23	33

Standard is 1 µg/ml Mg in H$_2$O   Blank is H$_2$O

signal by almost 50%. The reason for the suppression of the magnesium signal is not certain but silicon may inhibit the complete ionization of magnesium.

The ratio of Si:Mg in the analyte solutions resulting from the alkali fusion of solid Si$_3$N$_4$ samples is typically 1000:1. The silicon concentration in such solutions is in the range of 120-240 µg/ml while concentration of various impurities is in the 1 ng/ml to 100 ng/ml range. The silicon suppression of the magnesium in these alkaline solutions is evidenced by the calibration line in Figure 1 derived from the data points marked by triangles. These data were obtained on solutions containing 24 g/l of the alkaline flux, 120 mg/l of silicon and doped with varying amounts of magnesium in the

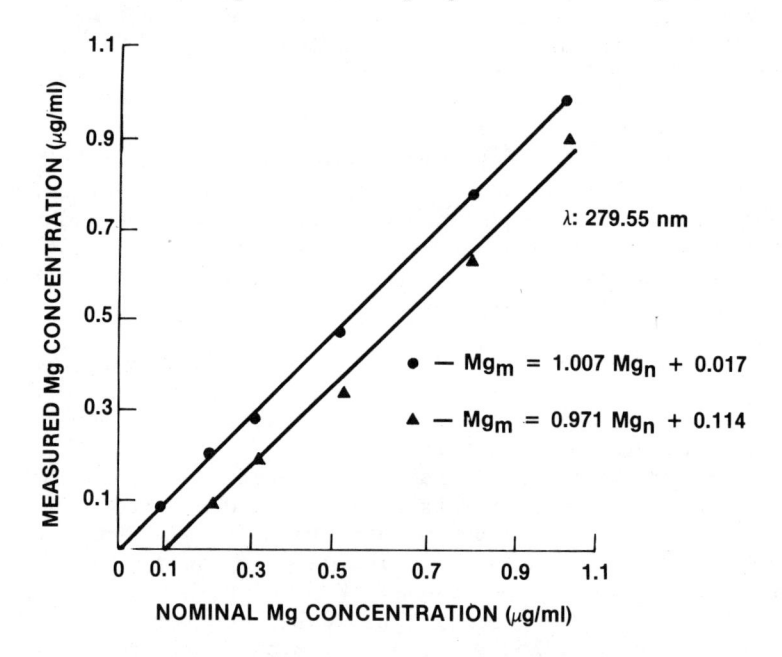

λ: 279.55 nm

● — Mg$_m$ = 1.007 Mg$_n$ + 0.017

▲ — Mg$_m$ = 0.971 Mg$_n$ + 0.114

Figure 1. Magnesium Calibration Curve

0.1-1 µg/l concentration range. In this measurement sequence the Mg reference standard was dissolved in the 24 g/l alkaline flux solution which was also used as the blank. The emission signal from the 279.55 nm Mg line was monitored. The data show a poor correlation between the measured (Mg_m) and nominal (Mg_n) Mg concentration values, the suppression of the Mg signal evidenced by the negative y-intercept and the shift of the calibration line to the right.

The silicon interference can be compensated by adding to both the reference standard and the blank solution an amount of silicon equal to that in the analyte solution, in this case 120 mg/l. The calibration line obtained under such conditions (circles in Figure 1) passes through the origin, has improved data point scatter and provides good correlation between the measured and nominal Mg concentrations. Thus, it is possible to obtain reliable Mg determinations down to concentrations of 0.1 µg/ml. It is noteworthy that the two calibration lines obtained with and without silicon compensation in the standard and blank solutions, respectively, have virtually identical slopes.

Similar problems are encountered in the measurement of the calcium impurity in Si_3N_4 samples because of the suppressing effect of silicon. The lower calibration curve of Figure 2 is the one obtained without a compensating amount of Si in the standard and blank solutions while the upper one was derived from data on samples where the Si was included in the standard and blank solutions. It can be seen that satisfactory correlation is obtained only in the latter case. In this system the slopes of the two calibration lines are substantially different suggesting that the mechanism of the Si suppression effect may be different in the case of Ca than Mg. But regardless of the cause for the observed interference, it can be corrected with proper matrix matching.

Other elements found as impurities in Si_3N_4 powder and sintered ceramics are Al, B, Cr, Fe, Mo and W. These impurities can be introduced during the actual synthesis of the powders, in the milling process, or in a number of other processing steps. Detectability of the various elements in solution ranges from 1 ng/ml for Cr to 0.1 µg/ml for B, Fe, and W. The results of this survey are summarized in Table II. These detection limits directly correlate to the emission intensity of the respective spectral lines. All determinations were made in a matrix of 24 g/l of alkaline flux. Suppression of the analyte signal intensity was not observed for these elements even when Si was present in ratios of more than 1000 times that of the element of interest. In extrapolating these results to impurity levels in the solid, detection limits are increased by three orders of magnitude because the concentration of the Si_3N_4 sample in the analyte solution was 1 mg/ml. Thus, a sensitive technique for determining trace impurities in the range of 1-100 µg/g was established in Si_3N_4 sintered pieces and Si_3N_4 powders.

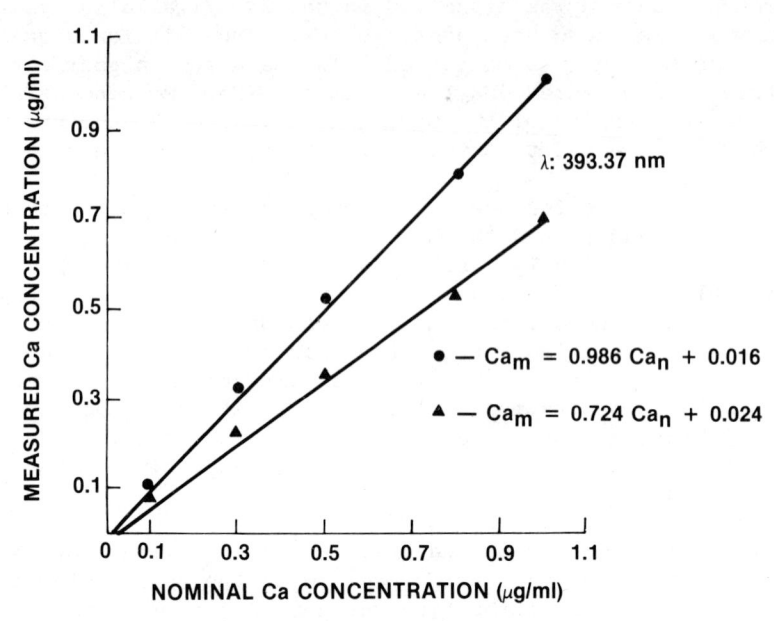

Figure 2. Calcium Calibration Curve

λ: 393.37 nm

$Ca_m = 0.986\ Ca_n + 0.016$

$Ca_m = 0.724\ Ca_n + 0.024$

Table II. Impurity Measurements in Si_3N_4

Element	Wavelength (nm)	Measured in Solution (µg/ml)	*Determinable in Solid (µg/g)
Al	396.15	0.01	10
B	249.77	0.10	100
Ca	393.37	0.010	10
Cr	425.43	0.001	1
Fe	371.99	0.10	100
Mg	279.55	0.01	10
Mo	379.82	0.010	10
W	400.87	0.10	100

*Based on 1 mg/ml solution of solid Si_3N_4

CONCLUSION

The analysis of Si_3N_4 powders and sintered pieces by dc plasma emission spectrometry is feasible. The dissolution of Si_3N_4 powder and sintered pieces by alkaline fusion is simple and rapid. Resulting interferences encountered during analysis can be completely compensated for by careful matrix matching. The analysis of Si_3N_4 sintered products using a dc plasma emission spectrometer is a sensitive and accurate technique for determining trace impurities as low as 1 ng/ml in solution and 1 µg/g in the solid.

REFERENCES

1. J. Reednick, A Unique Approach to Atomic Spectroscopy, High Energy Plasma Excitation and High Resolution Spectrometry, Am. Lab. 11:53 (1979).

2. W.F. Davis and E.J. Merkle, Dissolution of Bulk Specimens of Silicon Nitride, Anal. Chem. 53:1139 (1981).

3. S. Natansohn and G. Czupryna, Determination of Impurities in Industrial Products by DC Plasma Emission Spectrometry, accepted for publication in Spectrochim. Acta. Part B.

4. A.T. Zander, DCP Spectrometry Gives Reliable R&D and QA Results, IR&D 24:145 (1982).

5. D.C. Bankston, S.E. Humphris, and G. Thompson, Major and Minor Oxide and Trace Element Determinations in Silicate Rocks by Direct Current Plasma Optical Emission Echelle Spectrometry, Anal. Chem. 51:1222 (1979).

ACID-BASE PROPERTIES OF CERAMIC POWDERS

Alan Bleier

Massachusetts Institute of Technology
77 Massachusetts Avenue, Room 12-011
Cambridge, MA 02139

ABSTRACT

The fundamental aspects of potentiometric titration, electrokinetics, and conductometric titration in evaluating surface and interfacial thermodynamic behavior are addressed. The characterization of aqueous systems which are pertinent to the processing of ceramic powders is emphasized.

GENERAL BACKGROUND

It is becoming increasingly appreciated that colloidal and surface properties of powders significantly influence, if not determine, the processing of aqueous or polar organic ceramic suspensions, particularly, that of finely divided powders. Surface properties that contribute to the chemical nature of ceramic powders include acid or base strength relative to the solvent medium, the tendency of surface groups to form complexes with solute species, solubility, and the tendency of specific components in the suspending medium to react chemically, for instance, to hydrolyze or oxidize.

The role of electrochemical phenomena in generating the acid-base properties of relevant ceramic powders and related experimental techniques for characterizing powders are emerging as important areas in which ceramists must become knowledgable if the full potential of ceramic components is to be realized.

Various mechanisms have been developed to account for the generation of surface charge on particles in suspension. These are[1,2] (a) dissociation of surface groups via chemical reactions analogous to those which occur in bulk solution for carboxylic acids, amine-containing molecules, and metal-hydroxy species, (b) unequal dissolution of ions comprising the solid, (c) adsorption of charged species from bulk solution, and (d) adsorption of dipolar molecules leading to the subsequent adsorption in a second layer consisting of ions that normally do not adsorb onto the solid in the absence of the adsorbed, dipolar species. Although the generation of fixed surface charge for clay-containing systems, such as montmoillonite, kaolinite, and other clays often encountered by ceramists, seems to require a fifth mechanism because of isomorphic substitution[3,4], this is not the case since such fixed charges are balanced in the dry solid by counterions which are solubilized and, thereby, constitute part of the electrical double layer when these clays are placed in aqueous media. Consequently, the generation of surface charges is usually via mechanism (b) when fixed charges dominate.

Irrespective of the mechanism by which surface charge is generated, the magnitude of the corresponding surface potential largely determines the state of agglomeration in a suspension. For suspensions of ceramic powders, important physical properties relate, most often, to changes in the state of agglomeration. Some of these properties are diffusion, Brownian motion, osmosis, rheology, flowability, and optical and electrical behavior.

Since minimal agglomeration is usually desired for the reproducible and reliable packing of ceramic powders and in view that mechanism (a) often operates, it is desirable to understand the development of surface charge and, possibly more importantly, to measure acid-base and complexing properties of ceramic powders. This is specifically accomplished by identifying chemical surface groups and evaluating intrinsic thermodynamic data that reflect the reactivity and equilibration of powder surfaces. These goals constitute a critical step toward predicting and controlling ceramic powder processing.

While some powders behave according to the Nernst equation, others do not.[5] In many practical systems, however, soluble additives ultimately determine the surface properties. Thus, identification of surface groups is best accomplished using many techniques. Common ones are infrared spectroscopy[6-8], isotopic exchange[9], and nuclear magnetic resonance[10]. Surface thermodynamic quantities, namely intrinsic ionization and complexation constants[5], are evaluated using potentiometric techniques[11]. Electrokinetic[12-16] and conductometric titration[16-18] methods can also be used to augment potentiometric and coagulation studies. Although these techniques have not been widely applied to ceramics processing, they are suitable for studying such systems, particularly aqueous and polar organic suspensions.

ACID-BASE PROPERTIES OF CERAMIC POWDERS

DESCRIPTION AND PROCEDURE OF PERTINENT TECHNIQUES

Theoretical Considerations

Acid-Base Concepts. Many types of chemical equilibria which have been identified for bulk solution also apply in principle to oxide powders in aqueous media. These are hydrolysis, dissociation, dissolution, and complexation. Examples of these solution reactions, along with those of their surface reaction analogues, are summarized in Table I. Of these types of reactions, two which are of principal concern here are hydrolysis and dissociation, particularly acid-base dissociation. Although dissolution and complexation are not treated in detail in this

TABLE I

SAMPLE SOLUTION AND SURFACE EQUILIBRIA

Solution	Surface Analogue
Hydrolysis	
$CH_3C(O)OCH_3 + H_2O \rightleftharpoons CH_3C(O)OH + CH_3OH$	$\equiv MOM\equiv + H_2O \rightleftharpoons 2 \equiv MOH$
$CO_3^{2-} + H_2O \rightleftharpoons HCO_3^- + OH^-$	$\equiv MO^- + H_2O \rightleftharpoons \equiv MOH + OH^-$
Dissociation	
$Al(OH)_2^+ \rightleftharpoons Al^{3+} + 2\ OH^-$	$\equiv MOH_2^+ \rightleftharpoons \equiv MO^- + 2\ H^+$
$CH_3C(O)OH \rightleftharpoons CH_3C(O)O^- + H^+$	$\equiv MOH \rightleftharpoons \equiv MO^- + H^+$
Dissolution	
$BaSO_4(s) \rightleftharpoons Ba^{2+} + SO_4^{2-}$	--
$Al(OH)_3(s) + OH^- \rightleftharpoons AlO_2^- + H_2O$	--
Complexation	
$Cu^{2+} + 4OH^- \rightleftharpoons Cu(OH)_4^{2-}$	$\equiv MO^- + Na^+ \rightleftharpoons \equiv MO^-Na^+$
$R_3N + HCl \rightleftharpoons R_3NH^+Cl^-$	$\equiv MOH + HCl \rightleftharpoons \equiv MOH_2^+Cl^-$

paper, the theoretical treatment for these reactions and the appropriate experimental techniques are easily derived in a manner that is similar to the following arguments.

For aqueous media an acid is a substance which is usually defined as a chemical capable of releasing a proton and a base, therefore, is a substance which is capable of reacting with an acid via proton transfer.[5] These definitions follow from the works of Bronsted[19] and Lowry[20]. Thus, the inverse functions of acidity and basicity in aqueous media are related by Eq. (1),

$$\text{Acid} \rightleftarrows \text{Base} + H^+ , \qquad (1)$$

such that one increases as the other decreases. Neutrality corresponds to the condition pH = $pK_w/2$ which is, of course, at pH ~7, depending on ionic strength and temperature[5,21]. Details of aprotic and nonionizing substances which react with Bronsted-Lowry acids or bases are discussed elsewhere[22-24].

Although these concepts were derived to explain chemical reactions and equilibria ocurring in bulk solution, they apply also to interfacial reactions, such as those encountered in suspensions containing oxidic, ceramic powders. As the reactions summarized in Table I suggest, surface equilibria can be described thermodynamically using the concept of an intrinsic surface equilibrium constant.

The bulk solution equilibrium constant (K_{eq}) for the generalized, reversible reaction[5,25],

$$p_1 A_1 + p_2 A_2 + p_3 A_3 + \ldots \rightleftarrows q_1 B_1 + q_2 B_2 + q_3 B_3 + \ldots , \qquad (2)$$

where p_1, p_2, p_3 and q_1, q_2, q_3 are the stoichiometric numbers of reacting molecules, is given by the expression,

$$K_{eq} = \frac{(a_{B1})^{q_1}(a_{B2})^{q_2}(a_{B3})^{q_3}\ldots}{(a_{A1})^{p_1}(a_{A2})^{p_2}(a_{A3})^{p_3}\ldots} . \qquad (3)$$

Here, the a_i-terms represent molecular or molar activities[26], such that the activity of species 'i' is related to the concentration of 'i', c_i, by the activity coefficient γ_i.

$$a_i \equiv \gamma_i c_i \qquad (4)$$

Whereas Eqs. (2) through (4) describe bulk solution equilibria, surface equilibria are described analogously by replacing the a_i-, γ_i-, and c_i-terms with the corresponding solid-liquid interfacial values, using surface excess concentrations, (Γ_i) and surface activity coefficients (f_i). The expression for the surface region, corresponding to Eq. (3), is, therefore,

$$K_{eq}^{int} = \frac{\Pi (f_{Bi}\Gamma_{Bi})^{q_i}}{\Pi (f_{Ai}\Gamma_{Ai})^{p_i}} . \quad (5)$$

If the f_i-values are approximately unity,

$$K_{eq}^{int} \cong \frac{(\Gamma_{B1})^{q_1}(\Gamma_{B2})^{q_2}(\Gamma_{B3})^{q_3} \ldots}{(\Gamma_{A1})^{p_1}(\Gamma_{A2})^{p_2}(\Gamma_{A3})^{p_3} \ldots} . \quad (6)$$

Note that Eq. (6) represents a valid equality if

$$\Pi(f_{Bi})^{q_i}/\Pi(f_{Ai})^{p_i} = 1 . \quad (7)$$

Moreover, it is anticipated that apparent K_{eq}^{int}-values for real aqueous oxide suspensions depends on temperature.

Although these equilibria refer to surface processes that involve acids and bases, notation used in Eqs. (2) through (7) is general and can be adapted to the specific equilibria from Table I which operate under a given set of conditions. Such concepts may also extend to complicated phenomena that involve soluble polymers and surface active agents[5].

<u>Potentiometric Titration</u>. Potentiometric titration involves the change in the electromotive force, emf, measured across a chemical cell containing reference and indicator electrodes; the change in emf is induced by the addition of a titrant. A common electrode arrangement for aqueous systems consists of a calomel, reference electrode, represented by the chemical half-cell, $Hg|Hg_2Cl_2$ (saturated), KCl (saturated, 0.1 or 1.0 M), and an indicator electrode, which is often made of lithium silicate glass containing lanthanum or barium ions.

For acid-base reactions which occur in bulk solution and are generalized by Eq. (1), the equivalence point is the amount of titrant required to adjust the reacting acidic and basic species to the same activity (concentration). This condition corresponds to pH ~7 when the titration involves strong acids and bases and to a pH different from 7 if either the acid or the base is incompletely dissociated. The equivalence point and its corresponding pH are experimentally determined using pH-titration curves. Usually, the most direct procedure is to measure and plot pH (or emf) versus volume of titrant. The inflection point in such a plot corresponds to the equivalence point. Difference plots, representing the first and second derivatives, $\partial pH/\partial V$ and $\partial^2 pH/\partial V^2$, respectively, may also be used to locate the equivalence point. Such derivative plots yield the most precise evaluation of the inflection point. Knowledge of the this value yields, in turn, the pK_a.

Similar data may be obtained using colloidal suspensions to evaluate the acid-base properties of oxide powders. See Figures 1 and 2. These types of surfaces tend to contain weakly acidic and basic sites which have pK_a^{int}-values between 2.0 and 12.5. The titration procedure is similar to that developed for determining the equivalence points and the corresponding pK_a-values for the binding of H^+ by amino acids[26].

Gavels and Christ[27] and Blackmon[28] extended the interpretation of potentiometric titration data to estimate the density of surface sites. This information leads directly to the surface charge of powder as a function of pH. However, it was recently suggested[29] that these investigators neglected the possible effect of a powder's surface charge on the proper

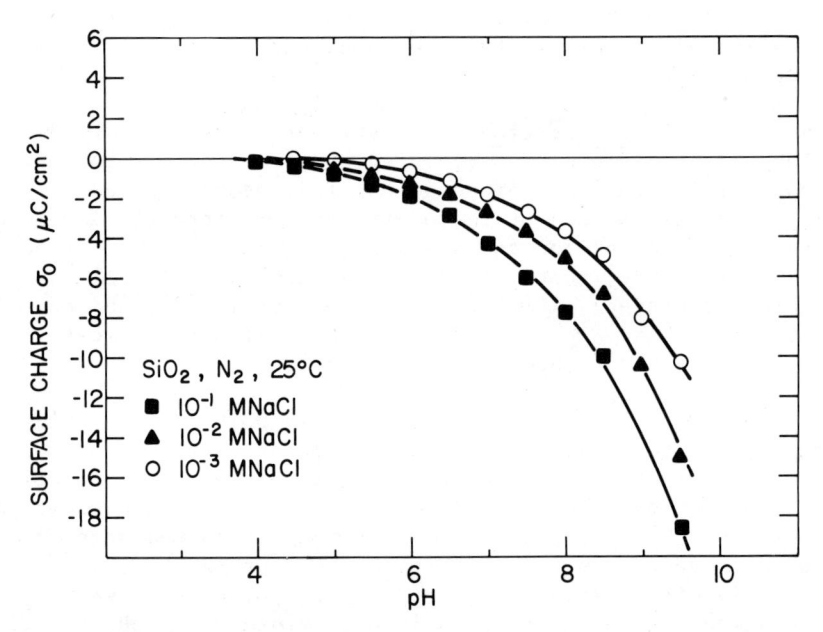

Figure 1. Surface charge of QUSO 30 silica vs pH at 25 °C. Data were obtained with a $N_2(g)$-purge to exclude $CO_2(g)$ from the system. Estimated pK_a^{int}-values are: -2.7 (pK_{a1}) and 6.7 (pK_{a2}). Isoelectric point is 2.0.

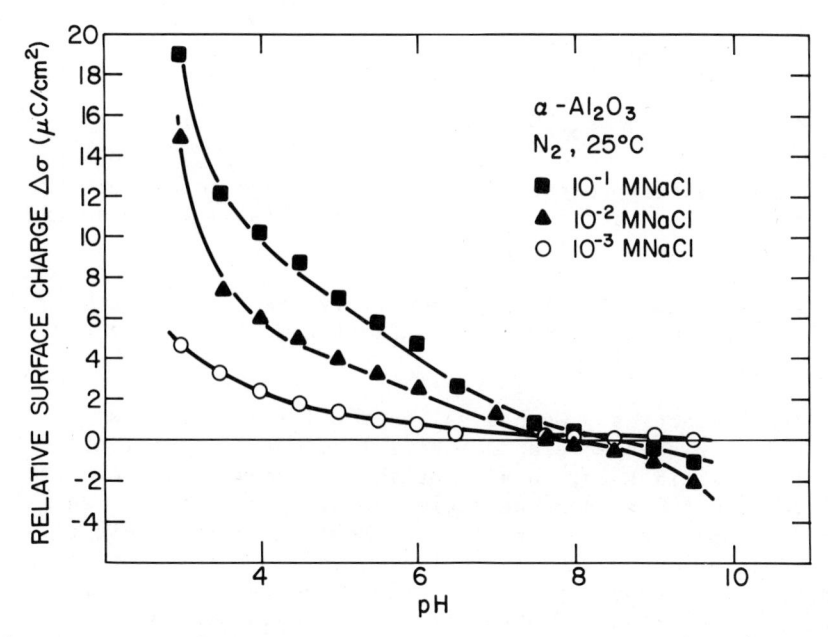

Figure 2. Relative change in surface charge vs pH for Meller's α-Al$_2$O$_3$. Plot is not corrected for the apparent isoelectric point since the intersection of the curves is close to Δσ = 0. Purging conditions are the same as those for data in Figure 1.

surrounding ceramic powders is required in order to use either of the methods described by James and Parks. For example, Schindler interpretation of potentiometric titration data. This oversight leads to significant error in evaluating pK_a^{int}. James and Parks[29] describe, comprehensively, procedures for properly evaluating pK_a^{int}-values for oxidic and nonoxidic suspensions. Basically, two graphical techniques are described. The first is a method to estimate surface potential (ψ_o) and charge (σ_o). This procedure is used when the point of zero charge (see the next section), the acid-dissociation constants for the surface sites, ≡MOH$_2^+$ and ≡MOH, and their surface densities are known. The second procedure for determining ψ_o and σ_o entails a double extrapolation method that accounts for ion-binding or complexation phenomena, e.g., the formation of surface groups such as ≡MO$^-$Na$^+$ and ≡MOH$_2^+$NO$_3^-$, in addition to the simpler, acid-base species already discussed. The double extrapolation method also yields the intrinsic pK_a-values for these processes.

A mathematical model describing the electrical double layer

and Gamsjager[30] give an equation relating an apparent K_a for a charged surface and K_a^{int}. This is

$$K_a = K_a^{int} \exp(z_o e_o \psi_o / kT) . \qquad (8)$$

Thus, experimental procedures are also required to measure the "effective" electrostatic interaction potential ψ_o. These are most commonly electrokinetic techniques.

Tables II and III give, respectively, samplings of the pK_a^{int}-values and complexation constants compiled by James and Parks[29] and others[31].

<u>Electrokinetics</u>. When a solid-liquid suspension is subjected to either an electromotive force or a mechanical one which induces, in turn, in a relative displacement between the particles and the supporting fluid, an electrokinetic potential, ζ, is developed at the plane (or surface) of shear. This plane is a microscopic one and is considered to be located within the fluid but close to the particle. Liquid between the particle surface and the plane of shear is considered stationary with respect to the particle and, thus, is quantitatively treated as

TABLE II

pK_a-VALUES FOR OXIDES IN AQUEOUS ELECTROLYTE SOLUTIONS
(AFTER JAMES AND PARKS[29])

Solid	pK_a
SiO_2	7.2 (2nd)[a]
TiO_2 (Rutile)	2.7 (1st), 9.1 (2nd)
TiO_2 (Anatase)	3.2 (1st), 8.7 (2nd)
α-Fe_2O_3	6.7 (1st), 10.3 (2nd)
α-FeOOH	4.2 (1st), 10.3 (2nd)
γ-Al_2O_3	5.2 (1st), 11.8 (2nd)
Putnam Clay	6.8 (1st), 11.0 (2nd)

[a]1st Dissociation step is strong; i.e., $\Gamma_{SiOH_2^+} \ll \Gamma_{SiOH}$.

TABLE III

pK-VALUES FOR COMPLEXATION OF OXIDES IN AQUEOUS ELECTROLYTE SOLUTIONS[29,31]

Solid	Other
SiO_2	K^+, 6.7
TiO_2 (Rutile)	NO_3^-, 4.1; ClO_4^-, 4.5; K^+, 7.1; Li^+, 7.2; Cd^{2+}, 9.3
TiO_2 (Anatase)	Cl^-, 4.6; Na^+, 7.1
α-Fe_2O_3	Cl^-, 7.5; K^+, 9.5
α-FeOOH	NO_3^-, 7.5; K^+, 9.5
$Fe_2O_3 \cdot H_2O$ (Amorphous)	Cd^{2+}, 4.8; Co^{2+}, 11.3; Zn^{2+}, 4.8; SeO_4^{2-}, 13.4, (1st), 17.3 (2nd); SeO_3^{2-}, 12.9 (1st), 21.4 (2nd); CrO_4^{2-}, 13.2 (1st), 17.4 (2nd); AsO_4^{3-}, 27.7 (1st), 33.5 (2nd)
γ-Al_2O_3	Cl^-, 7.9; Na^+, 9.2; Co^{2+}, 3.3 (1st), 7.3 (2nd)
Putnam Clay	Na^+, 3.2 (1st), 7.4 (2nd); K^+, 2.5 (1st), 6.8 (2nd)

part of the mobile unit in electrophoretic and sedimentation phenomena and part of the stationary surface in electrosmotic and streaming phenomena. References (32) and (33) should be consulted for experimental details.

Figure 3 gives an example of the electrophoretic data that may be obtained. The point of zero charge, i.e., pH at which ζ = 0 mV, apparently depends to some extent on the experimental conditions, possibly owing to nonequilibrium for some of the systems represented in this figure.

An essential concept is that the proper characterization of acid-base properties and the reproducible and reliable processing of ceramic powders in aqueous media depend, respectively, on knowledge of the detailed structure of the electrical double layer surrounding suspended powder and the interactions between

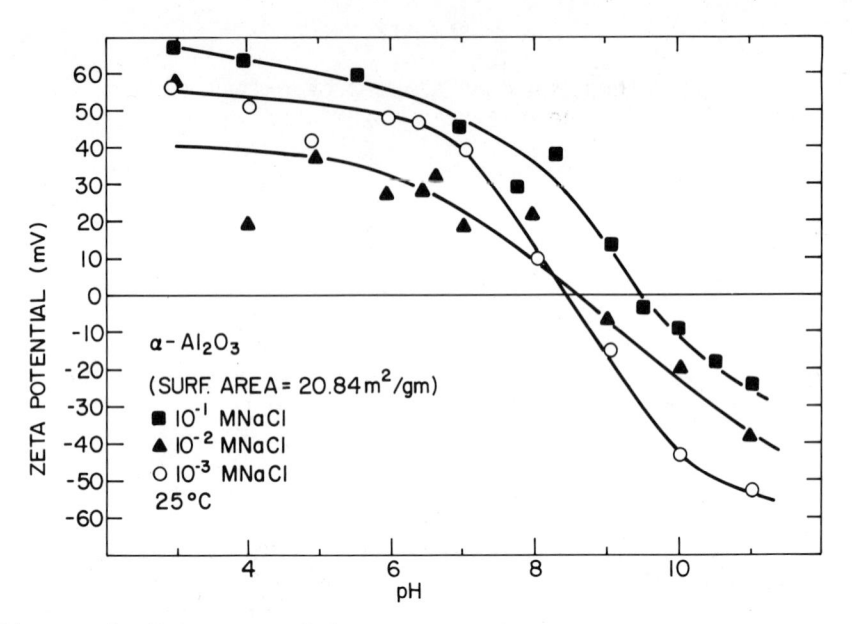

Figure. 3. Zeta potential as a function of pH for a sample of α-Al$_2$O$_3$, corresponding to the alumina represented in Figure 2.

neighboring particles. Consequently, proper measurement of ζ is helpful, if not necessary, for a consistent evaluation of surface pK_a-values and related suspension properties. Since the intrinsic constants describing surface chemical equilibria ultimately determine ψ_o and σ_o, it is logical that electrokinetic data have been used to estimate critical parameters in various proposed models[34] of the electrical double layer.

Specifically, if ζ is low and a suspension does not contain counterionic species that adsorb, a simplified model indicates that $\psi_o \approx \zeta$. The Gouy-Chapman relation then provides an expression for the surface charge, σ_o, in terms of the ζ-potential. That is,

$$\sigma_o \cong 2 \ \varepsilon kT \ \Sigma_i n_i(\infty)[\exp(-z_i e_o \zeta/kT) - 1]^{1/2} \ , \qquad (9)$$

where ε is the dielectric constant of the solvent; k is Boltzmann's constant; T is the absolute temperature; $n_i(\infty)$ is the bulk concentration of species 'i' having valence charge, z_i; and e_o is the electronic unit charge. Eq. (9) applies to planar interfaces such as those of platelike clays. Surface charge density of spherical particles with radius, a, can be approximated when ζ is low by using Eq. (10).

$$\sigma_o \equiv Q/4\pi a^2 \cong \varepsilon[a^{-1} + \kappa]\zeta \qquad (10)$$

The term, κ, is the Debye-Huckel reciprocal length parameter[33] and is given by

$$\kappa = [e_o^2 \Sigma_i n_i(\infty) z_i^2 / \varepsilon kT]^{1/2} . \qquad (11)$$

Eqs. (9) and (10) are approximations, but provide guidelines for more detailed models[34] for the electrical double layer. All of the more complicated models requires a ψ_δ-value which is either considered identical to ζ or can be approximated using electrokinetic measurements.

Figure 4 illustrates the relationship between a spherical particle surface, the Stern layer containing adsorbed counterions, the Gouy-Chapman or diffuse layer, and the bulk solution. Consult References (29), (34), and (36) for detailed explanation of the mathematical relationships among these various portions of a suspension.

Detailed models for the electrical double layer consider the region between the surface and the outer Helmholtz or Stern plane, the potential of which is designated as ψ_δ, to behave electrically as a single parallel-plate condenser or a series of them.[1,5,33] Under these conditions, $\psi_o \neq \zeta$ and a knowledge of ζ (i.e., ψ_δ) is required to evaluate pK_a^{int}-values properly. This can be done using a modified version of the computer program developed by Westall et al.[37] and used by Davis and coworkers in a series of studies[11,38-40]. A review[41] of computer programs which have been developed for evaluating solution equilibria should be consulted for a discussion of the advantages and disadvantages associated with the various numerical techniques and chemical concepts. Recently, Westall[42] updated Reference (37) to account for simultaneous equilibria in bulk solution and at electrically charged surfaces.

<u>Conductometric Titration</u>. The variation of electrical conductivity of a solution or suspension is followed during the course of a titration in this method. Any property which is proportional to the specific conductance may be measured for evaluation of acid-base properties of powders, as long as the proportionality constant is known. The shape and the equivalence point of a conductometric titration curve indicate, respectively, the nature of the acidic and basic surface sites and their number, N_s.[29] This latter value can be used to evaluate potentiometric data via a graphical technique outlined by James and Parks[29].

In a suspension of high ionic strength, the contribution of

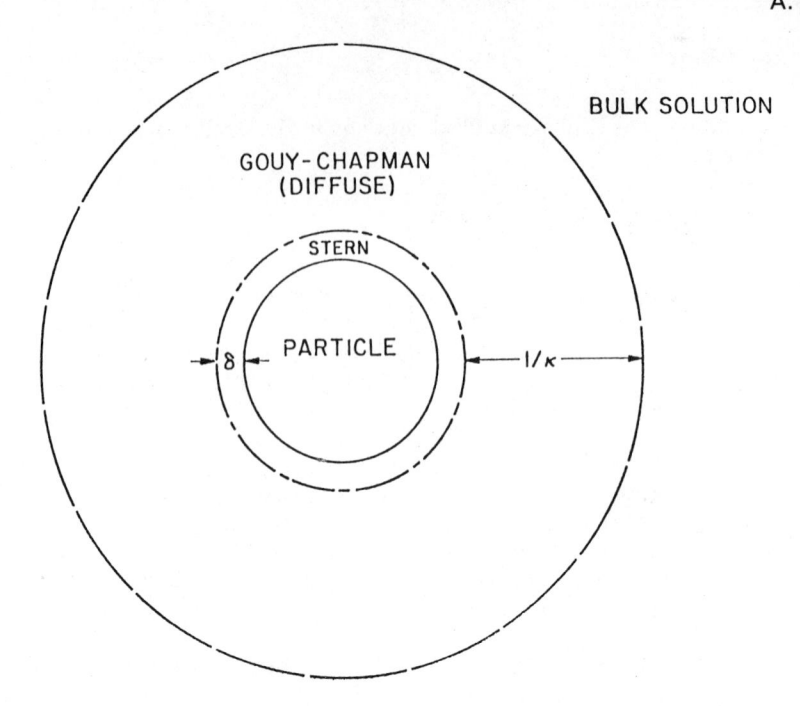

Figure 4. Schematic diagram illustrating the solution surrounding a suspended colloidal particle.[36] The thickness of the Stern layer is δ and that of the Gouy-Chapman layer is κ^{-1}.

the electrical double layers surrounding particles, Λ_p, is negligible compared to that of the supporting electrolyte, Λ_f. Overbeek[32] showed that Eq. (10), when combined with the appropriate expression for the electrophoretic velocity, leads to an expression for the colloidal contribution to the specific conductance, namely,

$$\Lambda_p = (2/3)\varepsilon^2\zeta^2 a(1 + \kappa a)/\eta , \qquad (12)$$

where η represents the suspension viscosity. If the solids concentration is low, η approaches the viscosity of the supporting medium, η_o. Bower[43] has modified Overbeek's treatment and has partly investigated the modified form of Eq. (12) experimentally. Bower's treatment applies if $\Lambda_f \gg \Lambda_p$.

Using a computer[37], James et al.[11] have calculated the total specific conductance of suspensions, Λ_T ($\equiv \Lambda_f + \Lambda_p$), based on experimentally derived potentiometric titrations and electrokinetic data. Interestingly, their calculations are in accord with experimentally derived conductometric titrations. However, the theoretical model for Λ_T neglects surface conductance and is consequently limited to dilute suspensions[29].

O'Brien[44] has recently investigated theoretically the electrical conductivity of dilute suspensions which contain nonconducting, charged particles and derives an approximate solution for the case of spherical particles with low ζ-potential. Comparison of the approximate formula for Λ_T with numerically derived values using the exact equations suggests that the conductivity may be estimated to within a few percent if $\zeta < 50$ mV.[44] It is evident from O'Brien's analysis that many experimental tests of the theory are needed to draw reliable conclusions concerning calculation of a suspension's conductivity.

Other Useful Characterization Methods. Other pertinent analyses for properly evaluating acid-base properties of ceramic powders include surface area analysis, spectrophotometry, isotopic exchange, and nuclear magnetic resonance.

Surface area is usually determined using electron microscopy for particle size, shape, and size distribution, gas adsorption for size and degree of porosity, and adsorption from solution for size and porosity. These techniques are well-documented and will not be described here.

Infrared spectrophotometry has been used by many investigators[7,8,45,46] to elucidate the nature and number density of chemical groups on oxide surfaces. This technique, when combined with isotopic exchange[9] and nmr[10] procedures can provide insight into the chemical structure of acidic and basic surface groups on oxide ceramic powders in aqueous media.

Table IV, adapted from Reference (29), provides an quantitative overview of the types of data that can be obtained. It is evident that many techniques used in conjunction help to establish an accurate profile of oxide surfaces.

CONCLUDING REMARKS

This discussion attempts to clarify the role of novel analytical techniques that will increasingly contribute to the advanced characterization of ceramic powders. Cited references are chosen to guide the reader in the evaluation of recently developed acid-base and complexation concepts and their applications to the processing of oxide ceramics. Other pertinent references concerning the thermodynamics of oxide surfaces can be found in References (5) and (47).

TABLE IV

SURFACE DENSITIES OF IONIZABLE PROTONS FOR MATERIALS SIMILAR TO THOSE, Number per nm^2 (AFTER JAMES AND PARKS[29])

Solid	$\alpha-SiO_2$ and SiO_2 Gels	$\chi-Al_2O_3$ (Corrundum)	$\lambda-AlOOH$ (Boehmite)
Surface Crystal Analysis	4.4-5.9	-	-
Isotopic Exchange	11.4	-	-
IR and H_2O (De)Sorption	4.4-5.9 ($\alpha-SiO_2$) 10.8 (Gels)	25	16.5
Chemical Reaction[a]	4.2 ($\alpha-SiO_2$) 7.9 (Gel)	-	-

[a] Not including acid-base titrations; see Reference (29).

ACKNOWLEDGEMENTS

The author thanks W. Hasz for gathering portions of the experimental data reported herein and K. S. Henchey for preparing the manuscript. This work was funded by the Department of Energy, Contract No. DE-AC02-81ER1053.

REFERENCES

1. A. L. Smith, Electrical Phenomena Associated with the Solid-Liquid Interface, in: "Dispersion of Powders in Liquids", G. D. Parfitt, Ed., Applied Science Publ., Ltd., Barking, England (1981).
2. A. L. Smith, Private communication, Potsdam, N. Y., September 7, 1970.
3. H. van Olphen, in: "Introduction to Clay Colloid Chemistry", Interscience, New York (1977).
4. D. W. Kingery, H. K. Bowen, and D. D. Uhlmann, "Introduction to Ceramics", John Wiley & Sons, New York (1970).
5. A. Bleier, "Relationships Between Solution Chemistry and Surface Properties of Ceramic Powders", To be submitted for publication, Am. Ceram. Soc.

6. G. D. Parfitt and C. H. Rochester, Surface Characterization: Chemical, in: "Characterization of Powder Surfaces", G. D. Parfitt and K. S. W. Sing, Eds., San Francisco (1976).
7. M. L. Hair, in: "Infrared Spectroscopy in Surface Chemistry", Marcel Dekker, New York (1967).
8. L. H. Little, "Infrared Spectra of Adsorbed Species", Academic Pr., New York, (1966).
9. Y. G. Bérubé, G. Y. Onoda, and P. L. de Bruyn, Surf. Sci. 8:448 (1967).
10. J. J. Fripiat, Catalysis Revs. 5:269 (1971).
11. R. O. James, J. A. Davis, and J. O. Leckie, J. Colloid Interface Sci. 65:331 (1978).
12. R. H. Ottewill and J. N. Shaw, J. Electroanal. Chem. 37:133 (1972).
13. D. N. Furlong and G. D. Parfitt, J. Colloid Interface Sci. 65:548 (1978).
14. G. R. Wiese, R. O. James, and T. W. Healy, Discuss. Faraday Soc. 52:302 (1971).
15. A. Bleier and E. Matijevic, J. Colloid Interface Sci., 55:510 (1976).
16. D. E. Yates, R. A. Ottewill, and J. W. Goodwin, J. Colloid Interface Sci. 62:356 (1977).
17. A. Homola and R. O. James, J. Colloid Interface Sci. 59:123 (1977).
18. J. Stone-Masui and A. Watillon, J. Colloid Interface Sci. 52:479 (1975).
19. J. N. Bronsted, Rec. Trav. Chim. Pays-Bas 42:718 (1923).
20. T. M. Lowry, Chem. Ind. (London) 42:43:1048 (1923).
21. L. G. Sillen and A. E. Martell, Stability Constants of Metal-Ion Complexes, The Chemical Society Special Publications Nos. 17 and 25, Burlington House, London (1964) and (1971).
22. G. N. Lewis, J. Franklin Inst. 226:293 (1938).
23. W. B. Jensen, "Lewis Acid-Base Concepts", Wiley-Interscience, New York (1980).
24. H. L. Finston and A. C. Rychtman, "A New View of Current Acid-Base Theories", Wiley-Interscience, New York (1982).
25. A. Vogel, "A Textbook of Quantitative Inorganic Analysis", Longman Group, Ltd., London (1978).
26. T. V. Parke and W. W. Davis, Anal. Chem. 26:642 (1954).
27. R. M. Garrels and C. C. Christ, Amer. J. Sci., 254:372 (1958).
28. P. D. Blackmon, Amer. J. Sci., 256:733 (1958).
29. R. O. James and G. A. Parks, Characterization of Aqueous Colloids by their Electrical Double-Layer and Intrinsic Surface Chemical Properties, in: "Surface and Colloid Science", Vol. 12, E. Matijevic, Ed., Plenum Pr., New York (1982).

30. P. W. Schindler and H. Gamsjager, Disc. Faraday Soc., 52:286 (1971).
31. "Adsorption from Aqueous Solutions", P. H. Tewari, Ed., Plenum Pr., New York (1981).
32. J. Th. G. Overbeek, Electrokinetic Phenomena, in: "Colloid Science", H. R. Kruyt, Ed., Elsevier, Amsterdam, Chapter 5 (1952).
33. R. J. Hunter, "Zeta Potential in Colloid Science", Academic Pr., London (1981).
34. J. C. Westall and H. Hohl, Adv. Coll. Interface Sci. 12:265 (1980).
35. P. Debye and E. Huckel, Phys. Z. 24:185 (1923).
36. A. Bleier, "Stability of Mixed Colloidal Dispersions", Ph. D. Thesis, Clarkson College of Technology, Potsdam, N. Y. (1976).
37. J. C. Westall, J. L. Zachary, and F. M. M. Morel, "MINEQL- A Computer Program for the Calculation of Chemical Equilibrium Composition of Aqueous Solutions", M.I.T. Technical Note No. 18, Massachusetts Institute of Technology, Cambridge, MA. (1976).
38. J. A. Davis, R. O. James, and J. O. Leckie, J. Colloid Interface Sci. 63:480 (1978).
39. J. A. Davis and J. O. Leckie, ACS Symp. Ser. 93:299 (1979).
40. J. A. Davis and J. O. Leckie, J. Colloid Interface Sci. 74:32 (1980).
41. D. K. Nordstrom et al., ACS Symp. Ser. 93:857 (1979).
42. J. C. Westall, "FITEQL", Computer Program, Oregon State University, Corvallis, OR. (1982).
43. C. A. Bower, Soil Sci. Soc. Amer. Proc. 25:196 (1958).
44. R. W. O'Brien, J. Colloid Interface Sci. 81:234 (1981).
45. K. Klier and A. C. Zettlemoyer, J. Colloid Interface Sci. 58:216 (1977).
46. J. H. Anderson and G. A. Parks, J. Phys. Chem. 72:3662 (1968).
47. A. Bleier, The Science of the Interactions of Colloidal Particles and Ceramics Processing, in: "Emergent Process Methods for High Technology Ceramics", Mat. Sci. Res. Ser., R. F. Davis, H. Palmour III, and R. L. Porter, Eds., Plenum Pr., New York, 1983, in press.

RECENT ADVANCES IN COMPUTERIZED HIGH TEMPERATURE

DIFFERENTIAL THERMAL ANALYSIS

>   Charles M. Earnest*, W.P. Brennan, and
>   M.P. DiVito
>   Perkin-Elmer Corporation
>   Main Avenue - M/S 131
>   Norwalk, CT   06856

INTRODUCTION

It was in 1887 when Le Chatelier,[1] published heating curves of clays (including kaolinite and halloysite) using a photographic recording of the reflection of a galvonometer mirror which was activated by a thermocouple in the sample. Although this study was not conducted in a differential fashion, DTA originated twelve years later in 1899,[2] and the study of clays and silicate minerals propagated the technique for the next 40 years. According to Wendlandt,[3] with few exceptions, DTA was used exclusively for the studies of clay materials by geologists, ceramicists, soil scientists and mineralogists until the late 1940's and early 1950's.

The equipment used by these workers was of the "home built" variety. Commercial DTA instrumentation became available in the late 1950's and the quality of DTA instrumentation changed very little until the introduction of the microcomputer-based DTA-1700 High Temperature Differential Thermal Analysis System in 1980. This instrument, shown in Figure 1 along with the Perkin-Elmer Thermal Analysis Data Station (TADS), has since won the IR-100 award as one of the major advances in analytical instrumentation and has found its way into numerous laboratories for use in both research and quality control applications.

COMPUTERIZATION OF THERMAL ANALYSIS INSTRUMENTATION

The recent addition of microcomputer control to thermal analysis instrumentation as well as the addition of versatile

---

* To whom correspondence should be addressed.

microcomputers for enhanced and automated data reduction has greatly improved the thermal analysis techniques. The incorporation of computer "firmware" chips into the thermal analysis hardware not only improves the reproducibility, but such programmable ROM chips are also utilized for conditioning the thermocouple responses delivered for both the ordinate and abscissa signals for DTA thermal curves.

Figure 2 shows a block diagram which describes the microcomputer signal conditioning utilized in the Perkin-Elmer DTA-1700 High Temperature Differential Thermal Analysis System. In the diagram, "S" and "R" refer to the sample and reference Pt/Pt, 10%Rh thermocouples utilized by this system. A platinum shunt is shown below these thermocouple junctions for connecting the two Pt, 10%Rh (negative) portions of each thermocouple pair. This allows a differential thermocouple signal to be connected directly to a ΔT amplifier as is shown in the diagram. The amplified ΔT response is the voltage signal utilized by most commercial, as well as "home built", instruments as the ordinate signal in obtaining DTA thermal curves.

Through the use of computer firmware components, the DTA-1700 System allows the operator a choice of two types of signal conditioning for further improvement of the amplified ΔT ordinate signal. By selecting the proper switch position, $\otimes$ in the diagram, this signal is adjusted versus sample temperature by either Mode 1 (used for routine DTA characterizations) or Mode 2 (used for quantitative assignment of heats of transitions) as described in Figure 2. Since thermocouple responses are known to become nonlinear with increasing temperature, and especially so with high temperature systems, the signal conditioning in Mode 1 is one of linearization of the ΔT ordinate signal versus the sample temperature. Thus, this mode of operation is used for routine DTA (ΔT versus T) characterizations and does not allow a declining ΔT sensitivity versus temperature.

Historically, differential thermal analysis has been considered somewhat less quantitative, with respect to the assignment of the energy absorbed or liberated by a thermal event of interest, than that desired. It was this fact which lead to the development of Mode 2 of the microcomputer conditioning which is available in the DTA-1700 System. In the development of this conditioning program, numerous metal fusion standards representing the entire temperature range of operation of the instrument were employed. The melting profiles of these metals using the amplified ΔT response of the thermocouples were monitored versus time as they were heated through the appropriate melting regions. A separate experiment was performed for each metal standard employed. Knowing the heats of fusion of these metals from published literature, the peak areas from these experiments were compared to the ideal peak area re-

COMPUTERIZED HIGH TEMPERATURE DIFFERENTIAL THERMAL ANALYSIS 517

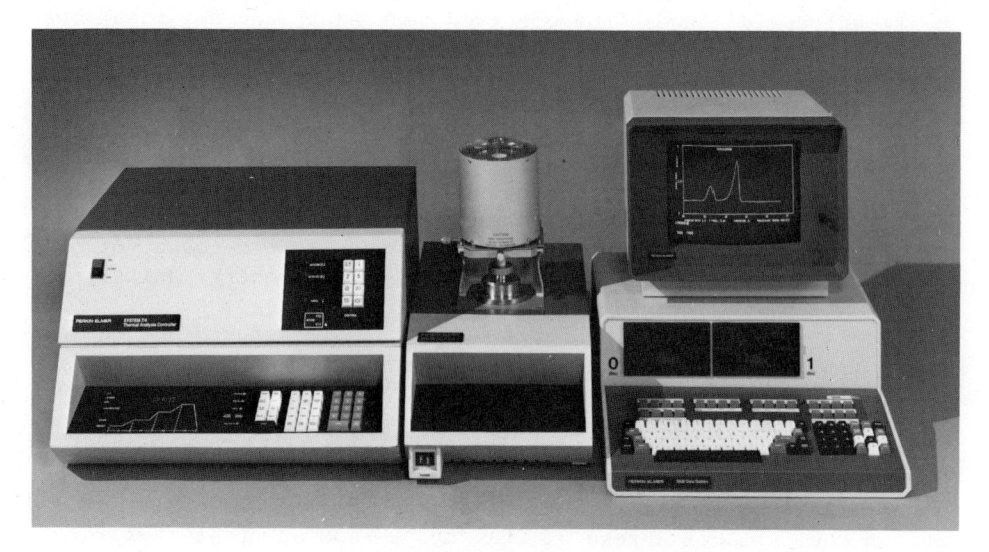

Figure 1. Perkin-Elmer DTA-1700 High Temperature
Differential Thermal Analysis System
with Thermal Analysis Data Station.

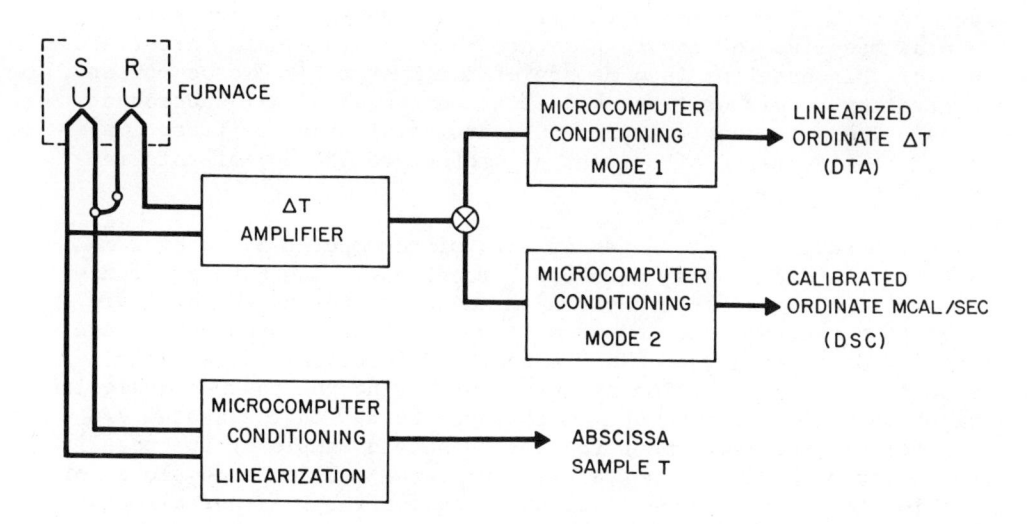

Figure 2. Block Diagram of Computerized Ordinate
and Abscissa Signal Conditioning
of the DTA-1700 System.

sponse for a true calibrated heat flow ordinate scale. Based on the relationship between the observed peak areas and the ideal peak areas, a complex equation was generated and programmed into the ROM chip which is now employed in Mode 2 of the ordinate signal conditioning of the DTA-1700 System. The net effect of the ordinate conditioning of this firmware is a calorimetric factoring of the ordinate response versus temperature. As in the measurement of heats of transition in any differential scanning calorimeter (DSC), thermal curves obtained by the operation of the DTA-1700 System in the DSC Mode (Mode 2) should be obtained using a time base, or program temperature, if areas are to be assigned in energy units.

As may be observed in Figure 2, the sample temperature thermocouple is utilized for monitoring the sample temperature on the abscissa of the DTA thermal curves obtained from the DTA-1700 System. As was mentioned earlier, due to the nonlinearity of thermocouple responses additional firmware components were entered into the system for the purpose of achieving a linear temperature axis in the final DTA thermal curve. Thus, a ROM chip was programmed to contain the National Bureau of Standards "look up" table for Pt/Pt, 10%Rh thermocouples. Therefore, on each temperature measurement by the sample thermocouple, the temperature corresponding to the generated emf is displayed on the microcomputer programmer screen display, for driving an x-y recorder, or sent through an RS-232 port for use with computerized data handling systems.

Additional computerized features of the DTA-1700 System which are not included in the diagram in Figure 2, include a scanning autozero (SAZ) function which allows the storage of the instrumental baseline for the temperature range of interest. After storage of this baseline into memory of the System 7/4 Microcomputer Controller, a mirror image signal is generated on all subsequent runs. Thus, a greatly improved instrumental baseline is obtained over the wide range of operation (ambient to 1500°C) of this instrument.

Two additional features of this microcomputer based system are the "Autozero" and "Autoslope" functions. The Autozero function offers a rapid (pushbutton) means of restoring the ordinate signal to recorder zero (0.00 millivolts) at any point during the heating or cooling run. The "Autoslope" function offers a rapid baseline slope correction at any point during the thermal analysis experiment. On pushing the "Autoslope" button on the System 7/4 controller, the instrument takes a 30 second sample of the baseline; then, after the 30 second delay, applies an extrapolated slope correction to all subsequent data obtained for the DTA thermal curve. Furthermore, this "Autoslope" function may be used even though a scanning autozero baseline is being applied to the instrumental baseline.

Figure 3 shows a DTA thermal curve, obtained using a Perkin-Elmer XY_1Y_2 Thermal Analysis Recorder with the DTA-1700 System, of the Chambers, Arizona "Cheto" type montmorillonite specimen. In this case, the author, through his own negligence, forgot to center the DTA-1700 furnace prior to starting the heating program. As one can observe in Figure 3, the negative trend of the DTA thermal curve not only was affecting the peak shape but would have intersected the temperature axis long before completion of the heating program. Instead, the slope was corrected and the run salvaged by the mere push of the "Autoslope" and "Autozero" buttons on the System 7/4 keyboard of the DTA-1700 System.

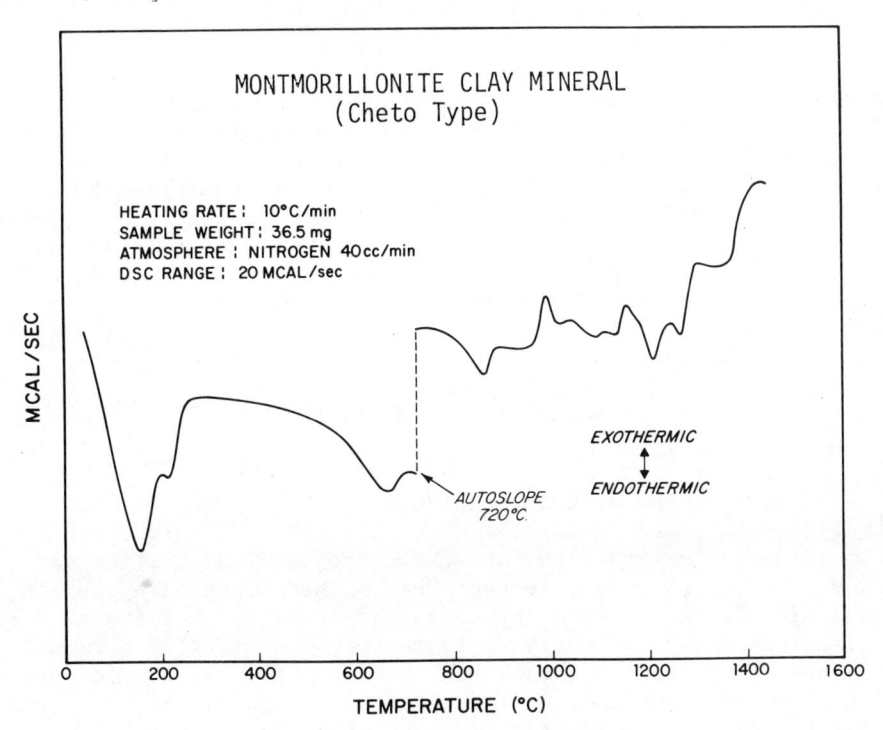

Figure 3. DTA Scan Showing the Use of "Autoslope" Feature to Salvage a Thermal Curve.

ADDITION OF MICROCOMPUTERS FOR AUTOMATED AND ENHANCED DATA REDUCTION

The recent addition of versatile microcomputers to thermal analysis instrumentation has led to many time-saving advantages for improved thermal curves, push button data manipulation, and storage of both raw data and calculated results. The Perkin-Elmer Thermal Analysis Data Station (TADS), shown in Figure 1 along with the DTA-1700 High Temperature Differential Thermal Analysis System, is the most recent addition to the total computerization of the

DTA-1700 System. The TADS combines two 8 bit registers to give
64K of memory and employs an RS-232 data bus for serial transfer
of data to the central processing unit (CPU).

Two microfloppy disk drives are used with the TADS. One disk
drive is used for application program storage and the other for data
storage. A multipurpose CRT display is employed for both observing
the thermal curves, in real time, and data manipulations. The high
resolution TADS printer/plotter is used for hard copy printout of
the DTA thermal curves and/or data analysis results.

It has recently been pointed out,[4] that a computer without
software is nothing but another piece of laboratory furniture. After much work and engineering, the TADS DTA Standard Software Package was recently completed and introduced,[5] at the 84th Annual
Meeting of the American Ceramic Society in Cincinnati, Ohio. The
DTA Standard Software Package was designed to handle both the DTA
and DSC modes of operation of the DTA-1700 System. The DTA Standard Software may be considered to contain two major programs. One,
the "SET-UP Program", for setting up and running the DTA experiment
and the other, the "ANALYSIS Program", for both optimizing the DTA
thermal curves and data handling routines. For convenience and
space considerations, Tables 1 and 2 list the major functions of
each program. In the following discussion of the DTA analysis of
selected carbonates, some of these functions will be demonstrated.

DIFFERENTIAL THERMAL ANALYSIS OF LIMESTONES AND DOLOMITES

Limestones and dolomites play an important role in the ceramics industry. These naturally occurring raw materials are generally heated or calcined to form the corresponding alkaline earth
oxides CaO and CaO, MgO, respectively. These limes are used in
pottery bodies and glazes, glasses, and enamels. Large quantities
of limestone are also consumed in the manufacturing of cements.
Limestones are also important raw materials for the chemical industry and find use in such things as soil conditioners, whiting
and whitewash, as a flux for smelting ores, and as a filler for
polymeric materials. The naturally occurring $CaCO_3$ specimens are
calcite and aragonite.

Aragonite is the rhombic crystal form and calcite is the
trigonal (rhombohedral) variety. Aragonite is unstable with time
and transforms into calcite at ordinary temperatures at a very
slow rate,[6]. The rate of this transformation increases with temperature and is endothermic. Thus, most calcium carbonate specimens
are found in nature as calcite.

Calcite exhibits a single decomposition to calcium oxide
with the liberation of $CO_2(g)$ in both DTA and TG thermal curves.

Table 1. Functions Contained in the SET-UP Program
of the DTA STANDARD SOFTWARE PACKAGE.

SET-UP PROGRAM

- MODIFY PARAMETERS
- CONDITIONS
- ZERO
- START
- STOP
- QUICK COOL
- OVERRIDE T

SPECIAL NOTES:

- All temperatures displayed are actual sample thermocouple temperatures (in K or $^\circ$C).

- Peak Search Option is included.

- Entry of Purge Gas Identity and Flow Rate.

- Entry of Calibration Factor for assignment of peak areas in energy units for DSC Mode.

- Choice of English or Metric Units for DSC Mode.

Table 2. Functions and Analytical Routines Contained
in the Analysis Program of the
DTA STANDARD SOFTWARE PACKAGE.

ANALYSIS PROGRAM

- Slope
- Rescale Y
- Rescale T
- Y Shift
- Peak
- Tg
- Derivative
- Results
- Plot Screen
- Plot Calc
- Save
- Delete
- Content
- Plot Content
- Recall
- Restore Original
- Options
- Go To Set Up

The peak temperature will depend on both the sample size and atmosphere to which the sample is experimentally subjected. The calcite decomposition peak typically falls between 710 and 960°C and has an endothermic energy of approximately 1.7 KJ/g. Figure 4 shows the DTA thermal curve obtained in dynamic nitrogen purge with the computerized DTA-1700/TADS System using the DSC mode of operation for a relatively pure blue calcite specimen from New York State (U.S.A.). This figure is a photograph of the hard copy printout of the DTA thermal curve obtained using the TADS/Printer Plotter. Figure 5 shows the same DTA thermal curve after "peak" analysis from 600°C to 930°C. One can see that the DTA-1700 Standard Software assigns the extrapolated onset temperature (760.9°C), the endothermic energy of decomposition (+1675 joules/gram), and the peak minimum (867.5°C) for this 17.15 milligrams of sample. Furthermore, the sample identity, sample weight, heating rate, purge gas identity, and flow rate of dynamic purge are all listed at the top of this hard copy. At the bottom, the operator is identified along with the date and time of data acquisition. The file name (BC1.DT) is also listed which is the operator's own code for storage and recalling this data set from the floppy disk employed by the TADS.

A characteristic of all DTA thermal curves of calcite or calcite bearing materials is that the rate of return to the baseline is much more rapid than that of the leading edge of the DTA peak. This is exemplified in Figure 6 where the first derivative trace is displayed simultaneously with the DTA thermal curve. In this case, the first derivative of the DTA thermal curve was generated, after displaying of the DTA thermal curve, by a single keystroke on the TADS keyboard.

If the calcium oxide formed from the decomposition of the calcite is cooled in the presence of $CO_2(g)$, a portion of the $CaCO_3(s)$ will be regenerated. This fact has been utilized in our laboratories by Culmo and Fyans,[7] to determine the amount of $CaO(s)$ in both the fly ash and bed ash from a fluidized bed combustion furnace using thermogravimetry. Figure 7 showing the DTA cooling curve obtained when 18.56mg of the same blue calcite specimen was cooled from 1020°C to 570°C in dynamic $CO_2(g)$ purge (P_{CO_2}= 1 atm) at 10°C/min using the DTA-1700/TADS System. An important fact to be pointed out here is that the peak analysis as performed by the TADS DTA Standard Software is that for a cooling curve. The peak onset temperature is extrapolated using the cooling onset - an important consideration for any DTA or DSC software package. One will observe that the exothermic peak area is given a negative sign (-659.41 J/gram by the TADS "Peak" analysis routine. When this exothermic reformation energy is compared with the endothermic decomposition energy for this blue calcite specimen given in Figure 5, one may calculate that the exothermic peak area is only 39.35% of that for the endothermic decomposition of the calcite specimen.

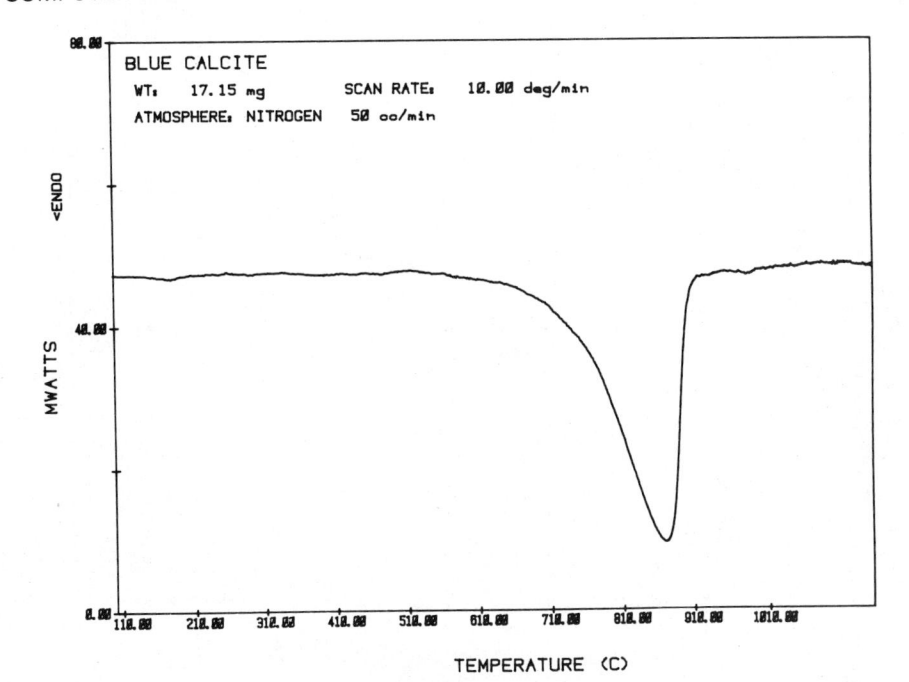

Figure 4. DTA Thermal Curve for Calcite using DTA-1700/TADS System.

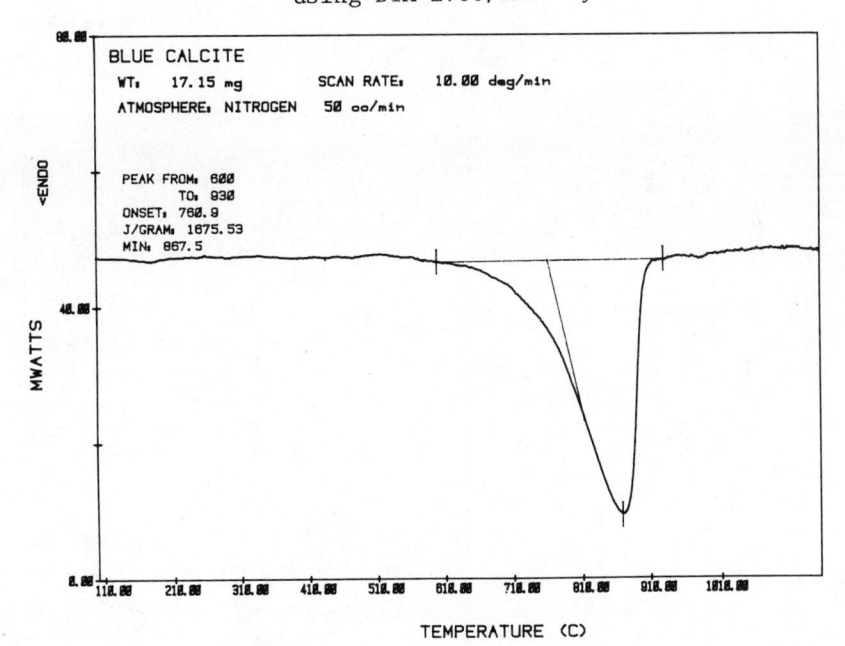

Figure 5. DTA Thermal Curve for Calcite with TADS PEAK Analysis.

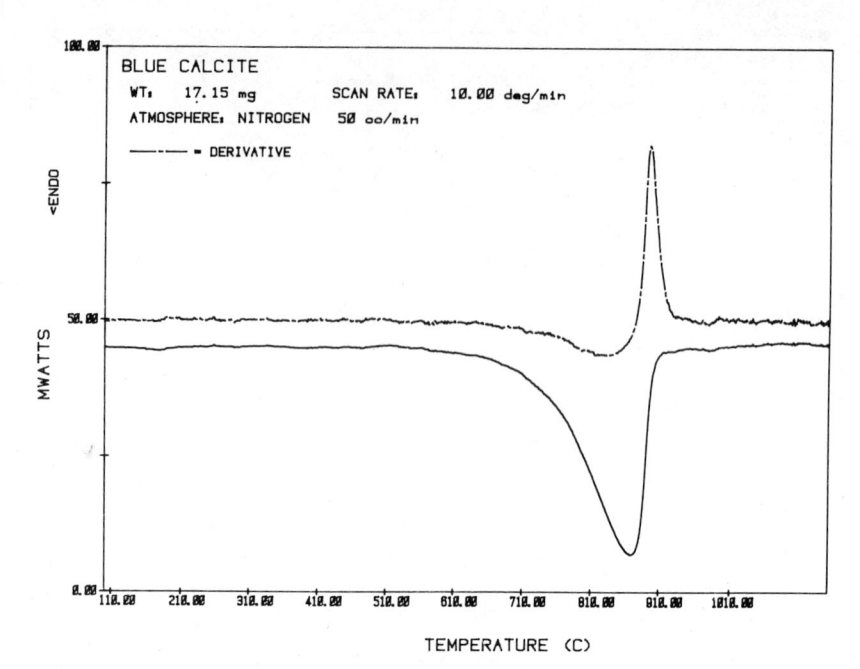

Figure 6. Simultaneous Display of Both DTA and First Derivative Thermal Curves for Calcite.

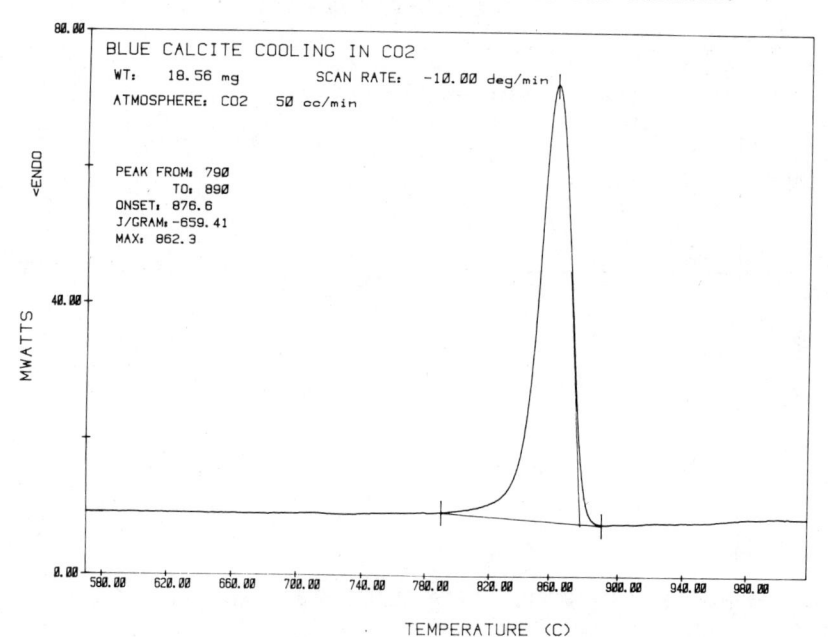

Figure 7. DTA-1700/TADS, Cooling Curve for Calcite with PEAK Analysis.

Figure 8 shows the DTA thermal curve for the National Bureau of Standards SRM 88a dolomitic limestone specimen. This DTA thermal curve was obtained using a heating rate of 10°C/min with the DTA-1700/TADS System. In a true, unsubstituted dolomite, Ca,Mg(CO3)2, the ratio of gram atoms of Ca to Mg is 1.00. In this specimen, the calcium ions are in excess, therefore, the term "dolomitic limestone" has been applied. This specimen also contains 1.20% SiO_2. As will be observed in Figure 8, the dolomitic specimen decomposes in two steps. These may be described by the following equations:

$$Ca,Mg(CO_3)_2 \longrightarrow CaCO_3(s) + MgO(s) + CO_2(g)$$

then

$$CaCO_3(s) \longrightarrow CaO(s) + CO_2(g)$$

As can be seen by the thermal curve in Figure 8, these two events are not completely separated in dynamic nitrogen atmospheres. The peak minima for these two events are separated, however, and for this 10.99 milligram sample are observed at 765°C and 820°C.

As was first shown by Rowland and Lewis,[8] the two decompositional events may be clearly separated by the use of dynamic $CO_2(g)$ atmospheres.

The separation of these two events in $CO_2(g)$ purge is due to the greater effect of the $CO_2(g)$ on the $CaCO_3(s) \longrightarrow CaO(s) + CO_2(g)$ equilibrium than that for the less reversible magnesium carbonate, magnesium oxide, $CO_2(g)$ system. Hence, the second thermal event in the decomposition of unsubstituted dolomitic specimens is shifted to higher temperatures and the peak is both sharper and intensified. Figure 9 shows the DTA curves obtained for this dolomitic limestone specimen obtained on both heating and cooling in dynamic CO_2 purge. As can be seen in this figure, the second stage of the dolomite decomposition is shifted approximately 120°C up the temperature scale and complete separation of the two events is achieved. On cooling, as was observed with the blue calcite specimen, a portion of the calcium oxide is converted back to $CaCO_3(s)$ while no such reformation exotherm is observed for the MgO.

The use of several atmospheres to affect such shifts in carbonate peak temperatures, as well as metal oxide reactions in some cases, has been utilized by Warne,[9,10,11] as a diagnostic tool for the recognition and identification of numerous carbonate mineral species. Thus, Warne has used the term "Variable Atmosphere DTA" for such studies. Substitution of the cations in dolomitic specimens can lead to several isomorphous series. Substitution of Fe^{+2} for Mg^{+2} leads to the dolomite - ferroan dolomite - ankerite series of trigonal carbonates. This series has been extensively studied by Warne,[9,10,11] and Warne, et. al,[13] using variable atmosphere DTA

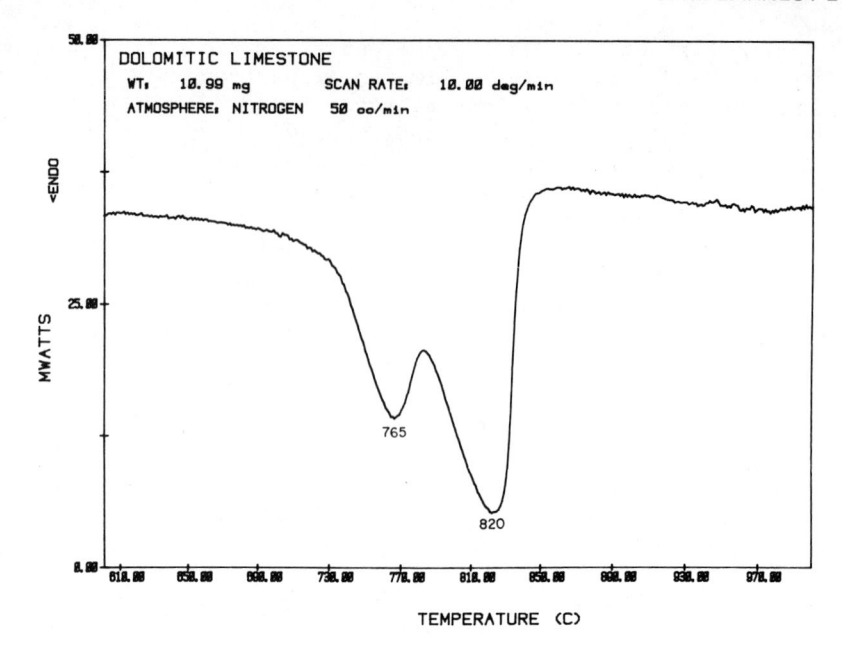

Figure 8. DTA Thermal Curve for NBS 88a Dolomitic Limestone in Dynamic Nitrogen Purge.

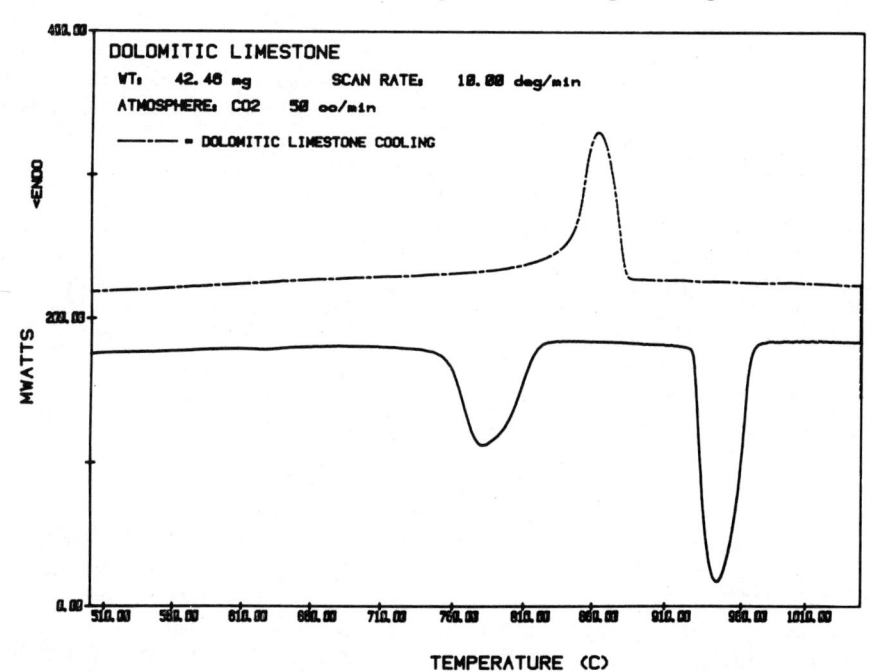

Figure 9. DTA Heating and Cooling Curves for NBS 88a Dolomitic Limestone in Dynamic CO_2 Purge (P_{CO_2} = 1 atm).

techniques. Figure 10 shows the thermal curve obtained for an ankerite specimen obtained in dynamic CO_2 purge using the DTA-1700/ TADS system in the DTA mode of operation. This ankerite specimen was originally collected at Blue Rock Tunnel, Pennsylvania (U.S.A.) and the carbonate component was analyzed to contain 51.63% $CaCO_3$, 19.05% $MgCO_3$, 28.23% $FeCO_3$, and 1.09% $MnCO_3$. As has been previously shown by Warne,[11] the peaks are clearly separated when the DTA thermal curves are obtained in dynamic CO_2 atmosphere. The identities of the peaks are not as straightforward as those for the unsubstituted dolomite specimens. Warne,[11] has shown, using specimens of varying Fe^{+2} levels, that the peak at 852°C is definitely related to the iron content. Milodowski and Morgan,[13] using x-ray diffraction as well as evolved gas analysis along with DTA, TG, and DTG, have concluded that the peak at 658°C is due to the decomposition of all of the $FeCO_3$ component and most of the $MgCO_3$ forming $CaCO_3$, MgO, and magnesioferrite ($MgO \cdot Fe_2O_3$). The small amount of remaining $MgCO_3$ was believed to exist in solid solution with the $CaCO_3$ and is gradually eliminated with increasing temperature. Most of the iron (II) oxide formed in this decomposition is oxidized by $CO_2(g)$ to iron (III) oxide (hematite) liberating carbon monoxide, $CO(g)$. The remaining FeO (wustite) combines with Fe_2O_3 forming the spinel form Fe_3O_4 (magnetite).

The peak at 852°C, according to Milodowski and Morgan,[13] is due to the reaction of the $CaCO_3$ with magnesioferrite to form non-magnetic dicalcium ferrite ($2CaO \cdot Fe_2O_3$) and MgO. Also, the magnetite is converted to hematite as the remaining iron (II) oxide is oxidized during this step by $CO_2(g)$. The last peak at 917°C is due to the decomposition of unreacted $CaCO_3$.

Figure 11 shows the DTA-1700/TADS thermal curve after three successive peak analyses performed by the TADS DTA Standard Software. One can see that when the data is acquired in the DTA mode of operation, the DTA Standard Software assigns peak heights rather than peak areas as was observed for the DSC mode. Again, the extrapolated onset temperatures, and peak minima are assigned for each individual peak analyzed. Furthermore, on exiting this analysis routine, the TADS CRT screen prompt will ask the operator if he/she wants to save the results. If so, the operator may reply "yes" with the keyboard and the TADS CRT screen will then ask for a file name.

CONCLUSION

The total computerization of thermal analysis instrumentation has led to improved hardware performance as well as many time-saving and convenient data handling features. Probably the greatest advantage of this total computerization is that one will seldom have to repeat the DTA experiment due to either improper choice of ordinate scale sensitivity or sloping baseline characteristics. Furthermore, the raw data may now be stored on microfloppy disks

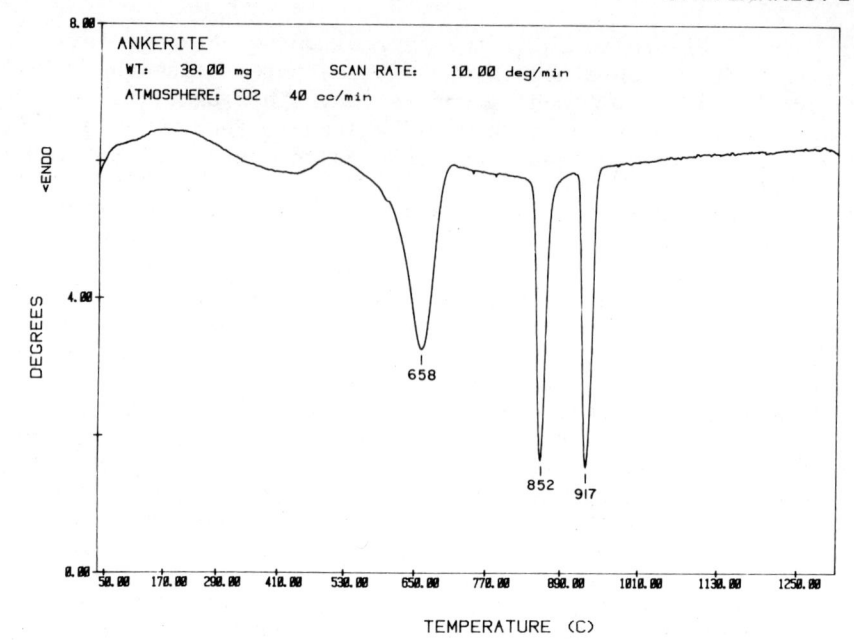

Figure 10. DTA Thermal Curve for Ankerite (Blue Rock Tunnel, Pa.) Specimen in Dynamic CO_2 Purge.

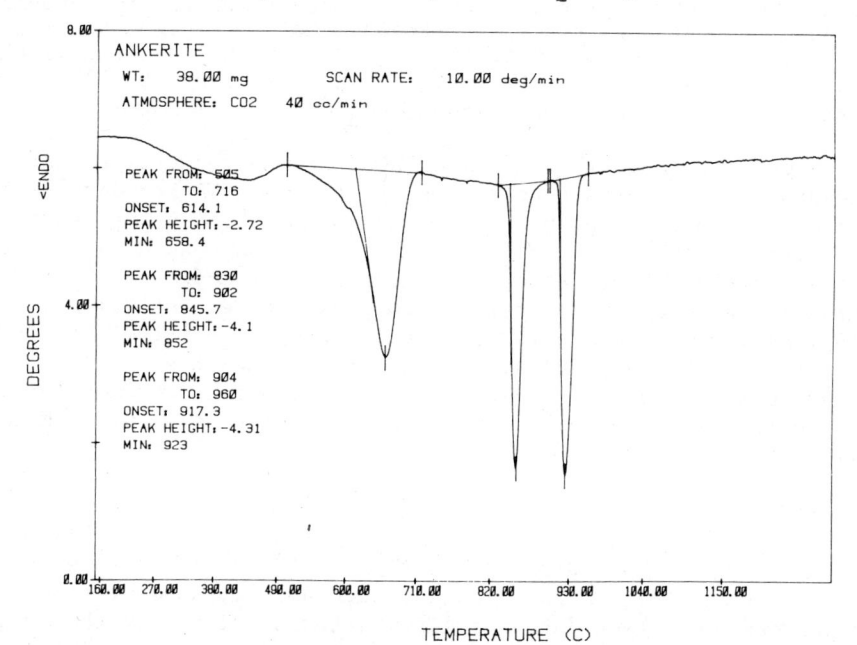

Figure 11. Hard Copy of Ankerite DTA Thermal Curve After Three Successive "Peak Analyses" by the TADS DTA STANDARD SOFTWARE.

as a permanent record and may be recalled at a later date for analysis.

REFERENCES

1. H. LeChatelier, Bull. Soc. Fr. Mineral Cristallogr, 10: 204 (1887).
2. W.C. Roberts-Austen, Proc. Mech. Eng., 35 (1899).
3. W.W. Wendlandt, "Development of Differential Thermal Analysis in the U.S., 1887-1976," Amer. Lab., 9 (#1): 59 (1977).
4. W.P. Brennan and M.P. DiVito, "Computers in the Plastics Laboratory: Trends and Prospects," presented at the 1982 Pittsburgh Conference on Analytical Chemistry and Applied Spectroscopy, Atlantic City, N.J., March, 1982, Paper #484.
5. C.M. Earnest, "Recent Advances in Computerized Thermal Analysis of Ceramic Raw Materials," presented at the 84th Annual Meeting of the American Ceramic Society, Cincinnati, Ohio, May, 1982, Paper No. 1-JXIV-82.
6. D.N. Todor, "Thermal Analysis of Minerals," Abacus Press, Kent (1976).
7. R.F. Culmo and R.L. Fyans, "The Determination of Combustion Efficiency and Calcium Utilization of a Fluidized Bed Combustion Furnace," presented at the 1981 Pittsburgh Conference on Analytical Chemistry and Applied Spectroscopy, Atlantic City, New Jersey, March, 1981, Paper No. 496.
8. R.A. Rowland and D.R. Lewis, Am. Miner., 36: 80-91 (1951).
9. S. St. J. Warne, "Predictable Curve Modifications and Variable Atmosphere DTA: Uses in Diagnostic Mineralogy, Particularly for Carbonates," in: "Thermal Analysis: Proc. of the Sixth ICTA," Birkhauser Verlag, Basel, 1980.
10. S. St. J. Warne, "An Improved Differential Thermal Analysis Method for the Identification and Evaluation of Calcite, Dolomite, and Ankerite in Coal," J. Inst. of Fuel 48: 142-145 (1975).
11. S. St. J. Warne, "The Detection and Elucidation of the Iron Components, Present as Carbonates, in Coal, by Variable Atmosphere Differential Thermal Analysis," in "Thermal Analysis: Proc. of the Fifth ICTA," Heyden, London, 1977.
12. S. St. J. Warne, D.J. Morgan, and A.E. Milodowski, "Thermal Analysis of the Dolomite, Ferroan Dolomite, Ankerite Series, Part 1. Iron Content Recognition and Determination By Variable Atmosphere DTA," Thermochim. Acta 51: 105 (1981).
13. A.E. Milodowski and D.J. Morgan, "Thermal Decomposition of Minerals of the Dolomite - Ferroan Dolomite - Ankerite Series in CO_2 Atmosphere," Proc. of the Second ESTA," Heyden, London, 1981.

REFLECTANCE TECHNIQUE FOR MEASUREMENT OF BOTTOM SURFACE TIN CONCENTRATION ON CLEAR FLOAT GLASS

Thomas O. LaFramboise

Ford Motor Company - Glass Division
25500 West Outer Drive, Lincoln Park, MI 48146

In the float glass process, some tin is absorbed into the bottom surface of the ribbon. If the glass is to be reheated for a bending or tempering operation; a surface haze, commonly called bloom, can occur when the tin concentration is sufficiently high. The degree of bloom is dependent upon the amount of stannous tin available for oxidation in the surface of the glass. Although the tin penetrates to a depth of about 10 microns, the stannous ions in the first micron appear to control the formation of bloom.

The common procedure for estimating the total surface tin is to ratio the tin peak intensity (by x-ray flourescence) of the unknown to that of a standard. This procedure yields a value referred to as the "tin count". In our laboratory, glass with tin counts above .45 is not considered suitable for heat treatment. This number does not accurately predict bloom susceptibility, since x-rays penetrate several microns into the surface and examine an integral volume. An x-ray procedure cannot analyze the tin in the critical surface micron that is the cause of bloom.

Tin oxide has a significantly higher refractive index than glass; therefore, the presence of tin will increase the index and surface reflectance of the glass - the greater the concentration of tin, the higher the surface reflectance. A reliable method has been developed to predict the bloom potential of a clear float glass by measuring this increase in reflectance. This is accomplished by coating the atmosphere surface of the glass with a black paint (to eliminate reflectance) and measuring the reflectance of the tin surface using a commercial glossmeter - an instrument that measures light reflected specularly by non-metallic surfaces.

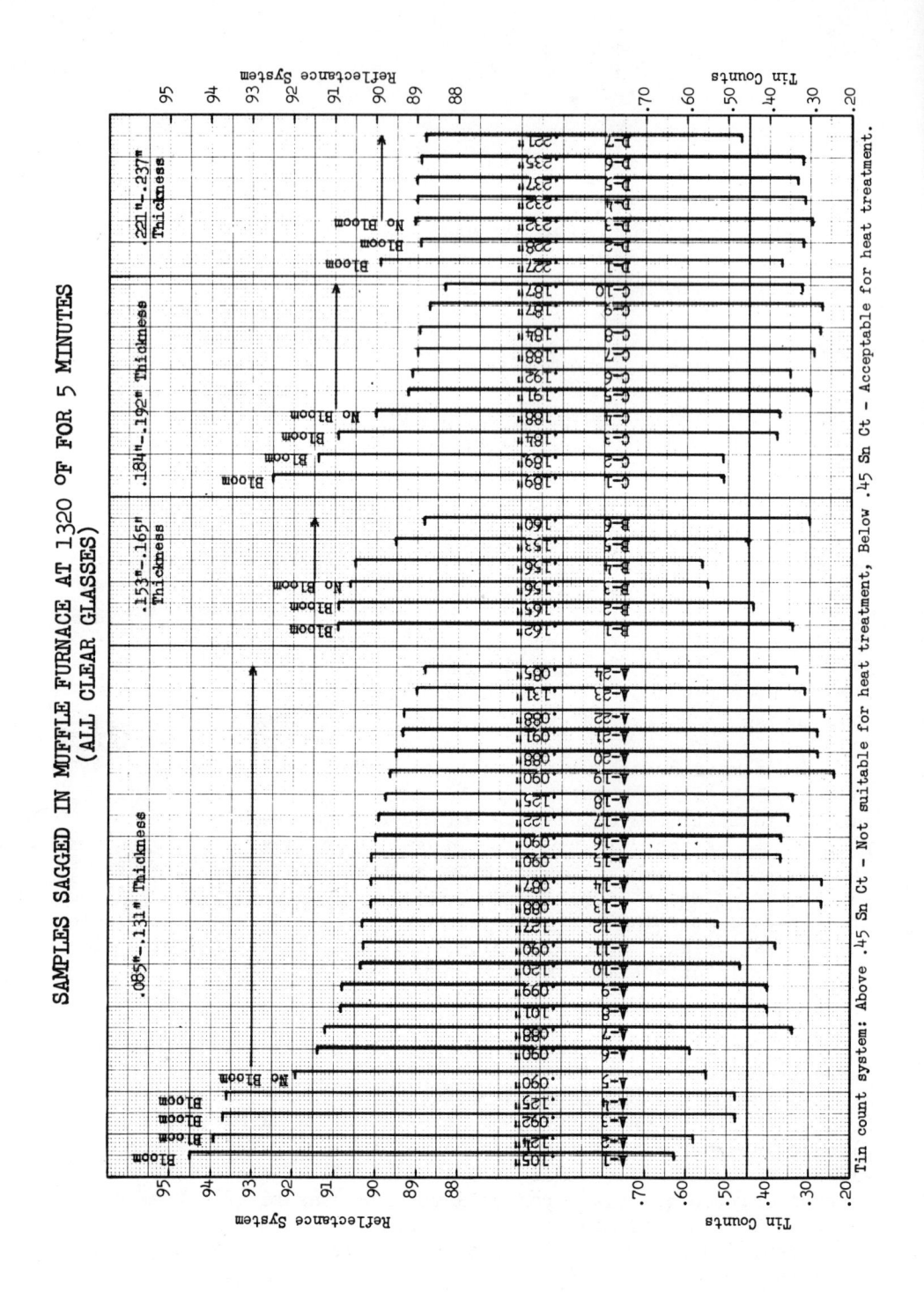

Fig. — SAMPLES SAGGED IN MUFFLE FURNACE AT 1320 °F FOR 5 MINUTES (ALL CLEAR GLASSES)

Tin count system: Above .45 Sn Ct - Not suitable for heat treatment, Below .45 Sn Ct - Acceptable for heat treatment.

In order to simulate bloom, a laboratory test was devised in which production float glass samples were cut into 1" x 3-1/2" strips, placed tin side up across a 2-1/2" x 2-1/2" x 5/8" deep platinum dish, and heated in a muffle furnace at 1320 °F for five minutes to induce sagging. The sagged samples facilitated visual observations of bloom.

Fourty-seven representative production glasses were selected to determine the correlation between the bloom furnace sag test, the tin count method, and the reflectance readings. The results of the experiments are illustrated in the accompanying graph.

The data indicate a 98% correlation between reflectance and laboratory-induced bloom for the thickness range of the samples tested. These experimental observations indicated that thicker glass has a greater tendency to bloom. For example .221" - .237" thick glass (Samples D-1 through D-7) is likely to bloom when the reflectance of the surface is only 89.5, while the critical reflectance readings for .153" - .192" (B and C Series Samples) glass is 90.7, and glass in the .085" - .131" (A Series) range did not bloom until a reflectance of above 92.5 was achieved. Using the tin count system the correlation with bloom formation was only 72%.

This technique is easily instituted into a Q.C. laboratory operation, as little training time is required to achieve consistent test results.

CHARACTERIZATION OF REINFORCEMENTS FOR INORGANIC COMPOSITES

S. W. Bradstreet

Nevada Engineering & Technology Corp., MMCIAC Division
Long Beach, CA 90806

ABSTRACT

Data analysis for structural composites inevitably depends upon the careful application of the Law of Mixtures,- the mathematical identity of a property value of a composite with the sum of the volume-weighted values of that property for the several constituents. Most important among characterizing parameters is the real density. Using for illustration a currently available polycrystalline Al_2O_3 fiber in a well-characterized but fictitious borosilicate glass matrix, the simple calculations necessary to reconcile calculated with observed values for real density, longitudinal elastic modulus and linear coefficient of thermal expansion are shown, and the extreme importance of accurately measured and fully reported real densities emphasized. A modest change in the current zero-porosity density of polycrystal alpha-alumina is suggested.

INTRODUCTION

The earliest parameter for characterizing a solid body must have been its weight, and the simple balance, fortuitously subjecting the body and balancing weights to the same gravity field, was essential to the development of coinage. By the time Euclid established the principles of mensuration and geometry, the fact that many solid materials exhibited different and characteristic densities can be regarded as a first step in what we now call materials science.

The public spectacle of a moist philosopher rushing from the public baths of Syracuse, shouting "I've got it!" (in Greek, of course) has been amply chronicled, and often used to illustrate the excitement of the flash of insight which solves a problem. Archimedes' problem would be suitable for a crime-detection story: his relative, King Hieron, had commissioned a golden crown and provided a weighed gold ingot for its working. The ornate crown exactly matched the king's expectations and the weight of the ingot. The problem arose from the kingly suspicion,- could the wily goldsmith have extracted a modicum of gold and added an equal weight of silver, or even copper? So he instructed his gifted relative to prove or disprove his suspicion without in any way harming the crown, which was too irregular in shape and surface for volume by measurement. And Archimedes, in that insightful moment in the bath, discovered what is still, more than twenty centuries later, the most accurate of nondestructive tests for measuring the volume of a solid body.

REAL AND OTHER DENSITIES

The distinction made in distinguishing the double-weighing method based on the buoyancy of a specimen immersed in a fluid of known density and all other methods is not that they are unreal, but that they are less accurate. For most laboratory specimens, the mass determined on a modern balance in air can easily and quickly be determined to one part in 10,000. Submerged in a liquid (such as absolute ethanol) for which the density is as exactly known, it can be weighed with equal exactitude. Obviously, the temperature of measurements should be the same. Refinements in technique (Ref 1) permit accuracy of the W/V ratio to better than 30 ppm.

It has been frquently pointed out to the writer that in most ceramics and composites, the density of each specimen is appreciably different. This criticism neglects the point that since the Archimedean density is completely nondestructive, it permits the use of the same specimen for several nondestructive property tests, and from these it is usually possible to identify nonconforming specimens and learn the reasons for individual behavior.

Second in accuracy to the Archimedean density measurement is one which depends upon Archimedean principles. This involves the fact that a solid will become weightless in a liquid of identical density, and the use of a density-gradient column has become popular, particularly for particulate and short-fiber specimens (Ref 2). Here the mixing tendency of such columns leads not only to inexact isodensity levels but may require such relatively large local gradients for stability as to lead to lower discriminating power.

The problems involved with __bulk density__ measurements in which the volume is determined by mensuration are, of course, simplified when the specimen is reasonably hard and tough enough for cutting, machining, and polishing (an example will be given later). For laboratory sized specimens, an accuracy of more than four significant figures in the calculated density will be found to require optical polishing of flat surfaces and extreme care to avoid scratches and edge damage. It must be noted that the error in the volume measurement will be the product of the three measurement errors, usually with the largest error in the smallest dimension. The general practice of obtaining bulk density values for thin composite sheets and calculating the specimen volume from its dimensions and the average sheet thickness is suspect.

The <u>theoretical density</u> is usually applied to materials in which the crystal(s) are assigned a chemical identity by analysis and a spatial structure (lattice constants) from X-ray or electron diffraction patterns. The term "theoretical" arises from the assumed ideality of atomic packing in the crystal, from which its density can be calculated. For most single crystals (whiskers) the deviation from ideality may be small (less than a percent or so), but the usual practice of "characterizing" a polycrystalline body as having a density which is "x percent of the theoretical (or X-ray) density" is of no value to the analyst unless that density is reported. The remainder of the body is regarded as porosity, where the pore fraction is given by: $P = (\rho_0 - \rho)/\rho_0$, where ρ_0 is the X-ray, and ρ the measured, density. This will be discussed later in some detail.

<u>Specific gravity</u> is a term originally meant to describe the ratio of the specimen mass to that of an identical volume of water; with the discovery that the density of water does not change linearly with temperature, the term usually is taken to mean the real density of the solid component in a porous body, obtained following pulverization or grining sufficient to eliminate closed pores.

For small or powder samples, the <u>pycnometer density</u> is frequently used; the calibrated holding vessel is filled with a liquid of known density to a fixed level, the preweighed sample is added and displaces a volume of liquid which can be measured with reasonable accuracy. With proper attention to temperature control, accuracies of at least three significant figures can be obtained.

For solids having interconnected pores, the <u>porosimeter</u> is a valuable instrument; in it the preweighed specimen is immersed in a liquid which will not ordinarily enter the pores (mercury is most frequently used with ceramic materials), and the change in volume with increasing pressure can be used not only to determine the open porosity but also pore size (an advantage of mercury is that X-radiography can be used for observing the pore pattern. For systems in which the pores are extremely fine, other fluids (alcohols, oils, glycerine, water, argon, air, hydrogen, and helium have been used.

The graphitic fibers appear always to comprise either planes or scroll-like fibrils of essential ideal graphite bonded together by a "disorganized" carbon, less dense and chemically more active. Moderately dense polycrystalline graphites are not usually penetrated by glycerine, but the less dense carbon-carbon composites are first weighed, thinly coated with volatizable polymer of known density, reweighing to obtain the coating volume, with a final weighing in a non-reactive liquid. The same technique can usually be applied to porous ceramic bodies.

This brief review of densitometry may be redundant at Alfred University, where several of the techniques were pioneered and more have been refined. Nevertheless, few modern reports or papers provide density values of the accuracy or precision with which individual measurements can readily be made. The reasons for this are not far to seek; the natural conservatism of the scientific observer makes "rounding off" data easy, and much of today's literature shows it to be common. The 20C densities of the pure metals of interest to Archimedes are given (Ref 3) as: Au = 19.3, Ag = 10.5, and Cu = 8.933 gm/cc. It seems highly improbable that copper can be tenfold more accurately determined than the more noble metals. A decided improvement can be found (Ref 4) for which, at 25C, the respective densities are given as 19.302, 10.847, and 8.933 gm/cc. Since lattice dimensions are also provided, the X-ray densities are readily calculated to be: 19.282_8, 10.501_3, and 8.933_1 gm/cc. A method for reconciling X-ray and real densities for some pure metals will be shown (Appendix 1).

The analyst, of course, cannot modify the observed data, but in some instances it is possible to provide insight regarding the interactions between the constituents of a composite. We are, of course, familiar with the Rule of Mixtures and frequently use it to forecast a number of composite properties.

THE RULE OF MIXTURES (RoM) AND THE LAW OF MIXTURES (LoM)

The RoM expresses the incontrovertible statement that the whole is the sum of its parts. Mathematically it is expressed:
$$P_c = P_1 V_1 + P_2 V_2 + \ldots + P_n V_n \tag{1}$$
, where P is the value of a property in the composite and in the constituents 1 through n, and V represents the volume fraction of each component.

There are three requirements implicit in this expression. The first is that the composite is large enough and the distribution of unit volumes even enough to provide a statistically valid population.

The second is that the test conditions do not change either P or V values.

The third is that all of the constituents must be included, and their volume fractions add to unity. Mass fractions cannot be used.

For a number of particulate composites and for some of the "advanced" fiber-reinforced polymer composites, the RoM is well enough obeyed to serve for preliminary design. For the advanced fiber-reinforced inorganic matrix combinations, both constituent properties and volumes are changed through interactions which occur during consolidation or later treatment. For analysis of these, the writer has found useful the Law of Mixtures. This is mathematically identical with the RoM, but includes two more requirements.

Fourth, the volume fraction for each constituent must be that which obtains in the mixture. For simple unidirectional composites with coarse filaments the volume fractions can be closely approximated from magnified cross-section photographs.

Fifth, the property values for the constituents must be those which obtain in the composite. These are not readily measurable, and the LoM is valuable simply because it provides a method for verifying or rejecting calculated values.

Of the several properties amenable to treatment by the LoM, the real density is most valuable. From the RoM we have, for a binary composite of fibers and matrix: $\rho_c = \rho_f V_f + \rho_m V_m$, where ρ_f is the fiber density and ρ_m that of the matrix.

With very few exceptions, advanced inorganic matrix composites exhibit density values appreciably different from that calculated by the RoM. Any of three mechanisms can be responsible:
 a) there has been a net gain or loss of mass,
 b) there has been a net gain in volume through the development of porosity, and/or
 c) there has been a change in volume because of a chemical or physical interaction between the constituents.

A change in mass can be descried by quantitative chemical analysis of a specimen whose density has been accurately determined. Since this necessarily destroys the specimen, it is recommended that the analysis be deferred in favor of obtaining supporting data from it.

If porosity exists, it is usually the result of entrapment of dissolved gases or volatile species, and can be observed through metallographic examination. A second source of porosity is the simple failure of the matrix to encapsulate and bond to the reinforcement or to itself, indicating that higher temperatures, greater pressures, and/or longer consolidation times are necessary.

In virtually all fiber-reinforced inorganic matrix composites there is an interaction responsible for a change of volume following consolidation. This involves the formation of an interfacial zone as the result of diffusion, and the development of residual stresses caused by cooling materials having different thermal expansivities.

ANALYTICAL EXAMPLE: A UNIDIRECTIONAL FIBROUS FP*-GLASS COMPOSITE

To illustrate some of the problems (and their potential solutions) encountered in composite data analysis, this example involves two constituents: parallel and evenly distributed fine fibers of high-purity polycrystalline Al_2O_3* in a matrix of essentially non-porous glass. It will be assumed that bonding between fiber and glass is complete, that the interfacial zone is infinitesimally thick and exhibits properties intermediate between those of fiber and matrix and can thus be neglected. Because the Al_2O_3 fibers are stronger, more rigid, and are higher in coefficient of thermal expansion (CTE) than the glass, cooling from the consolidation temperature will result in the development of residual tensile stresses in the fibers and balancing compressive stresses in the matrix. The composite is thus a longitudinally prestressed ceramic material (Ref 5). It is also assumed that both filament and glass are elastically isotropic in the unstresses condition.

Preliminary Analysis of FP Fiber

Vendor data for the FP fibers are limited to the following: RT density 3.90 gm/cc, Young's modulus (longitudinal) 50-55 Msi, ultimate tensile strength (UTS) more than 200 ksi. X-ray diffraction shows the somewhat diffuse pattern of fine-grained polycrystalline alpha-Al_2O_3 of at least 99.9% purity. Average fiber diameter is 20 um.

Particularly with regard to density and elastic modulus, these data are insufficient to characterize the fibers adequately. Fortuneately, a great deal of careful and occasionally elegant work has been carried out and reported.(Ref 6, 7, 8)

Ryshkewitch (Ref 9) observed that tensile failure in fine-grained polycrystalline alumina at ordinary temperatures was as apt to be transgranular as intergranular at ordinary temperatures, and postulated that both strength and elastic moduli could be related to those of sapphire if the porosity is taken into account.

For hot-pressed and cold-pressed and sintered polycrystalline alumina, Spriggs et al. (Ref 10) developed for the Young's modulus:

$$E = E_o \exp -3.95 P \qquad , \text{where} \qquad (2)$$

E is the calculated elastic modulus, E_o has a value of 410.2 GPa (59.49 Msi) and P is the volume pore fraction: $P = (\rho_o - \rho)/\rho_o$. It should be noted that Eq (2) must be modified for elastically anisotropic aluminas.

The density of sapphire used is the X-ray density obtained from

$$\rho_o = N \cdot M / V \cdot Av, \text{ where N is the number of molecules per unit} \quad (3)$$

---
* Product of E. I. DuPont Adv. Laboratories, Wilmington, Del.

cell, M the molecular mass in amu, V the cell volume in $Å^3$, and Av is Avogadro's number, then taken to be 6.024 x 10^{23} amu/gm, the X-ray density of sapphire at RT was calculated to be 3.986 gm/cc, and the Young's modulus for the fiber is calculated to be 54.6 Msi. It can be reported to but three significant figures because that is the limiting accuracy of the fiber density.

For its lowest and highest bounds, the fiber density might be 3.895 to 3.904 gm/cc, and the calculated modulus limits are 54.3_6 and 54.8_5 Msi respectively. This difference is larger by nearly five times than the discrimination norm for the stress/strain ratio in static tension, and many times that for dynamic longitudinal resonance.

A second refinement is necessary. The earlier value for Av. has now been superseded by the more accurate 6.02294 x 10^{23} amu/gm, measured by the NBS using hyperpure, fully dense silicon. Using this value, ρ_o becomes 3.9873 gm/cc, and the upper and lower modulus bounds are 54.2_9 and 54.7_8 Msi.

We have seen that X-ray density is not necessarily more accurate than bulk density. Lang (Ref 11), using both hot-pressed and cold-pressed and sintered rods of rectangular cross-section, took great care to measure their dimensions and even adjusted their masses for the buoyancy of air. He estimated a coefficient of variation of less than one percent for the density measurement and for the dynamic longitudinal modulus E_L calculated from the expressions:

$$c_L = 2 L F_L \quad \text{and} \quad E_L = \rho c^2$$

, where c is the speed of sound in cm/sec, L is the resonating length of the specimen in cm, F_L is the fundamental frequency of dynamic resonance in Hz, and ρ is the bulk density in gm/cc. The unit of E_L are in kilobars, or 10 GPa.

Lang shows a nearly linear ratio between E_L and ρ, but careful examination of his selected data points show the curve to be slightly concave upward and the linear extension of his most dense specimens (E_L = 317.797 ρ -856.954 GPa) yields, at $\rho = 3.9872_7$ gm/cc, a value for E_0 of 410.188 GPa (59.49_3 Msi).

This exact agreement with Spriggs et al. may be fortuitous, but is generally conceded that the principal reason for slightly higher dynamic modulus values than static ones in polycrystalline ceramics is that the static tests are so slowly loaded as to permit some stress relaxation, presumably by boundary sliding. There appears, then, no reason not to use Eq (2) for calculating the bounds given above for E_L of the FP fiber, or a mean value of 54.57 Msi, using it as compared to 55.47 Msi from interpolation in the Lang data. Until the average density of the FP fibers is known to four significant figures, one must use, for static measurements, E_f = 54.54 ± 0.24 Msi.

Other needed data for the fiber can be closely estimated from Ref 8, esp. chapters by R. Kirby and J. Wachtman.

Preliminary Description of Glass Matrix

The glass matrix used in this example may not exist; its properties are synthesized from three alkali metal and alkaline earth metal modified borosilicate glasses, and its property values of interest are chosen for illustrative purposes only. Perhaps wishfully, one finds it to be unusually well characterized with regard to density, Young's and bulk moduli, and CTE at RT. Unlike polymer matrices, which are easily compressed, or metals capable of plastic flow above their yield stress, the glass is presumed to be wholly elastic below its fictive temperature (about 415C). Other pertinent property values for it are shown in Table 1.

Analysis of Calculated and Observed Values of RT ρ, E_L, and CTE_L

The composite is charged with parallel fibers and finely powdered glass in a thin metal envelope, evacuated at a temperature sufficient to remove the fugitive binder and adsorbed water. The envelope is then sealed and hot isostatically pressed at a pressure and temperature sufficient to insure full densification of the glass and firm attachment to the fibers. The envelope and its contents are cooled slowly below the fictive temperature of the glass to insure even stress distribution.

All measurements are presumed accurate to four significant digits, except that the density of the fibers is taken as 3.90 $\pm$ .005 gm/cc.

Density and elastic modulus are volume-controlled properties and exactly measurable by the LoM, but CTE, being a linear strain, cannot be so used. Attachment of fibers to matrix, however, requires that a common strain exists at the interface, resulting in a residual stress in the fiber direction given by: (Ref 12)

$$CTE_L \, E_L = CTE_f \, E_f \, V_f + CTE_m \, E_m \, (1-V_f) \quad \text{in a binary system.}$$

The nominal volume fraction of the composite, made by weighing the components in proportion to their densities, lies between the bounds of 3.215 and 3.220 gm/cc if the weights used were exactly in proportion of 3.900:2,535; in either case the volume fractions of fiber and matrix are 0.5000 $\pm$ .00184 and the mean density 3.2175 gm/cc. Tobserved density, however, is 3.2250 gm/cc, and the RoM value is in error between the bounds of +0.3101% and 0.1550%; the composite has either undergone a net increase in mass or a decrease in volume; chemical analysis rules out mass change and porosity.

Again using the RoM, we find that the measured value of Young's modulus is from 4.88 to 4.17% higher than predicted The values shown are static, but one could as well have used dynamic and hence non-destructive measurement. In the same way, the nondestructively meas-

REINFORCEMENTS FOR INORGANIC COMPOSITES

CTE_L value is larger than predicted by from 3.00 to 2.46%. Note that all of these measurements can be performed on the same specimen, so that sample variation can be ruled out.

It is quite clear that in cooling from the consolidation temperature both volume fractions and property values have changed. The linear shrinkage of the fibers over the 25-415C range is .00205, and the glass .00096; the differential between them .001092, producing a net longitudinal tensile stress of 59.56 ksi in the fibers and a balancing compressive stress of 11.42 in the glass (hoop stresses being neglected). With these and the bulk moduli of the constituents the densities of FP can be calculated using the Spriggs expression (but noting that the tensed fibers occupy a larger volume, and the compressed glass a smaller, as the result of these stresses.

It will be observed that the RoM errors for the longitudinal modulus are about 6 times the CTE errors, and on the average more than 20 times the density errors. Clearly the accuracy of the density measurement is of paramount importance.

Table 1

Data Pertinent to RoM and LoM Calculations for 0.50 FP-Gl Composite

Property P_c (Obs.)	P_f (FP)	P_m (Gl)	RoM	error(%)
ρ_{25} (gm/cc) 3.225$_0$	3.895	2.535	3.215	-0.31
	3.905		3.220	-0.155
E_L (Msi) 34.05	54.296	10.45	32.37$_3$	-4.933
	54.782		32.61$_6$	-4.211
$CTE_{25-415C}$ (ppm/C)	5.21	2.46		
CTE_{25-100} 4.71$_0$	5.15	2.458	4.67$_4$	-.764$_3$
			4.67$_7$	-.700$_6$
B_{25}	35.0	6.85	--	--

Lin. stress (Msi)$_{100}$	.047690	-.009935
	.048160	-.009542
Lin. stress $_{25}$	.060464	-.012533
	.061081	-.012254

ρ_{25} (ρ_{100}) gm/cc	3.875$_2$ (3.889$_9$)	2.536$_6$ (2.536$_2$)	3.225	0.0
	3.885$_0$ (3.899$_9$)			
V_f	.51426$_8$ (.50740$_9$)	1-V_f		
	.51054$_9$ (.50371$_0$)			
E_L 25(100) 34.05(-)	55.370 (55.095)	11.477$_6$ (11.212$_8$)*	34.05	0.0
	55.935 (55.639)	11.212$_8$ (11.023$_8$)*		
$CTE_{25-100} \cdot E_L$ 160.38	285.16	28.206	4.707$_0$	-0.06$_4$
	288.07	27.555	4.713$_6$	0.07$_6$

Probable density of FP 3.900 gm/cc

* The increased glass modulus is attributable to the physical constraint exerted by the more rigid fiber through attachment.

SUMMARY AND CONCLUSIONS

The thermomechanical interactions between the rigid and strong advanced reinforcements with the matrix cannot be ignored and should not be approximated with inorganic matrices. At this time, the most accurate method for calculating the volume fractions of the constituents requires measurement of the real density of the composite. While X-ray and bulk density measurements are of supportive value, the greater precision and lower cost of Archimedean measurement lends to its advantage in nondestructively characterizing a composite specimen prior to other tests and providing for easy determination of the changes in constituent properties and volumes.

The unusual characteristics of a semi-commercial polycrystalline fibrous Al_2O_3 make it useful for this purpose because it appears to exhibit a Young's modulus calculable from its density. Whether its apparent "porosity" actually consists of submicroscopic pores in the crystals or, more probably stem from lattice vacancies and from the presence of a less-dense Al_2O_3 at crystallite boundaries remains to be determined.

For a unidirectional fibrous composite, the density, the initial longitudinal modulus, and the linear coefficient of thermal expansion are all nondestructive and applicable without correction to the Rule of Mixtures. Deviations from RoM-calculated values of these properties are valid indicators that either properties or unit volumes (or both) have changed, and permit the direction and magnitude of change to be calculated. In this example, the deviations observed are determined to be the result of: 1) residual stresses in both fiber and matrix resulting from cooling from the consolidation temperature, and 2) modification of the longitudinal modulus of the matrix as a result of its firm attachment to the more rigid, stronger fibers.

A necessary part of the technical analysis is a review of the technical data for the components. A modest change in the density of pore-free polycrystalline Al_2O_3 is recommended on the basis of the use of 6.022094 amu/gm for Avogadro's number; it should be noted that this will slightly increase X-ray densities in much of the literature, and modify some of the small differences observed between X-ray-derived and real density measurements in single crystals.

It is concluded that for typical laboratory specimens of advanced inorganic matrix composites and their constituents, mean densities can easily be reported to four significant figures. Without larger expenditure of time or effort, but careful control of temperature and correction for air buoyancy, accuracy of at least five significant figures can be obtained. It is strongly recommended that such density values not be rounded off in reporting them.

APPENDIX I

Technical data points initially stand alone. In the foregoing paper the point is made that density values derived from X-ray diffraction are dubious only to the extent that while interatomic distances can thus be measured with adequate precision, their transfer into density values depends upon a number of assumptions which are dubious. Take, for example, the pure solid metals Au, Ag, and Cu. All crystallize in the face-centered cubic mode, and if the atoms are identical spheres, will each be in contact with its twelve neighbors; from spherical packing theory one calculates that PE, the packing efficiency, says that 74.04805 v/o of space will be occupied by them. Unfortunately, packing theory states that it is impossible for one sphere to mutually touch twelve others if any one of the thirteen differs from the others in size.

For the pure metals there are at least two possible causes of packing variation. The most notable of these is partial ionization through removal of a valence electron by a single atom to join the "cloud" of electrons enveloping a number of atoms and are a major factor in electrical conductivity. The second is usually rejected on the basis that although the atoms of polyisotopic elements must certainly have slightly different masses, mass effects are negligible in modifying the lattice vibrations which determine the mean excursion of an atom from its "normal" lattice site.

The real density measurement, whether bulk or Archimedean, is subject to criticism in the sense that it reflects a volume in which occasional cancancies, dislocations, and even pores are a part. Carefully annealed specimens of Ag, for example at 25C, should exhibit a density of 10.501_3 gm/cc, but the best real density (Metals Handbook) is only 10.487, or 0.133 % porous. Such an objection can be countered by observing that even purer Cu, which at the same temperature has an X-ray density of 8.933_1, in good agreement with the real density of 8.933_0 gm/cc. With Au, on the other hand, the X-ray density is 19.283, or .099 % less than the 19.302 gm/cc obtained by Archimedean measurement. Clearly, a negative porosity is not tenable.

For other reasons, the possibility that Na should, in the Table of the Elements, be considered to be the lightest member of the Cu, Ag, Au family is of interest. Like Au it is a monoisotope, Like Au its X-ray density (0.9657) is almost ½% smaller than the real density of almost 0.9702 gm/cc. The outstanding electrical conductivity of Na as compared to the other alkali metals suggests an interesting, if not conclusive possibility: that both the real density and the observed interatomic distances are correct for all four elements; the flaw lies in the assumption that PE will be ideal for all of them. (The ideal PE for body-centered cubic Na is 68.01748 v/o.)

Table 2

Data Used for Comparison of Au, Ag, Cu, and Na with respect to PE$_{25C}$

Ele	($_{25C}$ gm/cc	W amu	N$_{25}$ 10^{22}at/cc	R$_{25}$ Å	V$_{at}$ Å^3	PE v/o	100C ohm-cm^{-1}	mf*
Au	19.302	196.9665	5.90143	1.44200$_3$	12.55993	74.1215	42.55$_6$	1.0
Ag	10.487	107.868	5.85472	1.44469$_3$	12.63034	73.9472	62.89$_3$	1.9725$_7$
Cu	8.933	63.54	8.4663$_8$	<u>1.27812</u>7	8.746023	74.047$_2$	59.73$_3$	1.6204*
Na	.97016	22.9898	2.54130	1.85849$_0$	26.88867	68.332$_2$	19.157	1.0

mf* is a mass factor, calculated on the basis of $(1+2_{1e})(W+W_h)/(W+W_l)$ where 1_e is the mass fraction of the lesser isotope, W_h and W_l the atomic masses of the heavier and lighter isotopes, and W the atomic mass of the element. In the Table of the Isotopes the values of mass given for ^{63}Cu and ^{65}Cu do not accord with the accepted atomic mass of 63.54 amu and the proportions of each were readjusted on the basis of W_{Cu} = 63.540 amu. The above data for Au, Ag, and Na were used to determine k, z, and b in the expression: #

$$PE/PE_{ideal} = k\,(mf)^z\,(W/Z)^b \qquad \text{and}$$

PE$_{Cu}$ for an exact fit is 73.9999$_4$, from which R$_{Cu}$ = 1.27786 Å . On this basis it is reluctantly suggested that the unit cell for Cu is 0.064% too large or that additional characterizing parameters beyond mf* and W/Z are required for reconciling Na with the nominally univalent metals Cu, Ag, and Au.

The above example illustrates the lengths to which the analyst is sometimes forced to go. It is perhaps of interest that in the search for Cu-Cu interatomic distances the values have ranged between 2.5550 and 2.55582 Å. There appear, however, to be instances in which a missing or doubtful value can be approximated by using a related property value as a characterizing parameter.

Ref. "Correlation of Materials Properties with the Atomic Density Concept", Contract NAS8-28517 to NETCO, NASA-Marshall SFC, 1972-76 In it the simple expressions for the atomic density, N, are equated:

$$\frac{(\cdot Av}{W} = N = \frac{PE}{V_{at}}$$

References

1   Smakula, A. and Sils, V., "Precision Density Determination of Large Single Crystals by Hydrostatic Weighing", Phys. Rev. 99(6) 1955

2   Oster, G. and Yamamoto, M., "Density Gradient Techniques", Chem. Rev. 63 257 1963

3   "Thermophysical Properties of Matter" Touloukian and Dewitt ed., Vol. 7, TEPIAC-CINDAS, West Lafayette, Ind. 1970

4   Metals Handbook, Vol. 2, revised 1978

5   Tinklepaugh, J.R., Metals Eng. Quarterly 3 56 1963, also Corbett, W.J. and Walton, J.D., "Metal Fiber Reinforced Ceramics", ASM Symposium on Fiber Composite Materials" ASM, OH 1965

6   Gitzen, W.H. "Alumina Ceramics", AFML TR-66-13, WP AFP, OH 1966

7   Mong, L.E. and Pendergast, W.L. "Control of Factors Affecting Reproducibility of Mechanical Properties of Refractory Semidry Press Specimens", J. Am. Ceram. Soc. 39 (9) 301 1956

8   "Mechanical and Thermal Properties of Ceramics" Symposium Proc., Wachtman, J.B. Jr. ed., NBS Sp. Publ. 303, 1968

9   Ryshkewitch, E., "Tensile Strength of Several Ceramic Bodies of a One-component System", Ber. deut. keram. Ges 22, 363 1941

10   Spriggs, R.M., "Expression for Effect of Porosity on Young's Modulus of Polycrystalline Refractory Materials, Particularly Aluminum Oxide", J. Am. Ceram. Soc. 44 628 1961 Also see "Elastic Deformation of Ceramics and Other Refractory Materials" in Ref. 8

11   Lang, S. M., "Properties of High-Temperature Ceramics & Cermets: Elasticity and Density at Room Temperature" NBS Monograph 6 1960

12   Turner, P.S., "Thermal Expansion Stresses in Reinforced Plastics" J. Res. NBS 37 239 1943. The expression shown simply converts Turner's ratios of mass fractions/densities to volume fractions, and deduces its basis on the more easily understood concept of a common strain at the interface. Because there is an elastic continuum in the fiber direction, the residual stress pattern can be considered to persist throughout the length of the specimen. A similar constraint in particulate composites and the importance of attachment is shown in: Fulrath, R.M., "Controlled Microstructure of Refractory Bodies" Proc. Int. Union of Pure & Applied Chem. 1964

NUCLEAR REACTION ANALYSIS OF GLASS SURFACES: THE STUDY OF THE REACTION BETWEEN WATER AND GLASS[+]

W.A. Lanford[++] and C. Burman

Department of Physics, SUNY/Albany
Albany, New York 12222

R.H. Doremus, Y. Mehrotra and T. Wassick

Materials Engineering, R.P.I.
Troy, New York 12181

ABSTRACT

Analysis for hydrogen and other light elements is notoriously difficult by most traditional analytic methods. Because of this, techniques based on nuclear reactions have been developed which can determine quantitatively the concentration profiles of hydrogen, oxygen and other elements. The use of the ^{15}N hydrogen profiling, resonant oxygen scattering, and backscattering spectroscopy to study the changes in elemental composition near glass surfaces as a result of exposure to aqueous solution will be discussed as will the importance of these data to our understanding of the mechanisms of the reaction between water and glass. Combining these nuclear reaction techniques with electromigration or ion implantation to introduce "markers" makes it possible to study dissolution rates directly and to study the effects of ion implantation on both the ionic exchange and dissolution rate of glasses exposed to water.

[+]Research supported by grants from the Office of Naval Research at Albany and from the NSF at RPI.
[++]An Alfred P. Sloan Foundation Fellow.

I. INTRODUCTION

When a glass surface is exposed to water or aqueous solution, a number of chemical reactions occur resulting in changes in the composition both of the near surface region of the glass and of the solution. Because of the difficulty in directly analyzing for changes in the elemental composition of the glass surface using traditional methods, much of the work on the study of the reaction between water and glass has relied on analysis of what went into solution, with these results used to infer what changes occurred in the glass surface.

In the past ten years or so, a number of techniques for measuring the concentration <u>vs</u> depth of hydrogen, oxygen, sodium, ... have been developed based on nuclear reactions. In addition, over approximately the same period Rutherford backscattering spectrometry became widely used in the development of thin film materials used in microelectronics(3). In the present paper, we wish to illustrate the use of both nuclear reaction analysis and Rutherford backscattering in the study of the reaction between water and glass.

For convenience, we have drawn illustrative examples from the work by the Glass Study Group at SUNY/Albany. While these techniques are not yet widely used in glass science, there are a number of other groups in the U.S. and Europe using them, most notably G. Arnold <u>et al</u>. at Sandia(2).

In order to emphasize the basic principles and utility of these methods for glass and ceramic science, we have not included much "technical detail." Such detail is available in the literature.

II. NUCLEAR REACTION ANALYSIS OF HYDROGEN CONCENTRATION <u>VS</u> DEPTH

Because it has no x-ray or Auger transitions, analysis for hydrogen is notoriously difficult by traditional analytic techniques. Because of this difficulty and because hydrogen is a common contaminant which can have important effects on the physical, chemical and electrical properties of many materials, techniques for hydrogen analysis based on nuclear reactions have been developed and successfully applied to many problems(2-8). While a number of different nuclear reactions have been used as probes for hydrogen, the reaction with ^{15}N(6) is widely considered the most generally useful. We will discuss only the ^{15}N hydrogen profiling technique. However, all the other resonant nuclear reaction methods are based on the same principles.

The nuclear reaction used is $^{15}N + ^{1}H \rightarrow ^{12}C + ^{4}He + 4.4$ MeV gamma-ray.

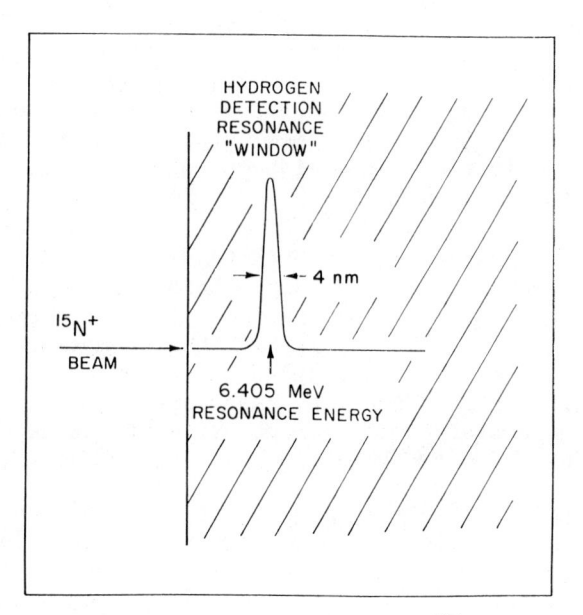

Fig. 1. A schematic representation of the ^{15}N hydrogen profiling techniques.

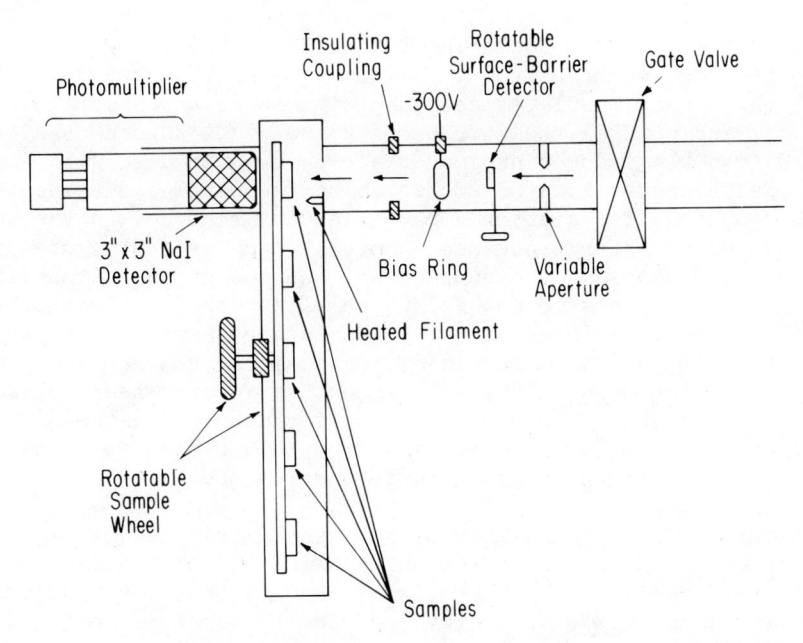

Fig. 2. A schematic diagram of the chamber used at SUNY/Albany for nuclear reaction analysis and Rutherford backscattering.

To use this reaction as a probe for hydrogen, the sample to be analyzed is bombarded with ^{15}N ions from a nuclear accelerator, and the yield of characteristic 4.4 MeV gamma-rays is measured with a scintillation detector. Because there is a narrow (width=6 keV), isolated resonance with a large peak cross-section (0.45 barns) in this reaction, this reaction will occur only when the ^{15}N ions are at the resonance energy, E_r=6.405 MeV. If a sample is bombarded with ^{15}N ions at $E=E_r$, the yield of characteristic gamma-rays is proportional to hydrogen on the surface of the sample. If the beam energy is raised, no reactions with surface hydrogen occur because $E>E_r$, but the ^{15}N ions lose energy as they penetrate the sample, reaching $E=E_r$ at some depth. Now the measured yield of characteristic gamma-rays is proportional to the hydrogen concentration at this depth. Hence, by measuring gamma-ray yield vs ^{15}N energy, the concentration of hydrogen vs depth is determined(6). This situation is shown schematically in Figure 1.

Because this method relies on a nuclear reaction, it is generally insensitive to the chemical state of the hydrogen within the sample. While this is a disadvantage in the sense that we can learn nothing directly about the chemical bonding, it is also an advantage in the sense that the technique is inherently easily made quantitative.

Figure 2 shows the analysis chamber routinely used at SUNY/Albany for both nuclear reaction analysis for hydrogen (and other elements) and for backscattering (to be discussed below). The ion beam enters the chamber from the right through some variable slits, passes through an insulated coupling with an electron suppressing bias ring and strikes the sample being analyzed. Gamma-rays emitted from the induced nuclear reaction are measured with a 3x3 (inch) NaI detector located 2 cm. behind the samples outside the vacuum chamber. Backscattered particles are recorded in the surface barrier detector. Because we often study sets of systematically prepared samples, the samples are mounted on the circumference of a rotatable sample wheel which allows the samples to be rapidly changed. In order to stop insulating samples from charging to high voltage as a result of the positive charge deposited by the analyzing beam, a heated filament is located near the sample. The filament is powered by batteries mounted on the analysis chamber in order that the electrons from this heated filament do not interfere with the beam current integration.

Figure 3 shows an example of the use of this technique in the analysis of the hydrogen content of a thin film of plasma deposited amorphous silicon(9). This figure shows both the raw data (gamma-ray counts vs beam energy) and the hydrogen concentration profile. The analysis of the data is very simple, consisting of multiplying the raw data axes by known constants(9). Another feature illustrated by Figure 3 is that, because the technique

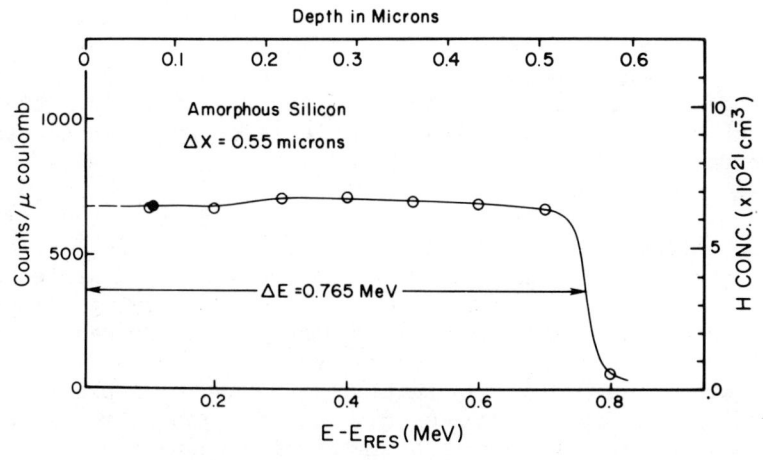

Fig. 3. A hydrogen profile of a thin film of hydrogenated amorphous silicon. This figure shows both the raw data (counts vs energy) and final results (hydrogen concentration vs depth). The solid datum point was measured after completing the profile.

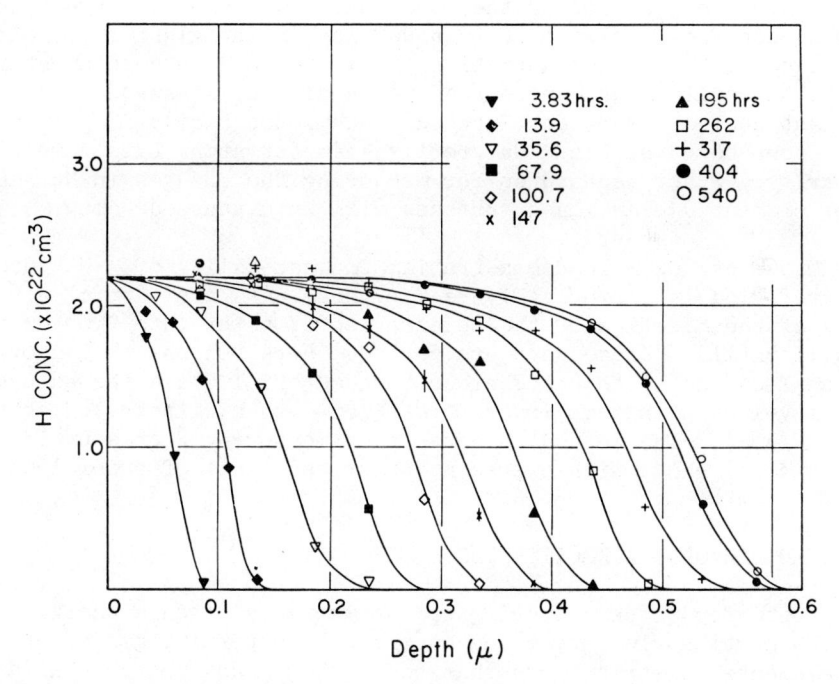

Fig. 4. Hydrogen profiles of a series of soda-lime glasses exposed to water at 90°C for varying times. From reference 10.

does not "destroy" the sample, one can check the results by repeating data points. For example, the solid datum point was measured after completing the other measurements in order to check to see if hydrogen was leaving the sample during the analysis. It typically takes 1-2 minutes of beam time for each hydrogen concentration measurement.

An example of the use of this technique in the study of the reaction between water and glass surfaces is shown in Figure 4, which shows the hydrogen concentration profiles of a series of soda-lime glasses exposed to water at 90°C for varying times(10). Figure 5 shows the depth of penetration of the hydrogen (to the one-half concentration depth) vs $\sqrt{T}$ for the data in Figure 4. From Figure 5, one sees that initially the hydrogen penetration is governed by a diffusion process but that for long times an equilibrium is reached. This equilibrium results when the incremental diffusion is cancelled by the dissolution of the glass surface. Hence, by measuring a data set such as that shown in Figures 4 and 5, both the diffusion rate and etch rate of glass in water are determined(10).

In the above example--the exposure of a soda-lima glass to water--the first process to occur is not diffusion of molecular water into the glass but rather the ionic exchange of a hydrogen bearing ion from the water with a Na^+ ion in the glass. If this is the case, the Na concentration profile should be complementary to the H profile. Na concentration can also be measured by a resonant nuclear reaction (induced by a proton beam). Figure 6 (10) shows both the H and Na profile, demonstrating this complementarity. Other techniques for measuring the changes in Na (and other elements) near glass surfaces will be discussed below.

One other example demonstrating the capability of ^{15}N hydrogen profiling is shown in Figure 7. These data are part of a study of the kinetics of the penetration of water into SiO_2 grown on silicon(11). These data are included here because they show the application of this technique to a material where the saturation level of H is much lower ($\sim 10^{20} H/cm^3$) than illustrated above ($\sim 10^{22} H/cm^3$). The method is capable of measuring H present in the ppt (atomic) range with a present practical limit of order 100 ppm.

III. RUTHERFORD BACKSCATTERING

Rutherford backscattering spectrometry is often a useful complement to nuclear reaction analysis in determining changes in the elemental distributions that occur near glass surfaces as a result of reaction with water. Rutherford backscattering spectometry relies on the fact that at sufficiently low energy (typically 2 MeV He ions), the scattering of ions from nuclei is (1) elas-

NUCLEAR REACTION ANALYSIS OF GLASS SURFACES

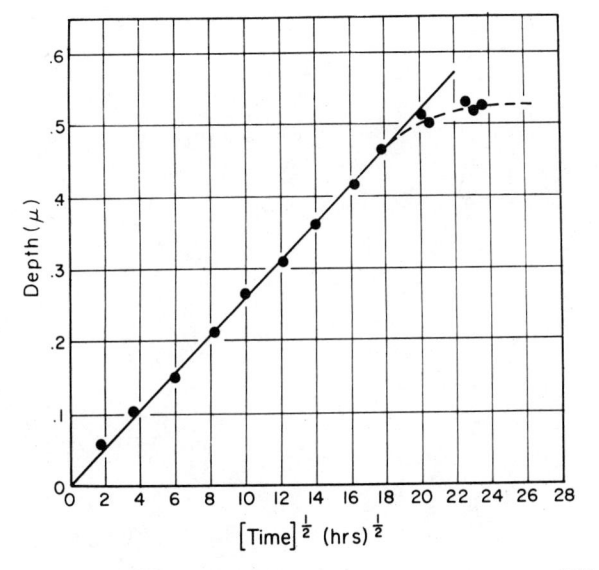

Fig. 5. The depth of hydrogen penetration vs $\sqrt{T}$ for the data in Figure 4.

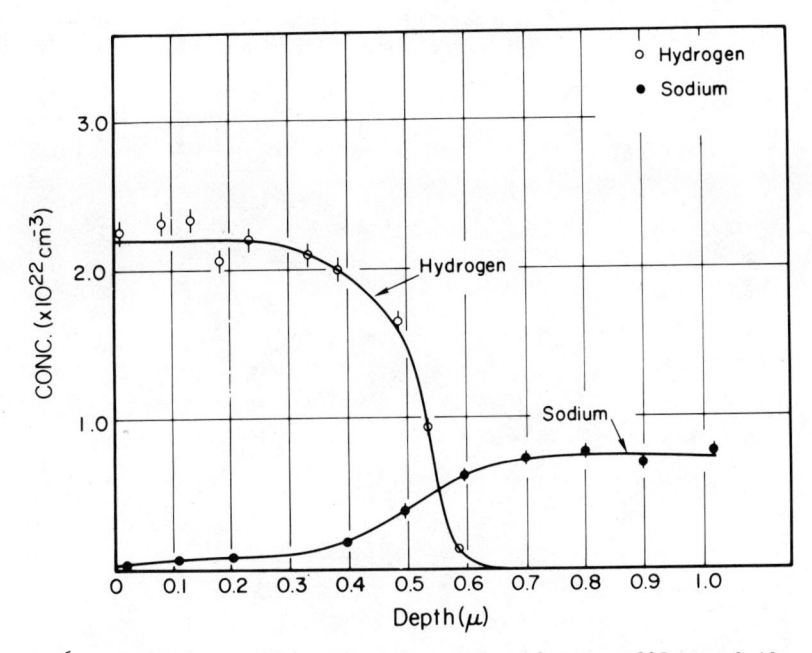

Fig. 6. Hydrogen and sodium concentration profiles of the surface of a soda-lime glass exposed to water. Both profiles were measured by nuclear reaction analysis. From reference 10.

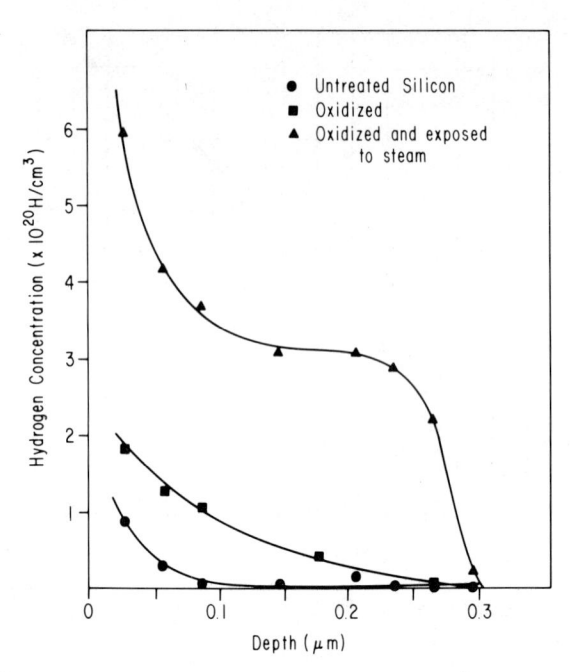

Fig. 7. Hydrogen profiles of a thin film (0.27 μm) of SiO_2 grown on silicon (dry) then exposed to steam at 320°C for 167 hours. See reference 11.

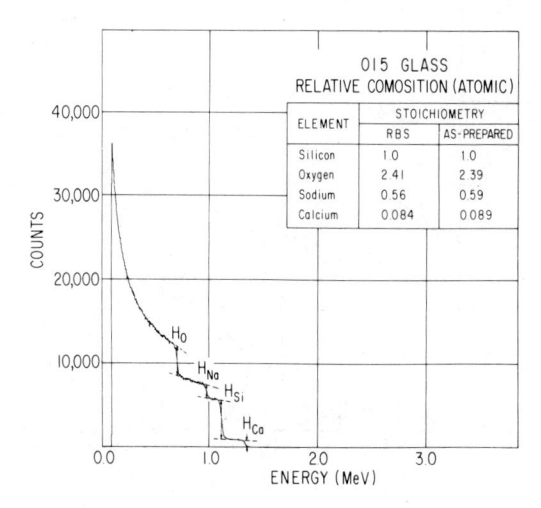

Fig. 8. A Rutherford backscattering spectrum of a soda-lime glass. A comparison of the stoichiometry determined by backscattering with the known composition is given. From reference 12.

NUCLEAR REACTION ANALYSIS OF GLASS SURFACES

tic and (2) governed purely by the Coulomb force, and, hence, the scattering cross-section is given by the Rutherford formula. Because the scattering is elastic, two-body kinematics uniquely give the energy of a backscattered ion as a function of the mass of the target atom. For example, if an ion of mass m_1 is scattered to 180° (relative to the incident direction) by a target atom of mass m_2, the energy of the backscattered ion will be

$$E_f = E_i (m_2 - m_1)^2 / (m_2 + m_1)^2$$

where E_i is the energy of the incident ion before scattering. Hence, by measuring the energy of the backscattered ion, the mass of the target atom is determined.

The above brief description is appropriate for Rutherford backscattering analysis of thin targets where the measurement of the number of backscattered ions <u>vs</u> their energy immediately determines (quantitatively) the mass (and hence elemental) composition of the target. Backscattering from a thick target sample is slightly more complicated because one has to consider also the energy loss of the ions both before and after scattering. Figure 8 shows a typical Rutherford backscattering spectrum(12) from a simple soda-lime glass.

As indicated in Figure 8, because the ion can lose energy penetrating the solid, the yield of backscattered particles <u>vs</u> energy shows steps, with each step occurring at the kinematic maximum energy for the various elements present. The height of each step (e.g. H_{Ca}, H_{Si}, etc., shown in Figure 8) is proportional to the concentration of that element at the surface of the glass. By analyzing these step heights, the stoichiometry of the glass can be determined(1). A comparison of this procedure with the known composition is also shown in Figure 8.

While one would not normally use backscattering for a simple elemental analysis of a glass, it is often a useful extra benefit of the use of backscattering for another purpose. We have found the compositional check useful in a number of cases.

Figure 9 shows a number of backscattering spectra of related glasses prepared and analyzed as part of Y. Mehrotra's Ph.D. thesis(13). These include the 015 glass also shown in Figure 8, a glass with the same composition except that Sr has been substituted for some of the Ca (004), a glass the same as 015 except Cs has been substituted for some of the Na, and a commercial soda-lime glass where Sb has been added as a fining agent.

One of the features illustrated by the backscatter spectra shown in Figure 9 is the high sensitivity of the method to small amounts of heavy elements present in the dominantly light silicate

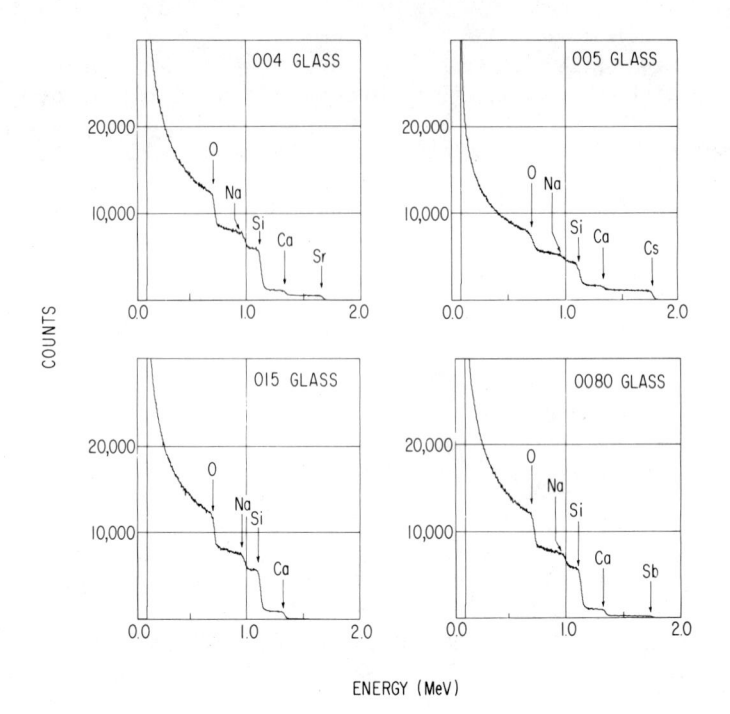

Fig. 9. Rutherford backscattering spectra from a series of soda-lime glasses. See reference 13.

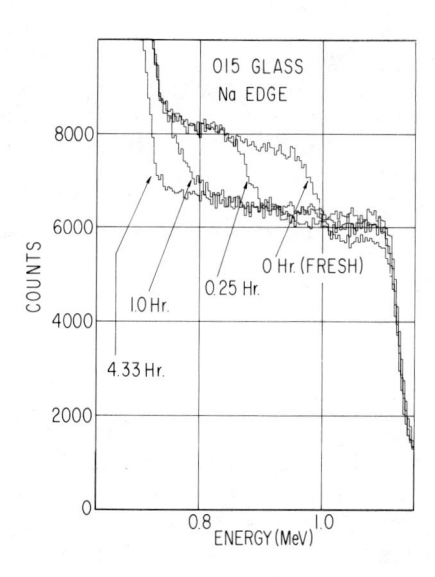

Fig. 10. An enlargement of the Na step for a series of glasses exposed to water for varying times. The shift of the step shows depletion of Na.

matrix. This sensitivity results both from the fact that the heavy elements are kinematically separated in energy from the lower mass elements and from the fact that the cross-section for Rutherford backscattering increases with the atomic number of the element (proportional to Z^2).

The converse is also true: It is very difficult to use Rutherford backscattering to analyze for low mass elements in a matrix containing significant amounts of heavy elements.

Rutherford backscattering is very useful for measuring the changes in the concentration of various elements in the near surface region of glasses as a result of some surface treatment. Above (see Figure 6), the depletion of Na near the surface of a commercial soda-lime glass exposed to water was illustrated by use of a resonant nuclear reaction for Na. Such a depletion should also be evident in the backscattering spectra.

Figure 10 shows an enlargement of the region of the Na step in a backscatter spectrum for a series of glasses exposed to water at 90°C for various times(12). As can be seen in Figure 10, even a 0.25 hour exposure causes the Na step to shift to lower energy by a substantial amount. This shift implies that Na is depleted from the surface into the depth corresponding to the measured energy of the Na edge. In the present case, the shift for 0.25 hours represents a depletion depth of 0.21 µm.

The depletion of other elements can also be measured easily by backscattering. For example, Figure 11 shows an enlargement of the Ca edge for a series of samples of the same glass exposed to water at 90°C for varying lengths of time(12). As can be seen, Ca is also preferentially removed from the surface of the glass but _very_ much more slowly than is the Na. In 5 days exposure at 90°C, the Ca depletion depth is only 0.053 µm. Figure 12 shows the region of the Sr edge for the glass made where some of the Ca was replaced with Sr. Analysis of this shift and the corresponding shift for Ca in the same glass shows that both these elements are depleted at essentially the same rate(12).

These examples of backscattering are intended to illustrate its use in easily obtaining quantitative information about the composition of the near surface region of glass. In these and many other cases it works extremely well. However, in some cases the backscatter spectra can become so complex as to make analysis difficult. In principle, one cannot uniquely determine the concentration profiles of a number of elements from the backscatter spectrum alone. However, there is a unique backscatter spectrum for any set of concentration profiles. For example, the backscatter spectrum from a very complex glass (e.g. a rad-waste glass) with many different heavy elements may be too complicated for

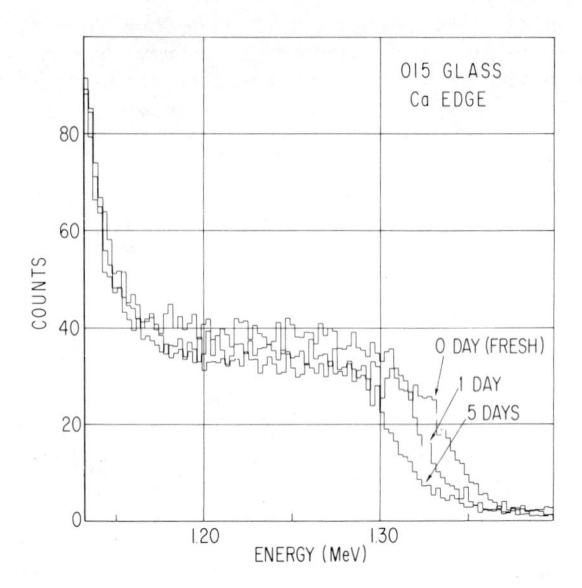

Fig. 11. An enlargement of the Ca step showing depletion of Ca on exposure to water.

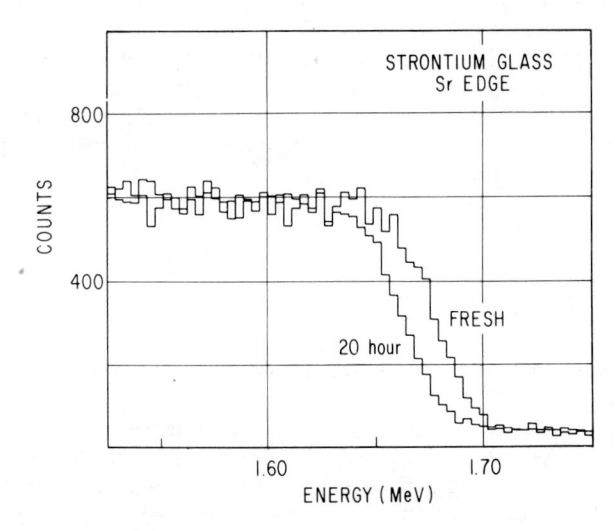

Fig. 12. An enlargement of the Sr step showing depletion of Sr on exposure to water.

NUCLEAR REACTION ANALYSIS OF GLASS SURFACES

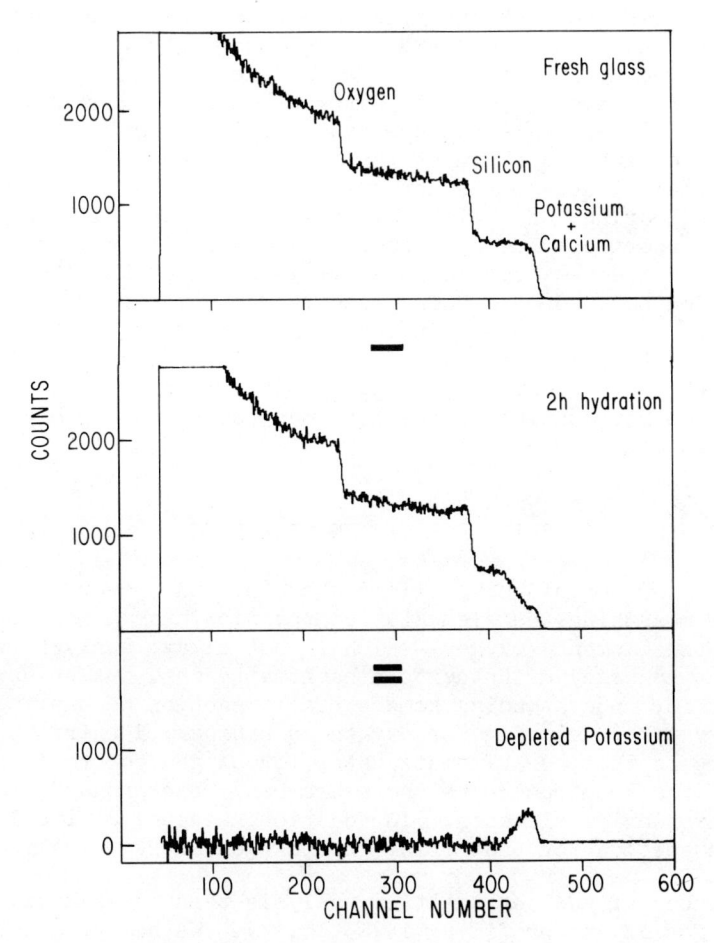

Fig. 13. The backscatter spectra of samples of potash glass--
one fresh, one exposed to water at 90°C for 2 hours.
The bottom shows the difference between these spectra.

detailed analysis. However, looking at differences in spectra recorded from systematically treated glasses may still give useful information about concentration changes.

The use of this subtraction procedure is illustrated in Figures 13 and 14 for a simple potash glass. Here the complication is the similar masses for K (39 amu) and Ca (40 amu). The top in Figure 13 shows the backscattering spectrum from an untreated sample. As can be seen, the K and Ca edges occur too close in energy to be resolved by our detector. The middle in Figure 13 shows the spectrum for a glass exposed to water at 90°C for 2 hours. The difference between these two spectra (obtained simply by numerically subtracting) is shown at the bottom. The difference is almost certainly due to the depletion of K and its replacement by a hydrogen bearing ion. With this assumption, the data in Figure 13 can be interpreted as a K concentration profile and compared with a H profile measured on the same glass using the ^{15}N hydrogen profiling method. Such a comparison is shown in Figure 14.

The point we are trying to emphasize is that backscattering can still be very useful in _practical_ applications, even in cases where the interpretation of the data may not be completely free from ambiguity.

IV. RESONANT SCATTERING FOR OXYGEN PROFILING

While the techniques outlined above provide powerful analytic tools for measuring the elemental composition near glass surfaces, there is one element--oxygen--which is not always analyzed with sufficient accuracy by Rutherford backscattering, but which is important for understanding some surface reaction mechanisms. The difficulty with analyzing for oxygen in ordinary Rutherford backscattering is that the kinematic step for oxygen occurs in a region of the spectrum where there are many "background" counts from other heavier elements. In addition, because of its low Z, the cross-section for scattering from oxygen is relatively small.

A number of nuclear reaction analysis techniques have been proposed for oxygen profiling(14,15,16,17). We have found that resonant oxygen backscattering is a convenient method(14). The basis of this method is a large resonance in the cross-section of ^{4}He scattering at E=3.05 MeV, i.e. at this energy the scattering cross-section is not Rutherford but an order of magnitude larger. This greatly enhanced cross-section provides the additional sensitivity needed in certain problems.

Figure 15 shows a backscattering spectrum recorded at E=3.06 MeV from a soda-lime glass (15% Na_2O, 10% CaO, 75% SiO_2). The large peak that occurs at and below the kinematic step for oxygen

NUCLEAR REACTION ANALYSIS OF GLASS SURFACES

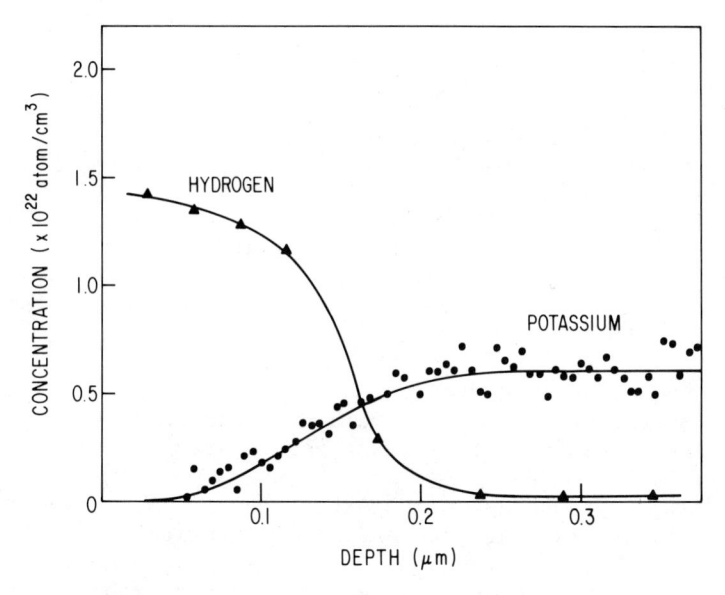

Fig. 14. The difference spectra from Figure 13 interpreted as resulting from K depletion plotted along with the hydrogen profile for the same glass.

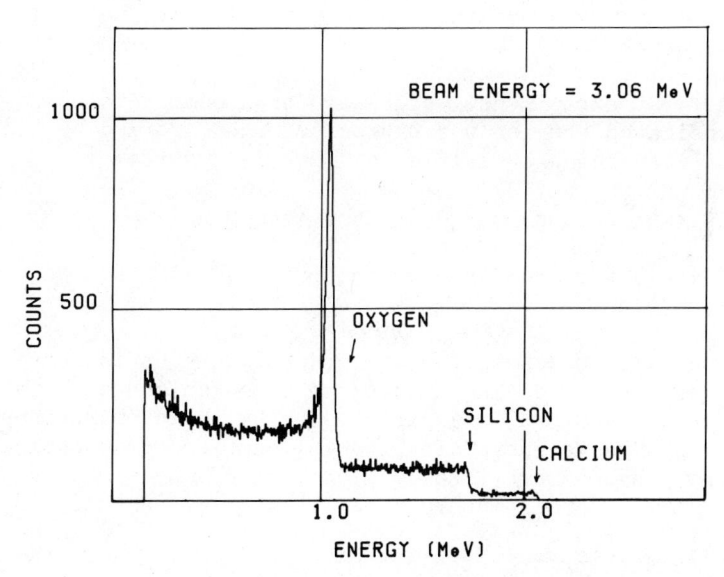

Fig. 15. A backscattering spectrum of a soda-lime glass recorded above the energy of resonant oxygen backscattering. The large peak results from resonant oxygen scattering.

results from resonant scattering. Note that except at the energy of this resonance, the scattering is nearly Rutherford so that the backscatter spectrum appears much as before except for this additional peak.

To use this resonant scattering to measure oxygen profiles, we have integrated the net counts in the resonance peak and plotted this resonant yield <u>vs</u> He energy. This procedure is fundamentally the same as the resonant nuclear reaction profiling method described in section II above except that we are using enhanced resonant scattering as a probe instead of a resonant nuclear reaction.

This technique is convenient because it requires precisely the same experimental setup as ordinary Rutherford backscattering spectrometry, except for the higher beam energy.

An example of the use of this technique is in the confirmation that, in the ion exchange and interdiffusion reaction that occurs between water and a soda-lime glass, oxygen as well as hydrogen is transported into the glass to replace the out diffusing Na^+. Above we have demonstrated that, for both soda-lime and potash-lime glasses, the alkali ion concentration and the hydrogen concentration are complementary, indicating ionic substitution. However, as seen in Figure 6 or 14, there are always more H atoms entering the glass than alkali ions leaving. The question remains: What is the hydrogen bearing ion responsible for the ionic transport of hydrogen in glass?

The first experiment, which was for a commercial soda-lime glass, indicated the ion was hydronium, H_3O^+, formed by the following ion exchange reaction at the glass surface:

$$2H_2O + Na^+(glass) \rightarrow H_3O^+(glass) + NaOH.$$

The H_3O^+ would then continue to interdiffuse with the Na^+ ions leaving the glass.

If this is the case, one expects not only an increase in the hydrogen concentration in a glass, but also an increase in the oxygen content. Shown in Figure 16 is the yield of resonant scattering counts <u>vs</u> He energy for several samples of soda-lime glass. Two of the glasses were freshly etched in HF and two were hydrated in water at 90°C (one 24 hours, one 274 hours). As seen in Figure 16, near the surface there is approximately a 17% increase in the resonant scattering from the two hydrated samples over the two fresh ones.

To convert the measured enhancement in resonant counts to the oxygen concentration near the surface requires a taking into

NUCLEAR REACTION ANALYSIS OF GLASS SURFACES

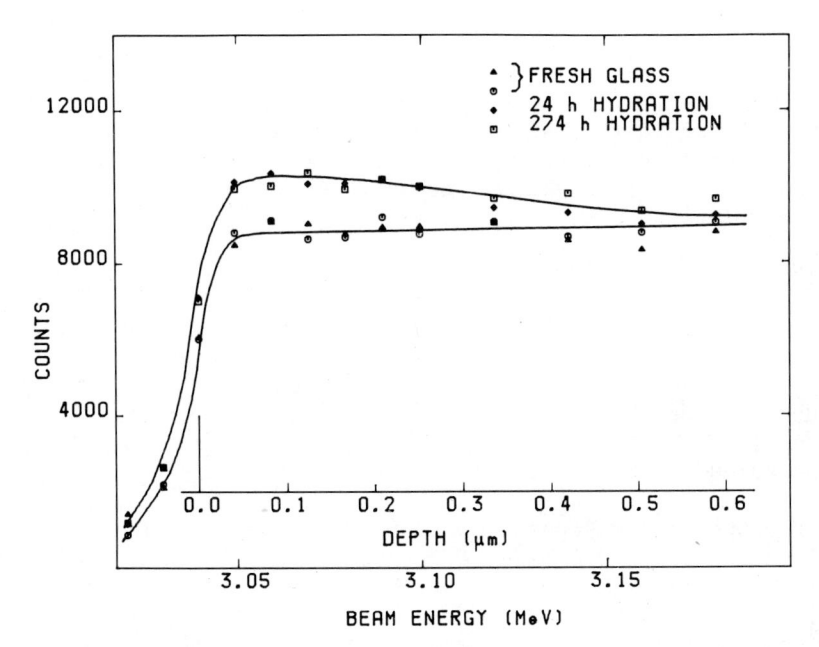

Fig. 16. The integrated yield of resonant scattering from a soda-lime glass vs He energy. Two samples were fresh, two were hydrated (24 hours, 274 hours) at 90°C. The increased yield in the hydrated samples results from both an increase in oxygen and the decrease in Na. See text.

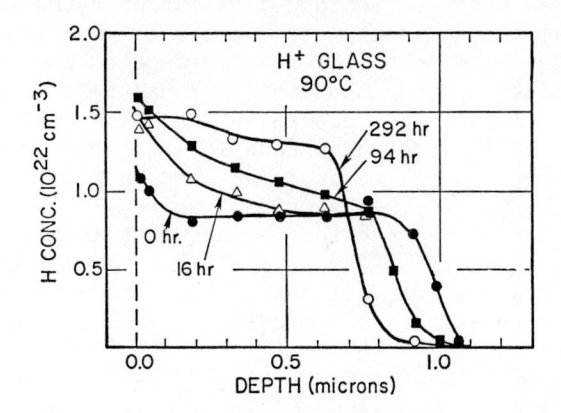

Fig. 17. Hydrogen profiles of a commercial soda-lime glass electrolyzed to replace the Na^+ ions with hydrogen ions to a depth of one micron. These samples were then exposed to water at 90°C for varying times. From reference 18.

account the other known changes in composition of this glass surface, i.e. the removal of the Na and the addition of ~ 2.6H for each Na removed. (The Ca depleted region is thin enough that it can be ignored in this analysis.) The effects of these compositional changes enter through changes in the rate of energy loss (dE/dx) of the He ions in the glass. The oxygen concentration (ρ_o) is related to the count rate through:

$$\rho_o = K \, (dE/dx) \, (counts)$$

where K is a constant, taking into account the resonant scattering cross-section, detector efficiency, etc. Hence, a self-consistent analysis is needed where the deduced oxygen concentration is included in the (dE/dx) evaluation. A simple procedure for making this analysis is to calculate and graph counts vs ρ_o and then use the measured counts and this graph to deduce ρ_o.

For the case discussed above (Figure 16), this analysis implies that 0.8 ± 0.2 oxygens are added to this glass for every Na that diffuses out.

Note that in this case, while the H or O concentrations are consistent with simply $Na^+ \leftrightarrow H_3O^+$ exchange, an exchange of $Na^+ \leftrightarrow H \cdot (H_2O)_n$ with n=0.8 is favored.

V. USE OF BACKSCATTERING AND NUCLEAR REACTION ANALYSIS IN CONJUNCTION WITH MARKERS

It may appear that it is difficult to use the ion beam techniques discussed above to measure dissolution rates directly because these methods measure depth profiles with the depth scale always measured relative to the real surface. In section II above, we demonstrate that the etch rate can be deduced by measuring both the diffusion of hydrogen into a glass and the equilibrium thickness of a hydrated surface layer. This procedure is somewhat indirect and relies on being able to reach the equilibrium condition. However, if a marker can be placed deep into the glass, either Rutherford backscattering or nuclear reaction analysis can be used to measure the distance from the marker to the surface. Dissolution will change this distance and, consequently, etch rates can be measured quickly and directly(18).

This type of procedure has been used with markers driven into a soda-lime glass by an applied electric field and by markers made by ion implantation of noble gases.

For example, Figure 17 shows the hydrogen profile of a series of soda-lime glasses which have been electrolyzed to sweep hydrogen ions into the glass replacing the Na^+ ions normally present(18).

NUCLEAR REACTION ANALYSIS OF GLASS SURFACES 567

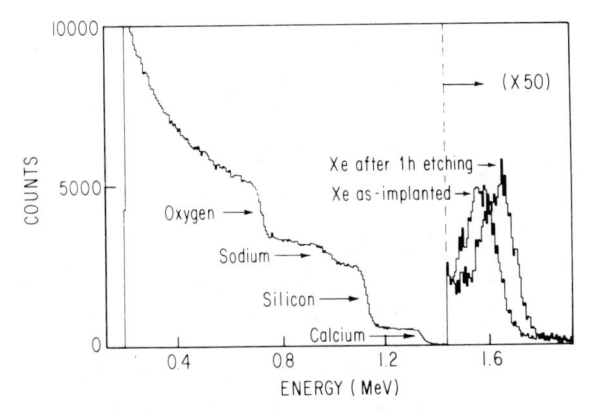

Fig. 18. Rutherford backscattering spectra from two samples of soda-lime glass implanted with 2.35×10^{15} Xe/cm^2 at 1 MeV. One sample was exposed to saline at 90°C for 1 hour. From reference 19.

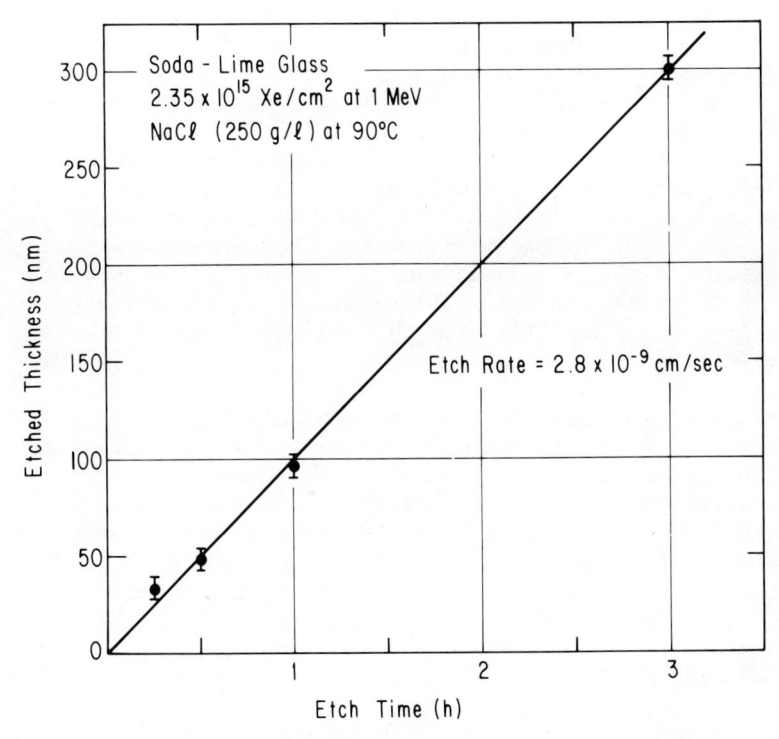

Fig. 19. The etched thickness, determined by Rutherford backscattering, for ion implanted glasses vs etch time. From reference 20.

These samples were then exposed to water at 90°C for varying lengths of time. This exposure to water both increases the hydrogen content (by water diffusing into the glass) and decreases the thickness of the electrolyzed layer by etching away the glass surface. By measuring the change in thickness of the hydrogen bearing layer with exposure time, the etch rate is deduced(18).

In principle, one may object that this method measures the etch rate of electrolyzed glass not untreated glass. However, since the first reaction that occurs when a soda-lime glass is exposed to water is the substitution of a hydrogen bearing ion for the Na^+ ion, the composition and structure of glass surface in contact with water is probably the same whether it was electrolyzed or not.

Figure 18 illustrates the use of markers made by ion implanting noble gas ions into the glass, with the distance of the marker from the surface measured by Rutherford backscattering. Figure 18 shows the backscatter spectra from two samples of commercial soda-lime glass implanted with 2.35×10^{15} Xe/cm^2 at 1 MeV(19). One sample was exposed to saline for 1 hour; one was not. The left side of the spectra is essentially the same for both and represents the backscattering from the glass. The peaks in the high energy end of the spectra are from backscattering from the implanted Xe marker. The width of these peaks results from fluctuations in the stopping distance of 1 MeV Xe ions.

As seen in Figure 18, the Xe peak occurs at higher energy for the sample exposed to the saline. This implies the marker is closer to the surface (by approximately 100 nm) as a result of etching away of the glass surface. This shift is not due to diffusion of the Xe, which would result in a broadening of the Xe peak. Figure 19 shows the etch thickness vs time of a series of such measurements(20).

The use of ion implantation to introduce a marker necessarily results in radiation damage to the glass, and this damage may change the etch rate. Within limits, the amount of damage introduced can be controlled by varying the dose and mass of the implanted ion. For studies of the effects of radiation damage on etch rate of glasses, the combination of ion implantation and backscattering is very powerful. However, for this procedure to be of general utility, a procedure of annealing out the radiation damage, without allowing the marker to diffuse away, needs to be developed. We are presently investigating such procedures.

VI.   DISCUSSION AND CONCLUSIONS

Above we have tried to review the use of nuclear reaction

analysis and backscattering for measuring elemental profiles near glass surfaces. These procedures can be used to study the kinetics of both the diffusion (ionic interdiffusion) processes and the dissolution processes. Combining these methods with ion implantation techniques provides a natural means for studying radiation damage effects(19,20).

We have not emphasized the problems associated with the application of these techniques. The most important concern is the effect of the analyzing beam on the distribution of elements in the sample being analyzed. In most cases this is not a problem. Further, unlike techniques which consume the sample during analysis, one is always free to check for beam effects by repeat measurements.

There are two situations where the analysis procedure can have an effect on the sample. One case is when working with any nondurable glasses. Even the process of putting these glasses in a vacuum can result in loss of hydrogen (water), and the analyzing beam can enhance this loss. This problem can be avoided by freezing the samples before evacuation. The second situation where beam effects can be important is in the nuclear reaction profiling for Na (Figure 6). The Na concentration near the surface was seen to increase with proton bombardment, probably as a result of motion of Na^+ carrying the charge deposited by the beam deep in the sample. This was minimized by frequently changing the analysis point on the glass sample.

We have not tried to compare these techniques with other analytical methods. Perhaps the most important comment to make is that these techniques are complementary (rather than competitive) to many of the more traditional methods. There are a number of situations where the use of nuclear reaction analysis to measure changes in the elemental composition of the surface should be used in combination with other analytic techniques such as IR-absorption (to determine bonding) or wet chemical analysis (to determine what went into solution) or electrolysis measurements (to determine total charge flows), etc.

Finally, because these methods rely on nuclear reactions and scattering, they are completely insensitive to poorly understood matrix phenoma (such as those which govern sputtering of compounds or light emission during sputtering), and hence these methods are easily made quantitative.

REFERENCES

1. W.K. Chu, J.W. Mayer and M.A. Nicolet
   Backscattering Spectrometry, Academic Press, New York (1978).

2. For example, see G.W. Arnold, P.S. Peercy and B.L. Doyle
   Nuclear Instruments and Methods 182/183 (1981) 733.
3. P.N. Adler, E.A. Kánykowski and G.M. Padwar
   Hydrogen in Metals, American Society for Metals (1974) 623.
4. B.L. Cohen, C.L. Fink and J.D. Degan
   Journal of Applied Physics 43 (1972) 19.
5. B.L. Doyle and P.S. Peercy
   Applied Physics Letters 34 (1979) 811
6. W.A. Lanford, H.P. Trautvetter, J. Ziegler and J. Keller
   Applied Physics Letters 28 (1976) 566.
7. D.A. Leich and T.A. Tombrello
   Nuclear Instruments and Methods 108 (1973) 67.
8. E. Ligeon and A. Guivarch
   Radiation Effects 22 (1974) 101.
9. W.A. Lanford
   Solar Cells 2 (1980) 351.
10. W.A. Lanford, K. Davis, P. LaMarche, T. Laursen, R. Groleau and R.H. Doremus
    Journal of Non-Crystalline Solids 33 (1979) 249.
11. R. Pfeffer, R. Lux, H. Berkowitz, W.A. Lanford and C. Burman
    Journal of Applied Physics 53 (1982) 4226.
12. R.H. Doremus, Y. Mehrotra, W.A. Lanford and C. Burman
    Journal of Material Science, to be published.
13. Y. Mehrotra, Ph.D. Thesis, R.P.I. 1981.
14. G. Megey, J. Gyulai, T. Nagy, E. Kotai and A. Manuba
    Ion Beam Surface Layer Analysis eds. O. Meyer, G. Linker and F. Kappeler, Plenum Press, New York (1976) 303.
15. G. Amsel, J.P. Nadai, E.D. Artemare, D. David, E. Girard and J. Moulin, Nuclear Instruments and Methods 92 (1971) 481.
16. G. Dearnaley, P.D. Goode, N.S. Miller and J.F. Turner
    Ion Implantation in Semiconductors ed. B.L. Crowder.
17. A. Turos, L. Wielunski and A. Barcz
    Nuclear Instruments and Methods 111 (1973) 605.
18. W.A. Lanford
    IEEE Trans. NS 26 (1979) 1795.
19. C. Burman, Wang Ke-Ming and W.A. Lanford
    Scientific Basis for Nuclear Waste Management Vol. VI ed. S.V. Topp, North-Holland (1982) 641.
20. C. Burman and W.A. Lanford
    to be published.

A STUDY OF WATER IN GLASS BY AN AUTORADIOGRAPHIC METHOD

THAT UTILIZES TRITIATED WATER

S. H. Knickerbocker, S. B. Joshi, and S. D. Brown

University of Illinois at Urbana-Champaign
Department of Ceramic Engineering
Urbana, IL 61801

Abstract

   Water concentration and spatial distribution in glass were determined by an autoradiographic method that makes use of tritiated water as the tagged species. The method is described and some typical results are presented. Advantages and disadvantages associated with the method are listed and discussed vis-a-vis other methods that might be used for the study of water in glass.

Introduction

   The characterization of water in glass--its content, distribution and movement, and the ways in which it exists and reacts relative to the glass structure--is important. Small quantities (in some instances, as little as a few hundredths of a wt. %) of dissolved and/or adsorbed water can have vivid effects upon such key glass properties and processes as melt viscosity and surface tension, durability, density, electrical conductivity, and devitrification (e.g., Refs. 1-14). Under certain conditions, water uptake by a glass melt (e.g., from the melter atmosphere or steam bubblers) can speed processing steps such as homogenization, firing and forming, possibly cutting costs (Ref. 15). On the other hand, water vapor in the melter atmosphere can enhance sodium vaporization from soda-lime-silicate glass melts to increase particulate emission and the corrosion of refractories in the stack, crown, and regenerator (Ref. 16). Plainly, there is a trade-off situation. Finally, the long term, possibly deleterious effects of water leaching on glasses that may be used to encapsulate radioactive wastes are of vital concern (Refs. 17,18).

   Several experimental methods are available for investigating water and/or hydrogen in materials including glass: for example, infra-red spectroscopy (IR) (Refs. 19-27); measurement of the quantity of water vaporized into a vacuum or dry carrier gas from a glass at elevated temperature (Refs. 28, 29); nuclear resonance reaction analysis (NRRA) (Ref. 30-33); secondary ion mass spectroscopy (SIMS)(Refs. 34,35); electron spin resonance spectroscopy (ESR) (Ref. 36); nuclear magnetic resonance spectroscopy (NMR) (Ref. 37); sputter-induced photon spectrometry (SIPS) (Refs. 35,38); incorporation of tritiated water as a tagged species in the glass, followed by the use of a scintillation counter to ascertain the water content of the glass, generally from grinding swarf (Refs. 39-41); incorporation of ^{18}O-tagged water in the glass followed by reaction of ground sections of the glass with $KBrO_4$ at 450°C to release the oxygen which is subsequently analyzed for isotopic make-up by mass spectrometry (Ref. 42); and a method in which the water is released from the glass into a dry O_2 carrier gas by the reaction of the glass with metallic Cu, then reacted with CaC_2 to form C_2H_2 which is subsequently measured by gas chromatography (Ref. 43). However, each method has its unique advantages and drawbacks. Each is deficient in that it provides only partial information.

The technique dealt with in this paper is an autoradiographic method that utilizes tritiated water as the tagged species. It is simple and straightforward, and works well for glasses and ceramics. It can provide quantitative two-dimensional maps of the water content within a micron or so of a specimen surface. It can be performed with safety provided the necessary precautions are taken.

Before the nature of radioactivity was recognized, St.-Victor (Ref. 44) observed blackening of AgCl and AgI emulsions by uranyl nitrate and tartrate, thinking it was a luminescence phenomenon. Becquerel (Refs. 45-47) reported the same phenomenon, considering it to be fluorescence. Lacassagne, et al. (Refs. 48-49) made systematic investigations of the response of photosensitive emulsions to ionizing radiation, and it is this work that marks the beginning of autoradiography. With the development of radioisotopes, autoradiography became practical. As early as 1940, for instance, Leblond (Ref. 50) prepared autoradiographs showing the distribution of iodine in thyroid glands. Since then, autoradiographic techniques have been used widely, especially in the life sciences (Ref. 51). Applications in materials research were reviewed by Condit (Ref. 52). More recently, Caskey (Ref. 53) examined tritium autoradiography as it is used to investigate hydrogen in metals.

An emulsion subjected to autoradiographic exposure develops a cumulative and permanent record of the radiation incident thereon. When obtained by correct procedures, this record enables one to determine accurately the spatial distribution of the radioactive species in the specimen material. Moreover, since the background development in the emulsion can be standardized, the method can be used to detect the presence of small levels of activity by making long exposures. Autoradiographic techniques are also valuable because the detector (i.e., the emulsion) can usually be placed sufficiently near the source that even radioisotopes having comparatively weak emissions with limited ranges can be used. Furthermore, many of the available emulsions are suitable for the detection of low energy emissions. Selection of the particular type to use depends upon the nature of the radiation involved, and entails a compromise between sensitivity and resolution.

Sometimes the radiation emitted has such limited range that adequate exposures of the photographic film are not obtained. The small air gap between the sample and the photographic film with which it is placed in macroscopic contact is enough to prevent acheivement of satisfactory results by ordinary autoradiographic procedures. An example is the beta radiation emitted from tritiated water that has been incorporated in an inorganic glass, the case with which this paper deals. We have found that use of a carefully selected and applied fluorescent coating on the specimen to convert the short-range beta radiation into visible light or longer range is a satisfactory means of overcoming the problem in many situations.

Procedure

Overview. In a nutshell, the technique involves the following five sequential steps: (1) An isothermal, high-temperature treatment is given a dry glass in a water vapor/inert gas mixture of known tritiated water content; then, the glass is quenched to room temperature. (2) The glass thus treated is sectioned, and the surface or surfaces of interest, are carefully polished flat. (3) A fluorescent intensifier, usually scintillation grade 2,5 diphenyloxazole (PPO),* is applied uniformly to the polished surface and a photosensitive film is placed over the intensifier. The sandwich thus obtained is wrapped snugly in plastic film and aluminum foil and banded so as to preclude adventitious moisture and light, and movement of the film with respect to the glass. (4) Depending upon the film used and the resolution and/or sensitivity required, exposure times may range from a day or so to a few weeks (at dry ice temperatures to reduce thermal noise to an acceptable level and increase the intensifier efficiency). (5) The density of the film exposure is compared quantitatively with a similar film exposure made in contact with a standard; e.g., a single crystal of alum appropriately doped with tritiated water during the crystal growth process. From this comparison, the absolute water content of the glass is calculated. Procedures used in connection with a specific case are detailed in the following paragraphs.

Dry Glass Preparation. A soda-lime silicate glass (composition, Table 1) was prepared from a pelletized batch. A 1.0 kg sample was melted in a mullite crucible in a SiC-element furnace. Dry N_2 (dew point = 77.2°K) was passed through the furnace chamber in contact with

---

*Eastman Kodak Company, 2400 Mt. Read Blvd., Rochester, NY 14650.

Table 1. Make-Up of the Soda-Lime-Silicate Glass

Batch Composition	wt %	Oxide Composition As Calculated From Batch	wt %	Wet Chemical Analysis of the Soda-Lime-Silicate Glass (Accuracy = ± 2%)	wt %
SiO_2	59.87	SiO_2	71.80	SiO_2	61.0
Na_2CO_3	19.30	Na_2O	13.99	Na_2O	12.0
Na_2SO_4	0.88	CaO	9.93	CaO	8.0
$CaCO_3$	14.77	MgO	2.00	MgO	2.0
$MgCO_3$	3.49	Al_2O_3	2.04	Al_2O_3	17.0
Al_2O_3	1.70	As_2O_3	0.24		
As_2O_3	0.20				

the sample during melting. The sample was held 48 hrs. at 1500°C, then annealed. Glass from the center of the glass mass was used for testing. A sample of the glass was analyzed by wet chemical methods, mass spectroscopy, and emission spectroscopy. The glass density was determined to be 2.50 gm/cc by immersion techniques. Using an optical microscope, no bubbles, cords or other defects were found in the glass.

Apparatus. The furnace used for the experiments is shown in Figure 1. It was heated by eight 1.3 cm diameter SiC elements. The 99.9% Al_2O_3 muffle was a closed end tube 12.7 cm I.D., 14.0 cm O.D., by 61.0 cm in length. Heat shields (fabricated from 50% clay - 50% grog, fired to a high density, and capable of withstanding temperatures somewhat greater than 1400°C) were fitted snugly into the furnace chamber. The sample holder was slip cast from high purity alumina and lined with platinum foil. Gas inlet and outlet tubes were 1.3 cm O.D. alumina. The furnace cap was water cooled stainless steel with a Teflon gasket. Pt-10% Rh thermocouples were spaced to read the temperature from one end of the sample holder to the other. The furnace was contained in a stainless steel and glass glove box. All water hoses, gas hoses and electrical connections were passed into the box through an air-tight port. A steady stream of fresh, filtered air circulated through the glove box, and a negative gauge pressure was maintained inside so that tritium-tagged water from the furnace would not escape to the laboratory.

Figure 2 illustrates the furnace atmosphere control system. The gas was continually circulated through the furnace by means of a nickel bellows attached to a variable speed motor. All tubing was maintained at a constant temperature by the use of asbestos heat tape. The humidity controller was filled with approximately one pint of water prior to each experiment. Condensation of water vapor on

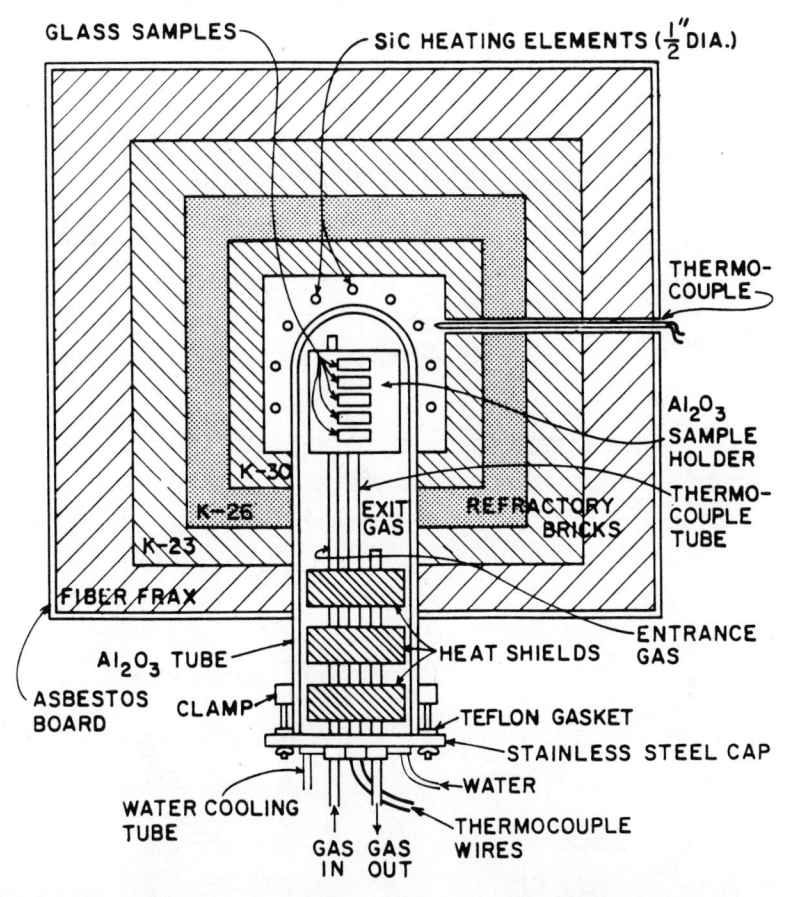

Fig. 1. Cut-away view of the furnace.

tubing walls was eliminated by maintaining the temperature of the humidity controller approximately 10°C lower than that of the surrounding tubing.

In these particular experiments, the humidity controller was kept at 50°C, the tubing, 60°C. A syringe was used to inject tritiated water into the system through a septum covered tube. A gas of known composition could be bled into the system prior to operation.

A series of U-tubes filled with granulated $CaCl_2$ desiccant was used to clear the furnace and atmosphere control system of tritiated water at the end of the experiment. Liquid nitrogen cold traps were

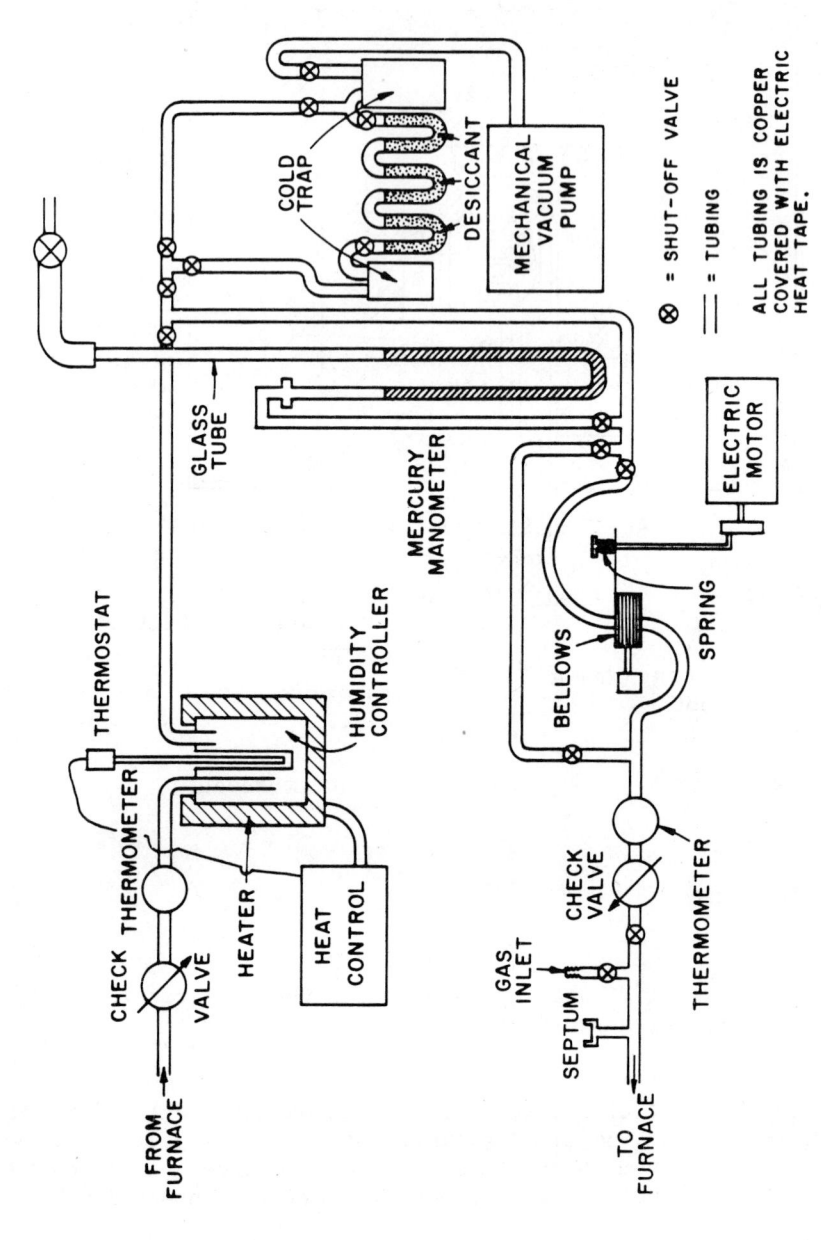

Fig. 2. Furnace atmosphere control system.

placed on each side of the U-tubes. A mechanical vacuum pump was used to evacuate the system after the furnace was opened. A system of valves allowed sections of the atmosphere control system to be preferentially opened or closed. The entire atmosphere control system was located under a hood.

Addition of Tritiated Water to Glass. Five 6 gm. pieces of dry glass were placed in the alumina sample holder as indicated in Figure 1. A tank of gas of known composition was connected to the gas entry valve shown in Figure 2. Approximately 1.42 m^3 of gas was flushed through the system and then the system was sealed off. The power to the heating elements was then turned on and the temperature was increased to 1000°C. At 1000°C, 0.5 ml of 75 mC/ml tritiated water was injected into the system, and the circulation of the furnace atmosphere was started. The temperature of the hot zone of the furnace was raised to 1350°C and maintained for 48 hrs. following which the glass samples were removed from the furnace and quenched to room temperature. The gas present in the atmosphere control unit and the furnace chamber was drawn through the cold traps and desiccant.

Autoradiographic Techniques. Following the foregoing treatment, each glass sample was mounted in quick-setting epoxy resin as an aid in handling during the polishing procedure. The surface of each glass sample was exactingly polished for a set time period with 200, 400, then 600 mesh SiC. Dry mineral oil was used as a medium. If either water or moist oil is used, non-tritiated water exchanges with tritiated water in the glass surface, compromising the eventual accuracy of the measurement. Since the beta radiation from tritium penetrates approximately 1 μm into a solid with the density of glass and only several thousand μm into air, the surface finish on the glass must be exceedingly smooth. When film is later placed over the surface to detect and record the radiation, a rough surface may cause the low areas on the surface to exclude detection. After polishing, the samples were washed in acetone, then stored in a desiccator until needed for the next stage of the procedure.

The polished surface of each glass sample was covered uniformly with a thin layer ($\approx$ 0.2 mm thick) of PPO, a fluorescent intensifier. PPO application was best accomplished by rubbing the powder across the polished glass surface. Kodak No-Screen X-ray film* was placed in contact with the PPO. The film and samples were surrounded by foam rubber. Rubber bands were used to hold the foam rubber around the film and samples. The foam rubber kept the film from moving in relation to the glass and also applied a uniform pressure to the film. This ensured an intimate film-PPO-sample contact. All of this was surrounded by a plastic film to keep moisture out and by aluminum foil to preclude adventitious light. The bundles thus prepared were placed inside an opaque plastic bag and over dry ice (at -77°C) for an expos-

*Eastman Kodak Company, 2400 Mt. Read Blvd., Rochester NY 14650.

sure time of 72 hrs. Liquid nitrogen temperature was not used because the sensitivity of the emulsion would decrease. Ions cannot diffuse as rapidly to trapped electrons to form the latent image at this temperature. Standard procedures were used to develop the film.

A densitometer was used to analyze the developed film. The incident intensity I_0 of a one mm diameter beam of light was compared with the transmitted intensity I and the optical density D was calculated from $D = \log(I_0/I)$. The average value of exposure density was used to calculate water content. Sometimes, higher resolutions than that achievable with an ordinary densitometer were required; for example, when diffusion profiles were considered. In such instances, an optical microscope was used. It was fitted with a simple mechanical scanning device and a photocell with the electronic circuitry necessary to ascertain the optical density as a function of position in the plane of the film (Figure 3).

After analyzing the exposures from each glass sample, comparisons of exposure density were made and relative water contents determined. Without a suitable standard, however, absolute water contents could not be obtained. A crystal of alum, $AlK(SO_4)_2 \cdot 12H_2O$, was grown in tritiated water to serve as a standard. Alum was selected because the water of hydration is highly stable at 12. A seed crystal $\approx$ 3.2 mm long was grown from a saturated aqueous solution of alum by the evaporation method. A second saturated solution was prepared with 0.5 ml of 75 mC/ml tritiated water per 473.2 ml of distilled water. The seed was grown to a length of 12.7 mm in this solution.

The alum crystal surface was polished with SiC grit and coated with PPO in the same manner as were the glass samples. Exposure techniques and film analysis also were identical.

Results and Discussion

Glass Preparation. The wet chemical analysis (Table I) indicates that the Al_2O_3 content of the glass increased by $\approx$ 15% during melting and that the Na_2O and SiO_2 contents decreased by similar amounts. Corrosion of the mullite crucible and volatilization of the melt from its exposed surface account for these changes. If control of the dry glass composition is important, it is clear that crucible materials less vulnerable to attack by the melt would need to be used; for example, a suitable platinum alloy. Moreover, appropriate measures to compensate for and/or manage volatilization during melting would need to be taken.

Drier and more homogeneous glass than that studied in the case reported here could be obtained were the melt stirred and the desiccated N_2 bubbled through it during melting. However, these measures would worsen the volatilization problem.

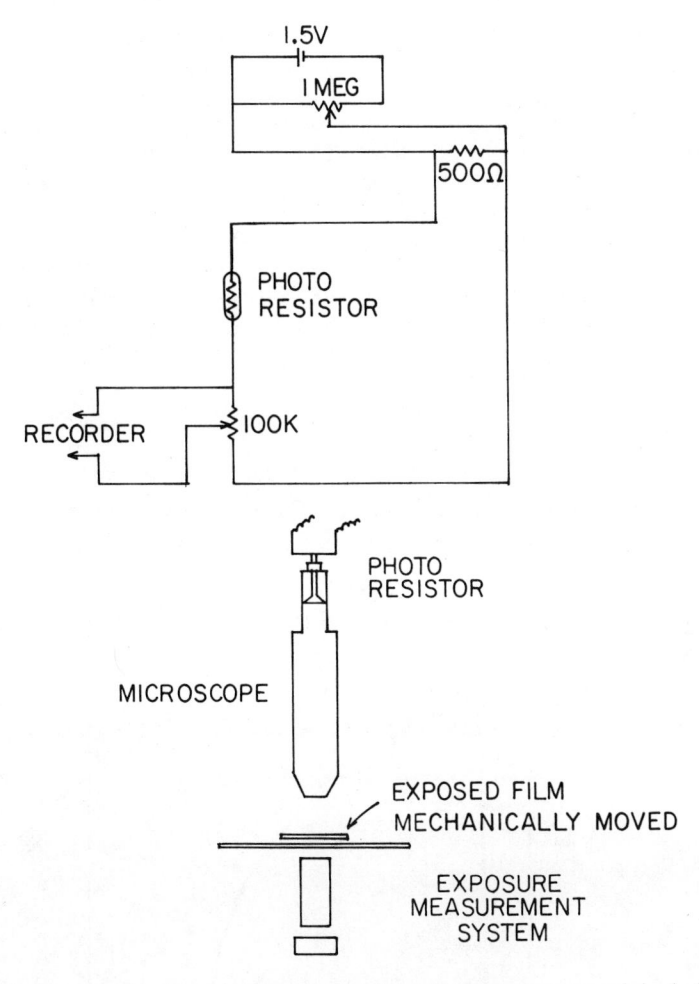

Fig. 3. Schema of the microscopic densitometer assembled for the determination of diffusion profiles.

It was necessary to add 0.88% Na_2SO_4 and 0.20% As_2O_3 to the batch as fining agents. Otherwise, the resultant glass was found to have an unacceptable content of bubbles and other heterogeneities.

No particular problem was encountered during the addition of tritiated water to the glass. Necessary safety measures were followed scrupulously. Small quantities of tritiated water having appropriately low levels of radioactivity were used. Adequate ventilation was maintained. While H^3_2O may enter and accumulate in the human body more readily than H^3_2, its rate of permeation through apparatus walls and seals is orders of magnitude less. Moreover, water vapor is more easily condensed and trapped than hydrogen gas. Therefore, while the matter of safety was not trivial in these experiments, the elaborate equipment required for the safe handling of H^3_2 was unnecessary.

Film Selection. The Kodak No-Screen X-ray film used contains larger silver halide crystals but with higher crystal density than visible photography films. The larger crystal size decreases resolution but the greater density increases the sensitivity. Films with

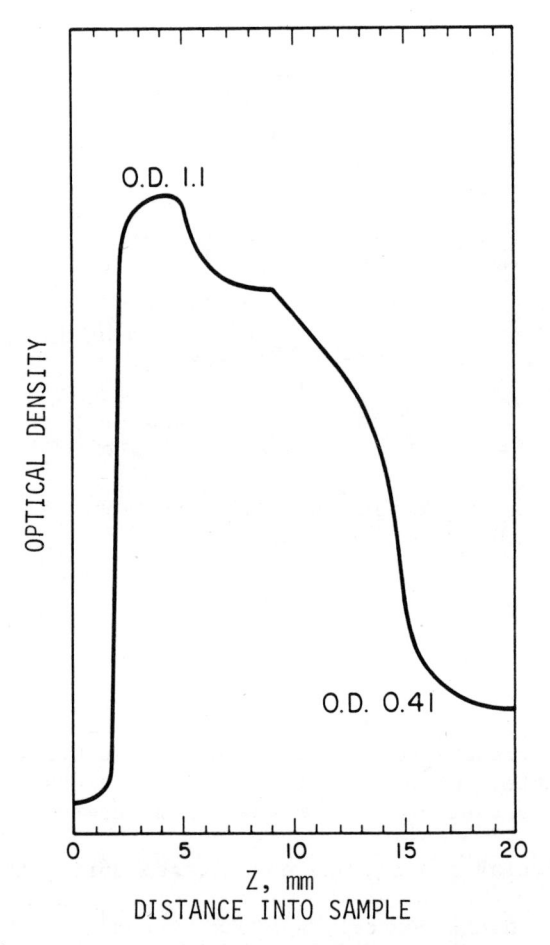

Fig. 4. Densitometer trace of diffusion profile.

greater density exist but were found to be extremely pressure sensitive which complicated exposure and handling. Less dense films were too insensitive.

Lateral resolutions better than 10^{-7} m are possible with special fine-grained photographic emulsions when an electron microscope is used to measure the exposure density (Refs. 56,57). No fluorescent intensifier is used in these cases; the emulsion is applied directly and intimately to the polished surface of the specimen. Although the sensitivity is increased by the use of an appropriate intensifier, the resolution is impaired.

<u>Film Analysis</u>. The densitometer was capable of resolving a 1% difference in optical density. The concentration of water in the glass was calculated from the known water concentration in the alum crystal and the measured optical densities of the autoradiographs. Concentrations of tagged water in the solution from which the crystal was grown and in the furnace atmosphere to which the glass was exposed were taken into account. Also, the effects of sample density on the beta particle range were factored into the calculations. In the specific case we report, the water concentrations of the five samples ranged from 0.025 to 0.039 wt % for a 1350°C furnace atmosphere of 50°C dew point. The range involved sample-to-sample variations and occurred primarily because of temperature changes with sample location within the furnace. While these water concentrations are not inconsistent with those reported for soda-lime-silica glasses on the basis of IR data (Refs. 22,25), they are higher by as much as 30%. This was expected because of the relatively high Al_2O_3 content of the present glass. Moreover, the IR values may be low (Refs. 24,33,58).

Diffusion profiles were obtained from quenched samples that had been exposed to the moist, active furnace atmospheres for periods of time too short for the chemical potentials of water in the gas and melt phases to equilibrate. These samples were sectioned parallel to the water concentration gradient and subjected to the autoradiographic procedures already described. One of the densitometer traces obtained, which shows the optical density of the film vs. the diffusion distance, is shown in Figure 4. Figure 5, taken from reference 53, is another example of a kind of diffusion profile that can be achieved by tritium autoradiography. In that case gaseous tritium rather than water was the diffusing species. The brief initial rise in optical density with distance into the sample which precedes the expected decrease stems from the fact that the surface had been polished in ordinary water, facilitating the exchange of tritiated water in the glass for the untagged water of the polishing medium.

<u>Advantages and Disadvantages of the Method</u>. Autoradiography in which tritiated water is the tagged species has been proved to be an effective, quantitative procedure for investigating water in inorganic glasses. We see no reason why the method cannot be extended success-

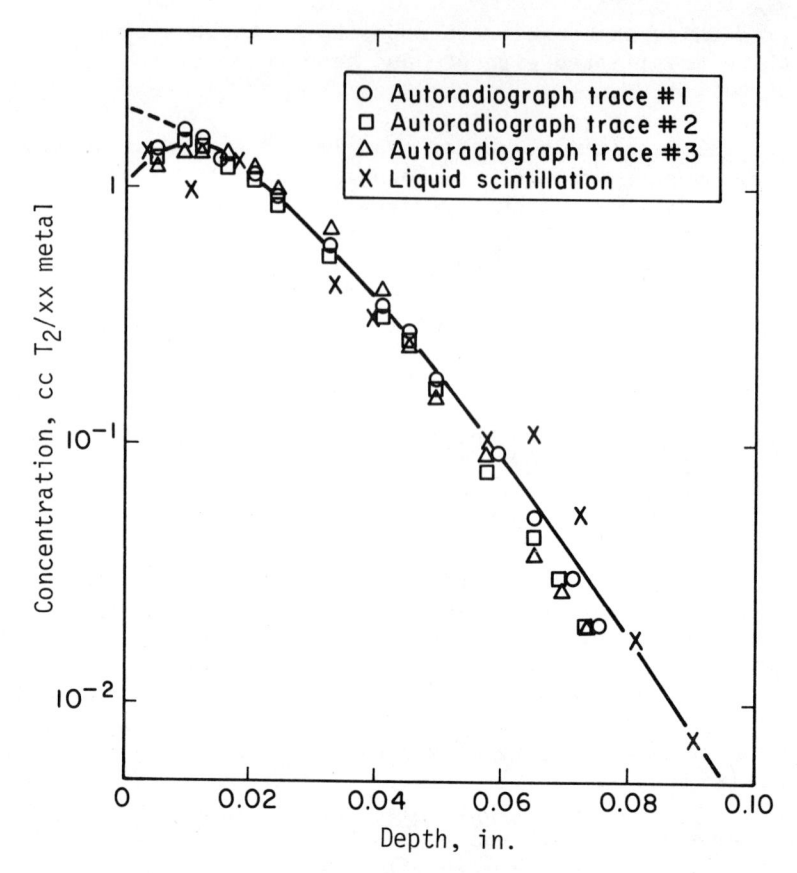

Fig. 5. Comparison of autoradiographic and liquid scintillation techniques for diffusion of tritium into gas 304 type stainless steel (Ref. 53, by permission of AIME).

fully to the study of water in many of the crystalline ceramics, organic polymers, and composite materials. Tritiated water is readily available and is an ideal species for the purpose: Its specific activity is 2150 Curie/ml. Tritium is a low-energy beta emitter (maximum energy = 18.6 KeV, most probable energy = 3.6 KeV) with a half-life of 12.26 years. The low energy of the radiation makes the volume of material that is active in developing grains in the film's emulsion very small ($\approx 10^{-12}$ to 10^{-11} cm^3 in glass). The radiation penetration range (≈ 0.05 to 0.06 mg/cm^2) centers around 1 μm in many solids. For all practical purposes, this eliminates the background problem ordinarily associated with the autoradiography of thick samples, making the procedure convenient for obtaining surface water content and diffusion profiles (from sections cut perpendicularly to surfaces through which the tagged species entered the glass). Much of the power of the method lies in the fact that a two-dimensional map of the spatial distribution of water can be obtained from a single section. The long half-life (the activity decreases less than 0.016% in a day) enables performance of experiments according to reasonable schedules without having serious decay problems.

Lateral resolutions of several µm are possible on a routine basis. Finer resolutions (better than 10^{-7} m) may be feasible with an especially fine-grained emulsion, e.g., Kodak SP 129-01 of grain size 5×10^{-8} m (Refs. 56,56); however, an electron microscope would need to be used to ascertain exposure density and some sacrifice of sensitivity would be expected. There is always a trade off between resolution and sensitivity. Water concentrations as low as 0.001 wt % ($\approx$ 10 ppm or $\approx 10^{18}$ water molecules/cc) are measured routinely by the method outlined herein. Lower concentrations can be dealt with by using longer exposure times (up to 6 wks) and/or tritiated water of a higher activity.

Sample preparation involves standard polishing techniques. The entire process of preparing water-free glass, tritiated water exposure, sample preparation, and film exposure can be accomplished in 8 days for the typical case. A densitometer can be used to analyze the autoradiographs. Quantitative values of the water concentration can be obtained by comparing sample autoradiographs with a suitable standard; e.g., the autoradiograph of a single crystal of alum, $KAl(SO_4)_2 \cdot 12H_2O$, grown from a tritiated solution of known activity.

The major disadvantage of tritium autoradiography lies in the fact that it requires a radioactive tagged species. Ordinary water is not detected. Moreover, the user must deal with the rigors necessarily involved in the safe handling of radioactive substances. For instance, the tritiated water must be adequately contained and inventoried. Use of small quantities of low activity tritiated water and sufficient ventilation of the laboratory area does much to reduce risk to acceptable levels. Of course, the radioactive character of the tagged species precludes the use of the technique for experiments in industrial glass-melters.

The starting glass must be dry. Polishing must be done in dry media and techniques must be virtually identical from one sample to the next. Application of the fluorescent intensifier must be such that the layer formed is in intimate contact with the glass surface and of sensibly constant thickness and density. Exposure times greater than about a month must be avoided owing to the latent image fading in the film.

Some degree of backscattering of the radiation will occur leading to a loss in resolution. Because the penetration of a material by a beta particle is dependent upon the particle energy, which varies from one particle to the next, the beta particles penetrate the film different distances. This too compromises resolution in some measure. Due to their difference in isotopic mass, tagged and untagged water participate in various kinetic processes at different rates. However, this problem is easily resolved by calculation. Unfortunately, no information is provided about the bonding between the glass structure and the water.

IR, ESR, and NMR have the very important advantage that they provide information regarding how water is bonded to the glass structure. Transmission IR, in particular, has found widespread use for determining the content and structural character of water in glasses, and has sensitivity of a few ppm. However, there is much evidence to suggest that IR may not detect all of the water present in glass (Refs. 24,26,33). ESR and NMR, while especially useful for determining the nature of the structure vicinal protons and certain other cations, may not be sufficiently sensitive to be useful in some of the lower concentration ranges that often are of concern. Finally, these methods cannot provide the resolution that is possible with the tritium autoradiographic method reported herein, or NRRA, SIMS and SIPS.

Those methods that rely upon vaporizing the water from the glass (Refs. 28,29) or releasing it by means of some chemical reaction (Ref. 43) are quantitative. However, they provide no information as to how the water is bonded to the glass structure or what its spatial distribution is within the glass. Moreover, they are less convenient to use than most of the more modern techniques.

The non-autoradiographic methods that involve water tagged with either ^{3}H or ^{18}O (Refs. 39-42) can be made sensitive to a few ppm. And, with repeated sectioning of the sample, diffusion profiles can be obtained. However, the value obtained for the water concentration from the grinding swarf necessarily represents an average for the removed section. Therefore, the resolution of these methods cannot compare with the submicron resolutions possible with autoradiography, NRRA, SIMS, or SIPS. Furthermore, no information regarding the role of water in the glass structure is obtained.

NRRA (Refs. 30-33) is a particularly powerful analytical tool that can be used to study water in glasses. It is simple experimentally and provides a quantitative value for the H content (and hence that of water) without reference to standards. The sensitivity is approximately 100 ppm. It is capable of near surface depth resolution of about 40 Angstroms and a lateral resolution of a few millimeters. It can profile H to a depth of about 4 µm, and the time required to obtain one point in the profile is approximately one minute. There is no background radiation to complicate the experiment. Unfortunately, the method provides no information about how the water is bonded to the glass structure. The cost of the equipment required is considerable and may preclude use of the method by many laboratories.

SIMS and SIPS (Refs. 34,35,38) have been used successfully to determine hydrogen concentrations at and/or near the surfaces of various materials; e.g., in the depth-profiling of water in leached glass. We note, that the water concentrations in the bulk regions of glasses often are less than those in the leached surfaces by an order of magnitude or more. The sensitivities of these methods, while satisfactory for many investigations, may prove to be borderline in some cases of interest. Problems arise from various sources: residual hydrogenous species from the sample chamber can contaminate the freshly exposed sample surface to give an erroneous signal and corrective measures may not always be possible with the apparatus at hand (Ref. 59). Electromigration of species at or near the surface under the influence of the excitation beam can alter the local chemistry as can the localized vaporization of water. Furthermore, as pointed out by Pantano (Ref. 35), complications that arise owing to the accumulation of excess positive charge at the specimen surfaces severely impair the use of SIMS and SIPS for glasses and ceramics. The methods do not provide information as to how the water is bonded to the glass structure.

Summary

Tritium autoradiography has been shown to be a valuable technique for measuring the content and spatial distribution of water in inorganic glasses. The technique yields unique information, particularly in regard to spatial distribution, when compared with techniques of IR spectroscopy, SIMS, SIPS, NRRA, ESR and NMR. If necessary, large areas (e.g., several square inches) of sample can be mapped in a single exposure. No extremely expensive equipment is necessary and results can be obtained within eight days. The spatial resolution of water in the glass network can be 10^{-7} m, so very accurate diffusion profiles are obtainable. Absolute water contents are possible with an alum crystal standard. A sensitivity of 0.001 wt % (or $\approx$ 10 ppm) can be realized.

ACKNOWLEDGEMENTS

This work was supported by the Department of Energy, Division of Materials Sciences, under Contract DE-AC02-76ER01198 through the Materials Research Laboratory at the University of Illinois at Urbana-Champaign. Important early support was received from the Columbia Gas System Service Corporation (Columbus, Ohio) and the Southern California Gas Company (Los Angeles, California). We acknowledge the counsel of J. R. Schorr, J. P. Hummel, L. S. Cook, and J. Rice. We are also grateful to P. McGuire for help in constructing equipment.

REFERENCES

1. F. R. Bacon and G. L. Calcamuggio, "Effect of Heat Treatment in Moist and Dry Atmospheres on Chemical Durability of Soda-Lime Glass Bottles," *Am. Ceram. Soc. Bull.*, 46 [9] 850 (1967).
2. F. E. Wagstaff, S. D. Brown, and I. B. Cutler, "The Influence of Water and Oxygen Atmospheres on the Crystallization of Vitreous Silica," *Phys. Chem. Glasses*, 5, 76 (1964).
3. C. R. Whitworth, L. R. Bunnell, and S. D. Brown, "Viscosity of Fused Silica Doped with Alumina," p. 87 in *Proceedings of the 1969 International Commission on Glass*, Canadian Ceramic Soceity, Don Mills, Ontario, Canada (1971).
4. M. E. Bullard, *The Effect of Gaseous Atmospheres on the Surface Tension of Soda-Lime-Silicate Glasses*, M.S. Thesis, Dept. of Ceramic Engineering, University of Illinois at Urbana-Champaign, Urbana, IL (1972), 38 pp.
5. Tran Bao Van, *Effect of Furnace Atmosphere on the Viscosity and Homogeneity of Soda-Lime-Silica Glass*, M.S. Thesis, Dept. of Ceramic Engineering, University of Illinois at Urbana-Champaign, Urbana, IL (1972), 69 pp.
6. C. J. Koenig, *Effect of Atmospheres in Firing Ceramics*, Columbia Gas System Service Corp., Columbus, Ohio; Southern California Gas Co., Los Angeles, California (1971), 40 pp.
7. C. J. Koenig, *Water Vapor in High-Temperature Ceramic Processes, Literature Abstracts*, Columbia Gas System Service Corp., Columbus, Ohio; Consolidated Natural Gas Service Co., Inc., Cleveland, Ohio; Corning Glass Works, Corning, New York; Frazier-Simplex, Inc., Washington, Pennsylvania; Southern California Gas Co., Los Angeles, California (1973).
8. W. E. Martinsen and T. D. McGee, "Effect of Water Content on Electrical Resistivity of Na_2O-SiO_2 Glasses," *J. Am. Ceram. Soc.*, 54 [3] 175 (1971).
9. J. F. Shackelford, J. S. Masaryk, and R. M. Fulrath, "Water Content, Fictive Temperature, and Density Relations for Fused Silica," *ibid.*, 53 [7] 417 (1970).
10. S. B. Joshi, T. B. Van, M. E. Bullard, and S. D. Brown, "The Effects of Gaseous Atmospheres on Factors Affecting Melting and Refining of Glass," *Collected Papers, 33rd Annual Conference on Glass Problems*, Ohio State Univ., Columbus, Ohio (1972), pp. 92-113.
11. G. L. McVay and E. H. Farnum, "Anomalous Effects of H_2O on Na Diffusion in Glass," *J. Am. Ceram. Soc.*, 57 [1] 43 (1974).
12. M. Takata, M. Tomozawa, and E. B. Watson, "Electrical Conductivity of $Na_2O-3SiO_2$ Glasses with High Water Content," *ibid.*, 63 [11-12] 710 (1980).
13. M. Takata, J. Acocella, M. Tomozawa, and E. B. Watson, "Effect of Water Content on the Electrical Conductivity of $Na_2O-3SiO_2$ Glass, *ibid.*, 64 [12] 719 (1981).

14. M. Tomozawa, C. Y. Erwin, M. Takata, and E. B. Watson, "Effect of Water Content on the Chemical Durability of $Na_2O-3SiO_2$ Glass," ibid., 65 [4] 182 (1982).
15. S. D. Brown and L. S. Cook, Furnace Atmosphere Control for Improved Glass Manufacturing, Columbia Gas System Service Corp., Columbus, OH; Southern California Gas Co., Los Angeles, CA (1977), 20 pp.
16. D. M. Sanders, M. E. Wilke, S. Hurwitz, and W. K. Haller, "Role of Water Vapor and Sulfur Compounds in Sodium Vaporization During Glass Melting," J. Am. Ceram. Soc., 64 [7] 399 (1981).
17. G. L. McVay and L. R. Pederson, "Gamma Irradiation Effects on Glass Leaching," Paper 14-JXII-81, presented at the 83rd Annual Meeting of the American Ceramic Society, 3-6 May 1981, Washington, D. C.; ibid.
18. B. O. Barnes and D. M. Strachan, "Leachability of Some Nuclear Waste Forms," Paper 16-JXII-81, presented at the 83rd Annual Meeting of the American Ceramic Society, 3-6 May 1981, Washington, D. C.; Am. Ceram. Soc. Bull., 60 [3] 368 (1981).
19. A. J. Harrison, "Water Content and Infra-red Transmission of Simple Glasses," J. Am. Ceram. Soc., 30 [12], 362 (1947).
20. R. Bruckner and H. Scholze, "On Infra-red Bands of Freshly Made and Weathered B_2O_3 Glass," Glastech. Ber., 31 [11], 417 (1958).
21. H. Scholze, "The Incorporation of Water in Glasses. I. The Influence of Water Dissolved in Glass on the Infra-red Spectrum, and the Quantitative Determination of Water in Glasses by the Aid of Infra-red Spectroscopy," ibid. 32 [3], 81 (1959).
22. H. Scholze, "Water in the Glass Structure," Glass Ind., 40 [6], 301-3, 338, (1959).
23. I. Simon, "Infra-red Studies of Glass," Modern Aspects of the Vitreous State, Vol. 1, J. D. Mackenzie, editor, Butterworths, Washington, D.C. (1960), pp. 120-51.
24. A. Kats, Y. Haven, and J. M. Stevels, "Hydroxyl Groups in α-Quartz," Phys. Chem. Glasses, 3 [3] 69 (1962).
25. H. Scholze, "Gases and Water in Glass," Glass Ind., 47 [10], 546; [11], 622; [12], 670 (1966).
26. G. Kocsis, "Study of Structural Water in Glasses," Epitoanyag., 20 [1], 18 (1968).; Ceram. Abstr., [3], 66 (1970).
27. F. M. Ernsberger, "Molecular Water in Glass," J. Am. Ceram. Soc., 60, [1] 91 (1977).
28. B. J. Todd, "Outgassing of Glass," J. Appl. Phys., 26 [10], 1238 (1955).
29. B. J. Todd, "Equilibrium Between Glass and Water at Bake-Out Temperatures," J. Appl. Phys., 27 [10] 1209 (1956).
30. W. A. Lanford, "Characterization of the Reaction Between Water and Soda-Lime and Reactor Waste Glasses," IEEE Trans. on Nuclear Science, 26 1795 (1979).
31. W. A. Lanford, "Glass Hydration: A Method for Dating Glass Objects," Science, 196, 975 (1977).
32. C. J. Altstetter, "Applications of Nuclear Reaction Analysis for Determining Hydrogen and Deuterium Distribution in Metals,"

pp. 211-232 in Advanced Techniques for Characterizing Hydrogen in Metals, N. Fiore and B. J. Berkowitz, editors, TMS-AIME, Warrendale, PA (1982).
33. G. J. Clark, C. W. White, D. D. Allred, B. R. Appleton, and I. S. T. Tsong, "Hydrogen Concentration Profiles in Quartz Determined by a Nuclear Reaction Technique," Phys. Chem. Minerals, 3, 199 (1978).
34. P. Williams, C. A. Evans, Jr., M. L. Grossbeck, and H. H. Birnbaum, "Ion Microprobe Analysis for Niobium Hydride in Hydrogen-Embrittled Niobium," Analytical Chem., 48, [7] 964 (1976).
35. C. G. Pantano, "Surface and In-Depth Analysis of Glass and Ceramics," Am. Ceram. Soc. Bull., 60 [11], 1154 (1981).
36. R. A. Weeks and M. M. Abraham, "Electron Spin Resonance of Irradiated Quartz: Atomic Hydrogen," J. Chem. Phys., 42, 68 (1965).
37. F. Meyer and W. Spalthoff, "Ermittlung des Wassergehaltes in Glasern mit Hilfeder Magnetischen Kernresonanz und Vergleich der Ergebnisse mit Messungen der infraroten OH-Banden," Glastech.Ber. 34, 184 (1961).
38. I. S. T. Tsong, C. A. Houser, W. B. White, and G. L. Power, "Glass Leaching Studies by Sputter-Induced Photon Spectrometry (SIPS)," J. Non-Cryst. Solids, 38 & 39, 649 (1980).
39. T. Drury and J. P. Roberts, "Diffusion in Silica Glass Following Reaction with Tritiated Water Vapour," Phys. Chem. Glasses, 4 [3], 79 (1963).
40. G. J. Roberts and J. P. Roberts, "Influence of Thermal History on the Solubility and Diffusion of 'Water' in Silica Glass," Phys. Chem. Glasses, 5 [1] 26 (1964).
41. E. W. Shaffer, A. R. Cooper and A. H. Heuer, "Solubility and Diffusion of 'Water' in Quartz," Paper 74-B-72, the 74th Annual Meeting of the American Ceramic Society, Washington, D.C., 6-11 May 1972; Bull. Am. Ceram. Soc. 51 [4], 335 (1972).
42. G. J. Roberts and J. P. Roberts, "An Oxygen Tracer Investigation of the Diffusion of 'Water' in Silica Glass," Phys. Chem. Glasses, 7 [3], 82 (1966).
43. M. Noshiro and T. Yarita, "Determination of Water in Glass by Gas-chromatography," J. Ceram. Soc. Japan, 90, [5] No. 1041, 215, (1982).
44. N. de St.-Victor, "On a New Action of Light," Compt. Rend, 65, 505 (1867).
45. H. Becquerel, "On the Invisible Radiations Emitted by Phosphorescent Solids," Compt. Rend., 122, 501 (1896).
46. H. Becquerel, "On the Invisible Radiations Emitted by Salts of Uranium," Compt. Rend., 122 689 (1896).
47. H. Becquerel, "Emission of Novel Radiations by Metallic Uranium," Compt. Rend., 122, 1086 (1896).
48. A. Lacassagne and J. Lattes, Bull. Histol. Appl. Physiol. et Pathol. et Tech. Microscopique, 1, 279 (1924) [cited in Ref. 47].
49. A. Lacassagne, J. Lattes, and J. Lavedan, J. Radiol. et Electrol. 9 [1] 67 (1925) [cited in Ref. 47].
50. C. P. Leblond, "Localization of Newly Administered Iodine in the

Thyroid Gland as Indicated by Radio-Iodine," *J. Anat.* 77, Part 2, 149 (1943).
51. G. A. Boyd, Autoradiography in Biology and Medicine, Academic Press, New York (1955), 399 pp.
52. R. H. Condit, "Autoradiographic Techniques in Metallurgical Research," Chapt. 20 in Techniques of Metals Research, Vol. II, Part 2, R. F. Bunshah, editor, Interscience, New York, N. Y. (1969), pp. 877-952.
53. G. R. Caskey, "Tritium Autoradiography," pp. 61-76 in Advanced Techniques for Characterizing Hydrogen in Metals, N. Fiore and B. J. Berkowitz, editors, TMS-AIME, Warrendale, PA (1982).
54. S. B. Joshi, Development of a Fluoro-Autoradiographic Technique for Studying Water-Glass Interactions at High Temperatures, M. S. Thesis, University of Illinois at Urbana-Champaign, Urbana, IL (1974), 76 pp.
55. S. A. Huffsmith, The Use of Tritium Autoradiography to Determine the Concentration and Spatial Distribution of Water in Soda-Lime-Silicate Glass, M. S. Thesis, University of Illinois at Urbana-Champaign, Urbana, IL (1980) 54 pp.
56. M. M. Salpeter and M. Szabo, "An Improved Kodak Emulsion for Use in High Resolution Electron Microscope Autoradiography," *J. Histochem. Cytochem.*, 24 [11] 1204 (1976).
57. Data Release, "Kodak Special Product, Type 129-01," Kodak Pamphlet No. P-65; Eastman Kodak Company, Health Sciences Markets Div.; Rochester, NY (1980).
58. I. S. T. Tsong, A.C. McLaren, and B. E. Hobbs, "Determination of Hydrogen in Silicates Using the Ion Beam Spectro-Chemical Analyzer Application to Hydrolytic Weakening," *Am. Mineral.*, 61, 921 (1976).
59. P. Williams, private communication to S. D. Brown, July, 1981.

CHARACTERIZATION OF BOROSILICATE GLASS CONTAINING SAVANNAH RIVER PLANT RADIOACTIVE WASTE

Ned E. Bibler and P. Kent Smith

E. I. du Pont de Nemours & Co.
Savannah River Laboratory
Aiken, South Carolina 29808

ABSTRACT

Vitrification is the reference process for immobilization of radioactive waste from the Savannah River Plant. The waste, consisting mostly of hydrous oxides of Fe, Al, and Mn contaminated with fission products and alpha-emitting radionuclides, is mixed with glass-forming chemicals and melted at 1150°C to produce a durable borosilicate glass. In this paper, the results of studies characterizing glass containing actual radioactive waste are presented. The glass was produced in a small-scale joule-heated melter in a shielded facility at the Savannah River Laboratory. Feed for the melter was a mixture of 35 wt % waste (primarily Fe_2O_3 and Al_2O_3) and 65 wt % frit (primarily SiO_2, Na_2O, and B_2O_3). The melt was poured into 600-cc stainless steel beakers and allowed to cool. The glass was intensely radioactive. Dose rates at the surface of the glass were approximately 10^6 rad/hr. Specific activities (dpm/g glass) of the principal radionuclides were: 1.2×10^{10} for Cs-137, 7.9×10^9 for Sr-90, and 1.0×10^8 for alpha activity (primarily Pu-238 and Cm-244). Examination by optical and electron microscopy indicated the presence of 1-5 wt % spinel crystals - primarily $NiFe_2O_4$. Leach tests indicated good durability in deionized water, brine, and silicate water. These latter two leachants simulate groundwater from possible geologic repositories. Release rates at 40°C based on gamma, beta, and alpha activity in a 28-day test were approximately 0.01 g glass/m^2-day. This result agrees with results of tests with glass containing simulated waste and indicates that the borosilicate glass effectively immobilizes SRP radioactive waste. Also, during the leach tests, the pH of the leachates did not decrease even though they were exposed to the intense gamma, beta, and alpha radiation from this glass.

INTRODUCTION

Savannah River Plant (SRP) has been producing nuclear materials for national defense for many years. Several nuclear reactors and two chemical separations plants are involved in the production of these materials. As a result of the chemical separations and other processes, approximately 28 million gallons of high-level radioactive waste has been produced at SRP. The waste is stored as a caustic slurry in underground tanks at SRP. Currently, a process is being developed to immobilize this waste for eventual transportation to a geologic repository for permanent storage.

A great deal of effort has been directed toward developing a suitable matrix and process for immobilizing SRP waste. The reference process produces borosilicate glass in a continuous joule heated melter.[1] Borosilicate glass was selected because of its durability, its ability to tolerate the varied compositions of the waste at SRP, and its excellent radiation stability. In addition, the technology for large-scale glass production can be adapted for remote operation.

The final product will be a cylinder of glass in a stainless steel canister, 10 ft tall by 2 ft. To immobilize all the waste, more than 6000 canisters will be necessary. Full-scale production of the glass and laboratory-scale production of test specimens have been performed with nonradioactive, simulated waste that has a chemical composition similar to that of the radioactive waste. Many studies have been performed to characterize this nonradioactive glass.[2] Glass containing actual radioactive waste has been made on a small scale in shielded facilities. An earlier study characterized radioactive glass prepared in that facility by crucible and in-can melting.[3] That study showed that the glass contained only a small amount of iron oxide type crystals and had a low leach rate, i.e., a high chemical durability. This study characterizes radioactive glass made in a small, joule-heated melter in the shielded facility. This melter[4] simulates more closely the reference process for full-scale melting. The results of this study indicate that the glass prepared in the small melter is similar to that prepared in the nonradioactive full-scale melter. Tests measuring the leaching of gamma, beta, and alpha radioactivity from the glass by water indicate that the glass is suitable for immobilizing radioactive SRP waste.

GLASS PREPARATION

Feed for the joule-heated melter was a dry, blended mixture of glass-forming chemicals and radioactive waste sludge from SRP storage Tank 13. The dried waste was obtained by washing the radioactive waste slurry to remove soluble salts. The resulting sludge was then dried at approximately 250°C. Aliquots of the blended mixture were then continuously fed to the melt which was at 1150°C. Residence time in the melter was approximately 19 hours. Small cans of glass were prepared by tilting the melter assembly so the glass poured into 600-cc stainless steel beakers. After a beaker was full, it was removed from the pour chamber and allowed to cool to ambient temperature. No attempt was made to control the cooling rate. Samples of glass for characterization were obtained by sectioning the can. Recently this melter has been successfully modified so that a slurry rather than a dry powder can be fed.[5]

GLASS AND WASTE COMPOSITION

The feed to the melter was 65 wt % glass-forming chemicals and 35 wt % waste sludge. Major components of the waste are listed in Table 1. The calculated composition of the final glass is in Table 2. The waste also contained approximately 1 wt % of zeolite loaded with Cs-137. This zeolite resulted from a demonstration test for removal of soluble Cs from the SRP waste supernate. The specific activities of the major radionuclides in the glass were determined by dissolving a known amount of glass by a Na_2O_2 fusion followed by HNO_3 dissolution of the resulting salts. Specific activities (disintegrations per minute per gram of glass, dpm/g glass) were: 1.2×10^{10} for Cs-137 gamma-rays, 7.9×10^9 for Sr-90 beta-particles, and 1.0×10^8 for alpha-particles from Pu-238 and Cm-244. The values for beta and alpha activity were in reasonable agreement with those expected from analysis of the waste sludge. The gamma activity is approximately 100 times higher than accounted for by the waste sludge because of the added Cs-137. A chemical dosimeter (Fricke dosimeter) was used to estimate the dose rate at the surface of the glass. This dose rate was approximately 10^6 rad/hr and primarily due to Sr-90, Y-90 beta particles.

GLASS MICROSTRUCTURE AND CRYSTALLINITY

Glass microstructure and crystallinity were characterized by examining polished samples of the glass by optical or electron microscopy. Figure 1 shows a typical distribution of crystals in a polished section. The shape, size, and distribution of the crystals are typical of those observed in nonradioactive glass samples. Examination of many samples from the radioactive glass at various positions in the canister indicated that the weight percent crystallinity varied from 0 to 5. The size of the individual crystals varied from 5 to 20μm as indicated by scanning electron microscopy. X-ray energy spectroscopy of a typical crystal (Figure 2) indicated that the crystals contained primarily Fe and Ni along with smaller amounts of Cr and Mn. These crystals are probably ferrite spinels,

Table 1. Major Components of Tank 13 Waste Sludge

Tank 13 Waste Sludge[a]

Composition as Oxides		Radionuclide Composition	
Oxide	Wt %	Radionuclide	mCi/g
Fe_2O_3	39.9	Sr-90	15.5
MnO_2	13.9	Ru-106	0.40
Al_2O_3	13.4	Eu-154	0.30
U_3O_8	4.7	Cs-137	0.30
Na_2O	4.2	Pu-238, Pu-239, Cm-244	0.28
CaO	3.2		
HgO	2.3		
CeO_2	1.2		
NiO	0.6		
Mg_2O	0.5		
Cr_2O_3	0.1		
K_2O	0.1		

[a]Taken from Reference 6.

Table 2. Calculated Composition of the Final Glass[a]

Component	Wt %	Component	Wt %
SiO_2	34.1	MnO_2	4.9
Fe_2O_3	14.0	Al_2O_3	4.7
Na_2O	13.5	CaO	4.4
TiO_2	6.5	Li_2O	2.6
B_2O_3	6.5	U_3O_8	1.6
		Other Oxides	7.2

[a]Based on 35 wt % waste (Table 1) and 65 wt % glass-forming chemicals of the composition (wt %); SiO_2, 52.5; Na_2O, 18.5; TiO_2, 10.0; B_2O_3, 10.0; CaO, 5.0; Li_2O, 4.0 (taken from Reference 7).

Fig. 1. Typical crystal distribution in SRL radioactive waste glass

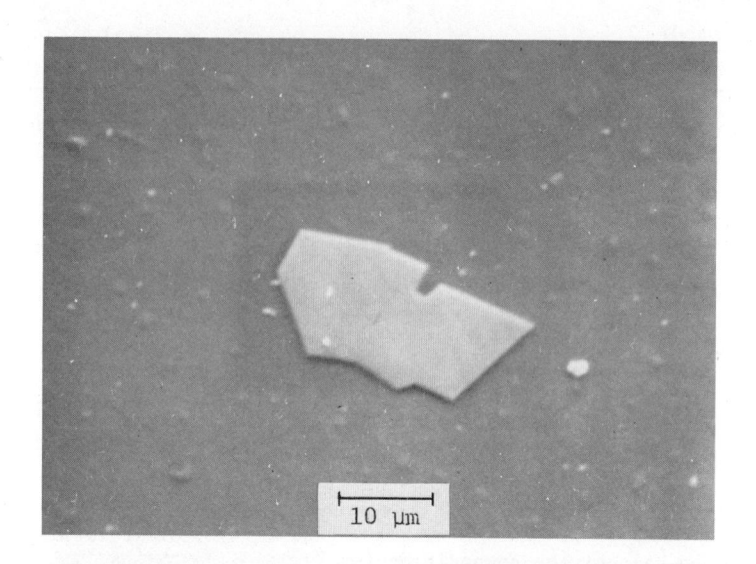

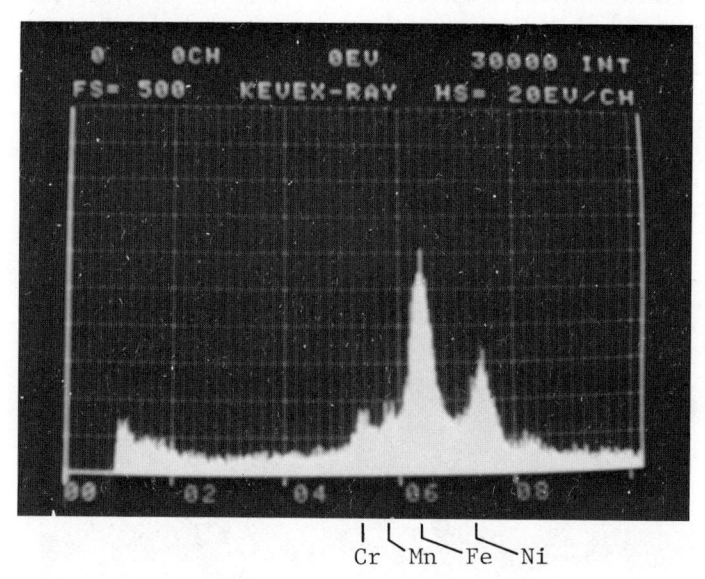

Fig. 2. Scanning electron micrograph and x-ray energy analysis of a typical crystal

in this case trevorite ($NiFe_2O_4$). Cr may have resulted from corrosion of the refractory or electrodes of the melter since Cr is a very minor component of the waste. Analysis of the glass in the vicinity of the crystals by x-ray energy spectroscopy showed no depletion of Ni or Fe from the glass, indicating that the crystals had probably formed while the glass was still fluid and not as a result of devitrification after the glass had hardened.

An earlier study of radioactive glass found that the glass surrounding spinel crystals leached faster than the crystals themselves.[3] This observation suggests that formation of ferrite spinels should not significantly decrease the durability of the glass for immobilizing nuclear waste. Leach tests based on both radioactive and nonradioactive components of the glass confirm this.[2,3]

GLASS DURABILITY

The leaching of radionuclides from the glass is a measure of its durability; a low leachability indicates a high durability. To determine leachability, tests were performed in deionized water and in two leachates representing groundwaters of possible geologic repositories. These groundwaters were a brine representing a salt repository and a silicate water representing a basalt repository. For the tests, the glass was supended in the respective leachate for 28 days. The leachate was at 40°C because this temperature is that expected in a repository after long storage times. After 28 days, the glass was removed. The amount of Cs-137, Sr-90, and alpha activity leached from the glass was then measured. Leach rates in terms of grams of glass leached per day per unit surface area were then calculated. Results are presented in Table 3. The specific activities of the Cs-137, Sr-90 and alpha activity in the glass were used to convert from radioactivity leached to grams of glass leached. The detailed equation for calculating the leach rate is:

$$L_i = \frac{A_i}{(Sp.\ Act.)_i\ (SA)_g\ t}$$

where:

L_i = leach rate (g/m²-day) based on radionuclide i

A_i = total activity of radionuclide i leached in units of disintegrations per minute

$(Sp.\ Act.)_i$ = disintegrations per minute of radionuclide i per gram of glass

$(SA)_g$ = geometric surface area of the glass leached, m²

t = time leached (days)

Table 3. Leach Rates of Actual SRP Radioactive Glass at 40°C[a]

Leachate	Leach Rates (g/m^2-day)[b]			pH Change
	Cs-137	Sr-90	Gross Alpha[c]	Initial - Final
Deionized Water	0.012	0.016	0.050	6.5 - 7.7
Brine[d]	0.014	0.0094	0.010	6.5 - 6.5
Silicate Water[e]	0.010	0.0012	0.0054	7.5 - 8.3

[a] Glass is 65% Frit 21, 34% Tank 13 sludge, and 1% Cs-137 loaded zeolite. Specific activity of the glass (dpm/g glass): Cs-137, 1.2×10^{10}; Sr-90, 7.9×10^9; gross alpha, 1.0×10^8.
[b] Approximately 10 grams of glass (surface area approximately 20 cm^2) leached in 200 mL leachant for 28 days. Results based on acidified leachates from duplicate tests. Relative precision is nominally 20%.
[c] Activity was 23% Pu-239, 44% Pu-238, and 33% Cm-244.
[d] MCC brine (48 g/L KCl, 90 g/L NaCl, 116 g/L MgCl$_2$; pH = 6.5).
[e] MCC silicate water (0.18 g/L NaHCO$_3$, 0.058 g/L SiO$_2$; pH = 7.5).

Standard procedures proposed by the Materials Characterization Center (MCC)[8] for leaching nonradioactive waste forms were followed as closely as possible. However, because of the intense radioactivity of the glass, a stainless steel basket (Figure 3) was used to suspend the sample rather than a plastic filament as recommended by MCC. The plastic filament would have failed due to radiation damage. Polypropylene leach containers were used (Figure 3). This plastic was not damaged by radiation because the leachate absorbed most of the radiation (beta and alpha particles). Also, the absorbed dose rate from gamma radiation was too small to damage the container during the leach tests.

Because waste glass will probably be in the form of fractured monoliths in a repository, fractured shards were used for leaching. The surface area of the glass being leached was estimated by placing each surface of that piece against a calibrated grid and calculating its area. The individual surface areas were then summed. The error introduced by this procedure was estimated to be approximately 20%. The ultrasonic cleaning procedures recommended by MCC were found sufficient to remove surface contamination from the glass. Finally, control tests confirmed that radioactive contamination on the cell walls, floors, etc. and on the master slave manipulators was not a problem.

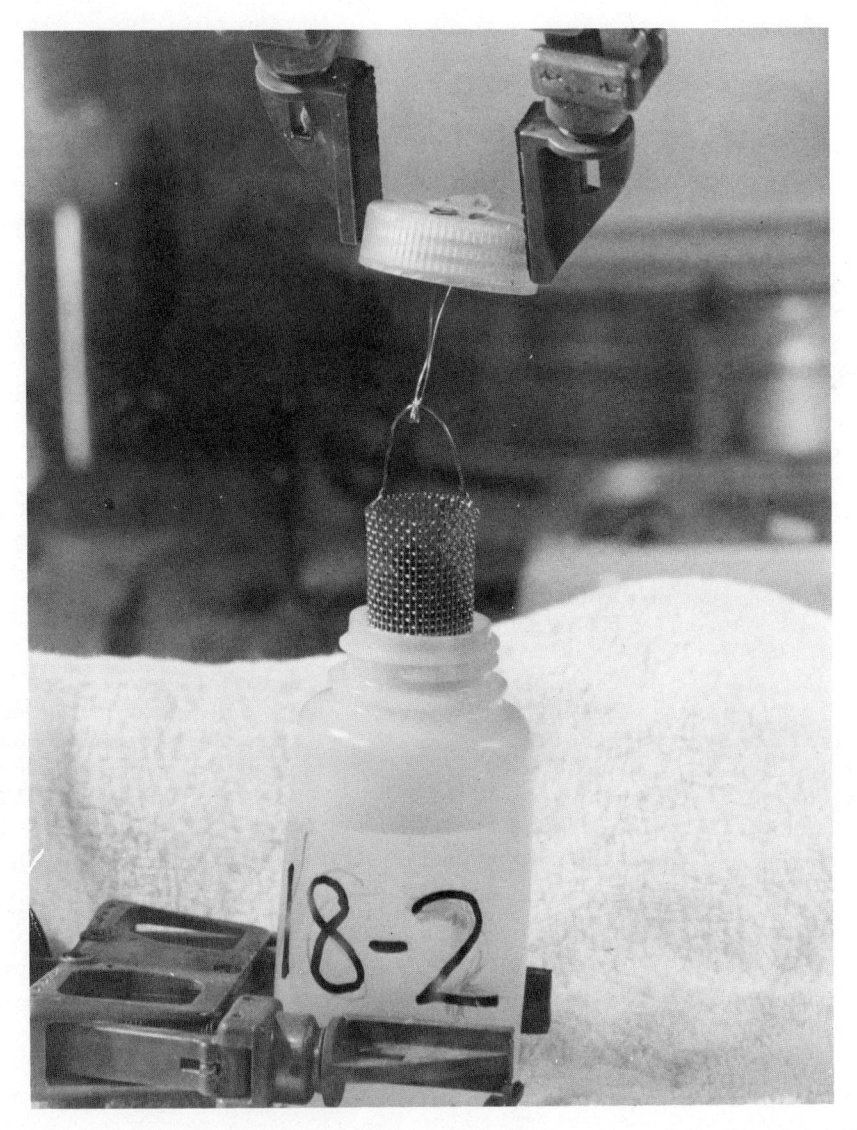

Fig. 3. Leaching container and stainless steel basket for holding the glass in leach tests with SRP radioactive waste glass

The leach rates in Table 3 agree with those for nonradioactive, simulated waste glass tested under similar conditions.[2] This agreement indicates that the glasses prepared to test vitrification for immobilizing nuclear waste have similar durabilities. Also, the leach rates in Table 3 in the simulated groundwaters are equal to or less than the respective values in deionized water. Thus, repository groundwaters will probably not have an adverse effect on the durability of the glass.

Another significant result in Table 3 is the pH change that resulted from the leaching process. A pH increase is expected during leaching as Na^+ ions from the glass exchange with H_3O^+ (or H^+) ions in the solution. As a result, the OH^- ion concentration increases and thus the pH increases. This behavior has been observed in many leach tests.[2] However, radiolysis of the leachant from the radioactive glass may cause an opposite effect, a pH decrease if dissolved air is present. This has been observed in leach tests in the presence of Co-60 gamma radiation.[9] In these tests, the pH decreased to approximately 4 during leaching. Radiation caused HNO_3 to form from the dissolved air in the leachate. Failure of radiation from the radioactive glass to cause a decrease in the pH, supports the proposal that radiolytic nitric acid formation will not be significant during geologic storage of the glass.[10] The reason that a pH decrease was not observed with the radioactive glass is currently being investigated.

A longer leach test was performed in deionized water under the same conditions. This test was for 134 days. Results indicated that leach rates after 134 days are lower than those after 28 days for all three of the radionuclides. The rates (in g/m^2-day) in the 134 day test were 0.0070 for Cs-137, 0.0047 for Sr-90, and 0.012 for alpha activity. This decrease at longer times has been observed in all leach tests with glass.[2] Such a decrease results from slower diffusion of the radionuclides from the glass as the leach time increases. Slower diffusion may be due to the formation of a gel layer on the glass or to the precipitation of the leached radionuclides on the surface of the glass. Such a decrease improves the performance of borosilicate glass as a matrix for immobilizing defense nuclear wastes. The above rates can be used in calculations to estimate the amount of radioactivity released from a geologic repository over long storage times.[13]

ACKNOWLEDGMENT

The information contained in this article was developed during the course of work under Contract No. DE-AC09-76SR00001 with the U.S. Department of Energy.

REFERENCES

1. R. G. Baxter, "Description of DWPF Reference Waste Form and Canister," USDOE Report DP-1606, E. I. du Pont de Nemours & Company, Savannah River Laboratory, Aiken, South Carolina (1981).
2. M. J. Plodinec, G. G. Wicks, and N. E. Bibler, "Performance of Borosilicate Glass in the Repository Environment," USDOE Report DP-1629, E. I. du Pont de Nemours & Co., Savannah River Laboratory, Aiken, SC (1982).
3. W. N. Rankin and J. A. Kelley, "Microstructures and Leachability of Vitrified Radioactive Wastes," Nucl. Technol. 41, 373 (1978).
4. M. J. Plodinec and P. H. Chismar, "Design and Operation of Small-Scale Glass Melters for Immobilizing Radioactive Waste," Conference Record of the IAS Annual Meeting, p. 20, Cincinnati, OH (Sept. 23 - Oct. 3, 1980).
5. G. G. Wicks, "Vitrification of Simulated High-Level Radioactive Waste by a Slurry-Fed Ceramic Melter," Nucl. Technol. 55, 601 (1981).
6. J. A. Stone, J. A. Kelley, and T. S. McMillan, "Sampling and Analysis of SRP High-Level Waste Sludges," USERDA Report DP-1399, E. I. du Pont de Nemours & Company, Savannah River Laboratory, Aiken, SC (1976).
7. M. J. Plodinec, J. R. Wiley, "Evaluation of Glass as a Matrix for Solidifying Savannah River Plant Waste: Properties of Glasses Containing Li_2O," USDOE Report DP-1498, E. I. du Pont de Nemours & Co., Savannah River Laboratory, Aiken, SC (1979).
8. "Materials Characterization Center Test Methods, Preliminary Version," Prepared by Material Characterization Center, J. E. Mendel, Manager, USDOE Report PNL-3990, Pacific Northwest Laboratory, Richland, WA (1981).
9. G. L. McVay and L. R. Pederson, "Effect of Gamma Radiation Glass Leaching," J. Am. Ceram. Soc. 64, 154 (1981).
10. W. G. Burns, A. E. Hughes, J. A. C. Marples, R. S. Nelson, and A. M. Stoneham, "Radiation Effects and The Leach Rates of Vitrified Radioactive Waste," Nature, 295, 130 (1982).
11. C. M. Koplik, M. F. Kaplan, and Ross, B., "The Safety of Repositories for Highly Radioactive Wastes," Rev. Mod. Phys. 54, 269, (1982).

MICROSTRUCTURE OF PHASE-SEPARATED SODIUM BOROSILICATE GLASSES

Peter Taylor[1], Allan B. Campbell[1], Derrek G. Owen[1], David Simkin[2] and Pierre Menassa[2]

[1]Atomic Energy of Canada Limited, Whiteshell Nuclear Research Establishment, Pinawa, Manitoba, Canada ROE 1L0
[2]Chemistry Department, McGill University, 801 Sherbrooke Street W., Montreal, Quebec, Canada H3A 2K6

ABSTRACT

The microstructures of various phase-separated glasses in the systems $Na_2O-B_2O_3-SiO_2$ and $Na_2O-MnO-B_2O_3-SiO_2$ have been examined by scanning electron microscopy. Examination of glasses lying just within the miscibility gap in the system $Na_2O-B_2O_3-SiO_2$ at 650, 700 and 750°C has permitted estimation of plait point compositions at these temperatures. Results are consistent with the tie-lines proposed by Mazurin and Streltsina [J. Non-Cryst. Solids, 11, 199 (1972)] on the basis of glass transition temperature measurements. X-ray microanalysis of some MnO-containing glasses has aided the description of tie-line orientation in the system $Na_2O-MnO-B_2O_3-SiO_2$.

INTRODUCTION

Many authors have investigated liquid immiscibility (amorphous phase separation) in glass-forming systems; this subject was most recently reviewed by Tomozawa[1]. The sodium borosilicate system has received particularly close attention[2-11]. The topography of the miscibility gap in this system is well established[2,3], but some uncertainty remains about the locations of tie-lines[3-7]. Mazurin and Streltsina[5], and Scholes[6], have discussed the limitations of various methods that can be applied to tie-line determination, with special reference to the sodium borosilicate system.

We have taken a slightly different approach from earlier microstructural studies, by examining many phase-separated glasses lying just within the boundary of the miscibility gap. The most detailed examination was near the 750°C isotherm, with less detailed work at 650 and 700°C. This work forms part of a general study of phase separation in borosilicate glass systems of potential use for the incorporation of high-level radioactive waste for permanent disposal[12-13]. Although many authors have published a few micrographs of phase-separated sodium borosilicate glasses, or have studied the microstructures of a few glass compositions in detail[2-10], we are aware of only one paper, by Baron and Wey[11], in which the microstructures of many glasses are reported. The discussion in that paper is limited to a qualitative description of observed microstructures, and does not include any correlation with the position of compositions within the miscibility gap.

THEORETICAL PRINCIPLES

The theoretical basis of our work is illustrated in Fig. 1.

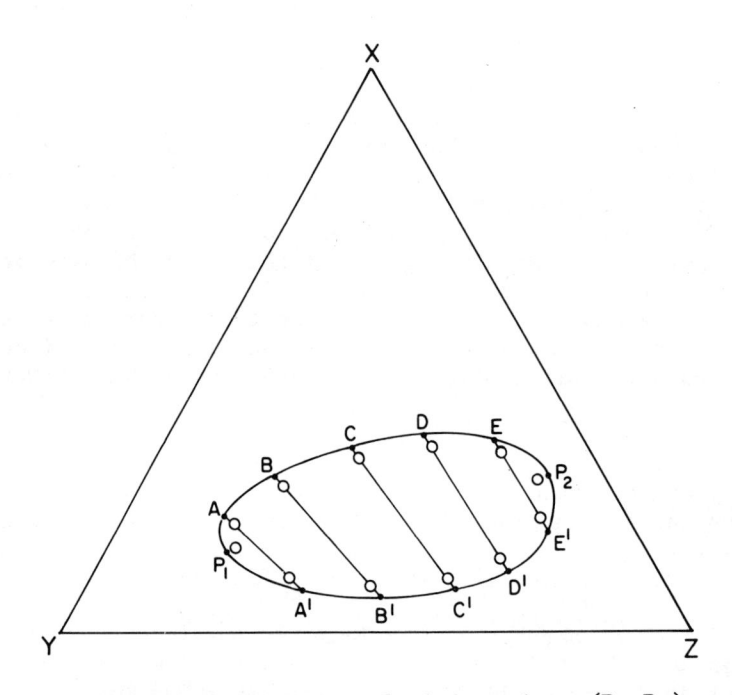

Fig. 1. Generalized arrangement of plait points (P_1, P_2) and tie-lines (AA', BB', etc.) in an "island" miscibility gap in a ternary system X-Y-Z at temperature T. Open circles represent compositions just within the miscibility gap at this temperature.

Here, the miscibility gap in the ternary system X-Y-Z at temperature T is a closed loop, $P_1...B...P_2...B'...P_1$. This resembles the behaviour of the sodium borosilicate system at 750°C[2,3]. Tie-lines join conjugate liquid compositions, A-A', etc. The tie-lines merge into the perimeter of the miscibility gap at the plait points, P_1 and P_2.

Consider the series of compositions shown just within the miscibility gap, assuming that the liquids are too viscous to separate completely into discrete layers, as is the case with most glass-forming systems. Those compositions lying near the upper portion of the loop, $P_1....B...P_2$, will contain a small quantity of X-deficient phase (A',B', etc.), dispersed within an X-rich matrix (A, B, etc.). Conversely, those compositions lying near the lower portion of the loop, $P_1...B'...P_2$, will contain the X-rich phase dispersed within an X-deficient matrix. Near the plait points, the material should consist of near-equal volumes of near-identical liquids. Our experiments were done in anticipation of this type of behaviour in the sodium borosilicate system.

EXPERIMENTAL

The methods of glass melting, heat treatment, clearing-temperature determination and scanning electron microscopy have been described elsewhere[13]. Glasses were prepared with compositions lying near the 650, 700 and 750°C isotherms of the sodium borosilicate miscibility gap, as determined by Haller et al.[3]. Scanning electron microscopy was performed on samples that had been heated at or near these temperatures to induce phase separation, then quenched to room temperature. Some MnO-doped glasses were prepared by the addition of manganese (II) sulfate or acetate to the batch composition. These were prepared primarily for luminescence-spectroscopic studies, to be described elsewhere, but the microstructure was examined in detail in some cases.

RESULTS AND DISCUSSION

Behaviour Near the 750°C Isotherm

We examined 24 glass compositions lying near or within this isotherm; the compositions are shown in Fig. 2. The phases denoted "silica-rich" and "silica-poor" in the following discussion were those showing greater and lesser resistance, respectively, to the etching in water or dilute HF, which was performed on freshly fractured surfaces in preparation for electron microscopy. Micrographs of seven of these glasses are shown in Figures 3 to 9.

The clearing temperature, T_c, of glass 1 (Fig. 2) was 745 ± 5°C. A sample heated for 1 h at 735°C and etched for 10 s in 1% HF showed

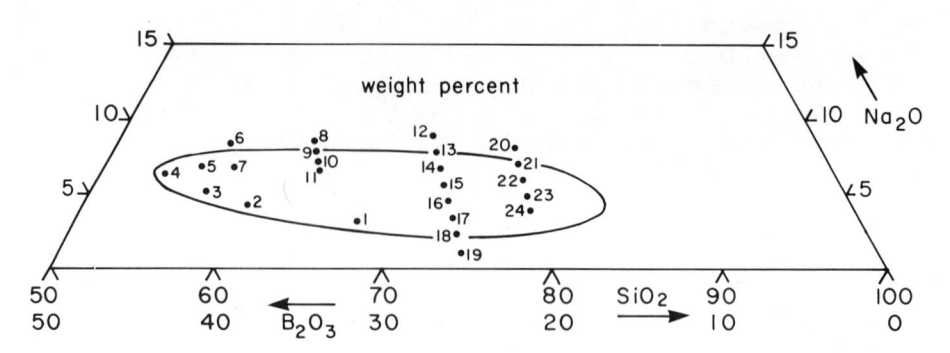

Fig. 2. Compositions examined near the 750°C isotherm in the sodium borosilicate miscibility gap (isotherm from ref. 3).

a non-connective porous microstructure, indicating selective etching of discrete silica-poor particles from a silica-rich matrix. The microstructure was similar but more clearly developed after 16 h at 735 C, as depicted in Fig. 3a. Fig. 3b shows a micrograph of the same glass, etched in water rather than HF; here the silica-poor phase is still visible. This composition evidently lies toward the silica-rich end of a tie-line.

Glass 2 also had a T_C = 745 ± 5°C. The micrograph of this glass after 16 h at 735°C (Fig. 4) shows larger silica-poor phase domains than in glass 1, with a larger volume fraction and substantial connectivity. This indicates that we are approaching the plait point, although probably still toward the silica-rich side.

Glass 3, after 16 h at 745 C, showed a high degree of connectivity in both phases, with approximately equal volume fractions (Fig. 5). This composition thus appears to lie very close to the plait-point locus.

Glass 4 had a T_C = 745 ± 5°C. After 16 h at 735°C the microstructure consisted of discrete silica-rich droplets, about 0.5 μm in diameter, dispersed in a silica-poor matrix (Fig. 6). This composition evidently lies on the silica-poor side of the plait-point locus. Glass 5 showed a very similar microstructure after 16 h at 745 C. Glass 6 had a T_C = 725 ± 5°C and was not examined further. Glass 7 had a T_C = 745 ± 5°C; after 16 h at 735°C its microstructure resembled those of glasses 4 and 5, but with a higher volume fraction of silica-rich particles.

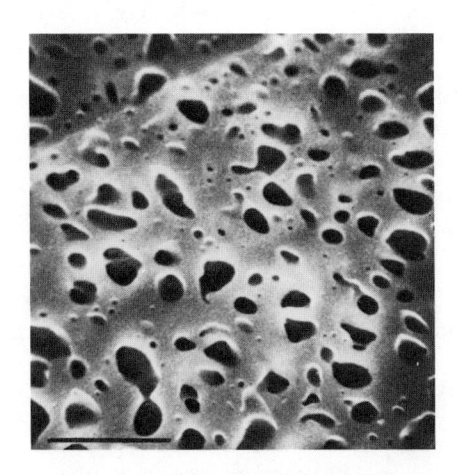

PHASE-SEPARATED SODIUM BOROSILICATE GLASSES 607

Fig. 3a. Glass composition 1 after 16 h at 735°C. Bar = 10 µm. (This and all other micrographs are of HF-etched, freshly fractured surfaces, unless otherwise stated).

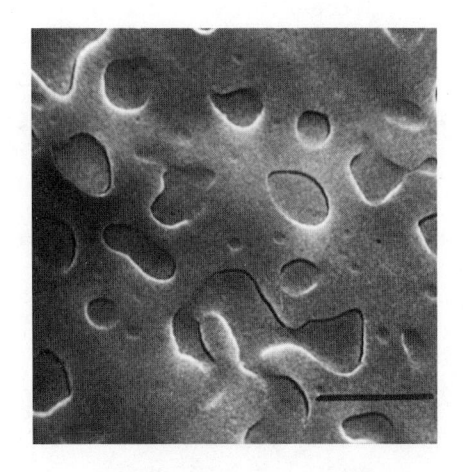

Fig. 3b. As Fig. 3a, but with water etch. Bar = 2 µm.

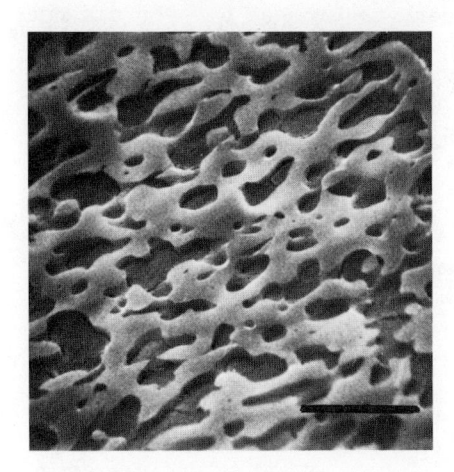

Fig. 4. Glass composition 2 after 16 h at 735°C. Bar = 10 µm.

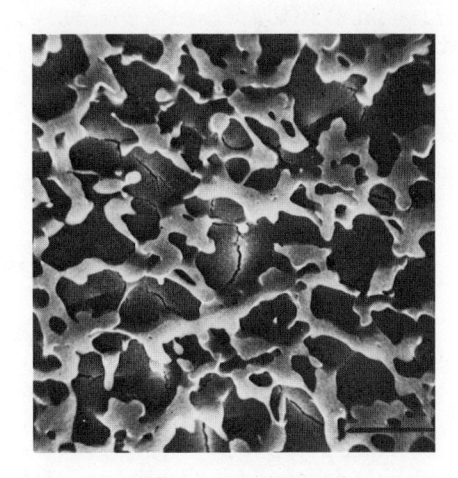

Fig. 5. Glass composition 3 after 16 h at 745°C. Bar = 10 µm.

Glasses 8 and 9 had $T_C < 730°C$, and showed no observable phase-separation microstructure after 16 h at 745°C, even following a 4 h nucleation step at 700°C. Glass 10 was slightly opalescent after this treatment, and the micrograph (Fig. 7) shows a low volume fraction of silica-rich droplets within a silica-poor matrix. Glass 11 had a $T_C = 750 \pm 5°C$. After 4 h at 700°C and 16 h at 745°C, it contained a much higher density of silica-rich droplets than glass 10 (Fig. 8). This higher particle density is expected as the composition progresses toward the centre of the miscibility gap.

Observations of phase separation became more difficult with the more silica-rich compositions, glasses 12 to 24. This was partly due to the difficulty of preparing satisfactory homogeneous samples of these viscous glasses, partly due to their slow kinetics of phase separation, and partly due to extensive devitrification occurring in competition with amorphous phase separation. Devitrification was too rapid with glasses 12, 13, 22 and 23 to determine whether phase separation had also occurred. Satisfactory homogeneous glasses could not be prepared with compositions 18 and 19. With glasses 14, 15, 20, 21 and 24, it was possible to examine the cores of samples, within a thick crust of devitrified material, after 72 h at 745°C. There was no indication of phase separation in glasses 20 and 21. Glasses 14, 15 and 24 all showed similar microstructures, with both phases highly interconnected (Fig. 9). Samples 15 and 24 had a somewhat higher volume fraction of silica-rich material than 14. Contrast between the phases was poor, indicating a rather small disparity between the phase compositions. These results are not conclusive, in view of the experimental conditions, but they are consistent with the second plait point being in this region of composition at 750°C. It is possible, however, that these glasses did not achieve the metastable equilibrium phase compositions because of their high viscosity, as discussed by Mazurin and Streltsina[5]. Glasses 16 and 17, after 16 h at 745°C, showed still higher volume fractions of silica-rich phase than glasses 15 and 24, with microstructures resembling those of samples 2 and 1, respectively. This confirms that composition 17 is is well removed from either plait point.

Behaviour Near the 650 and 700°C Isotherms
──

Glasses were prepared with compositions near these isotherms, as shown in Fig. 10. The experimental difficulties with silica-rich glasses, described above, were still more pronounced at these lower temperatures, so work was restricted to the B_2O_3-rich end of the miscibility gap. A gradation of microstructures, similar to that described above for glasses 1 to 5, was observed when glasses 25 to 33 were heated at 685 to 695°C for 3 to 4 h, and when glasses 34 to 37 were heated at 635°C for 16 h. From these results we estimated the plait-point positions shown in Fig. 11. These results are

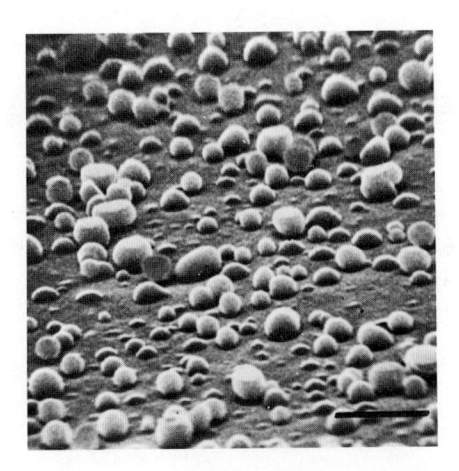

Fig. 6. Glass composition 5 after 16 h at 735°C. Bar = 2 μm.

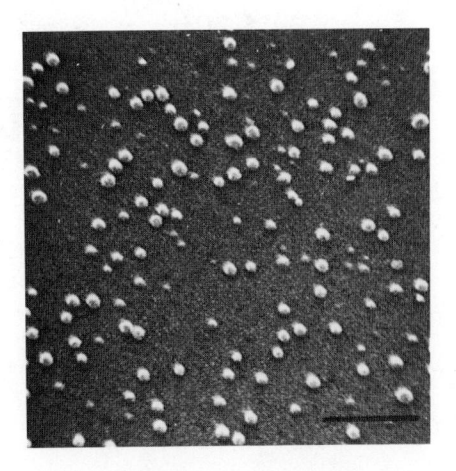

Fig. 7. Glass composition 10 after 4 h at 700°C, then 16 h at 745°C. Bar = 2 μm.

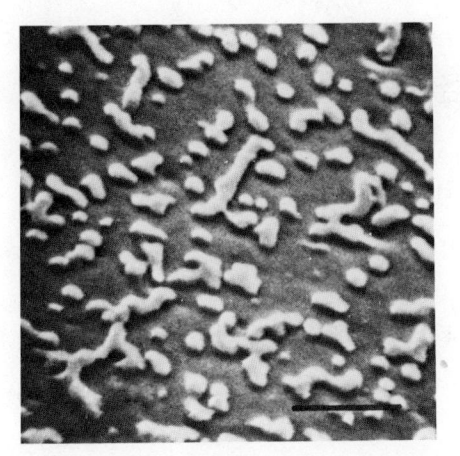

Fig. 8. Glass composition 11 after 4 h at 700°C, then 16 h at 745°C. Bar = 2 μm.

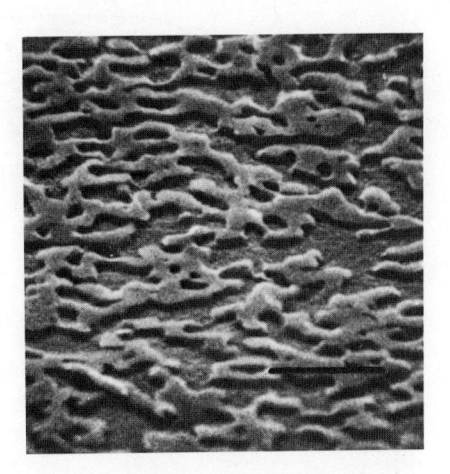

Fig. 9. Glass composition 14 after 72 h at 745°C. Bar = 2 μm.

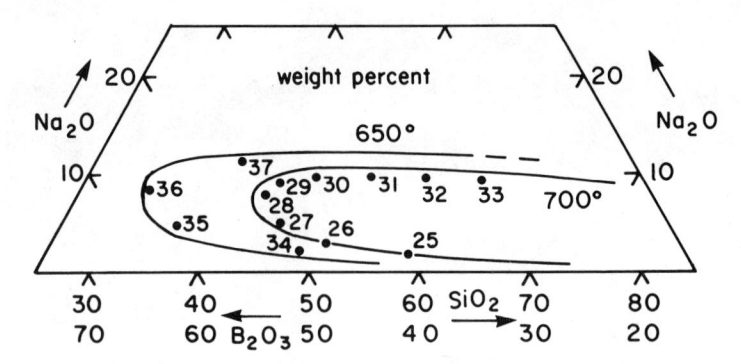

Fig. 10. Compositions examined near the 650 and 700°C isotherms in the sodium borosilicate miscibility gap (isotherms from ref. 3).

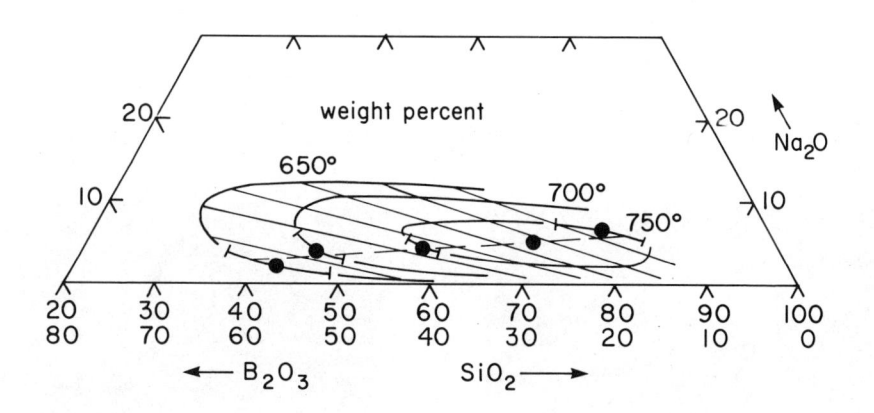

Fig. 11. Estimated plait points in the sodium borosilicate system at 650, 700 and 750°C, superimposed on isotherms from ref. 3 and tie-lines from ref. 5.

consistent with the tie-lines proposed by Mazurin and Streltsina, on the basis of glass-transition temperature measurements, and with the results of some earlier workers they cite[5]. The latter authors concluded that substantial rotation of the tie-lines does not occur with changing temperatures within the observable miscibility gap. Our estimate of the plait-point locus tends to confirm this, but our technique is relatively insensitive in this regard.

Progression of Microstructures Along a Tie-Line

We have extended this microstructural study to a series of MnO-doped glass compositions lying parallel to a tie-line, as proposed by Mazurin and Streltsina[5], and shown in Fig. 12. A more complete account of phase separation in the sodium manganese (II) borosilicate system will be given elsewhere (Menassa, Simkin and Taylor, unpublished work) but a brief description of microstructural observations is appropriate here.

Glasses with compositions 38 to 47 (Fig. 12), doped with 0.1 to 6.0 weight percent of MnO, were prepared and phase-separated by heating them near 700°C. A progressive change in microstructure was observed with these compositions; the microstructure was relatively insensitive to the presence of MnO. Glasses 38 to 43 all consisted of fairly large (2 - 10 μm) silica-rich particles in a silica-poor matrix. The volume fraction of the silica-rich phase increased between compositions 38 and 43. Between compositions 44 and 47, the more viscous silica-rich phase was continuous, with dispersed silica-poor particles a few tenths of a micrometre in diameter.

The relatively large phase domains which occurred in compositions 38 to 43 permitted us to perform some energy-dispersive X-ray microanalysis in conjunction with the scanning electron microscopy. A glass with composition 40, doped with 4% MnO (introduced as the acetate), was phase-separated by heating for 16 h at 695°C, and

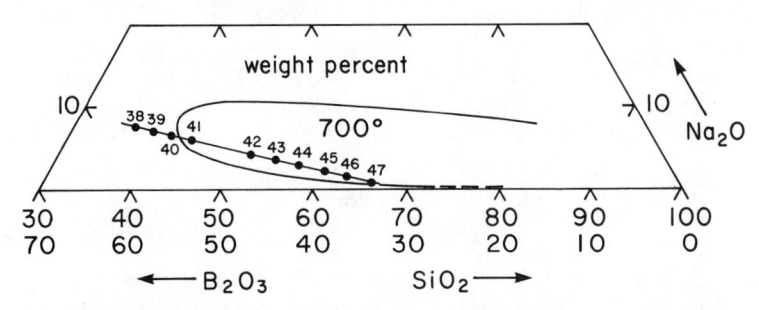

Fig. 12. Compositions of MnO-doped glasses lying along a tie-line in the sodium borosilicate system (0.1 to 6.0 wt % MnO added).

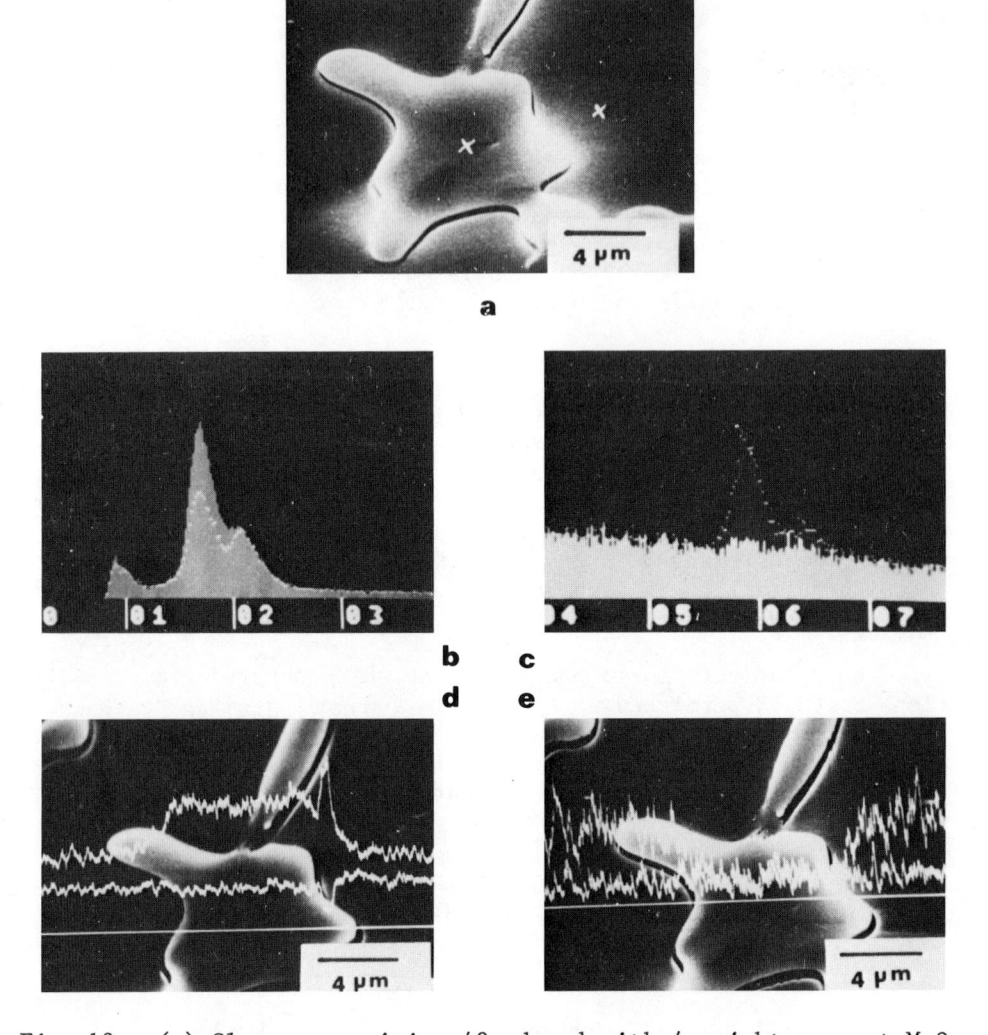

Fig. 13. (a) Glass composition 40, doped with 4 weight percent MnO, after 16 h at 685°C; water-etched fracture surface. Crosses represent points analyzed by energy-dispersive X-ray (EDX) "spot analysis". (b) and (c) Spectra obtained from EDX "spot analyses", demonstrating the concentration of Mn in the Si-poor phase (dotted spectrum) (Peak identity: 1.74 keV, Si; 2.12 keV Au coating of sample; 5.89 keV, Mn). (d) and (e) EDX "line scans" showing the distribution of Si and Mn, respectively. Lower scan in (d) is on the Au peak, that in (e) is on the background adjacent to the Mn peak. "Spikes" in (d) appear to be due to irregularity in the gold layer at the phase boundary.

examined in detail, as shown in Fig. 13. "Spot analyses" (Figs. 13b and 13c) and "line scans" (Figs. 13d and 13e) confirmed the identity of the "silica-rich" and "silica-poor" phases, and showed that manganese (II) is strongly partitioned into the latter phase. The ratio of Si K_α X-ray intensities obtained from the two phases was about 1.65, uncorrected for matrix effects. This is reasonably consistent with the mole-percentage ratio of about 1.55 at 700°C, indicated by the tie-line in Fig. 11. Fuller interpretation of these results requires care, since we are now dealing with the complexities of phase separation in a quaternary system[13]. These results are comparable to microprobe studies of some similar CoO-doped glasses by Ehrt et al.[14], and studies of two undoped sodium borosilicate glasses by Scholes and Wilkinson[7].

CONCLUSIONS

There are wide variations in microstructure with relatively small compositional changes in phase-separated borosilicate glasses. These variations can be rationalized in terms of the tie-line orientations reported by Mazurin and Streltsina, and variations in glass viscosity. There is a small "window" of glass compositions around the molar composition $5\pm2Na_2O-45\pm6B_2O_3-50\pm6SiO_2$, where phase domains can be grown sufficiently large for microanalytical probes to be applied.

REFERENCES

1. M. Tomozawa, Phase separation in glass, pp 71-113 in: "Treatise on Materials Science and Technology, vol. 17," M. Tomozawa and R.H. Doremus, eds., Academic Press, New York, 1979.
2. F.Ya. Galakhov and O.S. Alekseeva, Metastable liquation in the sodium oxide-boron oxide-silicon dioxide system, Dokl. Akad. Nauk SSSR 184:1102 (1969).
3. W. Haller, D.H. Blackburn, F.E. Wagstaff and R.J. Charles, Metastable immiscibility surface in the system $Na_2O-B_2O_3-SiO_2$, J. Am. Ceram. Soc. 53:34 (1970).
4. G.R. Srinivasan, I. Tweer, P.B. Macedo, A. Sarkar and W. Haller, Phase separation in $SiO_2-B_2O_3-Na_2O$ system, J. Non-Cryst. Solids 6:221 (1971).
5. O.V. Mazurin and M.V. Streltsina, Determination of tie-line directions in the metastable phase-separation regions of ternary systems, J. Non-Cryst. Solids 11:199 (1972).
6. S. Scholes, Tie lines in the metastable immiscibility region of the $Na_2O-B_2O_3-SiO_2$ system, J. Non-Cryst. Solids 12:266 (1973).
7. S. Scholes and F.C.F. Wilkinson, Glassy phase separation in sodium borosilicate glasses, Disc. Faraday Soc. 50:175 (1970).

8. T.H. Elmer, M.E. Nordberg, G.B. Carrier and E.J. Korda, Phase separation in borosilicate glasses as seen by electron microscopy and scanning electron microscopy, J. Am. Ceram. Soc. 53:171 (1970).
9. W. Haller, Rearrangement kinetics of the liquid-liquid immiscible microphases in alkali borosilicate melts, J. Chem. Phys. 42:686 (1965).
10. A. Makishima and T. Sakaino, Scanning electron micrographs of phase-separated glasses in the system $Na_2O-B_2O_3-SiO_2$, J. Am. Ceram. Soc. 53:64 (1970).
11. J. Baron and R. Wey, Electron microscope study of phase separation in sodium borosilicate glasses, Verres et Réfractaires 28:16 (1974).
12. P. Taylor and D.G. Owen, Liquid immiscibility in the system $K_2O-B_2O_3-SiO_2$, J. Am. Ceram. Soc. 64:C-158 (1981), and references therein.
13. P. Taylor and D.G. Owen, Liquid immiscibility in the system $Na_2O-ZnO-B_2O_3-SiO_2$, J. Am. Ceram. Soc. 64:360 (1981).
14. D. Ehrt, H. Reiss and W. Vogel, Microstructural study of CoO-containing $Na_2O-B_2O_3-SiO_2$ glasses, Silikattechnik 28:359 (1977).

THE MEASUREMENT OF THERMAL DIFFUSIVITY OF SIMULATED

GLASS FORMING NUCLEAR WASTE MELTS

>   James U. Derby, L. David Pye
>   and M. J. Plodinec*
>
>   New York State College of Ceramics
>   Alfred University
>   Alfred, N.Y.
>   *E. I. duPont de Nemours & Co.
>   Savannah River Laboratory
>   Aiken, S.C.

INTRODUCTION

High-level nuclear waste is generated during reprocessing of nuclear reactor fuels. At present, these wastes are stored at various locations in the United States until a final waste form (i.e., glass, SYNROC, Supercalcine, etc.) is chosen as the best method to immobilize these wastes. For borosilicate glasses to be used in this way it is important to determine both its low and high temperature properties. Especially important in the latter category is thermal diffusivity since this parameter can be used in the design of glass melting units needed in this process. Accordingly, the present study was directed at measuring the high temperature diffusivity of three melts of simulated nuclear waste glasses. Their compositions are given in Table 1. All melted easily at moderately low temperatures (900-1200°C).

EXPERIMENTAL PROCEDURES

In this investigation the measurement of thermal diffusivity was based on the modified Angstrom method used by VanZee and Babcock in earlier work on soda-lime-silica glasses.[1] A crucible of glass, shown in Fig. 1-a was positioned at the center of a cylindrically-shaped furnace. Heat was produced from electric heating elements which were positioned at equal distances from the

Table 1. Nominal Compositions of Glass Forming Melts
Studied in this Investigation

	Glass		
Component	A	B	C
SiO_2	41.3	41.0	41.0
Al_2O_3	14.2	3.0	0.4
Fe_2O_3	4.0	14.4	17.6
MgO	1.4	1.4	1.4
CaO	0.3	1.8	1.2
NiO	0.6	1.7	3.0
MnO_2	3.3	3.8	1.2
Na_2O	16.1	14.5	14.2
B_2O_3	10.5	10.5	10.3
Li_2O	4.0	4.0	4.0
TiO_2	0.7	0.7	0.7

sides of the crucible. A cam-type controller was used to input a sinusoidal voltage to the heating elements producing an inward radial heat flow which varied sinusoidally above and below the median temperature at which the thermal diffusivity was being measured.

Two thermocouples were placed in the melt such that the first was positioned at the outside of the melt and the second was at the center. Thermocouple No. 1 sensed changes in heat prior to thermocouple No. 2 due to the time required for the heat to travel through the extra melt. Figure 1-b shows how the sinusoidal wave from thermocouple No. 2 falls behind that of thermocouple No. 1 by an angular lag θ. A large angular lag is characteristic of glasses with low thermal diffusivity values whereas small lags represent large values of thermal diffusivity. In this work, data from each thermocouple were analyzed by a Xerox 560 computer which assigned constants to the sine wave of the form

$$y = A + B \sin(C + DX).$$

The constants were then used to find the angular lag, θ, which in turn was used through another program to find the Z value given in the following expression.

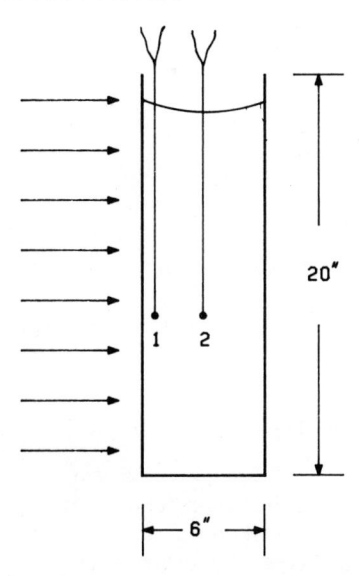

Fig. 1a. Crucible and thermocouple arrangement used to make thermal diffusivity measurements.

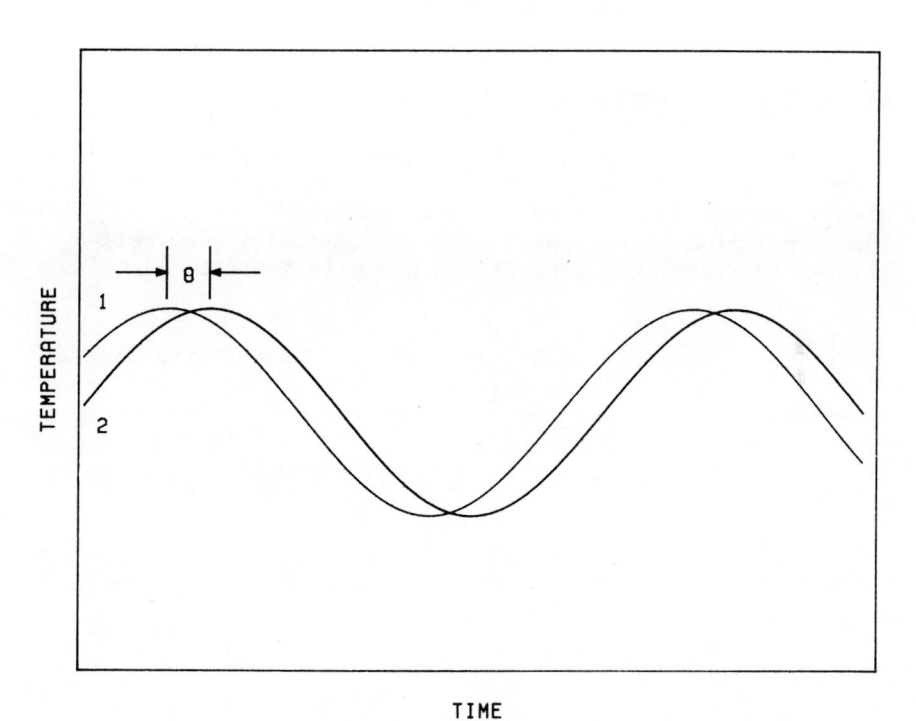

Fig. 1b. Anticipated temperature variation of thermocouples shown in Fig. 1a when the furnace temperature is varied sinusoidally.

$$\tan \theta = \frac{\text{bei } Z}{\text{ber } Z}$$

The imaginary component of the Bessel function of Z, bei, Z, is given by the expression

$$\text{bei } Z = \frac{Z^2}{2^2} - \frac{Z^6}{2^2 \cdot 4^2 \cdot 6^2} + \frac{Z^{10}}{2^2 \cdot 4^2 \cdot 6^2 \cdot 8^2 \cdot 10^2}$$

and the real component of the Bessel function of Z is

$$\text{ber } Z = 1 - \frac{Z^4}{2^2 \cdot 4^2} + \frac{Z^8}{2^2 \cdot 4^2 \cdot 6^2 \cdot 8^2} \cdot$$

By knowing the distance between the thermocouple beads, Ω_o, and the angular velocity of the heat wave, ω, the thermal diffusivity, h, can be calculated from the expression

$$h = \frac{\Omega_o^2}{Z^2} \omega$$

as was shown by VanZee and Babcock.

As a method of proving experimental procedures in this work, measurements were first done on a standard soda-lime-silica glass with a composition similar to that studied by VanZee and Babcock. The composition of this glass and the composition of melt No. 5 studied by VanZee and Babcock are given in Table 2.

Table 2. Nominal Compositions of Soda-Lime-Silica Glasses Used in this Work

	VanZee/Babcock*	Present Work
SiO_2	72.3	73.5
Al_2O_3	0.35	1.82
Fe_2O_3	0.03	0.05
MgO	0.0	0.5
CaO	14.4	9.93
Na_2O	12.5	13.7
K_2O	0.0	0.5

*Composition No. 5

Fig. 2 shows the wave forms obtained on the clear soda-lime silicate glass at an average temperature of 1080°C. The waves are very close to each other representing a time lag of 2 minutes and 42 seconds, which is an angular lag of 8 degrees. Heat transfer took place so rapidly in this melt that data had to be recorded using a voltmeter with a precision of ± 0.001 millivolts, corresponding to a sensitivity of about 0.1°C. Without such precision, it was difficult to differentiate the temperatures between thermocouples 1 and 2.

Fig. 3 is a graph of the log of the diffusivity (cm^2/sec) versus temperature for the results obtained in this study in comparison to those of VanZee and Babcock for the soda-lime-silica glass. The results of this study correlate closely to those of VanZee and Babcock with a deviation of approximately 5%. The thermal diffusivity for the soda-lime-silica glass increases exponentially with temperature in the temperature range being studied. Here heat flow consists of a conductive and a radiative component. At room temperatures the primary heat transfer process is that of conduction, but at elevated temperatures the radiative component becomes more important. This radiative or photon conductivity of a material is based on its ability to absorb radiation at a certain wavelength and to reradiate it. The ability of a material such as glass to absorb radiation is given by its optical absorption coefficient, γ, in the equation

$$I_x = I_o e^{-\gamma x}$$

where I_o is the intensity of the monochromatic radiation entering a transparent medium normal to its surface and I_x is the residual intensity after the radiation has penetrated some distance, x, into the medium. The diathermancy of a body, such as glass, is temperature dependent in that at low temperatures, long wavelength radiations are traveling through the medium but as temperature increases the frequency of the radiation also increases with a corresponding decrease in wavelength. The wavelength dependence of the absorption coefficient and the spectral composition of the emitted radiation are significant factors in determining heat transfer.[2]

Fig. 4 shows the wave forms obtained on one of the simulated nuclear waste melts. The angular lag is 53.7 degrees as opposed to the 8 degrees for the clear glass. The high iron content of the nuclear waste glass and the corresponding dark color causes the radiative component of heat transfer to be greatly diminished. The linear temperature dependence shown in Fig. 5-a indicates that in the temperature range studied, the simulated nuclear waste melts fail to exhibit the same exponential increase of thermal diffusivity with temperature as that shown by the soda-lime-silica glass.

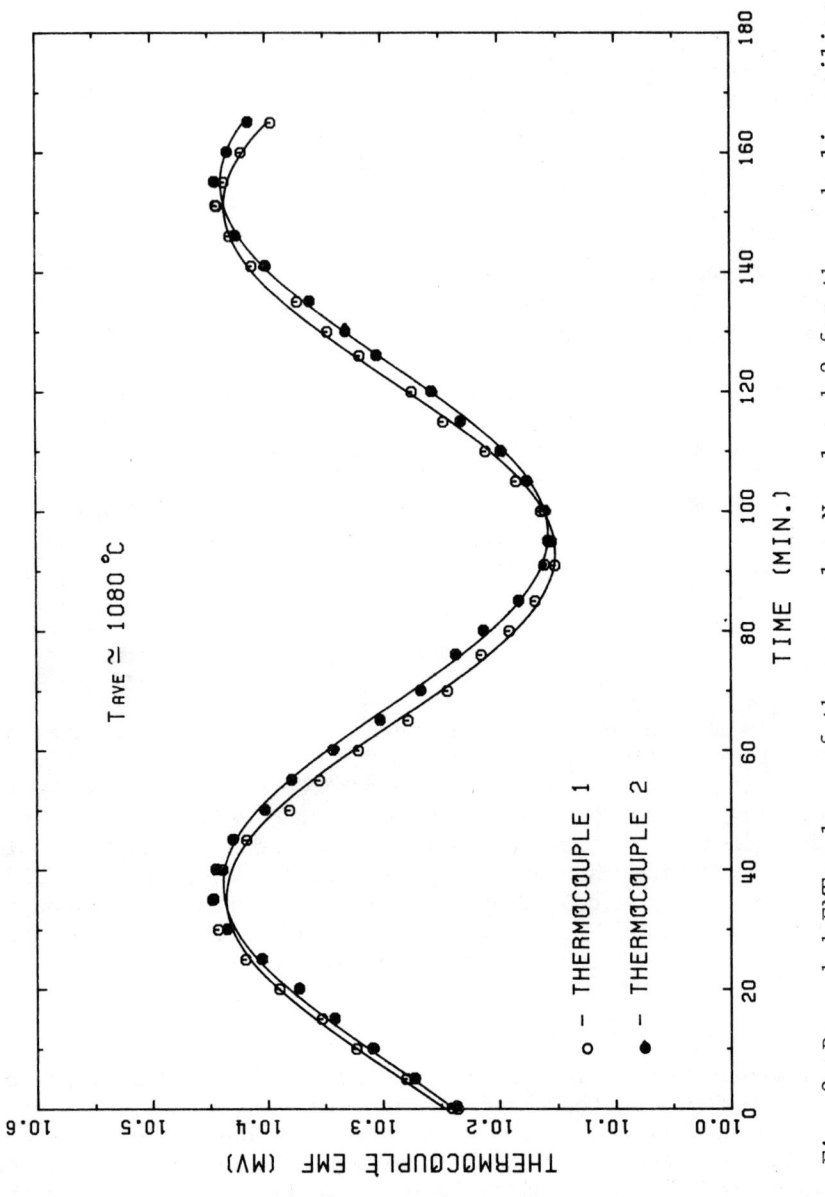

Fig. 2. Recorded EMF values of thermocouples No. 1 and 2 for the soda-lime-silica melt at 1080°C.

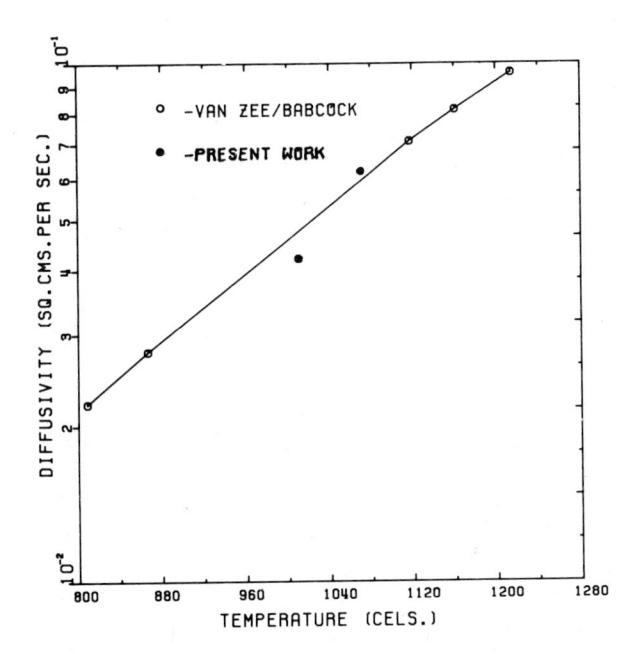

Fig. 3. The thermal diffusivity of a soda-lime-silica glass as determined by VanZee/Babcock and in this work.

In work by Lucks et al.[3] it was shown that the thermal conductivity of vitreous silica, in the range of 700°K to 1070°K, consisted of a radiative and conductive component. The thermal conductivity increased linearly with temperature without radiation present but showed rapid increase when both components were present. For the melts studied here, the strong absorption coefficient of the Fe_2O_3 probably controls the radiative component. Glass A which has the lowest Fe_2O_3 content has the highest thermal diffusivity, while glass C having the highest Fe_2O_3 content shows the lowest thermal diffusivity. The relationship between thermal diffusivity and Fe_2O_3 content at three different temperatures is shown in Fig. 5-b.

The following empirical formula can be used to calculate the thermal diffusivity of the simulated nuclear waste melts in the temperature range of 950°C to 1150°C given the weight percent of Fe_2O_3 and the temperature in °C.

$$h = 6.45 \times 10^{-5} (T) - 0.027257 (\beta) - 0.0542$$

T is the average temperature of the glass and β is the weight percent of Fe_2O_3. This empirical formula was used to calculate the values of h at the various temperatures producing the results shown in Table 3.

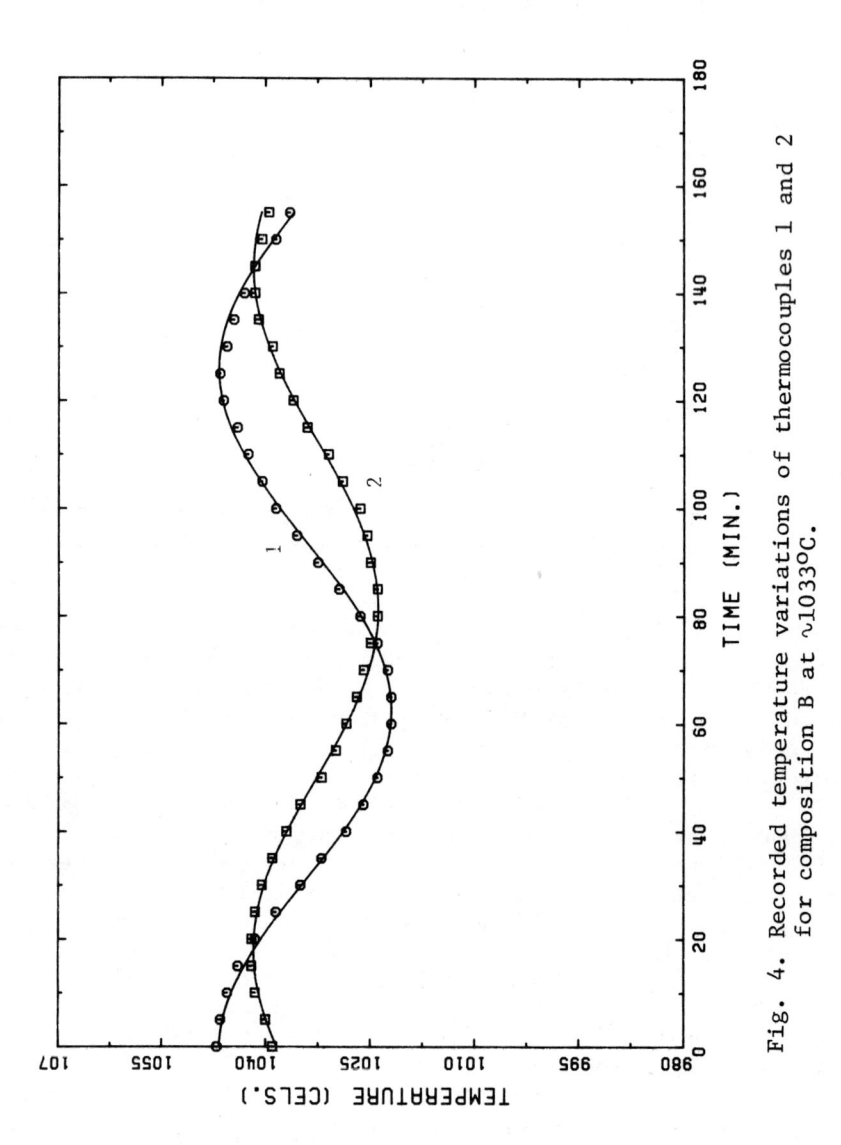

Fig. 4. Recorded temperature variations of thermocouples 1 and 2 for composition B at ~1033°C.

MEASUREMENT OF THERMAL DIFFUSIVITY

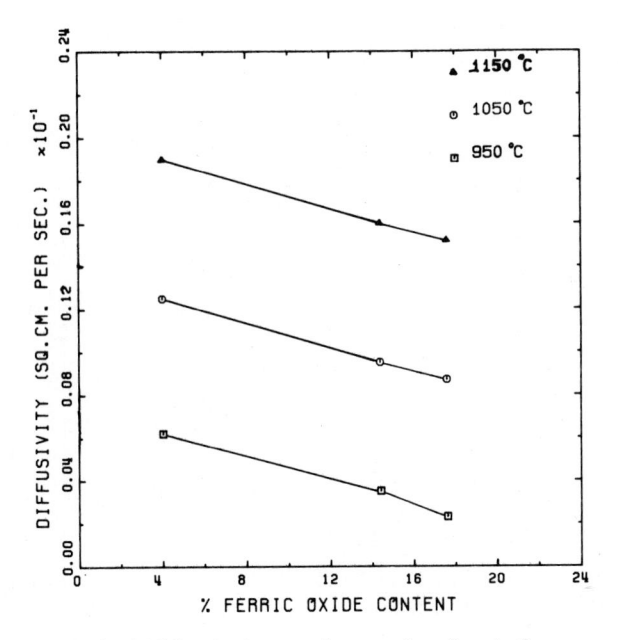

Fig. 5a. Thermal diffusivity values obtained for compositions A, B and C as a function of temperature.

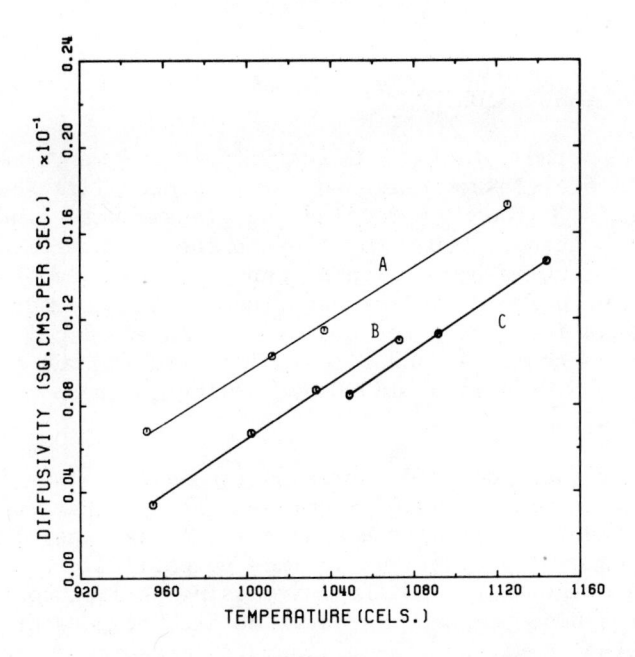

Fig. 5b. Values of thermal diffusivity as a function of temperature and ferric oxide content.

Table 3. Calculated and Measured Values of Diffusivity

	T	β	$h \times 10^{-3}$ (cm²/sec.) (calculated)	(measured)
Glass A	1012°C	.04	10.0	10.3
	1037°C	.04	11.6	11.5
	1125°C	.04	17.27	17.3
Glass B	955°C	.1441	3.5	3.4
	1002°C	.1441	6.5	6.7
	1033°C	.1441	8.5	8.7
	1073°C	.1441	11.1	11.0
Glass C	1020°C	.1761	6.8	7.2
	1049°C	.1761	8.7	8.5
	1092°C	.1761	11.4	11.3
	1144°C	.1761	14.8	14.7

SUMMARY AND CONCLUSIONS

In this study, thermal diffusivity measurements were made on borosilicate glasses designed for the immobilization of radioactive wastes. Three compositions of glasses were studied in the temperature range of 900°C to 1150°C using the modified Angstrom radial heat flow method of VanZee and Babcock. The thermal diffusivities were found to increase linearly with temperature and to decrease with increasing Fe_2O_3 content. An empirical formula was derived which could be used to calculate the thermal diffusivity in the range of 950°C to 1150°C by knowing the temperature and Fe_2O_3 content.

Plummer[4] has reported thermal diffusivity values for a synthetic tektite glass of 0.0053 cm²/sec. at a temperature of 950°C. The iron oxide content of the synthetic tektite was 4.9%. When the empirical formula derived for the borosilicate glass is used to calculate the thermal diffusivity using an iron content of 4.9%, a value of 0.0057 cm²/sec. is obtained at a temperature of 950°C. Therefore the results of the diffusivity measurements on these borosilicate glasses are comparable with those obtained for other glasses with similar iron content.

ACKNOWLEDGMENTS

The help of S. C. Cherukuri, B. Bowen, G. Cartledge, I. Joseph, J. Gesner, L. Burzycki, D. Snowden and Brockway Glass Company in the course of conducting this work is gratefully acknowledged.

REFERENCES

1. A. F. VanZee and C. Babcock, A method for the measurement of thermal diffusivity of molten glass, J. Amer. Ceram. Soc., 34:244 (1951).
2. R. Gardon, A review of radiant heat transfer in glass, J. Amer. Ceram. Soc., 44:305 (1961).
3. Lucks, Matalich, Van Welzon, "The Experimental Measurement of Thermal Conductivity, Specific Heats and Densities of Metallic Transparent and Protective Materials," Pt. III: WADC AFTR6145 (1954).
4. Plummer, W. A., "A Thermal Diffusivity Measurement Technique," Proc. Third Conference on Thermal Conductivity, Corning Glass Works, Corning, N.Y.

VOLUME-TEMPERATURE RELATIONSHIPS IN

SIMULATED GLASS FORMING NUCLEAR WASTE MELTS

L.D. Pye, R. Locker and M.J. Plodinec*

New York State College of Ceramics
Alfred University, Alfred, N.Y.

*I.E. duPont de Nemours & Co.
Savannah River Laboratory, Aiken, S.C.

INTRODUCTION

The classical description of a melt cooling to form a solid glass is expressed by the volume-temperature behavior shown in Fig. 1. The point where a change of slope takes place in this graph is defined as the glass transition temperature, T_g. Its exact value is moderately dependent on the rate of cooling, faster rates giving larger values. That the volume change at T_g is not discontinuous (as for example, during melting or crystallization), is often taken to mean that the glass transition is a quasi second-order thermodynamic transformation of the type described by Ehrenfest.[1] In this classification, properties defined by the first derivative of free energy, i.e., enthalpy, entropy, and volume are expected to change gradually during a second-order transformation, whereas properties defined by the second derivative of free energy (heat capacity, thermal expansion), are expected to show a discontinuous change. Thus, extension of the Ehrenfest criteria to the glass transition would predict a discontinuous change of the coefficient of thermal expansion at T_g. We shall examine below the extent to which this prediction is fulfilled for several simulated nuclear waste glasses.

Beyond these theoretical considerations, the change of specific volume of a glass forming melt with temperature is important in many aspects of glass manufacture: melting, annealing, sealing, tempering, and others. During melting it is this change of volume (or density) with temperature that gives rise to convention currents necessary for the satisfactory mixing

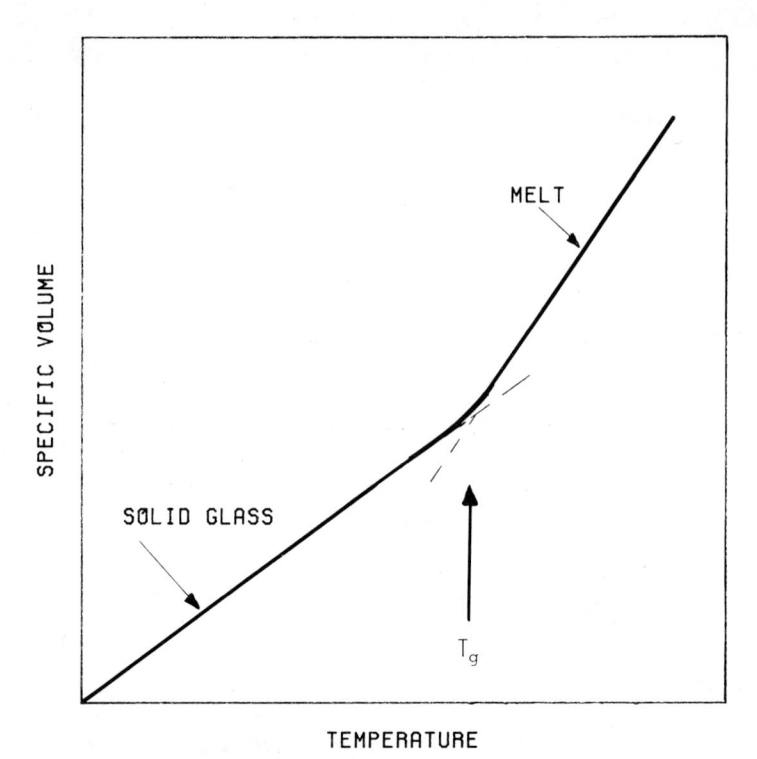

Fig. 1. Classical volume-temperature behavior of a glass forming when cooled to a solid glass.

of a melt. In this sense it becomes an especially important parameter in the design of glass melting units. In annealing, volume-temperature relationships, expressed as the coefficient of thermal expansion, allow the prediction of cooling rates necessary to avoid fracture during cooling.

It was for these reasons, primarily, that the present investigation involving simulated glass forming nuclear wastes melts was undertaken. We also note that the number of glass forming systems for which expansivity data are available for both the molten and solid state is relatively small. At the same time, it is easier by far to measure volume (expansion) changes of a glass as it is heated from room temperature through its transition range. Since this measurement gives low temperature information on the volume-temperature behavior shown in Fig. 1, it is often inquired as to what error can be expected by extrapolating these measurements to predict melt density, e.g. density of the molten state. This question is examined in some detail below since in this work, both high and low temperature densities and expansions were measured thereby allowing a comparison to be made.

EXPERIMENTAL PROCEDURES

Compositions Studied

The chemical compositions of the melts studied here are the same as those reported by Derby, et al. in this volume. Their designation of Glass A, B, and C is carried forward in this communication. Also a melt of nitrogen dried B_2O_3 was examined in our high-temperature density studies for reasons outlined below.

Low Temperature Thermal Expansion

The linear thermal expansion of each glass (above and below T_g) was measured with a dilatometer made by Theta Industries (Dilamatic 1)*. By making this same measurement on an NBS standard reference material (SRM 731), it was found that the maximum error of the coefficient of thermal expansion was less than 10% for a sample of length ~5.0 cm. Both expansion and temperature were recorded in a digital form so as to allow (1) the direct calculation of the coefficient of thermal expansion above and below T_g (HIGH CTE and LOW CTE respectively) and (2) an estimation of the dilatometric softening point (DSP). These analyses were carried out with a computer program developed by S. C. Cherukuri of Alfred University. The results obtained for glass A are shown in Fig. 2. The glass transition temperature can be determined from this graph by mechanically estimating the point where the transition from low to high expansion regions occurs.

High Temperature Density Measurements

Melts of each glass studied were placed in a doubly-wound resistance furnace, shown schematically in Fig. 3. A platinum bob was weighed in air (W_a) and its volume (V_o) determined by calibration against liquids of known density using Archimedes' principle. At the melt temperature (T) of interest (>900°C), the bob was allowed to sink in a melt contained by a 250cc platinum-iridium crucible. The new weight of the bob (W_m) was determined by use of an analytical balance (sensitivity ± 0.1 mg) placed over the furnace. The melt density, d, was then calculated by the following equation:

$$d = \frac{W_a - W_m}{V_o(1 + \alpha_{vp} \Delta T)} \quad 3$$

In this equation, α_{vp} is the volume expansion coefficient of

---
*Port Washington, N.Y.

HIGH CTE = 92.00 X 10^{-6}/·C
LOW CTE = 9.81 X 10^{-6}/·C

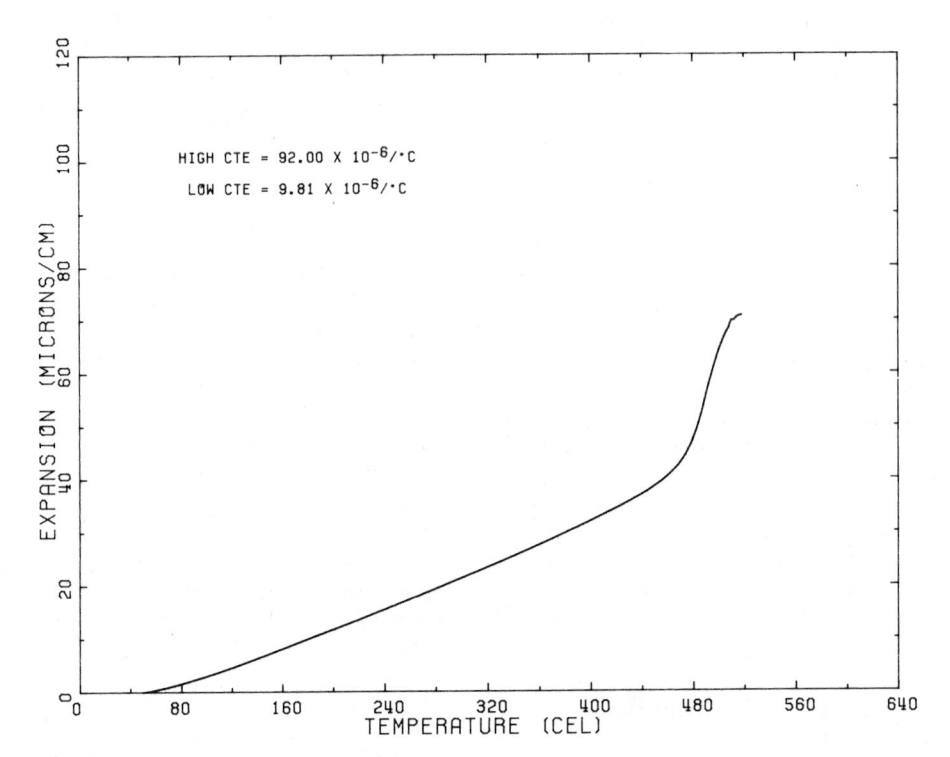

Fig. 2. Linear expansion of glass A with temperature.

platinum. No correction was made for surface tension forces exerted by the melt on a 10 mil platinum wire used to connect the platinum bob to the analytical balance. A correction for the amount of platinum wire submerged in the melt, however, was taken into account in the final calculation of d. Finally, no differences in the weight of the platinum bob, when freely suspended in air, could be detected over the range of 100-1200°C. This was taken to mean that convection currents in the furnace were minimal.

In an effort to verify our procedures, the density of molten B_2O_3 was measured at two different temperatures. Prior to this measurement, the melt was "dried" by bubbling nitrogen through it. Our results, and those reported by Napolitano et al.[2] are presented as follows:

Temperature	This Work	Napolitano et al.[2]
900°C	1.553 (g/cc)	1.541 (g/cc)
1000°C	1.542 (g/cc)	1.528 (g/cc)

For each temperature the difference between these values is less than 0.6%, which was regarded as satisfactory agreement for the present purposes.

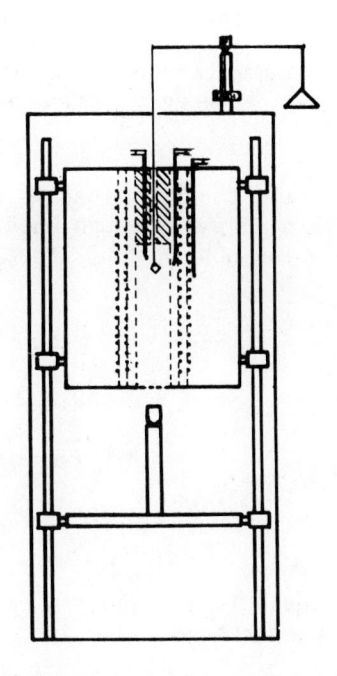

Fig. 3. Cross section of furnace used to make density measurements. The crucible was inserted into the furnace by raising the lower platform.

Results

Our low temperature dilatometry data are summarized in Table I.

Table I

Relevant Expansivity Parameters

Parameters	Glass Type A	Glass Type B	Glass Type C
LOW CTE[+][*]	9.81	11.1	11.3
T_g (°C)	465	460	450
HIGH CTE[+][#]	92.0	112	111
DSP (°C)	518	508	488

[+] ppm/°C
[*] over the temperature range 80-400°C
[#] linear range between T_g and the DSP (see Fig. 2.)

The measured densities for the glass melts studied here are given in Fig. 4. The line drawn through each set of data was that calculated from an equation obtained by least squares analyses:

Glass	Equation
A	$d = 2.682 - 2.491 \times 10^{-4} T$
B	$d = 2.777 - 2.961 \times 10^{-4} T$
C	$d = 2.723 - 2.191 \times 10^{-4} T$

From these equations, the volume expansion coefficient, α_v, could be calculated for the temperature range 1000-1200°C by the following relationship:

$$\alpha_v = \frac{d_1 - d_2}{d_2} \cdot \frac{1}{T_2 - T_1}$$

TABLE II

High Temperature Density and Expansivity Data

Glass	density (g/cc) measured[●]	calculated[●]	%Δd	α_v [■]	$\alpha_\ell(X)$ [■]	$\alpha_\ell(Y)$ [▲]	Y/X	wt% Fe_2O_3
A	2.433	2.241	7.8	102	34	92	2.7	4.0
B	2.481	2.285	7.8	119	39	112	2.9	14.4
C	2.504	2.287	8.6	87	29	111	3.8	17.6

expansion coefficient[+]

● at 1000°C
■ 1000 - 1200°C
▲ from Table I, T_g - DSP

Here d_2, d_1 are densities at temperatures T_2 and T_1 respectively ($T_2 > T_1$). α_v was converted to the linear expansion coefficient, α_ℓ by dividing by three, e. g., $\alpha_v \approx 3\alpha_\ell$. These calculations yielded the data summarized in Table II.

As in Table I, expansion coefficients in Table II are expressed as ppm/°C. Perusal of these data indicate that (1) as expected, an increase in wt. % Fe_2O_3 raised melt density by a slight amount, (2) there is a wide difference between the linear expansion coefficients derived from high-temperature density measurements and those obtained from low-temperature dilatometry (above T_g) and (3) the ratio of the linear expansion coefficient obtained from low temperature dilatometry to that obtained by density measurements (Y/X) tends to be larger than for those glasses containing greater amounts of Fe_2O_3. Since the high temperature volume measurements yield equilibrium values of density (and by extension, equilibrium expansion coefficients) it is possible that the low temperature coefficients are too high because of non-equilibrium behavior of glass between T_g and the DSP. Thus, although Fig. 1 is correct for melts cooling to the solid state, upon re-heating, non-equilibrium effects may develop. It may be recalled that high-iron glasses have lower thermal diffusivity. Therefore, in the course of carrying out these dilatometry experiments, the lag of the sample temperature behind the furnace temperature during heating would be greater for

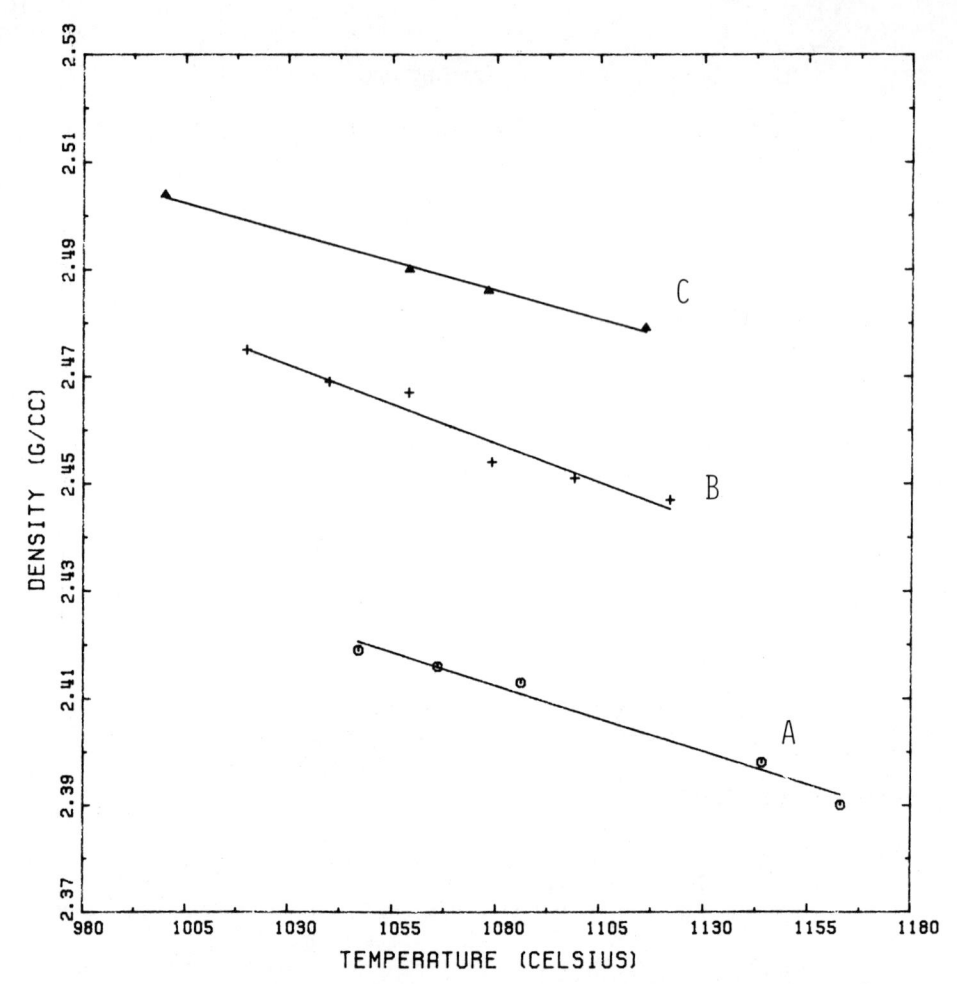

Fig. 4. Density-temperature relationships for melts of composition A, B, and C.

glasses containing larger amounts of Fe_2O_3. Thus, non-equilibrium transition range behavior might be exaggerated in such cases. As a way of testing this concept, samples of composition B (1 x 1 x 0.2 cm) were annealed at their dilatometric softening point for one hour and then quenched to room temperature. A portion of these samples were then annealed for eighteen hours at 480°C, another at 489°C for the same time and then quenched as before. It was anticipated that a long-term anneal at these temperatures would allow volume equilibration to occur. After measuring their room temperature densities by the sink-float method, the coefficient of expansion was calculated as above and gave a value of 73 ppm/°C. This value is between the low-temperature and high-temperature values reported in Table II. While this result is encouraging, more work on the transformation range behavior of this composition would be useful for confirming this interpretation.

In any event, it is clear that the expansion coefficients between T_g and the dilatometric softening point (DSP) bear little relation to those obtained by high-temperature density measurements. This inference is heightened by plotting α_ℓ vs temperature for the glasses studied here (Fig. 5). It is seen that a different high-temperature and low expansion coefficient is obtained in accordance with the Ehrenfest criteria, but near the glass transition temperature, the expansion coefficients rise to an anomously high value. This behavior was also observed by Shartiss and Spinner in earler work on borosilicate melts[3] and is discussed briefly by Kingery et al.[4]

Despite the above differences, it was found that by using the expansion coefficients derived by low temperature dilatometry, an error no greater than 10% would be obtained by extrapolating room temperature densities to 1000°C (see Table II). In making this extrapolation, the LOW CTE was used over the range of 80°C to the glass transition temperature and the HIGH CTE was used from there to 1000°C. As might be expected, because of the anomously high expansion coefficients between T_g and the DSP, the predicted densities are all low, e.g., the calculated volume changes were too large. For higher temperatures, the difference between the calculated and measured densities would be larger.

SUMMARY/CONCLUSIONS

This investigation has shown that volume-temperature relationships in simulated glass forming nuclear waste melts can be understood satisfactorily. Reasonable predictions of density as a function of temperature can be made by dilatometric measurements on solid glass samples although the predicted values will probably be low for high temperatures. Additional characterization of the transformation range behavior of these glasses would be desirable in providing confirmation of non-equilibrium values of expansion coefficients obtained from solid state dilatometry.

ACKNOWLEDGEMENTS

The authors acknowledge the considerable help of S. C. Cherukuri, I. Joseph, B. Bowen, G. Cartledge, L. Burzycki, J. Gesner, and J. Derby in carrying out this work. The help of L. Hanks is especially appreciated.

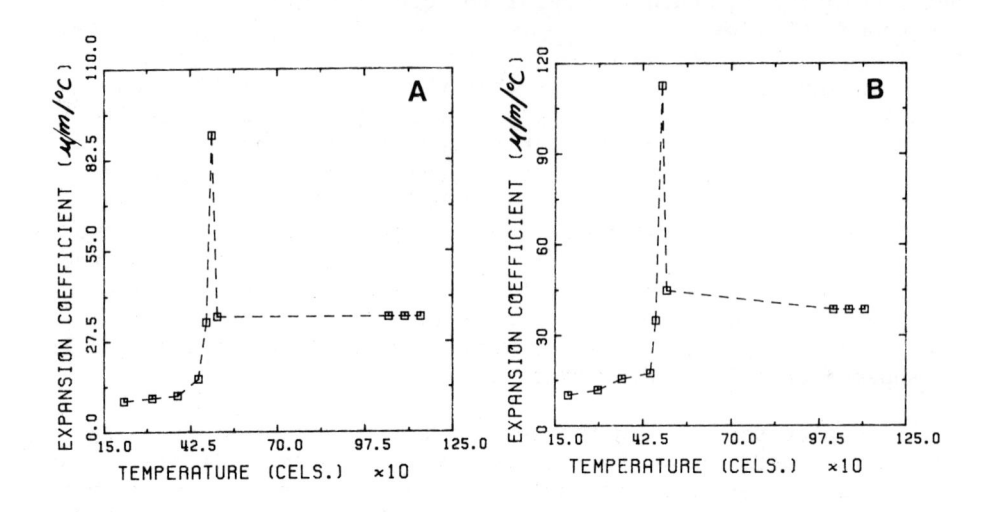

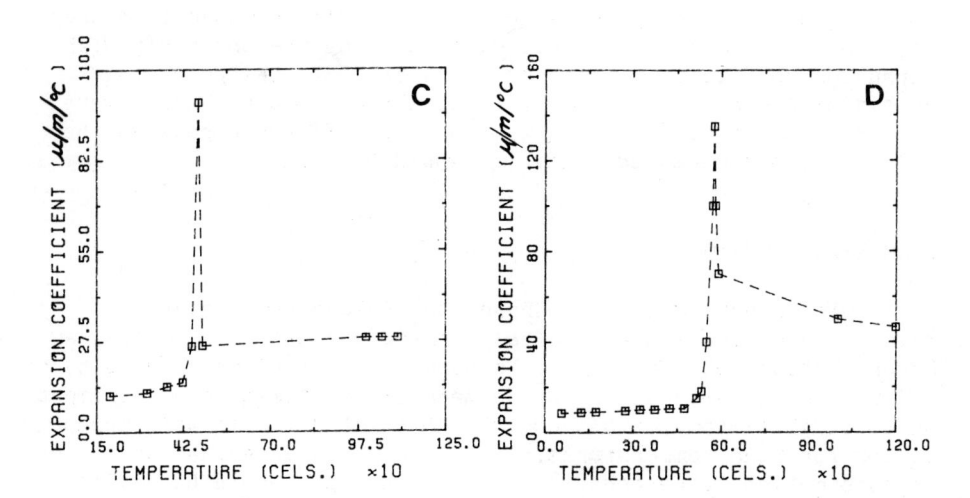

Fig. 5. Expansion coefficients of glasses studied in this work (A,B,C) and the borosilicate melt (D) investigated by Shartiss and Spinner[3]. Coefficients are expressed as microns/meter/°C which converts directly to ppm/°C.

REFERENCES

1. Ehrenfest, Proc. Amsterdam Akad, 36, 153 (1933).

2. A. Napolitano, P. B. Macedo, E. G. Hawkins, "Viscosity and Density of Boron Trioxide," J. Am. Ceram. Soc., 48, 12 613 (1965).

3. L. Shartsis, S. Spinner, "Viscosity and Density of Molten Optical Glasses," J. Res. Nat. Bur. Standards, 46 3 176 (1951).

4. W. D. Kingery, K. Bowen, D. Uhlmann, Introduction to Ceramics, John Wiley and Sons, 2nd Edition, 598 (1976).

PROPERTY/MORPHOLOGY RELATIONSHIPS IN GLASSES

J. E. Shelby

New York State College of Ceramics
at Alfred University
Alfred, N.Y.

INTRODUCTION

Many properties of glasses are sensitive to the morphology of the particular samples studied[1-4]. These properties can be subdivided into three classes: those which are only slightly sensitive to the presence of phase separation and insensitive to the details of the microstructure (density, refractive index, thermal expansion coefficient), those which are very sensitive to the presence of phase separation, but are relatively insensitive to the details of the microstructure (light scattering, glass transformation temperature), and those which are sensitive indicators of both the presence and the details of the microstructure (dilatometric softening temperature, viscosity, chemical durability, diffusion, and electrical conductivity). Shelby[3] has recently shown that the proper choice of property measurements will allow the determination of both the phase separation boundaries and the limits of interconnectivity of each phase for a binary glassforming system such as $PbO-B_2O_3$. This paper will review the technique described by Shelby[3] and present results of the application of this method for determining morphology for other glassforming systems.

DETERMINATION OF PHASE SEPARATION BOUNDARIES

Since the effects of morphology on the bulk properties of glasses (density, refractive index, thermal expansion coefficient, elastic modulus) are very small, study of these properties does not lend itself to the accurate determination of phase separation boundaries and/or interconnectivity limits. On the other hand, measurements of the glass transformation temperatures in binary glassforming systems has proven especially useful for establishing both the presence of phase separation and the immiscibility boundaries. Under ideal conditions, differential scanning calorimetry (DSC) or differential thermal analysis (DTA) measurements of a phase separated glass sample will indicate the existence of two glass transformation temperatures[4]. An example of this type of curve for a lead borate glass is shown in Figure 1. Curves exhibiting the presence of two glass transformation temperatures therefore conclusively establish the presence of two vitreous phases in the sample, and hence, if care has been taken to insure good mixing in the original glass melt, the existence of phase separation. Furthermore, since the compositions of the two co-existing phases on any tie line should be independent of the bulk glass composition[3], the glass transformation temperatures observed should be constant over the entire phase separation region for glasses lying along a tie line. This effect is shown in Figure 2 for the lead borate glasses, which exhibit two constant glass transformation temperatures over the compositional range 0-20 mol% PbO.

Although DSC curves for phase separated borate and germanate glasses frequently exhibit two glass transformation temperatures,

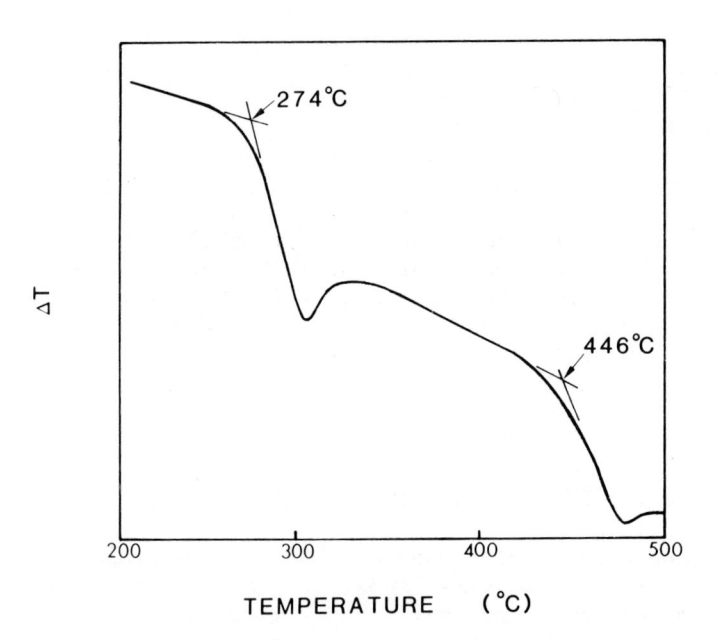

Figure 1. A DSC curve for a 10PbO-90B$_2$O$_3$ glass. Note the existence of two glass transformation temperatures.

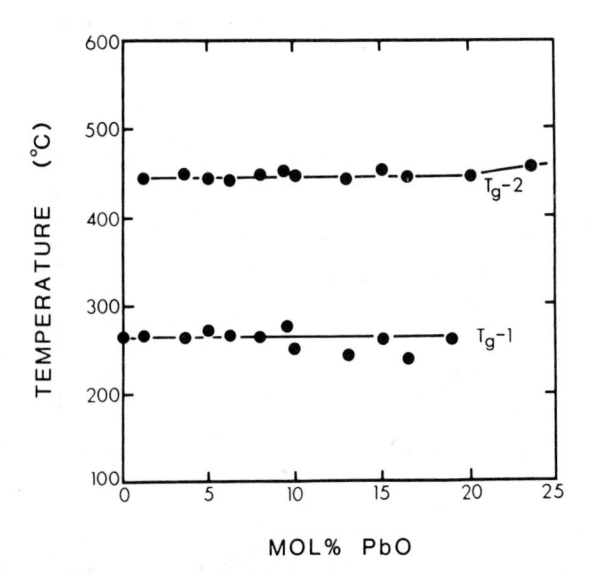

Figure 2. Effect of glass composition on the glass transformation temperatures observed for the lead borate glasses.

similar curves for phase separated silicate glasses often exhibit only one such transition. This is usually the result of metastable immiscibility, where the glass transformation temperature of the silica-rich phase is above the miscibility temperature for the composition studied. In this case, the sample reverts to a homogeneous glass during heating at a temperature below the upper glass transformation temperature. Since the T_g for the homogeneous glass would be below the miscibility temperature, only the T_g for the lower viscosity phase is observed. The value of this T_g, however, should still be a constant for all glasses lying along the same tie line. It follows that the observation of a single, constant T_g for a series of glasses is strongly suggestive of phase separation, e.g. as in the data for the lithium silicate glasses shown in Figure 3. (It should be noted that the existence of a region of constant T_g for a series of glasses is not, in itself, proof of phase separation.)

Glass transformation temperatures can also be obtained from thermal expansion curves. However, in this case, the observation of two glass transformation temperatures requires that the phase having the higher transformation temperature also must have a continuous, rather than droplet, microstructure. If this is not the case, the glass will begin to flow at a temperature slightly (30-50°C) above the T_g for the less viscous phase and will not sustain the load

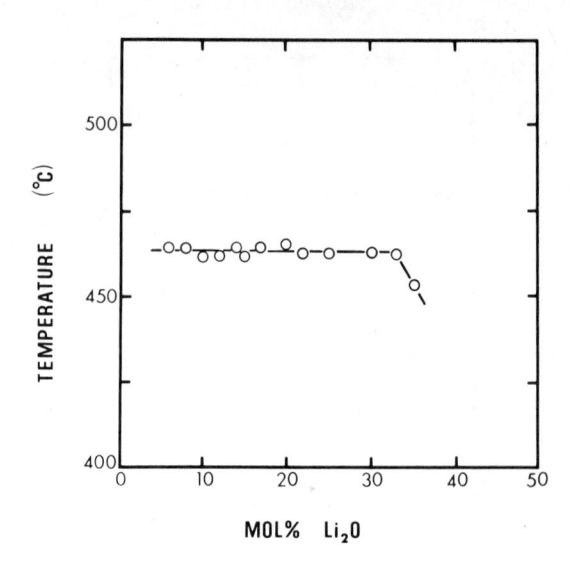

Figure 3. Effect of glass composition on the glass transformation temperature of lithium silicate glasses.

necessary for continued measurement in a standard dilatometer. It follows that, if two glass transformation temperatures are observed in a thermal expansion curve, the glass is indeed phase separated. However, the presence of only one T_g in a thermal expansion curve is not necessarily an indication of a homogeneous glass. In either case, the T_g measured for a series of glasses lying on the same tie line must be constant.

DETERMINATION OF CONNECTIVITY LIMITS

Most of the properties of glasses which are dependent upon the presence of phase separation are also dependent upon the specific morphology of the sample. As mentioned above, thermal expansion curves will only indicate the existence of two glass transformation temperatures if the more viscous phase has a connected microstructure. It follows that the sudden appearance or disappearance of a second glass transformation temperatures in a series of thermal expansion curves for glasses known to lie within a miscibility gap indicates the composition of the connectivity limit for the more viscous phase. This can be illustrated by the lead borate glasses (Figure 4), where glasses containing <10 mol% PbO exhibit only one T_g, whereas glasses containing between 10 and 20 mol% PbO exhibit two glass transformation temperatures. This sudden change in behavior indicates that the connectivity limit for the $20PbO-80B_2O_3$ phase lies at approximately 10 mol% PbO.

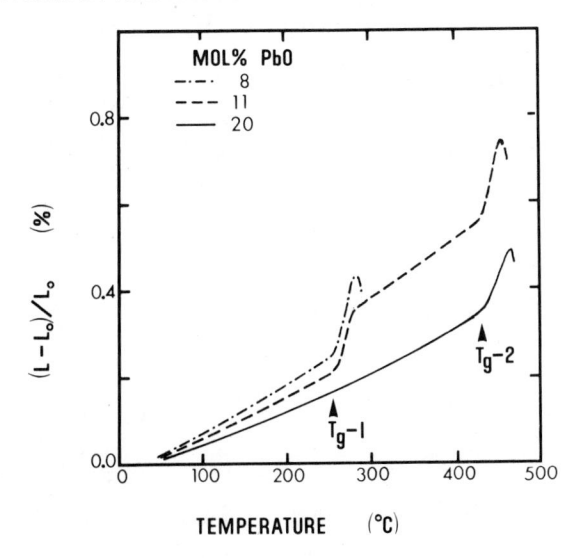

Figure 4. Thermal expansion curves for phase separated lead borate glasses.

As discussed earlier, thermal expansion curves for phase separated glasses frequently exhibit only one glass transformation even when the more viscous phase is continuous. Under these conditions, the dilatometric softening temperature is often a good indicator of the degree of connectivity of the more viscous phase, since this phase will prevent softening of the sample up to near the miscibility temperature so long as it remains continuous. In general, T_d is usually within 20 to 50°C of the glass transformation temperature for a homogeneous glass, regardless of the particular glassformer present. If the difference between the glass transformation and dilatometric softening temperatures exceeds 100°C, the sample is almost certainly either phase separated or partially crystallized. It follows that a plot of T_d-T_g as a function of glass composition can be used to determine the connectivity limit for the more viscous phase, even if the thermal expansion curves exhibit only one transformation temperature. An example of this approach is shown in Figure 5 for the lithium silicate glasses, where only one T_g is observed in the thermal expansion curves. The sharp change in T_d-T_g at about 23 mol% Li_2O is an indication of the presence of a connectivity boundary at this composition.

Of course, plots of T_d-T_g versus glass composition can also be used to determine connectivity limits even if two glass transformation temperatures are evident in the thermal expansion curve. The difference used should be that between the lower T_g and the

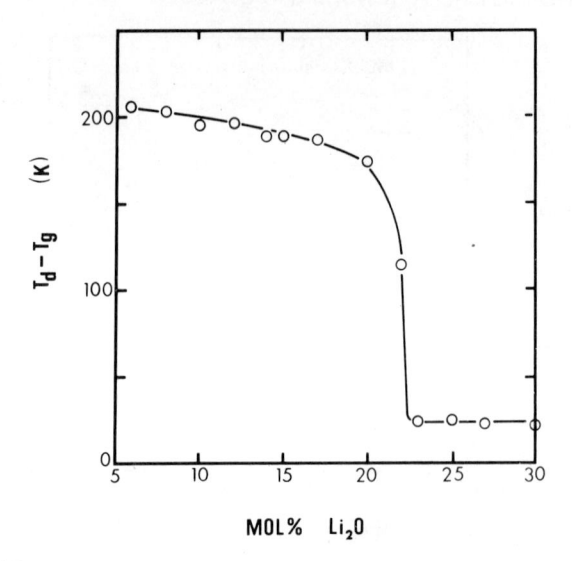

Figure 5. Effect of glass composition on the difference between the glass transformation and dilatometric softening temperatures for lithium silicate glasses. Loss of continuity of the silica-rich phase is indicated by the sharp change in T_d-T_g at about 23 mol% lithium oxide.

dilatometric softening temperature. Under these conditions, a sharp change will also be observed upon the disappearance of the lower T_g, i.e. upon passing the miscibility boundary, as shown in Figure 6.

A number of other properties also exhibit significant changes at connectivity boundaries[2,3]. Shelby[3,5] has shown that the helium permeability coefficient is extremely sensitivity to changes in the connectivity of the high permeability vitreous phase in two phase glasses. Since this phase is usually the modifier-poor, or glassformer-rich, phase, measurements of helium mobility can be used to determine the connectivity limit for this phase. Other diffusion-controlled phenomena can also be used to determine the limit of connectivity of a given vitreous phase for specific systems. Electrical conductivity measurements can detect changes in the connectivity of an alkali-rich phase. Chemical durability measurements can be used to detect the presence of a connected, low durability phase. While each of these measurements can be particularly useful under certain circumstances, they are all relatively time-consuming and/or require special apparatus. It follows that measurement of the glass transformation and dilatometric softening temperatures of a series of glasses is usually the easiest technique commonly available for the determination of both phase separation

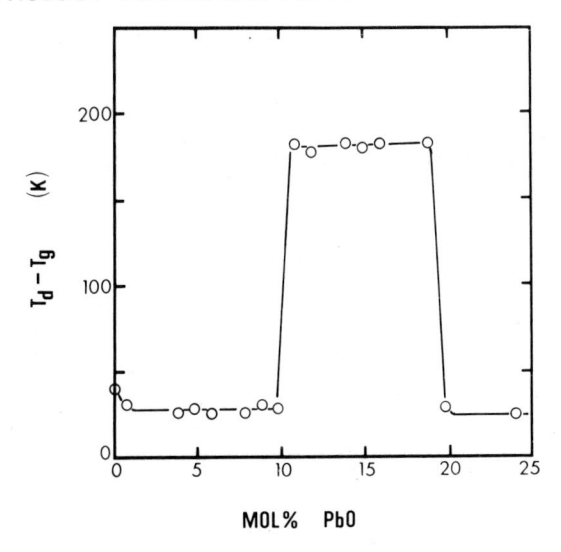

Figure 6. Effect of glass composition on the difference between the glass transformation and dilatometric softening temperatures for lead borate glasses.

and connectivity boundaries. Of course, if only the phase separation boundary is desired, DSC measurements can provide the desired information in a minimum of time.

REFERENCES

1. D. R. Uhlmann, "Microstructure of Glasses: Does it Really Matter," J. Non-Cryst. Solids 49:439 (1982).
2. M. Tomozawa, "Phase Separation in Glasses," in Treatise on Materials Science and Technology; Vol. 17: Glass II, M. Tomozawa and R. H. Doremus, eds., Academic Press, (1979).
3. J. E. Shelby, "Characterization of Glass Microstructure by Physical Property Measurements," J. Non-Cryst. Solids 49:287 (1982).
4. O. V. Mazurin, G. P. Roskova, and V. P. Kluyev, "Properties of Phase-Separated Soda-Silica Glasses As A Means of Investigation of Their Structure", Faraday Soc. London: Discussions 50:191 (1970).
5. J. E. Shelby, "Molecular Solubility and Diffusion," in: Treatise on Materials Science and Technology; Vol. 17: Glass II, M. Tomozawa and R. H. Doremus, eds., Academic Press, (1979).

THE CHARACTERIZATION OF INDIVIDUAL REDOX IONS IN GLASSES

Henry D. Schreiber

Department of Chemistry
Virginia Military Institute
Lexington, Virginia 24450

INTRODUCTION

The presence of particular redox states of multivalent elements in glasses {and in glass forming melts} control important physical as well as chemical properties of the glasses {and their corresponding melts}. Although the multivalent elements in glass-forming systems are typically the same as those encountered in aqueous solutions, the characterization of these elements in their various redox states may be quite different in the two solvent systems.

Some of the elements that can exist in more than one redox state in a glass are listed in the following table:

Transition Metals: Ti(IV)-Ti(III), V(V)-V(IV)-V(III)-V(II), Cr(VI)-Cr(V)-Cr(III)-Cr(II), Mn(III)-Mn(II), Fe(III)-Fe(II)-Fe(0), Co(III)-Co(II)-Co(0), Ni(III)-Ni(II)-Ni(0), Cu(II)-Cu(I)-Cu(0), as well as Mo, Ru, Rh, Ag, W, Re.

Other Metals: Sn(IV)-Sn(II), Pb(IV)-Pb(II)

Lanthanides: Ce(IV)-Ce(III), Eu(III)-Eu(II)

Actinides: U(VI)-U(V)-U(IV)-U(III), Pu(IV)-Pu(III), as well as Np and others.

Metalloids: As(V)-As(III), Sb(V)-Sb(III)

Non-Metals: $S\{SO_4^{2-}, S^{2-}, S^0\}$, $Se\{SeO_4^{2-}, Se^{2-}, Se^0\}$, $C\{CO_3^{2-}, "CO", C\}$, $N\{NO_3^-, N_2\}$, $O\{O_2, O^{2-}\}$

The chemistry of a multivalent element in a glass-forming melt is dependent on the prevailing redox state or states of that element within the melt. Synthesis conditions which include melt temperature, base composition, oxygen fugacity, equilibration time, concentration, and quenching procedure control the concentrations of particular redox ions. The multivalent elements, with the exception of iron, are typically not major components of the glass matrix but are introduced in trace to minor quantities as additives or impurities.

Glass {or melt} properties as diverse as color and foaming can be attributed to the occurrence of certain redox ions in the glasses {or melts}. Various properties of redox ions in glasses and the applicability of these properties to such fields as geochemistry, glass science, nuclear waste immobilization, slag recycling, and materials science among others have been reviewed previously.[1]

OBJECTIVES

The purpose of this paper is to survey the analytical methods available for the characterization of individual redox ions in glasses. In particular, emphasis will be placed on their characterization in terms of concentrations, coordination chemistries, and mutual interactions. Glasses which contain only one multivalent element will be considered initially because of the ease of characterization with a minimum of interferences. The applicability of the procedures to glasses containing two or more multivalent elements will then be discussed.

METHODS FOR CHARACTERIZATION

"Wet" Chemistry

The classic procedure to measure the concentration of a particular redox ion in glass is to first digest the glass in acidic medium and then to analyze the quantity of that redox ion present by redox titrations, colorimetric methods, or selective vaporization. The key step is to dissolve the glass in H_2SO_4, $HClO_4$, and/or HF solutions without disturbing the concentrations of the redox ions present in the glass.

Redox Titrations: After dissolution in an aqueous medium, the concentration of one redox ion of the redox couple can be determined by a standard redox titration with ceric sulfate, potassium permanganate, iodine-thiosulfate, etc. The end point for the titration can be monitored by a visual indicator {for example, ferroin} or by potentiometric means. In the cases where the redox state to be determined is not stable in the aqueous solution, an indirect titration can be done whereby either the glass is dissolved in the presence of excess Fe(III) in the case of reductants such as Cr(II) or Eu(II) in the glass, or in the presence of excess Fe(II) in the case of oxidants such as Cr(VI) or Ni(III). The resulting amount of Fe(II) can then be related to the concentration of the particular redox ion in the glass.

Colorimetric: Instead of the measurement of the redox ion concentration via redox titration after glass digestion, a complexing agent which is specific for a certain redox state can be added under controlled conditions to form a colored adjunct. For example, o-phenanthroline forms bright red complexes with Fe(II) but weakly blue complexes with Fe(III). Upon proper calibration with standard solutions, the absorbance of the peak maximum for the Fe(II)-phenanthroline complex will determine the initial Fe(II) concentration. Ferrozine is an example of another complexing agent which is selective for Fe(II), Co(II), and Ru(III).

Selective Vaporization: In special situations, the volatility of a compound formed from a particular redox state can be employed in the analytical scheme after glass digestion. For determinations of tri- and penta-valent arsenic in glass, the analysis usually relies on the volatility of As(III) in HF and H_2SO_4. Upon evaporation of the sample, only As(V) remains in the sample so that the concentration of the two redox states can be measured separately.

Characterizations: Redox ion concentrations

Advantages: Absolute concentration measurements, applicability to many redox couples, adaptation to microanalyses. Disadvantages: Potential disturbance of redox ion concentrations in the digestion step, usually non-selective if more than one multivalent element is present.

References: Redox Titration: W. D. Johnston, Oxidation-reduction equilibria in molten $Na_2O \cdot 2SiO_2$ glass, J. Amer. Ceram. Soc. 48:184 (1965). W. P. Close & J. F. Tillman, Chemical analysis of some elements in oxidation-reduction equilibria in silicate glasses, Glass Technol. 10:134 (1969). Colorimetric: C. R. Gibbs, Characterization and application of ferrozine iron reagent as a ferrous iron indicator, Anal. Chem. 48:1197 (1976). D. R. Jones, W. C. Jansheski, & D. S. Goldman, Spectrophotometric determination of reduced and total iron in glass with 1,10-phenanthroline, Anal. Chem. 53:923 (1981). Selective Vaporization: P. Close, H. M. Shepherd, & C. H. Drummond, Determination of several valences of iron, arsenic and antimony, and selenium in glass, J. Amer. Ceram. Soc. 41:455 (1958).

Optical Spectra

Most redox ions dissolved in glass possess characteristic absorptions in the near-ultraviolet, visible, and near-infrared regions. The origin of these electronic spectral bands is the d and f electron splittings of an ion in a particular chemical environment. The absorption bands for each ion are characteristic in terms of number, position, shape, and intensity. The positions and shapes of the absorptions can be correlated to the ionic state and its local electronic environment (ligands and coordination site), whereas the intensities of the absorptions can be correlated to the concentration of that ion in the glass structure. The absorptions of these redox ions have also been employed as probes into the basicities of glass melts.

Characterizations: Redox ion concentrations, redox states present, color and chromophores, structural chemistry (coordination, ligands), crystal-field stabilization energy, solubility.

Advantages: Ease of comparison to aqueous systems, sensitivity for many ions, unique spectra for each ion. Disadvantages: Broad bands and interferences from Fe(II) "contaminations" may hinder resolution.

References: C. R. Bamford, "Colour Generation and Control in Glass", Elsevier, Amsterdam (1977). D. S. Goldman & J. I. Berg, Spectral study of ferrous iron in Ca-Al-borosilicate glass at room and melt temperatures, J. Non-Cryst. Sol. 38:183 (1980).

Electron Paramagnetic Resonance (EPR) Spectroscopy

Ions having partially filled inner electron shells, as in the transition and rare earth element groups, with non-zero total angular momentum are those of interest that will show EPR absorptions. Typically, those transition and rare earth ions that have unpaired electron(s) can possess such absorptions. The EPR spectral pattern {in terms of g-factor, line-shape, fine structure, and intensity} of the individual ions provides a direct description of the ground state and of the effects of the nearest neighbor on the energy level of the paramagnetic center. For example, ions commonly investigated by EPR spectra include Fe(lII), Ti(III), Cr(III), Mn(II), V(IV), Cu(II), and Eu(II) among others; but possible equilibrium species such as Fe(II), Ti(IV), etc. respectively do not contribute absorptions to the spectra.

Characterizations: Presence of certain redox states, coordination of ion, concentration of that ion.

Advantages: Highly specific for certain ions, very narrow

bandwidths are common {e.g.; Fe(III)}, quite sensitive for many ions. Disadvantages: Possible line broadening for certain ions in glasses, interferences at some g-values, care in ascertaining concentrations of ions.

Reference: H. D. Schreiber and G. B. Balazs, Mutual interactions of Ti, Cr, and Eu redox couples in silicate melts, Phys. Chem. Glasses. 22:99 (1981).

Moessbauer Spectroscopy

Moessbauer spectroscopy makes use of a nuclear resonance phenomenon which occurs in certain isotopes and is apparent in the gamma-ray region of the electromagnetic spectrum. Isotopes of iron, tin, antimony, and europium in a glass matrix have all been investigated by Moessbauer spectra. Parameters such as chemical shift, quadrupole coupling/splitting, and linewidth in the spectra are all dependent on the short range order about the isotope investigated. For example, the chemical shift of Fe(II) is greater than that for Fe(III) in glass, while the chemical shift for either ion in an octahedral site is greater than that in a tetrahedral site.

Characterizations: Concentrations and coordinations of redox ions.

Advantages: Specific for certain elements and their ions. Disadvantages: Resolution of absorptions limits sensitivity and reliability, lack of applicability to most redox ions.

Reference: G. H. Frischat & G. Tomandl, Moessbauer investigation of the valence state and coordination of iron in silicate glasses, Glastechn. Ber. 42:182 (1969).

Raman Spectroscopy

Laser raman spectra of crystalline and vitreous compounds are very characteristic, well-resolved, and suitable for quantitative analysis in the identification of structures. Structures in the vitreous phases, in particular anionic units, can be determined by comparison of the Raman spectra to those of known crystalline compounds. Differential spectra have been used to extract the contribution of arsenic to the Raman spectra of arsenic-containing silicate glasses. Most spectral bands are due to silicate network contributions, but arsenate and arsenite structures were ascertained. As(V) showed up as AsO_4^{3-} and $As_2O_7^{4-}$, while As(III) existed as AsO_2^-. The anionic structure of iron-containing glasses were also used to aid in deducing of the coordination of the iron redox states.

of oxygen accompanying the reduction or oxidation respectively. TGA thus monitors the redox ions indirectly by the gaseous species in the redox equation; in cases where the redox component can be a gas {e.g.; S as SO_2 or Ru as RuO_4}, the redox equilibria can be directly monitored.

Characterizations: The kinetics involved in the establishment or reestablishment of redox equilibria in glass-forming melts.

Advantages: In situ analysis {on the melt and not the quenched state}, determines dynamic redox properties. Disadvantages: Only monitors the particular redox ions indirectly in most cases, a relatively untested method in this application.

Reference: R. J. Williams & I. -M. Chou, Some new directions for research on oxidation-reduction phenomena, EOS (Trans. Amer. Geophys. Union). 59:399 (1978).

Electrochemistry

Electrochemical techniques such as chronopotentiometry and potential sweep voltammetry can be applied to glass-forming melts at high temperatures. Such measurements have been employed to determine electrode reactions/processes and reversible electrode potentials in these melts, and subsequently to ascertain the presence of certain redox states, an effective redox series of multivalent elements, equilibrium constants for internal redox reactions, and diffusion coefficients. In addition, solid electrolyte electrochemical cells have been used to measure the activities of Co(II), Ni(II), and Fe(II) in silicate melts.

Characterizations: Concentration, reducing power, interactions, diffusion properties, activity coefficients.

Advantages: In situ analysis {on the melt, not the vitreous state}, thermodynamic characterizations. Disadvantages: Background noise at high temperature and other experimental complications.

References: H. W. Jenkins, G. Mamantov, D. L. Manning, & J. P. Young, EMF and voltammetric measurements on the U(IV)/U(III) couple in molten $LiF-BeF_2-ZrF_4$, J. Electrochem. Soc. 116:1712 (1969). A. M. Lacy & J. A. Pask, Electrochemical studies in glass: III, the system $CoO-Na_2Si_2O_5$, J. Amer. Ceram. Soc. 54:236 (1971). K. Takahashi & Y. Miura, Electrochemical studies on diffusion and redox behavior of various metal ions in some molten glasses, J. Non-Cryst. Sol. 38:527 (1980).

Electrical Conductance

The effect of multivalent element additives on the electric conductivity or the resistivity of a glass {as a function of temperature} can be used to understand the character of the ions dissolved within the glass. For example, resistivities of silicate glasses containing antimony systematically varied with the Sb(V)/Sb(III) ratio. Electrical conductance has also been used to investigate the possibility of conduction pairs such as Cu^+-O-V^{5+}, $Fe^{3+}-O-V^{4+}$, etc. in the glass structure. The formation of such complexes depends on the participation of the ions therein in the glass network as network formers and modifiers.

Characterizations: Redox ion ratios, nature of complexes composed of redox ions {coordination sites}.

Advantages: Characterization of certain complexes. Disadvantages: Limited sensitivity and the need for relatively high concentrations.

References: D. Chakravorty, D. Kumar, & G. V. S. Sastry, Electrical conduction in silicate glasses containing antimony oxide, J. Phys. D. 12:2209 (1979). T. Yoshida & Y. Matsuno, The electronic conduction of glass and glass ceramics containing various transition metal oxides, J. Non-Cryst. Sol. 38:341 (1980).

Viscosity

In an iron-containing glass melt {silicate}, Fe(II) tends to decrease the melt viscosity while Fe(III) tends to increase the viscosity. This observed effect is due to the different sites that the ions frequent in the melt structure.

Characterizations: Viscosity should be able to be related to the redox ratio in the melt if all other parameters such as temperature and composition are held constant. Inferences about the size and coordination of the redox ions may be made if the redox ratio is held constant.

Advantages: Direct measurement on the melt. Disadvantages: Viscosity is a bulk property and would be relatively insensitive to trace redox components.

Reference: L. C. Klein & B. V. Fasano, Viscous flow behavior of four iron-containing silicates with alumina, Lunar Planet. Sci. XIII. 395 (1982).

Magnetic Susceptibility

Magnetic susceptibility measurements provide an estimate of the number of unpaired electrons associated with a metal ion in a complex. As such, the method can be used to ascertain chemical structure, stereochemistry, and the degree of ionic/covalent bonding in the characterization of transition metals in glasses. These measurements are especially appropriate for elements such as iron, cobalt, and nickel, since as metals they are ferromagnetic but in their various redox states they are either diamagnetic or paramagnetic. The actual amount of paramagnetism can then be related to the amount of that specific ion, its coordination site within the glass structure, and the presence of other neighboring ions that may perturb its local environment.

Characterizations: Concentration and coordination of redox ions.

Advantages: Sensitivity to unpaired electrons and thus ionic state. Disadvantages: Difficulty in interpretation when more than one paramagnetic element is present.

Reference: A. N. Thorpe, F. E. Senftle, & F. Cuttitta, Magnetic and chemical investigations of iron in tektites, Nature. 197:836 (1963).

DISCUSSION

In the previous section, the methods of analysis for the characterization of redox ions in glasses and glass-forming melts have been discussed. Although each method was only briefly outlined, one or more representative references listed for that method will supply more detail. The procedures are more direct in the investigation of the quenched state {glass} under ambient conditions, but recent advances in instrumentation have simplified the usage of optical spectrophotometry, electrochemistry, and thermogravimetric analysis on systems at high temperature in the molten state. Each method provides certain characterizations of the redox ions; typically, concentration and coordination are the most common properties desired and measured by the procedures. In order to characterize fully a particular redox ion in a glass, one should never rely exclusively on one instrumental method -- other methods are available to supplement and to confirm identifications in this sense.

The choice of analytical procedures in the characterization of redox ions in glass depends largely on the availability of instrumentation and on the desired sophistication. Prior studies have focused on "wet" chemistry, optical absorption spectroscopy, and EPR spectroscopy because of their applicability to a wide range of redox ions and their relative ease of operation/understanding. Fluorescence, Raman, and Moessbauer spectroscopy are limited in the range of redox ions measureable; while methods such as XPS and the electron microprobe remain relatively untested for many systems.

For simplified systems in which only one multivalent element is present in the glass, the characterization of one redox ion or one redox couple by the methods discussed is fairly straightforward. Most potential interferences are eliminated, and the interpretation of the collected data follows directly.

However, "real" systems most likely do not contain just one multivalent element in the glass matrix, as several multivalent elements are dissolved simultaneously in the glass. This expands the number of possible redox ions in the glass and, thus, the number of interferences in the analytical procedures to characterize a particular redox state. Because of mutual interactions with other multivalent elements[2] in glasses or melts, the character of a certain redox ion may change. Its equilibrium concentration might be changed by internal oxidation-reduction reactions, and its structural site may be perturbed by coupling, complex formations, associations, etc.

In the characterization of a redox ion in a glass where several multivalent elements may be present, it becomes imperative to employ more than just one instrumental method to minimize the chance of systematic error. The degree of mutual interaction, that is the concentration of a redox ion in the presence of other possible redox couples, can be determined by first calibrating a signal {e.g.; EPR, optical absorption peak} specific to that redox state to its concentration when that element is the sole multivalent element in the glass and then monitoring the intensity of that signal. The choice of signal depends on the system. Interferences are to be minimized and sensitivities maximized. In the Cr-Ti system at reducing conditions where the redox couples Cr(III)-Cr(II) and Ti(IV)-Ti(III) are operational in the glass, EPR spectra would be chosen over optical absorption spectra because Cr(III), Cr(II), and Ti(III) all have overlapping absorptions in the visible region of the optical spectra but Cr(III) and Ti(III) have well-separated resonances in their EPR spectra.[3,4] Changes in the shapes, positions, and/or pattern of the monitored signal could then be used to understand coordination changes in the nature of mutual interactions. The analytical procedures can be combined in creative ways to obtain complementary information and to obtain a better characterization of redox ions in glasses.

ACKNOWLEDGMENT

This research in the characterization of redox ions in glasses and glass-forming melts has been supported by the Savannah River Laboratory (subcontract AX522039) administered by E. I. DuPont de Nemours & Co. for the U. S. Department of Energy, NASA (contract NSG-7355), and Research Corporation. The author expresses his gratitude to Charlotte Schreiber for her aid in manuscript preparation. L. M. Minnix, B. E. Carpenter, and G. B. Balazs also helped in the organization of this manuscript.

REFERENCES

1. H. D. Schreiber, Properties of redox ions in glasses: An interdisciplinary perspective, J. Non-Cryst. Sol. 38:175 (1980).

2. J. Wong and C. A. Angell, "Glass: Structure by spectroscopy", Marcel Dekker, New York (1976).

3. H. D. Schreiber, T. Thanyasiri, J. J. Lach, and R. A. Legere, Redox equilibria of Ti, Cr, and Eu in silicate melts: Reduction potentials and mutual interactions, Phys. Chem. Glasses. 19:126 (1978).

4. H. D. Schreiber and G. B. Balazs, Mutual interactions of Ti, Cr, and Eu redox couples in silicate melts, Phys. Chem. Glasses. 22:99 (1981).

AUTHOR INDEX

Abraham, M. M., 588
Acocella, J., 586
Adams, D. M., 197
Adams, R. N., 107
Adamson, A. W. 130
Adar, F., 199, 214, 215
Adler, P.N., 570
Ahearn, A. J., 69
Ahlberg, L. A., 411
Ahn, W. S., 157
Aigeltinger, E., 293
Akashi, T., 358
Akinc, M., 159, 170
Albee, A., 295
Alekseeva, O. S., 613
Allred, D. D., 588
Almond, D. P., 489
Alper, A. M., 349
Altstetter, C. J., 587
Amsel, G., 570
Anderson, H. H., 38
Anderson, J. H., 514
Andersson, S., 307
Angell, C. A., 658
Angerstein-Kozlowska, H., 108
Antoon, M. K., 197
Appleton, B. R., 57, 588
Apps, R. L., 489
Arashi, H., 198
Amstrong, N. R., 89
Arnold, G. W., 57
Arnold, G. W., 37, 570
Artemare, E. D., 570
Aust, K. T., 319
Avouris, Ph., 252, 270
Azuma, K., 396

Babcock, C., 625
Bach, H., 38
Bacon, J. R., 69
Bagley, R. D., 319

Baitinger, W. E., 37
Bailey, G. W., 293
Bailey, N. A., 292
Baker, B. G., 107
Baker, R. F., 292
Balau, J. R., 384
Balazs, G. B., 651, 658
Bamford, C. R., 650
Bando, Y., 222
Bankston, D. C., 497
Barcz, A., 570
Bard, A. J., 398
Barnard, R. S. 293,
Barnes, B. O., 587
Barnes, K. K., 107
Baron, F. R., 586
Baron, J., 614
Barrer, R. M., 131, 132
Barrett, E. P., 130
Barten, H., 464
Basile, L. J., 197
Baszyk, P. E., 237
Bates, J. B., 197
Bauer, E., 57
Bayman, A., 265
Baxter, R. G., 600
Bazan, F., 384, 385
Beahm, E. C., 69
Becquerel, H., 588
Beechan, C. R., 156
Beirens, L. C. M., 57
Belin, M., 264
Bell, A. T., 268
Bell, R. J., 197
Bence, A. E., 295
Bennett, H. E., 276
Bennett, J. M., 277
Benninghoven, A., 107
Bennison, S. J., 309
Bentley, J., 350
Bently, J., 293
Berard, M. F., 170

Berg, J. I., 650
Berger, H., 423
Berkowitz, B. J., 588, 589
Berkowitz, H., 570
Berlin, E., 132
Berndt, C. C., 465, 473, 489
Berthod, J., 69
Berube, Y. G., 513
Bettman, M., 307
Bibler, N. E., 590, 600
Biggers, J. V., 277
Binns, D. B., 336
Birgeneau, R. J., 89
Birks, L. S., 292, 295
Birnbaum, H. H., 588
Black, P. E., 237
Blackburn, D. H., 613
Blackmon, P. D., 513
Blanc, M., 307
Bleier, A., 499, 511, 513
Bliss, M., 464
Bockris, J. O'M., 107
Boder, E. E., 130, 131
Boldish, S. I., 277
Bongers, P. F., 358
Booker, G. R., 365
Books, W. C., 155
Borders, J. A., 37
Borodulenko, G. P., 278
Bovin, J. O., 307
Bowen, H. K., 319, 320, 512
Bowen, K., 637. 638,
Bower, C. A., 514
Bowser, W. M., 265
Boyd, G. A., 589
Boyde, G. E., 197
Bradt, R. C., 336, 358
Bradley, J. A., 265

659

Brakel, J. V., 156
Brasch, J. W., 197, 198
Bratt, P. W., 441
Bratton, R. J., 488
Bremser, A. H., 433
Brennan, W. P., 515, 529
Bright, E., 387, 399
Bril, T. W., 237
Bristow, J. R., 336
Broers, A. N., 291
Brongersma, H. H., 57
Bronsted, J. N., 513
Brook, R. J., 319, 320
Brown, A., 464
Brown, G. E., 384
Brown, J. K., 265
Brown, R. H., 198
Brown, S. D., 571, 586, 587, 589
Brownell, W. E., 464
Bruckner, R., 587
Brunauer, S., 130
Bucher, E., 278
Buchner, S. P., 221
Budiansky, B., 336
Budworth, D. W., 320
Buessem, W. R., 294
Bulko, J. B., 109, 131
Bullard, M. E., 586
Bunnell, L. R., 586
Bunshah, R. F., 107, 589
Bunting, E. M., 197, 198
Burke, J., 320
Burman, C., 549, 570
Burns, W. G., 600
Burr, K., 384
Busch, H., 291
Butterworth, B., 156
Buyukozturk, O., 440, 433
Burge, D. K.,
Buster, J.H.J.M., 237
Byer, N. E., 221
Bykovskii, 68

Byron, K., 237

Cachard, A., 57
Cadenhead, D. A., 130
Calcemuggio, G. L., 586
Calderwood, F. W., 336
Calvert, L. D., 464
Campbell, A. B., 603
Campbell, J. H., 384, 385
Campbell, S. S., 387
Campion, A., 265
Cannon, R. M., 276, 319, 320
Capellen, J., 69
Carbone, T. G., 157
Carlson, T. A., 89, 107
Carpenter, R. W., 293, 350, 383
Carrier, G. B., 614
Carriere, B., 38
Carslaw, H. S., 108
Carter, C. B., 297
Case, E. D., 337
Caskey, G. R., 589
Castaing, R., 292
Castro, J. H. C., 156
Chakraborty, I. M., 223
Chakraborty, D., 655
Chalmers, B., 411
Champion, P., 38
Chang, C. C., 107
Chang, R. K., 265
Chappell, R. A., 37
Charles, R. J., 276, 613
Charlier, H., 37
Chess, C., 277
Chess, C. A., 277
Chess, D., 277
Chesters, M. A., 38
Cheveigne, S. De., 264
Chiang, C-H., 197
Chimenti, D. E., 424
Chismar, P. H., 600

Chen, H., 653
Cheremisinov, V. P., 228
Chou, C. H., 411, 412, 424
Chou, I-M., 654
Chowdhryand, U., 319
Christ, C. C., 513
Christian, J. W., 411
Christie, J. M., 293
Chu, W. K., 57, 570
Clancy, F. K., 198
Clarebourgh, L. M., 294
Clark, G. J., 588
Clark, R. J. H., 214
Clarke, C. P., 242
Clarke, D. R., 199, 214, 292, 295, 323, 337, 348, 349, 367, 383, 384
Claussen, N., 336, 337
Cleveland, J. J., 336
Cline, J., 464
Close, W. P., 649
Coble, R. L., 307, 399
Coburn, J. W., 38
Cohen, B. L., 570
Cohen-Tenoudji, F., 412
Colby, J. W., 295
Coleman, R. V., 264
Colombin, L., 37
Condit, R. H., 589
Condrate, R. A., Sr., 223
Connes, J., 197
Connes, P., 197
Conway, B. E., 107, 108
Conzemius, R. J., 59, 69
Cook, L. S., 587
Cooley, J. W., 197
Cooper, A. R., 588
Cooper, W. D., 294
Cornilsen, B. C., 239, 248

AUTHOR INDEX

Costley, J. L., 292
Cox, R. L., 489
Craig, R. G., 156
Cranston, R. W., 131
Criddle, E. E., 108
Cross, L. E., 384
Crowder, B. L., 570
Cubicciotti, D. D., 384
Culmo, R. F., 529
Cunnion, J. P., 441
Cushing, J. R., 108
Ctuler, I. B., 586
Cutler, R. A., 436
Cuttitta, F., 656
Czupryna, G., 490, 496

Dallaire, S., 472
Danforth, S. C., 214
Dennheim, V. H., 38
Das, G., 293
David, D., 570
Davis, G. D., 221
Davis, J. A., 513, 514
Davis, K., 57, 570
Davis, L. E., 89, 358
Davis, R. F., 514, 295
Davis, W. F., 497
Davis, W. W., 513
Davisson, C. J., 291
Dearmaley, G., 570
de Billy, M., 412
Debras, G., 37
DeBoer, J. H., 130, 132
DeBruyn, P. L., 170, 513
Debye, P., 514
Defourneau, D., 264
Degan, J. D., 570
Dehoff, R. T., 294
Delhaye, M., 214
Delahay, P., 107
Della Mea, G., 37
DeMeyer, K. M., 266
Deming, L. S., 130
Deming, W. E., 130
Dempster, A. J., 68

Dench, W. A., 38
Derby, J. V., 615
Dernier, P. D., 278
de.St.Victor, N., 588
Devlin, J. P., 198
Dhamelincourt, P., 214
Diamond, S., 156
Dietz, R. E., 248
Dietz, V. R., 132
Dimaria, D. J., 266
Din, F., 131
Dinger, T. R., 339
DiVito, M. P., 515, 529
Dole, S. L., 170
Domange, L., 277, 278
Domarkas, V., 411
Dong, D. W., 266
Doremus, R. H., 57, 549, 613, 645
Dosch, R. G., 384
Dove, D. B., 255
Doveren, H. V., 38
Doyle, B. L., 57, 570
Drake, L. C., 145, 156
Dranova, G. N., 277
Dresselhaus, G., 220
Drew, P., 350
Drigo, A. V., 37
Drummond, C. H., 649
Drury, T., 588
Dubinin, M. M., 130
Duigou, L. E., 320
Dulka, C. E., 276

Earnest, C. M., 515, 529
Echlin, P., 294, 295
Economou, E. N., 265
Edington, J. W., 293
Edmonds, J. W., 464
Edwards, R., 294
Ehrenfest, 637
Ehrt, D., 614
Eliseev, A. A., 278
Ellialtioglu, R., 265
Elmer, T. H., 614
Elsley, R.K., 411,412

Emmett, P. H., 130, 131
Endriz, J. G., 265
Ernsberger, F. M., 587
Erwin, C. Y., 587
Eshelby, J. D., 336
Evans, A. G., 358, 411, 412, 433
Evans, C. A., 107
Evans, C. A., Jr., 588
Everett, D. H., 131, 132
Eyring, L., 69, 277

Fahlman, A., 108
Farmer, R. L., 425
Fasano, B. V., 655
Fellgett, P. B., 197
Felter, T. E., 108
Ferraro, J. R., 171, 197, 198
Fertig, K., 412
Few, I. S., 237
Fink, C. L., 570
Fink, W., 198
Finn, B. P., 131
Finston, H. L., 513
Fiore, N., 412, 588, 589
Fiore, N. F., 423
Fiori, C., 294, 295
Firestone, R. F., 293
Fischbach, D. B., 156
Fisher, I. A., 488
Flahaut, J., 277, 278
Fleischman, M. P., 265
Fletcher, J., 365
Forward, C. J., 294
Franks, F., 131
Frech, R., 197
Frechette, V. D., 237
Freiman, S. W., 431
French, P. W., 37
Frevel, L. K., 157, 463
Friedman, H., 292
Fripiat, J. J., 513
Frischat, G. H., 651

Fritz, G. S., 294
Fullman, R. J., 320
Fulrath, R. M., 320, 547, 586
Funkenbusch, E. F., 239, 248
Furtak, T. E., 265
Furlong, D. N., 513
Furnam, E. H., 586
Furukawa, T., 237
Fyans, R. L., 529

Galakhov, F. Y., 613
Gale, L. H., 89
Galeener, F. L., 57
Gallagher, P. K., 169
Gamsjager, H., 513
Gardon, R., 625
Garrels, R. M., 513
Garwood, G. A.,Jr., 91, 108
Gatti, A., 276
Gauckler, L. J., 350
Gebhardt, J. J., 276
Geick, R., 214
Geiss, R. H., 292, 294, 350
Gentilman, R., 277
Gerassimora, I., 220
Gerdeman, D. A., 472
Gere, J. M., 439
Gerlach, R. L., 107
German, R. M., 252
Germer, L. H., 291
Ghez, R., 266
Gibbs, C. R., 645
Gibbs, J. W., 157
Gileadi, E., 107
Girard, E., 570
Gitzen, W. H., 547
Glasser, A. M., 320
Gleiter, H., 319
Gobel, H. E., 464
Godwin, L. M., 265
Goehner, R. P., 463
Golabi, S. M., 277
Golay, M., 464
Goldman, D.S., 649,650

Goode, P. D., 570
Goodsel, A. J., 157
Goldstein, J. I., 292, 293, 294, 295, 350, 358
Gomer, R., 291
Goo, E. K., 358
Goo, E. K. W., 358
Goodwin, J. W., 513
Gordon, S. G., 463
Goringe, M. J., 358
Gorte, R. J., 130
Goss, S. A., 424
Gossink, R. G., 38, 89
Gottardi, V., 37
Graham, L. J., 411
Graham, T. P., 264
Grainger, J., 292
Grant, J. T., 108
Green, D. J., 323, 336
Green, H. W., 294
Greenler, R., 107
Greenler, R. G., 198
Gregg, S. J., 130
Greil, P., 350
Grescovich, C., 399
Greve, D., 336
Griffiths, P. R., 198
Grizik, A. A., 278
Grizzle, V. M., 265
Groleau, R., 57, 570
Grossbeck, M. L., 588
Gryzinski, M., 108
Gschneidner, K. A., 67
Gschneidner, K. A. 277
Guinan, M., 384
Guittard, M., 277
Guivarch, A., 570
Gupta, T. K., 399
Gyulai, J., 570

Haas, et. al., 69
Haas, T. W., 108
Hair, M. L., 270, 513
Halenda, P. P., 130
Hall, J. T., 265
Hall, P. M., 358
Haller, W., 613, 614

Haller, W. H., 587
Halsey, G. D., 148, 157
Hammond, J. S., 107
Hanada, T., 653
Hanawalt, J. D., 463
Hannah, R. W., 198
Hansma, P. K., 264, 265
Hare, T. M., 295
Harmer, M., 319
Harmer, M. P., 309, 320
Harrick, N. J., 198
Harrison, A. J., 587
Harrison, W. A., 276
Harrington, W. L., 57
Hartstein, A., 266
Hasegawa, Y., 348
Hasselman, D. P., 336
Hasselman, D.P.H.,350
Hatfield, W. T., 463
Hattori, H., 132
Haven, Y., 587
Hawkins, D. T., 107
Hawkins, E. G., 637
Hawkins, G. F., 412
Hayes, W., 214
Head, A. K., 294
Headley, T. J., 384
Hecht, N. L., 472
Heggarty, D., 157
Heinrich, K. F. J., 296
Hench, L. L., 251, 296
Hendra, P. J., 265
Henoc, J., 295
Herman, J., 465, 472, 473, 489
Herman, J. S., 57
Herman, R. G., 109, 131
Hermanek, F. J., 489
Herron, R. H., 156
Hershberger, J.F., 108
Hester, R. E., 214
Heuer, A. H., 293, 383, 588

AUTHOR INDEX

Hibberson, W., 384
Hildenbrand, D.L., 384
Hill, S. L., 198
Hillert, M., 319
Hillier, J., 291, 292
Hinman, D. C., 157
Hinz, W., 37
Hirsch, P. B., 294, 358
Hirschfeld, T., 198
Hlava, P. F., 384
Hobbs, B. E., 384, 589
Hoenig, C., 384
Hohl, H., 514
Hojiund-Nielsen, P.E. 107
Holleran, L. M., 295
Hollmagel, H., 198
Homola, A., 513
Horiuchi, S., 307
Hoshino, Y., 278
Houser, C. A., 57, 588
Houston, J. E., 107
Howard, A., 464
Howe, R. G., 293
Howie, A., 294
Hren, J. J., 292, 293, 294, 348, 358
Huang, T. C., 464
Hubbard, A. T., 91, 107, 108
Hubbard, C. R., 463, 464
Huckel, E., 514
Huffsmith, S. A., 589
Hughes, A. E., 600
Humble, P., 294
Humphris, S. E., 489
Hunter, O., Jr., 170, 336
Hunter, R. J., 514
Hurwitz, S., 587
Huseby, I. C., 276
Hyde, B. G., 307

Ibach, H., 107
Iida, S., 358
Inkley, F. A., 131

Inomata, Y., 350
Irven, J., 237
Isaacson, M., 292, 293
Isakasaa, S., 214
Ishida, H., 197, 198
Ishigame, M., 214
Ishii, T., 214
Ishikawa, R. M., 108

Jack, K. H., 349, 350
Jacquinot, P., 198
Jaeger, J. C., 108
Jakobsen, R. J., 198
James, R. O., 513, 514
Jansen, W. T., 89
Jansheski, W. C., 649
Jeanmaire, D. L., 265
Jelli, A., 37
Jenkins, E. J., 293, 295
Jenkins, H. W., 651
Jennings, H. M., 214
Jenson, W. B., 513
Jha, S. S., 265
Joannopoulos, J.D., 38
Johari, O., 293
Johari, O. H. M., 294
Johnson, D., 293
Johnson, D. W., Jr., 169
Johnson, G., 424
Johnson, G. E., Jr., 294
Johnson, P. F., 281, 293, 295
Johnson, O. C., 464
Johnson, W. C., 306, 307
Johnston, W. D., 645
Jonca, A., 465
Jones, A. J., 439
Jones, D. R., 649
Jordan, G., 170
Jordan, G. W., 170
Joshi, A., 89
Joshi, S. B., 571, 586, 589

Joy, D. C., 292, 293, 294, 295, 350, 358
Joyner, L. G., 130
Jungman, A., 412

Kahara, E., 463
Kahn, F. J., 89
Kamagnaito, O., 349
Kamarzin, A. A., 277
Kametani, H., 398
Kane, P. F., 57
Kanykowski, E. A., 570
Kaplan, M. F., 600
Kappeler, F., 570
Kariyama, M., 350
Kaska, W. C., 265
Katekaru, I., 108
Kats, A., 587
Kay, E., 38
Keil, R. G., 264
Keller, J., 570
Kelley, J. A., 600
Kelley, R., 38
Kelso, J. F., 1, 89
Kenik, E. A., 293
Keramidas, V. G., 214
Kessler, L., 413, 423
Kessler, L. W., 412, 424
Kesson, S. E., 384
Khuri-Yakub, B. T., 401, 411, 412, 424
Kijima, K., 214
Kim, K. S., 38
Kim, Y. S., 169
Kingery, W. D., 512, 637, 638
Kino, G. S., 411, 412, 424
Kirby, E. M., 237
Kirk, R. E., 220
Kirtley, J. R., 252, 264, 270
Kiselev, A. V., 131
Kishii, T., 37
Kissl, W., 653
Klaeboe, P., 198
Klein, J., 264

Klein, L. C., 655
Klier, K., 131, 514
Kluyev, V. P., 645
Knapp, G. S., 653
Knickerbocker, S.H., 571
Kobylinski, T. P., 131
Kocsis, G., 587
Koenig, C. J., 586
Koenig, J. L., 197, 198
Kolipari, R., 465
Konijnendijk, W. L., 237
Konynenberg, R.V., 384
Koplik, C. M., 600
Korda, E. J., 614
Kossell, W., 292
Kotai, E., 570
Kotanigawa, T., 132
Kozkukhavov, V., 220
Krautkramer, J., 424
Kreidl, N. J., 237
Kressley, L. J., 157
Krezhov, K., 220
Krishnan, K., 198
Krivanek, O. L., 349, 350, 358
Kroeker, R. M., 265
Kruyt, H. R., 514
Kucheria, C. S., 156
Kuczynski, G. C., 156, 383
Kyjima, T., 350
Kumar, D., 655
Kupperman, D. S., 412, 423, 424
Kuszyk, J. A., 336
Kuwana, T., 89
Kuzuba, T., 214
Kvaas, R. E., 215

Lacassagne, A., 588
Lach, J. J., 658
Lacharme, J. P., 38
Lacy, A. M., 654
Ladell, J., 464
LaFramboise, T.O., 531

Laine, R. M., 265
Lam, D. J. 653
LaMarche, P., 57, 570
Lambe, John, 265
Landon, T. G., 411
Lane, R. F., 108
Lanford, W. A., 37, 57, 549, 570, 587
Lang, B., 38
Lang, S. M., 547
Lange, F. F., 350, 383
Langmuir, I., 130
Lapinski, N. P., 412, 423
Larrabee, G. B., 57
Lattes, J., 588
Lau, A. L. Y., 108
Lau, S. K., 488
Lauer, J. L., 198
Laursen, T., 57, 570
Lavedan, J., 588
Lay, K. W., 251, 399
Lebiedzik, J., 294
Leblond, C. P., 588
Lechatelier, H., 529
Leckie, J. O., 513, 514
Lee, D. H., 38
Lee, J. H., 387
Lee, R. J., 294
Lee, S. Y., 488
Lee, W. M., 248
Leger, A., 264
Leger, D., 38
Legere, R. A., 658
Leich, D. A., 570
Levinson, L. M., 358
Lewis, D. R., 529
Lewis, G. N., 513
Liang, K., 412
Lifshin, E., 294, 295
Lifshitz, I. M., 399
Ligeon, E., 570
Lin, A. W. C., 89
Lin, I. N., 351, 358
Linker, G., 570
Lippens, B. C., 130, 132

Lippincott, E. R., 198,
Litster, J. D., 89
Little, L. H., 513
Locker, R., 627, 640
Lommen, T. P. A., 38, 89
Longinotti, L. D., 278
Lorimer, G. W., 292
Lorprayoon, V., 239, 248
Lou, L. K. V., 383
Loudon, R., 214
Low, M. J. D., 198
Lowe-Ma, C., 267
Lowe-Ma, C. K., 277
Lowell, S., 133, 146
Lowenheim, F. A., 108
Lowry, T. M., 513
Lucks, 625
Lucovsky, G., 57
Lumby, R. J., 350
Lux, R., 570
Lynch, A. W., 384

MacCrone, R. K., 465
MacDonald, N. C., 89
Macedo, P. B., 613, 637
Mackenzie, J. D., 587
Mackintosh, W. D., 57
Macleod, D. M., 132
Madey, T. E., 107
Madix, R. J., 107
Magee, C. W., 57
Maggs, F. A. P., 132
Maiman, T. H., 68
Major, A., 384
Makishima, A., 614
Malissa, H., 653
Mallett, G. R., 295
Mallory, C. L., 463
Malovitsky, Yu.N., 277
Mamantov, G., 654
Mann, C. K., 107
Manning, D. L., 654
Manning, W. R., 336
Manuba, A., 570

AUTHOR INDEX

Sadana, D. K., 359, 365
Safai, S., 472
Sakaino, T., 614
Sakuri, T., 214
Salganik, R. L., 336
Salko, S., 319
Salpeter, M. M., 589
Sanders, D. M., 587
Sarkar, A., 613
Sastry, G. V. S., 655
Sather, A., 423
Sato, M., 278
Sato, T., 214
Saunders, K., 277
Savitzky, A., 464
Scalapino, D. J., 265
Scheidecker, R.W., 170
Schilling, R. F., 57
Schindler, P. W., 513
Schmidt, F. A., 69
Schmidt, L. D., 107
Schoeffel, J. A., 108
Scholes, S., 613
Scholze, H., 587
Schreiber, H. D., 647, 651, 658
Sciammarella, C., 412, 423
Scruby, C. B., 489
Seager, R. E., 169
Seah, M. P., 38
Seifert, F., 652
Sekiawa, Y., 214
Sellers, D. J., 306
Senftle, F. E., 656
Serrano, C. M., 266
Shackelford, J.F., 586
Shaffer, E. W., 588
Shankar, N. R., 489
Sharma, S. K., 197
Sharp, W. B., 108
Shartsis, L., 637, 638
Shaw, J. N., 513
Shaw, T. M., 349, 350, 358
Shelby, J. E., 639, 645

Shepherd, H. M., 649
Sheppard, N., 198
Sherren, A. T., 198
Shields, J. E., 133, 146
Shiers, L. E., 170
Shii, K., 214
Shuman, H., 292
Shyne, J. C., 412
Siebeneck, H. J., 336, 350
Siegbahn, K., 108
Sieger, J. S., 37
Sih, G. C., 412
Sillen, L. G., 513
Sils, V., 547
Simkin, D., 603
Simmons, G., 337
Simmons, G. W., 131
Simon, I., 587
Simonsen, M. G., 264
Sing, K. S. W., 130, 131, 513
Singh, P., 239, 248
Singh, J. P., 336
Singhal, S. C., 350
Sipe, J. J., 156
Six, H. A., 89
Skalny, J. P., 131
Slack, G. A., 276
Sleight, A. W., 277
Sloane, H. J., 198
Slyoz, V. V., 399
Smakula, A., 547
Smets, B. M. J., 38
Smith, A. L., 512
Smith, D. W., 237
Smith, F., 198
Smith, G. S., 463
Smith, K. L., 71
Smith, P. K., 590
Smith, S. T., 464
Smolyo, A. P., 292
Smolyo, A. V., 292
Smothers, W. J., 156
Snow, J. H., 293
Snyder, R. L., 291, 449, 463, 464

Soga, N., 653
Sokolov, V. V., 277
Solin, S. A., 214
Somiya, S., 319
Somorjai, G. A., 107, 108
Somoyai, G. A., 38
Soriaga, M. P., 108
Spalthoff, W., 588
Speake, J. H., 489
Spence, J., 358
Spicer, W. E., 265
Spinner, S., 637, 638
Spivack, B. D., 441
Spriggs, R. M., 547
Springer, C., 337
Srinivasan, G. R., 613
Stacy, D. W., 336
Stanchell, F. W., 89
Stecura, S., 488
Steeds, J. W., 292, 358
Steele, W. A., 130
Stein, D. F., 306
Stevels, J. M., 237, 587
Stewart, G. H., 336
Stoddart, C. T. H., 37
Stone, F. S., 131, 132
Stone, J. A., 600
Stoneham, A. M., 600
Stone-Masui, J., 513
Strachan, D. M., 587
Strathman, M., 365
Streptsina, M. V., 613
Strudel, J. L., 307
Strunk, H., 337
Suchicital, C.T.A., 156
Sugimoto, M., 358
Sullivan, D. A., 156
Sun, T. S., 221
Suscavage, M.J., 1, 37
Svata, M., 156
Svec, H. J., 59, 69
Swanson, S. R., 439
Swarts, E. L., 37
Swindlehurst, W., 489

Sydzhimov, B., 220
Sykes, L.J., 294
Szabo, M., 589
Szalkowski, F.J., 107

Takahashi, K., 654
Takata, M., 586, 587
Tanabe, K., 132
Tanzilli, R. A., 276
Taylor, A. J., 350
Taylor, J. A., 89
Taylor, M., 384
Taylor, N. J., 107, 108
Taylor, P., 603, 614
Teller, E., 130
Templeton, T., 465
Terwilliger, G.R., 383
Tewari, P. H., 514
Thanyasiri, T., 658
Theis, T. N., 265
Thepaine, Y., 252
Thomas, G., 292, 339, 349, 350, 351, 358, 497
Thomas, H. E., 157
Thomas, J. P., 57
Thomas, R. L., 412
Thompson, D. O., 424
Thompson, V. S., 156
Thorpe, A. N., 656
Tien, J., 412
Tien, T. Y., 350
Tighe, N. J., 306, 383
Tillman, J. F., 649
Tinklepaugh, J.R., 547
Titchmarsh, J. M., 365
Tittmann, B. R., 411, 412
Todd, B. J., 587
Todor, D. N., 529
Toide, T., 278
Tolk, N. H., 57
Tolstova, V. A., 278
Tomandl, G., 38, 651
Tombrello, T. A., 570
Tomozawa, M., 586, 587, 613, 645

Tong, S. S. C., 57
Trautvetter, H.P., 570
Tressler, R. E., 358
Tretyakov, Y. D., 399
Truell, R., 424
Tsang, J. C., 252, 266
Tsong, T. M., 447
Tsong, I. S. T., 39, 57, 588, 589
Tsuji, S., 57
Turkey, J. W., 197
Tumoshenko, S. P., 439
Turk, D. H., 131
Turner, J. F., 570
Turner, L. L., 307
Turner, P. S., 547
Turos, A., 570
Tustison, R. W., 277
Tweer, I., 613
Tyburezy, J. A., 384

Uematsu, K., 319
Uhlmann, D., 637, 638
Ulhmann, D. D., 512
Uhlmann, D. R., 645
Ulmer, G. G., 156
Underwood, E. E., 294, 398
Ure, A. M., 69
Utiyama, M., 132
Utlavt, M., 295
Utsunomiya, T., 278

Valverde, N., 398
VanDerLigt, G.C.J., 57
VanDerMeulen, Y.J., 57
van Duyne, R. P., 265
Van, T. B., 586
VanValkenburg, H.,198,
Van Welzon, 625
Van Zee, A. F., 625
Varadan, V. J., 412
Varadan, V. V., 412
Varela, J.A., 156, 157
Vasilos, T., 306
Vasil`yeva, I. G., 277
Veal, B. W., 653
Vekshina, N. V., 277

Verbist, J., 37
Verhoeven, J.A.T., 38
Vermilyea, D.A., 398
Verweij, H., 237, 652
Vest, R. W., 248
Virgo, D., 652
Vogel, A., 513
Vogel, W., 614
Volynets, F. K., 277
Votava, W. E., 295

Wachtman, J.B.,Jr. 547
Wada, N., 214
Wadley, H. N. G., 489
Waff, H. S., 384
Wagner, C., 398
Wagner, C. D., 89
Wagstaff, F. E., 586, 613
Walling, P. L., 198
Wang, K., 411
Wang, Y. H., 412
Ware, N. G., 384
Warman, G. P., 198
Warman, M. O., 320
Warne, S. St. J., 529
Warren, J. B., 292
Washburn, E. W., 145, 155
Washburn, J., 365
Wassick, T., 549
Watanake, H., 358
Watillon, A., 513
Watson, A., 156
Watson, E.B., 586, 587
Weber, R. E., 89
Weeks, R. A., 588
Wefers, K., 321
Weinberg, W. H., 264
Weinstein, B. W., 439
Weir, C. E., 198
Weiss, J., 350
Welsh, F. E., 198
Wendlandt, W. W., 529
Wentdorf, R. H., 214
Werner, H. W., 37
Westall, J. C., 514
Wey, R., 614

AUTHOR INDEX

Weyl, W. A., 652
Whelan, M. J., 294
White, C. W., 57, 588
White, E. W., 294
White, H. W., 265
White, W., 276
White, W. B., 57, 214, 237, 277, 278, 588
Whittemore, O. J., 147, 156, 157
Whitworth, C. R., 586
Wicks, G. G., 600
Wielunski, L., 570
Wiese, G. R., 513
Wiles, D. B., 464
Wiley, J. R., 600
Wilke, M. E., 587
Wilkinson, F.C.F., 613
Williams, D. P., 427
Williams, K. A., 387
Williams, P., 57, 588, 589
Williams, R. J., 654
Willis, J. R., 336
Wilson, R. G., 365
Wilson, W. I., 349
Winograd, N., 38, 107
Wintenberg, A.L., 57
Wittmack, K., 37
Woldback, T., 198
Wolfram, T., 264
Wong, J., 214, 658
Wood, E. A., 108
Wood, R. W., 198
Worthington, R., 237

Yakowitz, H., 295
Yamamoto, M., 132, 547
Yan, M. F., 319, 320
Yarita, T., 588
Yates, D. E., 513
Yates, J. T.,Jr., 107
Yeates, A. T., 157
Ying, C. F., 424
Yoshida, T., 655
Young, J. P., 655
Young, R. A., 464
Youngblood, G. E., 336

Yuhas, D. E., 412, 413, 423, 424
Yund, R. A., 384

Zachariasen, W.H., 277
Zachary, J. L., 514
Zagofsky, A., 464
Zaluzec, N.J., 293, 350, 383
Zander, A. T., 497
Zangvila, A., 350
Zatorski, R. A., 465
Zeller, M. V., 89
Zettlemoyer, A.C., 131, 514
Ziegler, J., 570
Ziegler, J. F., 57
Zlomanov, V. P., 220
Zworykin, V. K., 291

SUBJECT INDEX

Absorbance-substraction
  techniques, 189
Absorption coefficient, 273
Acetone-toluene-acetone (ATA)
  treatment, 160
Acid-base reaction
  concepts, 501
  properties of ceramic
    powders, 499
  properties of ceramic
    powders, conductometric
    titration, 509
  properties of ceramic
    powders, electrokinetics,
    506
  properties of ceramic
    powders, infrared
    spectroscopy, 511
  properties of ceramic
    powders, potentiometric
    titration, 503
  conductometric titration, 509
Acoustic amplitude image, 415
  amplitude micrograph,
    delaminated ceramic
    capacitor, 419
  amplitude micrograph, silicon
    carbide, 416
  amplitude micrograph, silicon
    nitride, 416
  characterization, 401
  defect detection techniques,
    403
  emission (AE), coatings, 474
  interferogram, 415
  microscopy, 413
  velocity measurement, 403
Adhesion strength, 47
Adipic acid, 255
Adsorption-desorption
  isotherms, 124
Agglomeration, 163
β-Ag I, 192
AgI-type ionic conductors, 192
Al-Al$_2$O$_3$ tunnel junctions, 258
Al-Al$_2$O$_3$-Pb junction, 252
Alumina, 144, 149
Al$_2$O$_3$, 117
Al$_2$O$_3$ fibers in glass, 540
Alumina, abnormal grain growth,
  309
  acoustic migrograph, 421
  chemical inhomogeneity, 318
  density-grain size
    trajectories, 317
  electron energy loss
    spectroscopy, 302
  energy dispersive analysis,
    299
  formation of micro-dense
    island structures, 312
  K-∞^{111}-Al$_2$O$_3$ secondary
    particles, 300
  Mg-Al spinel secondary
    particles, 300
  MgO sintering additive, 309
  Ni-containing secondary
    particles, 299
  physical inhomogeneity, 318
  second phase particles, 297
  transmission electron
    microscopy, 298
  zero-porosity density, 535
Analytical electron microscopy
  (AEM) applications, 290
  historical development, 282
  interpretation of data, 288
  limiting factors, 288
  nuclear waste ceramics, 367
Analytical scanning electron
  microscopy (ASEM), 286
Analytical scanning
  transmission electron
  microscopy (ASTEM), 287
Ankerite, 527
Antiferromagnetic material, 239
  ordering, 242, 248
Aqueous dissolution kinetics,
  392
  precipitation, 390

SUBJECT INDEX

Archimedean density, 536
Argon, 109
  adsorption, 125
Argon desorption, 125
Argon trapped in secondary
  particles, 302
Arsenic-containing glasses,
  Raman spectroscopy, 651
A-Scan (low frequency-high
  frequency), 405
Atomic aborption spectroscopy,
  269
Attenuated total reflectance
  (ATR), 175
Auger analyses, 4
Auger electron spectroscopy
  (AES), 72, 79, 92, 106, 273
Auger spectroscopy, 215, 220
Auger electron spectroscopy,
  MnZn ferrite, 352
Automated x-ray phase
  identification, 461

B_2O_3, 223
β 118^1 sialons, electron
  micrographs, 343
  Fe-rich inclusions, 347
  x-ray energy dispersive
    spectroscopy, 343
Backscattering spectroscopy,
  549
BET surface area, 155
Biaxial compression testing at
  various temperatures, 440
Binary $K_2O \cdot B_2O_3$ system, 225
Birefringence scattering, 267
Birks methods, 290
Bloom potential, 531
Boron nitride, 201, 202
Borosilicate glasses for
  radioactive waste
  immobilization, 590
  glass and waste compositions,
    592
  glass durability, 596
  glass microstructure and
    crystallinity, 592
  glass preparation, 591
  high temperature density
    measurements, 629
  leaching Cs-137 and Sr-90
    from glasses, 596
  low temperature thermal
    expansion, 629
  pH change during leaching
    process, 599
  scanning electron microscopy
    (SEM), 592
  thermal diffusivity, 615
  volume-temperature
    relationships, 627
  x-ray energy spectroscopy,
    592
Boundary breakaway phenomena,
  310
  by solute breakaway, 319
Bulk Density, 536
  bulk density of composite,
    542
Buried flaws, 421

Cr-Ti glass system, 657
Cr_2O_3, 117
CuO, 118
CO, 255
CaF_2, 258
CaO-NiO solid solutions, 239
Calcined ZnO, 129
CoO, 240
$CdTeO_3$, 216
CdO, 215
Ca, Mg $(CO_3)_2$, 525
C-D vibrational intensity, 255
CaO in fly ash and bed ash, 522
Catalysis, 249
Calcite, 520
Cathodoluminescent stain, 332
Causes for abnormal grain
  growth, 310
Ceramic characterization, 273
Ceramic fabrication, 271
Characterization technique, 276
Chemical durability, 644
Chemical vapor deposition, 201
Chemisorption, 249, 254
"Cheto" type montmorillonite,
  519
Chronopotentiometry,

SUBJECT INDEX

multivalent elements in
 glasses, 654
Clearfloat glass, 531
Coefficient of thermal
 expansion of composite, 542
Colorimetric, multivalent
 elements in glasses, 649
Commercial titanate ceramics
 (CTC), 369, 378
 elemental analysis, 378
 glassy phases, 378
 hollandite, 378
 magneto plumbite, 378
 perovskite, 378
 zirconolite, 378
Composites, 535
Conchoidal fracture, 165
Condensation - polymerization
 type bonding, 169
 reaction, 194
Connectivity limits, 642
Connes advantage, 172
Constant displacement method,
 426
 rate method, 426
 load method, 426
Convergent beam electron
 diffraction (CBD)
 instrumentation, 287
 MnZn ferrite, 355
Corrosion kinetics, 388
Counter metal electrode, 253
Coupling agents, 194
Crack detection, 419
Crack growth of partially
 stabilized zirconia, 431
Crack velocity, 429
 stress intensity
 relationship, 429
 growth rate, 426
 cracks in refractory linings,
 440
Critical angle, 176, 189
Cross-sectional transmission
 electron microscopy (XTEM),
 359
C-scan imaging, 407
dc plasma atomic emission
 spectroscopy, 491
 dissolution of silica nitride
 using alkali fusion, 491
 factors affecting
 quantitative analysis, 493
 silicon interferences, 493,
 495
Defect structure, 268
Defense titanate ceramic (DTC)
 elemental analyses, 372
 glassy and minor phases, 377
 hercynitic spinel, 372
 nepheline, 372
 perovskite, 372
 ulvo spinel, 372
 uranium, 376
 x-ray diffraction analyses
 (XRD), 372
 zirconolite, 372
Degree of vacancy ionization,
 244
Delaminated ceramic capacitor,
 419
 in a ceramic material, 416
Delayed fracture, 425
Dense refractory concrete, 445
Densities, 163, 223
 fibers plus matric, 539
Depth of penetration, 176
 profiles, 36
Diamond anvil cell (DAC), 187
Dielectric constants, 260
Differential scanning
 calorimetry (DSC), 239
Diffuse reflectance technique,
 179
Differential thermal analysis,
 calcite, 520
 Ca, Mg $(CO_3)_2$, 525
 "Cheto" type montmorillonite,
 519
 dolomite ferroan, dolomite-
 ankevite series, 525
 limestones and dolomites, 520
 magnesium carbonate,
 magnesium oxide, CO_2
 system, 525
Diffusion profiles for quenched
 glass exposed to moist
 atmospheres, 581
Digital filter, 454
Dilatometric softening point

(DSP), 637, 643
Dimethylsulfoxide, 103
Dispersive IR, 172
Dissolution kinetics of TiO_2, 393
Dissolution of NiO agglomerates, 393
Dolomites, 520
Double countilever beam method, 425
Double torsion method, 425
Double weighing method, 536
Dynamic resistance, 252

Effective angle, 176
Effective electrostatic interaction potential, 506
Ehrenfest criteria, 637
Elastic module measurements, 324
Elastic recoil detection (ERD), 49, 56
Electrical conductance, multi-valent elements in glasses, 655
Electrical conductivity of dilute suspensions, 511
Electrical double layer, 500, 509
Electrochemistry, multivalent elments in glasses, 654
Electrode surfaces, 91
Electrokinetics, 506
Electromigration markers, 549
Electron microprobe (EMP), 286
Electron energy loss spectroscopy, 199
Electron-microprobe, multivalent elements in glasses, 659
Electron energy loss spectroscopy (EELS), 287
Electron probe microanalysis (EMPA), nuclear waste ceramics, 367
Electron-paramagnetic resonance (EPR), multivalent elements in glasses, 650
Electron spin resonance (ESR), water in glass, 584
Electron tunneling spectra, 252
Electronic absorption, 268
Electrostatic precipitator, 269
Elemental analyses, 269
Emission spectroscopy, 181
Emission techniques, 181
Energy dispersive analysis (EDA), plasma sprayed coatings, 475
EPR spectra, 464
Etch rate of electrolyzed glass, 568
Ethylidene ($CHCH_3$), 255
Evaporation decomposition of solution technique (EDS), 269
Explosive compaction techniques, 205
Extended x-ray absorption fine structure, multivalent elements in glasses, 652

FeO, 240
Failure mechanisms, 473
Fellgett advantage, 172
Ferri-magnetic resonance, 468
Fermi surface, 251
Fermi levels, 250
Ferro-magnetic resonance, 468
Fiber density, 539
Fibers, graphitic, 537
Field emission source, 282
Figure of merit, F_N, 450
Flaw detection, 413
Flaw image characteristics, 416
Float glass process, 531
Fluorescence phenomena, 205
spectroscopy, multivalent elements in glasses, 652
Fourier component, 259
transformation, 171
transform interferometry (FT-IR), 171, 271
Fracture of ceramics, 401
surfaces, 4, 207
FT-IR spectroscopy, 196
GeO_2, 223
Gas adsorption, 113
Gas phase hydrolysis, 388

SUBJECT INDEX

Glass hollow microspheres, 433
Glass-reinforced composites, 194
Glass transformation temperatures, lead borate glasses, 640
Gouy-Chapman layer, 509
Gouy-Chapman relation, 508
Grescovich-Lay mechanism, 398
Group analysis, 202

H_2 - ZnO system, 395
H_3BO_3, 223
$HC^0-Fe_2O_3$ system, 395
HgO, 215
Hg_2O, 216
Heavy-element staining, 332
Helium mobility, 644
Helium permeability coefficient, 644
High frequency sound waves, 413
Higher order Laue zone (HOLZ) lines, 355
High resolution electron microscopy (HREM), 287
High resolution spectroscopy, 196
High temperature differential thermal analysis, 515
  computerization, 525
  enhanced data reduction, 519
  microcomputers, 519
  diffusivity, 515
High voltage electron microscopy (HVEM), Mn Zn ferrite, 356
Hollandite, 369
Holographic inversion techniques, 423
Hot isostatic press (HIP), 271
Hot press, 271
Hot-pressed alumina, no Na-β^{111}-Al_2O_3 secondary particles, 302
Hot-pressed alumina, spinel secondary particles, β^{111}-Al_2O_3 secondary particles, 300
Hydrogen concentration profiles of soda-lime glasses, 554
Hydrogen content of amorphous silicon film, 552
Hydrogen in metals, 570
Hydrogenation reaction, 255
Hydrolysis/oxidation, 273
Hydroxylated silica, 116
Hysteresis, 133

Immersion technique, 161
Immobilization of radioactive waste, 590
Imperfections, 273
Inclusion analysis, 207
Index matching liquid (IML), 224
Inelastic electron tunneling spectroscopy (IETS), 249
Inelastic tunneling process, 251
Inhomogeneous densification, 310
Infrared diffuse reflectance, 271
Infrared spectroscopy, soda-lime glasses, 581
Infrared spectroscopy, water in glass, 484
Infrared transmission, 269
Injection mechanism, 264
Inorganic composites, 535
Insulating refractory concrete, 445
Interferometric measurements of glass sphere wall thicknesses, 433
Intermediate bond coat, 473
Internal standards, 270
Intrinsic defect detection, 403
Ion beam techniques, 39
Ion bombardment, 34
Ionic exchange of hydrogen bearing ion from water with Na^+ in glass, 554
Ion implantation markers, 549
Ion implanting noble gas into glass, 568
Ising model theory, 242
Isolated sub-surface flaws, 421

Isostatic compression of hollow glass microspheres, 434
Iron-containing glasses, Moessbauer spectroscopy, 651
Raman spectroscopy, 651
Isoelectric point, 164
Isothermal adsorption, 109
Isothermal desorption, 109
ISS analyses, 5, 39, 45, 56

Jacquinots advantage, 172

K_2CO_3, 223

LaB_6 source, 282
Laser mass spectrometry, 59
Laser mass spectrometry of solids, 61
Latent image fading, 583
Lattice parameter, 269
Law of mixtures (LOM), 538
Leaching of radio-nuclides from glass, 596
Lead borate glasses
 differential scanning calorimetry (DSC), 640
 glass transformation temperature, 641
 thermal expansion curves, 642
Leed, 93, 106
Light emitting tunnel junction experiments, 259
Limestone, 520
Liquid doping technique, 254
Liquid mix technique (LMT), 240
Liquid phase consolidation, 201
Lithium drifted silicon detecta, 286
Lithium silicate glasses, dilatometric softening temperature, 643
Load control, 477
Longitudinal modulus of composite, 542
Lorentz TEM microscopy, Mn Zn ferrite, 352
Low Al β^1 sialons, 340

Low temperature thermal expansion, 629
Lubrication, 249

MnO, 240
Magnesio ferrite, 527
Magnesium carbonate, magnesium oxide, CO_2 system, 525
Magnetic susceptibility, multivalent elements in glasses, 656
Magnified acoustic images of ceramics, 413
Matrix density, 539
Mechanism of reinforcement, 194
Mercury-cadmium telluride (MCT) detector, 176, 180
Mercury porosimetry, 133, 147
Microcrack detection acoustic emission, 329
 elastic modulus measurements, 324
 porosimeter measurements, 327
 scanning electron microscopy, 330
 small angle neutron scattering (SANS), 329
 thermal diffusivity measurements, 327
 transmission electron microscopy, 333
 differential thermal contraction, 326
 zirconia based ceramics, 324
Microstructure development in porous ceramic powder compacts, 395
Microtechnique, 187
Miscibility gap, 605
MnZn ferrite, auger electron spectra, 352
 convergent beam electron diffraction (CBD), 355
 domain wall-grain boundary interaction under magnetic field, 353
 intergranular CaO containing phase, 353
 Lorentz TEM microscopy, 352

SUBJECT INDEX

high voltage electron
 microscopy (HVEM), 356
 sintering temperature
 effects, 356
Moesbauer spectroscopy,
 iron-containing glasses, 651
Molecular optical laster
 examiner (MOLE), 215
Morphology, 161
Muconic acid, 255
Molecular orientations, 252
Monochromator, 172
Multichannel analyzer, 260
Multivalent elements in
 glasses, 647
 chronopotentiometry, 654
 colorimetric, 649
 electrochemistry, 654
 electron microprobe, 653
 electron paramagnetic
  resonance (EPR), 650
 extended x-ray absorption
  fine structure (EXAFS), 652
 fluorescence spectroscopy,
  652
 magnetic susceptibility, 656
 Moesbauer spectroscopy, 651
 potential sweep voltammetry,
  654
 redox titrations, 648
 selective vaporization, 649
 thermogravimetric analysis
  (TGA), 653
 viscosity, 655
 wet chemistry, 648
 x-ray photoelectron
  spectroscopy (XPS), 652
 glasses on melts, 657

N_2 adsorption, 161
Nepheline, 370
^{15}N hydrogen profiling, 549
NiCrAlY bond coat, 474
Nickel hydroxide, 388
 oxide, dissolation, 393
 powders, 388
Non-autoradiographic methods
 involving isotope-tagged
 water, 584

Non-electrochemical cell, 258
Non-linear dissulution
 kinetics, 392
Nonstoichiometry, 239
Nuclear magnetic resonance,
 (NMR) water in glass, 584
Nuclear reaction analysis
 (NRA), 53, 56
 glass surfaces, 549
 hydrogen concentration
  versus depth, 550
 measure distance from marker
  to surface, 566
Nuclear waste ceramics, 367
 analytical electron
  microscopy (AEM), 367
 electron probe microanalysis
  (EMPA), 367
 scanning electron microscopy
  (SEM), 367
 transmission electron
  microscopy (TEM), 367
 wavelength dispersive
  spectroscopy (WDS), 370
 x-ray diffraction analyses
  (XRD), 370
 redox ions in glasses, 648
 thermal diffusivity, 615
 volume-temperature
  relationships, 627

Optical microscope, 161, 199
 multichannel detector, 261
 quality, 273
 reflectance measurements, 259
 spectra, multivalent elements
  in glasses, 650
 transmission, 268
Ostwald ripening theory, 395
Oxygen nonstoichiometry, 244,
 247
 profiling, 562

P_2O_5, 224
Particle size analysis, 387
 size effect, 271
Pb counter electrode, 255
Penetration of water into SiO_2

SUBJECT INDEX

on silicon, 554
Perovskite, 369
Phase equilibria, 239
  separation boundaries, 639
  separation of glasses,
    chemical durability, 644
  differential thermal
    calorimetry, 640
  dilatometric softening
    temperature, 643
  electrical conductivity
    measurements, 644
  glass transformation
    temperature, 644
  heliim permeability
    coefficients, 644
  microstructure, 603
  properties relationships, 639
  thermal expansion curves, 642
Photoacoustic FT-IR, 186
  spectroscopy (PAS), 186
  spectroscopy, 407
Photoionization cross-sections, 8
Photomicrograph, 210
Photoresist type spinner, 250
pK_a^{int} - values for ceramic suspensions, 505
Plasma sprayed coatings, 473
  bond strength, 487
  stainless steel coatings, 473, 478
  spraying, 465
  steel coatings, 473
  cohesive fracture, 481
  sprayed titania, 465
  titania adhesion, 470
  titania EPR spectra, 466
  titania dc magnetization measurements, 471
  titania, the defect state, 472
  titania SCEM, 468
  yttria stabilized zirconia coatings, 473
  yttria stabilized zirconia coatings, as sprayed, 481, 487
  yttria stabilized zirconia coatings, AE/TAT data, 479
  yttria stabilized zirconia coatings, 473
  yttria stabilized zirconia coatings, fracture surfaces, 483
  yttria stabilized zirconia coatings, heat treated, 481, 487
  yttria stabilized zirconia coatings, SEM fractography, 483
  yttria stabilized zirconia, coatings, with NiCrAlY, 484
Platinum surfaces, 98
Pollucite, 370
Point defect, 239
Point defect dependence, 244
Point defect sensitive mode, 247
Point of zero charge, 161
Pore structure characterization, 147
Porosimeter measurements, 327
Porosity, 416
Porosities, 163
Porosity determinations, 115
Porous chromia gel, 116
Potash glass, 561
Potassia silica system, 31
Potential sweep voltammetry, multivalent elements in glasses, 654
Potentiometric titrations, 161
Potentiometric titration, 503
Powder analysis techniques, 388
Powder morphology, 388
Precipitation technique, 160
Preferred orientation, 458
Processing techniques, 207
Propagation loss measurements, 403
Property/morphology relationships in glasses, 639
Pulse-echo techniques (high frequency, low frequency), 408
Pure blue calcite, 522
Pycnometer density, 536
Pyridinium (Ag_5I_6), 192

Quantitative energy dispersive
 x-ray analysis (EDX), 270
Quantitative microscopy (QM),
 288
 stereology, 388
 x-ray analysis, 462

Radioactive waste, borosilicate
 glass, 590
Raman scattering technique, 263
 spectra of
  potassium-borogermanate
  glasses, 223
 spectroscopy, 186, 239, 242
 spectroscopy,
  arsenic-containing glasses,
  651
 iron-containing glasses, 651
Rare earth sulfides, 267
Redox ions in glasses, 647
Redox titrations, multivalent
 elements in glasses, 648
Reflection losses, 273
Refractive index, 223
Refractory concretes, 440
Reinforcements for inorganic
 composites, 535
Reitveld refinement, 462
Resonance angle, 260
Resonant oxygen scattering, 549
Resonant scattering for oxygen
 profiling, 560
Resulfurizing-cleaning, 273
Role of electrochemical
 phenomena on acid-base
 properties of ceramics, 499
Rotating anode generator, 461
Rule of mixtures (ROM), 538
Rutherford backscattering
 spectroscopy, 549
 spectrometry (RBS), 47, 56
 analyzing for low mass
  elements in matrix
  containing heavy elements,
  559
 analyzing for oxygen, 560
 measure distance from marker
  to surface, 566
 potash glass, 561
 soda-lime glass, 557

surface treatments, 559
water reaction with glass
 containing Ca, 559

Si_3N_4-SiO_2-Al_2O_3-AlN system,
 339
Si_2N_2O, 211
SiC/Si_3N_4 refractory, 211
Savitsky-Golay polynominal
 smoothing procedures, 454
Scanning electron micrograph,
 210
 electron microscope, 199, 201
 transmission electron,
  microscope, 161
 borosilicate glasses for
  radioactive waste
  immobilization, 592
 SEM, 286
 microcrack detection, 330
 nuclear waste ceramics, 367
 plasma sprayed coatings, 473
 laser acoustic microscope
  (SLAM), 414
 SLAM, 404
 computer holographic
  reconstruction of images,
  423
 images of inclusions, 422
 transmission electron
  microscope (STEM), 286
 secondary electron images,
  288
Secondary ion mass spectroscopy
 (SIMS), water in glass, 585
Second phase precipitates, 273
Seeding, 201
Selection preference, 252
Selection vaporization,
 multivalent elements in
 glasses, 649
Semiautomatic image analysis,
 387
Semiconductors, 359
 corrosive aqueous media, 392
Semiquantitative emission
 spectrographic analyses
 (EMS), 270

SUBJECT INDEX

Shape anisotropy, 387
Silicon, arsenic implantation, 361
Silicate glasses, 1
Silicon, Rutherford
  backscattering (RBS), 362
  secondary ion mass spectroscopy, 362
  silver implantation, 362
  structural defects, 362
  transmission electron microscopy (TEM), 362
  x-ray powdered diffraction pattern, 460
  XTEM, 361
Silicon carbide, 201
  acoustic amplitude micrograph, 416
  acoustic characterization, 400
  interferences, 495
Silicon nitride, 201
  acoustic amplitude micrograph, 416
  acoustic characterization, 401
  acoustic micrograph, 409
  chemical analysis, 491
  dc plasma atomic emission spectroscopy, 491
  Fe inclusions, 406, 422
  Si defects, 406
  silicon inclusion, 422
  surface acoustic waves (SAW), 409
  x-ray energy dispersive spectroscopy, 339
Silicon oxynitride, 201
SIMS analysis, 4, 39, 41, 55
  depth profiles, 25
Single crystal structure x-ray analysis, 462
Sinterability, 169
Sintering, 159, 271
Slow crack growth, 425
  automated test techniques, 419
  software development, 421
Soda-lime glass, 557
Soda-lime glass thermal
  diffusivity, 619
Sodium borosilicate glasses,
  behaviour near the 650 and 700°C isotherms, 608
  Co-O doped glasses, 613
  microstructure of phase separation, 603
  miscibility gap, 605
Sodium manganese (II) borosilicate system, 611
Solid inclusions, 422
Solution-reprecipitation process, 202
Spark source, 59
  mass spectrometry, 59
Spatial resolution, 200
Specific conductance of suspensions, 510
Specific gravity, 330
Specific surface areas, 221, 133
Specular reflectance, 181
Spray drying, 458
Sputter-induced photon spectrometry (SIPS), 30, 40, 55
  water in glass, 585
Stable Hillert criterion, 310
Standard adsorption data, 116
State of agglomeration analysis, 387
Stroke control, 477
Stokes shifted radiation, 260
Stray-light advantage, 172
Structure-property relationships, 239
Structural defects profiling, 365
Surface and interface studies, 71
Surface area determinations, 161
  surface area, 151
  by gas adsorption, 110
  calculations, 114
  characterization, 109
  enhanced Raman spectroscopy (SER), 249, 257
  plasmon polaritons, 259
  Raman scattering, 257

resonances, 260
studies, 194
Subcritical crack growth, 425
Surface acoustic waves (SAW), 408
  densities of ionizable protons for oxides, 512
  equilibria, 502
  flaws, 421
  oxide on aluminum alloys, distribution of elements, 321
  oxides on aluminum alloys, elemental composition, 321
  oxies on aluminum alloys, film thickness, 321
  reflectance of float glass, 531
Synchrotron x-ray sources, 461
Synroc, 369

TeO_2, 215
Tecktite glasses, 624
TEM images, 288
Temperature-pressure phase diagram, 202
Tensile adhesion test (TAT), 474
Theoretical density, 536
Thermal conductivity of vitreous silica, 621
Thermal diffusivity, measurements, 327
  nuclear waste melts, 619
  soda-lime-silicate glass, 610
Thermal expansion curves, lead borate glasses, 642
Thermal shock technique, 427
Thermionic source, 282
Thermogravimetric analysis (TGA), multivalent elements in glasses, 653
  evaluation, 244
Tin-oxide doped silicate, 7
  on float glass, bloom potential, 531
  surface reflectance, 531
  x-ray fluorescence, 531
Titanate-based nuclear waste ceramics, 369
Titania, 466
Titanium dioxide dissolution kinetics, 396
Transmission electron microscopy, microcrack detection, 333
  nuclear waste ceramics,(TEM) 367
Transmittance, 267
Trevorite ($NiFe_2O_4$), 596
Tritium autoradiography, 571
  advantages and disadvantages, 581
  apparatus, 574
  film analysis, 581
  film selection, 580
  hydrogen in metals, 572
  latent image fading, 583
  lateral sample resolutions, 581
  techniques, 571
  tritium-doped alum crystal standards, 578
  water detection limits, 583
  water in glass, 572

Ultra-high vacuum, 257
Ultrasonic techniques, 404
Unidirectional fibrous FP* glass composite, 540

Vapor-phase mechanism, 211
Vapor transport, 395
Variable atmosphere DTA, 525
Vibrational oscillator strengths, 252
Vibrational spectroscopies of molecular monolayers, 249
Viscosity, multivalent elements in glasses, 655
Voltage dependent differential conductance, 252
Water in glass, tritium autoradiography, 571
Water in materials, electron spin resonance spectroscopy (EPR), 572

SUBJECT INDEX

infrared spectroscopy (IR), 572
nuclear magnetic resonance spectroscopy (NMR), 572
nuclear resonance reaction analysis (NRRA), 572
secondary ion mass spectroscopy (SIMS), 572
sputter induced photon spectroscopy (SIPS), 572
Wave-guide system, 223
Wavelength dispersive spectroscopy (WDS), nuclear waste materials, 370
Wet chemical analysis, 161
Wet chemistry, multivalent elements in glasses, 648

X-ray analysis of rapid events, 462
  density, 545
  diffraction analyses (XRD), nuclear waste ceramics, 370
  diffraction, 200
  fluorescence (XRF), 239
  microanalysis technique, 199
  photoelectron spectroscopy (XPS), 215, 220, 273
  dye-enhanced radiography, 403
  energy dispersive spectroscopy, 339
  microphase analysis, 462
  photoelectron spectroscopy (XPS), multivalent elements in glasses, 652
X-ray powder diffraction, 449
  accuracy of diffraction intensities, 459
  background determination, 452
  computer automated powder diffraction, 450
  data smoothing, 454
  improved computer procedures, 461
  inaccuracy in the measurement of 2θ, 450
  limit of phase detectability, 458
  locating peaks, 452
  profile fitting, 456
  spectral stripping, 454
  impure silicon, 460
  wollastonite, 459
X-ray position, sensitive detectors, 460

Yttria, 159
Yttria-aluminum-garnet, 65
Yttrium hydroxynitrate hydrates, 159

ZnO, 118
ZAF corrections, 290
Zero-porosity density, 535
Zeta potential, 162, 164
Zincblende semiconductors, 204
Zirconia, 201, 206
Zirconolite, 369